AF389473

Foodborne Pathogens Management: From Farm and Pond to Fork

Editor
Frans J.M. Smulders
University of Veterinary Medicine
Wien, Austria

Editorial Office
MDPI
St. Alban-Anlage 66
4052 Basel, Switzerland

This is a reprint of articles from the Special Issue published online in the open access journal *Foods* (ISSN 2304-8158) (available at: https://www.mdpi.com/journal/foods/special_issues/foodborne_pathogen_management).

For citation purposes, cite each article independently as indicated on the article page online and as indicated below:

LastName, A.A.; LastName, B.B.; LastName, C.C. Article Title. *Journal Name* **Year**, *Volume Number*, Page Range.

ISBN 978-3-0365-8104-0 (Hbk)
ISBN 978-3-0365-8105-7 (PDF)

Foodborne Pathogens Management: From Farm and Pond to Fork

Editor

Frans J.M. Smulders

MDPI • Basel • Beijing • Wuhan • Barcelona • Belgrade • Manchester • Tokyo • Cluj • Tianjin

Contents

About the Editor

Frans J.M. Smulders

Frans J.M. Smulders, Ph.D., Dr. med. vet, Dr. h.c., Dipl.ECVPH, Professor emeritus, has been active in the field of meat quality and meat safety since his graduation as a vet in 1978. He became Full Professor and Chair of the Institute of Meat Hygiene, Meat Technology and Food Science at the University of Veterinary Medicine Vienna in 1996 and held this position until his retirement in 2020. He was not only active in research, but also in competence building, postgradual education and knowledge transfer, in particular in the European College of Veterinary Public Health (ECVPH) and as editor of a book series on Veterinary Public Health. At European Union level, he served in the Scientific Committee on Veterinary Measures Related to Public Health (SCVMPH), and, after the foundation of the European Food Safety Authority (EFSA), he was member of EFSA's Scientific Committee and various expert panels. Since the focus of his work was on fresh meat quality and physical-chemical treatments to ensure fresh meat safety as well as food-chain approaches for food safety, it was not surprising that he took the chance to combine scientific expertise on various food commodities or food chains and on various foodborne hazards into a Special Issue, which is now presented to a broad readership.

Preface to "Foodborne Pathogens Management: From Farm and Pond to Fork"

Dear Colleagues,

In the 1930s, the US food microbiologist Samuel Cate Prescott (1872–1962), his Swiss colleague Karl Friedrich Meyer (1884–1974), and the UK microbiologist Sir Graham Selby Wilson (1895–1987) first suggested to follow a more active intervention strategy against food-transmitted diseases of microbial aetiology. In the early 1960s, the US National Aeronautics and Space Administration (NASA) introduced the basics of a novel food safety assurance system (Ross-Nazal, 2007), which evolved, in the 1970s, into the Hazard Analysis Critical Control Point (HACCP) concept (Lachance, 1997; Weinroth et al., 2018).

Thus, the fundaments of the Longitudinally Integrated Safety Assurance (LISA) approach were created (Mossel, 1989). Over the past few decades, the latter concept has inspired many (veterinary) food microbiologists to stress the longitudinal character of this approach by suggesting more 'jazzy' terms such as: 'From Conception to Consumption', 'From Production to Consumption', 'From Stable to Table' and 'From Farm to Fork' (or variants such as 'From Pond to Fork' or 'From Forest to Fork' when one wants the reader to concentrate on particular foods such as fish or game). In essence, the researchers took the same path as epidemiologists would have taken when investigating outbreaks of foodborne disease, the only difference being that epidemiologists would follow the 'top-down' route, and scientists working to reveal a functioning LISA principle would take a 'bottom-up' direction in order to prevent conditions that would render food unsafe for consumption.

'From Farm to Fork' was recently chosen by European authorities as the title of a document released in May 2020. This document describes the declared EU policies aiming at reducing the environmental/climate impact of primary production, while at the same time ensuring fair economic returns for farmers and striving to meet the 'Green Deal' objectives—that is, achieving Sustainable Development Goals (SDGs) such as improving food security by reducing food loss and waste without impairing food safety. The big question is whether or not the resource footprints of future food production in terms of land, water, energy and resources will be within our common planetary boundaries (Steffen et al., 2015). This means that resolving conflicting goals will become a key challenge. A 'Farm-to-Fork' framework might be indispensable for meeting our future challenges in terms of food safety, security and sustainability (Hanning et al., 2012).

Obviously, this interpretation of 'Farm to Fork' goes far beyond its original purpose of stressing the longitudinal approach to safety assurance rather than the traditional end-product-oriented control of foods of animal origin.

This Special Issue indeed makes use of such a broad definition of 'Farm to Fork'. Consequently, it contains contributions on the state of the art in food safety assurance, with the ambition to contribute to the evidence-based trade-offs that our future food safety, security and sustainability necessitate.

Since the majority of the contributions focus on specific food commodities, the articles are arranged according to the thematic foci "Safety of meat and dairy products from primary production to primary processing", "Fish and seafood", "Wild game", "Insects", followed by papers on "Epidemiology of parasites and pathogenic bacteria in various food chains" and "Food technology and food safety".

References

Hanning, I.B., O'Bryan, C.A., Crandall, P.G., Ricke, S.C. (2012). Food Safety and Food Security. Nature Education Knowledge 3(10):9.

Lachance, P.A. (1977). How HACCP started. Food Technol., 51:3.

Mossel, D.A.A. (1989). Adequate protection of the public against food-transmitted diseases of microbial aetiology. Achievements and challenges half a century after the introduction of the Prescott-Meyer-Wilson strategy of active intervention. Int. J. Food Microbiol., 1989, Dec. 19 (4) 271–94 doi: 10.1016/0168-1605(89)90097-4.

Ross-Nazal, J. (2007). "From Farm to Fork": how space food standards interpreted the food industry and changed food safety standards. Chapter 12, p 219-236. In: Steven J. Dick, Roger D. Launius (Eds.): Societal impact of Space Flight, NASA History Division, Office of External Relations, SP-2007-4801.

Steffen, W.K., Richardson, K., Rockström, S.E., Cornell, S.E., Fetzer, I., Bennett, E.M., Biggs, R, Carpenter, S.R., de Vries, W., de Wit, C.A., Folke, C., Gerten, D., Heinke, J., Mace, G.M., Persson, L.M. Ramanathan, V., Reyers, B., Sörlin, S. (2015). Planetary boundaries: Guiding human development on a changing planet. Science 347: 736, 1259855, DOI: 10.1126/science 125985.

Weinroth, M.D. Belk, A.D., Belk, K.E. (2018). History, development and current status of food safety systems worldwide. Anim. Front. 2018 Aug 30;8(4):9–15. doi: 10.1093/af/vfy016. eCollection 2018 Oct.

Frans J.M. Smulders
Editor

Review

Disease Occurrence in- and the Transferal of Zoonotic Agents by North American Feedlot Cattle

Osman Y. Koyun [1], Igori Balta [2,3], Nicolae Corcionivoschi [2,3] and Todd R. Callaway [1,*]

[1]   Department of Animal and Dairy Science, University of Georgia, Athens, GA 30602, USA
[2]   Bacteriology Branch, Veterinary Sciences Division, Agri-Food and Biosciences Institute, Belfast BT4 3SD, UK
[3]   Faculty of Bioengineering of Animal Resources, University of Life Sciences King Mihai I from Timisoara, 300645 Timisoara, Romania
*   Correspondence: todd.callaway@uga.edu

Abstract: North America is a large producer of beef and contains approximately 12% of the world's cattle inventory. Feedlots are an integral part of modern cattle production in North America, producing a high-quality, wholesome protein food for humans. Cattle, during their final stage, are fed readily digestible high-energy density rations in feedlots. Cattle in feedlots are susceptible to certain zoonotic diseases that impact cattle health, growth performance, and carcass characteristics, as well as human health. Diseases are often transferred amongst pen-mates, but they can also originate from the environment and be spread by vectors or fomites. Pathogen carriage in the gastrointestinal tract of cattle often leads to direct or indirect contamination of foods and the feedlot environment. This leads to the recirculation of these pathogens that have fecal–oral transmission within a feedlot cattle population for an extended time. *Salmonella*, Shiga toxin-producing *Escherichia coli*, and *Campylobacter* are commonly associated with animal-derived foods and can be transferred to humans through several routes such as contact with infected cattle and the consumption of contaminated meat. Brucellosis, anthrax, and leptospirosis, significant but neglected zoonotic diseases with debilitating impacts on human and animal health, are also discussed.

Keywords: feedlot cattle; zoonoses; STEC O157:H7; *Salmonella*; *Escherichia coli*; *Campylobacter*; *Cryptosporidium*; *Brucella*; *Bacillus anthracis*; *Leptospira*

Citation: Koyun, O.Y.; Balta, I.; Corcionivoschi, N.; Callaway, T.R. Disease Occurrence in- and the Transferal of Zoonotic Agents by North American Feedlot Cattle. *Foods* **2023**, *12*, 904. https://doi.org/10.3390/foods12040904

Academic Editor: Frans J.M. Smulders

Received: 30 November 2022
Revised: 13 February 2023
Accepted: 15 February 2023
Published: 20 February 2023

1. Introduction

Cattle, along with other ruminants, have provided humanity a stable supply of meat and dairy products since their domestication. In 2021, the per capita consumption of beef was approximately 26.7 kg in the United States [1], 16.9 kg in Canada [2], and 14.8 kg in Mexico [3]. North America is a large producer of beef for both domestic and export purposes, with more than 119 million heads of cattle, which represents approximately 12% of the world's cattle inventory [4,5]. The United States has the largest cattle inventory (approximately 98.8 million cattle and calves in both beef and dairy operations) in North America [4,5]. Feedlots have been an integral part of modern beef cattle production in North America for more than 60 years, producing wholesome, highly desirable and marketable carcasses throughout the course of the year at a low cost to produce a high-quality protein food for humans [6,7]. Feedlots are typically located in the Great Plains region of North America and are located near both grain production and stocker/backgrounding regions. Cattle are fed in feedlots to take advantage of the economies of scale related to having many cattle located in one facility.

Readily digestible, high-energy rations are provided to cattle through communal feed bunks or troughs (Figure 1) at feedlots (i.e., a confined area for growing or fattening cattle) during their final stage of growth, which is also known as finishing. It is at this point that marbling (i.e., intramuscular fat) is deposited in muscular tissues [6,7]. Feedlot rations mostly rely on corn (*Zea Mays* L.) supplemented with a protein source and often include by-products from other industries (e.g., dried distiller's grains, brewer's yeast) [6,8]. Cattle are usually fed 2–3 times per day in order to maximize feed consumption and growth efficiency. Feedlot cattle typically gain 1–2 kg/d and have a feed efficiency of approximately 5 to 6 kg feed/kg gain [9]. Commonly, these feedlot rations contain less than 10% forage (e.g., corn silage), and the feeding of such high-energy density rations can lead to the development of ruminal acidosis (low ruminal pH) [10]. When we feed cattle, we are actually feeding the microbial population of the rumen and hindgut (i.e., cecum, colon, and rectum), which ferment feedstuffs to produce Volatile Fatty Acids (VFAs) that cattle utilize for energy, and Microbial Crude Proteins (MCP), which ruminants use as their primary dietary protein source [11]. Feeding with starch has an advantage, as the microbial fermentation produces a greater proportion of propionate than when cattle are fed forage-based rations. Propionate is glucogenic and leads to intramuscular fat deposition (i.e., marbling) [12].

Figure 1. Cattle share communal feed bunks or troughs.

Despite ground-breaking advancements in the animal production and animal health aspects of feedlot systems, cattle can still have certain conditions and diseases that impact their health, growth performance, and carcass characteristics, and some of these can also impact human health [13–16]. Diseases are mostly transferred between cattle in a fecal–oral or direct contact fashion; however, they can originate from the environment and be spread by vectors (e.g., animals, rodents, or insects) or fomites (e.g., water, feed, surfaces, and soil), and pathogen carriage in the gastrointestinal tract (GIT) of cattle often leads to the direct or indirect contamination of feeds and the feedlot environment (e.g., water troughs and feed bunks, and feedlot pen surfaces) [14,15,17,18]. The circulation (and re-circulation) of pathogenic bacteria between different hosts, vectors, and the feedlot environment is ripe for the development of an on-farm endemic pathogen population that can impact both animal and human health.

Amongst zoonotic pathogenic bacteria, foodborne pathogens such as *Salmonella* spp., Shiga toxin-producing *Escherichia coli* (STEC), and *Campylobacter* spp. are commonly associated with animal-derived foods and can be transferred to humans through several routes: (i) contact with positive cattle or carcasses, (ii) the consumption of contaminated or infected meat, and/or (iii) the consumption or irrigation of crops with water contaminated with cattle manure [13,19]. In addition, other zoonotic pathogenic agents with public health relevance such as *Cryptosporidium*, *Brucella*, *Bacillus anthracis*, and *Leptospira* and the diseases that they cause in humans are also discussed in this review.

2. Structure of the North American Beef Industry

Beef cattle production in the United States is inextricably linked with the founding mythos of the Great Plains, or the "Old West". Cattle ranchers from the frontier are often portrayed in movies and stories as independent and self-reliant heroes. Today's North American cattle producers are heirs to this image and remain very independent and self-reliant. While increasing corporatization has impacted some segments of the cattle industry at the cow-calf production level, the beef industry of North America currently remains largely comprised of small producers. The beef industry has traditionally been highly decentralized and fragmented into five basic segments: cow-calf producers, stocker/backgrounder, feedlots, packers, and retail. The packer and retail segments are largely beyond the scope of this review, yet it is important to remember their role in the industry, which drives the production decisions made by ranchers for years before cattle reach the market. The beef production continuum is shown in Figure 2 and is best visualized as a pyramid in terms of the number of producers involved at each phase. However, an increasing degree of consolidation and vertical integration at the packing and retail levels has entered the beef production industry because there are fewer participants who can implement the required/suggested practices on the farms. This means that many of the practices that can implemented at larger, more well-funded production locations may not be implemented due to the economic and logistical constraints faced by the small producers. In the present review, we primarily focus on the live animal phases of beef production (Figure 2).

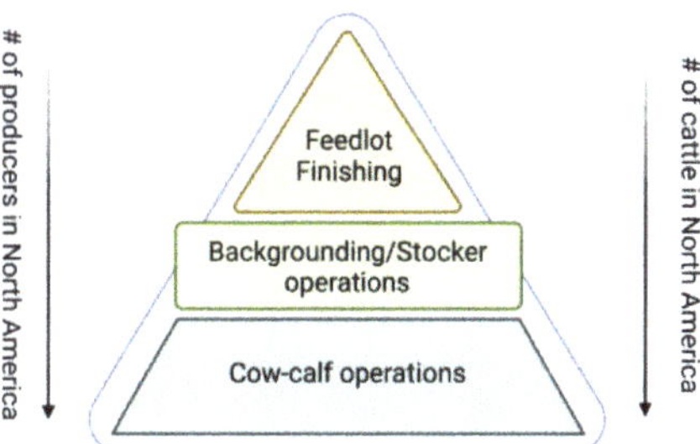

Figure 2. The beef production continuum visualized as a pyramid in terms of the number of producers involved at each phase.

2.1. Cow-Calf Producers

Cow-calf producers are the foundation of cattle in the U.S. and are the most decentralized phase of cattle production with thousands of producers scattered across the country, raising approximately 30 million calves each year. Cow-calf producers are often not able to be full-time cattle producers but must often work a "traditional job" (i.e., off-farm/ranch employment to generate a stable, consistent income) and must perform all of the farm tasks in their in their spare time, and as a result, many of their production decisions are driven by necessity, time availability, and logistics. This often limits the type of animal care procedures, as well as the procedures aimed at improving production efficiency, that can be implemented on any single farm. A typical beef producer in the southeastern United States is almost 60 years of age and works cattle on weekends and evenings when the weather and day-length allow. While most producers desire to maximize their profitability, many do not use the most up-to-date production methodologies (e.g., artificial insemination and estrus synchronization) due to the expense, time, and lack of skills and/or facilities involved. In general, producers attempt to calve in the spring and some use artificial insemination to improve their herd genetics and have a calf crop within a specified time window, with the majority utilizing a herd bull for ease of breeding.

Most cow-calf herds contain fewer than 50 cows, and these producers maintain a fairly stable herd size over the course of the year, marketing their calves themselves (from 180–240 d of age, see Figure 3), often through local auction markets or sale barns [20,21]. When calves leave their farm of origin, they bring an "internal record of exposure and vaccination" with them in the form of their immune systems, which means that that while the calves are less susceptible to pathogens that they have previously been exposed to, they remain susceptible to novel pathogens (bacterial, protozoal, and viral). Stress acts as an immunosuppressant and is cumulative in its impacts. Calves at auction markets can undergo multiple simultaneous stresses from weaning and transport, as well as social stresses, and can therefore be moderately to severely immunosuppressed when commingled with calves from other farms. Collecting calves from multiple farms in a close-quarters environment is a recipe for disease amplification in a population of susceptible calves, including the spread of zoonotic pathogens within these calves, commingled with calves that originated from across broad geographic origins.

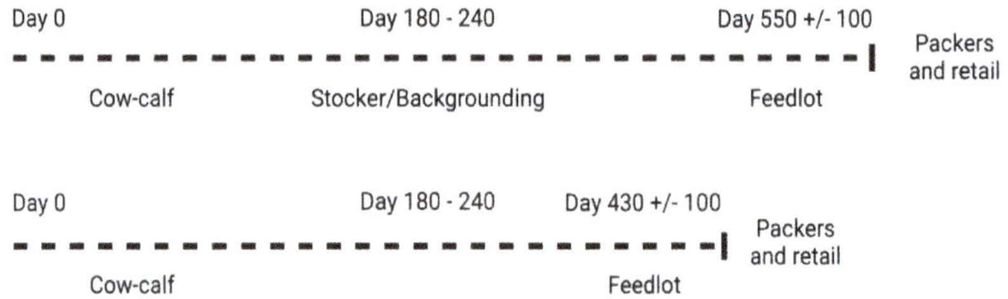

Figure 3. The beef industry has traditionally been fragmented into five segments: cow-calf producers, stockers/backgrounders, feedlots, packers, and retail. Created with BioRender.com.

Calves (weighing approximately 120–360 kg) typically remain at an auction market for 24 to 48 h before they are shipped to either a backgrounder/stocker facility or directly to a feedlot. The decision as to which pathway is utilized depends on calf size/age, breed, owner marketing strategies, and packer demands for quality or type of beef to be produced. Larger and older calves may be sent straight to a feedlot instead of to a background/stocker facility in order to begin the finishing process, but smaller calves may instead be sent to backgrounding/stocking to allow for slower growth and development.

2.2. Backgrounders/Stockers and Feedlots

A tractor-trailer load of stressed and newly commingled calves is often transported for an additional 12 to 24 h (frequently transiting more than 1500 km in this time frame, whilst undergoing feed and water withdrawal, and often profound temperature changes) to a stocker or feedlot facility, which further exacerbates the susceptibility of these calves to disease exposure from cohorts. Upon arrival at either the feedlot or stocker facility, calves are typically rapidly vaccinated, identified, and allowed to rest and recuperate from the stresses of transport. These first days upon arrival are critical in setting cattle up for success as stresses can accumulate and result in the development of shipping fever in calves, which can impact morbidity and mortality among animals. Thus, it is critical to ensure that calves receive a ration designed to tempt them into beginning feed consumption quickly, in order to begin the supply of glucose, protein, and minerals to the immune system. Calves that are classified as "high risk" often require special care and added nutritional metaphylaxis and prophylaxis in the first few days after arrival in a stocker facility or feedlot. Stocker operators commonly feed native forages or crop residues (e.g., corn or wheat stubble) to cattle for 2–6 months in order to increase their growth and develop their frame (Figure 3). During backgrounding/stocking, cattle may consume protein or energy concentrates in their ration to increase their energy or protein intake; however, the amount of grain consumed in the stocker phase is typically much lower than that used in feedlots. The rations of stocker producers often contain by-products such as distiller's grains, but mostly contain corn, with varying levels of processing (e.g., cracking or flaking) to improve its digestibility. When calves reach feedlot market weight (typically 270–370 kg), they are shipped to the feedlot for finishing or fattening prior to slaughter.

In the feedlot, cattle are segregated in pens based on body weight, breed, sex, and special program enrollment (e.g., No Antibiotics Ever) and eat from communal feed bunks at the front of each pen. Cattle often enter the feedlot at approximately 350 kg and are fed diets containing a high Net Energy for Gain (NEG) concentration, which is achieved by feeding them diets rich in starch until they reach approximately 625 kg, the current market weight. The feeding/finishing period can last 90–300 d, depending on the size of the cattle when they enter the feedlot.

Typically, according to the United States Department of Agriculture (USDA) animal census, there are more than 12 million cattle in U.S. feedlots at any time. While the vast majority of feedlot operations have a capacity of under 1000 heads, they only market a small percentage of the fed cattle to consumers. Feedlots with a capacity of more than 32,000 heads provide more than 40% of the fed cattle marketed [22]. Feedlots in the U.S. can reach a capacity of over 100,000 heads, which—assuming a 450 kg average weight for feedlot steers that consume 2% of their body weight (as dry matter (DM))—would require 9 kg (DM)/hd/d of feed, and a 50,000-head feedlot would require approximately 450,000 kg DM or 642,000 kg (as fed) of feed per day (approximately 7–8 train cars, or 20–25 tractor-trailer loads of feed). This typically requires feedlots to be largely self-contained facilities with an on-site feed mill (Figure 4). This means that many trucks bringing feed to each feedlot may take feed to other lots, and this represents a potential vector for zoonotic pathogens to be transmitted between feedlots. In addition, manure is often composted on site to mitigate the environmental impact and potentially generate a revenue stream by the sale of soil amendment for gardens; however, this can also carry zoonotic pathogens that can be transmitted to humans and other animals. It is clear that the infrastructure and activities needed to operate feedlots offer numerous opportunities for zoonotic pathogens to colonize and proliferate in cattle.

3. Zoonotic Agents with Public Health Relevance

There is a variety of pathogenic bacteria that are commonly found in cattle across North America. Most of these pathogens can (i) impact animal health; (ii) pose a threat to human health, such as foodborne pathogenic bacteria; and (iii) live in the GIT and are often undetected, as they may not cause illness in the host animal. This means that these pathogens may only be detected during the specific surveillance of cattle populations housed in a specific feedlot. Furthermore, many of these pathogenic bacteria can exist simultaneously in cattle, but little information currently exists on this issue of multiple pathogen colonization. Herein, we endeavor to discuss the most well-known human/animal threatening zoonotic agents of cattle with public health relevance.

3.1. Salmonella spp.

Salmonella enterica serovars are one of the most important foodborne pathogens in North America, comprising more than 2500 serotypes that are often harbored in the GIT of a variety of animals such as mammals, birds, reptiles, and amphibians, as well as in a variety of different environments [14,23–27]. The major *Salmonella enterica* serovars associated with clinical infections in both cattle and humans are Dublin, Enteritidis, Heidelberg, Kentucky, Montevideo, Newport, and Typhimurium, and it should be noted that several of these serotypes can colonize the same animal simultaneously [14,25,28,29]. *S. Montevideo* was the most frequently reported serotype in North American cattle, while it was not one of the most frequently reported serotypes in other continents [23,28]. Moreover, *Salmonella* prevalence varies considerably by geographical region; a lower prevalence was recorded in the northern U.S. states and Canada than in southern states [30].

Figure 4. Aerial image of randomly chosen commercial feedyard. Feedmill is indicated by 1; silage pits are depicted by 2; manure/pen surface composting is tagged 3; 4 denotes water retention pond; 5 indicates cattle pens; and 6 highlights cattle working facilities. Image selected from Google Maps.

In the United States, non-typhoidal *Salmonella* is one of the most common bacterial foodborne diseases, resulting in an estimated 1.2 million domestically acquired foodborne infections, along with 450 deaths from approximately 130 outbreaks every year [19,29,31]. The infective dose for non-typhoidal *Salmonella* is reported at 10^3 bacterial cells [30]. Salmonellosis in humans is often localized and self-limiting; however, severe cases require antimicrobial therapy and hospitalization [19,24,29,31]. Salmonellosis in humans is less associated with the consumption of beef or dairy products than compared to pork and poultry products [19,23,30]. However, certain cases have been traced back to cattle [19,23,30]. The contamination of lymph nodes that are processed into ground beef is one of the main ways for *Salmonella* spp. to enter the food chain [32,33].

The key transmission route of *Salmonella* in cattle is fecal–oral, and the prevalence of the pathogen in cattle varies, with reported estimates of 2–42% between-herd prevalence and

0–73% within-herd prevalence [14,34,35]. Cattle are asymptomatic carriers of *Salmonella* (i.e., a commensal of their GIT microbiota) [17,28] and can shed it at 10^3 to 10^5 CFU/g of feces, contaminating the farm environment and equipment [30,36]. It is believed that exposure to transport and lairage stress can increase the fecal excretion of *Salmonella* in feedlot cattle prior to slaughter [28,37]. The fecal shedding of *Salmonella* is subject to seasonal variation, reaching higher rates in the summer and early fall, declining in the winter months, and it has been reported that there is a correlation between shedding by animals and outbreaks in humans [14,17,32,38,39]. Although a physical correlation to ambient temperature has been observed, the internal temperature of the GIT is mostly stable; thereby, it seems that temperature is not the only source of the seasonality of *Salmonella* shedding through feces. Moreover, antimicrobial-resistant *Salmonella* (represented by varied serotypes such as *Salmonella* Newport, *Salmonella* Typhimurium, and *Salmonella* Reading) were detected in over 5000 individual fecal samples collected from multiple feedlots in the United States [40]. In Canada, the *Salmonella* prevalence in manure from feedlot cattle, beef carcasses, ground beef, and environmental samples is often reported to be low [13].

3.2. Shiga Toxin-Producing Escherichia coli (STEC)

Shiga toxin-producing *E. coli* (STEC), also known as enterohemorrhagic *E. coli* (EHEC) or Vero toxin-producing *E. coli* (VTEC), are a family of zoonotic foodborne pathogens that can be naturally present in the GIT of cattle [41,42]. STEC that infect the human GIT are able to cause clinical symptoms ranging in severity from mild diarrhea to hemorrhagic colitis and life-threatening hemolytic uremic syndrome (HUS), a critical cause of acute renal failure in children [41,42]. STEC is characterized by a very low infective dose (<100 bacterial cells) in humans; however, hosts can asymptomatically harbor these pathogens as part of their GIT microbiota [43]. The frequency of STEC O157:H7 infections has been on the decline in North America over the past two decades due to improvements in meat safety, especially the implementation of "Test and hold" procedures for ground beef prior to shipment to consumers [44,45]. While most STEC-related illnesses have been often associated with the consumption of undercooked ground beef or through contaminated produce, pathogen transmission to humans can occur through contaminated drinking or recreational water, contact with cattle, pen surface contamination, and human-to-human contact [46,47].

Among STEC strains, enterohemorrhagic *E. coli* serotype O157:H7 has become one of the most important and well-studied pathogens as it frequently colonizes the GIT of cattle in North America [48–50]. While this is the most well-known and common STEC in North America, it is becoming clear that other STEC serotypes are impactful and play a role in human health. In the United States, along with O157, the top six non-O157 STEC serogroups (e.g., O26, O45, O103, O111, O121, and O145) have been recognized as adulterants in raw and non-intact ground beef [42,48]. This provides an economic incentive in addition to the ethical and moral incentives to reduce STEC contamination. The colonization and re-colonization of cattle with STEC occurs through fecal–oral contamination or the consumption of contaminated drinking water sources, or contaminated feeds, and the lower GIT of cattle, particularly the mucosa of the recto–anal junction (RAJ), is considered the major region for persistent colonization by *E. coli* O157:H7 [48,51,52]. STEC infections in cattle are usually asymptomatic, as they lack vascular receptors for the Shiga toxins (*Stx*), allowing this potent pathogen to thrive in the GIT while not causing damage to the host intestinal tissue or stimulating immune host defenses [42,47,53].

The levels of STEC O157:H7 in the GIT, digesta, and on hides of cattle prior to entering the commercial abattoir play a crucial role in the occurrence of carcass contamination during slaughter and processing [41,48,54]. Higher levels of STEC in cattle were correlated with higher carcass contamination levels. The previous literature showed that grain feeding increased the number of acid-resistant *E. coli* in feces of cattle, which has critical implications for food safety as the acid-resistance of the pathogen seems to be a factor in the transmission of this pathogen from cattle to humans [55]. In addition, STEC O157:H7 prevalence was increased in hide samples of cattle during transport (i.e., a common stressor to animals) from the feedlots to the abattoir and/or during lairage prior to slaughter [37,54,56]. Cattle that shed STEC O157:H7 at a rate of greater than 10^3 or 10^4 CFU/g of feces have been defined by the term "super-shedder", and these high-shedding cattle remains the main vector of animal-to-animal transmission and production environment contamination [44,47,48,57]. STEC (*E. coli* O157 and non-O157) have been found in feedlot cattle feces and in feedlot environmental sources such as water troughs, lagoons, and soils in Canada [13,58]. Fecal prevalence rates of 0–79% have been reported for *E. coli* O157:H7 and 7–94% for the other 'top six STEC' (O26, O45, O103, O111, O121, and O145), and the prevalence is often higher during spring/summer than fall/winter [13,47,48,59,60]. It was reported that feedlot cattle farms can disseminate *E. coli* O157:H7 in the environment and that other animal vectors (e.g., feral swine), as well as flies, can contaminate leafy green vegetables on farms located in close proximity [46,61–63]. In North America, European starlings (*Sturnus vulgaris*) are considered a high-risk invasive bird species associated with the environmental dissemination of antimicrobial-resistant *E. coli* as these birds utilize feedlots during winter months for food sources [64]. Other studies have demonstrated that there is a potential spread of zoonotic pathogens to nearby fields and humans through dust spread from feedlot surfaces [61].

3.3. *Campylobacter spp.*

Campylobacter is one of the leading causative agents of bacterial foodborne gastro-enteritis in humans in the United States and Canada and can be transmitted to humans through human–animal contact (often via petting zoos), occupational exposure, the consumption of contaminated dairy (e.g., unpasteurized milk) and meat products, and contact with environment) [19,65–68]. *Campylobacter* is estimated to cause 1.3 million human illnesses every year in the United States [68], and the infection is often accompanied by abdominal pain and in some cases may lead to the development of the more severe Guillain–Barré syndrome in patients [69]. *Campylobacter* can also cause serious diarrhea in humans and has a very low infectious dose of as few as 500 organisms [67,68]. *Campylobacter jejuni* is the leading agent of reported human infections [65,67]. While poultry products are considered to be the leading source of human infections with *Campylobacter* in North America, cattle can serve as a vehicle for the transmission of this pathogen to humans [19]. Foodborne *Campylobacter* outbreaks in the United States (during 1998–2016) were attributed to dairy products (32%), chicken products (17%), and vegetables (6%), and more human outbreaks were reported during the summer (35%) than the spring (26%) and fall (22%) [67].

The colonization of *Campylobacter*, as a common commensal, in the GIT of cattle is significant not only regarding the potential for the contamination of the carcass at slaughter, but also regarding the environmental burden on farm and in transport through fecal shedding. It was reported that *Campylobacter* shedding by cows was 1.1×10^2 CFU/g of feces, while shedding in calves was approximately 250-fold (2.7×10^4 CFU/g of feces) more [30]. Studies conducted across the United States reported a *Campylobacter* prevalence ranging from 20 to 60% in feedlot and dairy cattle feces [70]. In particular, *C. jejuni* was detected in fecal samples collected from feedlots in the United States and Canada at a prevalence of 72–82% [13,65,66,70]. Up-to-date studies from Alberta, Canada, reported an increased antibiotic-resistant profile of fluoroquinolone-resistant *C. jejuni* isolates from around 1300 diarrheic patients connected to domestically acquired infections from cattle reservoirs [65]. Moreover, other researchers showed that, from 320 *C. jejuni* and 115 *C. coli* isolates collected from feedlot cattle farms in multiple states of the U.S., 35.4% of *C. jejuni* and 74.4% of *C. coli* isolates displayed increased fluoroquinolone resistance, which was remarkably higher than previously documented in United States reports [71]. *Campylobacter* species from feedlot manure runoff contaminates water supplies through agricultural runoff (due to rain events), posing serious human health concerns and increasing the risk of a waterborne outbreak [70]. Another important route of transmission is through migratory birds (e.g., European Starlings), and *Campylobacter jejuni* has been widely detected and identified in their feces [70].

3.4. Cryptosporidium spp.

Cryptosporidiosis is a disease in humans and cattle caused by a ubiquitous opportunistic enteric protozoan of the genus *Cryptosporidium*, is a global disease and one of the most common causes of diarrhea in both humans and livestock, and can be spread to humans from food animals and vice versa [72–74]. In cryptosporidiosis, parasite invasion and epithelial destruction in the small intestine by this causative agent results in crypt hyperplasia and apoptosis, villus atrophy and fusion, and physiological changes that impair intestinal nutrient absorption and cause diarrhea in the host [75,76]. Children, neonatal animals, and immunocompromised individuals are most susceptible to this parasitic disease, which is transmitted primarily through the fecal–oral route [74]. Contact with cattle, particularly with infected pre-ruminant calves, has been implicated as the root cause of many outbreaks in humans (e.g., veterinarians, researchers, and children attending agriculture-based activities and petting zoos) [74]. Moreover, food or water (e.g., lakes, rivers, and municipal drinking water without treatment) that is contaminated by cattle manure has been identified as a source of cryptosporidiosis outbreaks in humans [74,77,78]. The predominant *Cryptosporidium* species infecting humans are *C. parvum* and *C. hominis*, while *C. bovis*, *C. ryanae*, and *C. anderseni*, in addition to *C. parvum*, are the causative agents of bovine cryptosporidiosis [73].

In the United States and Canada, pre-weaned calves are considered important sources of zoonotic cryptosporidiosis transmission to humans. The previous literature documented that the prevalence of *Cryptosporidium* spp. between pre-weaned and post-weaned calves is age-related [79–81]. *Clostridium parvum*, the only prevalent zoonotic species in cattle, caused 85% of the *Cryptosporidium* infections in pre-weaned calves, while only 1% of the *Cryptosporidium* infections in post-weaned calves was due to this species [81]. In addition, a lower prevalence of cryptosporidiosis in 1–2-year-old dairy cattle (post-weaned calves and heifers) was found compared to younger (pre-weaned) calves [79,80]. Neonatal calves, which are not functional ruminants during the first 3–4 weeks of life, that are infected by *C. parvum* can suffer from serious scours (i.e., diarrhea with yellow pasty to watery feces) which can last up to 2 weeks and cause serious dehydration [72,82]. Infected calves can shed large numbers of infective oocysts in their feces, leading to environmental contamination and posing a threat to susceptible calves as well as humans [72,83]. Economic losses due to *Cryptosporidium* infections in neonatal calves are mostly associated with the cost of managing diarrheic animals, as well as mortalities [72,75]. Dehydration, weight loss, retarded growth performance, decreased feed efficiency, and losses due to mortality and morbidity are other repercussions of cryptosporidiosis, all of which leads to considerable economic losses [72,75].

3.5. Brucella spp.

Brucellosis, caused by *Brucella* spp., is a significant but neglected widespread bacterial zoonotic disease present around the world with debilitating impacts on human and animal health [84–87]. Humans are commonly infected through consuming adulterated unpasteurized/raw milk or dairy products [88–91]. However, direct contact with infected animals or their contaminated biological secretions (e.g., fetal or vaginal fluids and aborted fetuses or placentae), and exposure to anti-*Brucella* vaccines are other transmission routes of this occupational disease among animal handlers, veterinarians, and laboratory and abattoir personnel [90,92]. The inhalation of airborne agents was also reported as another transmission route of brucellosis in humans [87]. The use of personal protective equipment (PPE) to reduce the risk of brucellosis transmission is an effective measure among occupations that directly handle animals or their products [91]. Approximately 500,000 human brucellosis cases are reported each year to the World Health Organization (WHO), of which *Brucella melitensis* is the common causative agent [87,93]. The human brucellosis, also known as undulant fever or Malta fever, is characterized by non-specific clinical symptoms such as arthralgia, myalgia, sweats, miscarriage, abdominal pain, back pain, headache, profuse sweating, chills, and hepatomegaly [87,88,90]. Several countries in the world (located in the developed parts of Western and Northern Europe, Canada, Japan, Australia, and New Zealand) are free from the infectious agent [87,93]. Brucellosis is still endemic in Mexico, certain parts of Central and South America, the Mediterranean basin, the Middle East, India, and North Africa [89]. Nowadays, brucellosis in the United States is relatively rare (100–150 cases per year) and occurs more commonly in states that border Mexico (e.g., Texas and California) and in states where raw milk sale is legal [89–91,94]; a total of 75% of U.S. states allow different types of raw milk sales [89–91,94]. The incidence of human brucellosis in the United States has declined considerably over the years due to the successful U.S. State-Federal Cattle Brucellosis Eradication Program, as well as milk pasteurization [89,90].

Bovine brucellosis, caused by *B. abortus*, is a disease that occurs globally and causes substantial production loss along with a serious financial burden on producers [95]. The cattle farm environment is a convenient niche for brucellosis introduction, proliferation, and spread; improper biosecurity and management practices exacerbate the brucellosis progression in livestock animals [95]. The bacteria can live in soil, water, pasture, and manure for an extended time [96]. Therefore, the excretion of *Brucella* into the environment poses a risk to animal health [96]. In pregnant females, the primary symptom of brucellosis is abortion; however, the disease progression is often asymptomatic in young animals and non-pregnant females [97]. The bacterial agent can spread to multiple animals or herds through contaminated biological secretions such as fetal or vaginal fluids and aborted fetuses or placenta [92,97].

The smooth strain S19 and the rough strain RB51 vaccines are used in livestock for epidemiological control, yet both vaccines have disadvantages [90]. The RB51 strain, which has been used in the United States to vaccinate cattle against *B. abortus*, is virulent for humans (the infectious dose for *B. abortus* is 10–100 bacteria) and resistant to rifampin, a commonly used antibiotic used for treating human brucellosis [84,90,91]. Vaccinated animals can shed the strain into their milk; therefore, the presence and persistence of *Brucella* spp. in dairy products remain critical public health and food safety issues worldwide [90,91]. The contamination of the raw milk typically occurs either during milking or from the blood of infected animals being transferred to the milk [98]. Reportedly, animals infected with *B. abortus* can shed 10^3 CFU/mL from blood to raw milk, yet supper-shedder hosts can shed even more (10^4 CFU/mL) [97].

Brucella infections have been detected in varied terrestrial wild animals living in distinct environments (i.e., subtropical and temperate regions to arctic regions) [95]. The epidemiology of brucellosis in wildlife is often linked to the occurrence of the disease in livestock animals. Wild species can contribute to the re-introduction of *Brucella* agents along with infections in livestock (i.e., spillback) even in regions that are brucellosis-free or have had a successful eradication program [95]. Focusing on North America, bison, elk, and wild boars can become *Brucella* spp. reservoirs, and the latter two can spread the pathogenic agent to nearby cattle farms [95,98]. Brucellosis-impacted elk and bison populations from the Yellowstone Area in the United States have been shown to have a prevalence in the range of 35–60% [99].

3.6. Bacillus anthracis

Anthrax, known to humankind since ancient times, is a serious, naturally occurring, global zoonotic disease that affects domestic and wild animals, and directly/indirectly affects humans [100,101]. Anthrax is no longer considered a concern in developed countries due to effective control measures (e.g., vaccination, carcass disposal, and decontamination practices), yet it still occurs sporadically [101–103]. Anthrax is often found in agricultural regions of Central and South America, sub-Saharan Africa, central and southwestern Asia, southern and eastern Europe, and the Caribbean [101]. Over the years, there have been periodic outbreaks of anthrax in North America [102,103].

The causative agent of anthrax is *Bacillus anthracis*, an aerobic, Gram-positive, spore-forming, rod-shaped bacterium belonging to the *Bacillaceae* family [104,105]. In addition to causing naturally occurring anthrax, *B. anthracis* has been known to be a bioterrorism/agroterrorism weapon; therefore, surveillance systems have sought early detection of the disease [18,103]. The (dormant) spores produced by *B. anthracis* can persist in varied environments (e.g., soil, water, and animal hosts) for an extended time and are resistant to chemical and physical treatments such as radiation, desiccation, and heat application [104–106]. The spores enter the human body through varied routes and turn into active growing cells once the conditions are favorable, yet anthrax is not contagious [104,105,107,108]. The inhalation of spores from the hide or wool of infected animals, the ingestion of undercooked contaminated meat, skin abrasion, and, rarely, insect vectors (e.g., biting flies) are the main routes [104,105,107,108]. Reportedly, as few as 10 spores for herbivores and 200 to 55,000 spores for humans can be sufficient to cause an infection [109,110].

Anthrax in humans caused by the cutaneous transmission route accounts for approximately 95% of cases worldwide, due to the handling of carcasses and *B. anthracis*-contaminated by-products (e.g., hair, hides, and wool) of animals that were sick or died from the disease [18,105,107,108,111]. Animals often contract the disease through an oral ingestion of soil that is contaminated with spores [107,112]. It was reported that *B. anthracis* spores can survive in a soil environment for 300 years [107,112]. The most common clinical sign is a few sudden deaths in the herd without premonitory signs; bloating and hemorrhage from natural orifices (e.g., the nostrils, mouth, vulva, and anus) can be seen in dead animals [104,105,107,108].

In the United States, it was reported that *B. anthracis* spores can persist in alkaline soils present in the geographical corridor from Texas through Colorado, North and South Dakota to Montana, posing a primary risk for cattle and other herbivores [113–115]. In particular, a total of 63 anthrax cases in animals were confirmed in reference laboratories in Texas, the United States, during 2000–2018, and the last naturally occurring human case of cutaneous anthrax due to livestock exposure in Texas was reported in 2001 [111]. Texas experienced an increase in the number of animal anthrax cases in 2019 and state agencies suggested that more than 1000 animal losses might be attributed to the outbreak [111]. In Canada, repeated outbreaks in the wild bison populations still lead to concerns in the Northwest Territories, Northern Alberta, Manitoba, and Saskatchewan [116]. In 2006, an outbreak occurred in Saskatchewan and resulted in the loss of 804 livestock [117].

3.7. Leptospira spp.

Leptospirosis, caused by the spirochetal bacteria of the genus *Leptospira*, is considered one of the most widespread but neglected bacterial zoonotic diseases, affecting over 1 million humans globally every year with approximately 60,000 cases resulting in death [118–120]. Leptospirosis can cause a range of symptoms in humans, ranging from a mild fever, headache, and myalgia to more severe symptoms such as jaundice, renal failure, and multi-organ failure (i.e., known as Weil's disease) that is primarily characterized by kidney and liver damage [118–120]. The disease is often misdiagnosed or even not recognized in humans as leptospirosis causes a myriad of symptoms that are also commonly displayed in many other diseases such as influenza and dengue fever, hampering the diagnosis accuracy of the disease in humans [118–120].

Leptospirosis is transmitted to humans by varied species of animals (e.g., cattle, sheep, pigs, horses, rodents, and dogs) through their infected urine as the bacteria can persist in the renal tubules of the host and are then excreted into (soil or water) environment through urination [121,122]. The bacteria can live in soil or water for an extended period of time, and humans can contract the disease through open wounds, conjunctiva, and mucous membranes when they are exposed to urine-contaminated soil or water [123,124]. Therefore, working in an abattoir or animal farms (i.e., occupational exposure) and swimming or wading in water bodies contaminated with urine (i.e., recreational exposure) are considered the main high-risk activities affecting the transmission course of leptospirosis in humans [118,119]. Approximately 100–150 human leptospirosis cases are reported every year in the United States, with Puerto Rico reporting the majority of the cases, followed by Hawaii [125]. In Mexico, during 2000–2010, there were over 1500 human leptospirosis cases reported (with

198 mortalities), and the majority of the cases were reported during the rainy season of the country [126].

Leptospirosis is a ubiquitous disease found in varied species of animals (e.g., cattle, sheep, pigs, horses, rodents, and dogs) and differs from human leptospirosis in terms of epidemiology, pathogenesis, clinical presentation, diagnosis, and control measures [122,127]. In particular, cattle are a common livestock reservoir and significantly impacted by varied *Leptospira* spp. that can cause abortion, neonatal illness, and reduced milk production in the hosts [122,127]. Bovine leptospirosis is commonly caused by three different serovars of *Leptospira*: *Leptospira borgpetersenii* serovar Hardjo (Hardjobovis), *Leptospira interrogans* serovar Hardjo (Hardjoprajitno), and *Leptospira interrogans* serovar Pomona [128–130]. Exposure to *Leptospira*-contaminated water sources, co-grazing with sheep, and the preference of natural service over artificial insemination are some of the major risk factors for leptospirosis disease in cattle [122,127]. Due to the colonization ability of *Leptospira* spp. in the renal tubes of cattle, bacterial shedding through urination into the environment can continue for an extended period of time and can also occur through semen and uterine discharges [128,131]. Vaccination strategies are used to prevent the shedding of leptospires in cattle urine [132,133]. According to a report by the USDA, based on a study conducted by National Animal Health Monitoring System (NAHMS), approximately one in five feedlots use vaccination to provide protection against leptospirosis in cattle [134].

4. Conclusions

Overall, there are many challenges that face producers of beef cattle in North America, including zoonotic pathogens that threaten both human and animal health. Zoonotic diseases are often transferred amongst pen-mates, but they can also originate from the environment and be spread by vectors (e.g., wild birds and insects) or fomites (e.g., animal contacting surfaces and airborne dust). Zoonotic pathogens such as *Salmonella*, Shiga toxin-producing *Escherichia coli*, and *Campylobacter* are commonly harbored in the GIT of cattle and are all too often associated with animal-derived foods as they can be transferred to humans through contact with infected cattle or carcasses, the consumption of contaminated or infected meat, and the consumption of water that is contaminated with cattle manure. The challenges posed by the presence of these pathogens as undetected passengers in the GIT of cattle are extensive and must be addressed in a holistic fashion. Furthermore, neglected but significant zoonotic agents such as *Cryptosporidium*, *Brucella*, *Bacillus anthracis*, and *Leptospira* still cause debilitating diseases in North American human populations that come in direct or indirect contact with cattle, cattle-surrounding environments, or cattle-originated biological materials, although relatively rarely compared to other parts of the world.

The beef cattle industry of North America has implemented numerous post-harvest pathogen reduction strategies, and has recently focused on on-farm or pre-harvest pathogen reduction strategies to improve human and animal health. It must be emphasized that these strategies must include non-antibiotic activities to avoid the development of antimicrobial/antibiotic resistance and improve the production efficiency or sustainability in order to ensure adoption by the industry. In addition, vaccination strategies have been used to provide protection against zoonotic diseases for several decades by the North American beef cattle industry.

Author Contributions: Conceptualization, O.Y.K., I.B. and T.R.C.; writing—original draft preparation, O.Y.K. and I.B.; writing—review and editing, N.C. and T.R.C.; supervision, T.R.C. All authors have read and agreed to the published version of the manuscript.

Funding: This research was funded by University of Georgia Foundation (UGA 20-400).

Data Availability Statement: Not applicable.

Conflicts of Interest: The authors declare no conflict of interest.

References

1. National Chicken Council. Per Capita Consumption of Poultry and Livestock, 1965 to Forecast 2022, in Pounds. Available online: https://www.nationalchickencouncil.org/about-the-industry/statistics/per-capita-consumption-of-poultry-and-livestock-1965-to-estimated-2012-in-pounds/ (accessed on 28 November 2022).
2. Shahbandeh, M. Consumption of Beef Per Capita in Canada from 1980 to 2023, by Type (in Pounds). Available online: https://www.statista.com/statistics/735166/consumption-of-milk-per-capita-canada/ (accessed on 28 November 2022).
3. Estévez-Moreno, L.X.; Miranda-de la Lama, G.C. Meat consumption and consumer attitudes in México: Can persistence lead to change? *Meat Sci.* **2022**, *193*, 108943. [CrossRef] [PubMed]
4. Moyer, R.; Solano, P. 9. Commission for Environmental Cooperation (CEC). *Yearb. Int. Environ. Law* **2015**, *26*, 592–601. [CrossRef]
5. Trenda, E. Number of Cattle in Mexico from 2010 to 2021 (in Million Heads). Available online: https://www.statista.com/statistics/992638/catttle-number-heads-mexico/ (accessed on 28 November 2022).
6. Wagner, J.J.; Archibeque, S.L.; Feuz, D.M. The modern feedlot for finishing cattle. *Annu. Rev. Anim. Biosci.* **2014**, *2*, 535–554. [CrossRef] [PubMed]
7. United States Department of Agriculture (USDA); Economic Research Service (ERS). Topics, Animal Products, Cattle & Beef. Sector at a Glance. Available online: https://www.ers.usda.gov/topics/animal-products/cattle-beef/sector-at-a-glance/ (accessed on 28 November 2022).
8. Hales, K.; Lourenco, J.; Seidel, D.S.; Koyun, O.Y.; Davis, D.; Welch, C.; Wells, J.E.; Callaway, T.R. The use of feedlot/cereal grains in improving feed efficiency and reducing by-products such as methane in ruminants. In *Improving Rumen Function*; McSweeney, C., Mackie, R., Eds.; Burleigh Dodds Science Publishing Limited: Cambridge, UK, 2020; pp. 693–726. [CrossRef]
9. Lamb, G.C.; Maddock, T.; Feed Efficiency in Cows. Florida Beef Cattle Short Course. 2009; pp. 35–42. Available online: https://animal.ifas.ufl.edu/beef_extension/bcsc/2009/pdf/lamb.pdf (accessed on 28 November 2022).
10. Nagaraja, T.G.; Lechtenberg, K.F. Acidosis in feedlot cattle. *Vet. Clin. North Am. Food Ani. Prac.* **2007**, *23*, 333–350. [CrossRef]
11. Hungate, R.E. *The Rumen and Its Microbes*; Academic Press: New York, NY, USA, 1966. [CrossRef]
12. Tokach, R.J.; Ribeiro, F.R.; Chung, K.Y.; Rounds, W.; Johnson, B.J. Chromium Propionate Enhances Adipogenic Differentiation of Bovine Intramuscular Adipocytes. *Front. Vet. Sci.* **2015**, *2*, 26. [CrossRef]
13. Pogue, S.J.; Kröbel, R.; Janzen, H.H.; Beauchemin, K.A.; Legesse, G.; de Souza, D.M.; Iravani, M.; Selin, C.; Byrne, J.; McAllister, T.A. Beef production and ecosystem services in Canada's prairie provinces: A review. *Agric. Syst.* **2018**, *166*, 152–172. [CrossRef]
14. Rukambile, E.; Sintchenko, V.; Muscatello, G.; Kock, R.; Alders, R. Infection, colonization and shedding of *Campylobacter* and *Salmonella* in animals and their contribution to human disease: A review. *Zoonoses Public Health* **2019**, *66*, 562–578. [CrossRef]
15. Rahman, M.T.; Sobur, M.A.; Islam, M.S.; Ievy, S.; Hossain, M.J.; El Zowalaty, M.E.; Rahman, A.T.; Ashour, H.M. Zoonotic diseases: Etiology, impact, and control. *Microorganisms* **2020**, *8*, 1405. [CrossRef]
16. Jones, B.A.; Grace, D.; Kock, R.; Alonso, S.; Rushton, J.; Said, M.Y.; McKeever, D.; Mutua, F.; Young, J.; McDermott, J.; et al. Zoonosis emergence linked to agricultural intensification and environmental change. *Proc. Natl. Acad. Sci. USA* **2013**, *110*, 8399–8404. [CrossRef]
17. Callaway, T.R.; Edrington, T.; Anderson, R.C.; Byrd, J.A.; Nisbet, D.J. Gastrointestinal microbial ecology and the safety of our food supply as related to *Salmonella*. *J. Anim. Sci.* **2008**, *86*, E163–E172. [CrossRef]
18. McDaniel, C.J.; Cardwell, D.M.; Moeller, R.B.; Gray, G.C. Humans and cattle: A review of bovine zoonoses. *Vector-Borne Zoonotic Dis.* **2014**, *14*, 1–19. [CrossRef] [PubMed]
19. Dewey-Mattia, D.; Manikonda, K.; Hall, A.J.; Wise, M.E.; Crowe, S.J. Surveillance for foodborne disease outbreaks—United States, 2009–2015. *MMWR Surveill Summ.* **2018**, *67*, 1–11. [CrossRef] [PubMed]
20. McBride, W.D.; Kenneth, M., Jr. The Diverse Structure and Organization of U.S. Beef Cow-Calf Farms. United States Department of Agriculture (USDA), Economic Research Service (ERS). 2011. Available online: https://www.ers.usda.gov/publications/pub-details/?pubid=44532 (accessed on 28 November 2022).
21. Feuz, D.M.; Umberger, W.J. Beef cow-calf production. *Vet. Clin. N. Am.* **2003**, *19*, 339–363. [CrossRef] [PubMed]
22. United States Department of Agriculture (USDA), National Agricultural Statistics Service (NASS), Census of Agriculture. 2017. Available online: https://www.nass.usda.gov/AgCensus/ (accessed on 28 November 2022).
23. Ferrari, R.G.; Rosario, D.K.; Cunha-Neto, A.; Mano, S.B.; Figueiredo, E.E.; Conte-Junior, C.A. Worldwide epidemiology of *Salmonella* serovars in animal-based foods: A meta-analysis. *Appl. Environ. Microbiol.* **2019**, *85*, e00591-19. [CrossRef]
24. Ruby, T.; McLaughlin, L.; Gopinath, S.; Monack, D. *Salmonella*'s long-term relationship with its host. *FEMS Microbiol. Rev.* **2012**, *36*, 600–615. [CrossRef]
25. Boore, A.L.; Hoekstra, R.M.; Iwamoto, M.; Fields, P.I.; Bishop, R.D.; Swerdlow, D.L. *Salmonella enterica* infections in the United States and assessment of coefficients of variation: A novel approach to identify epidemiologic characteristics of individual serotypes, 1996–2011. *PLoS ONE* **2015**, *10*, e0145416. [CrossRef]
26. Boyle, E.C.; Bishop, J.L.; Grassl, G.A.; Finlay, B.B. *Salmonella*: From pathogenesis to therapeutics. *J. Bacteriol.* **2007**, *189*, 1489–1495. [CrossRef]
27. Issenhuth-Jeanjean, S.; Roggentin, P.; Mikoleit, M.; Guibourdenche, M.; De Pinna, E.; Nair, S.; Fields, P.I.; Weill, F.-X. Supplement 2008–2010 (no. 48) to the white–Kauffmann–Le minor scheme. *Res. Microbiol.* **2014**, *165*, 526–530. [CrossRef]
28. Gutema, F.D.; Agga, G.E.; Abdi, R.D.; De Zutter, L.; Duchateau, L.; Gabriël, S. Prevalence and serotype diversity of *Salmonella* in apparently healthy cattle: Systematic review and meta-analysis of published studies, 2000–2017. *Front. Vet. Sci.* **2019**, *6*, 102. [CrossRef]

29. Laufer, A.S.; Grass, J.; Holt, K.; Whichard, J.M.; Griffin, P.M.; Gould, L.H. Outbreaks of *Salmonella* infections attributed to beef–United States, 1973–2011. *Epidemiol. Infect.* **2015**, *143*, 2003–2013. [CrossRef]

30. Conrad, C.C.; Stanford, K.; Narvaez-Bravo, C.; Callaway, T.R.; McAllister, T. Farm fairs and petting zoos: A review of animal contact as a source of zoonotic enteric disease. *Foodborne Pathog. Dis.* **2017**, *14*, 59–73. [CrossRef] [PubMed]

31. Scallan, E.; Hoekstra, R.M.; Angulo, F.J.; Tauxe, R.V.; Widdowson, M.A.; Roy, S.L.; Jones, J.L.; Griffin, P.M. Foodborne illness acquired in the United States—Major pathogens. *Emerg. Infect. Dis.* **2011**, *17*, 7–15. [CrossRef] [PubMed]

32. Wottlin, L.R.; Edrington, T.S.; Anderson, R.C. *Salmonella* Carriage in Peripheral Lymph Nodes and Feces of Cattle at Slaughter Is Affected by Cattle Type, Region, and Season. *Front. Anim. Sci.* **2022**, *3*, 859800. [CrossRef]

33. Nickelson, K.J.; Taylor, T.M.; Griffin, D.B.; Savell, J.W.; Gehring, K.B.; Arnold, A.N. Assessment of *Salmonella* prevalence in lymph nodes of US and Mexican cattle presented for slaughter in Texas. *J. Food Prot.* **2019**, *82*, 310–315. [CrossRef] [PubMed]

34. Semenov, A.M.; Kuprianov, A.A.; van Bruggen, A.H. Transfer of enteric pathogens to successive habitats as part of microbial cycles. *Microb. Ecol.* **2010**, *60*, 239–249. [CrossRef] [PubMed]

35. Hoelzer, K.; Moreno Switt, A.I.; Wiedmann, M. Animal contact as a source of human non-typhoidal salmonellosis. *Vet. Res.* **2011**, *42*, 34. [CrossRef]

36. Gopinath, S.; Carden, S.; Monack, D. Shedding light on *Salmonella* carriers. *Trends Microbiol.* **2012**, *20*, 320–327. [CrossRef]

37. Rostagno, M.H. Can stress in farm animals increase food safety risk? *Foodborne Pathog. Dis.* **2009**, *6*, 767–776. [CrossRef]

38. Edrington, T.S.; Callaway, T.R.; Ives, S.E.; Engler, M.J.; Looper, M.L.; Anderson, R.C.; Nisbet, D.J. Seasonal shedding of *Escherichia coli* O157: H7 in ruminants: A new hypothesis. *Foodborne Pathog. Dis.* **2006**, *3*, 413–421. [CrossRef]

39. Naumova, E.N.; Jagai, J.S.; Matyas, B.; DeMaria, A.; MacNeill, I.; Griffiths, J. Seasonality in six enterically transmitted diseases and ambient temperature. *Epidemiol. Infect.* **2007**, *135*, 281–292. [CrossRef]

40. Mollenkopf, D.F.; Mathys, D.A.; Dargatz, D.A.; Erdman, M.M.; Habing, G.G.; Daniels, J.B.; Wittum, T.E. Genotypic and epidemiologic characterization of extended-spectrum cephalosporin resistant *Salmonella enterica* from US beef feedlots. *Prev. Vet. Med.* **2017**, *146*, 143–149. [CrossRef] [PubMed]

41. Karmali, M.A.; Gannon, V.; Sargeant, J.M. Verocytotoxin-producing *Escherichia coli* (VTEC). *Vet. Microbiol.* **2010**, *140*, 360–370. [CrossRef] [PubMed]

42. Sapountzis, P.; Segura, A.; Desvaux, M.; Forano, E. An overview of the elusive passenger in the gastrointestinal tract of cattle: The Shiga toxin producing *Escherichia coli*. *Microorganisms* **2020**, *8*, 877. [CrossRef] [PubMed]

43. Caprioli, A.; Morabito, S.; Brugère, H.; Oswald, E. Enterohaemorrhagic *Escherichia coli*: Emerging issues on virulence and modes of transmission. *Vet. Res.* **2005**, *36*, 289–311. [CrossRef] [PubMed]

44. Karmali, M.A. Emerging public health challenges of Shiga toxin–producing *Escherichia coli* related to changes in the pathogen, the population, and the environment. *Clin. Infect. Dis.* **2017**, *64*, 371–376. [CrossRef]

45. Buck, P. The Role of Consumer Advocacy in Strengthening Food Safety Policy. In *Food Safety Economics: Incentives for a Safer Food Supply*; Roberts, T., Ed.; Springer: Cham, Switzerland, 2018; pp. 323–328. [CrossRef]

46. Berry, E.D.; Wells, J.E.; Bono, J.L.; Woodbury, B.L.; Kalchayanand, N.; Norman, K.N.; Suslow, T.V.; López-Velasco, G.; Millner, P.D. Effect of proximity to a cattle feedlot on *Escherichia coli* O157: H7 contamination of leafy greens and evaluation of the potential for airborne transmission. *Appl. Environ. Microbiol.* **2015**, *81*, 1101–1110. [CrossRef]

47. Ferens, W.A.; Hovde, C.J. *Escherichia coli* O157: H7: Animal reservoir and sources of human infection. *Foodborne Pathog. Dis.* **2011**, *8*, 465–487. [CrossRef]

48. Smith, D.R. Cattle production systems: Ecology of existing and emerging *Escherichia coli* types related to foodborne illness. *Annu. Rev. Anim. Biosci.* **2014**, *2*, 445–468. [CrossRef]

49. Majowicz, S.E.; Scallan, E.; Jones-Bitton, A.; Sargeant, J.M.; Stapleton, J.; Angulo, F.J.; Yeung, D.H.; Kirk, M.D. Global incidence of human Shiga toxin–producing *Escherichia coli* infections and deaths: A systematic review and knowledge synthesis. *Foodborne Pathog. Dis.* **2014**, *11*, 447–455. [CrossRef]

50. Lisboa, L.F.; Szelewicki, J.; Lin, A.; Latonas, S.; Li, V.; Zhi, S.; Parsons, B.D.; Berenger, B.; Fathima, S.; Chui, L. Epidemiology of Shiga toxin-producing *Escherichia coli* O157 in the province of Alberta, Canada, 2009–2016. *Toxins* **2019**, *11*, 613. [CrossRef]

51. Kudva, I.T.; Dean-Nystrom, E.A. Bovine recto-anal junction squamous epithelial (RSE) cell adhesion assay for studying *Escherichia coli* O157 adherence. *J. Appl. Microbiol.* **2011**, *111*, 1283–1294. [CrossRef]

52. Naylor, S.W.; Low, J.C.; Besser, T.E.; Mahajan, A.; Gunn, G.J.; Pearce, M.C.; McKendrick, I.J.; Smith, D.G.; Gally, D.L. Lymphoid follicle-dense mucosa at the terminal rectum is the principal site of colonization of enterohemorrhagic *Escherichia coli* O157: H7 in the bovine host. *Infect. Immun.* **2003**, *71*, 1505–1512. [CrossRef] [PubMed]

53. Pruimboom-Brees, I.M.; Morgan, T.W.; Ackermann, M.R.; Nystrom, E.D.; Samuel, J.E.; Cornick, N.A.; Moon, H.W. Cattle lack vascular receptors for *Escherichia coli* O157: H7 Shiga toxins. *Proc. Natl. Acad. Sci. USA* **2000**, *97*, 10325–10329. [CrossRef] [PubMed]

54. Arthur, T.M.; Bosilevac, J.M.; Brichta-Harhay, D.M.; Guerini, M.N.; Kalchayanand, N.; Shackelford, S.D.; Wheeler, T.L.; Koohmaraie, M. Transportation and lairage environment effects on prevalence, numbers, and diversity of *Escherichia coli* O157: H7 on hides and carcasses of beef cattle at processing. *J. Food Prot.* **2007**, *70*, 280–286. [CrossRef] [PubMed]

55. Diez-Gonzalez, F.; Callaway, T.R.; Kizoulis, M.G.; Russell, J.B. Grain feeding and the dissemination of acid-resistant *Escherichia coli* from cattle. *Science* **1998**, *281*, 1666–1668. [CrossRef]

56. Dewell, G.; Simpson, C.; Dewell, R.; Hyatt, D.; Belk, K.; Scanga, J.; Morley, P.; Grandin, T.; Smith, G.; Dargatz, D. Impact of transportation and lairage on hide contamination with *Escherichia coli* O157 in finished beef cattle. *J. Food Prot.* **2008**, *71*, 1114–1118. [CrossRef]

57. LeJeune, J.; Kauffman, M.B. E. coli O157 supershedders: Mathematical myth or meaningful monsters? In Proceedings of the 2006 VTEC Conference, Melbourne, Australia, 29 October–1 November 2006.

58. LeJeune, J.T.; Besser, T.E.; Hancock, D.D. Cattle water troughs as reservoirs of *Escherichia coli* O157. *Appl. Environ. Microbiol.* **2001**, *67*, 3053–3057. [CrossRef]

59. Ekong, P.S.; Sanderson, M.W.; Cernicchiaro, N. Prevalence and concentration of *Escherichia coli* O157 in different seasons and cattle types processed in North America: A systematic review and meta-analysis of published research. *Prev. Vet. Med.* **2015**, *121*, 74–85. [CrossRef]

60. Stanford, K.; Johnson, R.P.; Alexander, T.W.; McAllister, T.A.; Reuter, T. Influence of season and feedlot location on prevalence and virulence factors of seven serogroups of *Escherichia coli* in feces of western-Canadian slaughter cattle. *PLoS ONE* **2016**, *11*, e0159866. [CrossRef]

61. Besser, T.E.; Schmidt, C.E.; Shah, D.H.; Shringi, S. "Preharvest" Food Safety for *Escherichia coli* O157 and Other Pathogenic Shiga Toxin-Producing Strains. *Microbiol. Spectr.* **2015**, *2*, 419–436. [CrossRef]

62. Feng, P. Shiga toxin-producing *Escherichia coli* (STEC) in fresh produce—A food safety dilemma. *Microbiol. Spectr.* **2014**, *2*, 17. [CrossRef] [PubMed]

63. Hoff, C.; Higa, J.; Patel, K.; Gee, E.; Wellman, A.; Vidanes, J.; Holland, A.; Kozyreva, V.; Zhu, J.; Mattioli, M. Notes from the Field: An Outbreak of *Escherichia coli* O157: H7 Infections Linked to Romaine Lettuce Exposure—United States, 2019. *MMWR Morb. Mortal. Wkly. Rep.* **2021**, *70*, 689–690. [CrossRef] [PubMed]

64. Carlson, J.C.; Chandler, J.C.; Bisha, B.; LeJeune, J.T.; Wittum, T.E. Bird-livestock interactions associated with increased cattle fecal shedding of ciprofloxacin-resistant *Escherichia coli* within feedlots in the United States. *Sci. Rep.* **2020**, *10*, 1–8. [CrossRef]

65. Inglis, G.D.; Taboada, E.N.; Boras, V.F. Rates of fluoroquinolone resistance in domestically acquired *Campylobacter jejuni* are increasing in people living within a model study location in Canada. *Can. J. Microbiol.* **2021**, *67*, 37–52. [CrossRef]

66. Plishka, M.; Sargeant, J.M.; Greer, A.L.; Hookey, S.; Winder, C. The prevalence of *Campylobacter* in live cattle, Turkey, chicken, and swine in the United States and Canada: A systematic review and meta-analysis. *Foodborne Pathog. Dis.* **2021**, *18*, 230–242. [CrossRef]

67. Sher, A.A.; Ashraf, M.A.; Mustafa, B.E.; Raza, M.M. Epidemiological trends of foodborne *Campylobacter* outbreaks in the United States of America, 1998–2016. *Food Microbiol.* **2021**, *97*, 103751. [CrossRef]

68. Laughlin, M.E.; Chatham-Stephens, K.; Geissler, A.L. Campylobacteriosis. *CDC Yellow Book.* 2020. Available online: https://wwwnc.cdc.gov/travel/yellowbook/2020/travel-related-infectious-diseases/campylobacteriosis (accessed on 28 November 2022).

69. Bolton, D.J. *Campylobacter* virulence and survival factors. *Food Microbiol.* **2015**, *48*, 99–108. [CrossRef]

70. Tang, Y.; Meinersmann, R.J.; Sahin, O.; Wu, Z.; Dai, L.; Carlson, J.; Plumblee Lawrence, J.; Genzlinger, L.; LeJeune, J.T.; Zhang, Q. Wide but variable distribution of a hypervirulent *Campylobacter jejuni* clone in beef and dairy cattle in the United States. *Appl. Environ. Microbiol.* **2017**, *83*, e01417–e01425. [CrossRef]

71. Tang, Y.; Sahin, O.; Pavlovic, N.; LeJeune, J.; Carlson, J.; Wu, Z.; Dai, L.; Zhang, Q. Rising fluoroquinolone resistance in *Campylobacter* isolated from feedlot cattle in the United States. *Sci. Rep.* **2017**, *7*, 494. [CrossRef]

72. Thomson, S.; Hamilton, C.A.; Hope, J.C.; Katzer, F.; Mabbott, N.A.; Morrison, L.J.; Innes, E.A. Bovine cryptosporidiosis: Impact, host-parasite interaction and control strategies. *Vet. Res.* **2017**, *48*, 1–16. [CrossRef]

73. Feng, Y.; Ryan, U.M.; Xiao, L. Genetic diversity and population structure of *Cryptosporidium*. *Trends Parasitol.* **2018**, *34*, 997–1011. [CrossRef] [PubMed]

74. Checkley, W.; White Jr, A.C.; Jaganath, D.; Arrowood, M.J.; Chalmers, R.M.; Chen, X.-M.; Fayer, R.; Griffiths, J.K.; Guerrant, R.L.; Hedstrom, L.; et al. A review of the global burden, novel diagnostics, therapeutics, and vaccine targets for *Cryptosporidium*. *Lancet Infect. Dis.* **2015**, *15*, 85–94. [CrossRef] [PubMed]

75. de Graaf, D.C.; Vanopdenbosch, E.; Ortega-Mora, L.M.; Abbassi, H.; Peeters, J.E. A review of the importance of cryptosporidiosis in farm animals. *Int. J. Parasitol.* **1999**, *29*, 1269–1287. [CrossRef] [PubMed]

76. Di Genova, B.M.; Tonelli, R.R. Infection Strategies of Intestinal Parasite Pathogens and Host Cell Responses. *Front. Microbiol.* **2016**, *7*, 256. [CrossRef] [PubMed]

77. Efstratiou, A.; Ongerth, J.E.; Karanis, P. Waterborne transmission of protozoan parasites: Review of worldwide outbreaks-An update 2011–2016. *Water Res.* **2017**, *114*, 14–22. [CrossRef]

78. Ryan, U.; Hijjawi, N.; Xiao, L. Foodborne cryptosporidiosis. *Int. J. Parasitol.* **2018**, *48*, 1–12. [CrossRef]

79. Fayer, R.; Santín, M.; Trout, J.M.; Greiner, E. Prevalence of species and genotypes of *Cryptosporidium* found in 1–2-year-old dairy cattle in the eastern United States. *Vet. Parasitol.* **2006**, *135*, 105–112. [CrossRef]

80. Santín, M.; Trout, J.M.; Fayer, R. A longitudinal study of cryptosporidiosis in dairy cattle from birth to 2 years of age. *Vet. Parasitol.* **2008**, *155*, 15–23. [CrossRef]

81. Santín, M.; Trout, J.M.; Xiao, L.; Zhou, L.; Greiner, E.; Fayer, R. Prevalence and age-related variation of *Cryptosporidium* species and genotypes in dairy calves. *Vet. Parasitol.* **2004**, *122*, 103–117. [CrossRef]

82. Adkins, P.R. Cryptosporidiosis. *Vet. Clin. North. America. Food Anim. Pract.* **2022**, *38*, 121–131. [CrossRef]

83. Nydam, D.V.; Wade, S.E.; Schaaf, S.L.; Mohammed, H.O. Number of *Cryptosporidium parvum* oocysts or *Giardia* spp. cysts shed by dairy calves after natural infection. *Am. J. Vet. Res.* **2001**, *62*, 1612–1615. [CrossRef] [PubMed]

84. Corbel, M.J. *Brucellosis in Humans and Animals*; World Health Organization (WHO) Press: Geneva, Switzerland, 2006; pp. 4–10. Available online: https://apps.who.int/iris/handle/10665/43597 (accessed on 28 November 2022).

85. Sriranganathan, N.; Seleem, M.N.; Olsen, S.C.; Samartino, L.E.; Whatmore, A.M.; Bricker, B.; O'Callaghan, D.; Halling, S.M.; Crasta, O.R.; Wattam, A.R.; et al. Brucella. In *Genome Mapping and Genomics in Animal-Associated Microbes*; Nene, V., Kole, C., Eds.; Springer: Berlin/Heidelberg, Germany, 2009; pp. 1–64.

86. Dean, A.S.; Crump, L.; Greter, H.; Schelling, E.; Zinsstag, J. Global burden of human brucellosis: A systematic review of disease frequency. *PLoS Neglec. Trop. Dis.* **2012**, *6*, e1865. [CrossRef] [PubMed]

87. Lai, S.; Chen, Q.; Li, Z. Human Brucellosis: An Ongoing Global Health Challenge. *China CDC Wkly.* **2021**, *3*, 120–123. [CrossRef] [PubMed]

88. Baldi, P.C.; Giambartolomei, G.H. Pathogenesis and pathobiology of zoonotic brucellosis in humans. *Rev. Sci. Tech. Off. Int. Epiz.* **2013**, *32*, 117–125. [CrossRef] [PubMed]

89. Hassouneh, L.; Quadri, S.; Pichilingue-Reto, P.; Chaisavaneeyakorn, S.; Cutrell, J.B.; Wetzel, D.M.; Nijhawan, A.E. An outbreak of brucellosis: An adult and pediatric case series. *Open Forum. Infect. Dis.* **2019**, *6*, 384. [CrossRef]

90. Negrón, M.E.; Kharod, G.A.; Bower, W.A.; Walke, H. Notes from the field: Human *Brucella abortus* RB51 infections caused by consumption of unpasteurized domestic dairy products—United States, 2017–2019. *MMWR Morb. Mortal. Wkly. Rep.* **2019**, *68*, 185. [CrossRef]

91. Cossaboom, C.M.; Kharod, G.A.; Salzer, J.S.; Tiller, R.V.; Campbell, L.P.; Wu, K.; Negrón, M.E.; Ayala, N.; Evert, N.; Radowicz, J.; et al. Notes from the Field: *Brucella abortus* Vaccine Strain RB51 Infection and Exposures Associated with Raw Milk Consumption-Wise County, Texas, 2017. *MMWR Morb. Mortal. Wkly. Rep.* **2018**, *67*, 286. [CrossRef]

92. Pereira, C.R.; de Almeida, J.V.F.C.; de Oliveira, I.R.C.; de Oliveira, L.F.; Pereira, L.J.; Zangerónimo, M.G.; Lage, A.P.; Dorneles, E.M.S. Occupational exposure to *Brucella* spp.: A systematic review and meta-analysis. *PLoS Neglec. Trop. Dis.* **2020**, *14*, e0008164. [CrossRef]

93. Pappas, G.; Photini, P.; Nikolaos, A.; Leonidas, C.; Epameinondas, V.T. The new global map of human brucellosis. *The Lancet Infect. Dis.* **2006**, *6*, 91–99. [CrossRef]

94. Serpa, J.A.; Knights, S.; Farmakiotis, D.; Campbell, J. Brucellosis in Adults and Children: A 10-Year Case Series at Two Large Academic Hospitals in Houston, Texas. *South Med. J.* **2018**, *111*, 324–327. [CrossRef]

95. Rhyan, J.C.; Nol, P.; Quance, C.; Gertonson, A.; Belfrage, J.; Harris, L.; Straka, K.; Robbe-Austerman, S. Transmission of brucellosis from elk to cattle and bison, Greater Yellowstone Area, USA, 2002–2012. *Emerg. Infect. Dis.* **2013**, *19*, 1992–1995. [CrossRef] [PubMed]

96. Kaden, R.; Ferrari, S.; Jinnerot, T.; Lindberg, M.; Wahab, T.; Lavander, M. *Brucella abortus*: Determination of survival times and evaluation of methods for detection in several matrices. *BMC Infect. Dis.* **2018**, *18*, 259. [CrossRef] [PubMed]

97. Letesson, J.-J.; Barbier, T.; Zúñiga-Ripa, A.; Godfroid, J.; De Bolle, X.; Moriyón, I. *Brucella* Genital Tropism: What's on the Menu. *Front. Microbiol.* **2017**, *8*, 506. [CrossRef] [PubMed]

98. Dadar, M.; Shahali, Y.; Whatmore, A.M.; Whatmore, A.M. Human brucellosis caused by raw dairy products: A review on the occurrence, major risk factors and prevention. *Int. J. Food Microbiol.* **2019**, *292*, 39–47. [CrossRef]

99. Olsen, S.C.; Paola, B.; David, M.W.; Tonia, M. Biosafety concerns related to *Brucella* and its potential use as a bioweapon. *Appl. Biosaf.* **2018**, *23*, 77–90. [CrossRef]

100. Sternbach, G. The history of anthrax. *J. Emerg. Med.* **2003**, *24*, 463–467. [CrossRef]

101. World Health Organization (WHO). Anthrax in Humans and Animals. WHO. 2008. Available online: https://apps.who.int/iris/bitstream/handle/10665/97503/9789241547536_eng.pdf (accessed on 8 February 2023).

102. Ackerman, G.A.; Giroux, J. A history of biological disasters of animal origin in North America. *Rev. Sci. Tech. Off. Int. Epiz.* **2006**, *25*, 83–92. [CrossRef]

103. United States Department of Agriculture (USDA), Animal and Plant Health Inspection Service (APHIS), Veterinary Services. Differentiation of Naturally Occurring from Non-Naturally Occurring Epizootics of Anthrax in Livestock Populations. USDA, APHIS. 2007. Available online: https://www.aphis.usda.gov/animal_health/emergingissues/downloads/finalanthraxnaturalterror.pdf (accessed on 8 February 2023).

104. Fasanella, A.; Galante, D.; Garofolo, G.; Jones, M.H. Anthrax undervalued zoonosis. *Vet. Microbiol.* **2010**, *140*, 318–331. [CrossRef]

105. Bhunia, A.K. Bacillus cereus and *Bacillus anthracis*. In *Foodborne Microbial Pathogens*. Bhunia, A.K., Ed.; Food Science Text Series; Springer: New York, NY, USA, 2018; pp. 193–207. [CrossRef]

106. Driks, A. The *Bacillus anthracis* spore. *Mol. Asp. Med.* **2009**, *30*, 368–373. [CrossRef]

107. Shadomy, S.V.; Smith, T.L. Zoonosis update. Anthrax. *J. Am. Vet. Med. Assoc.* **2008**, *233*, 63–72. [CrossRef]

108. Salman, M.; Steneroden, K. Important Public Health Zoonoses Through Cattle. In *Zoonoses—Infections Affecting Humans and Animals*; Sing, A., Ed.; Springer: Berlin/Heidelberg, Germany, 2015; pp. 3–19. [CrossRef]

109. Smith, I.M. A brief review of anthrax in domestic animals. *Postgrad. Med. J.* **1973**, *49*, 571–572. [CrossRef] [PubMed]

110. Wallin, A.; Luksiene, Z.; Zagminas, K.; Surkiene, G. Public health and bioterrorism: Renewed threat of anthrax and smallpox. *Medicina* **2007**, *43*, 278–284. [CrossRef] [PubMed]

111. Sidwa, T.; Salzer, J.S. Control and Prevention of Anthrax, Texas, USA, 2019. *Emerg. Infect. Dis.* **2020**, *26*, 2815–2824. [CrossRef] [PubMed]
112. Nicholson, W.L.; Munakata, N.; Horneck, G.; Melosh, H.J.; Stelow, P. Resistance to *Bacillus* endospores to extreme terrestrial and extraterrestrial environments. *Microbiol. Mol. Biol. Rev.* **2000**, *64*, 548–572. [CrossRef] [PubMed]
113. Griffin, D.W.; Silvestri, E.E.; Bowling, C.Y.; Boe, T.; Smith, D.B.; Nichols, T.L. Anthrax and the geochemistry of soils in the contiguous United States. *Geosciences* **2014**, *4*, 114–127. [CrossRef]
114. Blackburn, J.K.; McNyset, K.M.; Curtis, A.; Hugh-Jones, M.E. Modeling the geographic distribution of *Bacillus anthracis*, the causative agent of anthrax disease, for the contiguous United States using predictive ecological [corrected] niche modeling. *Am. J. Trop. Med. Hyg.* **2007**, *77*, 1103–1110. [CrossRef]
115. Yang, A.; Mullins, J.C.; Van Ert, M.; Bowen, R.A.; Hadfield, T.L.; Blackburn, J.K. Predicting the geographic distribution of the *Bacillus anthracis* A1. a/Western North American sub-lineage for the continental United States: New outbreaks, new genotypes, and new climate data. *Am. J. Trop. Med. Hyg.* **2020**, *102*, 392. [CrossRef]
116. Xu, S.; Harvey, A.; Barbieri, R.; Reuter, T.; Stanford, K.; Amoako, K.K.; Selinger, L.B.; McAllister, T.A. Inactivation of *Bacillus anthracis* spores during laboratory-scale composting of feedlot cattle manure. *Front. Microbiol.* **2016**, *7*, 806. [CrossRef]
117. Himsworth, C.G.; Argue, C.K. Anthrax in Saskatchewan 2006: An outbreak overview. *Can. Vet. J.* **2008**, *49*, 235–237.
118. Haake, D.A.; Levett, P.N. Leptospirosis in humans. In *Leptospira and Leptospirosis*; Adler, B., Ed.; Springer: Berlin/Heidelberg, Germany, 2015; pp. 65–97. [CrossRef]
119. Costa, F.; Hagan, J.E.; Calcagno, J.; Kane, M.; Torgerson, P.; Martinez-Silveira, M.S.; Stein, C.; Abela-Ridder, B.; Ko, A.I. Global morbidity and mortality of leptospirosis: A systematic review. *PLoS Negl. Trop. Dis.* **2015**, *9*, e0003898. [CrossRef]
120. Bharti, A.R.; Nally, J.E.; Ricaldi, J.N.; Matthias, M.A.; Diaz, M.M.; Lovett, M.A.; Levett, P.N.; Gilman, R.H.; Willig, M.R.; Gotuzzo, E.; et al. Leptospirosis: A zoonotic disease of global importance. *Lancet Infect. Dis.* **2003**, *3*, 757–771. [CrossRef] [PubMed]
121. Nally, J.E.; Ahmed, A.A.; Putz, E.J.; Palmquist, D.E.; Goris, M.G. Comparison of Real-Time PCR, Bacteriologic Culture and Fluorescent Antibody Test for the Detection of *Leptospira borgpetersenii* in Urine of Naturally Infected Cattle. *Vet. Sci.* **2020**, *7*, 66. [CrossRef] [PubMed]
122. Ellis, W.A. Animal Leptospirosis. In *Leptospira and Leptospirosis*; Adler, B., Ed.; Springer: Berlin/Heidelberg, Germany, 2015; pp. 99–137. [CrossRef]
123. Casanovas-Massana, A.; Pedra, G.G.; Wunder Jr, E.A.; Diggle, P.J.; Begon, M.; Ko, A.I. Quantification of *Leptospira interrogans* survival in soil and water microcosms. *Appl. Environ. Microbiol.* **2018**, *84*, e00507-18. [CrossRef] [PubMed]
124. Ko, A.I.; Goarant, C.; Picardeau, M. *Leptospira*: The dawn of the molecular genetics era for an emerging zoonotic pathogen. *Nat. Rev. Microbiol.* **2009**, *7*, 736–747. [CrossRef] [PubMed]
125. Centers for Disease Control and Prevention (CDC). Leptospirosis. Fact Sheet for Clinicians. 2018. Available online: https://www.cdc.gov/leptospirosis/pdf/fs-leptospirosis-clinicians-eng-508.pdf (accessed on 9 February 2023).
126. Sánchez-Montes, S.; Espinosa-Martínez, D.V.; Ríos-Muñoz, C.A.; Berzunza-Cruz, M.; Becker, I. Leptospirosis in Mexico: Epidemiology and Potential Distribution of Human Cases. *PLoS ONE* **2015**, *10*, e0133720. [CrossRef]
127. Sykes, J.E.; Haake, D.A.; Gamage, C.D.; Mills, W.Z.; Nally, J.E. A global one health perspective on leptospirosis in humans and animals. *J. Am. Vet. Med. Assoc.* **2022**, *260*, 1589–1596. [CrossRef]
128. Putz, E.J.; Bayles, D.O.; Alt, D.P.; Nally, J.E. Complete genome sequence of four strains of *Leptospira borgpetersenii* serovar Hardjo isolated from cattle in the Central United States. *J. Genom.* **2022**, *10*, 45–48. [CrossRef]
129. Nally, J.E.; Hornsby, R.L.; Alt, D.P.; Bayles, D.; Wilson-Welder, J.H.; Palmquist, D.E.; Bauer, N.E. Isolation and characterization of pathogenic leptospires associated with cattle. *Vet. Microbiol.* **2018**, *218*, 25–30. [CrossRef]
130. Talpada, M.D.; Garvey, N.; Sprowls, R.; Eugster, A.K.; Vinetz, J.M. Prevalence of leptospiral infection in Texas cattle: Implications for transmission to humans. *Vector-Borne Zoonotic Dis.* **2003**, *3*, 141–147. [CrossRef]
131. Putz, E.J.; Nally, J.E. Investigating the immunological and biological equilibrium of reservoir hosts and pathogenic *Leptospira*: Balancing the solution to an acute problem? *Front. Microbiol.* **2020**, *11*, 2005. [CrossRef]
132. Sanhueza, J.M.; Wilson, P.R.; Benschop, J.; Collins-Emerson, J.M.; Heuer, C. Meta-analysis of the efficacy of *Leptospira* serovar Hardjo vaccines to prevent urinary shedding in cattle. *Prev. Vet. Med.* **2018**, *153*, 71–76. [CrossRef] [PubMed]
133. Rinehart, C.L.; Zimmerman, A.D.; Buterbaugh, R.E.; Jolie, R.A.; Chase, C.C. Efficacy of vaccination of cattle with the *Leptospira interrogans* serovar *hardjo* type *hardjoprajitno* component of a pentavalent *Leptospira* bacterin against experimental challenge with *Leptospira borgpetersenii* serovar *hardjo* type *hardjo-bovis*. *Am. J. Vet. Res.* **2012**, *73*, 735–740. [CrossRef] [PubMed]
134. United States Department of Agriculture. Feedlot 2011 "Part IV: Health and Health Management on U.S. Feedlots with a Capacity of 1000 or More Head" USDA–APHIS–VS–CEAH–NAHMS. Fort Collins, CO #638.0913. 2011. Available online: https://www.aphis.usda.gov/animal_health/nahms/feedlot/downloads/feedlot2011/Feed11_dr_PartIV_1.pdf (accessed on 11 February 2023).

Article

Symbiotic Husbandry of Chickens and Pigs Does Not Increase Pathogen Transmission Risk

Emma Kaeder [1,*], Samart Dorn-In [2], Manfred Gareis [1] and Karin Schwaiger [2]

[1] Chair of Food Safety and Analytics, Faculty of Veterinary Medicine, LMU Munich, Schoenleutnerstr. 8, 85764 Oberschleissheim, Germany
[2] Unit of Food Hygiene and Technology, Institute of Food Safety, Food Technology and Veterinary Public Health, University of Veterinary Medicine, Veterinaerplatz 1, 1210 Vienna, Austria
* Correspondence: emma.kaeder@ls.vetmed.uni-muenchen.de

Abstract: A symbiotic or mixed animal husbandry (e.g., pigs and chickens) is considered to have a positive effect for animal welfare and sustainable agriculture. On the other hand, a risk of infection and transmission of microorganisms, especially of zoonotic pathogens, between animal species may potentially occur and thus might increase the risk of foodborne illnesses for consumers. To prove these assumptions, two groups of animals and their environmental (soil) samples were investigated in this study. Animals were kept in a free-range system. In the first group, pigs and chickens were reared together (pasture 1), while the other group contained only pigs (pasture 2). During a one-year study, fecal swab samples of 240 pigs and 120 chickens, as well as 120 ground samples, were investigated for the presence of *Campylobacter* spp., *Salmonella* spp. and *E. coli*. Altogether, 438 *E. coli* and 201 *Campylobacter* spp. strains were isolated and identified by MALDI-TOF MS. *Salmonella* spp. was not isolated from any of the sample types. The prevalences of *Campylobacter coli* and *C. jejuni* in pigs were 26.7% and 3.3% in pasture 1 and 30.0% and 6.7% in pasture 2, while the prevalences of *C. coli* and *C. jejuni* in chickens from pasture 1 were 9.2% and 78.3%, respectively. No correlation between the rearing type (mixed vs. pigs alone) and the prevalence of *Campylobacter* spp. was observed. All swab samples were positive for *E. coli*, while the average prevalences in soil samples were 78.3% and 51.7% in pasture 1 and 2, respectively. Results of similarity analysis of the MALDI-TOF MS spectra (for *C. coli*, *C. jejuni* and *E. coli*) and FT-IR spectra (for *E. coli*) of the same bacterial species showed no recognizable correlations, no matter if strains were isolated from chickens, pig or soil samples or isolated at different sampling periods. The results of the study indicate that the symbiotic husbandry of pigs and chickens neither results in an increased risk of a transmission of *Campylobacter* spp. or *E. coli*, nor in a risk of bacterial alteration, as shown by MALDI-TOF MS and FT-IR spectra. In conclusion, the benefits of keeping pigs and chickens together are not diminished by the possible transmission of pathogens.

Keywords: *Campylobacter* spp.; *E. coli*; free-range rearing system; MALDI-TOF MS; FT-IR; animal welfare

Citation: Kaeder, E.; Dorn-In, S.; Gareis, M.; Schwaiger, K. Symbiotic Husbandry of Chickens and Pigs Does Not Increase Pathogen Transmission Risk. *Foods* **2022**, *11*, 3126. https://doi.org/10.3390/foods11193126

Academic Editors: Frans J.M. Smulders and Arun K. Bhunia

Received: 16 July 2022
Accepted: 30 September 2022
Published: 8 October 2022

Publisher's Note: MDPI stays neutral with regard to jurisdictional claims in published maps and institutional affiliations.

1. Introduction

In recent years, the meat industry has increasingly gained the interest of society. Partially triggered by scandals led by buzzwords such as zoonotic diseases (e.g., *Salmonella* spp., *Campylobacter* spp. and enterohemorrhagic *E. coli*), consumers are increasingly taking a critical look at primary production and the downstream stages. In addition to product quality and product safety, the social and ethical aspects of animal husbandry are a major concern [1–3].

In many ways, animal husbandry offers a high potential for improvement in animal welfare and sustainability, both ecologically and economically [4]. In many countries, a large part of conventional husbandry types is considered as unsustainable in the long run, such as that declared by the Federal Ministry of Food and Agriculture (Germany) [5]. This knowledge and a changed human-animal relationship have led to a critical rethinking [6]. Additionally, there is a broad support among the population demanding that animals are treated with care and respect and that they are given the opportunity to practice species-appropriate behavior [5].

Meat production takes up a large share in the food sector. This discrepancy between the demand for animal welfare and maximum economic value has led to an urgently needed review of animal welfare standards [7,8]. It is important to respond to this change in the society's perception by creating new opportunities in animal husbandry [9,10].

By keeping pigs and chickens together on the pasture, animal welfare-relevant symbiotic effects and the sustainability of animal husbandry systems can be optimally exploited. The benefits of keeping chickens and pigs together could include, for example, giving the chickens better access to earthworms and other food by having the pigs stir up the soil. For their part, the chickens could provide the pigs with protection from ectoparasites. Another benefit to the chickens could be that the pigs offer them protection from birds of prey such as the goshawk. However, at the same time, it raises the question as to whether this kind of animal husbandry leads to an increased exchange of pathogens and thus to a potentiation of the risk of disease transmission. Since *Campylobacter* spp., *Salmonella* spp. and *E. coli* are considered to be important pathogens in both pigs and chickens and are among the most common foodborne zoonoses in Europe [11], they were chosen as model microorganisms for the tracking investigations in this study.

Campylobacter spp. are gram-negative, microaerophilic bacteria. Campylobacteriosis caused by *Campylobacter* (*C.*) *jejuni* and *C. coli* is the most common bacterial diarrheal disease in humans [12]. They are considered as common zoonotic agents, with contaminated food being the main route of transmission, posing a high risk [13–15]. Although the two species mentioned above are not obligately host bound, *C. coli* are more frequently detected in pigs and *C. jejuni* in chickens [16,17].

After campylobacteriosis, the second most frequent, notifiable bacterial gastrointestinal disease in humans is salmonellosis [18]. Like *Campylobacter* spp., not all *Salmonella* serovars are obligately bound to the host. Nevertheless, there is a species-specific clustering of some serovars, e.g., *S.* Typhimurium in humans, pigs and chickens, *S.* Enteritidis in humans and chickens, *S.* Infantis and *S.* Gallinarum in chickens [19,20]. There are various possibilities for the transmission of *Salmonella* spp. within livestock. Depending on the serovar, it can be spread via latently infected animals, contaminated feed or other vectors, e.g., rodents, contaminated objects and birds [21,22]. The most common cause of human infection is the consumption of contaminated animal products [23].

The third investigated bacterial species in this study is *Escherichia coli*. They are gram-negative, facultatively pathogenic, flagellated rod-shaped bacteria that are commonly found in human and animal intestines [24,25]. Due to their ability to rapidly absorb and transfer genetic information, *E. coli* are considered as indicator and reservoir germs. Thus, they are particularly of interest for scientific studies dealing with epidemiological questions [26].

The aim of the study was to find out whether animal husbandry types (pigs and chickens vs. pigs alone) have an influence on the risk of shedding, and transmission of *Campylobacter* spp., *Salmonella* spp. and *E. coli*. Additionally, the isolated bacterial strains were investigated using MALDI TOF MS and FT-IR to see if the spectra are converging over time, which could indicate increased exchange between the animal species.

2. Materials and Methods

2.1. Study Design (Sampling)

2.1.1. Pre Sampling

A pre-sampling was performed to obtain the prevalence of investigated bacteria in animal and soil samples. Before starting the main experiment, rectal swabs were taken once from pigs ($n = 10$) and cloacal swabs were taken once from chickens ($n = 10$). At this point, the animals were each in their parent stocks and had no contact with each other. In addition, soil samples ($n = 10$) were taken once before the animals went out to pasture. The method of sample collection corresponds to the later applied study procedure (see sample collection, Section 2.2).

2.1.2. Forms of Husbandry

The animals were separated into two different groups, living on different pastures. Both pastures were not previously used for any agricultural purpose for the past ten years. For the study, pasture 1 was used for pigs (35) and poultry (about 250) as mixed husbandry and pasture 2 for pigs only (35; comparison group). Each pasture had an area of 2.5 ha. The distance between both pastures was two meters on each side separated by a double fence. Thus, direct contact between animals from both pastures can be ruled out. All investigated animals received feed from the same producer and the same source of water. Figure 1 shows the structure of each pasture. Pigs (3–5 months old) and chickens (4 weeks old) were obtained from the respective breeding stations of the same farm. They were kept in the pastures until reaching age of slaughtering, namely 12 months for pigs and 5 months for chickens. Then, new animals were continually introduced in the two pastures. Altogether, two pig and three chicken groups were introduced to the corresponding pastures. The whole study was localized in Upper Bavaria, Germany.

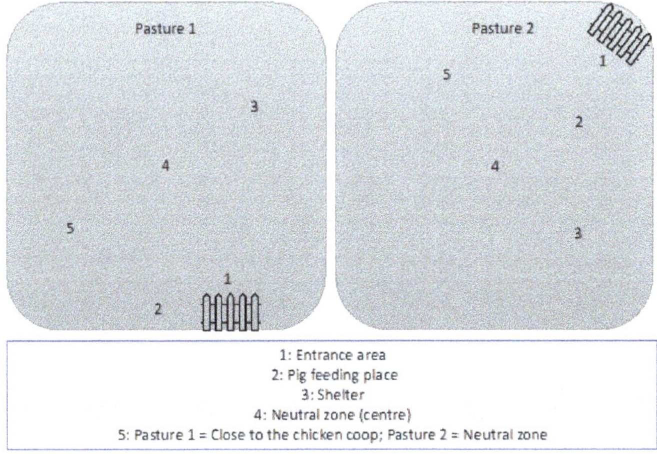

Figure 1. Schematic layout of the pastures and the soil sampling (1 to 5). Pasture 1: pigs and chickens; pasture 2: pigs alone.

2.2. Sample Collection

2.2.1. Animals

Rectal and cloacal swabs of 240 pigs and 120 chickens (from 12 monthly sampling runs with the exception of May and June 2020 due to the pandemic situation) were investigated between September 2019 and October 2020. For each sampling run, 10 rectal and 10 cloacal swabs were obtained from pigs and chickens from each pasture.

Two persons performed the swab sampling of animals. Sterile single-use swabs with Amies transport medium (Sarstedt, Germany) were inserted into the recta of pigs and the cloacae of chickens. The swabs were immediately put into the transport medium, individually labeled, packed in three different disposable bags (pasture 1—pig, pasture 1—chickens, pasture 2—pig), placed in a cooling box and transported to the laboratory within three hours. The animals were randomly selected. To assure that none of the animals was sampled twice, the pigs were marked using a marker pen immediately after the sampling was performed. As for chickens, the poultry coops were closed, and each chicken was released after the sampling procedure.

2.2.2. Soil

A total of 60 soil samples per pasture obtained from 12 sampling runs were investigated at the same time as the animal swab samples. The locations of five sampling sites from each pasture are shown in Figure 1. The soil sampling method was adopted from a procedure developed by the State Office for Nature, Environment and Consumer Protection in North Rhine-Westphalia, Germany [27]. The near-surface soil samples with a sampling depth of 2–4 cm were cut with a hole saw (Wolfcraft® GmbH, Kempenich, Germany), recorded with a diameter of 100 mm. For sampling, the metal cylinder was driven into the ground with a plastic hammer. After the excavation, the soil column in the cylinder (approximately 100 g) was transferred to a 200 mL sterile screw-type beaker (Sarstedt, Germany). Between the individual samples, the hole saw was freed from leftover soil with a knife and then disinfected with 70% alcohol. Samples were placed in a cooling box and transported to the laboratory within three hours.

2.3. Sample Preparation

The bacteriological analysis was started within 3 h after sample collection.

2.3.1. Animal Samples

The 20 rectal and 10 cloacal swabs from each sampling run were processed as individual samples under sterile conditions. In a first step, the swabs were streaked directly on a RAPID'*E. coli* agar (Bio-Rad, Feldkirchen, Germany). This agar is recommended for the enumeration of *E. coli* in water and food [28,29]. The protective cap of the swab was then removed using a sterile scissor, while the swab was put into a sterile disposable tube (Greiner Bio-One, Germany) that was previously filled with 5 mL of buffered peptone water. All 30 tubes containing swabs were closed and shaken for 25 min at 250 rounds/min (FL-3005 varioshake, GFL, Lauda, Lauda-Königshofen, Germany) at room temperature. The "peptone water sample suspension" (PSS) was used as the starting material for the subsequent culturing of *Campylobacter* spp. and *Salmonella* spp.

2.3.2. Soil Samples

Each of the ten screw cups (Sarstedt, Germany) containing soil samples was opened under a sterile laminar flow workbench. Soil was transferred into a sterile flask and weighed to 10 g, then mixed with 90 mL of peptone water by shaking at 250 rounds/min for 25 min at room temperature. This PSS of soil served as the starting material for the subsequent culturing of all target bacteria.

2.4. Bacteriological Investigation
2.4.1. Isolation of *Escherichia coli*

E. coli were isolated using the RAPID'*E. coli* 2 agar (Bio-Rad, Germany). While animal swabs were directly streaked on the selective agar, approximately 10 µg of the PSS of soil was transferred onto an agar plate and spread out using a sterile inoculation loop. The plates were aerobically incubated at 37 °C for 24 h. After that, one colony from each positive RAPID'*E. coli* 2 agar was subcultured on agar technical (Oxoid, Wesel, Germany) and incubated under the same conditions. The grown colonies proceeded to species identification/confirmation using MALDI-TOF MS (Bruker Daltoniks, Bremen, Germany) and to spectra analysis using FT-IR (Bruker Daltonik GmbH, Bremen, Germany).

2.4.2. Isolation of *Salmonella* spp.

A pre-enrichment procedure was applied in order to revive the potentially sublethally damaged cells of *Salmonella* spp. For this step, 1 mL of the PSS suspension was transferred to 5 mL of buffered peptone water (Thermo Scientific™, Waltham, MA, USA) and aerobically incubated at 37 °C for 16–20 h. From this pre-enrichment, 0.1 mL was dropped in triplicate onto the Modified Semisolid Rappaport-Vassiliadis (MSRV) medium (Oxoid, Germany) and incubated non-inverted at 42 °C for 24 h. Growth of *Salmonella* spp. on MRSV is indicated when a clear opaque halo has formed around the droplet. For further confirmation steps, material from the rim of the opaque halo was subcultured onto Xylose-Lysine-Tergitol 4 (XLT4) agar (Oxoid, Germany) and Brilliant-green Phenol-red Lactose Sucrose (BPLS) agar (Oxoid, Germany). The agar plates were aerobically incubated at 37 °C for 24 h.

2.4.3. Isolation of *Campylobacter* spp.

Enrichment of the thermophilic *Campylobacter* spp. was primarily performed, starting with transferring 1 mL of the PSS into 9 mL of a Preston selective broth (Carl Roth, Karlsruhe, Germany), followed by incubation under microaerobic conditions (5% O_2, 10% CO_2, Anaerocult™ C 2.5 l (Merck, Darmstadt, Germany)) at 42 °C for 48 h. The selective enrichment procedure was used in order to enhance the growth of *Campylobacter* spp. and at the same time to reduce or inhibit the growth of the accompanying microorganisms, which may be present in a high number in fecal swab and soil samples. After incubation, the suspension was filtered through a sterile membrane filter with a pore size of 0.65 μm (VWR, Hannover, Germany). Approximately 10 μL of the flow-through suspension was transferred to a Columbia blood agar containing sheep blood (CBA, Oxoid, Germany) with a disposable loop and was streaked using a 3-loop smear technique. The CBA plates were incubated under microaerobic conditions at 42 °C for 48 h. The grown colonies proceeded to species identification using MALDI-TOF MS.

2.4.4. Species Identification by MALDI-TOF MS

The colonies of bacterial cultures were identified to species level using Matrix Assisted Laser Desorption Ionization—Time of Flight Mass Spectrometry (MALDI-TOF MS). Colonies of pure cultures were extracted using the direct transfer method as described in the Bruker Daltonik User's manual [30]. An appropriate colony mass on the agar plate was taken using a toothpick and smeared on a ground steel BC target plate. Then, 1 μL of a low-molecular organic matrix solution (saturated solution of a cyano-4-hydroxycinnamic acid in 50% acetonitrile) was added. During the drying process at room temperature, a co-crystallization took place in which the analyte was incorporated into the matrix crystals. The MALDI-TOF MS measurements were performed using a Microflex LT (Bruker Daltoniks, Bremen, Germany). The analysis of the generated data was executed with the Software—Biotyper OC incl. Taxonomy (Version 3.1.66, Bruker Daltoniks, Bremen, Germany) and its automated settings.

2.4.5. FT-IR

The Fourier-transform infrared spectroscopy (FT-IR) measurement was applied to *E. coli* strains because of their role as indicator and reservoir bacteria as described in the introduction. Additionally, some studies related to the application of FT-IR have shown that the stability of the bacterial cell mass remains stable up to 24 h after subculturing [31]. For cell masses grown for shorter/longer periods or in other nutrient solutions, the FT-IR spectra sometimes differ considerably. Therefore, reproducible and meaningful information can only be expected from cell masses obtained under standardized conditions [32,33]. To ensure these standardized conditions during FT-IR measurement, all *E. coli* strains were cultured on the same medium, incubated at the same room temperature for exactly 24 h. The restrained growth of *Campylobacter* spp. did not allow this standardized measurement with the sample size, since the incubation time had to be extended if the growth rate was too slow or the colonies were too small.

For each *E. coli* strain, three biological replicates were prepared for FT-IR measurement. The material from each colony was removed from the agar technical after exactly 24 h of incubation using a 1-μL disposable loop. The amount was equivalent to an overloaded inoculation loop. It is important to note that the cell material was only removed from the confluent growth zone. The cell material was transferred to a 1.5-mL reaction tube that was prefilled with 50 μL ethanol (70%) and four inert metal cylinders (Bruker Daltonik GmbH, Bremen, Germany), then mixed by shaking at 250 rounds for 15 s. The 70% ethanol killed the microorganisms, thus stopping their ongoing metabolic activities. To increase the surface tension of the suspension, 50 µL of deionized water was added. Then, 15 µL of each isolate suspension was pipetted onto three spots (technical replicates) of the 96-well microtiter plate (Bruker Daltonik GmbH, Bremen, Germany). The spots on the plate had to be completely dried at 37 °C in an incubator (approximately 30 min) before they were subjected to FT-IR measurement.

Additionally, a quality control for each FT-IR measurement was required. This was carried out by pipetting 12 µL of each Bruker Infrared Test Standards Solutions (IRTS 1 and IRTS 2) on the same microtiter plate. These two standard solutions are part of the Bruker IR Biotyper kit (Bruker, Bremen, Germany). Finally, FT-IR spectroscopy was performed using an IR biotyper spectrometer (Bruker Daltonik GmbH, Bremen, Germany) according to the instructions of the producer [34]. Briefly, each *E. coli* strain was automatically scanned 64 times. Spectra were acquired up to 1500 cm^{-1} with a spectral resolution of 3 cm^{-1} and an aperture of 10 mm. All 64 spectra obtained from a single strain were automatically combined, resulting in a single spectrum. The analysis of the generated data was carried out using Biotyper software (Bruker Daltoniks, Bremen, Germany, version 1.5.0.90) and its automatic settings. The spectral data were automatically converted to dendrograms using the average mean spectra method that was further used for the statistical analysis (Chi-square test).

2.5. Analysis for Similarities

For each bacterial group, *Campylobacter* spp. and *E. coli*, the similarity of their protein spectra obtained by the MALDI-TOF MS, were analyzed using the clustering program BioNumerics (version 7.6, Applied Maths, Sint-Martens-Latem, Belgium).

Additionally, the similarity of the *E. coli* strains (isolated from animals, $n = 240$) was investigated by FT-IR spectroscopy. This involves comparing each spectrum within a species to all other spectra recorded using the same protocols and methods. The comparison of two spectra provides a spectral distance value. The more two spectra match, the smaller the spectral distance (Bruker Daltonik GmbH, 2017).

2.6. Statistical Analysis

2.6.1. Pearson´s Correlation

To evaluate the correlation between pasture types and the occurrence of the investigated bacteria in pigs/in soil samples, a Pearson correlation coefficient (r, Microsoft Excel, 2016) was computed. The strength of the correlation for absolute values of r is interpreted as follows; $r = 0$–0.19 is regarded as very weak, 0.20–0.39 as weak, 0.40–0.59 as moderate, 0.60–0.79 as strong and 0.80–1.0 as a very strong correlation (Evans, 1996). Additionally, the p-value was calculated based on a two-tailed t-test analysis in order to evaluate whether the correlation was statistically significant. In Microsoft Excel, the p-value was calculated using the formula = T.VERT.2S (t;df). The T.VERT.2S = two-tailed t-test, t = t-value and df = degree of freedom. The results were interpreted as statistically significant if the p-value was less than 0.05.

2.6.2. Chi-Square

The chi-square test (SPSS software, version 26.0) was used to examine the similarity of genotype identification of *E. coli* with FT-IR spectroscopy with respect to two research questions. First, whether the type of husbandry (mixed/symbiotic vs. control pasture) had a significant influence on the formation of the clusters and, second, whether the animal species had a corresponding influence. Pearson's chi-square test was calculated with calculation of a continuity correction. An asymptotic significance (two-sided), or p-value obtained by chi-square test less than 0.05 means that there is a statistically significant relationship between the factors and clusters. In addition, a likelihood-ratio test was performed. To exclude the possibility of inaccuracies in the chi-square due to small sample sizes, the frequencies to be observed were checked using Fisher's exact test and the linear correlation was also determined.

3. Results

The pre-sampling result showed that the prevalence of *Campylobacter* spp. was 10% in pigs, 20% in chickens, and 0% in soil samples. For *E. coli* it was 100% in all animal samples and 30% in soil samples.

In the main experiment, a total of 639 bacterial strains were isolated from 120 cloacal swabs from chickens, 240 rectal swabs from pigs, and 120 soil samples. These included 438 strains of *E. coli* and 201 strains of *Campylobacter* spp.

Salmonella spp. could not be isolated in any of the investigated samples.

3.1. Detection and Similarity Analysis of Campylobacter spp.

A total of 201 *Campylobacter* strains were isolated from 51.4% of all investigated animals and 12.5% of all soil samples. The prevalences of these bacteria were 87.5% in chickens and 33.3% (30.0% and 36.7% for pasture 1 and 2, respectively) in pigs. Species identification by MALDI-TOF MS revealed that 43.8% and 56.2% were *Campylobacter coli* and *C. jejuni*, respectively.

Figure 2 shows the distribution in detail and the prevalence of *Campylobacter* spp. in each animal group and in soil samples. The highest prevalence of *C. jejuni* was found in chickens (78.3%), while *C. coli* was mostly found in pigs (28.5% in total, and 27.0% and 30.0% of pigs from pasture 1 and 2, respectively). The prevalences of *C. jejuni* in pigs (3.3% and 6.7% for pasture 1 and 2, respectively) and *C. coli* in chickens (9.2%) were relatively low. The distribution of *C. coli* and *C. jejuni* in soil samples from pasture 1 was similar (12.0% and 10.0%, respectively), as well as in soil samples from pasture 2, where the prevalence was remarkably lower (3.0% and 2.0%, respectively) than pasture 1, but not statistically significant ($p > 0.05$).

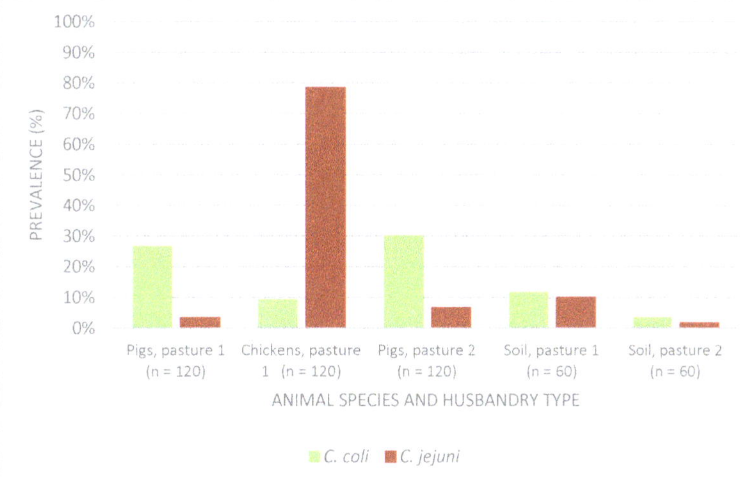

Figure 2. Prevalence of *Campylobacter* spp. in animal and soil samples from two husbandry types. Pasture 1: pigs and chickens were kept together (mixed husbandry). Pasture 2: pigs alone.

According to the Pearson correlation coefficient (*r* value), no correlation between husbandry types and detection of *C. coli* ($r = 0.03$, $p = 0.57$) as well as detection of *C. jejuni* in pigs ($r = 0.08$, $p = 0,24$) was found. For soil samples, a weak positive correlation was found between pasture type 1 and the contamination with *C. coli* and *C. jejuni* in soil ($r = 0.18$, $r = 0.16$, respectively). This means it was more likely to detect both *C. coli* and *C. jejuni* in ground samples from pasture type 1 than from pasture type 2. However, the correlation was evaluated as statistically not significant ($p = 0.05$, and $p = 0.08$, respectively).

Results of a similarity analysis of the protein spectra obtained by MALDI-TOF MS using the clustering program Bionumerics show that *Campylobacter* strains were classified into two major subgroups, *C. coli* and *C. jejuni*. The protein spectra of the same *Campylobacter* species were similar, regardless of their origin (chickens, pigs, or soil samples). Figure 3 shows the protein spectra of *C. coli* and *C. jejuni* isolated from chickens and pigs exemplarily. The peaks of the spectra within the same *Campylobacter* spp. (*C. coli*/*C. jejuni*) did not show any differences among isolates obtained from different samples (pigs/chicken/soil) and from different pastures. The differences of the peaks of MALDI-TOF spectra between *C. coli* and *C. jejuni* were indicated with arrows in Figure 2.

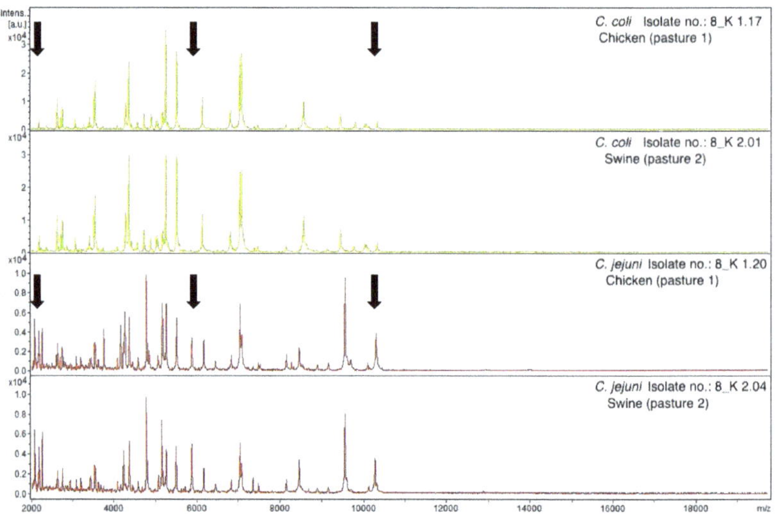

Figure 3. Examples of MALDI-TOF MS mass spectra of *C. coli* and *C. jejuni* isolated from chickens and pigs. Arrows indicate peaks that are absent or present in both species.

3.2. Detection and Similarity Analysis of Escherichia coli

As shown in Figure 4, 438 strains of *E. coli* were isolated from all animal swab samples, while in soil samples they were found in a wide range among sampling runs (between 0% and 100%) without recognizable influence of the duration of grazing. The average prevalence of *E. coli* in soil samples obtained from 12 sampling runs was 78.3% and 51.6% in pasture 1 and 2, respectively. The shedding of *E. coli* in ground samples was further analyzed using Pearson's correlation coefficient. A weak correlation was found between pasture types and the prevalence of *E. coli* ($r = 0.28$) in ground samples and shedding of *E. coli* on pasture 1 was evaluated as statistically significantly higher than on pasture 2 ($p = 0.002$).

Results obtained from similarity analysis (Bionumerics, Applied Maths) showed that the protein spectra of *E. coli* obtained by MALDI-TOF MS from all sample types have a high similarity (data not shown). The spectra were distributed randomly and were not grouped in sample types (pig/chicken swabs or soil samples) or husbandry types (pigs with chickens vs. pigs alone), but were rather grouped in sampling time (from September 2019 to October 2020). By comparing the spectra obtained from the same sampling run, it was observed that at the beginning of the study (sampling runs one to three) that there was a high diversity in the spectra of *E. coli*, resulting in a high number of clusters. Each cluster included isolates from both husbandry types and/or animal species. In the course of time (sampling runs 4–12), the number of clusters was reduced to one to three, since the spectra of the isolates became more similar, independent of whether they were isolated from chickens or pigs from pasture 1 or pasture 2. According to this analysis, a manifest transformation of a single *E. coli* isolate was not detected.

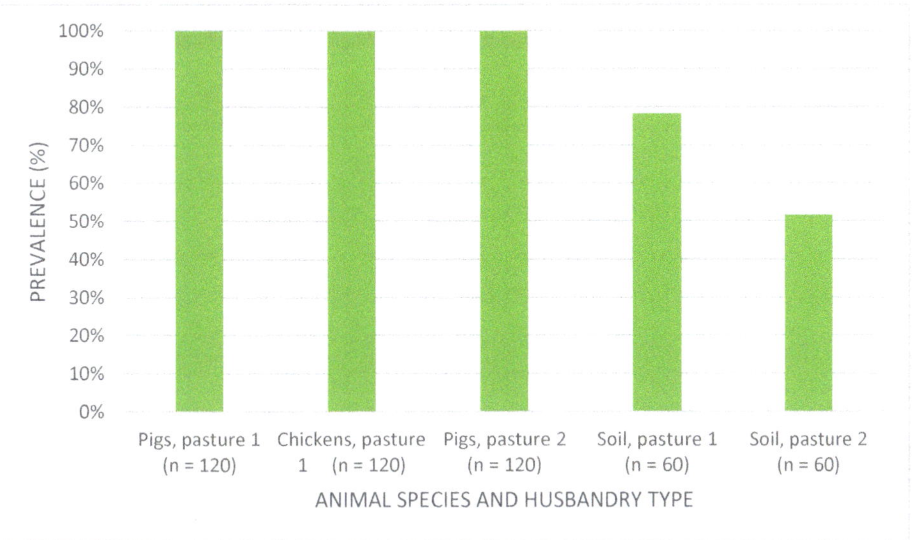

Figure 4. Prevalence of *Escherichia coli* in animal and soil samples from two husbandry types. Pasture 1: pigs and chickens were kept together (mixed husbandry). Pasture 2: pigs alone.

In addition, FT-IR spectroscopy was used to analyze whether the spectra of *E. coli* (isolated from animals, *n* = 240) converge over time or whether species-dependent differences persist. *E. coli* cultures that were used for FT-IR spectrometry always showed very uniform and brisk growth within the same cultivation period. Differences between the FT-IR spectra due to technical errors could be excluded by the three biological and three technical replicates or, if necessary, deviating spectra could be sorted out. The comparison of the three technical replicates and the three biological replicates showed that the spectra of one and the same biomass matched. After that, the dendrograms used for statistical analysis were generated as follows: for each sample run, one dendrogram contained the spectra of *E. coli* from the pigs kept in both husbandry types (pasture 1 and 2) and another dendrogram contained the spectra of *E. coli* from the chickens and pigs kept in pasture 1 (mixed husbandry).

Regarding the interpretation of the created dendrograms, the most important aspect was to find a reasonable cut-off value for distance to see which spectra belong to the same cluster. Since the cut-off value for differentiation at the strain level for bacteria varies slightly in each run, a stable cut-off value of 0.300 was set for differentiation. The cut-off value was set to be as low as possible to achieve a high discriminatory power, but also high enough for the technical replicates to not spread across multiple clusters. As a result, at least one major cluster occurred in all sampling runs, as shown in Figure 5.

The aim of the cluster evaluation was to find out whether the spectrum of the respective individual animal could be sorted into the corresponding cluster of its group. For this purpose, the largest cluster was determined and it was checked whether predominantly pig or chicken samples occurred in this cluster, and it thus was named the "pig cluster" or "chicken cluster". Subsequently, the number 1 or 0 was assigned for each individual animal sample. Number 1 meant that the animal sample could be sorted according to its cluster, while 0 meant that the animals were outside the assigned cluster.

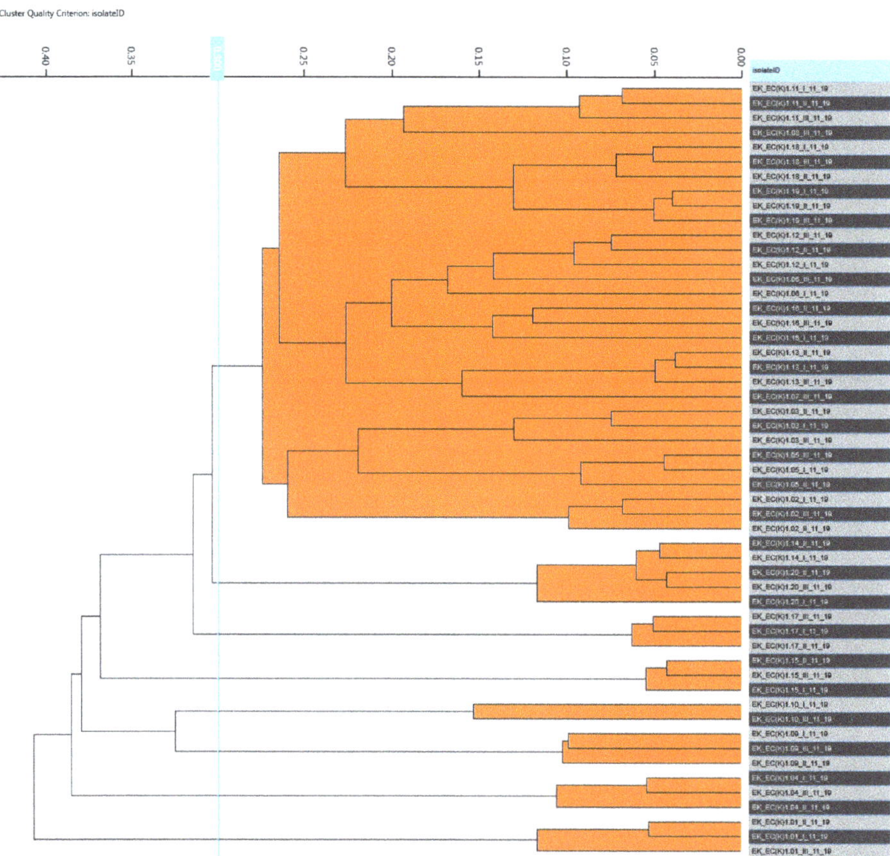

Figure 5. Example of a dendrogram of FT-IR spectra obtained from *E. coli* strains isolated from pigs and chickens kept in pasture 1 (mixed husbandry; third sampling run). The blue line indicates the stable cut-off value of 0.300. A main cluster is in the upper horizontal plane. The right side, highlighted in dark and light gray refers to *E. coli* strains coming from animals. The abbreviation, for example "EK_EC(K)_1.11_I_11_19" stands for: EK = name of author; EC(K) = *E. coli* (K = cloacal) 1.11 = Pasture 1, chicken no. 1 (no. 01–10 = pigs, no. 11–20 = chickens); I = first biological replicate; 11_19 = month November and year 2019.

The first statistical analysis aimed to find out whether the type of husbandry (mixed/ symbiotic vs. control pasture) had a significant effect on the formation of the clusters of *E. coli*. The statistical results revealed that no relationship between factors and clusters could be detected either within each sampling run or when comparing all 12 runs together (Pearson's chi-square test: asymptotic significance (two-sided) or p-value = 0.984, see Table 1). This means that the husbandry type had no influence on the cluster formation of *E. coli*.

The relationship between animal types (chicken/pig) and the formation of clusters was also statistically evaluated. The statistical result in Table 2 shows that no significant effect across all study time points was found (Pearson's chi-square test: asymptotic significance (two-sided) = 0.283). This indicates that the type of animal (chicken/pig) did not have any influence on the cluster formation of *E. coli* isolates.

Table 1. Chi-square test (FT-IR dendrograms). Influence of husbandry type on the cluster formation of *E. coli* isolated from pigs from pasture 1 (n = 120) and pasture 2 (n = 120).

Total	Value	Degree of Freedom	Asymptomatic Significance z (Two-Sided)	Exact Significance z (Two-Sided)	Exact Significance z (One-Sided)
Pearson's chi-square test	0.000	1	0.984		
Continuity correction	0.000	1	1.000		
Likelihood-ratio test	0.000	1	0.984		
Fisher's exact test				1.000	0.551
Linear correlation	0.000	1	0.984		
Number of valid cases	225				

Table 2. Chi-square test (FT-IR dendrograms): Influence of animal species on the cluster formation of *E. coli* isolated from pigs (n = 120) and chickens (n = 120) from pasture 1.

Total	Value	Degree of Freedom	Asymptomatic Significance z (Two-Sided)	Exact Significance z (Two-Sided)	Exact Significance z (One-Sided)
Pearson's chi-square test	1.153	1	0.283		
Continuity correction	0.868	1	0.351		
Likelihood-ratio test	1.154	1	0.283		
Fisher's exact test				0.321	0.176
Linear correlation	1.148	1	0.284		
Number of valid cases	231				

All isolates that did not pass the quality check during the FT-IR measurement were automatically sorted out so that the numbers of valid cases used for both statistical analyses were n = 225 (Table 1) and n = 231 (Table 2).

Furthermore, a multifactorial approach with the generalized linear model (GLM; distribution form of the dependent variable binomial) was applied to investigate the influence of animal species and husbandry type on the distribution of spectra. With the respective results, no statistically significant effects were found (animal species: p = 0.256, husbandry: p = 0.899).

4. Discussion

Topics related to animal welfare of livestock are increasingly discussed in society and have a high influence on consumer decisions regarding whether to buy meat and meat products. A symbiotic or mixed rearing system, in which, for example, two animal species are kept together in the same free ranging area, can significantly contribute to an increased animal welfare status [35]. Another major issue in the critical examination of agriculture is sustainability. Due to global issues such as the ever-growing global population, climate change and an increasing demand for animal protein, the need for more sustainable animal agriculture is more urgent than ever. The pressure to maximize the production of milk and meat has disturbed the equilibrium between feeding and yield, animal welfare, environmental impact and public acceptance [36,37]. More and more ways are being sought to make agriculture more sustainable in the long run and therefore more viable for the future [38]. If the food supply for the growing world population is to be secured in the long term, production systems and consumption patterns will have to change. The challenge is to increase yields on existing lands without leaching it out and losing its fertility [39]. Shared animal husbandry is an approach which is based on the same fundamental idea. By keeping two different species of animals together, only one pasture is needed instead of the usual two, thus increasing the capacity utilization of the space with positive effects on both sustainability and animal welfare. In addition, as observed as a side finding of this study, chickens always spread throughout the pasture and used all of the space for scratching and foraging. This may be a result of their positive feeling of being protected by the pigs from any of their foes such as birds of prey. On the contrary, many different studies have shown that even with a large free-range area, chickens stay very close to their coop out of fear [40,41], and only use the free-range area if they can find protection in the form of a shelter [42]. The findings of the present study clearly demonstrate the protective function of pigs in a mixed husbandry system.

However, the assumption that natural bacterial infection and disease transmission between animal species can increase when different animal species are kept together might impede the implementation of this rearing system for example due to veterinary authority reservations. Therefore, this study was conducted to prove whether the rearing system (pasture 1: chickens and pigs together; pasture 2: only pigs) has an influence on the prevalence of important zoonotic pathogens like *Campylobacter* spp., *Salmonella* spp. and *E. coli*, and whether there is an increased exchange of these isolates, as determined by MALDI-TOF MS and FT-IR spectra. For this purpose, a total of 240 pigs and 120 chickens were investigated between September 2019 and October 2020. Altogether, 438 *E. coli* and 201 *Campylobacter* strains were isolated and identified by MALDI-TOF MS.

In this study, *Salmonella* spp. could not be isolated in any of the investigated samples. With 8743 cases reported in 2019, salmonellosis is the second most common notifiable bacterial gastrointestinal disease in humans in Europe [11]. Farm animals (e.g., poultry, pigs and cattle) are considered to be the main reservoir, since almost all infected animals do not show any clinical symptoms [23]. A study conducted by the Federal Office of Consumer Protection and Food Safety in Germany (2020) showed that the prevalence of *Salmonella* spp. in caecal content samples of broiler was 2.6% and of broiler turkeys 2.4%, while 4.6% of fecal samples of wild boars and 4.0% of slaughtered fattening pigs were positive for this genus [43]. Although the prevalence of *Salmonella* spp. in farm animals in Germany is relatively low, they were included in the analysis for this study. Within livestock, there are several ways for *Salmonella* transmission, e.g., via latently infected animals, contaminated feed, or other vectors such as rodents, insects, wild birds and contaminated objects [21,22]. Free-range animals, such as in this study, could have a high risk of exposure to these vectors. Additionally, various studies have shown that free-range chickens have a higher prevalence of *Salmonella* spp. [44,45]. On the other hand, once *Salmonella* spp. entered the crops, the transmission rate was much lower in free range and especially in organic farming systems since there is more space available for each animal [46], and probably due to the better welfare aspects that could lead to a higher immune status of animal herds [47].

Thermophilic *Campylobacter* spp. could be detected in both pigs and chickens with a relatively similar prevalence to a study carried out in Bavaria (Germany) [48]. In this study, the detection rate of *Campylobacter* spp. in pigs (33,3% in total, 30% in pasture 1 and 36,7% in pasture 2) is slightly lower than in the above-mentioned study (36 %) and is considerably lower than the prevalence detected in other regions such as the Netherlands (46% [16] and 85% [49]). In a study from the United Kingdom, the prevalence of *Campylobacter* spp. is variable depending on the health status of animals, e.g., 77% for sick pigs compared to 44% for healthy pigs [3]. However, it should be noted that apart from ours and the Bavarian prevalence study, all the above-described studies collected the samples at the postmortem stage at the slaughterhouse. Stress and conditions during transport of animals to the slaughterhouse can increase the susceptibility of animals to the disease as well as the risk of disease transmission, possibly explaining the high prevalence of *Campylobacter* spp. in slaughtered pigs, as found in the mentioned studies. In addition to the moderate prevalence of *Campylobacter* spp. in pigs, a high colonization with thermophilic *Campylobacter* spp. (88%) in the chicken group was observed and is similar to data previously collected in Bavaria (75%, [48]). Regarding the bacterial species, *C. jejuni* and *C. coli* show a very different prevalence in the respective animal species in this study. The high prevalence of *C. jejuni* in poultry (over 78%) is consistent with previous reports, considering it as the most commonly detected *Campylobacter* species in chickens and as a natural gut inhabitant [16]. The low detection rate (5%) of *C. jejuni* and the predominance of *C. coli* in pigs are also consistent with the results of numerous studies [17,50,51].

The correlation of husbandry types (pasture 1 vs. pasture 2) and the risk of infection with *Campylobacter* spp. was analyzed. Pigs that were in close contact with chickens (pasture 1) have a risk of infection with *C. coli* similarly high to pigs that were kept alone (control group, pasture 2). However, pigs kept in pasture 2 showed a weak correlation to the risk of infection with *C. jejuni*, which is the species that is more frequently found in chickens. The prevalence of *C. jejuni* in the present study was higher in the pigs kept alone than in the pigs kept together with chickens (7% vs. 3%, respectively). Similar results were observed in Denmark, where pig herds kept alone or together with cattle have a tendency of increasing infection with *C. jejuni* than pig herds kept with poultry (i.e., 7.8%, 12.8%, and 4.4% of investigated pig herds, respectively) [50]. In this context, it may be possible that *C. jejuni* has adapted itself to invade other animal species when its specific host (poultry) is not present.

The shedding of *Campylobacter* spp. into soil/ground of pastures was additionally investigated. The prevalences of both *Campylobacter* species in soil samples from pasture 1 were higher than in soil samples from pasture 2. This may be due to the higher concentration of animals in the pasture (35 pigs and 250 chickens in 5 ha for pasture 1, and only 35 pigs for pasture 2). However, the difference was evaluated as statistically non-significant. According to the results of this study, it can be concluded that being kept on pasture 1 (pigs and chickens on mixed husbandry) did not increase the risk of infection of pigs with *Campylobacter* spp. compared to being kept on pasture 2 (pigs kept alone).

The cluster analysis of protein spectra of *Campylobacter* strains ($n = 201$) obtained by MALDI-TOF MS show that the strains were not sorted into groups based on husbandry, but solely into two groups according to the species *C. jejuni* and *C. coli*. The single spectra of the same *Campylobacter* species (*C. coli*/*C. jejuni*) show no differences between those of the pigs/chickens from pasture 1 (mixed husbandry) to the spectra of the pigs from pasture 2 (control group). Since there was no contact between the chickens (pasture 1) and the pigs of the control group during the project, transmission by direct contact can be ruled out. This result confirmed that no alteration regarding the protein composition of a single *Campylobacter* spp. was detected using this method, which does not indicate an increased exchange of these pathogens.

E. coli are mostly considered as harmless commensals, but this species also includes pathogenic variants that are associated with a variety of infections in humans and animals. They can be classified into non-pathogenic, commensal, intestinal pathogenic and extraintestinal pathogenic strains. *E. coli* exhibit a very flexible genome that quickly acquires genetic information horizontally. The genomic region contributes to the rapid evolution of variants [52]. Because of this resulting wide range of phenotypes, *E. coli* is a well-suited model organism for tracking studies. Pronounced genomic plasticity leads to a large variability. Other genomic changes such as DNA rearrangements and point mutations can also constantly alter the genome content and thus the fitness and competitiveness of individual variants in specific niches [53,54]. *E. coli* were isolated from all animal samples ($n = 360$). The shedding of *E. coli* in ground samples of pasture 1 (78.3%) was statistically significantly higher than of pasture 2 (51.6%), which may be the result of the higher concentrations of animals in pasture 1, as described in the discussion part for *Campylobacter* spp. By using protein spectrum analysis, the change of an individual strain and the formation of strain clusters can be recognized; thus, their spectra obtained by MALDI-TOF MS and from FT-IR proceeded to similarity analysis and the data was statistically evaluated. The mass spectrometry analysis was applied in this study, since previous studies have shown it to be highly reliable in terms of discriminatory power and the identification accuracy of microorganisms [33,55–57]. Additionally, it requires less material and cost and is rather easy to be conducted with a high number of samples. It may be noted that the results obtained could be extended in subsequent studies using next generation sequencing (NGS) or whole genome sequencing. One possibility would also be the combined and complementary NGS and MALDI-TOF MS techniques for bacterial characterization [58]. However, it was already mentioned in some studies that the 16S rRNA gene, which was often used for the NGS analysis, is rather insufficient at differentiating bacteria down to species level [59]. Thus, using this gene, the differentiation between *C. jejuni* and *C. coli* and between *E. coli* strains might also not be possible [60]. Therefore, specific gene sequences have to be properly selected for the genome analysis.

MALDI-TOF MS spectra of *E. coli* strains isolated within the same sampling run showed a high similarity. Subsequently, the spectra of all *E. coli* isolates ($n = 438$) were clustered according to the sampling time. Similar results were obtained by FT-IR analysis, indicating that the husbandry types (symbiotic living of chickens and pigs vs. pigs alone) and animal species (pigs vs. chickens) did not have any influence on the cluster formation of FT-IR spectra of *E. coli* isolates. Since an alteration of *E. coli* strains isolated from both animal species and husbandry types was not detected, an increased risk for pathogen exchange due to the symbiotic animal husbandry could not be observed in the one-year study period. However, it has to be mentioned that a methodological limitation of the study relates to the number of investigated colonies per plate. As described in the section material and methods, only one colony of *Campylobacter* spp./*E. coli* per culture plate was investigated by MALDI-TOF MS and FT-IR. In a single animal, there could be different bacterial strains. In this context, the observed effect might have been more pronounced if more colonies had been sampled.

Altogether, traditional culturing and state-of-the-art-methods (MALDI-TOF MS, FT-IR and similarity analysis) were applied to evaluate whether there was a risk of increasing disease transmission between two animal species that were kept together for one year. The results indicate that there is no species barrier regarding the transmission of *Campylobacter* spp. and *E. coli* between pigs and chickens. The prevalences of both *Campylobacter* spp. in both animal species are similar to the results of other studies conducted in the same region (Bavaria, Germany). Additionally, a high prevalence of *C. jejuni* in chickens did not result in a high infection rate of this bacteria in pigs raised in the same pasture. Furthermore, the characteristic alteration of *E. coli* was neither observed in the strains originally isolated from pigs or from chickens.

In terms of food safety, it can be concluded that keeping these animals together in free-ranging husbandry does not increase disease susceptibility and transmission regarding *Campylobacter* spp. and *E. coli*. Subsequently, meat and their products from mixed animal husbandry have no additional risk of being contaminated with pathogens (*Campylobacter* spp., *Salmonella* spp.) and indicator bacteria (*E. coli*). The most important factors when aiming to keep infection rates at a low level are the hygienic management of the animal herd, farm biosecurity, and the density of animals. This study was conducted under optimal conditions, where the animals had plenty of space (the legal requirements for access of chickens to open-air runs (broilers) are 4 m^2 (organic) or 2 m^2 (conventional) [61]), and were raised on pastures that have not been used for a long time. To verify the results obtained in this study, further investigations are required, for example, under the condition that stocking density is increased and/or when the pastures have been continually used for rearing animals.

5. Conclusions

This study was conducted to investigate the influence of symbiotic animal husbandry on the risk of bacterial transmission between pigs and chickens and the risk of the exchange of bacterial isolates between both animal species. The results do not indicate an increased risk of transmission for pigs when they are kept together with chickens in a mixed husbandry system (pasture 1) compared to a pasture with pigs alone (pasture 2). The prevalence of *Campylobacter* spp. in pigs was 30.0% in pasture 1 and 36.7% in pasture 2, and 0% regarding *Salmonella* spp. and 100% for *E. coli* for both pastures. Results obtained by similarity analysis of the MALDI-TOF MS and FT-IR spectra show that husbandry types and animal species did not have any influence on the cluster formation of *Campylobacter* spp. and *E. coli* strains, indicating that protein alteration of isolates of both bacterial species did not occur to a significant extent during the studied period. Therefore, in addition to the highly positive effects on animal welfare and sustainability associated with the symbiotic rearing system, a higher risk of transmission of the investigated pathogens was not ascertained. Neither the composition of the animal groups nor the duration of grazing rearing had a significant influence on the similarity or exchange of individual pathogens in this study. Thus, the advantages of keeping pigs and chickens together under good grazing conditions are not diminished by the possible transmission of pathogens.

Author Contributions: Conceptualization, M.G. and K.S.; methodology, E.K., S.D.-I. and K.S.; validation, E.K., S.D.-I. and K.S.; formal analysis, E.K. and S.D.-I.; investigation, E.K.; resources, K.S.; data curation, E.K. and S.D.-I.; writing—original draft preparation, E.K.; writing—review and editing, K.S. and S.D.-I.; visualization, E.K.; supervision, K.S.; project administration, K.S. and E.K.; funding acquisition, K.S. and M.G. All authors have read and agreed to the published version of the manuscript.

Funding: This work was funded by the Schweisfurth Foundation (Munich, Germany) and Software AG—Foundation (Darmstadt, Germany).

Institutional Review Board Statement: After consultation with the responsible authority in the Bavarian State (Germany), no ethical statement of approved animal trials was required for our study. Similar study designs in Germany were also conducted without ethical approvals [62].

Informed Consent Statement: Not applicable.

Data Availability Statement: Data is contained within the article.

Acknowledgments: We would like to thank the Team of Herrmannsdorfer Landwerkstätten (Bavaria State, Germany) for their active support during the sampling process. The authors' great thanks due to their skillful laboratory works are for Sebastian Schlef, Verena Hohenester and Erika Altgenug from the Chair of Food Safety and—Analytics (LMU, Munich). A special thank goes to Sven Reese from the Chair of Animal Anatomy (LMU, Munich) for the statistical data analysis. This article is dedicated to Karl Ludwig Schweisfurth († 2020), a pioneer in the welfare of farm animals and organic food production.

Conflicts of Interest: The authors declare that they have no conflict of interest.

References

1. Christoph-Schulz, I. SocialLab—Nutztierhaltung im Spiegel der Gesellschaft. *J. Consum. Prot. Food Saf.* **2018**, *13*, 145–236. [CrossRef]
2. Jochemsen, H. An ethical foundation for careful animal husbandry. *NJAS—Wagening. J. Life Sci.* **2013**, *66*, 55–63. [CrossRef]
3. Brumfiel, G. Animal-rights activists invade Europe. *Nature* **2008**, *451*, 1034–1036. [CrossRef] [PubMed]
4. Albernaz-Gonçalves, R.; Olmos, G.; Hötzel, M. My pigs are ok, why change?–Animal welfare accounts of pig farmers. *Animal* **2021**, *15*, 100154. [CrossRef] [PubMed]
5. Wissenschaftlicher Beirat Agrarpolitik beim BMEL. *Wege zu Einer Gesell-Schaftlich Akzeptierten Nutztierhaltung. Kurzfassung des Gutachtens*; Wissenschaftlicher Beirat Agrarpolitik beim BMEL: Berlin, Germany, 2015.
6. Eurobarometer, S. *Attitudes of EU Citizens towards Animal Welfare*; European Commission: Brussels, Belgium, 2007.
7. Bundesministerium für Ernährung und Landwirtschaft. Deutschland, Wie es Isst. Available online: https://www.bmel.de/SharedDocs/Downloads/Broschueren/Ernaehrungsreport2017.pdf?__blob=publicationFile (accessed on 25 February 2022).
8. Tallentire, C.W.; Edwards, S.A.; Van Limbergen, T.; Kyriazakis, I. The challenge of incorporating animal welfare in a social life cycle assessment model of European chicken production. *Int. J. Life Cycle Assess.* **2019**, *24*, 1093–1104. [CrossRef]
9. Bundesanstalt für Landwirtschaft und Ernährung. Ökologische Tierhaltung. Available online: https://www.praxis-agrar.de/tier/artikel/oekologische-tierhaltung/ (accessed on 25 February 2022).
10. Aerts, S.; Lips, D.; Spencer, S.; Decuypere, E.; De Tavernier, J. A new framework for the assessment of animal welfare: Integrating existing knowledge from a practical ethics perspective. *J. Agric. Environ. Ethics* **2006**, *19*, 67–76. [CrossRef]
11. European Food Safety Authority; European Centre for Disease Prevention and Control. The European Union One Health 2020 Zoonoses Report. *EFSA J.* **2021**, *19*, e06971.
12. European Food Safety Authority. *The European Union One Health 2019 Zoonoses Report*; EFSA: Parma, Italy, 2021.
13. Robert-Koch-Institut. Campylobacter-Enteritis. Available online: https://www.rki.de/DE/Content/Infekt/EpidBull/Merkblaetter/Ratgeber_Campylobacter.html (accessed on 1 April 2022).
14. Friedman, C.R.; Hoekstra, R.M.; Samuel, M.; Marcus, R.; Bender, J.; Shiferaw, B.; Reddy, S.; Ahuja, S.D.; Helfrick, D.L.; Hardnett, F.; et al. Risk Factors for Sporadic Campylobacter Infection in the United States: A Case-Control Study in FoodNet Sites. *Clin. Infect. Dis.* **2004**, *38*, S285–S296. [CrossRef]

15. Endtz, H.P. 50—Campylobacter Infections. In *Hunter's Tropical Medicine and Emerging Infectious Diseases*, 10th ed.; Ryan, E.T., Hill, D.R., Solomon, T., Aronson, N.E., Endy, T.P., Eds.; Elsevier: London, UK, 2020; pp. 507–511.

16. Møller Nielsen, E.; Engberg, J.; Madsen, M. Distribution of serotypes of Campylobacter jejuni and C. coli from Danish patients, poultry, cattle and swine. *FEMS Immunol. Med. Microbiol.* **1997**, *19*, 47–56. [CrossRef]

17. Alter, T.; Gaull, F.; Kasimir, S.; Gürtler, M.; Mielke, H.; Linnebur, M.; Fehlhaber, K. Prevalences and transmission routes of *Campylobacter* spp. strains within multiple pig farms. *Vet. Microbiol.* **2005**, *108*, 251–261. [CrossRef]

18. Robert-Koch-Institut. *Infektionsepidemiologisches Jahrbuch meldepflichtiger Krankheiten für 2020*; RKI: Berlin, Germany, 2021.

19. Duijkeren, E.V.; Wannet, W.J.B.; Houwers, D.J.; Pelt, W.V. Serotype and Phage Type Distribution of *Salmonella* Strains Isolated from Humans, Cattle, Pigs, and Chickens in The Netherlands from 1984 to 2001. *J. Clin. Microbiol.* **2002**, *40*, 3980–3985. [CrossRef] [PubMed]

20. Shivaprasad, H. Fowl typhoid and pullorum disease. *Rev. Sci. Tech.* **2000**, *19*, 405–424. [CrossRef] [PubMed]

21. Silva, C.; Calva, E.; Maloy, S. One Health and Food-Borne Disease: Salmonella Transmission between Humans, Animals, and Plants. *Microbiol. Spectr.* **2014**, *2*, 26082128. [CrossRef] [PubMed]

22. Ellington, C.; Hebron, C.; Crespo, R.; Machado, G. Unraveling the Contact Network Patterns between Commercial Turkey Operation in North Carolina and the Distribution of Salmonella Species. *Pathogens* **2021**, *10*, 1539. [CrossRef]

23. Robert-Koch-Institut. Salmonellose (Salmonellen-Gastroenteritis). *RKI-Ratgeber Ärzte* **2016**, *13*. [CrossRef]

24. Allocati, N.; Masulli, M.; Alexeyev, M.F.; Di Ilio, C. *Escherichia coli* in Europe: An Overview. *Int. J. Environ. Res. Public Health* **2013**, *10*, 6235–6254. [CrossRef] [PubMed]

25. Kayser, F.H. *Medical Microbiology*; Thieme: Stuttgart, Germany, 2005; p. 292.

26. Schwaiger, K.; Harms, K.; Hölzel, C.; Meyer, K.; Karl, M.; Bauer, J. Tetracycline in liquid manure selects for co-occurrence of the resistance genes tet(M) and tet(L) in Enterococcus faecalis. *Vet. Microbiol.* **2009**, *139*, 386–392. [CrossRef]

27. Landesamt für Natur, Umwelt und Verbraucherschutz Nordrhein-Westfalen. *Probenahme und Untersuchung am 21.9./3.11.2011 in der Umgebung der Deponie Eyller Berg*; LANUV/NRW: Essen, Germany, 2011.

28. Lauer, W.F.; Martinez, F.L.; Patel, A. Validation of RAPID'*E. coli* 2 for Enumeration and Differentiation of *Escherichia coli* and Other Coliform Bacteria in Selected FoodsPerformance-Tested MethodSM 050601. *J. AOAC Int.* **2007**, *90*, 1284–1315. [CrossRef]

29. BIORAD. *RAPID'E.coli 2 for Water Testing for the Enumeration of Escherichia coli and Coliforms in Drinking Water for Human Consumption*; Adgene Laboratoire: Le Hom, France, 2019.

30. Bruker Daltonik GmbH. *Compass 1.4 for FLEX Series User Manual, Doc No. 269834. User Manual—Volume 1*; Bruker Daltonik GmbH: Bremen, Germany, 2015.

31. Filip, Z.; Hermann, S.; Demnerová, K. FT-IR spectroscopic characteristics of differently cultivated Escherichia coli. *Czech J. Food Sci.* **2009**, *26*, 458–463. [CrossRef]

32. Naumann, D.; Helm, D.; Labischinski, H. Microbiological characterizations by FT-IR spectroscopy. *Nature* **1991**, *351*, 81–82. [CrossRef] [PubMed]

33. Ekruth, J.; Gottschalk, C.; Ulrich, S.; Gareis, M.; Schwaiger, K. Differentiation of *S. chartarum* (Ehrenb.) *S. Hughes* Chemotypes A and S via FT-IR Spectroscopy. *Mycopathologia* **2020**, *185*, 993–1004. [CrossRef] [PubMed]

34. Bruker Daltonik GmbH. *Biotyper IR. User Manual. Doc. No. 5022573 ed*; Bruker Daltonik GmbH: Bremen, Germany, 2017.

35. Herrmannsdorfer Landwerkstätten Glonn GmbH & Co., KG. Weideschweine und Symbiotische Landwirtschaft. Available online: https://www.herrmannsdorfer.de/landwirtschaft/symbiotisch/ (accessed on 11 March 2022).

36. Eisler, M.C.; Lee, M.R.; Tarlton, J.F.; Martin, G.B.; Beddington, J.; Dungait, J.A.; Greathead, H.; Liu, J.; Mathew, S.; Miller, H. Agriculture: Steps to sustainable livestock. *Nature* **2014**, *507*, 32–34. [CrossRef]

37. Elzen, B.; Bos, B. The RIO approach: Design and anchoring of sustainable animal husbandry systems. *Technol. Forecast. Soc. Change* **2019**, *145*, 141–152. [CrossRef]

38. Murgueitio, E.; Barahona, R.; Chará, J.; Flores, M.; Mauricio, R.; Molina, J. The intensive silvopastoral systems in Latin America sustainable alternative to face climatic change in animal husbandry. *Cuba. J. Agric. Sci.* **2016**, *49*, 541–554.

39. Bundesministerium für Wirtschaftliche Zusammenarbeit und Entwicklung. Nachhaltige Landwirtschaft. Available online: https://www.bmz.de/de/entwicklungspolitik/ernaehrungssicherung/nachhaltige-landwirtschaft#anc=id_51074_51074 (accessed on 27 April 2022).

40. Stadig, L.M.; Rodenburg, T.B.; Ampe, B.; Reubens, B.; Tuyttens, F.A.M. Effects of shelter type, early environmental enrichment and weather conditions on free-range behaviour of slow-growing broiler chickens. *Animal* **2017**, *11*, 1046–1053. [CrossRef] [PubMed]

41. Landwirtschaft, B.L.F. *Evaluierung Alternativer Haltungsformen für Legehennen*; LfL: Thüringen, Germany, 2004.

42. Dawkins, M.S.; Cook, P.A.; Whittingham, M.J.; Mansell, K.A.; Harper, A.E. What makes free-range broiler chickens range? In Situ measurement of habitat preference. *Anim. Behav.* **2003**, *66*, 151–160. [CrossRef]

43. Bundesamt für Verbraucherschutz und Lebensmittelsicherheit (BVL). *Zoonosen-Monitoring 2020*; BVL: Osceola, FL, USA, 2021.

44. Parry, S.; Palmer, S.; Slader, J.; Humphrey, T.; Group, S.E.W.I.D.L. Risk factors for salmonella food poisoning in the domestic kitchen–a case control study. *Epidemiol. Infect.* **2002**, *129*, 277–285. [CrossRef]

45. Leotta, G.A.; Suzuki, K.; Alvarez, F.; Nuñez, L.; Silva, M.; Castro, L.; Faccioli, M.; Zarate, N.; Weiler, N.; Alvarez, M. Prevalence of *Salmonella* spp. in backyard chickens in Paraguay. *Int. J. Poult. Sci.* **2010**, *6*, 533–536. [CrossRef]

46. Bailey, J.; Cosby, D. Salmonella prevalence in free-range and certified organic chickens. *J. Food Prot.* **2005**, *68*, 2451–2453. [CrossRef]

47. Staley, M.; Conners, M.G.; Hall, K.; Miller, L.J. Linking stress and immunity: Immunoglobulin A as a non-invasive physiological biomarker in animal welfare studies. *Horm. Behav.* **2018**, *102*, 55–68. [CrossRef] [PubMed]
48. Bayerisches Landesamt für Gesundheit und Lebensmittelsicherheit. Prävalenz von Thermophilen Campylobacter spp. in Kotproben von Rindern, Schweinen und Geflügel Sowie in Lebensmittelproben. Available online: https://www.lgl.bayern.de/forschung/forschung_interdisziplinaer/fp_campylobacter_kotproben_lebensmittelproben.htm (accessed on 16 March 2022).
49. Weljtens, M.J.B.M.; Bijker, P.G.H.; Van der Plas, J.; Urlings, H.A.P.; Biesheuvel, M.H. Prevalence of campylobacter in pigs during fattening; an epidemiological study. *Vet. Q.* **1993**, *15*, 138–143. [CrossRef] [PubMed]
50. Boes, J.; Nersting, L.; Nielsen, E.M.; Kranker, S.; Enøe, C.; Wachmann, H.C.; Baggesen, D.L. Prevalence and Diversity of Campylobacter jejuni in Pig Herds on Farms with and without Cattle or Poultry. *J. Food Prot.* **2005**, *68*, 722–727. [CrossRef] [PubMed]
51. Wang, Y.; Dong, Y.; Deng, F.; Liu, D.; Yao, H.; Zhang, Q.; Shen, J.; Liu, Z.; Gao, Y.; Wu, C.; et al. Species shift and multidrug resistance of Campylobacter from chicken and swine, China, 2008–14. *J. Antimicrob. Chemother.* **2015**, *71*, 666–669. [CrossRef] [PubMed]
52. Dobrindt, U. (Patho-)Genomics of Escherichia coli. *Int. J. Med. Microbiol.* **2005**, *295*, 357–371. [CrossRef] [PubMed]
53. Thomas, C.M.; Nielsen, K.M. Mechanisms of, and barriers to, horizontal gene transfer between bacteria. *Nat. Rev. Microbiol.* **2005**, *3*, 711–721. [CrossRef] [PubMed]
54. Cameron, A.D.; Redfield, R.J. Non-canonical CRP sites control competence regulons in Escherichia coli and many other γ-proteobacteria. *Nucleic Acids Res.* **2006**, *34*, 6001–6014. [CrossRef] [PubMed]
55. Dubois, D.; Leyssene, D.; Chacornac, J.P.; Kostrzewa, M.; Schmit, P.O.; Talon, R.; Bonnet, R.; Delmas, J. Identification of a Variety of *Staphylococcus* Species by Matrix-Assisted Laser Desorption Ionization-Time of Flight Mass Spectrometry. *J. Clin. Microbiol.* **2010**, *48*, 941–945. [CrossRef]
56. Ilina, E.N.; Borovskaya, A.D.; Serebryakova, M.V.; Chelysheva, V.V.; Momynaliev, K.T.; Maier, T.; Kostrzewa, M.; Govorun, V.M. Application of matrix-assisted laser desorption/ionization time-of-flight mass spectrometry for the study of Helicobacter pylori. *Rapid Commun. Mass Spectrom.* **2010**, *24*, 328–334. [CrossRef] [PubMed]
57. Nagy, E.; Maier, T.; Urban, E.; Terhes, G.; Kostrzewa, M. Species identification of clinical isolates of Bacteroides by matrix-assisted laser-desorption/ionization time-of-flight mass spectrometry. *Clin. Microbiol. Infect.* **2009**, *15*, 796–802. [CrossRef] [PubMed]
58. Sabença, C.; de Sousa, T.; Oliveira, S.; Viala, D.; Théron, L.; Chambon, C.; Hébraud, M.; Beyrouthy, R.; Bonnet, R.; Caniça, M.; et al. Next-Generation Sequencing and MALDI Mass Spectrometry in the Study of Multiresistant Processed Meat Vancomycin-Resistant Enterococci (VRE). *Biology* **2020**, *9*, 89. [CrossRef] [PubMed]
59. Fadeev, E.; Cardozo-Mino, M.G.; Rapp, J.Z.; Bienhold, C.; Salter, I.; Salman-Carvalho, V.; Molari, M.; Tegetmeyer, H.E.; Buttigieg, P.L.; Boetius, A. Comparison of two 16S rRNA primers (V3–V4 and V4–V5) for studies of arctic microbial communities. *Front. Microbiol.* **2021**, *12*, 637526. [CrossRef] [PubMed]
60. Hansson, I.; Persson, M.; Svensson, L.; Engvall, E.O.; Johansson, K.-E. Identification of nine sequence types of the 16S rRNA genes of Campylobacter jejuni subsp. jejuni isolated from broilers. *Acta Vet. Scand.* **2008**, *50*, 10. [CrossRef]
61. DLG. *DLG Merkblatt 406: Haltung von Masthühnern*; DLG Verlag: Frankfurt, Germany, 2017.
62. Schwaiger, K.; Storch, J.; Bauer, C.; Bauer, J. Development of selected bacterial groups of the rectal microbiota of healthy calves during the first week postpartum. *J. Appl. Microbiol.* **2020**, *128*, 366–375. [CrossRef] [PubMed]

Article

Prevalence and Persistence of Multidrug-Resistant *Yersinia enterocolitica* 4/O:3 in Tonsils of Slaughter Pigs from Different Housing Systems in Croatia

Nevijo Zdolec [1,*], Marta Kiš [1], Dean Jankuloski [2], Katerina Blagoevska [2], Snježana Kazazić [3], Marina Pavlak [1], Bojan Blagojević [4], Dragan Antić [5], Maria Fredriksson-Ahomaa [6] and Valerij Pažin [1]

[1] Faculty of Veterinary Medicine, University of Zagreb, 10000 Zagreb, Croatia; mkis@vef.hr (M.K.); mpavlak@vef.hr (M.P.); vpazin@vef.hr (V.P.)
[2] Faculty of Veterinary Medicine, Food Institute, 1000 Skopje, North Macedonia; djankuloski@fvm.ukim.edu.mk (D.J.); kblagoevska@fvm.ukim.edu.mk (K.B.)
[3] Ruder Boskovic Institute, 10000 Zagreb, Croatia; snjezana.kazazic@irb.hr
[4] Faculty of Agriculture, Department of Veterinary Medicine, University of Novi Sad, 21000 Novi Sad, Serbia; blagojevic.bojan@yahoo.com
[5] Faculty of Health and Life Sciences, Institute of Infection, Veterinary and Ecological Sciences, University of Liverpool, Leahurst, Neston CH64 7TE, UK; dragan.antic@liverpool.ac.uk
[6] Faculty of Veterinary Medicine, University of Helsinki, 00014 Helsinki, Finland; maria.fredriksson-ahomaa@helsinki.fi
* Correspondence: nzdolec@vef.hr

Citation: Zdolec, N.; Kiš, M.; Jankuloski, D.; Blagoevska, K.; Kazazić, S.; Pavlak, M.; Blagojević, B.; Antić, D.; Fredriksson-Ahomaa, M.; Pažin, V. Prevalence and Persistence of Multidrug-Resistant *Yersinia enterocolitica* 4/O:3 in Tonsils of Slaughter Pigs from Different Housing Systems in Croatia. *Foods* **2022**, *11*, 1459. https://doi.org/10.3390/foods11101459

Academic Editor: Frans J.M. Smulders

Received: 2 March 2022
Accepted: 11 May 2022
Published: 18 May 2022

Publisher's Note: MDPI stays neutral with regard to jurisdictional claims in published maps and institutional affiliations.

Abstract: *Yersinia enterocolitica* is one of the priority biological hazards in pork inspection. Persistence of the pathogen, including strains resistant to antimicrobials, should be evaluated in pigs from different housing systems for risk ranking of farms. In this 2019 study, tonsils were collected from 234 pigs, of which 69 (29.5%) were fattened on 3 big integrated farms, 130 (55.5%) on 10 medium-sized farms, and 35 (15%) on 13 small family farms. In addition, 92 pork cuts and minced meat samples from the same farms were tested for the presence of *Y. enterocolitica* using the culture method. Phenotypic and genetic characteristics of the isolates were compared with previously collected isolates from 2014. The overall prevalence of *Y. enterocolitica* in pig tonsils was 43% [95% CI 36.7–49.7]. In pigs from big integrated, medium-sized, and small family farms, the prevalence was 29%, 52%, and 40%, respectively. All retail samples of portioned and minced pork tested negative for pathogenic *Y. enterocolitica*, likely due to high hygienic standards in slaughterhouses/cutting meat or low sensitivity of culture methods in these matrices. The highest recovery rate of the pathogen from tonsils was found when alkali-treated PSB and CIN agar were combined. The biosecurity category of integrated and medium farms did not affect the differences in prevalence of *Y. enterocolitica* ($p > 0.05$), in contrast to family farms. Pathogenic *ail*-positive *Y. enterocolitica* biotype 4 serotype O:3 persisted in the tonsils of pigs regardless of the type of farm, slaughterhouse, and year of isolation 2014 and 2019. PFGE typing revealed the high genetic concordance (80.6 to 100%) of all the *Y. enterocolitica* 4/O:3 isolates. A statistically significant higher prevalence of multidrug-resistant *Y. enterocolitica* 4/O:3 isolates was detected in the tonsils of pigs from big integrated farms compared to the other farm types ($p < 0.05$), with predominant and increasing resistance to nalidixic acid, chloramphenicol, and streptomycin. This study demonstrated multidrug resistance of the pathogen in pigs likely due to more antimicrobial pressure on big farms, with intriguing resistance to some clinically relevant antimicrobials used in the treatment of yersiniosis in humans.

Keywords: *Yersinia enterocolitica* 4/O:3; pigs; slaughter; farm; antimicrobial resistance

1. Introduction

Yersiniosis is one of the leading zoonoses in Europe, caused by pathogenic *Yersinia enterocolitica* bioserotypes and mainly transmitted through contaminated food. The pooled global prevalence of *Y. enterocolitica* in cases of human gastroenteritis has been recently estimated to be 1.97% [95% CI 1.32–2.74%], dominated by serotype O:3 [1]. According to the latest data from EFSA and ECDC Zoonoses Report, reporting data for 2020, there were 5668 human cases of this disease reported in Europe, with very limited surveillance data in the meat production chain [2]. In addition, six European countries reported only 0.2% of pigs (out of 2351 tested) positive for *Y. enterocolitica*, but these data were most likely related to fecal testing on farms. A total of 12.5% of pork sold at retail and 4.7% of samples (carcass swabs, pork) from cutting plants and slaughterhouses were *Y. enterocolitica* positive [2].

The main carriers of pathogenic *Y. enterocolitica* are pigs, with their tonsils being the main predilection site [3]. The reported prevalence of the pathogen in pigs varies widely among the numerous studies, which is to be expected considering the many risk factors involved from farm to slaughterhouse. Virtanen et al. [4] reported that factors contributing to fecal shedding of *Y. enterocolitica* include carriage of pathogen on the tonsils, purchase of feed from different suppliers, fasting of pigs prior to transport to slaughter, and snout contacts. Furthermore, Vilar et al. [5] claimed that the prevalence of *Y. enterocolitica* in pigs can only be reduced by supplying water of municipal origin and applying the "all-in-all-out" method, while risk factors contributing to increase were a lack of bedding and sourcing piglets from multiple farms. Existing pig farming systems differ significantly in terms of biosecurity levels and could, therefore, pose differing animal health risks. For example, important aspects include the transmission of *Y. enterocolitica* at the interface between livestock and wildlife and the role that wild and peridomestic rodents play as a source of this zoonotic pathogen for pigs [6]. Regarding the possibility of meat contamination during slaughter, Vilar et al. [7] indicated that risk factors include the presence of *Y. enterocolitica* in the intestines (OR: 35.6, 95% CI 2.8–8285), tonsils (OR: 38.4, 95% CI 5.0–854), and offal (OR: 16.6, 95% CI 1.9–1111). Furthermore, differences between slaughterhouses, where different hygiene practices are applied during slaughter and dressing, could increase cross-contamination from tonsils to carcasses [8]. In addition to farm- and slaughterhouse-related risk factors, differences in reported prevalences among studies could also be due to pathogen isolation methods. Therefore, traditional isolation methods are supplemented with more sensitive and rapid techniques such as polymerase chain reaction (PCR) screening. Additionally, matrix assisted laser desorption ionization-time of flight mass spectrometry (MALDI-TOF MS), PCR, pulsed-field gel electrophoresis (PFGE), multiple locus variable number of tandem repeats analysis (MLVA), and sequencing have been widely used for identification and characterization of *Yersinia* isolates [9].

Y. enterocolitica biotypes and serotypes associated with pathogenicity occur in both pigs and infected humans, with bioserotype 4/O:3 being the most common in continental Europe [10]. Consumption of raw and inadequately heat-treated pork and untreated water are considered the main risk factors for human infection [10]. Although pork is considered the main source of human infection; many studies have shown that pathogenic *Y. enterocolitica* is rarely found in portioned pork on the market, except for carcass parts and organs that are more likely to be contaminated at slaughter (cheeks, head, tongue, throat) [11]. However, the pathways of contamination and persistence of pathogenic strains have been confirmed over years in the pork production chain, linking the farm and the pork produced [12,13]. In recent years, research on antimicrobial resistance in foodborne pathogens has intensified to reduce the spread of resistance in the food chain. *Y. enterocolitica* is generally sensitive to clinically relevant antibiotics, and similar resistance profiles persist over time, which is explained by the genetic stability of the bacterium [14]. However, recent reports warn of foodborne yersiniosis outbreaks associated with multidrug-resistant *Y. enterocolitica* 4/O:3, which possess resistance genes of major public health concern that are acquired by horizontal transfer [15].

Therefore, the aim of this study was to determine the prevalence of (multidrug-resistant) *Y. enterocolitica* in the tonsils of slaughtered fattening pigs raised in different housing systems: big integrated farms, cooperative farms (medium-sized farms), and small family farms in Croatia. In addition, the presence of the pathogen on the market was evaluated in portioned pork and minced meat that originated from the investigated farms. The study also aimed to determine the persistence of the pathogen in the pork production chain by comparing the phenotypic and genetic characteristics of *Y. enterocolitica* with previously collected isolates in Croatia [16].

2. Materials and Methods

2.1. Farms and Slaughterhouses Included in the Study

All pigs included in this study originated from fattening farms, and were slaughtered in the same slaughterhouses as in previous survey from 2014. Three types of pig farms were included in the study: big integrated farms (>10,000 pigs), medium-sized farms (300–10,000 pigs), and small family farms (<300 pigs). The biosecurity category of investigated farms was obtained from the national database of registered farms; category 3 contains the farms with the highest biosecurity level, category 2 indicates that some biosecurity improvements are needed, and category 1 contains the farms with a low biosecurity level. A survey of the farms regarding their biosecurity levels was not conducted as a part of this study.

The big integrated farms involved ($n = 3$) used a vertical management system, their own piglets from separated breeding farms, their own produced crops and feed, and high biosecurity standards. The number of fattening pigs (per year) in these farms ranged from 11,000 to 31,000. Medium-sized farms ($n = 10$) purchased piglets from different local farms and import. The level of biosecurity in the medium-sized farms was medium to high. The number of fattening pigs on the investigated medium-sized farms ranged from 600 to 3000. Small family farms ($n = 13$) had their own sows and piglets that were fattened for slaughter. These farms had lower biosecurity conditions. The number of pigs on these farms ranged from 6 to 300.

Selected characteristics of the slaughterhouses involved in the study are shown in Table 1. Slaughterhouses were categorized as low, medium, or high risk based on the following parameters: slaughterhouse capacity and size of meat distribution area (factor of 0.30), past non-compliance in terms of infrastructure, equipment and hygiene (factor of 0.40), and the degree of implementation of HACCP principles and animal welfare rules (factor of 0.30) [17].

Table 1. Characteristics of slaughterhouses included in this study.

Parameter	Slaughterhouse 1	Slaughterhouse 2	Slaughterhouse 3	Slaughterhouse 4
Number of slaughtered fattening pigs per year	308,000	174,000	4000	55,000
Number of slaughtered pigs/h	130	160	20	140
Risk category	High risk	High risk	Medium risk	High risk
Biosecurity of farms (sampled in this study)	3	2–3	1–3	2
Contact between pigs from different farms, lairage	No	Yes *	Yes *	Yes *
Scalding technology	Water (5 min/62 °C)	Steam (20 min/60 °C)	Water (10 min/62 °C)	Water (7 min, 61.5 °C)
Pluck set organ removal techniques and organ placement	Knife, conveyor belt	Knife, hanging hook	Knife, hanging hook	Knife, hanging hook
Head removal and processing on separate line	No	No	No	No

* The pens in the lairage are separated by a fence that allows contact between the pigs.

2.2. Sampling of Tonsils and Retail Meat

Tonsils from 234 fattening pigs were collected by simple random sampling after pluck set removal in four slaughterhouses during 12 sampling sessions (slaughterhouse 1—pigs from three big integrated farms (n = 69); slaughterhouses 2, 3, 4—pigs from 10 medium-sized farms (n = 130), and slaughterhouse 3—pigs from 13 family farms (n = 35); Table S1—Supplementary Materials).

A total of 92 samples of retail pork cuts (neck, thigh, loin, shoulder, bacon) and minced pork, originating from the investigated farms, were tested. These samples were obtained from local markets/supermarkets owned by the same companies that owned the slaughterhouses. In addition, 36 samples were obtained from other local producers and from import. Tonsil and meat samples were transported refrigerated to the laboratory and analyzed within 30 min of arrival. The maximum time from sample collection to analysis was 3 h.

2.3. Microbiological Analyses of Tonsils and (Minced) Pork

Ten grams of each tonsil (n = 234) and meat sample (n = 128) were homogenized in 90 mL of enrichment broth (peptone, sorbitol, and bile salts, PSB, Sigma Aldrich, St. Louis, MO, USA), of which 10 mL was transferred to 90 mL of selective enrichment broth (Irgasan^{TM} Ticarcillin and Potassium chlorate, ITC, Sigma Aldrich, St. Louis, MO, USA). Subsequently, both solutions were incubated at 25 ± 1 °C for 44 ± 4 h followed by streaking on Cefsulodin, Irgasan^{TM}, and Novobiocin agar (CIN, Merck, Darmstadt, Germany) and CHROMagar^{TM} *Y. enterocolitica* (Paris, France). Broths cultures were then treated with alkaline solution (0.5% KOH) for 20 s, and streaked again on the same selective agars, incubated for 24 ± 2 h at 30 ± 1 °C [16]. Characteristic colonies on CIN agar (small, round, smooth, with dark red center and transparent edge—"bull's eye") were retained and subcultured for further identification and characterization. Colonies that were CHROMagar^{TM} purple (presumptive pathogenic) were also retained and subcultured. The alkali treatment of broth cultures was considered a risk factor for unsuccessful isolation of *Y. enterocolitica* on selective media. The odds ratio of the events (isolation and failed isolation of *Y. enterocolitica*) was calculated in relation to the prevalence detected after alkali treatment.

2.4. Assessment of Y. enterocolitica Persistence

Selected isolates of *Y. enterocolitica* obtained from this study (n = 84) were compared for phenotypic and genetic characteristics with selected isolates (n = 49) from a previous survey conducted in the same slaughterhouses and in pigs originated from comparable housing systems [16]. A total of 84 isolates were selected from 101 positive tonsils in this study for further characterization, representing all positive batches and farms. All isolates from the tonsils of pigs kept on small family farms were retained for further analysis (1–3 positive tonsils per farm). For medium and big farms, a maximum of seven isolates from one farm were retained (2 to 12 positives per farm).

2.4.1. Identification of Isolates by MALDI-TOF MS and Real Time PCR

A total of 84 isolates of presumptive *Y. enterocolitica* were selected for matrix-assisted laser desorption/ionization time of flight mass spectrometry identification (MALDI-TOF MS, Bruker Daltonik, Bremen, Germany), with detailed description provided in a recent study [18].

A total of 65 isolates from this study (representing all positive batches/farms) and 32 isolates from a previous study [16] were selected for Real Time PCR to confirm the presence of the *ail* gene. The number of tested isolates (97 in total) was conditioned by test assays (n = 100) provided in the diagnostic kit. The positive control was a human isolate of *Y. enterocolitica* 4/O:3 and the negative controls were two atypical colonies selected from CIN agar and CHROMagar^{TM}. DNA isolation was performed using the Gene JET Genomic DNA Purification Kit (Thermo Fisher Scientific, Waltham, WA, USA). PCR amplification and detection was performed according to the protocol of VIASURE *Yersinia enterocolitica* Real Time PCR detection kit (Certest Biotec S.L., Zaragoza, Spain). The sample was positive if the threshold cycle (Ct) value was below 40 and the internal control showed an amplification signal.

2.4.2. Biotyping, Serotyping, and PFGE Typing of Isolates

Isolates from both surveys (this study: $n = 84$, previous study: $n = 49$) were biotyped according to the standard HRN EN ISO 10273: 2017 [19] using the reactions of esculin, xylose, pyrazinamidase, tween esterase/lipase, trehalose, and indole. Xylose and trehalose solutions, slant agar pyrazinamidase, and Tween esterase/lipase plates were purchased from the Croatian Veterinary Institute, Zagreb. Esculin and indole reactions were tested on Rapid 20E and API 20E, respectively (bioMérieux, Marcy l'Etoile, France). Serotyping was performed by agglutination of *Y. enterocolitica* O:3 antiserum (Statens Serum Institute, Copenhagen, Denmark). Human isolate *Y. enterocolitica* 4/O:3 was used as a positive control (courtesy of Višnja Kružičević, MD, Croatian Institute of Public Health).

Molecular profiles of isolates were compared by PFGE in order to evaluate the possible persistence of specific genotypes in pig tonsils. The PulseNet One-Day (24–28 h) Standardized Laboratory Protocol for Molecular Subtyping of *Yersinia pestis* was used [20]. One rare-cutting restriction enzyme, AscI (New England Biolabs, Beverly, MA, USA) was used for restriction endonuclease digestion. The gels were stained with ethidium bromide and visualized and digitally photographed with a Molecular imager GelDoc XR+ camera system (Bio-Rad Laboratories, Hercules, CA, USA). Fragment size was determined with a low-range CHEF DNA Size Standard Lambda Ladder marker (Bio-Rad Laboratories, Hercules, CA, USA). The PFGE typing results were analyzed with FPQuest software version 5.10 (Bio-Rad Laboratories, Hercules, CA, USA). Dice coefficient with optimization and tolerance set at 1% was used to identify similarities between the PFGE types. A dendrogram was constructed with the unweighted pair group method using arithmetic means showing genetic similarity (percent). The position tolerance was set to 1.5%, with the average optimization value at 1.0%. A down limit for band interpretation at 33kbp was used as recommended for *Salmonella* by Peters et al. [21].

2.4.3. Testing the Susceptibility of *Y. enterocolitica* to Antimicrobial Agents

All isolates (this study: $n = 84$, previous study: $n = 49$) were tested for susceptibility to antimicrobial agents by the disk diffusion method. A 0.5 McFarland cell solution (Densimat, bioMérieux, Marcy l'Etoile, France) was prepared prior to the application of the test isolate on Mueller-Hinton agar (Bio-Rad Laboratories). Eleven antibiotics (MASTDISKS® AST, Mast Group, Bootle, UK) were used: Levofloxacin (5 μg), Ciprofloxacin (5 μg), Ampicillin (10 μg), Cephalothin (30 μg), Cefotaxime (30 μg), Tetracycline (30 μg), Nalidixic acid (30 μg), Ceftazidime 30 μg), Trimethoprim/Sulfamethoxazole (25 μg), Chloramphenicol (30 μg), and Streptomycin (10 μg). Zones of inhibitions were measured by automated system Scan 1200 (Interscience, Saint-Nom-la-Bretèche, France) and interpreted according to CLSI criteria for *Enterobacteriaceae* [22].

2.5. Statistical Analysis

In data processing, descriptive statistics methods were used for the quantitative data and data distribution to estimate the curve. Since most of the data were non-parametric, non-parametric tests were used: Spearman's correlation, Mann—Whitney U test, Kruskal—Wallis test, and Fisher exact test. All data were correlated and tested for differences between slaughterhouses, farms, and years. Depending on the data, the χ^2 test was used for qualitative data and proportional estimates, the Student's *t*-test was used to analyze differences between quantitative data between two groups when the data were normally distributed, the Mann—Whitney U test was used for other data distributions, and the Kruskal—Wallis test with multiple rank comparison was used to test multiple groups simultaneously. Differences were significant at the $p < 0.05$ level. The Statistica 13.1 program (Stata Corp., Lakeway Drive, TX, USA) was used.

3. Results

3.1. Prevalence of Y. enterocolitica in Pig Tonsils and Retail Meat

The study revealed a prevalence of *Y. enterocolitica* in pig tonsils of 43% (Table 2). In pigs from big integrated, medium-sized, and small family farms, the prevalence was 29%, 52%, and 40%, respectively. The percentage of *Yersinia*-positive pigs from integrated farms ranged from 14% to 43%. Although the three integrated farms were in the highest biosecurity category (i.e., category 3), a statistically significant difference in prevalence was found between two of these integrated farms ($p < 0.05$).

Table 2. Prevalence of *Y. enterocolitica* in tonsils of pigs from different housing systems and slaughterhouses.

Slaughterhouse	Farm Type	Biosecurity	No. of Farms	YE + Farms	No. of Pigs	YE + Pigs (*n*)	YE + Pigs (%)
1	Big integrated	3	3	3	69	20	29%
2	Medium-sized	2 and 3	6	6	74	31	42%
		3	5	5	62	24	39%
		2	1	1	12	7	58%
3	Medium-sized	2 and 3	3	3	14	12	86%
		3	2	2	10	8	80%
		2	1	1	4	4	100%
	Small family farms	1, 2 and 3	13	8	35	14	40%
		3	2	0	5	0	0
		2	10	8	29	14	48%
		1	1	0	1	0	0
4	Medium-sized	3	1	1	42	24	57%
			26	21	234	101	43%

Pigs from medium-sized farms were slaughtered in three slaughterhouses (2, 3, and 4). When *Y. enterocolitica* prevalences were compared depending on the place of slaughter (42%, 86%, and 57% at slaughterhouses 2, 3, and 4, respectively), a significant difference was found between slaughterhouse 2 and slaughterhouse 3 ($p < 0.05$). Considering slaughterhouse 2, the prevalence of positive pigs ranged from 15.4% to 67%, and 39% *Yersinia*-positive pigs originated from medium-sized farms of the highest biosecurity category 3. Comparing this result with the medium farms of lower biosecurity category 2 (58% positive pigs), the difference was not statistically significant ($p = 0.2104$, $\chi^2 = 1.568$). Similarly, biosecurity category did not significantly affect the proportions of *Yersinia*-positive pigs from medium-sized farms slaughtered in slaughterhouse 3. Excluding the slaughterhouse factor, within pigs from medium-sized farms, 44% of *Yersinia*-positive pigs originated from the highest biosecurity farms, while 60% were from lower biosecurity farms. However, this difference was not significant ($p = 0.2482$; $\chi^2 = 1.333$). In addition, within biosecurity category 3, no statistically significant differences in *Yersinia* prevalences were found between medium-sized farms and big integrated farms. The majority of family farms (77%) were in lower biosecurity category 2, and 48% ($n = 29$) of the pigs from these farms were *Yersinia*-positive. Compared to the family farms in category 3, the difference was significant ($p = 0.0460$, $\chi^2 = 1.333$). All retail samples of portioned and minced pork were negative for pathogenic *Y. enterocolitica*.

3.2. Recovery Rates of Y. enterocolitica with Different Isolation Procedures

As presented in Table 3, the lowest number of positive samples (*Y. enterocolitica* isolated from pig tonsils) was detected when only PSB broth was used followed by streaking on selective agars. The type of agar (CIN or CHROMagar^{TM}) did not significantly affect the success of bacterial isolation (p = 0.288). Alkali treatment of PSB broth cultures significantly increased the frequency of isolation of pathogenic *Y. enterocolitica*, by 5.4-fold on CIN agar and 3.7-fold on CHROMagar^{TM}, respectively (p = 0.000; p = 0.022) (Table 4). The frequency of *Y. enterocolitica* isolation after alkali treatment of PSB broth cultures was not statistically different with respect to the selective agar used (p = 0.05). Compared to PSB broth, enrichment in ITC broth showed a significantly higher number of *Yersinia*-positive tonsils after inoculation on CIN agar or CHROMagar^{TM} (p < 0.05). There were no differences in pathogen growth on the selective agars used (p = 0.70). KOH treatment of ITC broth cultures also showed an increase in the number of *Yersinia*-positive tonsils detected using CIN agar, but without statistical significance compared to untreated ITC broth (p = 0.422). Similarly, the frequency of pathogen isolation on CHROMagar^{TM} was not altered by alkali treatment of ITC broth (p > 0.05). Thus, a significantly higher frequency of *Y. enterocolitica* isolation was observed on CIN agar than on CHROMagar^{TM} after alkali treatment of ITC broth cultures (p = 0.0002).

Table 3. Comparison of different methods regarding recovery rate and isolation of *Y. enterocolitica* from pig tonsils.

Method of Isolation (Broth + Agar)	Number of Positives (%); n = 234	*Y. enterocolitica* Recovery Rate (%); n = 101
PSB and CIN	14 (5.9)	13.9
PSB and CHROMagar^{TM}	18 (7.7)	17.8
PSB + KOH and CIN	75 (32.0)	74.3
PSB + KOH and CHROMagar^{TM}	66 (28.2)	65.3
ITC and CIN	50 (21.4)	49.5
ITC and CHROMagar^{TM}	43 (18.4)	42.6
ITC + KOH and CIN	58 (24.8)	57.4
ITC + KOH and CHROMagar^{TM}	42 (17.9)	41.6

Table 4. *Y. enterocolitica* odds ratio and prevalence ratio between alkali-treated and untreated broths.

Broth and Agar Combinations	Prevalence Ratio	Odds Ratio (OR)	Fisher Exact Test; p	Confidence Interval (95% CI)
PSB + KOH and CIN vs. PSB and CIN	5.42	7.41	<0.0001	4.07–13.47
PSB + KOH and CHROMagar^{TM} vs. PSB and CHROMagar^{TM}	3.66	4.71	<0.0001	2.71–8.19
ITC + KOH and CIN vs. ITC and CIN	1.15	1.21	0.44	0.78–1.86
ITC + KOH and CHROMagar^{TM} vs. ITC and CHROMagar^{TM}	0.97	0.97	1	0.60–1.55

3.3. MALDI-TOF MS and Real Time PCR Identification, Bio-, Sero-, and PFGE-Typing

Isolates (n = 84) were confirmed by MALDI-TOF MS with a very high probability (score 2.30–3.00) to be *Y. enterocolitica*, while atypical colonies were assigned to *Citrobacter* or *Serratia* species. All the isolates belonged to biotype 4, characterized by negative reactions of aesculin, xylose, pyrazinamidase, lipase, and indole, with a positive reaction of trehalose. Serotyping confirmed that all biotype 4 isolates belonged to serotype O:3, regardless of the year of isolation and the origin of the pigs, i.e., the type of fattening farm. All tested isolates were also positive for the *ail* gene by Real Time PCR. PFGE analysis showed low variability of pulse types within successfully typed (n = 66) pathogenic *Y. enterocolitica* 4/O:3 isolates (Figure 1).

Yersinia enterocolitica (67 entries)

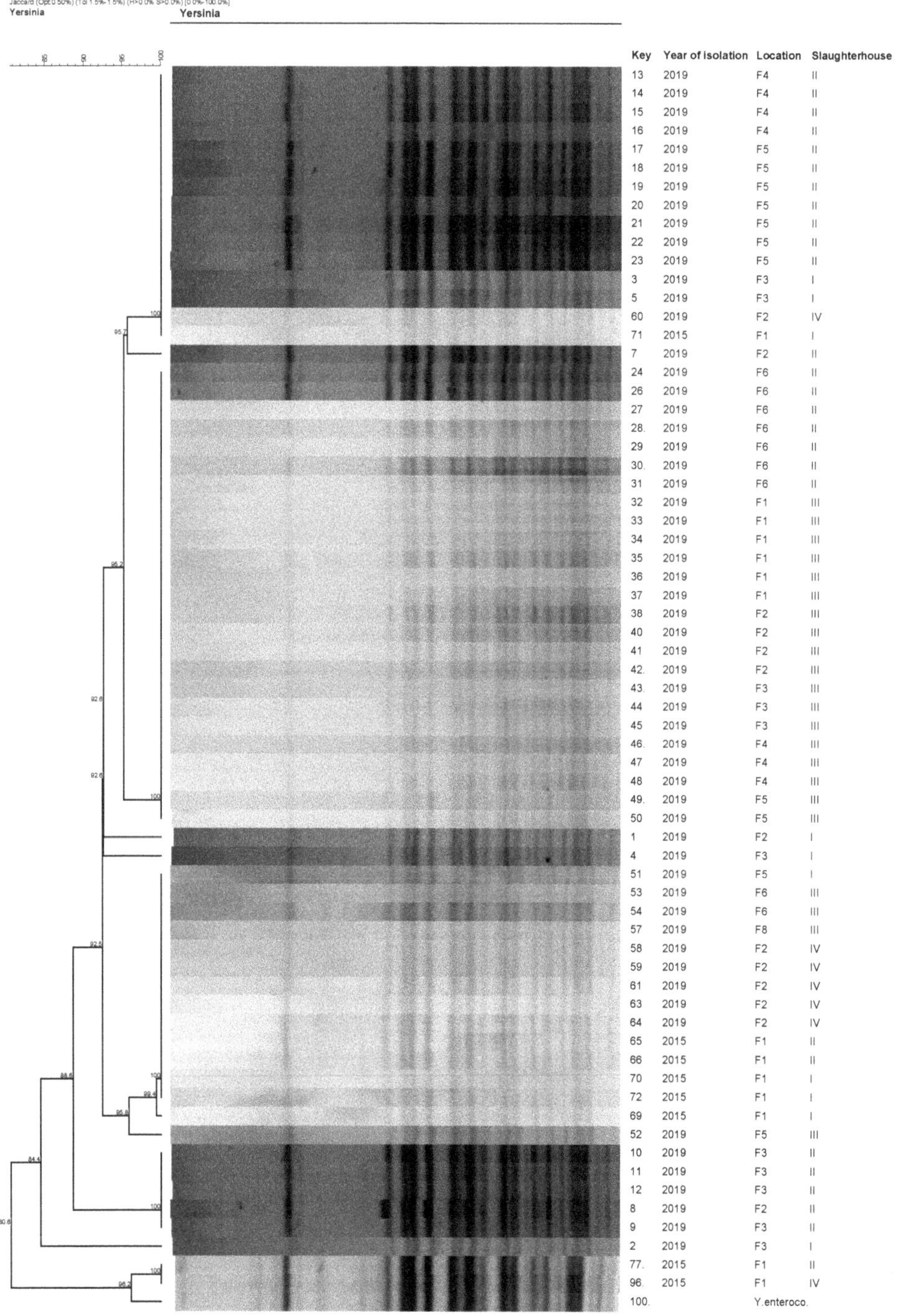

Figure 1. PFGE profiles of *Y. enterocolitica* 4/O:3 from different farm types, slaughterhouses, and years of isolation.

3.4. Susceptibility of Y. enterocolitica 4/O:3 Isolates to Antimicrobial Agents

In total (both surveys), 36 isolates of *Y. enterocolitica* 4/O:3 from big integrated farms, 84 isolates from medium-sized farms, and 13 isolates from small family farms were tested for susceptibility to 11 antimicrobial agents. Considering isolates from the previous survey (*n* = 49; integrated and medium farms), in addition to natural resistance to ampicillin (92% of isolates) and cephalothin (85%), resistance to chloramphenicol (31%), nalidixic acid (31%), streptomycin (27%), tetracycline (8%), and trimethoprim/sulfamethoxazole (2%) was observed. Only one isolate was sensitive to all antibiotics tested. Among *Y. enterocolitica* 4/O:3 isolates from medium-sized farms only one isolate showed multiresistance (nalidixic acid-chloramphenicol-cefotaxime). In contrast, isolates from big integrated farms were frequently resistant to chloramphenicol, nalidixic acid, and streptomycin. In total, 15 isolates of 24 tested from big integrated farms were multiresistant (Table 5).

Table 5. Prevalence and resistance patterns of multiresistant *Y. enterocolitica* 4/O:3 in pig tonsils from different farm types (2014).

Farm Type	Resistance Pattern	Number of Resistant Isolates	Number of Tested Isolates	% of Multiresistant Isolates/Patterns
Big integrated	NA-CHL-STR	13		54
	TET-NA-CHL-STR	1	24	4
	TET-NA-CAZ-TMP/SMX	1		4
Medium-sized	NA-CHL-CFX	1	25	4

NA: nalidixic acid, CHL: Chloramphenicol, STR: Streptomycin, TET: Tetracycline, CAZ: Ceftazidime, TMP/SMX: Trimethoprim/Sulfamethoxazole, CFX: Cefotaxime.

In this study, among the 84 isolates tested, resistance was detected, in addition to ampicillin and cephalothin, toward nalidixic acid (20% of isolates), streptomycin (18%), chloramphenicol (12%), ceftazidime (4.7%), levofloxacin (2.4%), and cephalotaxime (1.2%). Multiresistance was found in 10 isolates among 12 tested from big integrated farms. Nine of these isolates (75%) were simultaneously resistant to nalidixic acid, chloramphenicol, and streptomycin. One isolate was additionally resistant to cefotaxime. In contrast, only one isolate from a medium-sized farm was multiresistant (ceftazidime, trimethoprim/sulfamethoxazole, streptomycin). Similarly, among *Y. enterocolitica* 4/O:3 isolates from family farms, only one multiresistant isolate was found (Table 6).

Table 6. Prevalence and resistance patterns of multiresistant *Y. enterocolitica* 4/O:3 in pig tonsils from different farm types (2019).

Farm Type	Resistance Pattern	Number of Resistant Isolates	Number of Tested Isolates	% of Multiresistant Isolates/Patterns
Big integrated	NA-CHL-STR	9	12	75
	TET-NA-CHL-CFX	1		8
Medium-sized	CAZ-TMP/SMX-STR	1	59	2
Small	NA-CAZ-TMP/SMX	1	13	8

NA: nalidixic acid, CHL: Chloramphenicol, STR: Streptomycin, TET: Tetracycline, CAZ: Ceftazidime, TMP/SMX: Trimethoprim/Sulfamethoxazole, CFX: Cefotaxime.

Excluding the year of isolation, isolates of *Y. enterocolitica* 4/O:3 from integrated farms were more resistant to streptomycin, chloramphenicol, and nalidixic acid compared to isolates from the other two farm systems (Table 7). No significant differences were found with respect to the susceptibility/resistance of *Y. enterocolitica* isolates from big integrated farms and considering the year of isolation of the pathogen ($p > 0.05$). Similarly, no significant differences were found in the susceptibility/resistance of *Y. enterocolitica* isolates from medium-sized farms between both surveys ($p > 0.05$) (Tables 8 and 9).

Table 7. Antimicrobial susceptibility of *Y. enterocolitica* 4/O:3 isolates collected in two surveys of tonsils from pigs raised in different housing systems.

Antimicrobial Agent	Big Integrated Farms (*n* = 36)			Medium-Sized Farms (*n* = 84)			Small Family Farms (*n* = 13)			Total (*n* = 133)		
	S	I	R	S	I	R	S	I	R	S	I	R
Levofloxacin	36	0	0	81	3	0	13	0	0	130	3	0
Ciprofloxacin	36	0	0	84	0	0	13	0	0	133	0	0
Ampicillin	1	7	28	2	10	72	0	0	13	3	17	113
Cephalothin	2	5	29	16	2	66	0	0	13	18	7	108
Cefotaxime	35	0	1	81	2	1	13	0	0	129	2	2
Tetracycline	34	0	2	83	0	1	13	0	0	130	0	3
Nalidixic acid	9	0	27	76	3	5	12	0	1	97	3	33
Ceftazidime	35	0	1	79	2	3	12	0	1	126	2	5
Trimethoprim/ Sulfamethoxazole	34	1	1	83	0	1	12	0	1	129	1	3
Chloramphenicol	12	0	24	82	1	1	13	0	0	107	1	25
Streptomycin	11	3	22	72	8	4	10	1	2	93	12	28

S = sensitive, I = intermediate, R = resistant.

Table 8. Antimicrobial susceptibility/resistance of *Y. enterocolitica* 4/O:3 isolates from big integrated farms.

Antimicrobial Agent	Year 2014 (*n* = 24)			Year 2019 (*n* = 12)		
	S	I	R	S	I	R
Levofloxacin	24	0	0	12	0	0
Ciprofloxacin	24	0	0	12	0	0
Ampicillin	1	2	21	0	5	7
Cephalothin	1	5	18	0	0	12
Cefotaxime	24	0	0	11	0	1
Tetracycline	22	0	2	12	0	0
Nalidixic acid	9	0	15	0	0	12
Ceftazidime	23	0	1	12	0	0
Trimethoprim/Sulfamethoxazole	23	0	1	11	1	0
Chloramphenicol	9	0	15	3	0	9
Streptomycin	9	1	14	2	2	8

S = sensitive, I = intermediate, R = resistant.

Table 9. Antimicrobial susceptibility/resistance of *Y. enterocolitica* 4/O:3 isolates from medium-sized farms.

Antimicrobial Agent	Year 2014 (*n* = 25)			Year 2019 (*n* = 59)		
	S	I	R	S	I	R
Levofloxacin	25	0	0	56	0	3
Ciprofloxacin	25	0	0	59	0	0
Ampicillin	2	6	17	0	5	54
Cephalothin	0	2	23	0	0	59
Cefotaxime	22	2	1	59	0	0
Tetracycline	24	0	1	59	0	0
Nalidixic acid	23	0	2	53	3	3
Ceftazidime	24	1	0	55	1	3
Trimethoprim/Sulfamethoxazole	25	0	0	58	0	1
Chloramphenicol	24	0	1	58	1	0
Streptomycin	25	0	0	47	8	4

S = sensitive, I = intermediate, R = resistant.

4. Discussion

The study was based on the assumption that the overall prevalence of pathogenic *Y. enterocolitica* in the tonsils of pigs does not change significantly depending on the year, but that there are differences related to the type of husbandry, especially in the prevalence of resistant isolates. When pathogenic *Y. enterocolitica* is found in portioned and minced pork, the phenotypic and genetic characteristics of the isolates are expected to be identical to those obtained from the tonsils of pigs from the same farm/slaughterhouse.

4.1. Prevalence of Y. enterocolitica in Pig Tonsils at Slaughter and Retail Pork

Given the current lack of data on the prevalence of pathogenic *Y. enterocolitica* in pigs and pork in Croatia, this study aimed to map the production chain from farms to slaughterhouses and pork retail outlets to assess the risk of pathogen transmission to consumers. The relevance of the study stems from the fact that *Y. enterocolitica* is a priority biological hazard in pig meat inspection in Europe and a target of a new comprehensive meat safety assurance system [8,23]. This study builds on the preliminary results previously obtained from a smaller study conducted in 2014, which showed a *Y. enterocolitica* O:3 prevalence of 33% and 10% in tonsils and mandibular lymph nodes, respectively [16]. In comparison, the results of this study showed a higher prevalence of *Y. enterocolitica* in pig tonsils, i.e., 43% [95% CI 36.7–49.7]. The present results are in agreement with other European studies, such as Fredriksson-Ahomaa et al. [24] in Switzerland (prevalence of 34%), van Damme et al. [25] in Belgium (37%), and Martínez et al. [26] in Belgium (44%) and Italy (32%). On the other hand, Fredriksson-Ahomaa et al. [27] and Martínez et al. [26] warned of a high prevalence of pathogenic *Y. enterocolitica* in slaughtered pigs in Finland (62%) and Spain (93%), respectively. At the other extreme are the studies that found low prevalence: 2%, 4%, 8%, 9%, 11%, and 13% [28–33]. Several other studies conducted in Europe in recent years also show very different results and the prevalence of *Y. enterocolitica* ranges from 3% [34] (Sardinia), to 14% [35] (Central Italy), to 97% [36] (Finland).

When considering the relationship between *Y. enterocolitica* findings and biosecurity conditions, this study found that there were statistically significant differences in prevalence among integrated farms as well as among medium-sized farms, despite the same level of biosecurity. It is likely that prevalence was affected by slaughterhouse factors, such as possible contact between pig batches at lairage, or omitting sterilization of knife after pluck set removal, as reported before [16].

The opposite was true for family farms, where differences in prevalences were likely related to farm biosecurity levels. Pig farming systems vary among European countries, and comparisons of the prevalence of *Y. enterocolitica* as a function of the type of fattening pig farming system are rare in the literature. However, conventional and alternative (organic) housing systems have been compared, and Nowak et al. [37] found a higher number of positive pigs (29% vs. 18%) in conventional housing systems, with twice as many tonsils from conventionally housed pigs being positive for *Y. enterocolitica* (22% vs. 11%). Also of interest are the results of Novoslavskij et al. [38] in Lithuania, who linked the higher prevalence of *Y. enterocolitica* in pigs to lower farm biosecurity. However, detailed biosecurity factors used in farm categorization were not available in our study, which prevents us from correlating specific factors with observed prevalence.

In addition, practices at the harvest stage, such as lairage cross-contamination or removal of the pluck set, could influence the rate of contamination of tonsils with *Y. enterocolitica* [39]. All of this highlights the complexity of reporting the true prevalence (pre-harvest) of pathogenic *Y. enterocolitica* and the role of on-farm and slaughter practices in the spread of the pathogen to the consumer. In this context, the assessment of the prevalence of *Y. enterocolitica* based on tonsils as a predilection site needs to be complemented by other data, such as serological tests. In recent years, serological surveillance prior to slaughter has been recommended for risk management purposes in slaughterhouses [40]. Serological testing also showed significant differences in seroprevalence of *Y. enterocolitica* in pigs housed in different fattening systems [41]. Similar to *Salmonella*, data on seroprevalence and/or the presence of *Y. enterocolitica* in lymphoid tissues or intestines can help to reduce risk by implementing decontamination measures on pig carcasses [8].

No positive findings of pathogenic *Y. enterocolitica* were detected when marketed pork cuts and minced pork were examined (*n* = 128), indicating a low risk of *Y. enterocolitica* transmission to such meat. The same results were found in the study by Laukkanen-Ninios et al. [11]. Martins et al. [12] similarly isolated *Y. enterocolitica* from the tonsils and lymph nodes of pigs, but not from environmental samples or from pork cuts. Given slaughter techniques and possible hygiene deficiencies during processing, it is likely that contamination occurs first in the meat of the neck region, head, tongue, and throat, rather than on the carcass, as reported in other studies [11,42]. In contrast to our results, considerable contamination of minced meat with *Y. enterocolitica* was found in other studies [43–47].

Recovery Rate of *Y. enterocolitica* by Different Methods of Isolation and MALDI-TOF MS Determination

Another factor that may influence the outcome of determining the prevalence of *Y. enterocolitica* in pig tonsils is the methodology of sampling and isolation. The results obtained show that the success of isolating pathogenic *Y. enterocolitica* by enrichment of tonsils in selective ITC broth is higher than in PSB, but is vice versa after alkali treatment of PSB and ITC broths. Van Damme et al. [25] found that KOH treatment of broth, particularly PSB, was a key factor significantly affecting the success of isolating pathogenic *Y. enterocolitica* from pig tonsils. In our study, we also found that alkali treatment of PSB broth and inoculation on CIN resulted in a significantly higher number of positive samples compared to untreated samples (OR = 7.41, $p < 0.0001$). The same case was found with KOH treatment of PSB and inoculation on CHROMagar^{TM} (OR = 4.71, $p < 0.0001$).

MALDI-TOF MS identification of presumptive colonies demonstrated excellent selectivity of the agars used, especially in the case of CHROMagar^{TM} for screening pathogenic biotypes. This shortens the process for preliminary assessment of pathogenicity, which was determined at later stages by biotyping, serotyping, and detection of the *ail* gene. The use of other chromogenic media, such as YECA, has also been shown to be useful in shortening the process by direct detection of pathogenic biotypes in pig tonsil [48]. In addition, the combination of CHROMagar$^{®}$ and MALDI-TOF MS is less time consuming for the detection of pathogenic isolates compared to conventional isolation methods and biochemical tests. Moreover, MALDI-TOF MS can identify strains belonging to different *Y. enterocolitica* biotypes [49,50]. It is well known that isolation and identification of this bacterium is challenging. Therefore, more sensitive and rapid techniques than existing culture methods have been developed in recent years [9]. Peruzy et al. [51] generally believed that conventional isolation methods for *Y. enterocolitica* are not reliable enough, which they interpreted as due to competition with the background microbiota in tonsils.

4.2. *Y. enterocolitica* Biotyping, Serotyping, PCR, and PFGE Typing

The results obtained from both surveys show the persistence of the pathogenic bioserotype 4/O:3 in the tonsils of fattening pigs in Croatia. This pathogenic bioserotype is most commonly isolated from clinical cases of yersiniosis in humans as well as from carrier pigs in many European countries [52–56]. All *Y. enterocolitica* 4/O:3 isolates from this study carried the *ail* gene that is required for bacterial adhesion and invasion into the host cell as well as serum resistance. However, the gene is also sporadically present in nonpathogenic *Yersinia* species as well as in nonpathogenic *Y. enterocolitica* biotypes such as biotype 1A, so other tests are also needed to confirm the pathogenicity of *Y. enterocolitica* isolates [57]. Therefore, in our study, potential pathogenicity was assessed by colony morphology on chromogenic agar, detection of the *ail* gene, biotyping, and serotyping. The pathogenic bioserotype 4/O:3 is also the prevalent type among *Y. enterocolitica* isolates from fattening pigs sampled at the slaughter line (tonsils) in other European countries, such as Germany (99% of isolates, 2001, [58]), Switzerland (96% of isolates, 2007, [24]), or Finland, 2000, (100%, [3]). The persistence of this bioserotype of *Y. enterocolitica* has been confirmed in similar studies in later years in the same countries [36,59], which is in agreement with our results. In contrast, Bonardi et al. [40,60] reported lower prevalences (15% and 27%) of *Y. enterocolitica* 4/O:3 in two surveys conducted in Italy (2014, 2016). The persistence of the pathogenic *Y. enterocolitica* bioserotype 4/O:3 was recently confirmed in the Brazilian pork production chain (tonsils, oral cavity, head meat) by comparing the results of two studies two years apart, confirming the importance of slaughter hygiene and farming practices in the epidemiology of yersiniosis [13].

Persistence and epidemiology of pathogenic *Y. enterocolitica* is also assessed by molecular typing using methods such as PFGE, MVLA, or whole genome sequencing [9,15]. In our study, selected isolates (based on year of isolation and farm of origin) were subjected to restriction enzyme DNA fragment comparison by PFGE. We found the same pulsotypes occurred regardless of the year of isolation and the origin of the isolates, confirming the assumption of persistence of the pathogenic bioserotype 4/O:3 in pig tonsils. Although the analysis formed several clusters in the dendrogram, their agreement ranged from 80.6% to 100%, indicating low variability of this bioserotype (Figure 1). Similar results from pulsotyping *Y. enterocolitica* bioserotype 4/O:3 isolates were obtained by Martins et al. [13]. They compared pulsotypes of *Y. enterocolitica* bioserotype 4/O:3 isolates collected in 2016 and 2018 from tonsils, lymph nodes, and carcass swabs in the same slaughterhouses using macrorestriction enzymes (XbaI or NotI) and also found high agreement between isolates, ranging from 82.4 to 100%. The low variability of *Y. enterocolitica* 4/O:3 was also noted when comparing human and pig isolates, the pulsotypes of which were combined into a single cluster [61]. Despite the low genetic variability of the 4/O:3 bioserotype, Fredriksson-Ahomaa et al. [62] recommended the PFGE method for distinguishing genotypes present in pig farms using a combination of the restriction enzymes NotI, ApaI, and XhoI. However, the same genotype for bioserotype 4/O:3 isolates was found in most farms (71%).

4.3. Antimicrobial Susceptibility of Y. enterocolitica 4/O:3

In this work, the susceptibility of *Y. enterocolitica* isolates from pig tonsils to antimicrobials was investigated to gain insight into the variability of the resistance profile over time and the origin of the isolates (farm type). The presence of resistant *Y. enterocolitica* in slaughter pigs has been studied in many European countries in recent years [34,40], but not in Croatia. In Latvia [63], resistance to erythromycin and sulfamethoxazole was detected in all *Y. enterocolitica* tested. Bonardi et al. [40], in northern Italy, also reported a frequent prevalence of sulfonamide resistance in slaughtered pigs. In contrast, the prevalence of sulphonamide resistance in our study was rare, as was also reported by other authors from Switzerland and Germany [22,64]. In contrast to other studies [30,65], isolates from the current study were frequently resistant to chloramphenicol, nalidixic acid or streptomycin, and these multiresistant isolates were present in fattening pigs from big integrated farms. In addition, resistance to third generation of cephalosporins was detected in several isolates, which is of clinical relevance. The high public health relevance has been highlighted in recent reports [15] confirming *Y. enterocolitica* 4/O:3 as a novel multidrug-resistant pathogen possessing transmissible resistance determinants.

Therefore, our results show a significantly higher prevalence of multidrug-resistant isolates of *Y. enterocolitica* bioserotype 4/O:3 in big integrated pig farms, although the resistance profile has not changed significantly over the years of research (Table 8). The susceptibility/resistance of *Y. enterocolitica* to certain antimicrobials has also not changed significantly over the years in pigs from medium sized farms (Table 9). To our knowledge, no similar studies have been conducted in Croatia, so more accurate comparisons are not possible. For some bacterial species, resistance profiles can generally be observed with respect to the year of isolation to allow comparison, i.e., insight into an increase or decrease in resistance over time. An earlier study [66] (2007; Switzerland) found that isolates of *Y. enterocolitica* from pork, humans, and pig feces were highly resistant to ampicillin, cephalothin, and amoxicillin/clavulanic acid. In the same year, Fredriksson-Ahomaa et al. [24] found dominant resistance to ampicillin and erythromycin. Bonardi et al. [33] recorded the *Y.*

enterocolitica were resistant primarily to cephalothin, ampicillin, streptomycin, and then amoxicillin/clavulanic acid in Italian pig slaughterhouses (2013), and Sacchini et al. [35] reported resistance to ampicillin, streptomycin, sulfisoxazole, tetracycline, nalidixic acid, and chloramphenicol (2018). The resistance profiles of Y. *enterocolitica* have not changed significantly in recent years, likely due to the genetic stability of the pathogen [14]. Fredriksson-Ahomaa et al. [67] found no association between Y. *enterocolitica* genotypes and resistance profiles in pigs. In this context, although our Y. *enterocolitica* 4/O:3 isolates were all genetically similar by the methods used, isolates from the different housing systems showed significant variability in phenotypic antibiotic resistance. This likely reflects the greater exposure of the pathogens to antimicrobial agents on big integrated farms than on small farms.

5. Conclusions

Considering all the results presented in this work, the high prevalence of pathogenic Y. *enterocolitica* 4/O:3 in pig tonsils is an important risk factor for pig carcass contamination at slaughter. The pathogen was not isolated from pork cuts or minced meat placed on the market, likely due to good hygiene procedures in meat cutting and preparation, which indicates a low risk to consumers. The low recovery of pathogen from minced meat or pork cuts can also be affected by background microbiota and low sensitivity of culture method. The prevalence of the pathogen in pig tonsils did not depend on the biosecurity level of the farms, except in the case of family farms. Comparison of genetic profiles showed a high concordance of Y. *enterocolitica* isolates over the study years and in the investigated farm systems; the antimicrobial resistance patterns also did not change significantly by year or farm system. However, a significantly higher prevalence of multidrug-resistant isolates was found in pigs from big integrated farms, which could be due to greater pressure of antimicrobial agents used on such farms.

Further studies of this foodborne pathogen in the context of microbiological safety in pork production chain are needed to gain better insight into antimicrobial resistance and *Yersinia* epidemiology. In addition to culture methods, molecular and serological tests should be used to determine prevalence and distinguish natural infection or transmission from possible external contamination during carcass processing.

Supplementary Materials: The following supporting information can be downloaded at: https://www.mdpi.com/article/10.3390/foods11101459/s1, Table S1. Tonsil sampling scheme in slaughtered fattening pigs from different farm types.

Author Contributions: Conceptualization, N.Z. and V.P.; methodology, N.Z. and M.K.; software, D.J., S.K. and M.P.; formal analysis, M.K., K.B., N.Z. and V.P.; writing—original draft preparation, N.Z.; writing—review and editing, B.B., D.A. and M.F.-A.; project administration, N.Z.; funding acquisition, N.Z. All authors have read and agreed to the published version of the manuscript.

Funding: This research was funded by the University of Zagreb and the Ministry of Science and Education, Croatia, Programme Contracts 2015–2021. Proofreading and APC were funded by COST Action 18105 (Risk-based Meat Inspection and Integrated Meat Safety Assurance).

Institutional Review Board Statement: Ethical Committee Decision, Faculty of Veterinary Medicine University of Zagreb.

Informed Consent Statement: Not applicable.

Data Availability Statement: Data is contained within the article or supplementary material.

Acknowledgments: This publication is based on work from COST Action 18105 (Risk-based Meat Inspection and Integrated Meat Safety Assurance; www.ribmins.com, Accessed on: 12 April 2022) supported by COST (European Cooperation in Science and Technology; www.cost.eu, Accessed on: 12 April 2022).

Conflicts of Interest: The authors declare no conflict of interest. The funders had no role in the design of the study; in the collection, analyses, or interpretation of data; in the writing of the manuscript, or in the decision to publish the results.

References

1. Riahi, S.M.; Ahmadi, E.; Zeinali, T. Global prevalence of *Yersinia enterocolitica* in cases of gastroenteritis: A systematic review and meta-analysis. *Int. J. Microbiol.* **2021**, *2021*, 1499869. [CrossRef] [PubMed]
2. EFSA; ECDC. The European Union One Health 2020 Zoonoses Report. *EFSA J.* **2021**, *19*, e06971.
3. Fredriksson-Ahomaa, M.; Björkroth, J.; Hielm, S.; Korkeala, H. Prevalence and characterization of pathogenic *Yersinia enterocolitica* in pig tonsils from different slaughterhouses. *Food Microbiol.* **2000**, *17*, 93–101. [CrossRef]
4. Virtanen, S.E.; Salonen, L.K.; Laukkanen, R.; Hakkinen, M.; Korkeala, H. Factors related to the prevalence of pathogenic *Yersinia enterocolitica* on pig farms. *Epidemiol. Infect.* **2011**, *139*, 1919–1927. [CrossRef] [PubMed]
5. Vilar, M.J.; Virtanen, S.; Heinonen, M.; Korkeala, H. Management practices associated with the carriage of *Yersinia enterocolitica* in pigs at farm level. *Foodborne Pathog. Dis.* **2013**, *10*, 595–602. [CrossRef] [PubMed]
6. Arden, K.; Gedye, K.; Angelin-Bonnet, O.; Murphy, E.; Antic, D. *Yersinia enterocolitica* in wild and peridomestic rodents within Great Britain, a prevalence study. *Zoonoses Public Health* **2022**. [CrossRef] [PubMed]
7. Vilar, M.J.; Virtanen, S.; Laukkanen-Ninios, R.; Korkeala, H. Bayesian modelling to identify the risk factors for *Yersinia enterocolitica* contamination of pork carcasses and pluck sets in slaughterhouses. *Int. J. Food Microbiol.* **2015**, *197*, 53–57. [CrossRef]
8. Zdolec, N.; Kiš, M. Meat Safety from farm to slaughter—Risk-based control of *Yersinia enterocolitica* and *Toxoplasma gondii*. *Processes* **2021**, *9*, 815. [CrossRef]
9. Fredriksson-Ahomaa, M.; Joutsen, S.; Laukkanen-Ninios, R. Identification of *Yersinia* at the Species and Subspecies Levels Is Challenging. *Curr. Clin. Microbiol. Rep.* **2018**, *5*, 135–142. [CrossRef]
10. Guillier, L.; Fravalo, P.; Leclercq, A.; Thébault, A.; Kooh, P.; Cadavez, V.; Gonzales-Barron, U. Risk factors for sporadic *Yersinia enterocolitica* infections: A systematic review and meta-analysis. *Microb. Risk Anal.* **2021**, *17*, 100141. [CrossRef]
11. Laukkanen-Ninios, R.; Fredriksson-Ahomaa, M.; Maijala, R.; Korkeala, H. High prevalence of pathogenic *Yersinia enterocolitica* in pig cheeks. *Food Microbiol.* **2014**, *43*, 50–52. [CrossRef] [PubMed]
12. Martins, B.T.F.; Botelho, C.V.; Lopes Silva, D.A.; Lanna, F.G.P.A.; Libero Grossi, J.; Campos-Galvão, M.E.M.; Seiti Yamatogi, R.; Pfrimer Falcão, J.; dos Santos Bersot, L.; Nero, L.A. *Yersinia enterocolitica* in a Brazilian pork production chain: Tracking of contamination routes, virulence and antimicrobial resistance. *Int. J. Food Microbiol.* **2018**, *276*, 5–9. [CrossRef] [PubMed]
13. Martins, B.T.F.; Azevedo, E.C.; Yamatogi, R.S.; Call, D.R.; Nero, L.A. Persistence of *Yersinia enterocolitica* bio-serotype 4/O:3 in a pork production chain in Minas Gerais, Brazil. *Food Microbiol.* **2021**, *94*, 103660. [CrossRef] [PubMed]
14. Bonke, R.; Wacheck, S.; Stüber, E.; Meyer, C.; Märtlbauer, E.; Fredriksson-Ahomaa, M. Antimicrobial susceptibility and distribution of b-lactamase A (blaA) and b-Lactamase B (blaB) genes in enteropathogenic *Yersinia* species. *Microb. Drug Resist.* **2011**, *17*, 575–581. [CrossRef] [PubMed]
15. Karlsson, P.A.; Tano, E.; Jernberg, C.; Hickman, R.A.; Guy, L.; Järhult, J.D.; Wang, H. Molecular characterization of multidrug-resistant *Yersinia enterocolitica* from foodborne outbreaks in Sweden. *Front. Microbiol.* **2021**, *12*, 664665. [CrossRef]
16. Zdolec, N.; Dobranić, V.; Filipović, I. Prevalence of *Salmonella* spp. and *Yersinia enterocolitica* in/on tonsils and mandibular lymph nodes of slaughtered pigs. *Folia Microbiol.* **2015**, *60*, 131–135. [CrossRef]
17. Ministry of Agriculture. Multi-Annual National Plan of Official Controls of the Republic of Croatia. 2020. Available online: http://www.veterinarstvo.hr/UserDocsImages/Koordinacija%20slu%C5%BEbenih%20kontrola/VNPSK%20-%20final.pdf (accessed on 5 May 2022).
18. Zdolec, N.; Bogdanović, T.; Severin, K.; Dobranić, V.; Kazazić, S.; Grbavac, J.; Pleadin, J.; Petričević, S.; Kiš, M. Biogenic amine content in retailed cheese varieties produced with commercial bacterial or mold cultures. *Processes* **2022**, *10*, 10. [CrossRef]
19. *EN ISO 10273:2017*; Microbiology of the Food Chain—Horizontal Method for the Detection of Pathogenic *Yersinia enterocolitica*. ISO: Geneva, Switzerland, 2017.
20. CDC. *One-Day (24–28 h) Standardized Laboratory Protocol for Molecular Subtyping of Yersinia pestis by Pulsed Field Gel Electrophoresis (PFGE)*; Pulse Net: Atlanta, GA, USA, 2006.
21. Peters, T.M.; Maguire, C.; Threlfall, E.J.; Fisher, I.S.; Gill, N.; Gatto, A.J. The Salm-gene project—A European collaboration for DNA fingerprinting for food-related salmonellosis. *Euro Surveill.* **2003**, *8*, 46–50. [CrossRef]
22. CLSI. *Performance Standards for Antimicrobial Susceptibility Testing*, 29th ed.; CLSI Supplement M100; Clinical and Laboratory Standards Institute: Wayne, PA, USA, 2019.
23. Blagojević, B.; Nesbakken, T.; Alvseike, O.; Vågsholm, I.; Antic, D.; Johlerf, S.; Houf, K.; Meemken, D.; Nastasijevic, I.; Vieira Pinto, M.; et al. Drivers, opportunities, and challenges of the European risk-based meat safety assurance system. *Food Control* **2021**, *124*, 107870. [CrossRef]
24. Fredriksson-Ahomaa, M.; Stolle, A.; Stephan, R. Pravelence of pathogenic *Yersinia enterocolitica* in pigs slaughtered at a Swiss abattoir. *Int. J. Food Microbiol.* **2007**, *119*, 207–212. [CrossRef]
25. Van Damme, I.; Habib, I.; De Zutter, L. *Yersinia enterocolitica* in slaughter pig tonsils: Enumeration and detection by enrichment versus direct plating culture. *Food Microbiol.* **2010**, *27*, 158–161. [CrossRef] [PubMed]
26. Martínez, P.O.; Fredriksson-Ahomaa, M.; Pallotti, A.; Rosmini, R.; Houf, K.; Korkeala, H. Variation in the prevalence of enteropathogenic *Yersinia* in slaughter pigs from Belgium, Italy, and Spain. *Foodborne Pathog. Dis.* **2011**, *8*, 445–450. [CrossRef] [PubMed]
27. Fredriksson-Ahomaa, M.; Gerhardt, M.; Stolle, A. High bacterial contamination of pig tonsils at slaughter. *Meat Sci.* **2009**, *83*, 334–336. [CrossRef] [PubMed]

28. Kot, B.; Trafny, E.A.; Jakubczak, A. Application of multiplex PCR for monitoring colonization of pig tonsils by *Yersinia enterocolitica*, including biotype 1A, and *Yersinia pseudotuberculosis*. *J. Food Prot.* **2007**, *70*, 1110–1115. [CrossRef] [PubMed]

29. Kot, B.; Jakubczak, A.; Piechota, M. Isolation and characterization of *Yersinia enterocolitica* rods from pig tonsils. *Med. Wet.* **2008**, *64*, 283–287.

30. Simonova, J.; Borilova, G.; Steinhauserova, I. Occurrence of pathogenic strains of *Yersinia enterocolitica* in pigs and their antimicrobial resistance. *Bull. Vet. Inst. Pulawy* **2008**, *52*, 39–43.

31. Fondrevez, M.; Labbé, A.; Houard, E.; Fravalo, P.; Madec, F.; Denis, M. A simplified method for detecting pathogenic *Yersinia enterocolitica* in slaughtered pig tonsils. *J. Microbiol. Methods* **2010**, *83*, 244–249. [CrossRef]

32. O'Sullivan, T.; Friendship, R.; Blackwell, T.; Pearl, D.; McEwen, B.; Carman, S.; Slavić, Đ.; Dewey, C. Microbiological identification and analysis of swine tonsils collected from carcasses at slaughter. *Can. J. Vet. Res.* **2011**, *75*, 106–111.

33. Bonardi, S.; Bassi, L.; Briandini, F.; D'incau, M.; Barco, L.; Carra, E.; Pongolini, S. Prevalence, characterization and antimicrobial susceptibility of *Salmonella enterica* and *Yersinia enterocolitica* in pigs at slaughter in Italy. *Int. J. Food Microbiol.* **2013**, *163*, 248–257. [CrossRef]

34. Fois, F.; Piras, F.; Torpdahl, M.; Mazza, R.; Ladu, D.; Consolati, S.G.; Spanu, C.; Scarano, C.; De Santis, E.P.L. Prevalence, bioserotyping and antibiotic resistance of pathogenic *Yersinia enterocolitica* detected in pigs at slaughter in Sardinia. *Int. J. Food Microbiol.* **2018**, *283*, 1–6. [CrossRef]

35. Sacchini, L.; Garofolo, G.; Di Stefano, G.; Marotta, F.; Ricci, L.; Di Donato, G.; Miracco, M.G.; Perletta, F.; Di Giannatale, E. The prevalence, characterisation, and antimicrobial resistance of *Yersinia enterocolitica* in pigs from Central Italy. *Vet. Ital.* **2018**, *54*, 115–123. [PubMed]

36. Koskinen, J.; Keto-Timonen, R.; Virtanen, S.; Vilar, M.J.; Korkeala, H. Prevalence and dynamics of pathogenic *Yersinia enterocolitica* 4/O:3 among Finnish piglets, fattening pigs and sows. *Foodborne Pathog. Dis.* **2019**, *16*, 831–839. [CrossRef] [PubMed]

37. Nowak, B.; Mueffling, T.V.; Caspari, K.; Hartung, J. Validation of a method for the detection of virulent *Yersinia enterocolitica* and their distribution in slaughter pigs from conventional and alternative housing systems. *Vet. Microbiol.* **2006**, *117*, 219–228. [CrossRef] [PubMed]

38. Novoslavskij, A.; Šernienė, L.; Malakauskas, A.; Laukkanen-Ninios, R.; Korkeala, H.; Malakauskas, M. Prevalence and genetic diversity of enteropathogenic *Yersinia* spp. in pigs at farms and slaughter in Lithuania. *Res. Vet. Sci.* **2013**, *94*, 209–213. [CrossRef] [PubMed]

39. Laukkanen, R.; Ranta, J.; Dong, X.; Hakkinen, M.; Martínez, P.O.; Lundén, J.; Johansson, T.; Korkeala, H. Reduction of enteropathogenic *Yersinia* in the pig slaughterhouse by using bagging of the rectum. *J. Food Protect.* **2010**, *73*, 2161–2168. [CrossRef] [PubMed]

40. Bonardi, S.; Bruini, I.; D'incau, M.; Van Damme, I.; Carniel, E.; Brémont, S.; Cavallini, P.; Tagliabue, S.; Brindani, F. Detection, seroprevalence and antimicrobial resistance of *Yersinia enterocolitica* and *Yersinia pseudotuberculosis* in pig tonsils in Northern Italy. *Int. J. Food Microbiol.* **2016**, *235*, 125–132. [CrossRef] [PubMed]

41. Felin, E.; Jukola, E.; Raulo, S.; Fredriksson-Ahomaa, M. Meat Juice Serology and Improved Food Chain Information as Control Tools for Pork-Related Public Health Hazards. *Zoonoses Public Health* **2014**, *62*, 456–464. [CrossRef]

42. Meßelhäußer, U.; Kämpf, P.; Colditz, J.; Bauer, H.; Schreiner, H.; Höller, C.; Busch, U. Qualitative and quantitative detection of human pathogenic *Yersinia enterocolitica* in different food matrices at retail level in Bavaria. *Foodborne Pathog. Dis.* **2011**, *8*, 39–44. [CrossRef]

43. Vishnubhatla, A.; Oberst, R.D.; Fung, D.Y.C.; Wonglumsom, W.; Hays, M.P.; Nagaraja, T.G. Evaluation of a 5′-nuclease (TaqMan) assay for the detection of virulent strains of *Yersinia enterocolitica* in raw meat and tofu samples. *J. Food Protect.* **2001**, *64*, 355–360. [CrossRef]

44. Fredriksson-Ahomaa, M.; Korkeala, H. Molecular Epidemiology of *Yersinia enterocolitica* 4/O:3. In *The Genus Yersinia*; Skurnik, M., Bengoechea, J.A., Granfors, K., Eds.; Springer: New York, NY, USA, 2003; pp. 295–302.

45. Latha, C.; Anu, C.J.; Ajaykumar, V.J.; Sunil, B. Prevalence of *Listeria monocytogenes*, *Yersinia enterocolitica*, *Staphylococcus aureus*, and *Salmonella enterica* Typhimurium in meat and meat products using multiplex polymerase chain reaction. *Vet. World* **2017**, *10*, 927–931. [CrossRef]

46. Thong, K.L.; Tan, L.K.; Ooi, P.T. Genetic diversity, virulotyping and antimicrobial resistance susceptibility of *Yersinia enterocolitica* isolated from pigs and porcine products in Malaysia. *J. Sci. Food Agric.* **2017**, *98*, 87–95. [CrossRef]

47. Ferl, M.; Mäde, D.; Braun, P.G. Combined molecular biological and microbiological detection of pathogenic *Yersinia enterocolitica* in spiced ground pork, meat for production of ground pork and raw sausages. *JVL* **2020**, *15*, 27–35. [CrossRef]

48. Denis, M.; Houard, E.; Labbé, A.M.; Fondrevez, M.; Salvat, G. A selective chromogenic plate, YECA, for the detection of pathogenic *Yersinia enterocolitica*: Specificity, sensitivity, and capacity to detect pathogenic *Yersinia enterocolitica* from pig tonsils. *J. Pathog.* **2011**, *2011*, 296275. [CrossRef] [PubMed]

49. Fredriksson-Ahomaa, M. *Yersinia enterocolitica*. In *Foodborne Diseases*, 3rd ed.; Dodd, C.E.R., Aldsworth, T., Stein, R.A., Cliver, D.O., Riemann, H.P., Eds.; Academic Press: London, UK, 2017; pp. 223–232.

50. Morka, K.; Bystroń, J.; Bania, J.; Korzeniowska-Kowal, A.; Korzekwa, K.; Guz-Regner, K.; Bugla-Płoskońska, G. Identification of *Yersinia enterocolitica* isolates from humans, pigs and wild boars by MALDI-TOF MS. *BMC Microbiol.* **2018**, *18*, 86. [CrossRef] [PubMed]

51. Peruzy, M.F.; Aponte, M.; Proroga, Y.T.R.; Capuano, F.; Cristiano, D.; Delibato, E.; Houf, K.; Murru, N. *Yersinia enterocolitica* detection in pork products: Evaluation of isolation protocols. *Food Microbiol.* **2020**, *92*, 103593. [CrossRef] [PubMed]

52. Drummond, N.; Murphy, B.P.; Ringwood, T.; Prentice, M.B.; Buckley, J.F.; Fanning, S. *Yersinia enterocolitica*: A brief review of the issues relating to the zoonotic pathogen, public health challenges, and the pork production chain. *Foodborne Pathog. Dis.* **2012**, *9*, 179–189. [CrossRef] [PubMed]

53. Le Guern, A.S.; Martin, L.; Savin, C.; Carniel, E. Yersiniosis in France: Overview and potential sources of infection. *Int. J. Infect. Dis.* **2016**, *46*, 1–7. [CrossRef]

54. Centorame, P.; Sulli, N.; De Fanis, C.; Colangelo, O.V.; De Massis, F.; Conte, A.; Persiani, T.; Marfoglia, C.; Scattolini, S.; Pomilio, F.; et al. Identification and characterization of *Yersinia enterocolitica* strains isolated from pig tonsils at slaughterhouse in Central Italy. *Vet. Ital.* **2017**, *53*, 331–344.

55. Biasino, W.; De Zutter, L.; Woollard, J.; Mattheus, W.; Bertrand, S.; Uyttendaele, M.; Van Damme, I. Reduced contamination of pig carcasses using an alternative pluck set removal procedure during slaughter. *Meat Sci.* **2018**, *145*, 246–247. [CrossRef]

56. Moreira, L.M.; Milan, C.; Gonçalves, T.G.; Ebersol, C.N.; De Lima, H.G.; Timm, C.D. Contamination of pigs by *Yersinia enterocolitica* in the abattoir flowchart and its relation to the farm. *Ciênc. Rural* **2019**, *49*, 1–7. [CrossRef]

57. Joutsen, S.; Johansson, P.; Laukkanen-Ninios, R.; Björkroth, J.; Fredriksson-Ahomaa, M. Two copies of the ail gene found in *Yersinia enterocolitica* and *Yersinia kristensenii*. *Vet. Microbiol.* **2020**, *247*, 108798. [CrossRef] [PubMed]

58. Fredriksson-Ahomaa, M.; Bucher, M.; Hank, C.; Stolle, A.; Korkeala, H. High prevalence of *Yersinia enterocolitica* 4:O3 on pig offal in southern Germany: A slaughtering technique problem. *Syst. Appl. Microbiol.* **2001**, *24*, 457–463. [CrossRef] [PubMed]

59. Gürtler, M.; Alter, T.; Kasimir, S.; Linnebur, M.; Fehlhaber, K. Prevalence of *Yersinia enterocolitica* in fattening pigs. *J. Food Protect.* **2005**, *68*, 850–854. [CrossRef] [PubMed]

60. Bonardi, S.; Alpigiani, I.; Pongolini, S.; Morganti, M.; Tagliabue, S.; Bacci, C.; Briandini, F. Detection, enumeration and characterization of *Yersinia enterocolitica* 4/O:3 in pig tonsils at slaughter in Northern Italy. *Int. J. Food Microbiol.* **2014**, *177*, 9–15. [CrossRef]

61. Rusak, L.A.; Reis, C.M.F.; Barbosa, A.V.; Santos, A.F.M.; Paixão, R.; Hofer, E.; Vallim, D.C.; Asensi, M.D. Phenotypic and genotypic analysis of bio-serotypes of *Yersinia enterocolitica* from various sources in Brazil. *J. Infect. Dev. Ctries.* **2014**, *15*, 1533–1540. [CrossRef]

62. Fredriksson-Ahomaa, M.; Autio, T.; Korkeala, H. Efficient subtyping of *Yersinia enterocolitica* bioserotype 4/O:3 with pulsed-field gel electrophoresis. *Lett. Appl. Microbiol.* **1999**, *29*, 308–312. [CrossRef]

63. Terentjeva, M.; Berzinš, A. Prevalence and Antimicrobial Resistance of *Yersinia enterocolitica* and *Yersinia pseudotuberculosis* in Slaughter Pigs in Latvia. *J. Food Protect.* **2020**, *73*, 1335–1338. [CrossRef]

64. Bucher, M.; Meyer, C.; Grötzbach, S.; Wachneck, S.; Stolle, A.; Fredriksson-Ahomaa, M. Epidemiological data on pathogenic *Yersinia enterocolitica* in southern Germany during 2000–2006. *Foodborne Pathog. Dis.* **2008**, *5*, 273–280. [CrossRef]

65. Bonardi, S.; Paris, A.; Bassi, L.; Salmi, F.; Bacci, C.; Riboldi, E.; Boni, E.; D'incau, M.; Tagliabue, S.; Brindani, F. Detection, semiquantitative enumeration, and antimicrobial susceptibility of *Yersinia enterocolitica* in pork and chicken meats in Italy. *J. Food Protect.* **2010**, *73*, 1785–1792. [CrossRef]

66. Baumgartner, A.; Küffer, M.; Sutter, D.; Jemmi, T.; Rohner, P. Antimicrobial resistance of *Yersinia enterocolitica* strains from human patients, pigs and retail pork in Switzerland. *Int. J. Food Microbiol.* **2007**, *115*, 110–114. [CrossRef]

67. Fredriksson-Ahomaa, M.; Meyer, C.; Bonke, R.; Stüber, E.; Wacheck, S. Characterization of *Yersinia enterocolitica* 4/O:3 isolates from tonsils of Bavarian slaughter pigs. *Lett. Appl. Microbiol.* **2010**, *50*, 412–418. [CrossRef] [PubMed]

Article

Small Contaminations on Broiler Carcasses Are More a Quality Matter than a Food Safety Issue

Kacper Libera, Len Lipman and Boyd R. Berends *

Institute for Risk Assessment Sciences (IRAS), Utrecht University, 3508 TD Utrecht, The Netherlands
* Correspondence: b.r.berends@uu.nl; Tel.: +31-30-2535367

Abstract: Depending on the interpretation of the European Union (EU) regulations, even marginally visibly contaminated poultry carcasses could be rejected for human consumption due to food safety concerns. However, it is not clear if small contaminations actually increase the already present bacterial load of carcasses to such an extent that the risk for the consumers is seriously elevated. Therefore, the additional contribution to the total microbial load on carcasses by a small but still visible contamination with feces, grains from the crop, and drops of bile and grease from the slaughter line was determined using a Monte Carlo simulation. The bacterial counts (total aerobic plate count, *Enterobacteriaceae*, *Escherichia coli*, and *Campylobacter* spp.) were obtained from the literature and used as input for the Monte Carlo model with 50,000 iterations for each simulation. The Monte Carlo simulation revealed that the presence of minute spots of feces, bile, crop content, and slaughter line grease do not lead to a substantial increase of the already existing biological hazards present on the carcasses and should thus be considered a matter of quality rather than food safety.

Keywords: food safety; poultry; slaughter; carcass; contamination; Monte Carlo simulation; process hygiene criteria

Citation: Libera, K.; Lipman, L.; Berends, B.R. Small Contaminations on Broiler Carcasses Are More a Quality Matter than a Food Safety Issue. *Foods* **2023**, *12*, 522. https://doi.org/10.3390/foods12030522

Academic Editor: Frans J.M. Smulders

Received: 9 November 2022
Revised: 31 December 2022
Accepted: 19 January 2023
Published: 24 January 2023

1. Introduction

The muscles and internal organs of healthy slaughter animals are normally sterile, but during slaughtering, both the carcasses and internal organs invariably become contaminated with bacteria. Historically, the main factors affecting the final bacterial load of carcasses and consequently cuts of meat are driven by the cleanliness of the slaughterhouse environment and the skills of the slaughterhouse workers. However, increased mechanization has considerably reduced the human role in controlling the bacteriological quality and safety of meat. Today, the level of poultry carcass contamination is predominantly determined by the performance of the slaughterhouse machinery and the bacteriological status of the animals pre-slaughter. For example, not maintaining a constant high temperature (e.g., due to thermostat malfunction) in scalding machines increases the chances of carcass bacterial contamination by almost five times [1], and with respect to the bacteriological status of the animals pre-slaughter, *Campylobacter* spp.-positive flocks (positive caeca contents) are approximately four times more likely to cause *Campylobacter* spp. contamination of the carcasses at the end of the slaughter line compared to *Campylobacter* spp.-negative flocks [2]. Therefore, in a modern poultry slaughter line, the bacteriological safety and quality of the carcasses at the end of the line are ultimately determined by the number of bacteria present on and in the live animals as they arrive at the slaughterhouse in combination with the effectiveness and adjustment of the defeathering and evisceration equipment and that of the carcass washers [3–5].

Chicken carcass contamination continues to be a major food safety concern because broiler meat remains an important source of human campylobacteriosis. The latest data in the EU show that there were 120,964 confirmed cases of campylobacteriosis [6], and it is estimated that 20 to 30% of infections could be attributed to the handling, preparation, and consumption of broiler meat [7]. However, due to the self-limiting nature of this disease, the real prevalence is far higher. To design optimal interventions, it is crucial to understand how carcasses can become contaminated and which factors contribute to the contamination. Pacholewicz et al. [8] demonstrated that bacterial concentrations in the intestines of broilers are an important explanatory variable of carcass contamination because these were associated with fluctuations in *Campylobacter* spp. and *Escherichia coli* concentrations at various processing steps in the slaughter line. This is in accordance with a study performed by Tang et al., [9] who reported the highest prevalence of *Campylobacter* spp. contamination at the evisceration step due to the exposure of intestinal contents. In this study, the *Campylobacter*-positive carcass rate decreased from 53.4% during evisceration to 14.75% after cooling, which suggests that the cooling step is crucial for eliminating *Campylobacter* spp. on chicken carcasses. Furthermore, earlier research by Pacholewicz et al. [10] demonstrated that the changes in numbers of *Escherichia coli* and *Campylobacter* spp. on chicken carcasses during the various processing steps in a slaughter line are of a similar nature. However, a direct relationship between total bacterial load on chicken carcasses at the end of the slaughter line and the number of poultry-related cases of human disease have never been established, with the possible exception of a few risk assessment models for *Campylobacter* spp., such as the one described by Nauta et al. [11]. From the stable-to-table model of Nauta et al. [11] in particular, it was inferred that, in the slaughterhouse, compliance to a maximum threshold of about 1000 CFU/g of fresh chicken meat probably would halve the number of associated human cases of campylobacteriosis in the EU [7].

A strategy to reduce risk for consumers is to decrease the counts of *Campylobacter* spp. in the intestines of live birds with a range of control options available, including vaccinations, feed additives, or phage therapy. By lowering the concentration in the intestinal content, these control options aim at reducing *Campylobacter* spp. contamination during broiler processing and thus lead to lower concentrations on the broiler meat. A recent model suggests that a relative risk reduction (39%) could be obtained through a $2 \log_{10}$ reduction in caecal concentration ($9 \log_{10}\text{CFU/g}$ to $7 \log_{10}\text{CFU/g}$) [12]. However, it is important to note that the association between concentrations found in caeca and skin largely depend on the variation in hygiene practices between slaughterhouses and regions; consequently, the scale of potential risk-reducing effects may also vary greatly [12].

In almost entirely mechanised processes, the biological variation and physical condition of the animals are the most important factors with regard to the occurrence of slaughter defects, i.e., damaged intestines and/or gall bladders [4,13,14]. However, quantitative assessments of fecal contamination that are the result of this practice are limited. In an article from 1997, Russell and Walker [13] reported that the American inspection services found 0.8 to 5% fecally contaminated carcasses just before cooling. The study of Russell and Walker also demonstrated that, just after evisceration, 4–6% of the broiler carcasses showed evidence of fecal leakage on the inside and 5.2–8.4% on the outside [14]. Brizio et al. [15] investigated different types of carcass contamination and reported that 6% of carcasses were found to have fecal contamination, 1.45% of the carcasses were contaminated with bile, while 1.90% were contaminated with gastric content. In total, 9.35% out of 51,500 examined broiler carcasses were contaminated. Another field study found that, at the end of the slaughter line, just before cooling, 2–5% of the broiler carcasses were fecally contaminated [16]. It is important to highlight that there are significant differences in the prevalence of visibly contaminated carcasses between slaughterhouses representing different levels of compliance with food safety procedures [17].

The total bacterial load of chicken carcasses is often considerable, regardless of the presence of any visible contamination. Cibin et al. [18] reported, in an EU study, that carcasses visibly uncontaminated with feces and sampled just after evisceration showed *E. coli* loads ($\log_{10}$ CFU/g) that ranged from 1.30 to 7.38 and that visibly fecally contaminated carcasses showed loads from 2.40 to 7.04, respectively. Visibly uncontaminated carcasses sampled just after cooling showed *E. coli* loads that ranged from 1.00 to 6.95, whereas in visibly fecally contaminated carcasses, counts ranged from 2.65 to 5.28, respectively. With regard to the *Enterobacteriaceae*, after evisceration, the visibly clean carcasses had loads that ranged from 1.48 to 7.45, whilst counts on visibly fecally contaminated carcasses ranged from 2.45 to 7.26, respectively. After cooling, the loads with *Enterobacteriaceae* ranged from 1.00 to 7.08 for visibly clean carcasses and from 3.54 to 5.18 for fecally contaminated carcasses, respectively. Research by Jimenez et al. [19,20] reported comparable figures from Argentinean poultry slaughterhouses. In addition, they also demonstrated that there were no significant differences in numbers of *Enterobacteriaceae*, coliforms, and *Escherichia coli* per gram or cm^2 between visibly contaminated and uncontaminated carcasses. In contrast, however, *Campylobacter jejuni* and *Campylobacter coli* were detected in 58.8% and 11.6% (respectively) of broiler carcasses with visible fecal contamination, as compared to 17.6% and 9.8% in carcasses without visible fecal contamination [21]. However, the counts of *Campylobacter* spp. did not significantly differ between carcasses with and without contamination. At retail level, broiler carcasses are also characterized by an abundant microbiome, including pathogens as reported by Yu et al. [22], who found that 100% of organic carcasses were *Campylobacter*-positive compared to 8.33% in conventionally reared carcasses. Furthermore, 5 % of conventionally reared carcasses were contaminated with *Salmonella* spp., while the other most abundantly present bacteria included *Pseudomonas, Serratia* spp., and *E. coli*.

In 2022, the Association of Dutch Poultry Processing Industries (NEPLUVI) requested the Division of Veterinary Public Health of the Institute for Risk Assessment Sciences (IRAS) to estimate a) the total bacterial load of an 'average', visibly clean chicken carcass at the end of the slaughter line, b) to estimate what extent a small contamination would add to this 'average' total bacterial load, and c) to determine whether or not this would mean a substantial increase of any food risk already present that would make that carcass unfit for human consumption. Therefore, the aim of this study was to investigate the bacterial load on carcasses with different types of small but still visible contaminations with feces, crop content, and bile and grease from the line and compare these carcasses to those without any visible contamination with the use of a Monte Carlo simulation. We hypothesized that there are no significant differences in bacterial loads between carcasses with a small contamination and those with no visible contamination.

2. Materials and Methods

2.1. Data Used for Input in the Calculations

Bacterial counts (mean values of bacterial log_{10} counts ± standard deviation) used for the calculations were collected from peer-reviewed journals. The main selection criteria of articles included study design, performed laboratory analysis, sample size, year of publication, and parameters of the journal quality and impact. We aimed that the data from chosen articles should be representative and correspond as much as possible with a contemporary slaughterhouse environment.

2.2. Monte Carlo Simulation

A Monte Carlo simulation is a method used to predict the outcomes of events derived from multiple variations in their input [23]. It leads to insight into how ordinary or extraordinary certain final outcomes of these calculations are. The actual Monte Carlo simulation was performed using @Risk 8.0, which was part of the software package 'Decision Tools Suite' (Pallisade Corporation, 2020) and can be used as an add on to an Excel spreadsheet [24]. This method has already been successfully implemented to detect *Campylobacter* spp. presence and concentration using different chicken carcass samples [25].

In a Monte Carlo simulation, the variables that determine the outcome are repeatedly drawn from a range of values that follow a user-defined probability distribution [23,24]. @Risk was set to perform 50,000 iterations for each simulation.

The variables that were given a @Risk function were: (1) the total surface (weight) of the carcasses, (2) the number of bacteria per square centimeter (gram) already present on the skin surface of a clean carcass, (3) the total weight of a contamination, and (4) the total number of bacteria present per gram of contamination. A graphical explanation of the model design is given in Figure 1, and an example spreadsheet (Spreadsheet S1) model was uploaded in the Supplementary Materials.

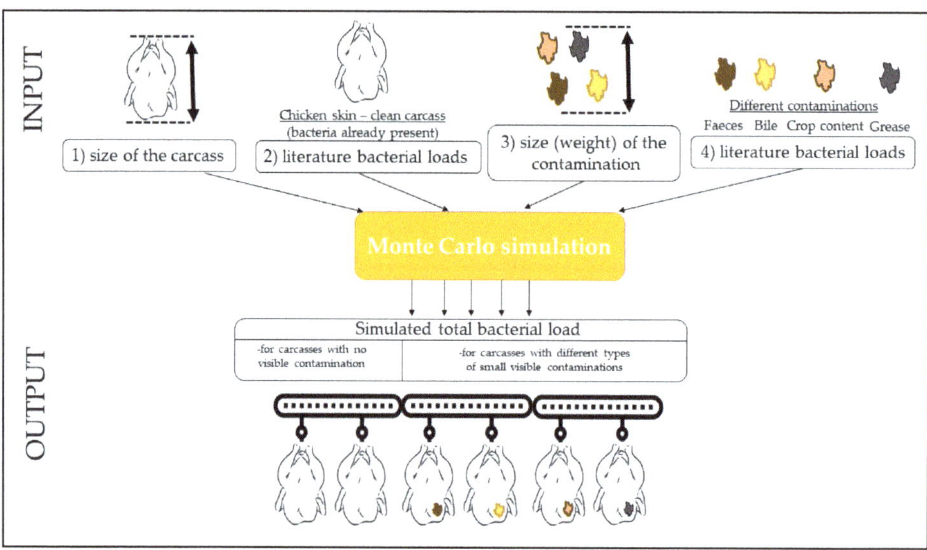

Figure 1. Data used for input in the Monte Carlo simulation.

The weights of the carcasses and the numbers of bacteria present on the carcasses and in the contaminations were processed with the @Risk function 'normal distribution' using values of mean and standard deviation. In order to avoid obtaining unrealistic results (e.g., lower than 0) of the total bacterial load prediction, we set the maximum and minimum value for the distribution of bacterial counts. For every bacterial count distribution, the minimum value was set to '0', while the maximum number of bacteria varied between materials and was set at 10^9 CFU/g for skin, 10^{10} for feces, 10^9 for crop content, 10^8 for bile, and 10^4 for grease. A similar procedure was followed by Nauta et al. [26]. The weights of the contaminations were processed with the @Risk function 'uniform distribution' using minimum and maximum values (see Table 1) because we had no knowledge about the real frequency distribution of the weights of small contaminations [23].

Table 1. Assumptions regarding small contaminations on broiler carcasses as input for the Monte Carlo simulation.

Type of Contamination	Minimal Amount (g)	Maximal Amount (g)
Feces	0.001	0.01
Bile	0.0375	0.15
Crop content	0.05	0.2
Grease	0.01	0.04

To determine the total load with bacteria on a 'typical' broiler carcass, a calculation was conducted with the aid of results from the study by Elfadil et al. [27]. From their study, it can be inferred that approximately one gram of body weight equals circa 1cm^2 surface. The spreadsheet model used an 'average' bird weight of 1600 g (i.e., 1600 cm^2), since this corresponded to the weight of the smaller animals both slaughterhouses confirmed to regularly process, and it is to be expected that a contamination has the biggest impact on a relatively small carcass. The total carcass load was then calculated by multiplying the bacterial counts per gram of the chicken skin with the total surface of the carcass.

All small contamination sizes were identified and described using a standardized number of grains or droplets (crop content and bile, respectively) or circle-shaped spots (feces and grease). In combination with the specific weights of the materials involved, the mass of a contamination could then be calculated. For the crop content and the bile, we used the generally acknowledged international standards of 0.065 g for a grain and 0.05 mL for a droplet. The specific weight in grams of the feces and bile fluid was estimated using Cussler et.al. [28] and Van der Meer [29]. When microscopically examined, the slaughter line grease turned out to be a mixture of chicken skin and feather material, minute metallic particles from the line, and the original food grade lubricant (see Figure 2). Therefore, we assumed that the specific weight would be in between that of the weight of feces, bile, and crop content.

Figure 2. The microscopic view of the grease.

The variation in carcass weights was approximated based upon the average weight of batches of animals sent to slaughter having a standard deviation of 5%, and that in a batch, the lightest animals weigh, on average, minus three times the standard deviation and that the heaviest animals weigh the average plus three times the standard deviation [24]. In this case, the 'average' weight was set at 1600 g, the minimum weight at 1350 g, and the maximum weight at 1850 g. The weights of the different contaminations that were used as model inputs are listed in Table 1.

3. Results

The microbial literature data that were used as the input for the Monte Carlo simulation can be found in Table 2. To the best of our knowledge, there are no bacteriological data available for grease; thus, we used our own (not published) data. Similarly, there are no studies describing bacterial counts in the bile after it has leaked from the gall bladder onto the machinery and/or the digestive tract before it drips onto the carcass during the evisceration. Therefore, we used bacterial counts from liver samples, assuming that the bacteriological loads of the liver correspond with bacteria potentially present in the bile after it has leaked from the gall bladder onto the machinery and the gut.

Table 2. Number of bacteria per gram in different types of contaminations expressed in $\log_{10}$.

	Total Aerobic Count	*Enterobacteriaceae*	*E. coli*	*Campylobacter* spp.
Skin				
mean $\pm$ sd ($\log_{10}/g$)	4.15 $\pm$ 0.46	3.77 $\pm$ 0.13	3.3 $\pm$ 0.6	2.99 $\pm$ 0.7
reference	[30]	[31]	[32]	[26]
Feces				
mean $\pm$ sd ($\log_{10}/g$)	3.36 $\pm$ 1.37	8.62 $\pm$ 0.58	8.44 $\pm$ 0.35	6.0 $\pm$ 1.52
reference	[33]	[34]	[34]	[26]
Bile				
mean $\pm$ sd ($\log_{10}/g$)	6.0 $\pm$ 0.7	3.1 $\pm$ 0.5	1.9 $\pm$ 1.1	2.795 $\pm$ 1.641
reference	[35]	[36]	[37]	[38]
Crop content				
mean $\pm$ sd ($\log_{10}/g$)	5.6 $\pm$ 0.1	4.2 $\pm$ 0.2	3.9 $\pm$ 0.2	3.63 $\pm$ 1.12
reference	[39]	[39]	[39]	[40]
Grease				
mean $\pm$ sd ($\log_{10}/g$)	3.40 $\pm$ 0.16	1.86 $\pm$ 0.41	0.86 $\pm$ 1.19	0.83 $\pm$ 0.67
reference	own data not published	own data not published	own data not published	own data not published

The simulated total bacterial loads on the broiler carcasses (mean 1600 g) with or without small visible contamination are given in Table 3.

Table 3. Monte Carlo simulation of the total bacterial load on the average broiler carcass (1600 g) with or without a contamination with a small amount of material (expressed in $\log_{10}$CFU).

Bacterial Species	Feces		Bile		Crop Content		Grease	
	No Contam.	With Contam.	No Contam.	With Contam.	No Contam.	With Contam.	No Contam.	With Contam.
Total aerobic count								
mean	7.3535	7.3536	7.3535	7.3625	7.3535	7.3552	7.3535	7.3535
sd	0.4604	0.4604	0.4606	0.4528	0.4603	0.4585	0.4606	0.4605
minimum	5.3536	5.3536	5.3394	5.5386	5.4267	5.4681	5.4143	5.4144
maximum	9.3717	9.3717	9.4324	9.4324	9.4114	9.4114	9.2766	9.2766
Enterobacteriaceae								
mean	6.9735	7.1181	6.9735	6.9735	6.9735	6.9736	6.9735	6.9735
sd	0.1316	0.1907	0.1318	0.1318	0.1319	0.1319	0.1320	0.1320
minimum	6.4232	6.4665	6.3800	6.3800	6.4164	6.4171	6.3775	6.3775
maximum	7.5920	8.0262	7.5122	7.5122	7.5480	7.5480	7.5612	7.5612
E. coli								
mean	6.5035	6.7640	6.5035	6.5036	6.5035	6.5039	6.5035	6.5035
sd	0.6003	0.4368	0.6003	0.6002	0.6005	0.6000	0.6004	0.6004
minimum	3.5649	5.1184	3.7284	3.7287	3.9146	4.0222	3.9726	3.9726
maximum	9.1898	9.1915	9.15620	9.1562	9.2393	9.2393	9.0849	9.0849
Campylobacter spp.								
mean	6.1936	6.2610	6.1936	6.2011	6.1936	6.2005	6.1936	6.1936
sd	0.7001	0.6784	0.7002	0.6946	0.7002	0.6931	0.7003	0.7003
minimum	3.3425	3.5869	3.3131	3.3245	3.3419	3.60291	3.2279	3.2287
maximum	9.2117	9.2117	9.26841	9.2684	9.10343	9.10343	9.1931	9.1931

The probability that a small visible contamination results in at least a 0.5 ($\log_{10}$CFU) increase in the total bacterial load of the average chicken carcass is given in Table 4. The value of ± 0.5 $\log_{10}$CFU is considered as the precision of classical microbiological methods [41]. From the practical point of view, differences below this value cannot be identified with classical microbiological culturing methods.

Table 4. The probability (%) that a small visible contamination results in at least a 0.5 ($\log_{10}$CFU) increase in the total bacterial load of the average chicken carcass (1600 g).

Type of Bacteria	Feces	Bile	Crop Content	Grease
Total aerobic count	0%	0%	0%	0%
Enterobacteriaceae	5.1%	0%	0%	0%
E. coli	16.7%	0%	0%	0%
Campylobacter spp.	4.1%	0%	0.2%	0%

It is important to note that the difference of 0.5 $\log_{10}$CFU is roughly equivalent to a three-fold increase in the total bacterial load (calculated based on CFUx10^x). The probability that small contaminations result in at least a three-fold increase in the total bacterial load of the average chicken carcass is given in Table 5.

Table 5. The probability (%) that a small visible contamination results in at least a three-fold increase (calculated based on CFUx10^x) in the total bacterial load of the average chicken carcass (1600 g).

Type of Bacteria	Feces	Bile	Crop Content	Grease
Total aerobic count	0%	0%	0%	0%
Enterobacteriaceae	5.6%	0%	0%	0%
E. coli	17.9%	0%	0%	0%
Campylobacter spp.	4.3%	0%	0.2%	0%

It is also important to determine what the contribution is of small contaminations to the total bacterial load compared to the already existing bacterial loads on the carcass. Therefore, the percentual contribution of a small contamination to the final total bacterial load on the carcass is given in Table 6. In the majority of the simulations, it was below 1%.

Table 6. The contribution (%) of small contaminations to the total bacterial load of the average chicken carcass (1600 g).

Type of Bacteria	Feces	Bile	Crop Content	Grease
Total aerobic count	0.001%	0.122%	0.02311%	<0.001%
Enterobacteriaceae	2.031%	<0.001%	0.00143%	<0.001%
E. coli	3.851%	0.002%	0.00615%	<0.001%
Campylobacter spp.	1.077%	0.121%	0.11128%	<0.001%

The highest probability that the bacterial count increase is higher than three-fold (17.9%) was obtained in the case of *E. coli* of a carcass contaminated with feces. This relationship is visualized in Figure 3.

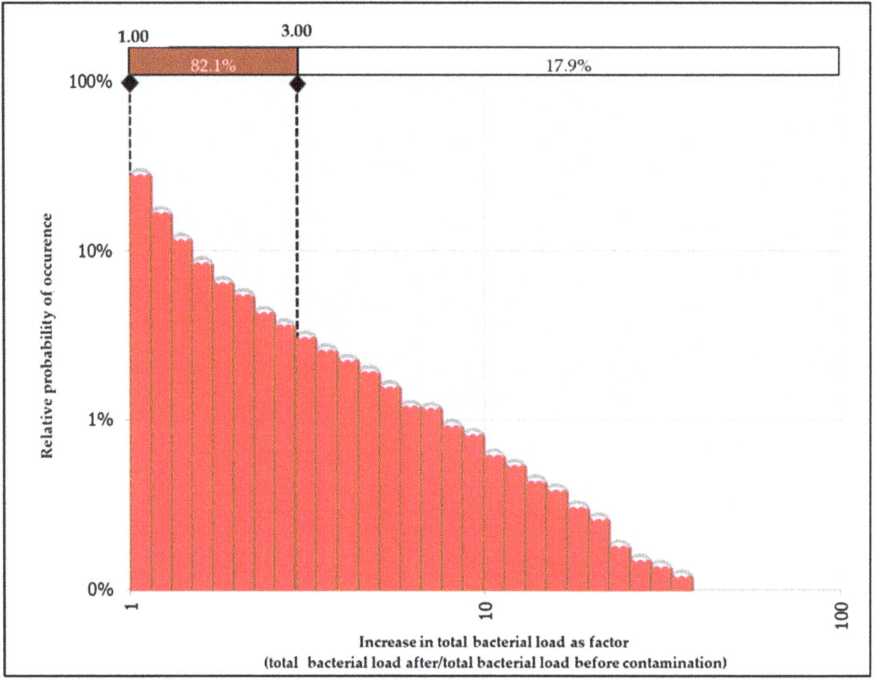

Figure 3. Relative probability of occurrence of different increases in total bacterial load expressed as a factor in the case of *E. coli* counts and contamination with feces. The red horizontal bar indicates the probability of increase less than three-fold (1.00–3.00), while the white horizontal bar indicates an increase more than three-fold.

4. Discussion

4.1. Bacterial Counts Used for Calculations

As expected, feces contained the most bacteria, and they far exceeded the counts on chicken skins. It is not surprising that, in the literature, there is a lack of data regarding microbial counts in the bile. Normally, bile fluids should contain zero to very few bacteria, since, otherwise, the animals would develop cholecystitis and become clinically ill and unfit for slaughter. It is challenging to either investigate or simulate bile bacterial counts found after the machinery has damaged the gall bladder. Specifically, it is particularly difficult to determine the bacterial counts present in bile itself before it reaches carcasses because the bile is usually mixed with gut content before it contaminates the carcasses. We assumed that microbial counts from chicken livers would approximate that of bile that contaminated the carcass via the machinery and the viscera. Crop content bacterial counts resembled an intermediate level between that of the skin and of the feces contents. The lowest counts of bacteria were observed in grease, since this material is mainly composed of lipids (almost no water), consequently creating a hostile environment for bacterial growth.

4.2. Monte Carlo Simulation

The results of the Monte Carlo Simulation after 50,000 iterations are summarized in Table 3. The high number of iterations ensures that the simulation included almost all of the possible combinations of carcass weight (surface), clean carcass bacterial numbers, and weights of the small contaminations with corresponding bacterial counts of the small contaminations. For modern computers, a simulation with 50,000 iterations is not a challenge and lasts for approximately one minute. As expected, the highest differences between bacterial loads in contaminated and non-contaminated carcasses were observed in the case of fecal contamination. In other cases, if the differences existed, they occurred at the second to the fourth decimal place of bacterial counts.

We also determined the probability that the increase in the total bacterial load would exceed the precision limit of classical microbiology methods (0.5 $\log_{10}$CFU). This value is important because increases below this number cannot be identified by classical microbiological methods and can thus be considered insignificant. Similarly, the highest values were observed in carcasses contaminated with feces, in particular in *E. coli* counts (Tables 4 and 5). For example, there was a probability of 16.7% that, after a small contamination with feces, *E. coli* counts would increase by 0.5 $\log_{10}$ (Table 4). In other words, roughly 1/6 of the carcasses with small visible contamination had significantly higher *E. coli* counts. For other types of contaminations, the probability of increasing the total bacterial load by at least 0.5 $\log_{10}$ was close to 0%. It is important to note that the increase by 0.5 $\log_{10}$CFU is approximately equal to a three-fold increase in bacterial load calculated based on CFUx10^x values. It might be useful to compare bacterial loads expressed in different units. In our study, the probability of total bacterial increase (*E. coli*, carcasses contaminated with feces) by at least 0.5 $\log_{10}$ was 16.7%, while the probability of occurrence of at a least three-fold increase was 17.9% (Figure 3).

These calculations confirmed that the majority of small contaminations have a negligible impact. Although the numbers of bacteria can be substantial on the spot where the small contamination has taken place (especially if feces are involved), when these numbers are related to the total bacterial load that is already present on a whole carcass, the impact of the small contamination becomes negligible (Table 6), consequently causing no extra threat to food safety. This is illustrated by a study by Giombelli and Gloria [21], who found that visible fecal contamination did not influence the counts of *Campylobacter* spp. on the carcasses *per se*, but that it did result in a higher prevalence of *Campylobacter*-positive carcasses, i.e., the number of positive carcasses was higher in the group of fecally contaminated carcasses than in the group of carcasses without any visible fecal contamination. This was also the case in a laboratory experiment on carcass contamination with 0.1 g of feces with cultured bacteria [42]. In addition, when the effects of cooling are taken into consideration, the effects will even be further diminished. As demonstrated by Cibin et al. [18], the cooling process reduced the overall bacterial levels to such an extent that, even in situations where there are significant differences between clean and visibly soiled carcasses at the end of slaughter, the cooling process renders these differences insignificant. Similar results were observed by Cason et al. [43], who reported no differences in bacterial counts post-chilling between carcass halves, from which one was not contaminated, while the other was artificially contaminated with fresh feces.

Many cases of visibly contaminated carcasses can be attributed to a faulty evisceration process. Machines can only be adjusted to work within a certain set of size ranges. Therefore, it would be highly desirable if the machines could be auto-adjusted in real time to the size of every single carcass processed, thus minimizing the risk of faulty evisceration (e.g., intestine or gall bladder disruption) and decreasing the prevalence of visibly contaminated carcasses. However, from the slaughterhouse's perspective, the reason for the carcass damage can be explained as the lack of uniformity of the delivered broiler flock. In other words, the birds do not meet the expected standard size, which should be the responsibility of the poultry producer. Maintenance of the equipment (or not appropriate maintenance) could also result in poultry carcass damage, including rupture of the gastrointestinal tract. Nevertheless, there are some interventions that aim at reducing bacterial counts on chicken carcasses. For example, the application of rapid surface cooling (immersion in liquid nitrogen) resulted in a reduction of counts of *Campylobacter* spp. by 1 $\log_{10}$ CFU/g on the chicken carcass skin [44]. Similar promising results were obtained when the combination of steam and ultrasound were used in the evisceration room, before the inside/outside carcass washer [45]. Carcass trimming or using water sprays to remove contaminations, however, offer no real solution. For example, a study by Giombelli and Gloria [21] showed that there were no differences in bacterial counts before and after the trimming of carcasses with visible fecal contamination, while the water spray (potable water) decreased bacterial loads by approx. 20% (i.e., a factor of 0.8, which in practice will not yield significant impact). Stefani et al. [46] also showed that the washing of carcasses with fecal contamination is more effective in reducing bacterial loads than trimming is. It is not surprising that, in practice, despite all of the actions taken, some carcasses will remain visibly contaminated until just before cooling. However, when seen in the light of our calculations regarding small contaminations, a zero-tolerance policy towards all visible contaminations by some food safety authorities in Europe can be seen as a mainly quality or politically driven and not a real food safety issue. As amply demonstrated by the results of Cibin et al. [18], high microbial counts of carcasses can also occur without visible contamination.

Inspecting chicken carcasses for small visible contaminations by the competent authorities is time-consuming, requires sharp eyesight, and can be highly subjective. Hence, more attention should be paid to robust hygiene criteria, which are far more effective than implementing a strict zero-tolerance policy towards small visible contaminations by the competent authorities. For example, the current process hygiene criterium for poultry production is a *Campylobacter* spp. count with an upper limit set at 1000 CFU/g [47]. For every batch, no more than 15 samples out of 50 should exceed this limit, but the aim is to reduce this number to 10 by 2025. As time progresses, adopting more stringent microbial criteria appears reasonable because producers have ample time to adjust and can at the same time claim to work actively with the competent authorities towards reducing the prevalence of food-borne diseases in humans. For example, a more stringent critical limit for *Campylobacter* spp. of 100 CFU/g could reduce consumer risk of campylobacteriosis via poultry by 98%, but currently, over 55% or more of the batches would not comply [48]; thus, this is currently an impossible criterium to adopt. Producers should still check for small contaminations since they might be the indicators that mechanical adjustments are required during the processing. While the meat inspectors should be also aware of this issue, they should at the same time try to focus on other indicators with well-established food safety implications, as stated above.

5. Conclusions

In conclusion, our calculations revealed that carcass contamination with minute amounts of feces, bile, grain from the crop, and grease from the lines will not lead to a significant increase of the already present food safety hazards. Maintaining a strict zero tolerance for these small contaminations on chicken carcasses does not improve the level of protection of the consumer. Instead, it would be far more effective to pay more attention to existing hygienic microbial criteria and a further improvement by a gradual tightening up of these regulations in the future. However, it is important to note that the biological hazards discussed above are best controlled at earlier stages of the production. Ensuring the highest possible animal health status as well as animal welfare standards should play a key role in minimizing food-borne risks for consumers.

Supplementary Materials: The following supporting information can be downloaded at: https://www.mdpi.com/article/10.3390/foods12030522/s1, Spreadsheet S1: The baseline model for used for Monte Carlo simulation.

Author Contributions: Conceptualization, B.R.B. and L.L.; methodology, B.R.B.; software, B.R.B. and K.L.; investigation, B.R.B. and K.L.; resources, K.L., B.R.B. and L.L.; data curation, B.R.B. and K.L.; writing—original draft preparation, K.L. and L.L.; writing—review and editing, B.R.B. and L.L.; visualization, B.R.B. and K.L.; supervision, B.R.B. and L.L.; project administration, L.L.; funding acquisition, B.R.B. All authors have read and agreed to the published version of the manuscript.

Funding: This study was financed by the Association of Dutch Poultry Processing Industries (NEPLUVI).

Institutional Review Board Statement: Not applicable.

Informed Consent Statement: Not applicable.

Data Availability Statement: Data is contained within the article or supplementary material.

Conflicts of Interest: The authors declare no conflict of interest. The funders had no role in the design of the study; in the collection, analyses, or interpretation of data; in the writing of the manuscript; or in the decision to publish the results.

References

1. Klaharn, K.; Pichpol, D.; Meeyam, T.; Harintharanon, T.; Lohaanukul, P.; Punyapornwithaya, V. Bacterial Contamination of Chicken Meat in Slaughterhouses and the Associated Risk Factors: A Nationwide Study in Thailand. *PLoS ONE* **2022**, *17*, e0269416. [CrossRef] [PubMed]
2. Emanowicz, M.; Meade, J.; Bolton, D.; Golden, O.; Gutierrez, M.; Byrne, W.; Egan, J.; Lynch, H.; O'Connor, L.; Coffey, A.; et al. The Impact of Key Processing Stages and Flock Variables on the Prevalence and Levels of *Campylobacter* on Broiler Carcasses. *Food Microbiol.* **2021**, *95*, 103688. [CrossRef] [PubMed]
3. Berends, B.R.; Snijders, J.M.A. De Hazard Analysis Critical Control Point benadering bij de productie van vlees. *Tijdschr Diergeneeskd* **1994**, *119*.
4. Rasekh, J.; Thaler, A.M.; Engeljohn, D.L.; Pihkala, N.H. Food Safety and Inspection Service Policy for Control of Poultry Contaminated by Digestive Tract Contents: A Review. *J. Appl. Poult. Res.* **2005**, *14*, 603–611. [CrossRef]
5. Rouger, A.; Tresse, O.; Zagorec, M. Bacterial Contaminants of Poultry Meat: Sources, Species, and Dynamics. *Microorganisms* **2017**, *5*, 50. [CrossRef] [PubMed]
6. European Food Safety Authority; European Centre for Disease Prevention and Control. The European Union One Health 2020 Zoonoses Report. *EFSA J.* **2021**, *19*, 20. [CrossRef]
7. European Food Safety Authority. EFSA Panel on Biological Hazards (BIOHAZ); Scientific Opinion on *Campylobacter* in broiler meat production: Control options and performance objectives and/or targets at different stages of the food chain. *EFSA J.* **2011**, *9*, 8. Available online: http://www.efsa.europa.eu/en/efsajournal/pub/2105 (accessed on 4 November 2022).
8. Pacholewicz, E.; Swart, A.; Wagenaar, J.A.; Lipman, L.J.A.; Havelaar, A.H. Explanatory variables associated with *Campylobacter* and *Escherichia coli* concentrations on broiler chicken carcasses during processing in two slaughterhouses. *J. Food Protect.* **2016**, *79*, 2038–2047. [CrossRef]
9. Tang, Y.; Jiang, Q.; Tang, H.; Wang, Z.; Yin, Y.; Ren, F.; Kong, L.; Jiao, X.; Huang, J. Characterization and Prevalence of *Campylobacter* spp. from Broiler Chicken Rearing Period to the Slaughtering Process in Eastern China. *Front. Vet. Sci.* **2020**, *7*, 227. [CrossRef]
10. Pacholewicz, E.; Swart, A.; Schipper, M.; Gortemaker, B.G.M.; Wagenaar, J.A.; Havelaar, A.H.; Lipman, L.J.A. A comparison of fluctuations of *Campylobacter* and *Escherichia coli* concentrations on broiler chicken carcasses during processing in two slaughterhouses. *Int. J. Food Microbiol.* **2015**, *205*, 119–127. [CrossRef]
11. Nauta, M.J.; Jacobs-Reitsma, W.F.; Havelaar, A.H. A Risk assessment Model for *Campylobacter* in Broiler Meat. *Risk Anal.* **2007**, *27*, 845–861. [CrossRef] [PubMed]
12. Nauta, M.; Bolton, D.; Crotta, M.; Ellis-Iversen, J.; Alter, T.; Hempen, M.; Messens, W.; Chemaly, M. An Updated Assessment of the Effect of Control Options to Reduce *Campylobacter* Concentrations in Broiler Caeca on Human Health Risk in the European Union. *Microb. Risk Anal.* **2022**, *21*, 100197. [CrossRef]
13. Russell, S.M.; Walker, J.M. The Effect of Evisceration on Visible Contamination and the Microbiological Profile of Fresh Broiler Chicken Carcasses Using the Nu-Tech Evisceration System or the Conventional Streamlined Inspection System. *Poult. Sci.* **1997**, *76*, 780–784. [CrossRef] [PubMed]
14. Russell, S.M. The effects of airsacculitis on bird weights, uniformity, fecal contamination, processing errors and populations of *Campylobacter* spp. and *Escherichia coli*. *Poult. Sci.* **2003**, *82*, 1326–1331. [CrossRef] [PubMed]
15. Brizio, A.P.; Marin, G.; Schittler, L.; Prentice, C. Visible Contamination in Broiler Carcasses and its Relation to The Stages of Evisceration in Poultry Slaughter. *Int. Food Res. J.* **2015**, *22*, 59–63.
16. Fletcher, D.L.; Craig, E.W.; Arnold, J.W. An evaluation of on-line "reprocessing" on visual contamination and microbiological quality of broilers. *J Appl. Poultry Res.* **1997**, *6*, 436–442. [CrossRef]
17. Pacholewicz, E.; Sura Barus, S.A.; Swart, A.; Havelaar, A.H.; Lipman, L.J.A.; Luning, P.A. Influence of Food Handlers' Compliance with Procedures of Poultry Carcasses Contamination: A Case Study Concerning Evisceration in Broiler Slaughterhouses. *Food Control* **2016**, *68*, 367–378. [CrossRef]
18. Cibin, V.; Mancin, M.; Pedersen, K.; Barrucci, F.; Belluco, S.; Roccato, A.; Cocola, F.; Ferrarini, S.; Sandri, A.; Lau Baggesen, D.; et al. Usefulness of *Escherichia coli* and *Enterobacteriaceae* as Process Hygiene Criteria in poultry: Experimental study. *EFSA J.* **2014**, *121*, 25. Available online: https://www.efsa.europa.eu/en/supporting/pub/en-635 (accessed on 4 November 2022). [CrossRef]
19. Jiménez, S.M.; Salsi, M.S.; Tiburzi, M.C.; Pirovani, M.E. A comparison between broiler chicken carcasses with and without visible fecal contamination during the slaughtering process on hazard identification of *Salmonella* spp. *J. Appl. Microbiol.* **2002**, *93*, 593–598. [CrossRef]
20. Jiménez, S.M.; Tiburzi, M.C.; Salsi, M.S.; Pirovani, M.E.; Moguilevsky, M.A. The role of visible fecal material as a vehicle for generic *Escherichia coli*, coliform, and other enterobacteria contaminating poultry carcasses during slaughtering. *J. Appl. Microbiol.* **2003**, *95*, 451–456. [CrossRef]

21. Giombelli, A.; Gloria, M.B.A. Prevalence of *Salmonella* and *Campylobacter* on Broiler Chickens from Farm to Slaughter and Efficiency of Methods to Remove Visible Fecal Contamination. *J. Food Prot.* **2014**, *77*, 1851–1859. [CrossRef] [PubMed]

22. Yu, Z.; Joossens, M.; Houf, K. Analyses of the Bacterial Contamination on Belgian Broiler Carcasses at Retail Level. *Front. Microbiol.* **2020**, *11*, 539540. [CrossRef] [PubMed]

23. Vose, D. *Risk Analysis: A Guide to Monte Carlo Simulation Modeling*; John Wiley and Sons: Chichester, NY, USA, 1996.

24. European Food Safety Authority. EFSA Panel on Biological Hazards (BIOHAZ); Scientific Opinion on Quantitative Assessment of The Residual BSE Risk in Bovine-Derived Products. *EFSA J.* **2005**, *307*, 34–35. [CrossRef]

25. Ellis-Iversen, J.; Gantzhorn, M.R.; Borck Høg, B.; Foddai, A.; Nauta, M. The Ability to Detect *Campylobacter* Presence and Concentration Using Different Chicken Carcass Samples. *Food Control* **2020**, *115*, 107294. [CrossRef]

26. Nauta, M.J.; Jacobs-Reitsma, W.F.; Evers, E.G.; van Pelt, W.; Havelaar, A.H. *Risk Assessment of Campylobacter in the Netherlands via Broiler Meat and Other Routes*; Report 250911006/2005; National Institute for Public Health and the Environment (RIVM): Bilthoven, The Netherlands, 2005.

27. Elfadil, A.A.; Vaillantcourt, J.P.; Duncan, I.J.H. Comparative study of body characteristics of different strains of broiler chickens. *J. Appl. Poult. Sci.* **1998**, *7*, 268–272. [CrossRef]

28. Cussler, E.L.; Fennell Evans, D.; DePalma, R. A Model for Gallbladder Function and Cholesterol Gallstone Formation. *Proc. Natl. Ac. Sci. USA* **1970**, *67*, 400–407. [CrossRef]

29. Van Der Meer, J. Het verschijnsel van de drijvende feces. *Ned. Tijdschr. Voor Geneeskd.* **1972**, *116*, 1662.

30. Seol, K.-H.; Han, G.-S.; Kim, H.W.; Chang, O.-K.; Oh, M.-H.; Park, B.-Y.; Ham, J.-S. Prevalence and Microbial Flora of Chicken Slaughtering and Processing Procedure. *J. Food. Sci. Anim. Resour.* **2012**, *32*, 763–768. [CrossRef]

31. Goksoy, E.O.; Kirkan, S.; Kok, F. Microbiological Quality of Broiler Carcasses during Processing in Two Slaughterhouses in Turkey. *Poult. Sci.* **2004**, *83*, 1427–1432. [CrossRef]

32. Altekruse, S.F.; Berrang, M.E.; Marks, H.; Patel, B.; Shaw, W.K.; Saini, P.; Bennett, P.A.; Bailey, J.S. Enumeration of *Escherichia Coli* Cells on Chicken Carcasses as a Potential Measure of Microbial Process Control in a Random Selection of Slaughter Establishments in the United States. *Appl. Environ. Microbiol.* **2009**, *75*, 3522–3527. [CrossRef]

33. Danek-Majewska, A.; Kwiecień, M.; Samolińska, W.; Kowalczyk-Pecka, D.; Nowakowicz-Dębek, B.; Winiarska-Mieczan, A. Effect of Raw Chickpea in the Broiler Chicken Diet on Intestinal Histomorphology and Intestinal Microbial Populations. *Animals* **2022**, *12*, 1767. [CrossRef] [PubMed]

34. Śliżewska, K.; Markowiak-Kopeć, P.; Żbikowski, A.; Szeleszczuk, P. The Effect of Synbiotic Preparations on the Intestinal Microbiota and Her Metabolism in Broiler Chickens. *Sci. Rep.* **2020**, *10*, 4281. [CrossRef] [PubMed]

35. Berrang, M.E.; Cox, N.A.; Meinersmann, R.J.; Bowker, B.C.; Zhuang, H.; Huff, H.C. Mild Heat and Freezing to Lessen Bacterial Numbers on Chicken Liver. *J. Appl. Poult. Res.* **2020**, *29*, 251–257. [CrossRef]

36. Dourou, D.; Grounta, A.; Argyri, A.A.; Froutis, G.; Tsakanikas, P.; Nychas, G.-J.E.; Doulgeraki, A.I.; Chorianopoulos, N.G.; Tassou, C.C. Rapid Microbial Quality Assessment of Chicken Liver Inoculated or Not with *Salmonella* Using FTIR Spectroscopy and Machine Learning. *Front. Microbiol.* **2021**, *11*, 623788. [CrossRef]

37. Stromberg, Z.R.; Johnson, J.R.; Fairbrother, J.M.; Kilbourne, J.; Van Goor, A.; Curtiss, R.; Mellata, M. Evaluation of *Escherichia Coli* Isolates from Healthy Chickens to Determine Their Potential Risk to Poultry and Human Health. *PLoS ONE* **2017**, *12*, e0180599. [CrossRef]

38. Whyte, R.; Hudson, J.A.; Graham, C. *Campylobacter* in Chicken Livers and Their Destruction by Pan Frying. *Lett. Appl. Microbiol.* **2006**, *43*, 591–595. [CrossRef]

39. Smith, D.P.; Berrang, M.E. Prevalence and Numbers of Bacteria in Broiler Crop and Gizzard Contents. *Poult. Sci.* **2006**, *85*, 144–147. [CrossRef]

40. Musgrove, M.T.; Berrang, M.E.; Byrd, J.A.; Stern, N.J.; Cox, N.A. Detection of *Campylobacter* Spp. in Ceca and Crops with and Without Enrichment. *Poult. Sci.* **2001**, *80*, 825–828. [CrossRef]

41. Food and Drug Administration Office of Regulatory Affairs. *Pharmaceutical Microbiology Manual, ORA.007*; Food and Drug Administration: Silver Spring, MD, USA, 2020.

42. Smith, D.P.; Northcutt, J.K.; Musgrove, M.T. Microbiology of Contaminated or Visibly Clean Broiler Carcasses Processed with an Inside-Outside Bird Washer. *Int. J. Poutry Sci.* **2005**, *4*, 955–958. [CrossRef]

43. Cason, J.A.; Berrang, M.E.; Buhr, R.J.; Cox, N.A. Effect of Prechill Fecal Contamination on Numbers of Bacteria Recovered from Broiler Chicken Carcasses Before and After Immersion Chilling. *J. Food Prot.* **2004**, *67*, 1829–1833. [CrossRef]

44. Burfoot, D.; Hall, J.; Nicholson, K.; Holmes, K.; Hanson, C.; Handley, S.; Mulvey, E. Effect of Rapid Surface Cooling on *Campylobacter* Numbers on Poultry Carcasses. *Food Control* **2016**, *70*, 293–301. [CrossRef]

45. Musavian, H.S.; Krebs, N.H.; Nonboe, U.; Corry, J.E.L.; Purnell, G. Combined Steam and Ultrasound Treatment of Broilers at Slaughter: A Promising Intervention to Significantly Reduce Numbers of Naturally Occurring *Campylobacters* on Carcasses. *Int. J. Food Microbiol.* **2014**, *176*, 23–28. [CrossRef] [PubMed]

46. Stefani, L.M.; Backes, R.G.; Faria, G.A.; Biffi, C.P.; de Almeida, J.M.; da Silva, H.K.; das Neves, G.B.; Langaro, A. Trimming and Washing Poultry Carcass to Reduce Microbial Contamination: A Comparative Study. *Poult. Sci.* **2014**, *93*, 3119–3122. [CrossRef] [PubMed]

47. European Commission. Commission Regulation (EC) No 2073/2005 of 15 November 2005 on Microbiological Criteria for Foodstuffs (Consolidated Version, Text with EAA Relevance). 2005. Available online: https://eur-lex.europa.eu/legal-content/EN/TXT/?uri=CELEX%3A02005R2073-20200308 (accessed on 4 November 2022).
48. Swart, A.N.; Mangen, M.J.J.; Havelaar, A.H. *Microbiological Criteria as a Decision Tool for Controlling Campylobacter in the Broiler Meat Chain*; RIVM Letter Report 330331008/2013; National Institute for Public Health and the Environment (RIVM): Bilthoven, The Netherlands, 2013.

Article

Systematic Review and Meta-Analysis of the Efficacy of Interventions Applied during Primary Processing to Reduce Microbial Contamination on Pig Carcasses

Nevijo Zdolec [1], Aurelia Kotsiri [2], Kurt Houf [3], Avelino Alvarez-Ordóñez [4], Bojan Blagojevic [5], Nedjeljko Karabasil [6], Morgane Salines [7] and Dragan Antic [2,*]

[1]   Faculty of Veterinary Medicine, University of Zagreb, 10000 Zagreb, Croatia; nzdolec@vef.hr
[2]   Institute of Infection, Veterinary and Ecological Sciences, University of Liverpool, Liverpool L69 3BX, UK; aurelia.kotsiri@liverpool.ac.uk
[3]   Faculty of Veterinary Medicine, Department of Veterinary and Biosciences, Ghent University, 9820 Merelbeke, Belgium; kurt.houf@ugent.be
[4]   Institute of Food Science and Technology, Department of Food Hygiene and Technology, Universidad de León, 24004 León, Spain; aalvo@unileon.es
[5]   Faculty of Agriculture, Department of Veterinary Medicine, University of Novi Sad, 21000 Novi Sad, Serbia; blagojevic.bojan@yahoo.com
[6]   Faculty of Veterinary Medicine, University of Belgrade, Department of Food Hygiene and Technology, 11000 Belgrade, Serbia; nedjakn@gmail.com
[7]   French Ministry of Agriculture, Office for Slaughterhouses and Cutting Plants, 75015 Paris, France; morgane.salines@agriculture.gouv.fr
*   Correspondence: dragan.antic@liverpool.ac.uk

Citation: Zdolec, N.; Kotsiri, A.; Houf, K.; Alvarez-Ordóñez, A.; Blagojevic, B.; Karabasil, N.; Salines, M.; Antic, D. Systematic Review and Meta-Analysis of the Efficacy of Interventions Applied during Primary Processing to Reduce Microbial Contamination on Pig Carcasses. *Foods* **2022**, *11*, 2110. https://doi.org/10.3390/foods11142110

Academic Editor: Frans J.M. Smulders

Received: 2 April 2022
Accepted: 5 July 2022
Published: 15 July 2022

Publisher's Note: MDPI stays neutral with regard to jurisdictional claims in published maps and institutional affiliations.

Abstract: Interventions from lairage to the chilling stage of the pig slaughter process are important to reduce microbial contamination of carcasses. The aim of this systematic review and meta-analysis was to assess the effectiveness of abattoir interventions in reducing aerobic colony count (ACC), *Enterobacteriaceae*, generic *Escherichia coli*, and *Yersinia* spp. on pig carcasses. The database searches spanned a 30 year period from 1990 to 2021. Following a structured, predefined protocol, 22 articles, which were judged as having a low risk of bias, were used for detailed data extraction and meta-analysis. The meta-analysis included data on lairage interventions for live pigs, standard processing procedures for pig carcasses, prechilling interventions, multiple carcass interventions, and carcass chilling. Risk ratios (RRs) for prevalence studies and mean log differences (MDs) for concentration outcomes were calculated using random effects models. The meta-analysis found that scalding under commercial abattoir conditions effectively reduced the prevalence of *Enterobacteriaceae* (RR: 0.05, 95% CI: 0.02 to 0.12, I^2 = 87%) and ACC (MD: −2.84, 95% CI: −3.50 to −2.18, I^2 = 99%) on pig carcasses. Similarly, significant reductions of these two groups of bacteria on carcasses were also found after singeing (RR: 0.25, 95% CI: 0.14 to 0.44, I^2 = 90% and MD: −1.95, 95% CI: −2.40 to −1.50, I^2 = 96%, respectively). Rectum sealing effectively reduces the prevalence of *Y. enterocolitica* on pig carcasses (RR: 0.60, 95% CI: 0.41 to 0.89, I^2 = 0%). Under commercial abattoir conditions, hot water washing significantly reduced ACC (MD: −1.32, 95% CI: −1.93 to −0.71, I^2 = 93%) and generic *E. coli* counts (MD: −1.23, 95% CI: −1.89 to −0.57, I^2 = 61%) on pig carcasses. Conventional dry chilling reduced *Enterobacteriaceae* prevalence on pig carcasses (RR: 0.32, 95% CI: 0.21 to 0.48, I^2 = 81%). Multiple carcass interventions significantly reduced *Enterobacteriaceae* prevalence (RR: 0.11, 95% CI: 0.05 to 0.23, I^2 = 94%) and ACC on carcasses (MD: −2.85, 95% CI: −3.33 to −2.37, I^2 = 97%). The results clearly show that standard processing procedures of scalding and singeing and the hazard-based intervention of hot water washing are effective in reducing indicator bacteria on pig carcasses. The prevalence of *Y. enterocolitica* on pig carcasses was effectively reduced by the standard procedure of rectum sealing; nevertheless, this was the only intervention for *Yersinia* investigated under commercial conditions. High heterogeneity among studies and trials investigating interventions and overall lack of large, controlled trials conducted under commercial conditions suggest that more in-depth research is needed.

Keywords: interventions; pig carcasses; aerobic colony count; *Enterobacteriaceae*; generic *E. coli*; *Yersinia*; abattoir; hot water washing; chilling

1. Introduction

Microbial contamination of pig carcasses (i.e., skin and meat) can arise from numerous sources and operations in abattoirs, from lairage to chilling. The level of contamination depends on the management of animal purchase, lairage conditions and slaughter technologies, which can vary significantly among abattoirs [1–3]. The level of hygiene during processing at slaughter and dressing is assessed based on process hygiene criteria (PHC), which includes testing for *Salmonella* presence, aerobic colony count (ACC) and *Enterobacteriaceae* count (EBC) on carcass surfaces before chilling [4]. Microbiological criteria are usually revised according to the current epidemiological status of animal production and new scientific knowledge. For example, the criteria for *Salmonella* proposed in European Food Safety Authority (EFSA) opinions on modernisation of meat inspection in pigs [5] are stricter and allow for only 6% *Salmonella*-positive pig carcasses in one sampling period of 10 weeks in order for an abattoir process to be considered as satisfactory [6]. On the other hand, PHC for *Yersinia enterocolitica* have not been envisaged in the legislation, although pigs are a common source of pathogenic strains causing yersiniosis in humans [7], and this is one of the priority hazards in pork [8]. *Campylobacter* spp., and particularly *C. coli*, is a frequent contaminant of prechilled pig carcass surfaces; however, given its sensitivity to drying and freezing when conventional dry or blast chilling is used, there is a significant decline of this pathogen on pig carcasses post-chilling [5]. Consequently, pig carcasses and pork are not considered an important source of *Campylobacter* in public health context, and it is not a priority hazard for control at the abattoir stage [5]. Common groups of indicator microorganisms, such as ACC, EBC, generic *Escherichia coli* count and total coliforms, are ideal for assessing the hygiene status of pig carcasses due to the fact of their existing higher levels and more uniform distribution on carcass surfaces compared to pathogens [9,10]. Indeed, the overall hygiene performance of pig abattoirs can be assessed by monitoring the ACC, EBC and generic *Escherichia coli* count before and after each specific slaughter operation. Many studies have shown that standard processing procedures, such as scalding, singeing or rectum sealing, reduce the number of indicator bacteria or the presence of pathogens, while dehairing, polishing and carcass splitting increase bacterial contamination [11–15].

Various interventions, usually hazard-based or good hygienic practice (GHP)-based in nature, are used in pig abattoirs to eliminate or reduce pathogens and spoilage bacteria from carcasses. GHP-based measures are prerequisites used at the preslaughter stage (e.g., lairage holding time and feed withdrawal) and during slaughter and carcass dressing (e.g., scalding, singeing, rectum sealing, head removal, knife trimming, carcass washing). More specific, hazard-based interventions, such as various thermal treatments for carcasses (hot water washing, steam pasteurisation), can be used in the prechilling phase, and do not require specific regulatory approval. On the other hand, chemical washes with organic acids and other chemicals undergo stringent risk assessment processes and regulatory approval [1,16]. Finally, carcass dry air chilling (conventional and blast) has some antimicrobial effect that is based on surface drying and can be complemented or replaced with spray chilling (with water or water plus organic acids or other approved chemicals) to increase the antimicrobial effect. However, the specific interventions used vary from country to country and are influenced by the regulatory framework, economic feasibility, seasonal variations, environmental impact, technical constraints and occupational health and safety [1,16].

Numerous studies using different experimental designs have been conducted over the last couple of decades with the aim of investigating the effectiveness of various interventions for pig carcasses. They usually produce different supporting evidence, depending on many factors (sample size, various study conditions, study design, etc.). One way to address the high heterogeneity between different study designs is to conduct a systematic literature review coupled with meta-analysis. This structured process enables the effectiveness of interventions to be measured with reduced bias and increased transparency and can be used to explain the differences in intervention effectiveness between different studies [17]. There is, however, a lack of meta-analysis studies on pig interventions during primary processing. Two meta-analysis studies, which investigated the effects of abattoir interventions and chilling on *Salmonella* only, found significant effects of organic acid washes, hot water washes, steam pasteurisation and chilling in reducing *Salmonella* on pig carcasses [18,19]. However, there are no meta-analysis studies to investigate interventions' effects in reducing indicator bacteria counts and *Yersinia* spp. on pig carcasses. Therefore, the aim of this study was to conduct a systematic review and meta-analysis of literature data reporting on the effectiveness of a range of interventions applied to pig carcasses during primary processing in abattoirs, on indicator bacteria (i.e., ACC, EBC and generic *Escherichia coli* count) and *Yersinia* spp.

2. Materials and Methods

2.1. Review Protocol and Research Question

A systematic review of the literature on the contribution of pig abattoir interventions to the reduction of bacterial load on pig carcasses was conducted, with a focus on the pre and post-slaughter production processes in abattoirs, up to and including primary chilling. The review considered evidence on pig interventions' efficacy available in the public domain, but only primary research studies were used for data extraction and reporting. The review question was: "What is the efficacy of all possible interventions to control microbial contamination on pig carcasses at any stage in the pork production chain from pigs received in the abattoir to the pig carcass chilling inclusive?" The review followed a structured, predefined protocol and PICO framework. The population studied was pigs produced for meat consumption, including their carcasses at primary processing. Relevant outcome measures for interventions were the effectiveness of each intervention in reducing log levels of indicator bacteria (aerobic colony count (ACC), *Enterobacteriaceae* count (EBC) and generic *E. coli* count) and log levels or prevalence of the foodborne pathogens *Salmonella* spp. and *Yersinia enterocolitica/pseudotuberculosis*. Subsequently, it was agreed to exclude data on *Salmonella* from further analysis, as it was found that an insufficient number of studies had been published since the previous systematic review by Young et al. [18] to justify data analysis. Any GHP- and hazard-based interventions applied from the stage of pigs being received in the abattoir lairage up to (and inclusive of) primary chilling in abattoirs were considered relevant.

2.2. Review Team and Search Strategy

Relevance screening, relevance confirmation, risk-of-bias assessment and data extraction were conducted by two review team members, and discrepancies were resolved through discussion or by judgment of a third reviewer. All developed protocols are provided in the Supplementary Material. A comprehensive search algorithm was developed and used for the search of peer-reviewed literature. The algorithm was developed by extracting key words from a selection of twenty known relevant primary research articles on pig interventions (different articles per intervention category), and by reviewing and adapting search strategies and key terms of previously published reviews and risk assessments on this and similar topics. Three databases were searched, Scopus, CAB Direct and SciELO. Key terms were combined using the Boolean operator "OR" into three categories: microorganism/outcome (*E. coli*, *Yersinia*, *Salmonella*, *Enterobacteriaceae*, aerobic colony count), intervention (intervention terms) and population (pig terms). The categories were combined using the "AND" operator. The algorithms were pretested using a list of twenty relevant articles (provided in the Supplementary Materials) in Scopus and CAB direct to ensure they could be sufficiently identified. The searched articles spanned a period of

30 years (1990–2021, except SciELO, which encompassed 2002–2021), with no language restrictions imposed. Search verification included reviewing the reference lists of ten relevant review and ten primary research articles (provided in the Supplementary Materials).

2.3. Relevance Screening and Eligibility Criteria for Prioritisation

All retrieved citations were first uploaded in Endnote X9.2 and duplicates removed. Remaining citations were then imported into the web-based systematic review platform Rayyan for subsequent relevance screening at the title and abstract level [20]. Each article was screened through its title and abstract using a prespecified relevance screening form, and then its relevance further confirmed after the full article was procured and using the prespecified checklist (see Supplementary Materials). All experimental and observational study designs were considered for data extraction (controlled, challenge and before-and-after trials, and cohort studies). These included studies measuring interventions' efficacy through the measurement of concentration (such as colony forming units, (CFU)/sample) and/or prevalence (absence or presence) of microorganisms. Intervention application settings were described as commercial (large or small) abattoirs and pilot plants (where industrial equipment was used in nonindustrial settings) as well as research conducted under laboratory conditions. "In vitro" studies (model broth system experiments) were excluded. The interventions were analysed and presented according to five intervention categories: (i) preslaughter, lairage interventions for live pigs; (ii) standard processing procedures for carcasses; (iii) pig carcass prechilling interventions; (iv) carcass chilling; (v) multiple interventions.

2.4. Risk of Bias Assessment and Data Extraction

The risk-of-bias (RoB) assessment was conducted for 25 primary research articles. It was performed using a prespecified tool that was adapted to suit the needs of the topic and study designs, from the Cochrane Collaboration's recommended tools for randomised and non-randomised study designs [21,22]. Two reviewers conducted RoB assessment independently and any disagreements between them were resolved by a third reviewer. The tool was structured into five domains through which bias might be introduced into the results: (1) bias arising from the randomisation process; (2) bias due to the presence of deviations from intended interventions; (3) bias due to the fact of missing outcome data; (4) bias in measurement of the outcome; (5) bias in selection of the reported result. The possible risk-of-bias judgements were: (1) low risk of bias; (2) some concerns; (3) high risk of bias.

Only articles assessed to be at low risk of bias were considered for detailed data extraction. The data extraction tool included targeted questions about intervention (category, specific intervention and detailed description about intervention parameters), population (i.e., live animal, skin and carcass surface), outcomes (microorganisms) measured, comparison group(s) and intervention efficacy results (concentration and prevalence data). Data were first stratified by study design and conditions, then into specific predefined intervention categories and, finally, by different outcome measures (*Yersinia*, ACC, EBC, generic *Escherichia coli* count). Where data in articles were presented only in visual form, such as graphs, and no other extractable data were present in the text, data on microbial reduction were not considered due to the reduced precision, and these articles were excluded. The detailed protocol followed for RoB assessment and data extraction is provided in the Supplementary Materials.

2.5. Random-Effect Meta-Analysis and Reporting

Data were first stratified by the study design and conditions (commercial abattoir or laboratory), then into specific groups for interventions and, finally, grouped together for different microbiological outcomes. If comparison groups had three or more trials that were eligible for meta-analysis, then the mean CFU/cm^2, $CFU/100\ cm^2$, and their respective standard deviations (SDs) or standard error of means (SEMs) were extracted from studies measuring concentration outcomes. For prevalence outcomes, only the number of positives in each group was extracted. If only the SEM was available, then a pooled SD was calculated. Trials without a direct comparison group were presented in a tabulated form. Random effects models were calculated using R (version 1.3.1093), including packages meta and metaphor [23–25]. These were pooled risk ratios (RRs) for prevalence outcomes and pooled log mean differences for concentration outcomes. If the RR was less than 1, this indicated a lower risk in the intervention group compared to the control one, whereas if the RR was greater than 1, it indicated an increased risk for the intervention group, suggesting the intervention may not be effective. Confidence intervals were also extracted. Weights in the random-effects meta-analysis were based on the size of each study (i.e., number of observations). Forest plots were created to summarise the effects and visualise heterogeneity measures. The results were then summarised and presented in a tabulated form with selected forest plots presented in the main text, while the remaining are available in the Supplementary Materials. Heterogeneity was assessed using I^2, which measures the percentage of variability in the effect size, which is not result of sampling error [26,27]. If I^2 values were greater than 50%, heterogeneity was considered as high, values between 25 and 50% were considered as moderate heterogeneity, whereas values less than 25% represented low levels of heterogeneity. A test for heterogeneity was performed (Cochran's Q-Statistic), which evaluates the null hypothesis that all studies evaluate the same effect. The resultant p-values were also presented; values less than 0.05 indicated that the studies were significantly heterogeneous. Therefore, the resultant forest plots can be split into three groups: those that were homogenous ($p > 0.05$ on the test for heterogeneity), those that were moderately heterogeneous ($p < 0.05$, $I^2 \leq 60\%$) and those that were highly heterogeneous ($p < 0.05$, $I^2 > 60\%$).

3. Results

3.1. Study Characteristics and Risk of Bias Assessment

The results from the systematic review, risk-of-bias assessment and data analysis are shown in Figure 1. Of the 17,340 articles retrieved in the database search and search verification, following the deduplication, 11,480 were screened at title and abstract levels for relevance. After screening, 152 articles were procured as full articles and checked for relevance using eligibility criteria, of which 74 reported interventions in pigs from lairage to chilling. For the purpose of this paper, articles reporting data on non-*Salmonella* outcomes (54 in total) were further checked for extractable data (i.e., data with measures of variability and excluding graphical format). The finalised list for subsequent risk-of-bias assessment included 25 articles (key characteristics shown in Table 1). These were twelve before-and-after trials, nine controlled trials, seven challenge trials and one cohort study.

Most studies on interventions in pigs, and the selected outcomes, were conducted in Europe (64%), followed by North America (24%). The majority of studies were conducted under commercial abattoir conditions (69.2%), followed by laboratory conditions (23.1%). Most of the studies investigated pig carcass prechilling interventions, chilling (air, spray and blast chilling) or standard processing procedures/GHP. Scalding and singeing were investigated in four studies each (10.3%) and lairage interventions were investigated in only two studies (5.1%). Among microorganisms, indicator bacteria (predominantly ACC) were investigated the most, and *Yersinia enterocolitica* in only six studies (13.3%) (Table 1).

Overall, 22 articles were judged to be at low risk of bias (and progressed to data extraction), two articles had some concerns, and one article was judged to be at high risk of bias. The main concerns for controlled trials, cohort trials and challenge trials were bias arising from the randomisation process, whereas only a limited number of before-and-after trials were associated with a similar risk of bias. The results from the RoB assessment process for the 25 articles are presented in Figure 2 in the form of weighted bar plots of the distribution of risk-of-bias judgements within each bias domain.

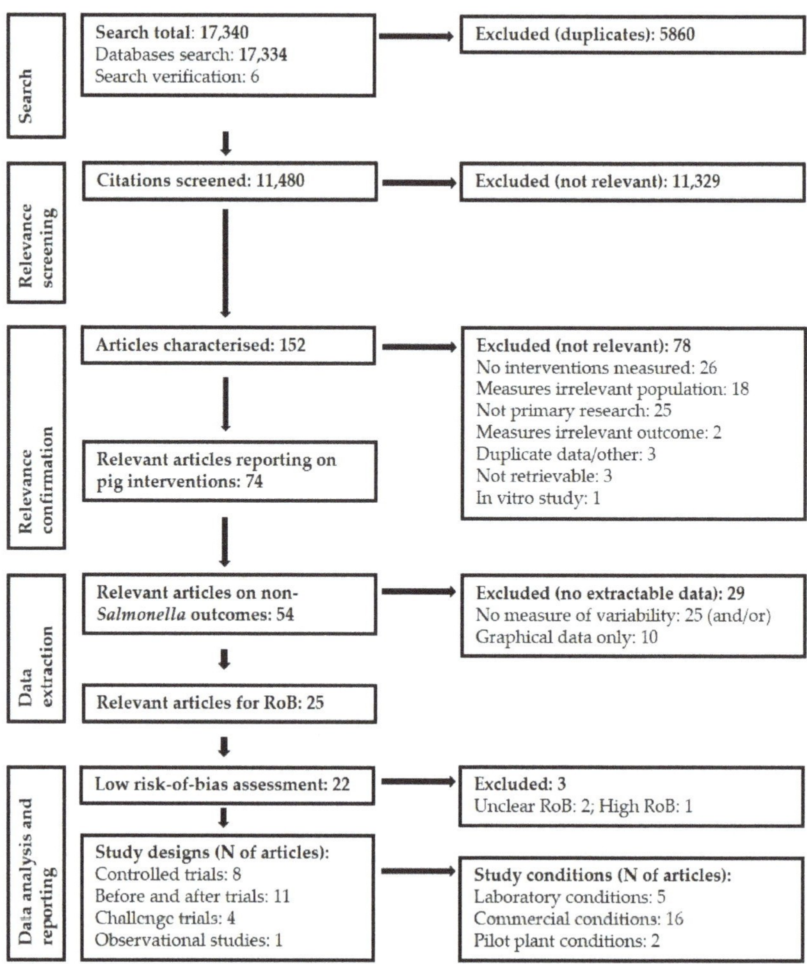

Figure 1. Flow chart of the systematic review and meta-analysis process.

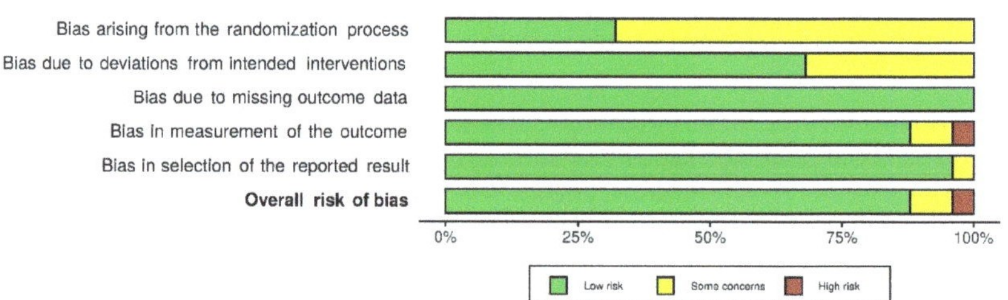

Figure 2. Distribution of risk of bias judgement within each bias domain for all 25 articles investigating pig interventions.

Table 1. Key characteristics of 25 relevant articles on pig interventions.

Article Characteristic	Number of Articles [1]	%
Region		
North America	6	24%
Europe	16	64%
Australia/South Pacific	1	4%
Asia/Middle East	2	8%
Central and South America/Caribbean	0	0
Africa	0	0
Document type		
Journal article	25	100%
Thesis	0	0
Conference paper	0	0
Government or research report	0	0
Study design		
Challenge trial	7	24.1%
Before-and-after trial	12	41.4%
Controlled trial	9	31%
Cohort study	1	3.4%
Study conditions		
Laboratory conditions	6	23.1%
Commercial abattoir conditions	18	69.2%
Research/pilot plant	2	7.7%
Intervention category/subcategory		
Pig handling in lairage	2	5.1%
Scalding	4	10.3%
Singeing	4	10.3%
Other standard processing procedures/GHP	8	20.5%
Carcass prechilling interventions	12	30.8%
Chilling, spray chilling, blast chilling	9	23.1%
Outcomes investigated		
Aerobic colony count	17	37.7%
Enterobacteriaceae count/prevalence	9	20.0%
Generic *E. coli* count/prevalence	12	26.6%
Yersinia enterocolitica count/prevalence	6	13.3%
Yersinia pseudotuberculosis prevalence	1	2.2%
Risk-of-bias concerns		
Low	22	88%
Some concerns	2	8%
High	1	4%

[1] Although the number of included articles was 25, the number of articles per category may not be equal, as often studies incorporated more than one study condition and/or intervention category and investigated multiple outcomes.

3.2. Random-Effects Meta-Analysis

For reasons of brevity, the results on the meta-analysis summary effects are shown below in tabulated form (Tables 2–5). Furthermore, three examples of forest plots are also given (Figures 3–5), and the remaining forest plots can be found in the Supplementary Materials. The results of the interventions for which there were not enough trials for a direct comparison of intervention effects are also presented in the Supplementary Materials.

3.2.1. Preslaughter and Lairage Interventions

Regarding the investigated outcomes, no studies were identified that reported logistic slaughter, and only two studies reported lairage holding time [28] or misting pigs with disinfectant [29]. Six trials from one study found that *Enterobacteriaceae* counts in pig caecal content increased with an increase in both feed withdrawal time and lairage holding time (MD: 0.48, 95% CI: −0.10 to 1.06, I^2 = 77%) [28]. Misting live pigs with disinfectant reduced

Enterobacteriaceae counts on pig skin significantly when compared to water misting alone in only one trial (MD: −1.36, 95% CI: −2.91 to −0.19) [29].

3.2.2. Standard Processing Procedures and GHP-Based Measures

Table 2 summarises the overall meta-analysis estimates of interventions' effects for standard processing procedures and GHP-based measures such as scalding, dehairing, singeing, polishing, water washing, rectum sealing, alternative pluck removal and standard fat trimming.

Several studies investigated the efficacy of scalding in reducing indicator bacteria counts, with sufficient data to calculate meta-regression summary effects. Eight before-and-after trials showed that scalding under commercial abattoir conditions effectively reduced *Enterobacteriaceae* prevalence on pig carcasses (RR: 0.05, 95% CI: 0.02 to 0.12, I^2 = 87%). In addition, 14 before-and-after trials from three studies showed that scalding significantly reduced ACC on pig carcasses by 2.84 $\log_{10}$ CFU/cm^2 (MD: −2.84, 95% CI: −3.50 to −2.18, I^2 = 99%). Another effective standard processing procedure for reducing *Enterobacteriaceae* prevalence and ACC on pig carcasses was singeing (RRL 0.25, 95% CI: 0.14 to 0.44, I^2 = 90% and MD: −1.95, 95% CI: −2.40 to −1.50, I^2 = 96%, respectively). In contrast, eight before-and-after trials investigating carcass water washing had a negligible effect in reducing *Enterobacteriaceae* prevalence (RR: 0.87, 95% CI: 0.80 to 0.94, I^2 = 19%), while it increased the risk of carcass contamination with generic *E. coli* (RR: 1.09, 95% CI: 0.94 to 1.27, I^2 = 26%). Water washing did not reduce ACC on pig carcasses as shown in 20 trials (MD: −0.12, 95% CI: −0.35 to 0.11, I^2 = 90%).

Furthermore, rectum sealing, investigated in two studies with 18 controlled trials, effectively reduced the prevalence of *Y. enterocolitica* on pig carcasses (RR: 0.60, 95% CI: 0.41 to 0.89, I^2 = 0%) (Figure 3). An alternative method with anal plugging prior to scalding and dehairing was investigated in only one study and reduced EBC around the anuses of plugged carcasses by 1.10 log CFU/cm^2 compared with unplugged carcasses [30].

Expectedly, other standard processing procedures for carcasses, such as dehairing, polishing and standard fat trimming, were ineffective in reducing the prevalence or log counts of indicator bacteria and more often led to increase in contamination (Table 2). Dehairing increased ACC by 1.94 $\log_{10}$ CFU/cm^2 (MD: 1.94, 95% CI: 1.67 to 2.11, I^2 = 97%), while also significantly increasing the prevalence of *Enterobacteriaceae* (RR: 17.36, 95% CI: 6.88 to 43.75, I^2 = 89%). Polishing at best did not change ACC or prevalence of *Enterobacteriaceae*, and similar results were reported with standard fat trimming (Table 2). One alternative pluck removal procedure, where the pluck set was partially removed, leaving the highly contaminated oral cavity, tonsils and tongue in place, did not meaningfully reduce the prevalence of *Y. enterocolitica*, *Enterobacteriaceae* and generic *E. coli*, and did not reduce ACC.

Table 2. A summary of the overall meta-analysis estimates of the interventions' effects for standard processing procedures and good hygiene practices on pig carcasses.

Intervention	Microorganism [a]	Study Design/ Conditions (No. of Studies/Trials) ‡	RR (95% CI) or MD (95% CI)	Heterogeneity I^2 (%) *	*p*-Value *	Reference(s)
Scalding	EBC	BA/Comm (1/8)	RR 0.05 (0.02, 0.12)	High (87%)	<0.01	[15]
Scalding	ACC	BA/Comm (4/14)	MD −2.48 (−3.50, −2.18)	High (99%)	0	[11,15,31,32]
Dehairing	EBC	BA/Comm (1/8)	RR 17.36 (6.88, 43.75)	High (89%)	<0.01	[15]
Dehairing	ACC	BA/Comm (3/12)	MD 1.94 (1.67, 2.21)	High (97%)	<0.01	[11,15,31]
Singeing	EBC	BA/Comm (1/4)	RR 0.25 (0.14, 0.44)	High (90%)	<0.01	[15]
Singeing	ACC	BA/Comm (3/9)	MD −1.95 (−2.4, −1.5)	High (96%)	<0.01	[11,15,32]
Polishing	EBC	BA/Comm (1/8)	RR 1.01 (0.8, 1.28)	High (86%)	<0.01	[15]
Polishing	ACC	BA/Comm (3/12)	MD 0.19 (−0.51, 0.89)	High (100%)	0	[11,14,15]

Table 2. *Cont.*

Intervention	Microorganism [a]	Study Design/ Conditions (No. of Studies/Trials) ‡	RR (95% CI) or MD (95% CI)	Heterogeneity I^2 (%) *	*p*-Value *	Reference(s)
Water washing	ACC	CT_BA/Comm (4/20)	MD −0.12 (−0.35, 0.11)	High (90%)	<0.01	[14,15,31,33]
Water washing	EBC	BA/Comm (1/8)	RR 0.87 (0.8, 0.94)	Low (19%)	0.28	[15]
Water washing	Generic *E. coli*	BA/Comm (1/8)	RR 1.09 (0.94, 1.27)	Low (26%)	0.22	[33]
Rectum sealing	*Yersinia pseudotuberculosis*	CT/Comm (1/5)	RR 1.33 (0.24, 7.49)	Low (38%)	0.17	[12]
Rectum sealing	*Yersinia enterocolitica*	CT/Comm (2/18)	RR 0.6 (0.41, 0.89)	Low (0%)	0.88	[12,34]
Pluck removal	EBC	CT/Comm (1/3)	RR 0.98 (0.94, 1.03)	Low (0%)	0.56	[13]
Pluck removal	*Yersinia enterocolitica*	CT/Comm (1/3)	RR 0.33 (0.03, 3.18)	Low (0%)	1.00	[13]
Pluck removal	Generic *E. coli*	CT/Comm (1/3)	RR 0.87 (0.68, 1.11)	High (71%)	0.03	[13]
Pluck removal	ACC	CT/Comm (1/3)	MD −0.04 (−0.3, 0.21)	Low (34%)	0.22	[13]
Standard fat trimming	EBC	BA/Comm (1/8)	RR 1.16 (1.01, 1.33)	High (71%)	<0.01	[15]
Standard fat trimming	ACC	BA/Comm (1/8)	MD 0.06 (−0.16, 0.27)	High (95%)	<0.01	[15]

‡ CT—controlled trial; BA—before-and-after trial; Comm—commercial abattoir conditions. [a] ACC—aerobic colony count; EBC—*Enterobacteriaceae* count. * Homogenous trials: $p > 0.05$ on the test for heterogeneity; moderately heterogeneous: $p < 0.05$, $I^2 \leq 60\%$; highly heterogeneous: $p < 0.05$, $I^2 > 60\%$.

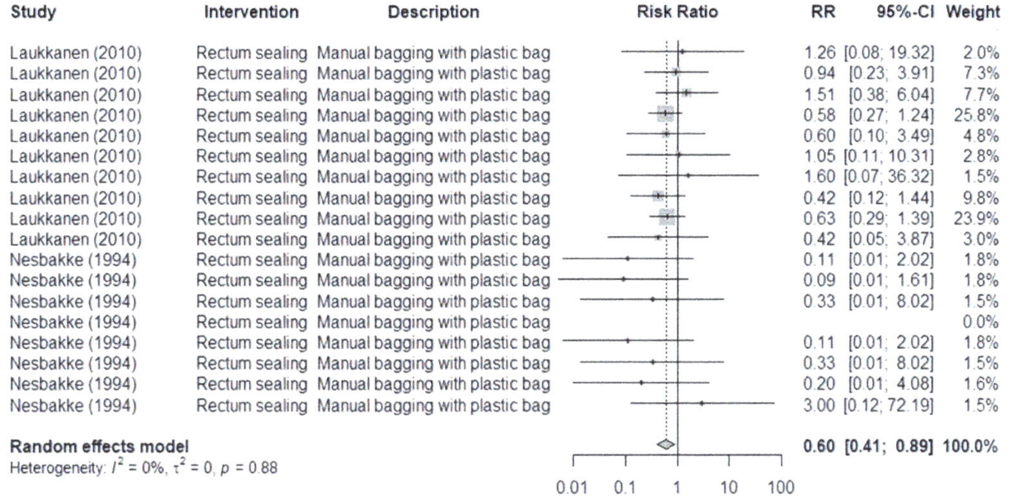

Figure 3. Forest plot of the results of controlled trials performed under commercial abattoir conditions to investigate the efficacy of rectum sealing in reducing *Yersinia enterocolitica* prevalence on pig carcasses [12,34].

3.2.3. Prechilling Carcass Interventions

Data for only four hazard-based interventions for pig carcasses applied at the prechilling stage were available from the literature; interventions were hot water washing, lactic acid or acidified sodium chlorite (ASC) washing and novel pulsed light treatment (Table 3). Hot water washing investigated under commercial abattoir conditions significantly reduced the prevalence of generic *E. coli* on pig carcasses (RR: 0.31, 95% CI: 0.15 to 0.64, $I^2 = 91\%$)

(Figure 4). It also significantly reduced ACC (MD: -1.32, 95% CI: -1.93 to -0.71, $I^2 = 93\%$) and generic *E. coli* count on pig carcasses (MD: -1.23, 95% CI: -1.89 to -0.57, $I^2 = 61\%$) (Table 3). Challenge trials conducted under laboratory conditions found that lactic acid wash reduced EBC by 0.72 $\log_{10}$ CFU/cm^2 (MD: -0.72, 95% CI: -1.40 to -0.05, $I^2 = 98\%$) and ACC by 1.07 $\log_{10}$ CFU/cm^2 (MD: -1.07, 95% CI: -1.33 to -0.81, $I^2 = 93\%$) on pig carcass meat. Another single study investigating prechilling lactic acid carcass spray efficacy after 24 h chilling found reductions of 0.49–1.05 $\log_{10}$ CFU/cm^2 for ACC and of 0.73–1.38 $\log_{10}$ CFU/cm^2 in generic *E. coli* count [35] (Supplementary Materials).

In 36 trials investigating pulsed light treatment, a significant reduction of 1.68 $\log_{10}$ CFU/cm^2 in *Y. enterocolitica* on pig carcass meat (MD: -1.68, 95% CI: -1.99 to -1.37, $I^2 = 97\%$) was demonstrated. ASC wash was investigated in only one study with two trials; therefore, meta-analysis summary estimates were not calculated. However, two trials found RRs of 0.13 and 0.43 in reducing the prevalence of generic *E. coli* and mean reductions of 0.47–1.30 $\log_{10}$ CFU/cm^2 for ACC and 1.05–1.64 $\log_{10}$ CFU/cm^2 for generic *E. coli* count [36] (Supplementary Materials).

Table 3. A summary of the overall meta-analysis estimates of the interventions' effects for pig carcass interventions: hot water washing, lactic acid washing and pulsed light treatment.

Intervention	Microorganism [a]	Study Design/ Conditions (No. of Studies/Trials) [‡]	RR (95% CI) or MD (95% CI)	Heterogeneity I^2 (%) *	*p*-Value *	Reference(s)
Hot water washing	Generic *E. coli*	CT_BA/Comm (3/6)	RR 0.31 (0.15, 0.64)	High (91%)	<0.01	[36–38]
Hot water washing	Generic *E. coli*	CT_BA/Comm (2/4)	MD -1.23 (-1.89, -0.57)	Moderate (61%)	0.05	[36,38]
Hot water washing	ACC	CT_BA/Comm (3/8)	MD -1.32 (-1.93, -0.71)	High (93%)	<0.01	[36–38]
Lactic acid washing	EBC	ChT/Lab (2/6)	MD -0.72 (-1.40, -0.05)	High (98%)	<0.01	[39,40]
Lactic acid washing	ACC	ChT/Lab (2/12)	MD -1.07 (-1.33, -0.81)	High (93%)	<0.01	[39,40]
Pulsed light treatment	*Yersinia enterocolitica*	ChT/Lab (1/36)	MD -1.68 (-1.99, -1.37)	High (97%)	<0.01	[41]

[‡] CT—controlled trial; BA—before-and-after trial; ChT—challenge trial; Comm—commercial abattoir conditions.
[a] ACC—aerobic colony count; EBC—*Enterobacteriaceae* count. * Homogenous trials: $p > 0.05$ on the test for heterogeneity; moderately heterogeneous: $p < 0.05$, $I^2 \leq 60\%$; highly heterogeneous: $p < 0.05$, $I^2 > 60\%$.

Study	Intervention	Description	Risk Ratio	RR	95%-CI	Weight
Hamilton, D (2010)	Hot water wash	83.5°C, 15 s		0.11	[0.04; 0.27]	16.7%
Hamilton, D (2010)	Hot water wash	83.5°C, 15 s		0.32	[0.23; 0.44]	21.9%
Gill (1997)	Hot water wash	Post-polishing, pre-evisceration, 85°C, 15 s		0.70	[0.51; 0.95]	22.0%
Gill (1997)	Hot water wash	Post-polishing, pre-evisceration, 85°C, 15 s		0.81	[0.59; 1.11]	21.9%
Gill (1998)	Hot water wash	85°C, 10 s, carcass split before		0.09	[0.01; 0.65]	8.7%
Gill (1998)	Hot water wash	85°C, 10 s, carcass split after		0.08	[0.01; 0.59]	8.7%
Random effects model				**0.31**	**[0.15; 0.64]**	**100.0%**

Heterogeneity: $I^2 = 91\%$, $\tau^2 = 0.6180$, $p < 0.01$

0.1 0.51 2 10

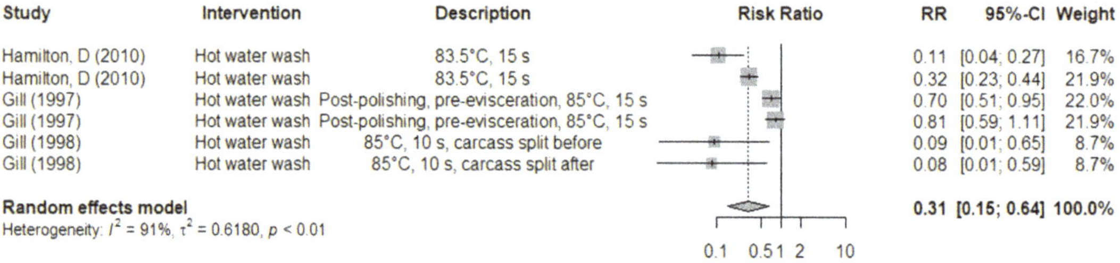

Figure 4. Forest plot of the results of combined controlled trials and before-and-after trials performed under commercial abattoir conditions to investigate the efficacy of hot water washing in reducing generic *E. coli* prevalence on pig carcasses [36–38].

3.2.4. Chilling

Three different methods of chilling were studied: conventional dry, blast and water spray chilling. Conventional dry chilling produced more consistent reductions in indicator bacteria counts, whereas other methods of chilling, such as combination of blast and conventional chilling, produced mixed results (Table 4). In four before-and-after trials under commercial conditions, conventional dry chilling effectively reduced *Enterobacteriaceae* prevalence on pig carcasses (RR: 0.32, 95% CI: 0.21 to 0.48, I^2 = 81%). Likewise, fifteen before-and-after trials showed a small but significant 0.36 $\log_{10}$ CFU/cm^2 reduction in ACC (MD: −0.36, 95% CI: −0.61 to −0.12, I^2 = 94%). Conventional chilling also significantly reduced ACC (MD: −1.77, 95% CI: −2.54 to −1.01, I^2 = 35%) and generic *E. coli* count (MD: −2.44, 95% CI: −3.93 to −0.95, I^2 = 89%) in four challenge laboratory trials on pig carcass meat.

Blast chilling followed by conventional dry chilling reduced prevalence of *Enterobacteriaceae* on pig carcasses (RR: 0.10, 95% CI: 0.02 to 0.47, I^2 = 78%), but not the prevalence of generic *E. coli* (RR: 0.61, 95% CI: 0.34 to 1.11, I^2 = 50%) or ACC (MD: −0.17, 95% CI: −0.47 to 0.12, I^2 = 93%) in four before-and-after trials conducted under commercial abattoir conditions. In four challenge trials, blast chilling produced similar reduction effects as conventional dry chilling for ACC (MD: −1.70, 95% CI: −2.81 to −0.59, I^2 = 57%) and generic *E. coli* count (MD: −2.64, 95% CI: −4.56 to −0.73, I^2 = 94%) on pig carcass meat.

Blast chilling followed by water spray chilling largely did not reduce the prevalence of *Enterobacteriaceae* (RR: 0.55, 95% CI: 0.34 to 0.90, I^2 = 46%) and actually led to increased ACC (MD: 0.01, 95% CI: −1.00 to 2.22, I^2 = 88%) on pig carcass meat in trials conducted under commercial abattoir conditions.

Table 4. A summary of the overall meta-analysis estimates of the interventions' effects for different chilling methods on pig carcasses.

Intervention	Microorganism [a]	Study Design/ Conditions (No. of Studies/Trials) [‡]	RR (95% CI) or MD (95% CI)	Heterogeneity I^2 (%) *	*p*-Value *	Reference(s)
Conventional dry chilling	EBC	BA/Comm (1/4)	RR 0.32 (0.21, 0.48)	High (81%)	<0.01	[15]
Conventional dry chilling	ACC	BA/Comm (4/15)	MD −0.36 (−0.61, −0.12)	High (94%)	<0.01	[11,15,33,42]
Blast and conventional chilling	EBC	BA/Comm (1/4)	RR 0.1 (0.02, 0.47)	High (78%)	<0.01	[15]
Blast and conventional chilling	Generic *E. coli*	BA/Comm (1/4)	RR 0.61 (0.34, 1.11)	Low (50%)	0.11	[33]
Blast and conventional chilling	ACC	BA/Comm (3/10)	MD −0.17 (−0.47, 0.12)	High (93%)	<0.01	[15,32,33]
Blast and water spray chilling	EBC	BA/Comm (2/3)	RR 0.55 (0.34, 0.9)	Low (46%)	0.16	[33,43]
Blast and water spray chilling	ACC	BA/Comm (2/3)	MD 0.01 (−1.0, 2.22)	High (88%)	<0.01	[33,43]
Blast chilling	Generic *E. coli*	ChT/Lab (1/4)	MD −2.64 (−4.56, −0.73)	High (94%)	<0.01	[44]
Blast chilling	ACC	ChT/Lab (1.4)	MD −1.7 (−2.81, −0.59)	Low (57%)	0.07	[44]
Blast vs conventional chilling	ACC	ChT/Lab (1.4)	MD −0.04 (−1.02, 0.94)	Low (30%)	0.23	[44]
Conventional dry chilling	ACC	ChT/Lab (1/4)	MD −1.77 (−2.54, −1.01)	Low (35%)	0.20	[44]
Conventional dry chilling	Generic *E. coli*	ChT/Lab (1/4)	MD −2.44 (−3.93, −0.95)	High (89%)	<0.01	[44]

[‡] BA—before-and-after trial; ChT—challenge trial; Comm—commercial abattoir conditions. [a] ACC—aerobic colony count; EBC—*Enterobacteriaceae* count. * Homogenous trials: $p > 0.05$ on the test for heterogeneity; moderately heterogeneous: $p < 0.05$, $I^2 \leq 60\%$; highly heterogeneous: $p < 0.05$, $I^2 > 60\%$.

3.2.5. Multiple Interventions

Several studies conducted under commercial abattoir conditions investigated the effects of multiple interventions sequentially applied on the slaughterline. The majority of these trials investigated the efficacy of sequential use of scalding, dehairing, singeing, polishing, trimming, water washing (with or without prechilling lactic acid spray) and blast and/or dry chilling (Table 5). Eight before-and-after trials investigating multiple interventions showed they effectively reduced *Enterobacteriaceae* prevalence on pig carcasses (RR: 0.11, 95% CI: 0.05 to 0.23, I^2 = 94%). Similarly, another fifteen before-and-after trials found significant reductions (2.85 $\log_{10}$ CFU/cm^2) of ACC on pig carcasses (MD: -2.85, 95% CI: -3.33 to -2.37, I^2 = 97%) (Figure 5). In only one study/trial that investigated the sequential use of scalding, dehairing, singeing and scraping, reductions achieved for ACC, EBC and generic *E. coli* were 0.87 $\log_{10}$ CFU/cm^2, 2.15 $\log_{10}$ CFU/cm^2 and 2.20 $\log_{10}$ CFU/cm^2, respectively [31].

Table 5. A summary of the overall meta-analysis estimates of the multiple intervention effects on pig carcasses.

Intervention	Microorganism [a]	Study Design/ Conditions (No. of Studies/Trials) [‡]	RR (95% CI) or MD (95% CI)	Heterogeneity I^2 (%) [*]	*p*-Value [*]	Reference(s)
Multiple **	EBC	BA/Comm (1/8)	RR 0.11 (0.05, 0.23)	High (94%)	<0.01	[15]
Multiple ***	ACC	BA/Comm (4/15)	MD -2.85 (-3.33, -2.37)	High (97%)	<0.01	[11,15,32,35]

[‡] BA—before-and-after trial; Comm—commercial abattoir conditions. [a] ACC—aerobic colony count; EBC—*Enterobacteriaceae* count. [*] Homogenous trials: $p > 0.05$ on the test for heterogeneity; moderately heterogeneous: $p < 0.05$, $I^2 \leq 60\%$; highly heterogeneous: $p < 0.05$, $I^2 > 60\%$. ** Interventions including scalding, dehairing, singeing, polishing, trimming, water washing and blast and/or dry chilling. *** Interventions including scalding, dehairing, singeing, polishing, water washing and/or lactic acid washing and blast/dry chilling.

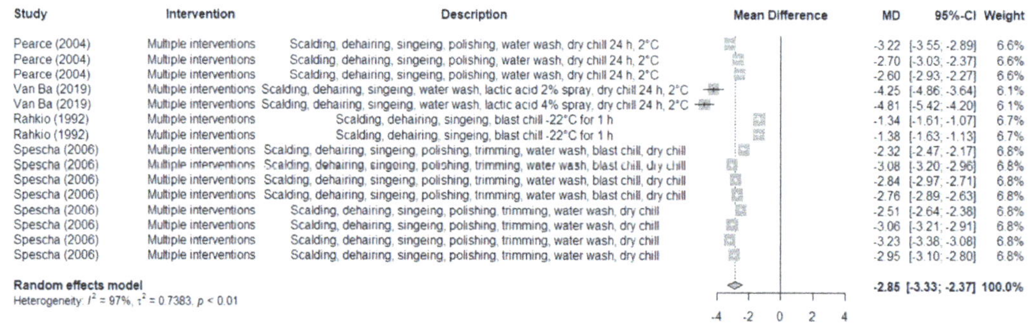

Figure 5. Forest plot of the results of before-and-after trials performed under commercial abattoir conditions to investigate the efficacy of multiple interventions in reducing aerobic colony count ($\log_{10}$ CFU) on pig carcasses [11,15,32,35].

4. Discussion

The aim of this study was to analyse a range of abattoir interventions and to identify those that have a significant reduction effect on microorganisms of concern (i.e., indicator bacteria and *Yersinia* spp.) using the statistical power of a meta-analysis tool. Overall, 30 years of literature were reviewed, and following a structured and stringent review process, 22 articles were found eligible to conduct a meta-analysis. The final outcomes were 48 forest plots and 40 meta-analysis summary effects generated. Data were included for interventions from the preslaughter stage (i.e., lairage holding time, feed withdrawal and misting pigs with disinfectant), standard processing procedures for pig carcasses, hazard-based prechilling interventions and multiple carcass interventions, to the final chilling stage. Despite the fact that this systematic review included such a large body of literature and investigated interventions for four microorganisms, the main findings and concerns are that pig interventions are not a well-researched area, and there are significant gaps in the literature. Furthermore, even when some studies existed for a given intervention/outcome, more than half of identified eligible studies either did not report measures of variability, which are essential for meta-analysis or had data presented in difficult-to-extract graph format. In line with the problems with methodological study design in some of the articles reviewed, the data reporting was a significant obstacle in obtaining more useful data for analysis purpose. Among 40 pooled meta-analysis summary effects (pooled risk ratios (RRs), for prevalence outcomes, or pooled log mean difference for concentration outcomes), only 13 were with low or moderate heterogeneity (and, therefore, we had better confidence in the results). Meta-analysis is a useful analytical tool for combining the results of multiple primary research studies into a weighted, average estimate for, in our case, intervention effect. The limitation of this analysis could be that, even though every effort was made to stratify data into the most similar subgroups, sometimes within-subgroup data likely resulted from studies/trials with recorded or unrecorded differences. This stratification approach was chosen for pragmatic reasons to combine a sufficient number of trials for meta-analysis, wherever it was possible, from a limited pool of data. As a consequence, details about intervention application parameters (e.g., acid concentration, temperature, duration) and differences between study sampling and laboratory methods were not investigated as possible sources of variation in intervention effects across studies. These and other study factors could well contribute to the heterogeneity in effects observed for many intervention categories, but it was beyond the scope of this systematic review and meta-analysis to investigate these factors in detail. However, the created forest plots contain sufficient information and description about analysed interventions. Overall, this systematic review clearly has identified a lack of large, controlled trials conducted under commercial conditions, with sound study design and adequate reporting of intervention protocols. This was particularly case with hazard-based, prechilling interventions at slaughter, and particularly for *Yersinia* spp. This was surprising given that carriage of *Yersinia enterocolitica* on pig tonsils and prevalence on pig carcasses at slaughter can be as high as 90% and 60%, respectively, and up to 30% in raw pork [7]. Inadequate reporting of protocols, lack of addressing any confounders, inappropriate choice of outcome measurement units when expected microbial counts were low and faults in reporting of results (e.g., lack of measures of variability) were common and reduced further the already sparse pool of scientific data in this area.

The microbial status of pig carcasses on the slaughterline depends on many factors, including preslaughter hygiene and animal cleanliness. Stress factors during transport and lairage can provoke the shedding of bacteria, including pathogens, increasing the risk of faecal contamination of carcasses during slaughter [3,28]. Lairage time and direct or indirect contact of groups of pigs during lairaging prior to slaughter influence the bacterial load of carcasses or the occurrence of pathogens in lymphoid tissues [45,46]. For example, a higher prevalence of *Y. enterocolitica* was found in the tonsils of pigs slaughtered in the slaughterhouses where pigs were held in the lairage pens separated by a fence that allowed contact between the pigs (40% and 52%), than in the tonsils of pigs slaughtered in the slaughterhouse that had a solid wall between lairage pens thus preventing contact between pigs (29%) [47]. Moreover, in one study, a higher prevalence of *Salmonella* was found in pigs in the lairage than in the farm of origin [6]. Considering the outcomes included in the present systematic review, only one cohort study on lairage interventions with six trials was eligible to conduct a meta-analysis [28]. This showed that feed withdrawal time and holding time in lairage have no significant effect on *Enterobacteriaceae* count in pig caecal content and likely no effect in their further spread on the slaughterline during slaughter and dressing. However, the lairage is known to be a source of contamination with *Salmonella* [48]; thus, it is important that further research is conducted to assess effective ways to reduce contamination before pigs enter the slaughterline.

Standard processing procedures and good hygienic practices in pig slaughtering are designed to maintain high levels of hygiene and produce final carcasses with low microbial load. Various slaughter operations affect the bacterial status of pig skin, offal and carcasses in a positive or negative way, i.e., they can increase contamination or reduce the microbial load [1]. Thermal treatments are a well-known hurdle used to reduce bacterial contamination and are used to varying degrees in pig carcass scalding (with warm water), singeing (open-flame gas burning) or spraying/washing (hot water, steam). The present meta-analysis identified that within standard processing procedures, scalding and singeing were the most effective in reducing *Enterobacteriaceae* prevalence and ACC on pig carcasses (by around 2 logs). Although their primary purpose is dehairing, they contribute to the reduction of microbial contamination of pig carcasses [1]. Scalding time and temperatures vary from abattoir to abattoir, and differences in these parameters produce different reductions of microbial contamination [15,31]. The current meta-analysis found dehairing and polishing, on the other hand, increased the counts and/or prevalence of aerobic bacteria and *Enterobacteriaceae*, as expected. Dehairing machines are always contaminated with bacteria and are washed with recirculated hot water only. A recent study reported that recycled water in the dehairing process is the main source of contamination of pig carcasses with *Salmonella* at the abattoir [49]. It is also generally accepted that subsequent polishing facilitates the redistribution of any surviving bacteria from the singeing process throughout the pig carcass [11]. In this meta-analysis, we found that polishing at best did not change ACC or prevalence of *Enterobacteriaceae*. Other GHP measures investigated provided mixed results. For example, water washing only negligibly reduced the prevalence of *Enterobacteriaceae* and ACC, and slightly increased prevalence of generic *E. coli*. Furthermore, combined effects of sequential use of several standard processing procedures (scalding, dehairing, singeing and scraping) achieved reductions in ACC, EBC and generic *E. coli* counts of up to 2 logs, although only one study/trial was eligible for this meta-analysis [31]. This was expected, as usually two or more interventions applied sequentially produce a larger effect than any individual intervention [16].

The evisceration procedure on the slaughterline is one of the most critical steps, which begins with the loosening and sealing of the rectum. The general purpose of this hygienic procedure is to avoid faecal contamination of the carcass and organs. Data analysis showed its efficacy in reducing *Y. enterocolitica* prevalence on carcasses, suggesting that this procedure should always be applied. In addition, the data analysis revealed strong evidence, derived from laboratory trials, of the efficacy of pulsed light to reduce *Y. enterocolitica* on pig carcasses. Pathogenic *Y. enterocolitica* is a priority hazard to control in pork production and more data are needed for its effective control in the meat chain [50,51]. The present systematic review did not identify any other published studies investigating other potentially relevant interventions to reduce *Y. enterocolitica* on carcasses. Thus, the effectiveness of interventions in reducing *Y. enterocolitica* on pig carcasses is an insufficiently researched area, and there is a serious lack of data in this respect.

In some abattoirs, carcass interventions are used with the aim of reducing bacterial loads and the carriage of pathogens detected at farm level. This includes hazard-based interventions, such as hot water washing. Combinations of controlled and before-and-after trials conducted under commercial abattoir conditions showed hot water washing effectively reduces the prevalence and counts of generic *E. coli* and ACC, by around 1 log. Hot water washing is also a very effective intervention commonly used for beef carcasses [16]. In Denmark, hot water washing is used on pig carcasses from batches originating from *Salmonella*-positive pig herds. It has been found to be more cost-effective than steam vacuum and lactic acid washing [52,53].

Lactic acid washing also significantly reduced EBC and ACC on pig carcass meat, but data eligible for meta-analysis came only from studies investigated under laboratory conditions. EFSA in 2018 issued a scientific opinion on the safety and efficacy of organic acids for pig carcasses [54]. In its review, EFSA found that spraying pig carcasses with lactic acid (2–5% solutions at temperatures of up to 80 °C) prior to chilling is of no safety concern (provided that the substances comply with the EU specifications for food additives) and was efficacious compared to untreated control. However, EFSA could not conclude whether lactic acid was more efficacious than water treatment when pig carcasses were sprayed at the prechilling stage. EFSA's review was systematic in nature and included 11 literature sources (16 eligible experiments) on lactic acid but without meta-analysis. Some of analysed literature sources were on pork meat cuts post-chill or ground pork (therefore, these were excluded from our study), and some of them did not report measures of variability that are needed for meta-analysis. Following a similar positive EFSA opinion from 2011 [55], lactic acid was permitted for use in EU abattoirs (Regulation EC 101/2013) for beef carcass washing [56]. Lactic acid washes are efficacious interventions for beef carcasses, usually reducing indicator bacteria counts by 1–1.5 logs under commercial abattoir conditions [16]. Studies investigating other organic acids (e.g., acetic acid) and other chemical agents for pig carcass washes were lacking or did not meet criteria for this meta-analysis.

Chilling is a procedure mandated by the legislation, and there are several methods of chilling with varying degrees of effectiveness with regard to reducing microbial contamination. Usually, dressed pig carcasses are blasted with air at approximately $-8\,^\circ\mathrm{C}$ to $-20\,^\circ\mathrm{C}$ for up to 1 h to quickly reduce carcass temperature, and then the carcasses are transferred to a conventional chiller at approximately $2\,^\circ\mathrm{C}$ for the remaining chilling time. Studies focusing on the effects of a combination of blast chilling followed by conventional chilling and/or each individual chilling method showed inconsistent results [15,33]. It is likely that the effectiveness of these interventions is influenced by temperature, air velocity, humidity, and duration [33]. Furthermore, it is likely that some microbial reductions are due to inactivation due to surface drying but also due to reduced viability of bacteria to recover from chilling for subsequent growth and/or inability of swabbing method to pick up bacteria cells from the dry surface. These factors hinder microbial detection, and therefore, proper study design and using specific media to enable microbial recovery is necessary when investigating the efficacy of chilling.

Multiple interventions when applied sequentially (scalding, dehairing, singeing, polishing, trimming, water washing (with or without prechilling lactic acid spray) and blast and/or dry chilling) produced the biggest reductions of up to 3 logs of ACC and significantly reduced the prevalence of *Enterobacteriaceae* on pig carcasses. Application of multiple slaughterline interventions is expected to improve the overall microbial status of carcasses and reduce risks further than do single interventions [16], particularly when they are extended in an overall multiple-hurdle strategy with decontamination of resulting portioned meat and pork trimmings [57]. Furthermore, use of interventions is necessary in high risk situations (e.g., when an abattoir is unable to sufficiently reduce risks arising from specific farms/animal batches by using process hygiene alone), to meet the targets on chilled carcasses [16,53]. As such, pig interventions at abattoir stage (preslaughter and slaughter) form an essential component of the meat safety assurance system.

5. Conclusions

This systematic review and meta-analysis were performed to assess the effectiveness of abattoir interventions in reducing indicator bacteria counts (i.e., ACC, EBC and generic *E. coli* count) and the count and/or prevalence of *Yersinia* spp. on pig carcasses. There were noticeable gaps in the literature spanning 30 years on studies investigating pig interventions. This was very clear, particularly with respect to interventions with proven efficacy in some other meat species (e.g., beef carcasses), such as carcass steam pasteurisation and organic acid washes (acetic acid and lack of data on lactic acid), and there is a distinct lack of sufficient data on hot water washing and blast chilling. Several commercial trials found that common standard processing procedures, such as scalding and singeing, are very effective in reducing indicator bacteria counts. This meta-analysis found that pig carcass scalding effectively reduces the prevalence of *Enterobacteriaceae* (RR 0.05) and ACC ($2.84\,\log_{10}\,\mathrm{CFU/cm^2}$), as does singeing (RR: 0.25, and $1.95\,\log_{10}\,\mathrm{CFU/cm^2}$, respectively). Rectum sealing effectively reduces the prevalence of *Y. enterocolitica* on pig carcasses (RR

0.60). A multiple hurdle approach that included the sequential application of carcass interventions significantly reduces *Enterobacteriaceae* prevalence (RR: 0.11) and ACC on carcasses (2.85 $\log_{10}$ CFU/cm^2). Nevertheless, most of the data were generated from highly heterogeneous studies and trials, likely due to the inherent differences between studies, but also from the small number of studies/trials eligible for this meta-analysis. This indicates that better designed research, with results presented numerically and with measures of variability, is needed. This is particularly the case for *Y. enterocolitica*, which is a priority pathogen for control in the pig meat chain. Overall, the results suggest that scalding, singeing, washing with hot water and/or lactic acid, and dry chilling effectively reduce the counts of indicator bacteria on pig carcasses. The meta-analysis also found evidence that pathogenic *Y. enterocolitica* on pig carcasses is effectively reduced by the standard procedure of rectum sealing; however, this was the only intervention for *Yersinia* investigated under commercial conditions. All these effective interventions should be recommended for commercial use in abattoirs and should form an essential part of integrated pig meat controls. Furthermore, the data generated in this meta-analysis can be used for further modelling and risk assessment work and for providing recommendations on the use of specific interventions in pig abattoirs.

Supplementary Materials: The following supporting information can be downloaded at: https://www.mdpi.com/article/10.3390/foods11142110/s1, Database search results; Detailed systematic review protocols; List of references for studies used in meta-analysis; Examples of intervention forest plots.

Author Contributions: Conceptualisation, D.A., K.H. and N.Z.; methodology, D.A., K.H. and A.K.; software, A.K.; validation, D.A. and A.K.; formal analysis, A.K. and D.A.; investigation, D.A., K.H., N.Z., A.A.-O., B.B., N.K. and M.S.; resources, D.A.; writing—original draft preparation, N.Z.; writing—review and editing, D.A., K.H., A.A.-O., B.B. and M.S.; supervision, D.A.; project administration, D.A.; funding acquisition, B.B. All authors have read and agreed to the published version of the manuscript.

Funding: This research received no external funding. The APC was funded by COST Action 18105 (Risk-based Meat Inspection and Integrated Meat Safety Assurance).

Institutional Review Board Statement: Not applicable.

Informed Consent Statement: Not applicable.

Data Availability Statement: Data is contained within the article.

Acknowledgments: This publication is based upon work from COST Action 18105 (Risk-based Meat Inspection and Integrated Meat Safety Assurance; www.ribmins.com (accessed on 2 May 2022)) supported by COST (European Cooperation in Science and Technology; www.cost.eu (accessed on 2 May 2022)).

Conflicts of Interest: The authors declare no conflict of interest. The funders had no role in the design of the study; in the collection, analyses, or interpretation of data; in the writing of the manuscript, or in the decision to publish the results.

References

1. Buncic, S.; Sofos, J. Interventions to control *Salmonella* contamination during poultry, cattle and pig slaughter. *Food Res. Int.* **2012**, *45*, 641–655. [CrossRef]
2. Houf, K. Control of *Salmonella* in pigs. *Wien Tierarztl. Monatsschr.* **2012**, *99*, 278–285.
3. De Busser, E.V.; De Zutter, L.; Dewulf, J.; Houf, K.; Maes, D. *Salmonella* control in live pigs and at slaughter. *Vet. J.* **2013**, *196*, 20–27. [CrossRef] [PubMed]
4. European Parliament and Council of the European Union. Regulation (EC) No 2073/2005 of 15 November 2005 on Microbiological Criteria for Foodstuffs. *Off. J. Eur. Union* **2005**, *338*, 1–26.
5. EFSA. Scientific Opinion on the public health hazards to be covered by inspection of meat (swine). *EFSA J.* **2011**, *9*, 2351. [CrossRef]
6. Bonardi, S. *Salmonella* in the pork production chain and its impact on human health in the European Union. *Epidemiol. Infect.* **2017**, *145*, 1513–1526. [CrossRef]

7. Laukkanen-Ninios, R.; Fredriksson-Ahomaa, M.; Korkeala, H. Enteropathogenic *Yersinia* in the pork production chain: Challenges for control. *Compr. Rev. Food Sci. Food Saf.* **2014**, *13*, 1165–1191. [CrossRef]

8. Buncic, S.; Alban, L.; Blagojevic, B. From traditional meat inspection to development of meat safety assurance programs in pig abattoirs–the European situation. *Food Control* **2019**, *106*, 106705. [CrossRef]

9. Cegar, S.; Kuruca, L.; Vidovic, B.; Antic, D.; Hauge, S.J.; Alvseike, O.; Blagojevic, B. Risk categorisation of poultry abattoirs on the basis of the current process hygiene criteria and indicator microorganisms. *Food Control* **2021**, *132*, 108530. [CrossRef]

10. Belluco, S.; Barco, L.; Roccato, A.; Ricci, A. Variability of *Escherichia coli* and *Enterobacteriaceae* counts on pig carcasses: A systematic review. *Food Control* **2015**, *55*, 115–126. [CrossRef]

11. Pearce, R.A.; Bolton, D.J.; Sheridan, J.J.; McDowell, D.A.; Blair, I.S.; Harrington, D. Studies to determine the critical control points in pork slaughter hazard analysis and critical control point systems. *Int. J. Microbiol.* **2004**, *90*, 331–339. [CrossRef]

12. Laukkanen, R.; Ranta, J.; Dong, X.; Hakkinen, M.; Martínez, P.O.; Lundén, J.; Johansson, T.; Korkeala, H. Reduction of enteropathogenic *Yersinia* in the pig slaughterhouse by using bagging of the rectum. *J. Food Protect.* **2010**, *73*, 2161–2168. [CrossRef] [PubMed]

13. Biasino, W.; De Zutter, L.; Woollard, J.; Mattheus, W.; Bertrand, S.; Uyttendaele, M.; Van Damme, I. Reduced contamination of pig carcasses using an alternative pluck set removal procedure during slaughter. *Meat Sci.* **2018**, *145*, 23–30. [CrossRef]

14. Yu, S.; Bolton, D.; Laubach, C.; Kline, P.; Oser, A.; Palumbo, S.A. Effect of dehairing operations on microbiological quality of swine carcasses. *J. Food Protect.* **1999**, *62*, 1478–1481. [CrossRef] [PubMed]

15. Spescha, C.; Stephan, R.; Zweifel, C. Microbiological contamination of pig carcases at different stages of slaughter in two Europian Union-approved abattoirs. *J. Food Protect.* **2006**, *69*, 2568–2575. [CrossRef]

16. Antic, D.; Houf, K.; Michalopoulou, E.; Blagojevic, B. Beef abattoir interventions in a risk-based meat safety assurance system. *Meat Sci.* **2021**, *182*, 108622. [CrossRef]

17. O'Connor, A.; Sargeant, J.; Wang, C. Conducting systematic reviews of intervention questions III: Synthesizing data from intervention studies using meta-analysis. *Zoonoses Public Health* **2014**, *61*, 52–63. [CrossRef]

18. Young, I.; Wilhelm, B.J.; Cahill, S.; Nakagawa, R.; Desmarchelier, P.; Rajić, A. A rapid systematic review and meta-analysis of the efficacy of slaughter and processing interventions to control nontyphoidal *Salmonella* in beef and pork. *J. Food Protect.* **2016**, *79*, 2196–2210. [CrossRef]

19. Gonzales-Barron, U.; Cadavez, V.; Sheridan, J.J.; Butler, F. Modelling the effect of chilling on the occurrence of *Salmonella* on pig carcasses at study, abattoir and batch levels by meta-analysis. *Int. J. Microbiol.* **2013**, *163*, 101–113. [CrossRef]

20. Ouzzani, M.; Hammady, H.; Fedorowicz, Z.; Elmagarmid, A. Rayyan—A web and mobile app for systematic reviews. *Syst. Rev.* **2016**, *5*, 210. [CrossRef]

21. McGuinness, L.; Higgins, J. Risk-of-bias VISualization (robvis): An R package and shiny web app for visualizing risk-of-bias assessments. *Res. Synth. Methods* **2020**, *12*, 55–61. [CrossRef]

22. Sterne, J.A.; Savović, J.; Page, M.J.; Elbers, R.G.; Blencowe, N.S.; Boutron, I.; Cates, C.J.; Cheng, H.-Y.; Corbett, M.S.; Eldridge, S.M. RoB 2: A revised tool for assessing risk of bias in randomised trials. *BMJ* **2019**, *366*, l4898. [CrossRef] [PubMed]

23. R Core Team. *A Language and Environment for Statistical Computing*; R Foundation for Statistical Computing: Vienna, Austria, 2021; Available online: https://www.R-project.org/ (accessed on 1 March 2022).

24. Balduzzi, S.; Rücker, G.; Schwarzer, G. How to perform a meta-analysis with R: A practical tutorial. *EBMH* **2019**, *22*, 153–160. [CrossRef] [PubMed]

25. Viechtbauer, W. Conducting meta-analyses in R with the metafor package. *J. Stat. Softw.* **2010**, *36*, 1–48. [CrossRef]

26. Lin, L. Comparison of four heterogeneity measures for meta-analysis. *J. Eval. Clin. Pract.* **2020**, *26*, 376–384. [CrossRef]

27. Higgins, J.P.; Thompson, S.G.; Deeks, J.J.; Altman, D.G. Measuring inconsistency in meta-analyses. *BMJ* **2003**, *327*, 557–560. [CrossRef]

28. Martín-Peláez, S.; Martín-Orúe, S.M.; Pérez, J.F.; Fàbrega, E.; Tibau, J.; Gasa, J. Increasing feed withdrawal and lairage times prior to slaughter decreases the gastrointestinal tract weight but favours the growth of cecal *Enterobacteriaceae* in pigs. *Livest. Sci.* **2008**, *119*, 70–76. [CrossRef]

29. Walia, K.; Lynch, H.; Grant, J.; Duffy, G.; Leonard, F.C.; Lawlor, P.G.; Gardiner, G.E. The efficacy of disinfectant misting in the lairage of a pig abattoir to reduce *Salmonella* and *Enterobacteriaceae* on pigs prior to slaughter. *Food Control* **2017**, *75*, 55–61. [CrossRef]

30. Purnell, G.; James, C.; Wilkin, C.A.; James, S.J. An evaluation of improvements in carcass hygiene through the use of anal plugging of pig carcasses prior to scalding and dehairing. *J. Food Protect.* **2010**, *73*, 1108–1110. [CrossRef]

31. Rivas, T.; Vizcaíno, J.A.; Herrera, F.J. Microbial contamination of carcasses and equipment from an Iberian pig slaughterhouse. *J. Food Protect.* **2000**, *63*, 1670–1675. [CrossRef]

32. Rahkio, M.; Korkeala, H.; Sippola, I.; Peltonen, M. Effect of pre-scalding brushing on contamination level of pork carcasses during the slaughtering process. *Meat Sci.* **1992**, *32*, 173–183. [CrossRef]

33. Gill, C.O.; Dussault, F.; Holley, R.A.; Houde, A.; Jones, T.; Rheault, N.; Rosales, A.; Quessy, S. Evaluation of the hygienic performances of the processes for cleaning, dressing and cooling pig carcasses at eight packing plants. *Int. J. Microbiol.* **2000**, *58*, 65–72. [CrossRef]

34. Nesbakken, T.; Nerbrink, E.; Røtterud, O.J.; Borch, E. Reduction of *Yersinia enterocolitica* and *Listeria* spp. on pig carcasses by enclosure of the rectum during slaughter. *Int. J. Microbiol.* **1994**, *23*, 197–208. [CrossRef]

35. Van Ba, H.; Seo, H.W.; Seong, P.N.; Kang, S.M.; Cho, S.H.; Kim, Y.S.; Park, B.Y.; Moon, S.S.; Kang, S.J.; Choi, Y.M.; et al. The fates of microbial populations on pig carcasses during slaughtering process, on retail cuts after slaughter, and intervention efficiency of lactic acid spraying. *Int. J. Microbiol.* **2019**, *294*, 10–17. [CrossRef]

36. Hamilton, D.; Holds, G.; Lorimer, M.; Kiermeier, A.; Kidd, C.; Slade, J.; Pointon, A. Slaughterfloor decontamination of pork carcasses with hot water or acidified sodium chlorite—A comparison in two Australian abattoirs. *Zoonoses Public Health* **2010**, *57*, 16–22. [CrossRef] [PubMed]

37. Gill, C.O.; Jones, T.; Badoni, M. The effects of hot water pasteurizing treatments on the microbiological conditions and appearances of pig and sheep carcasses. *Food Res. Int.* **1998**, *31*, 273–278. [CrossRef]

38. Gill, C.; Bedard, D.; Jones, T. The decontaminating performance of a commercial apparatus for pasteurizing polished pig carcasses. *Food Microbiol.* **1997**, *14*, 71–79. [CrossRef]

39. Van Netten, P.; Mossel, D.A.A.; Huis, J.H.J. Microbial changes on freshly slaughtered pork carcasses due to "hot" lactic acid decontamination. *J. Food Saf.* **1997**, *17*, 89–111. [CrossRef]

40. Van Netten, P.; Valentijn, A.; Mossel, D.A.A.; Huis, J.H.J. Fate of low temperature and acid-adapted *Yersinia enterocolitica* and *Listeria monocytogenes* that contaminate lactic acid decontaminated meat during chill storage. *J. Appl. Microbiol.* **1997**, *82*, 769–779. [CrossRef]

41. Koch, F.; Wiacek, C.; Braun, P.G. Pulsed light treatment for the reduction of *Salmonella typhimurium* and *Yersinia enterocolitica* on pork skin and pork loin. *Int. J. Microbiol.* **2019**, *292*, 64–71. [CrossRef]

42. Langkabel, N.; Großpietsch, R.; Oetjen, M.; Bräutigam, L.; Irsigler, H.; Jaeger, D.; Ludewig, R.; Fries, R. Microbiological status of pig carcasses in mobile chilling vehicles. *Arch. Lebensmittelhyg.* **2014**, *65*, 45–49.

43. Gill, C.O.; Jones, T. Assessment of the hygienic performances of an air-cooling process for lamb carcasses and a spray-cooling process for pig carcasses. *Int. J. Microbiol.* **1997**, *38*, 85–93. [CrossRef]

44. Chang, V.P.; Mills, E.W.; Cutter, C.N. Reduction of bacteria on pork carcasses associated with chilling method. *J. Food Protect.* **2003**, *66*, 1019–1024. [CrossRef]

45. Argüello, H.; Carvajal, A.; Álvarez-Ordóñez, A.; Jaramillo-Torres, H.A.; Rubio, P. Effect of logistic slaughter on *Salmonella* contamination on pig carcasses. *Food Res. Int.* **2014**, *55*, 77–82. [CrossRef]

46. De Busser, E.V.; Maes, D.; Houf, K.; Dewulf, J.; Imberechts, H.; Bertrand, S.; De Zutter, L. Detection and characterization of *Salmonella* in lairage, on pig carcasses and intestines in five slaughterhouses. *Int. J. Microbiol.* **2011**, *145*, 279–286. [CrossRef] [PubMed]

47. Zdolec, N.; Kiš, M.; Jankuloski, D.; Blagoevska, K.; Kazazić, S.; Pavlak, M.; Blagojević, B.; Antić, D.; Fredriksson-Ahomaa, M.; Pažin, V. Prevalence and persistence of multidrug-resistant *Yersinia enterocolitica* 4/O:3 in slaughter pigs from different housing systems in Croatia. *Foods* **2022**, *11*, 1459. [CrossRef]

48. Mannion, C.; Fanning, J.; McLernon, J.; Lendrum, L.; Gutierrez, M.; Duggan, S.; Egan, J. The role of transport, lairage and slaughter processes in the dissemination of *Salmonella* spp. in pigs in Ireland. *Food Res. Int.* **2012**, *45*, 871–879. [CrossRef]

49. Zeng, H.; Rasschaert, G.; De Zutter, L.; Mattheus, W.; De Reu, K. Identification of the source for *Salmonella* contamination of carcasses in a large pig slaughterhouse. *Pathogens* **2021**, *10*, 77. [CrossRef]

50. Zdolec, N.; Kiš, M. Meat safety from farm to slaughter—Risk-based control of *Yersinia enterocolitica* and *Toxoplasma gondii*. *Processes* **2021**, *9*, 815. [CrossRef]

51. Van Damme, I.; Berkvens, D.; Vanantwerpen, G.; Baré, J.; Houf, K.; Wauters, G.; De Zutter, L. Contamination of freshly slaughtered pig carcasses with enteropathogenic *Yersinia* spp.: Distribution, quantification and identification of risk factors. *Int. J. Microbiol.* **2015**, *204*, 33–40. [CrossRef]

52. Lawson, L.G.; Jensen, J.D.; Christiansen, P.; Lund, M. Cost-effectiveness of *Salmonella* reduction in Danish abattoirs. *Int. J. Microbiol.* **2009**, *134*, 126–132. [CrossRef] [PubMed]

53. Alban, L.; Sørensen, L.L. Hot-water decontamination is an effective way of reducing risk of *Salmonella* in pork. *Fleischwirtschaft* **2010**, *90*, 109–113.

54. EFSA Panel on Food Contact Materials, Enzymes and Processing Aids (CEP); Silano, V.; Barat Baviera, J.M.; Bolognesi, C.; Brüschweiler, B.J.; Chesson, A.; Cocconcelli, P.S.; Crebelli, R.; Gott, D.M.; Grob, K.; et al. Evaluation of the safety and efficacy of the organic acids lactic and acetic acids to reduce microbiological surface contamination on pork carcasses and pork cuts. *EFSA J.* **2018**, *16*, e05482. [PubMed]

55. EFSA Panel on Biological Hazards (BIOHAZ). Scientific Opinion on the evaluation of the safety and efficacy of lactic acid for the removal of microbial surface contamination of beef carcasses, cuts and trimmings. *EFSA J.* **2011**, *9*, 2317.

56. European Parliament and Council of the European Union. Regulation (EC) No 101/2013, Concerning the use of lactic acid to reduce microbiological surface contamination on bovine carcasses. *Off. J. Eur. Union* **2013**, *L 34*, 1–3.

57. Blagojevic, B.; Antic, D.; Adzic, B.; Tasic, T.; Ikonic, P.; Buncic, S. Decontamination of incoming beef trimmings with hot lactic acid solution to improve microbial safety of resulting dry fermented sausages—A pilot study. *Food Control* **2015**, *54*, 144–149. [CrossRef]

 *foods*

Article

The Value of Current *Ante Mortem* Meat Inspection and Food Chain Information of Dairy Cows in Relation to *Post Mortem* Findings and the Protection of Public Health: A Case for a More Risk-Based Meat Inspection

Pieter Jacobs [1], Boyd Berends [2,*] and Len Lipman [2]

[1] Netherlands Food and Consumer Product Safety Authority (NVWA),
    P.O. Box 43006, 3540 AA Utrecht, The Netherlands
[2] Institute for Risk Assessment Sciences, Department of Population Health Science,
    Faculty Veterinary Medicine, Utrecht University, P.O. Box 80175, 3508 TD Utrecht, The Netherlands
*   Correspondence: b.r.berends@uu.nl

Abstract: In this study, the contribution of the *ante mortem* (AM) inspection and the food chain information (FCI) to ensuring meat safety and public health was investigated, by evaluating the slaughterhouse findings of 223,600 slaughtered dairy cows in the Netherlands. The outcome of this study was that the *ante mortem* (AM) and *post mortem* (PM) inspections have a substantial overlap, and that with regard to food safety and public health in over 99% of cases the PM could even be omitted on the basis of the AM. In this study, the data provided by the dairy farmers on the current FCI forms contributed little to nothing with regard to the outcomes of AM and PM inspection. It is concluded that current meat inspection procedures need an update and a more risk-based approach needs to be adopted. Regarding this, the AM inspection of dairy cattle should remain, because it plays an important role in ensuring food safety (e.g., by preventing contamination of the slaughter line by excessively dirty animals, or animals with abscesses), monitoring animal welfare and in detecting some important notifiable diseases. The PM inspection, however, could in many cases be omitted, provided there is a strict AM inspection complemented by a vastly improved (automated) way of obtaining reliable FCI.

Keywords: meat safety; meat inspection; risk-based; legislation; veterinarian; official control

Citation: Jacobs, P.; Berends, B.; Lipman, L. The Value of Current *Ante Mortem* Meat Inspection and Food Chain Information of Dairy Cows in Relation to *Post Mortem* Findings and the Protection of Public Health: A Case for a More Risk-Based Meat Inspection. *Foods* **2023**, *12*, 616. https://doi.org/10.3390/foods12030616

Academic Editor: Frans J. M. Smulders

Received: 8 November 2022
Revised: 4 January 2023
Accepted: 16 January 2023
Published: 1 February 2023

1. Introduction

Current meat inspection was originally designed in Europe in the late 19th century and was almost entirely aimed at protecting the public's health [1]. With increasing international trade, detecting notifiable animal diseases soon became another important goal. The most recent addition to its goals is monitoring animal welfare. For all of these reasons, all animals destined for slaughter are subjected to a brief clinical examination by an official veterinarian before they enter the slaughter line (i.e., the *ante mortem* inspection, AM) and a concise pathological–anatomical examination after most of the internal organs have been removed and made accessible for close inspection (i.e., the *post mortem* inspection, PM). These two examinations together determine (i) whether an animal may be slaughtered for human consumption, and (ii) if that slaughtered animal is fit for human consumption, so that its meat and edible by-products may indeed enter the human food chain. The way AM and PM are performed in the EU is currently laid down in the Official Controls Regulation EU 2017/625 and follows a strict protocol regardless of the age of the animals or any other factor that may influence the possible outcomes and value of these procedures [2–4].

However, the threats to public health that can be associated with the slaughter of animals have changed during the last century, whereas the system of meat inspection has remained basically the same. Therefore, it seems that current meat inspection procedures are no longer adequate in protecting public health, and there is a need for a more risk-based form of meat inspection [5].

At the time meat inspection was designed, virtually all zoonotic diseases that were of primary concern had distinct clinical signs and/or caused distinct macroscopic pathological–anatomical abnormalities, as, for example, was the case for tuberculosis, anthrax and cysticercosis [1,6,7]. Currently, these zoonotic conditions do not play a significant role in modern western countries or are no longer even considered to be a major health risk anymore, such as bovine cysticercosis [8]. The currently important human health hazards remain practically always undetected during AM and PM inspection. Examples of these are animals that contain residues of veterinary drugs or environmental contaminants, animals that are infected with *Toxoplasma gondii*, and animals that are fecal carriers of *Salmonella spp.*, *E. coli* O157:H7 or Extended Spectrum Beta Lactamase (ESBL) producing *Enterobacteriaceae* [9–14].

Thus, the value of meat inspection with regard to its efficacy in protecting human and animal health in situations where animals are raised in modern systems of husbandry and provided with optimum health care may be seriously questioned [15]. Nowadays, the main function of the AM meat inspection appears to be (a) preventing the contamination of the slaughter line (e.g., by excessively dirty animals, or animals with an abscess), (b) monitoring animal welfare, and (c) acting as a last line of defense with regard to several notifiable animal diseases. *Post mortem* meat inspection, on the other hand, serves to detect abnormalities that are almost entirely related to food quality and on-farm (health) management issues [9,16–18].

This study was aimed at (a) assessing the value of the current EU meat inspection procedures with regard to the condemnation of whole carcasses declared 'not suitable for human consumption' (NHC) in the framework of the protection of public health, and (b) gauging whether data from official meat inspections—as a proof of principle—can potentially be used for determining which AM or PM procedures could, in a particular situation, be revised or even omitted (i.e., to transform our current system into a more flexible and risk-based one).

2. Materials and Methods

2.1. Population and Slaughterhouse

The slaughter of dairy cows was chosen as a test-case, because the slaughter of this group probably best resembles the situation for which meat inspection was originally designed. When compared to the slaughter of pigs or poultry, the number of animals from a farm sent to slaughter in one shipment is relatively low and the animals are also far more diverse with regard to the circumstances they were kept under, as well as their genetic make-up. Furthermore, dairy cows are currently the animals that, at the time of slaughter, display the largest variations in age and disease history [19].

For this study, the data from 223,600 animals that were slaughtered in 2014 and 2015 in the largest cattle slaughterhouse in the Netherlands were used. This slaughterhouse is considered to be representative for the entire Dutch situation, since the only difference with the other slaughterhouses was the scale of operations and not, for example, the type of breeds slaughtered or the regions from where the animals originated.

2.2. Data Sources and Management

The main data regarding the results of the AM and PM inspections came from the database for the Registration of Slaughter Findings (Registratie Slacht Gegevens, RSG) of the Netherlands Food and Consumer Product Safety Authority (NVWA) [19] and from the individual findings during the inspections of each animal as written by hand on official forms from the NVWA, also called "VOS forms" (Verzamelstaat Onderzoeksgegevens Slachtdieren, i.e., Summary Findings Meat Inspection) [20]. On these VOS forms, all relevant findings of the AM and PM inspection are briefly noted by the official veterinarians, including the final decision regarding the carcass and organs.

The Identification and Registration (I&R) data came from the official database for the registration of animals in the Netherlands. These data were needed for determining the location of the farms of origin, the age of the animals, their parity and their breed.

The individual Food Chain Information (FCI) forms were included in this study, when any relevant information was available. Food Chain Information is legally required for all animals to be slaughtered for human consumption (as laid down in EU regulation 853/2004, annex II, Section III) [5].

As a first step, all 223,600 VOS forms from 2014 and 2015 were—in a period of several months—thoroughly read. All the VOS forms of animals that were declared not suitable for human consumption (NHC) at the *post mortem* examination (PM) were used as a basis, and put in a spreadsheet together with the information about the AM inspection results, the information from the FCI forms, and information from the RSG and I&R databases. All data from 3933 individual NHC animals thus gathered were subsequently used for further analyses.

2.3. Definitions and Categorizations

The criteria determining whether an animal is suitable for human consumption are laid down in European legislation [4]. If an animal is declared 'NHC' by the official veterinarian, Regulation EU 2017/625 considers it by definition an unacceptable risk for food safety and/or public health, irrespective of the variety of underlying causes and diagnoses that can be made. Thus, we considered for this study the declaration of an animal as 'NHC' as our end point, too.

In this study, we have analyzed the number of animals considered suitable for human consumption (SHC) after *post mortem* inspection and the number of NHC animals as related to their *ante mortem* inspection (AM) data and/or their FCI forms, and whether there was a pattern to be seen between these findings and the PM results.

Animals with no clinical findings at *ante mortem* inspection were assigned to a group that was called *Ante Mortem*-1 (AM-1). If there were remarks on *ante mortem* inspection, such animals were placed either in a group of animals showing local deformities that was called *Ante Mortem*-2 (AM-2), or in a group we called *Ante Mortem*-3 (AM-3). The latter group consisted of animals with remarks about their habitus (e.g., abnormal postures, abnormal coats, general signs of discomfort, fatigue, emaciation, etc.), but which were not considered to be sick and/or otherwise unfit to be slaughtered for human consumption. Animals that were declared unfit to be slaughtered at AM inspection were by definition excluded from the spreadsheet.

Likewise, the origin (province), parity and the breed of the animals were included in the analysis. For this, the breeds were categorized as 'Holstein-Friesian' (HF), 'Meuse Rhine IJssel' (MRIJ) or 'other breeds'. Parity was considered as a proxy for age.

Finally, with the aid of simple 2 × 2 tables, we looked into some of the test characteristics and measures of agreement between the FCI forms and the AM results, using the PM results (NHC or SHC) as the "gold standard". This was because these two elements of the meat inspection procedures can also be considered as diagnostic (screening) tests for sieving out animals that pose a hazard to the consumers' health.

2.4. Statistical Analysis

All analyses were carried out in **R** (free software environment for statistical computing and graphics). The number of NHC cases was analyzed using logistic regression analysis with, as independent variables: province, *ante mortem* information, breed and number of calvings. To see if the *ante mortem* effect on the number of NHC depended on province, an "*ante mortem*–province interaction" was added to the model.

For similar reasons, a "breed–*ante mortem*" interaction and "number of calvings–*ante mortem* interactions" were added to the model. Akaike's Information Criterion (AIC) was used for model reduction. For the effects that were important according to the AIC odds-ratios and their profile (log-) likelihood confidence intervals were calculated. The log odds ratio for AM-2 (only local deformities) or AM-3 (aberrant habitus) were calculated against the AM-1 group (no abnormalities at the time of AM inspection).

For the number of calvings, a Poisson model was used with province and breed as independent variables. For important effects according to the AIC, odds-ratios and their 95% profile (log-) likelihood confidence intervals were calculated. (The complete results of this analysis and the R-script used are included in the Supplementary Results section as File S1 R-output and File S2 R-script).

3. Results

From the in-total 223,600 slaughtered dairy cows, 212,546 originated from 9500 farms throughout the whole of the Netherlands. The remaining 11,054 animals were imported from, mainly, Belgium (ca. 80%), Germany (ca. 18%) and France (ca. 2%). When important data were missing, imported animals were excluded from the analysis. Of the 223,600 slaughtered dairy cows, a total of 3933 animals (1.8%) were considered 'NHC' at PM. Table 1 summarizes the findings with respect to the categorization of the animals in groups with or without certain remarks during AM inspection, as written down on the official VOS forms.

Table 1. Overview of the number of animals slaughtered within each AM category and the number of animals that were declared 'not suitable for human consumption' (NHC) during PM.

AM Group	Number	NHC (%)
AM-1 (no remarks)	213,744	1700 (0.8%)
AM-2 (local abnormalities)	7195	1111 (15%)
AM-3 (aberrant habitus)	2661	1122 (42%)
Totals:	223,600	3933 (1.8%)

Note: all slaughtered animals in this table were not severely ill or otherwise unfit for slaughter; hence, it is stressed that animals with serious health problems, severe mastitis, inability to walk, severe pneumonia are NOT included.

The AM results differed between categories regarding the likelihood of an animal to be declared 'NHC' during PM. Animals from the AM-2 and AM-3 groups were significantly more likely to become declared 'NHC' than animals from the AM-1 group. On average, the calculated odds-ratios for an animal to become declared 'NHC' were 2.99 (2.42 < OR < 4.60; 95% confidence interval, ci) for the AM-2 group, and for the AM-3 group were 4.04 (2.71 < OR < 6.26; 95% ci).

The origin, i.e., the province where the animals came from, had a small but statistically significant effect on the likelihood that animals were being declared 'NHC' during PM (see File S1 R-output, and Table S1 Provinces-distances-NHC in the Supplementary Materials). However, these differences were inconsistent. It appeared that not the geographical distances, per se, but other factors that we did not consider when compiling the dataset played a role in this. The cause of these inconsistent differences in outcomes of certain particular regions should be further investigated, but a plausible explanation will be given in the discussion.

The differences between the parity of the slaughtered dairy cows and the number of carcasses declared 'NHC' during PM is shown in Table 2. From the total number slaughtered, 20.928 animals were excluded, because they were bulls or imported animals from which the parity was unknown. Almost 80% of the slaughtered animals had only calved four times or fewer. The percentage of animals declared 'NHC' more or less increased incrementally with parity, with a maximum percentage of almost 11 at a parity of ten calvings (OR = 1.032; 1.015 < OR < 1.049; 95% ci).

Table 2. Overview of the number of slaughtered animals, categorized by number of calvings and the number of animals declared 'not suitable for human consumption' ('NHC') during PM examination.

Number of Calvings	Total Number	NHC (%)
0	22,070	141 (0.6%)
1	35,220	462 (1.3%)
2	23,7089	472 (1.1%)
3	35,222	533 (1.7%)
4	28,920	572 (2.0%)
5	19,752	467 (2.4%)
6	11,791	289 (2.5%)
7	6447	365 (5.7%)
8	3296	265 (7.8%)
9	1573	44 (2.8%)
10	813	89 (10.9%)
11 or more	539	3 (0.6%)
Total:	202,732	3933 (1.8%)

The number of 'NHC' cases was also analyzed, using logistic regression analysis, and this demonstrated that the AM effect is dependent both on the number of calvings and on the breed (included in the supplementary materials).

The differences between breeds with regard to the number of animals declared 'NHC' is shown in Table 3. There was a significant difference between the Holstein Friesian and the other breeds. Imported animals were excluded from the calculations because their breed was often not known.

Table 3. Overview of the number of animals categorized by breed (HF: Holstein-Friesian; MRIJ: Meuse-Rhine-IJssel) and the number of animals that were declared 'not suitable for human consumption' ('NHC') during PM examination.

Breed	Total Number	NHC (%)
HF	145,182	2871 (2.0%)
MRIJ	60,714	718 (1.2%)
Other breeds	6650	59 (0.9%)
Total:	223,600	3648 (1.6%)

Animals from the Holstein-Friesian breed were significantly more likely to be declared 'NHC' than were cows from all other breeds. Depending on whether the animal was noted to display local abnormalities (AM-2), or more generalized signs of distress (AM-3), the OR varied from 2.4 to 3.0 (2.37 < OR < 3.03; 95% ci) for AM-2 animals, and between 1.3 and 2.3 for the AM-3 animals (1.32 < OR < 2.32; 95% ci).

From the 212,546 Dutch dairy cows brought to this slaughterhouse, only 7038 (3,3%) had one or more of the relevant questions answered with 'yes' on their Food Chain Information forms (designated either as 'FCI ok' or 'FCI Not ok'). These questions were about recent illness, the use of veterinary drugs and about withdrawal periods of the drugs used. There never was any 'yes' answers regarding the questions about the (disease) status of the holdings (e.g., *Salmonella*, paratuberculosis, etc.) or about relevant results from previous AM or PM inspections of animals from the same holdings. Of these 7038 animals, 380 (5%) were declared 'NHC' at PM. Table 4 summarizes the PM results with regard to each of the three AM categories.

Table 4. Numbers of animals that had one or more questions answered with 'yes' on the FCI-form (i.e., FCI-Nok = FCI Not OK) and the PM results per AM inspection category (1: no remarks, 2: local abnormalities, 3: aberrant habitus; 'SHC' is suitable- and 'NHC' is not suitable for human consumption).

	AM-1		AM-2		AM-3		Totals	
FCI-Form:	**SHC**	**NHC**	**SHC**	**NHC**	**SHC**	**NHC**	**SHC**	**NHC**
FCI-Nok	6148	319	246	19	264	42	6658	380 (5.4%)
FCI-ok	198,287	1280	2531	1001	1442	987	202,240	3268 (1.6%)
Total	204,435	1599	2777	1020	1686	1029	208,898	3648 (1.7%)

If the FCI is to be seen as a diagnostic (screening) test, with the PM results regarding 'NHC' as the gold standard, and calculated with a simple two by two table, the overall sensitivity (5.4%) and predictive value (10.4%) regarding animals being declared 'NHC' on the basis of any of the relevant questions being answered with 'yes' will be low. On the other hand, the negative predictive value seemed high (96%). In other words, in those cases that all FCI questions were answered with 'no', there was a 96% probability that the animal would also not be declared 'NHC' at PM. However, the likelihood ratio of a positive or negative test result, which indicates whether there is an increased probability of finding or not finding an 'NHC' animal at PM, was 3.4 and 0.96, respectively. That points towards the FCI being not useful as a quick test for sieving out likely 'NHC' or 'SHC' animals (the calculations are included in the uploaded Supplementary Materials).

The calculations on the overall properties of the AM inspection as a diagnostic test and the PM inspection as the gold standard showed that the overall sensitivity was 56.8% and the overall specificity 99.2%. In other words, if an animal had no remarks at the *ante mortem* inspection, there was a more than 99% probability that it would not be declared 'NHC' at *post mortem* inspection. In addition, the 'likelihood ratio' of a positive test result (LR+) was 16.3 and the LR- 0.46, a clear indication that the test results indeed lead to greater probabilities of finding or not declaring an animal NHC at PM (the calculations can be found in the uploaded Supplementary Materials).

When regarded as a set of parallel (screening) tests, the sensitivity of the combination of the FCI form and the AM inspection resulted in an overall sensitivity of the combined test results of about 59% and an overall specificity of 95%, which is lower than when the specificity of each test is considered separately. However, whether or not these tests can be considered as a useful combination that improves the performance of EU inspection procedures is entirely debatable. The determination of the measure of agreement between these two tests with Cohens' kappa showed that the agreement between the two tests was far from acceptable. The Cohen's kappa value for agreement between the two tests regarding animals declared 'NHC' at PM was −0.15 and for animals declared 'SHC' at PM about 0.07. This means that the two tests disagreed, and apparently measure different things and cannot be seen as a useful combination (calculations shown in the uploaded Supplementary Materials as File S3 Analysis with 2 × 2 tables).

4. Discussion

4.1. General Remarks

As far as we can conclude from the literature, there is little research into the value of current official EU meat inspection procedures in culled dairy cattle with respect to an efficient protection of the consumers' health on the basis of the data of large numbers of animals slaughtered. This study briefly looked into the relationships between the data of the Food Chain Information, AM inspection, PM inspection and the number of animals declared not fit for human consumption ('NHC') on the basis of a dataset that was derived from individual handwritten forms from over 223,000 slaughtered dairy cattle. The only recent study that used the data of large numbers of slaughtered bovines is a French study by Dupuy et al. in 2013 [21], which included the data of over 50,000 bovines that were slaughtered in 12 different slaughterhouses. However, that study was strictly aimed at assessing whether or not the meat inspection results as such could be used for (regional) syndrome surveillance, and did not allow for any inferences regarding the efficacy of inspection procedures as a diagnostic test used for the protection of the consumers' health.

4.2. Influence of Breed, Province, Parity and Slaughterhouse

The province the animals originated from had a small but statistically significant effect on the AM and PM inspection results and numbers of animals declared 'NHC', but these were inconsistent with the geographical distances to the slaughterhouse (Table S1 in the uploaded extra materials). An explanation could be that there were distinct differences in travel circumstances and/or the total duration of the journey to the slaughterhouse. Given that the Netherlands is a small country with maximum travel distances well below 400 km, a better explanation is that there are differences between provinces in the ways that the slaughterhouse obtained its animals. Later inquiries, made at the slaughterhouse and with different traders, revealed that there were distinct differences between the provinces and the number of animals purchased by agents of the slaughterhouse or bought via traders, whereby the agents appear to buy animals in a somewhat better condition. However, because these data could not be included in the dataset, this needs further investigation. Normally, these data are not a part of the FCI, or registered by the NVWA.

The slaughterhouse chosen can be considered fairly representative for the health situation in the Netherlands regarding the dairy cows brought to slaughter. This is in line with the results on slaughtered pigs reported by Harbers et al. [22], who also demonstrated that, in the Netherlands, the large slaughterhouses provided for the fairest representation of the nationwide health situation in a population of slaughter animals.

Inspection data can vary from inspector to inspector and from slaughterhouse to slaughterhouse. Regarding this, the use of a single, very large slaughter facility was considered an advantage. In the slaughterhouse that provided the data for this study, meat inspection was carried out by a stable group of experienced veterinarians that have worked there together for many years in a time-pressured, high-throughput environment, thus most likely ensuring optimum and uniform performance under stress. After all, studies by Harbers et al. [22] showed that the detection of clinical signs and pathological anatomical abnormalities differ greatly between meat inspectors, and that their performance is clearly influenced by, amongst other things, their working experience and their ability to work under time pressure.

When meat inspection is considered as a diagnostic (screening) test, it seems that, at least under circumstances resembling those under which dairy cows in the Netherlands are being kept, in over 99% of cases, a favorable result of the AM (no remarks at all) also meant a favorable result of the *post mortem* meat inspection (fit for human consumption; ok). In other words, in those cases the *post mortem* meat inspection procedures could just as well have been omitted.

The differences in outcome of the meat inspection between the two largest bovine breeds in the Netherlands can be explained by differences in robustness between the Meuse-Rhine-IJssel (MRIJ) and Holstein-Friesian breeds. Meuse-Rhine-IJssel cattle are still largely dual purpose animals and are generally considered robust, fertile and with firm, sturdy legs [23,24]. Holzhauer et al. [23] and Waag et al. 2005 [24] noted distinct differences in robustness between the Holstein-Friesian and Meuse-Rhine-Ijssel breeds with regard to a number of disease conditions. Not surprisingly, the conditions that were studied by Holzhauer et al. [23] and Waag et al. [24] make up of a large percentage of the conditions mentioned on the AM forms in this study (mastitis, lameness, vaginitis and other urogenital problems). In addition, also internationally, these health conditions (fertility problems, mastitis, lameness) comprise the main reasons for culling dairy cows [25–28].

4.3. Use of FCI

When the FCI is considered a means for pre-selecting the more 'risky' cows (i.e., a diagnostic test carried out independently from the AM inspection) it does not perform as well as it potentially could or should. In fact, the results of the FCI in this study on culled dairy cows in the Netherlands seems to bear little or no significance with regard to the results of the AM or PM inspection and, in the vast majority of cases, the FCI forms were not informative at all. Of the 212,000 forms that were specifically analyzed, only 7038 (3.3%) displayed any answer to questions that were related to the health and medicinal history of the animal. In these cases, the sensitivity of the FCI information with regard to animals being declared 'NHC' was approximately 5%. In contrast, the sensitivity of the AM inspection was approximately 57%. Furthermore, when looking at the measure of agreement between the FCI and the AM as a diagnostic test for sieving out 'NHC' cows, it appeared that the results of the FCI and the AM disagreed strongly and that these had little in common. At least in our study, in the Netherlands with culled dairy cows, the current FCI information and/or the way it is being used seems to be of little added value with regard to ensuring meat safety.

Nevertheless, this does not mean that the FCI as such is a bad instrument. What it does mean is that the competent authorities have to assess whether the FCI is indeed used as intended by the farmers and slaughterhouses. For example, in our study, a mere 3.3% of all the forms were filled out by the dairy farmers with a 'yes' on any of the relevant questions about recent illness, the use of veterinary drugs and about withdrawal periods of the drugs used. However, there were never any 'yes' answers regarding the relevant FCI questions about the (disease) status of the holdings the cows originated from (e.g., *Salmonella*, paratuberculosis, etc.) or about relevant results from previous AM or PM inspections of animals from the same dairy farm. From our own personal experience, Dutch dairy farmers seem to be foremost fixated on the questions about the recent use of veterinary drugs. Additionally, farmers were very reluctant to provide anyone with information that might harm their reputation or the outcome of the AM and PM, for example, the disease status of their dairy herd, or animals declared 'NHC' in the past. Additionally, farmers do not generally understand how many of the questions on the FCI forms relate to meat safety, possibly because meat production is not their core business and is often considered by them as an unavoidable necessity. Moreover, because the Dutch Food and Consumer Product Safety Authorities do not keep records of the herd histories, the dairy farmers are able to continue with this behavior. This aspect of the reliability of information given by farmers on the FCI forms certainly calls for further investigation.

That the percentage of 'NHC' animals rises with the parity (or age) of the animal was to be expected [23,24,27]. The sudden low percentage of animals declared 'NHC' after nine calvings in this study, however, is inexplicable and may be coincidence. The low percentage 'NHC' of the "very old" cows (i.e., >10 calvings) is possibly due to their already proven robustness by their long on-farm career.

4.4. Current AM/PM Examination, Use of FCI and Public Health

From earlier studies [29] and from the opinion on public health hazards in bovine meat of the EFSA (European Food Safety Authority), it can be inferred that, in fact, almost none of the hazards that they considered as currently important could be detected by our current AM and PM procedures [15]. However, when PM examination is conducted without any incisions ('vision-only'), the sensitivity of the detection of cysticercosis and bovine tuberculosis will drop [8]. Nevertheless, bovine tuberculosis is, nowadays, not an important threat anymore in countries with an optimally organized animal health care system and with effective eradication programs in place in practice, which is the case in countries such as the Netherlands, Germany or Denmark [8,15].

With regard to cysticercosis, sarcosporidiosis and toxoplasmosis, the question arises of whether omitting PM procedures does indeed lead to major increased risks to the consumers' health, and if there are alternative ways for preventing or mitigating any of these existing consumer health hazards.

In Europe, the prevalence of *Cysticercus bovis* in dairy cows is generally between circa 1 and 6%, and the sensitivity of detecting cysticercosis during PM examination in general 20% (10–30%) [8]. In other words, currently about 80% of cattle that are actually positive for cysticercosis will already pass PM unhindered and thus—in general—only about 0.25–1.50% of the slaughtered animals in Europe will, at PM, be labelled as positive for *Cysticercus bovis*. Furthermore, only 10% of the cysts found in these carcasses are viable and DNA sequencing of the cysts showed that about 20% of the viable and 50% of the non-viable cysts are not *Cysticercus bovis*. The probability of an infection with *Cysticercus bovis* increases with the age of the animals and the way they were grazed and housed. Therefore, a more risk-based approach, with surveillance in combination with ELISA testing of animals at risk of an infection, would provide a far better way of detecting animals with *Cysticercus bovis* than any current standard PM inspection could ever provide [8,15,21,30–34].

Studies in various European countries show that, in general, 80% or more of dairy cows are carriers of sarcosporidia. Bovines are an intermediate host for several sarcosporidia species, with probably the most important ones being *Sarcocystis bovihominis* and *S. cruzi*. In Western Europe, *S. bovihominis* is the most important zoonotic sarcocystis species carried by cattle and *S. cruzi* is a non-zoonotic species, because it has dogs as the final host. *S. cruzi* is also the most common *sarcocystis* species, and is carried by up to 75% of dairy cows. Current standard PM inspection procedures will identify only macroscopic lesions, which are mostly caused by *S. cruzi* and never by *S. bovihominis*. Moreover, the role of sarcosporidia in eosinophilic myositis in cows is still unclear [35–37]. Therefore, omitting PM inspection procedures for each individual carcass will be of little consequence to the already existing possible health hazards for consumers of beef.

In the case of echinococcosis, PM inspection procedures also have a low sensitivity when detecting smaller hydatid cysts. The Netherlands and many other parts of Europe are not considered endemic regions for echinococcosis, and most human cases of cystic echinococcosis are caused by eating raw vegetables or berries contaminated with feces from dogs or foxes (or other carnivores). In Western European countries where echinococcosis is sporadically found, meat inspection in cattle would suffice when the lungs and livers from imported cattle from countries where echinococcosis is endemic are condemned [38].

Toxoplasma gondii is one of the most important foodborne pathogens. Conventional PM inspection does not detect the tissue cysts of *Toxoplasma*. Most human infections occur after the ingestion of raw vegetables contaminated with cat feces, gardening without gloves and/or improper hand hygiene, or after cleaning the cat litter box and infection by the ingestion of tissue cysts in undercooked or unfrozen meat. Sheep are more often infected than cattle, but eating undercooked or raw beef is quite common. There is no detection of toxoplasmosis during the conventional PM inspection [28]. Again, omitting PM for a large number of carcasses will be of little consequence to the already exiting consumer health hazards.

Although, according to EFSA (31), drug residues are not considered a hazard for public health related to the consumption of bovine meat, it can be a reason for condemnation of the carcass. Food chain information should be a method of pre-selecting animals with suspected drug residue risk, but because of the limited number of FCI forms with information on health and medication (3.3% of FCI forms) and the pre-selection of animals to be slaughtered (no animals with serious health problems), whole carcass condemnation related to the risk of drug residues in this dataset was minimal. Improvement of the FCI is necessary for this specific risk.

4.5. Risk-Based Meat Inspection

It is clear that, at present, the protection of consumer health via the pre-slaughter collection of Food Chain Information followed by an AM and PM inspection does not function as it should, and that the system needs serious improvement to work properly again. When PM inspection procedures are omitted, *a priori* knowledge of the slaughter animals becomes even more important. Regarding this, the information collected via the Food Chain Information forms should be vastly improved, because the current forms contributed practically nothing to the PM decisions that were made. Additionally, the consequences of limiting or omitting PM inspection of animals without remarks should be investigated further [16,18,31].

In a future risk-based system, the AM inspection by an official veterinarian should remain in place. The PM inspection should only be performed when an animal is a hazard for the hygiene of the slaughter line (e.g., an excessively dirty animal, or an animal with an abscess and/or a hazard for food safety/public health). This would be the case if the results of the AM inspection gave reason to suspect this, when animals are of a certain age–breed combination, when they stem from a region or herd where, in the past, more than the usual numbers of animals were declared not suitable for human consumption [21], and/or when the Food Chain Information calls for it. The FCI should, therefore, always include important known risk indicators, for example those that were identified in this study. The FCI should, or could, for example, then contain information about parity (age), breed, and region, including the endemicity of certain diseases or environmental contaminants from the region the animal comes from. Other information may include the results of serological or other tests on the presence of certain diseases that were carried out (e.g., *Toxoplasma*, STEC), the herd history regarding diseases and treatments, and reports of animals from this farm that were declared 'NHC' in the past. With regard to these elements, and the possible lack of compliance, Dutch dairy farmers showed that, when filling in these forms, it is worth considering complete digitalization of the FCI. Thus, all the relevant information can be automatically retrieved by the slaughterhouse and the competent authorities, totally independent from FCI-forms that have to be filled out by hand by the farmers. Finally, by continuation of the AM inspection, animal welfare monitoring and the detection of notifiable diseases can still be carried out as intended [16,18,39].

5. Conclusions

This study has shown that, at least as a 'proof of concept', slaughterhouse data about culled dairy cows can be used for determining whether or not certain elements of our current set of fixed EU meat inspection procedures could be omitted and, thus, be changed into a more risk-based approach without negative consequences for public or animal health. With regard to Dutch dairy cows that are being brought to slaughter, the AM and PM inspections have a substantial overlap. With regard to food safety and public health, in over 99% of cases the PM could even be omitted on the basis of the AM, provided our current FCI is massively improved and all the risk factors that influence the inspection findings are known. To improve the reliability of the FCI, a transition to a fully automated system is worth considering. Such a system could prevent the information being unreliable due to incomplete or misleading information on forms filled out by the (dairy) farmers themselves.

However, what we found in our study on culled Dutch dairy cows that are being slaughtered in large scale slaughterhouses does not necessarily apply to smaller or other types of operations, other animal species, other countries or other regions throughout the EU. That is an integral part of risk-based meat inspection: for every situation, it should be determined—on the basis of identified risk factors—which elements of the meat inspection procedures should be improved or can be omitted. Thus, risk-based meat inspection will improve, in terms of the protection of public and animal health and welfare, while at the same time being as cost-effective as possible.

Supplementary Materials: The following supporting information can be downloaded at: https://www.mdpi.com/article/10.3390/foods12030616/s1, File S1 R-output docx.; File S2 R-script docx; Table S1 Provinces and Animals declared NHC, grouped by travel distances; File S3 Analyses with 2 × 2 tables

Author Contributions: Conceptualization, P.J. and B.B.; formal analysis in R, Jan van den Broek, P.J. and B.B.; analysis with 2 × 2 tables P.J. and B.B.; investigation, P.J.; data curation, P.J.; writing—original draft preparation, P.J.; writing—review and editing, B.B. and L.L.; visualization, P.J. and B.B.; supervision, B.B. and L.L. All authors have read and agreed to the published version of the manuscript.

Funding: The project received no external funding.

Institutional Review Board Statement: Not applicable.

Informed Consent Statement: Not applicable.

Data Availability Statement: The dataset that was compiled is formally owned by the Netherlands Food and Consumer Product Safety Authority (NVWA) and the slaughterhouse involved. It contains many data that fall under the Dutch and European privacy legislation and cannot be distributed freely. On a case by case basis, the authors may consider sharing parts of the dataset.

Acknowledgments: We wish to thank the slaughterhouse and the Netherlands Food and Consumer Product Safety Authority (NVWA) for their willingness to make the data available and their practical support during the study. We would also like to thank Jan van den Broek of the Division of Theoretical Epidemiology, Department of Population Health Science, Faculty Veterinary Medicine, Utrecht University, the Netherlands, for his invaluable support in the data analysis with the R-package.

Conflicts of Interest: The authors declare no conflict of interest. Neither the slaughterhouse nor the Netherlands Food and Consumer Product Safety Authority (NVWA) had a role in the design of the study; the analyses, or interpretation of data; in the writing of the manuscript; or in the decision to publish the results.

References

1. Schönberg, F. *Lehrbuch der Schlachttier- und Fleischuntersuchung Einschliesslich der Tierärztlichen Lebensmittelüberwachung für Tierärzte und Studierende der Veterinärmedizin*; Ferdinand Enke: Stuttgart, Germany, 1955.
2. European Parliament; Council of the European Union. Regulation (EU) 2017/625 of the European Parliament and of the Council of 15 March 2017 on official controls and other official activities performed to ensure the application of food and feed law, rules on animal health and welfare, plant health and plant protection products, amending Regulations (EC) No 999/2001, (EC) No 396/2005, (EC) No 1069/2009, (EC) No 1107/2009, (EU) No 1151/2012, (EU) No 652/2014, (EU) 2016/429 and (EU) 2016/2031 of the European Parliament and of the Council, Council Regulations (EC) No 1/2005 and (EC) No 1099/2009 and Council Directives 98/58/EC, 1999/74/EC, 2007/43/EC, 2008/119/EC and 2008/120/EC, and repealing Regulations (EC) No 854/2004 and (EC) No 882/2004 of the European Parliament and of the Council, Council Directives 89/608/EEC, 89/662/EEC, 90/425/EEC, 91/496/EEC, 96/23/EC, 96/93/EC and 97/78/EC and Council Decision 92/438/EEC (Official Controls Regulation). *Off. J. Eur. Union* **2017**, *95*, 1–142.
3. European Parliament; Council of the European Union. Regulation (EC) No 852/2004 of the european parliament and of the council of 29 April 2004 on the hygiene of foodstuffs. *Off. J. Eur. Union* **2004**, *47*, 1–54.
4. European Parliament; Council of the European Union. Regulation (EU) 853/2004 of the European Parliament and of the Council of 29 April 2004 laying down specific hygiene rules for food of animal origin. *Off. J. Eur. Union* **2004**, *47*, 55–205.
5. Blagojevic, B.; Nesbakken, T.; Alvseike, O.; Vågsholm, I.; Antic, D.; Johler, S.; Houf, K.; Meemken, D.; Nastasijevic, I.; Pinto, M.V.; et al. Drivers, opportunities, and challenges of the European risk-based meat safety assurance system. *Food Control* **2021**, *124*, 107870. [CrossRef]
6. Van Den Berg, P.M.A.; Hendriksen, R.J.M. Vleeskeuringswet en Destructiewet. In *Meat Inspection and Destruction Law*; Tjeenk Willink: Zwolle, The Netherlands, 1981. (In Dutch)
7. Van Den Berg, P.M.A.; Hendriksen, R.J.M. Uitvoeringsvoorschriften van de Vleeskeuringswet. In *Regulations on Implementation of the Meat Inspection and Destruction Law*; Tjeenk Willink: Zwolle, The Netherlands, 1981. (In Dutch)
8. Laranjo-González, M.; Devleesschauwer, B.; Gabriël, S.; Dorny, P.; Allepuz, A. Epidemiology, impact and control of bovine cysticercosis in Europe: A systematic review. *Parasites Vectors* **2016**, *9*, 81. [CrossRef]
9. Berends, B.R.; van den Bogaard, A.E.J.M.; van Knapen, F.; Snijders, J.M.A. Human health hazards associated with the administration of antimicrobials to slaughter animals. *Part 1. An assessment of the risks of residues of tetracyclines in pork. Vet. Q.* **2001**, *23*, 2–10. [PubMed]
10. EFSA (European Food Safety Authority); ECDC (European Centre for Disease Prevention and Control). The European Union summary report on trends and sources of zoonoses, zoonotic agents and food-borne outbreaks in 2015. *EFSA J.* **2016**, *14*, 4634. [CrossRef]
11. Opsteegh, M. Toxoplasma Gondii in Animal Reservoirs and the Environment. Ph.D. Thesis, Utrecht University, Utrecht, The Netherlands, 2011.
12. Persad, A.K.; LeJeune, J.T. Animal Reservoirs of Shiga Toxin-Producing Escherichia coli. *Microbiol. Spectr.* **2014**, *2*, EHEC-0027-2014. [CrossRef]
13. Vosough Ahmadi, B. Cost-Effectiveness of Escherichia Coli O157: H7 Control in the Beef Chain. Ph.D. Thesis, Wageningen University, Wageningen, The Netherlands, 2007.
14. Van Duijkeren, E.; Hengeveld, P.D.; Albers, M.; Pluister, G.; Jacobs, P.; Heres, L.; van der Giessen, A.W. Prevalence of methicillin-resistant Staphylococcus aureus carrying mecA or mecC in dairy cattle. *Vet. Microbiol.* **2014**, *171*, 3–4. [CrossRef]

15. EFSA BIOHAZ Panel (EFSA Panel on Biological Hazards). Scientific Opinion on the public health hazards to be covered by inspection of meat (bovine animals). *EFSA J.* **2013**, *11*, 3266. [CrossRef]
16. Stärk, K.D.C.; Alonso, S.; Dadios, N.; Dupuy, C.; Ellerbroek, L.; Georgiev, M.; Hardstaff, J.; Huneau-Salaün, A.; Laugiers, C.; Mateus, A.; et al. Strength and weakness of meat inspection as a contribution to animal health and welfare surveillance. *Food Control* **2014**, *39*, 154–162. [CrossRef]
17. Mousing, J.; Pointon, A.; World Association of Veterinary Food Hygienists (WAVFH). Liability of meat inspection. Should post-mortem meat inspection be abandoned? In Proceedings of the 12th International Congress, The Hague, The Netherlands, 27–29 August 1997; Wageningen Press: Wageningen, The Netherlands, 1997; pp. 1–3, ISBN 9074134459.
18. Hill, A.A.; Horigan, V.; Clarke, K.A.; Dewè, T.C.M.; Stärk, K.D.C.; O'Brien, S.; Buncic, S. A qualitive risk assessment for visual–only post-mortem meat inspection of cattle, sheep, goats and farmed/wild deer. *Food Control* **2014**, *38*, 96–103. [CrossRef]
19. Bruinier, E. Official Guidelines Regarding the Handling and Registration of RSG-Forms at Meat Inspection: Entry of the daily statement. In *Regulation Inspection and Testing-Document Code: VenI01*; Netherlands Food and Consumer Products Safety Authority (NVWA): Utrecht/The Hague, The Netherlands, 2014.
20. Kroeze, M.; Gerritsjans-Veenstra, R. Official Guidelines Regarding the Handling and Registration of VOS-Forms at Meat Inspection; Manual for completing the VOS form. In *Regulation Inspection and Testing-Document Code: VenI01RA-111*; Netherlands Food and Consumer Products Safety Authority (NVWA): Utrecht/The Hague, The Netherlands, 2014.
21. Dupuy, C.; Morignat, E.; Maugey, X.; Vinard, J.-L.; Hendrikx, P.; Ducrot, C.; Calavas, D.; Gay, E. Defining syndromes using cattle meat inspection data for syndromic surveillance purposes: A statistical approach with the 2005–2010 data from ten French slaughterhouses. *BMC Vet. Res.* **2013**, *9*, 88. [CrossRef] [PubMed]
22. Harbers, A.H.M.; Snijders, J.M.A.; Smeets, J.F.M.; Blocks, G.H.M.; Van Logtestijn, J.G. Use of information from Pig finishing herds for meat inspection purposes. *Vet. Q.* **1992**, *14*, 41–45. [CrossRef] [PubMed]
23. Holzhauer, M.; Hardenberg, C.; Bartels, C.; Frankena, K. Herd- and Cow-Level Prevalence of Digital Dermatitis in The Netherlands and Associated Risk Factors. *J. Dairy Sci.* **2006**, *89*, 580–588. [CrossRef]
24. van der Waaij, E.; Holzhauer, M.; Ellen, E.; Kamphuis, C.; de Jong, G. Genetic Parameters for Claw Disorders in Dutch Dairy Cattle and Correlations with Conformation Traits. *J. Dairy Sci.* **2005**, *88*, 3672–3678. [CrossRef]
25. De Vries, A.; Marcondes, M.I. Review: Overview of factors affecting productive lifespan of dairy cows. *Animal* **2020**, *14*, s155–s164. [CrossRef] [PubMed]
26. Ahlman, T.; Berglund, B.; Rydhmer, L.; Strandberg, E. Culling reasons in organic and conventional dairy herds and genotype by environment interaction for longevity. *J. Dairy Sci.* **2011**, *94*, 1568–1575. [CrossRef] [PubMed]
27. Pritchard, T.; Coffey, M.; Mrode, R.; Wall, E. Understanding the genetics of survival in dairy cows. *J. Dairy Sci.* **2013**, *96*, 3296–3309. [CrossRef] [PubMed]
28. Nor, N.M.; Steeneveld, W.; Hogeveen, H. The average culling rate of Dutch dairy herds over the years 2007 to 2010 and its association with herd reproduction, performance and health. *J. Dairy Res.* **2013**, *81*, 1–8. [CrossRef]
29. Berends, B.R.; Snijders, J.M.; Van Logtestijn, J.G. Efficacy of current EC meat inspection procedures and some proposed revisions with respect to microbiological safety: A critical review. *Vet. Rec.* **1993**, *133*, 411–415. [CrossRef] [PubMed]
30. Dupuy, C.; Morlot, C.; Gilot-Fromont, E.; Mas, M.; Grandmontagne, C.; Gilli-Dunoyer, P.; Gay, E.; Callait-Cardinal, M.-P. Prevalence of Taenia saginata cysticercosis in French cattle in 2010. *Vet. Parasitol.* **2014**, *203*, 65–72. [CrossRef] [PubMed]
31. Bonde, M.; Toft, N.; Thomsen, P.T.; Sørensen, J.T. Evaluation of sensitivity and specificity of routine meat inspection of Danish slaughter pigs using Latent Class Analysis. *Prev. Vet. Med.* **2010**, *94*, 165–169. [CrossRef] [PubMed]
32. Eichenberger, R.; Lewis, F.; Gabriël, S.; Dorny, P.; Torgerson, P.; Deplazes, P. Multi-test analysis and model-based estimation of the prevalence of Taenia saginata cysticercus infection in naturally infected dairy cows in the absence of a 'gold standard' reference test. *Int. J. Parasitol.* **2013**, *43*, 853–859. [CrossRef]
33. Abuseir, S.; Epe, C.; Schnieder, T.; Klein, G.; Kühne, M. Visual diagnosis of Taenia saginata cysticercosis during meat inspection: Is it unequivocal? *Parasitol. Res.* **2006**, *99*, 405–409. [CrossRef]
34. Geysen, D.; Kanobana, K.; Victor, B.; Rodriguez-Hidalgo, R.; DE Borchgrave, J.; Brandt, J.; Dorny, P. Validation of Meat Inspection Results for Taenia saginata Cysticercosis by PCR–Restriction Fragment Length Polymorphism. *J. Food Prot.* **2007**, *70*, 236–240. [CrossRef]
35. Dorny, P.; Vallée, I.; Alban, L.; Boes, J.; Boireau, P.; Boué, F.; Claes, M.; Cook, A.J.; Enemark, H.; van der Giessen, J.; et al. Development of harmonised schemes for the monitoring and reporting of Cysticercus in animals and foodstuffs in the European Union. *EFSA Support. Publ.* **2010**, *7*, 34E. [CrossRef]
36. Rosenthal, B.M. Zoonotic Sarcocystis. *Res. Vet. Sci.* **2021**, *136*, 151–157. [CrossRef]
37. Taylor, M.A.; Boes, J.; Boireau, P.; Boué, F.; Claes, M.; Cook, A.J.; Dorny, P.; Enemark, H.; Van Der Giessen, J.; Hunt, K.R.; et al. Development of harmonised schemes for the monitoring and reporting of Sarcocystis in animals and foodstuffs in the European Union. *EFSA Support. Publ.* **2010**, *7*, 33E. [CrossRef]

38. Hoeve-Bakker, B.; van der Giessen, J.; Franssen, F. Molecular identification targeting cox1 and 18S genes confirms the high prevalence of Sarcocystis spp. in cattle in the Netherlands. *Int. J. Parasitol.* **2019**, *49*, 859–866. [CrossRef]
39. Tourlomoussis, P.; Eckersall, P.; Waterson, M.; Buncic, S. A Comparison of Acute Phase Protein Measurements and Meat Inspection Findings in Cattle. *Foodborne Pathog. Dis.* **2004**, *1*, 281–290. [CrossRef] [PubMed]

Review

Transmission Scenarios of *Listeria monocytogenes* on Small Ruminant On-Farm Dairies

Dagmar Schoder [1,2,*], **Alexandra Pelz** [2] **and Peter Paulsen** [3]

1   Institute of Food Safety, Food Technology and Veterinary Public Health, Unit of Food Microbiology, University of Veterinary Medicine, Veterinaerplatz 1, 1210 Vienna, Austria
2   Vétérinaires sans Frontières Austria, Veterinaerplatz 1, 1210 Vienna, Austria
3   Institute of Food Safety, Food Technology and Veterinary Public Health, Unit of Food Hygiene and Technology, University of Veterinary Medicine, Veterinaerplatz 1, 1210 Vienna, Austria
*   Correspondence: dagmar.schoder@vetmeduni.ac.at; Tel.: +43-1-25077-3520

Abstract: *Listeria monocytogenes* can cause severe foodborne infections in humans and invasive diseases in different animal species, especially in small ruminants. Infection of sheep and goats can occur via contaminated feed or through the teat canal. Both infection pathways result in direct (e.g., raw milk from an infected udder or fresh cheese produced from such milk) or indirect exposure of consumers. The majority of dairy farmers produces a high-risk product, namely fresh cheese made from raw ewe's and goat's milk. This, and the fact that *L. monocytogenes* has an extraordinary viability, poses a significant challenge to on-farm dairies. Yet, surprisingly, almost no scientific studies have been conducted dealing with the hygiene and food safety aspects of directly marketed dairy products. *L. monocytogenes* prevalence studies on small ruminant on-farm dairies are especially limited. Therefore, it was our aim to focus on three main transmission scenarios of this important major foodborne pathogen: (i) the impact of caprine and ovine listerial mastitis; (ii) the significance of clinical listeriosis and outbreak scenarios; and (iii) the impact of farm management and feeding practices.

Keywords: *Listeria monocytogenes*; direct marketing; farm sales; cheese; dairy products; small ruminant; contamination routes; mastitis

Citation: Schoder, D.; Pelz, A.; Paulsen, P. Transmission Scenarios of *Listeria monocytogenes* on Small Ruminant On-Farm Dairies. *Foods* **2023**, *12*, 265. https://doi.org/10.3390/foods12020265

Academic Editor: Arun K. Bhunia

Received: 17 November 2022
Revised: 19 December 2022
Accepted: 22 December 2022
Published: 6 January 2023

1. Introduction

The bacterial genus *Listeria* comprises 21 species of Gram-positive, motile, facultative anaerobic, non-spore-forming rods up to 2 μm in length [1]. Of these, *Listeria monocytogenes* (*L. monocytogenes*) has been studied the most extensively. *L. monocytogenes* is a facultative intracellular bacterium, which can cause severe foodborne infection in humans and invasive diseases in different animal species, especially farm ruminants [2,3].

While ruminants, particularly small ruminants, are extraordinarily susceptible to *L. monocytogenes*, other vertebrate wild fauna and birds can excrete the bacteria without notice from their gastrointestinal tracts either continuously or intermittently and for weeks at a time [4]. The widespread occurrence of *L. monocytogenes* in rural environments and its strong association with domestic ruminants make eliminating the risk of listeriosis difficult. Indeed, it is well documented that *L. monocytogenes* is prevalent in the ruminant farm environment [5–12]. However, information on *Listeria* transmission dynamics on small ruminant on-farm dairies is scarce [13–16]).

Listeriosis outbreaks in ruminants have been repeatedly reported in the scientific literature and are often referred to as a "silage disease", as it is strongly associated with the ingestion of spoiled silage [17]. Infected farm ruminants and contaminated agricultural environments rarely appear to directly cause human infection. However, animal-derived food products that are not processed before consumption, such as raw milk, clearly represent a direct link [18]. Minimizing cases of human listeriosis is dependent upon improving our understanding of how to limit contamination of food with *L. monocytogenes*. This is a challenging task, as the pathogen is widely disseminated in nature, has successfully infiltrated both farm and food processing environments, and can enter the food chain at nearly every stage of production [19].

The extraordinary viability of *L. monocytogenes* over wide temperature and pH ranges and its ability to survive at high salt concentrations pose significant and ongoing challenges to the food industry and markedly affect the ultimate risk for the consumer [3]. Ready-to-eat foods can be contaminated post-processing (i.e., during portioning, slicing, and packaging). Intermediate moisture and non-acidic foods require refrigerated storage that actually favours the growth of this cold-tolerant pathogen [19,20].

In recent years, significant changes have been noted in the European food market. These essentially reflect increased production of regional and processed farm products. It is clear that, in a competitive marketing environment, farmers are seeking new niche products as prices decrease and various regulations place limits on production [21].

Due to the simplicity of the manufacturing procedure, many ewe and goat dairy farmers produce fresh cheese from raw milk. These milk products have recently become very popular [22]. Growing numbers of people now consume non-pasteurized sheep or goats' milk or cheese products for practical reasons (e.g., dairy farm families), medical reasons (allergies or intolerance to cow's milk), or the perceived health benefits of raw milk products. Nonetheless, direct marketers are operators of food businesses and therefore have a responsibility to ensure that the food they market is safe. Furthermore, direct marketers, and everyone involved in the industry, should have a high degree of "quality awareness" and must live by this ideal.

"Direct Marketing" or "Short Food (Supply) Chain" are terms used synonymously for direct sales from producers to final consumers. Direct marketing in the narrower sense is the sale of agricultural products directly from the farmer to the final consumer. For milk and dairy products, this covers the following distribution channels: (i) farm gate sales; (ii) farm markets or weekly markets; (iii) sales areas outside the farm; (iv) doorstep sales; (v) "new" sales channels (e.g., internet ordering and subsequent shipment of products); and finally (vi) delivery to private households.

Direct marketing in the broader sense of the term includes not only sales to the final consumer but also the following distribution channels: (i) delivery to large customers (canteens, restaurants, etc.); (ii) sales to individual retail outlets, natural food shops, delicatessens, farmers' shops; and (iii) consumer-producer communities (food co-ops, community-supported agriculture, etc.). However, most farmers operate several forms of direct marketing and there is often no clear distinction between the various forms of marketing [23].

Direct marketing farmers are food business operators and are therefore responsible for the safety of their products from primary production to delivery to the final consumer. Only "safe food" may be placed on the market. In contrast, "unsafe" is defined as harmful to health or unsuitable for consumption [24]. A major contribution to food safety is the implementation of a self-monitoring system. This means that the company must establish a self-control system for its operation [25]. Several manuals have also been developed to implement self-monitoring systems in order to align rules in-force more precisely with the actual requirements for direct marketers [26].

Since 2002, the European Union has issued general (Regulation (EC) No 178/2002, [24]; Regulation (EC) No 852/2004, [27]) and specific regulations concerning the hygiene of foods of animal origin (Regulation (EC) No 853/2004, [28]) and microbiological food safety criteria (Commission Regulation (EC) No 2073/2005, [29]). Subsequently, additional national laws or regulations were set in force for small enterprises with regional activities (e.g., in Austria, [30–32]). In principle, the same requirements for food safety apply to any undertaking in which foodstuffs are produced, manufactured, treated and/or placed on the market.

The following hygiene requirements apply to direct marketing farmers: (i) compliance with requirements for buildings and equipment (e.g., sanitary installations, lighting, ventilation equipment, flooring, walls, doors and windows); (ii) the use of appropriate raw material with known origin; (iii) safe handling of food (including packaging and transport); (iv) safe waste disposal; (v) pest control measures; (vi) cleaning and disinfection plans; (vii) water quality; (viii) compliance with the cold chain; (ix) personal hygiene; (x) training and, finally; (xi) application of Hazard Analysis and Critical Control Points (HACCP) principles, including good manufacturing practices and product tests [25,33,34].

Each farmer has his own registered number (for example, the farm and forestry operational information system number (LFBIS number)). Regardless of whether or not checks are carried out on farms under different brand programs, direct marketers—as any other food businesses—are subject to controls by food inspection bodies. Dairy direct marketing farmers are mainly inspected by veterinary control offices. The inspection bodies primarily check whether a suitable self-control system is in place and actually implemented. Food products are examined sensorially and microbiologically, as well as assessed for compliant labelling.

Generally, small scale on-farm dairies and their self-control systems almost exclusively spotlight their final product and merely identify problems passively as they occur, whereas food business operators, on the industrial scale, must implement a more complex HACCP-based system. Such a system embraces the entire production process proactively by prevention and ensures a consistent quality. In this way, a company can control or mitigate hazards that may arise at any point during the complete production process.

Despite the seemingly endless variety of food items, with more differing foodstuffs than ever being consumed, milk is still an essential basic food for the majority of consumers. However, it also remains a prime nutrient medium for a wide range of pathogens. Information from various large outbreaks importantly demonstrates that milk and milk products have served as the most common vectors for *L. monocytogenes* transmission [35].

Globally milk production is dominated by the dairy cattle sector, which, according to the FAO, accounts for 81% of worldwide production followed by 15% for buffalo and a combined 4% for goats (1.9%), sheep (1.3%), and camels [36]. Although there has been a decline of 8.9% in livestock ruminants across the EU within the last two decades, there was a significant production increase in raw milk on EU farms.

Another special feature is that in many rural or arid regions, particularly in the Mediterranean area, sheep and goats make an important contribution to the overall milk production. In 2020, according to the "key figures on the European food chain" from the annual report of sectoral and regional statistics, Eurostat, 589,000 to 684,000 tonnes of ewes' milk were produced in Spain and Greece, whereas the main producer of goats' milk in the EU was France with 523,000 tonnes milk per year. The report further states that the majority of raw milk production in the EU is delivered to dairies. Still, 10.6 million tonnes were used on farms, being consumed by the farmer's family, sold directly to consumers, used as feed or processed directly. With 78.7%, Romania holds the highest direct milk-marketing rate, followed by Bulgaria (55.9%). In all other member states, more than 70% of the total milk amount are delivered to dairy companies [37].

It is clear that with 450,000 goat and 850,000 sheep farms the small ruminant sector constitutes just a small share of the total output of the EU livestock sector. However, more than 1.5 million people work on these farms, and sheep and goat rearing takes place mostly on pastureland in remote and disadvantaged rural areas. Thus, the sheep and goat sectors actively contribute to landscape and biodiversity conservation [37].

As mentioned above, sheep and/or goat milk and their respective processed products may have many beneficial health impacts, which could appeal to modern consumers [38–40]. Sheep and goat milk, due to protein differences with cow milk, induce fewer allergy responses. The levels of minerals, vitamins, and essential fatty acids are also generally higher than in cow milk. Sheep milk has a much higher concentration of conjugated linoleic acid (CLA) than both goat and cow milk [38]. CLA is claimed, for example, to prevent obesity [41] and reduce triglyceride levels. It should, therefore, help to prevent coronary heart disease and atherosclerosis [39].

Due to its higher fat and protein content, sheep milk is also particularly beneficial for cheese production [40]. Less additives, such as calcium chloride and rennet, are needed in sheep milk curd production, compared to other ruminants' milk.

Taking all of these physical, nutritional, and health benefits into account, it is not surprising that there is a growing demand for milk from small ruminants. In Europe, the production of sheep milk accounted for around 2.8 million tonnes, whereas almost 33% (0.9 million tonnes) were processed directly on-farm into cheese [37].

As mentioned above, each direct marketer is a food business operator and, as such, is responsible for the safety of the food that he or she places on the market. Therefore, they must follow good hygiene practices and manage their operations in such a way as to monitor food safety hazards. However, the majority of on-farm dairies produce a high-risk product, which is cheese made of raw milk. The cheese production takes place in direct proximity to animals and the barn environment. Consequently, the microbial contamination pressure on the cheese production environment is classified as "very high". Yet, surprisingly, almost no scientific studies have been conducted dealing with the hygiene and food safety aspects of directly marketed dairy products. *L. monocytogenes* prevalence studies on small ruminant on-farm dairies are especially scarce. Therefore, it was our aim to focus on this sector and highlight various transmission scenarios of this important major foodborne pathogen.

2. Transmission Scenarios

In order to elucidate how contamination of dairy products with *L. monocytogenes* occurs on small ruminant on-farm dairies, we performed a systematic literature search in PubMed, SCOPUS and Web of Science databases, for the publication period from 1944 to 2022. The search strategy and the outputs are represented in Figure 1. Basically, we considered the epidemiology of subclinical listerial mastitis (scenario 1) and clinical listeriosis (scenario 2) and the impact of farming-, feeding- and milk processing practices (scenario 3).

2.1. Transmission Scenario 1: Impact of Ovine and Caprine Listerial Mastitis

L. monocytogenes can colonize the mammary complex of ruminants. Although *L. monocytogenes* is common in the faeces of ruminants and widespread in the environment [4], only a few cases of bovine [42–49] and ovine [50–56] listerial mastitis have been reported. We could retrieve merely a single study on caprine mastitis [57]. Interestingly, listerial mastitis has not yet been covered in any review article (Figure 1).

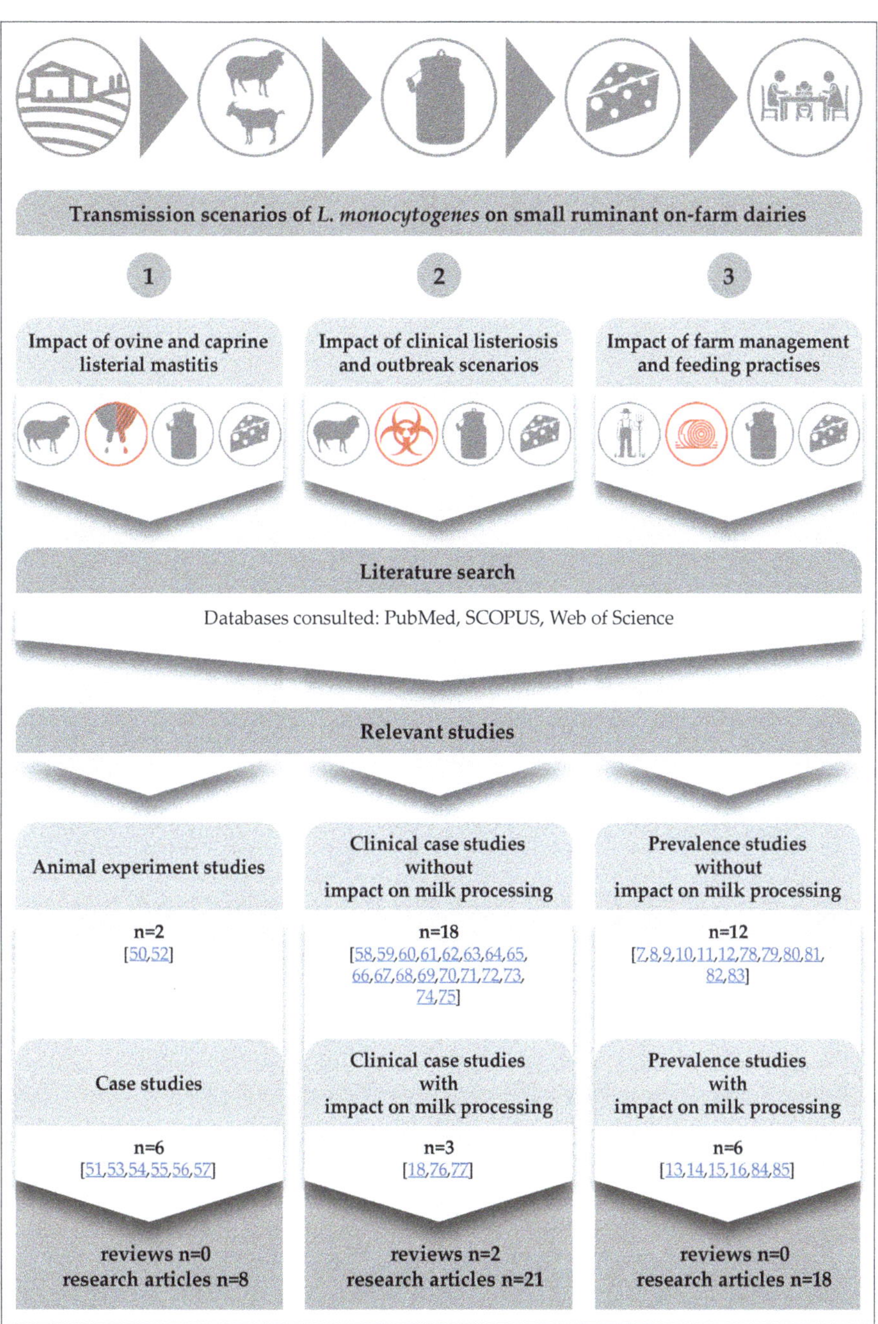

Figure 1. Results from a comprehensive literature search focusing on three main transmission scenarios on small ruminant on-farm dairies. The icons on the top of the figure depict the major steps in the field-to-table continuum [7–16,18,50–85].

Another surprising detail is that there is not a single study from America, Asia, Australia, or Africa; all listerial mastitis studies were performed in Europe. This fact is quite remarkable, if one considers that there was a dramatic increase especially in dairy goat production during the past decade, with Asia seeing the largest growth of 22%, followed by Africa (13%), Oceania (9%), and America (5%) [86].

A comprehensive search of the literature revealed two experimental studies on the course of listerial mastitis in small ruminants (Figure 1). They have shown that inoculations of 300 CFU to 1000 CFU of *L. monocytogenes* into the udder of ewes are sufficient to result in mastitis [45,52]. All inoculated ewes became infected and developed chronic subclinical mastitis, regardless of the serotype or origin of the strains used. According to Tzora et al. [52], only one single ewe out of 34 animals showed typical signs of acute clinical mastitis immediately after the inoculation. The gland was larger and hotter and its secretion contained clots. There was also an increase in the internal body temperature.

The somatic cell count of all infected sheep was always greater than 1.0×10^6 cells per ml and *L. monocytogenes* could be consistently isolated from the milk over a period of 88 days. *L. monocytogenes* was also detected from the mammary lymph nodes, but not from any internal organ of any inoculated ewe. Histologically, in the early stage of the infection, extra-alveolar neutrophilic infiltration and interstitial oedema were predominant. Subsequently, 25 days after inoculation, chronic inflammatory signs predominated, such as destruction of alveoli and fibrous tissue proliferation, with lymphocytes as the main cell type [52]. The findings of both studies provide clear evidence that *L. monocytogenes* is pathogenic for the ovine and caprine mammary gland.

With regard to naturally occurring cases of ovine mastitis (Figure 1), it is worthwhile to compare a Greek study from Fthenakis et al. [51] with the findings of an Austrian research team [53–55]. Fthenakis et al. [51] monitored the udder health, somatic cell count and the shedding of *L. monocytogenes* in 98 ewes. Half udder milk samples were collected at three separate time points during the lactation period, including: (i) 15 ± 30 days post- lambing; (ii) 6 ± 7 weeks after initial sampling; and (iii) 6 ± 7 weeks on from collection of the second sample. There were diagnoses of clinical mastitis in any of the ewes, though the prevalence of subclinical *L. monocytogenes* mammary infections was 3.1% during collection of the first and the second samples and this had increased to 6.2% by the third time point. Examination of the milk of ewes with mammary infection revealed somatic cell counts ranging from 1.8×10^6 to 3.0×10^6 cells/mL. Furthermore, *L. monocytogenes* could also be detected in the faeces of 19.4% of the animals. The authors concluded that infection of the mammary gland with *L. monocytogenes* had occurred via the bloodstream. Firstly, there was an 83% higher prevalence of bilateral mammary infection and, secondly, the pathogen was isolated from the liver of two of the four infected ewes. These findings are only partly in accordance with other case studies, which consider intramammary infection to be the most likely and emphasize that *L. monocytogenes* has to contaminate the teat before penetration into the udder [45,53–57].

Briefly, Schoder et al. [53,55] studied two cases of ovine *L. monocytogenes* mastitis over a period of 7 months. On a daily basis, the animals were clinically examined. After adspection and palpation of the mammary gland, the California mastitis test (CMT) was performed and half udder milk samples were collected. During the entire observation period, the animals continued to eat well and did not show any signs of distress or evidence of systemic reaction. The milk appeared to be normal, was not discoloured and did not contain any flakes or clots. However, CMT showed consistently thick gel (++) or thick and sticky gel (+++) reactions. Somatic cell counts averaged $\geq 10^6$ per ml milk. Both sheep shed *L. monocytogenes* at a mean concentration of 3.8×10^4 (range 9.0×10^1 to 4.0×10^5) and 2.2×10^4 (range 1.3×10^3 to 8.1×10^4) CFU/mL, respectively. Subclinical mastitis was diagnosed without palpable changes in the consistency of the udder parenchyma.

The histopathological and immunohistochemical findings revealed that chronic inflammatory features predominated [54,57]. There was a diffuse infiltration with lymphocytes, plasma cells and macrophages. Additionally, alveolar destruction and proliferation of fibrous tissue were recorded with a very strong immunoreactivity for CD5 cells.

Listeria could be cultivated from the mammary parenchyma of the infected halves and from the lymph nodes [52,54,57]. In contrast to the Greek study, all other internal organs showed no abnormalities, and no single *Listeria* could be isolated [54,57]).

Literature findings suggest that caprine and ovine mastitis are very much comparable. Furthermore, the typical listerial mastitis in small ruminants is defined: (i) by its subclinical nature; (ii) a high somatic cell count ($\geq 10^6$ SCC per ml); (iii) persistent shedding of the pathogen bacteria; (iv) by induration and atrophy of the mammary parenchyma in progredient stages of the infection and, finally; (v) the local invasion via the teat canal seems to be the most likely route of infection. Figure 2 illustrates the main clinical and pathological findings of listerial mastitis in small ruminants. A risk scenario was designed to highlight the dimension of the consumers' exposure.

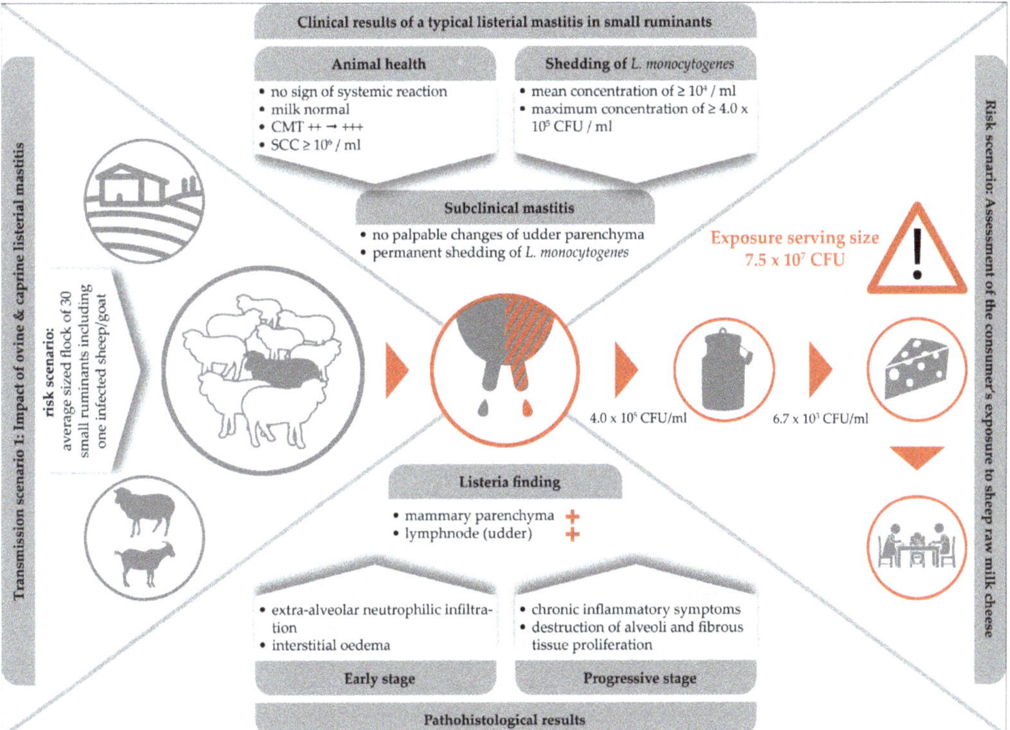

Figure 2. Clinical and histopathological findings in a typical listerial mastitis in small ruminants and consequences for the safety of cheese produced on-farm.

Clearly, mastitis attributed to *L. monocytogenes* is especially dangerous due to its subclinical nature. While milk from infected udders remains visually unchanged and the udders show no clinical signs, *L. monocytogenes* continues to be shed up to concentrations of 4.0×10^5 CFU/mL [53]. With respect to food safety, listerial mastitis has two main consequences: firstly, the direct contamination of bulk milk and raw milk cheese with high loads of the pathogen and, secondly, the increase of environmental colonization of the farm and the cheese processing environment. Furthermore, within the last decades, hypervirulent *L. monocytogenes* strains were found to be significantly associated with subclinical mastitis and were more commonly isolated from dairy products [87].

Remarkably, merely three single studies have been published demonstrating the consequences of ovine listerial mastitis on the further processing of milk to cheese [53,55,56]. Based on two cases of ovine mastitis, a risk scenario was designed in order to assess the consumer's exposure to *L. monocytogenes* per serving size of sheep raw milk cheese [55]. Various cheese-making procedures were performed. The results were alarming: the final level of contamination was up to 7.5×10^7 CFU/serving size. Certainly, such an extremely high dose qualifies the cheese to be hazardous for consumers (Figure 2).

Clearly, there is an urgent need to screen small ruminant farms for the presence of cases of subclinical *L. monocytogenes* mastitis by implementing CMT at least once per week [53]. With regard to caprine mastitis, however, milk SCC is a less reliable indicator of inflammation than in other dairy animals [88]. Therefore, the routine control of subclinical mastitis cases by SCC monitoring, such as with the CMT, is less meaningful than in cows or ewes [57]. In conclusion, Addis et al. [57] emphasized that the milk of all goats of a dairy farm should be screened for the presence of *L. monocytogenes* on a regular basis.

Together, these data suggest that small ruminant dairy farms, which sell milk and/or cheese made of raw milk directly to consumers or retailers, are in urgent need of an efficient monitoring program for the detection of *L. monocytogenes*.

2.2. Transmission Scenario 2: Impact of Clinical Listeriosis

L. monocytogenes is a globally distributed pathogen with the ability to cause disease in a wide range of animal species, though sheep are particularly susceptible to infection. In the northern hemisphere, infections are typically seasonal and most common sporadically in winter and early spring in association with silage feeding. Meanwhile, in the southern hemisphere, most listeriosis cases in ruminants occur during the warmest months of the year and the transition from rainy to dry season. It can be assumed, that not only silage, but also feedstuff and water generally play an additional role in the mode of infection [17].

Listeriosis of small ruminants is well documented and there is a numerous number of case studies, including two comprehensive review articles. However, case studies focusing on the impact on milk processing are scarce [18,76,77] (Figure 1). The disease is clearly and most commonly caused by oral infection, but other entrance sites, such as the conjunctiva, microlesions of the skin, buccal and genital mucosa, or the teat canal have also been described [89]. After oral infection, *L. monocytogenes* is able to colonize the gastrointestinal tract. Animals either become asymptomatic carriers or they develop mild symptoms of a self-limiting enteritis. In both cases, the bacterium is shed with the faeces and is able to heavily contaminate the farm and milk processing environment (Figure 3), [4]. The interplay of environmental reservoirs outside the farm and of vectors and the farm animals is shown in Figure 3. Notably, the excretion of *L. monocytogenes* by the farm animals is not only a food safety and herd health issue, but can also contribute to infection of wildlife.

Interestingly, there is a study describing the case of an orally infected sheep that carried *L. monocytogenes* in the spleen, liver and lymphoid organs without showing any clinical symptoms. The authors concluded that *L. monocytogenes* intestinal infection and translocation to visceral organs may occur asymptomatically [90]. Additionally, in the case of invasive listeriosis, the pathogen is able to cross the gastrointestinal barrier causing severe illnesses including abortion, septicaemia and rhombencephalitis—the so-called "circling disease"—which accounts for the vast majority of invasive clinical infections in small ruminants [17,89].

The incubation period in small ruminants varies according to pathogenesis. It can be as short as one to two days for septicaemia or gastrointestinal forms, two weeks for abortion, and between four and six weeks for the encephalitic form. The pathogen has a particular affinity for the central nervous system in sheep. The main clinical signs include apathy, fever, anorexia, head pressing or compulsive circling and unilateral or bilateral cranial nerve deficits. *L. monocytogenes* ascends the nervous system following peripheral traumatic lesions (e.g., ascending intra-axonal migration within the trigeminal nerve or other cranial nerves following small lesions of the buccal mucosa). Another route involves ascending infection via the sensory nerves of the skin [17].

Neurological symptoms leave little doubt as to their cause and affected animals can be removed from herds. The milk and meat of these affected animals is rather unlikely to enter the food chain [18,76,77]. However, especially during an outbreak event, massive contamination of the animal environment, both through contaminated feed and faecal shedding from exposed animals, may lead to cross-contamination of: (i) the milk processing and cheese-making environment; (ii) bulk tank milk; and (iii) subsequently, the cheese products themselves (Figure 3).

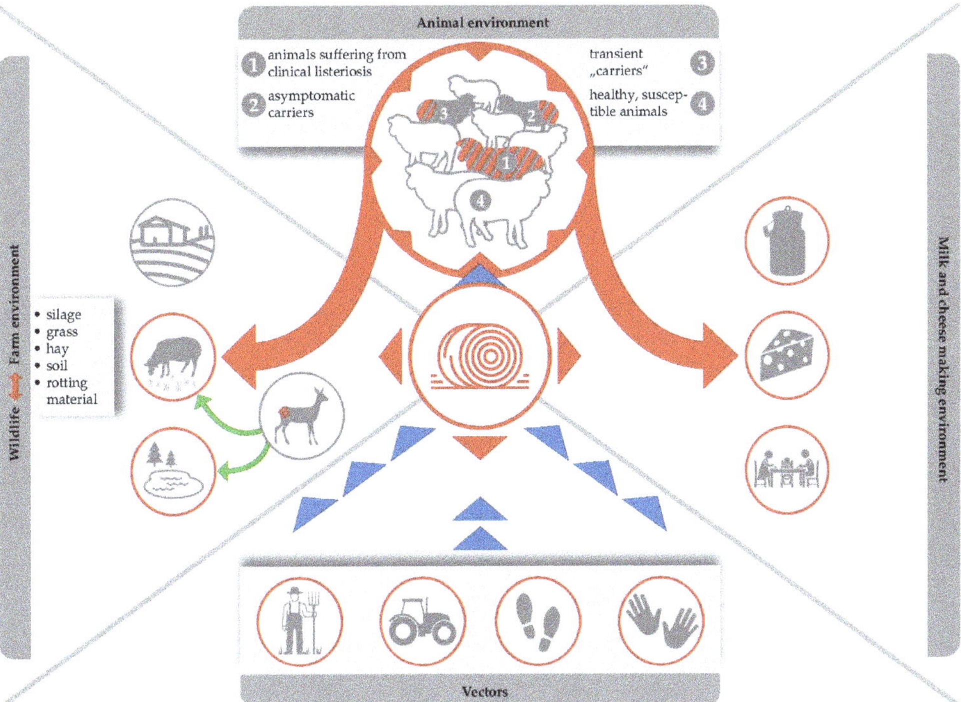

Figure 3. Silage serves as the most important *Listeria* reservoir. Ingestion of contaminated feed leads to the pathogen multiplying within animal hosts, and the bacteria are then excreted via faeces, which are in turn used as fertilizers, which forms a recurring cycle that favours the persistence of *L. monocytogenes* in both farm and natural environments.

Following an outbreak of clinical listeriosis in sheep, Wagner et al. [77] sourced the infectious agent to grass silage feed, which was contaminated with 10^5 CFU/g *L. monocytogenes*. The investigation took place on a dairy farm producing raw milk cheese made of 50% ewe and 50% cow milk. Dairy cows were not affected by this outbreak, reflecting the high susceptibility of sheep to listerial infection. Interestingly, the clinical manifestation within the flock of 55 sheep was also quite variable. Although they were all fed from the same batch of silage, only one ewe was affected by central nervous symptoms caused by rhombencephalitis, four ewes suffered from septicaemia and a further nine animals delivered a combined total of 20 stillborn mature foetuses.

From the animal that had developed central nervous symptoms, *L. monocytogenes* could neither be recovered from the visceral organs nor in the faeces, but was found in a blood sample taken directly from the heart, brain and the nasal mucosa. The authors concluded that the infection had originated within the nasal mucosa and spread to the brain, but that the liver, spleen and other visceral organs remained clear. The route of infection in the other animals was most likely via feedborne transmission. Those with septicaemia suffered from accumulation of *Listeria* in the liver, spleen, heart and lung, with a median concentration of 5.9×10^5 to 6.4×10^6 CFU/g. *L. monocytogenes* could also be detected in the foetal liver, spleen, lung, heart and brain with values ranging from 3.1×10^3 to 5.6×10^5 CFU/g.

Samples from both the farm environment and the cheese production chain, which were randomly taken from ewes, cattle and all individuals who lived on the farm, were positive for *L. monocytogenes,* including 62% of faecal samples and the bulk tank of the cows. Interestingly, one farm worker tested positive for an isolate that was so similar to the outbreak clone that it could not be distinguished genetically, which clearly occurred through them consuming contaminated raw bovine milk. Due to intensive consultation and the fact that the most important countermeasures were immediately taken (silage had been discarded, affected animals had been separated and cleaning and disinfection of the cheese making facilities were implemented), *L. monocytogenes* was not detected in the cheese samples [77].

2.3. Molecular Epidemiological Aspects of Listeria monocytogenes

Bagatella et al. [17] provides a comprehensive overview of epidemiological and experimental studies, which highlight the genetic heterogeneity of *L. monocytogenes* in humans and in ruminants. A heterogeneity, which is likely linked to the variability observed in virulence and in clinical manifestations, as well as to the environmental distribution of listeriosis [91–93]. Research is currently ongoing in an attempt to identify the bacterial determinants driving variability and niche adaptation in *L. monocytogenes*, as well as the principally associated mechanisms [94]. Several bacterial subtypes have been characterized and efforts made to associate them with particular niches and relative virulence. Of the 13 serovars identified, types 1/2a, 1/2b, and 4b were those most frequently found in clinical isolates from both humans and animals. Meanwhile, in cases of ruminant neurolisteriosis and in major outbreaks of listeriosis, serotype 4b was the most dominant [59,66,95–97].

All 3 serotypes, apart from being implicated in disease, were additionally isolated from food, food processing and farm environments, and animal faeces [46,98–100]. Isolates can be linked to clinical outcomes, the environment and foods through molecular typing methods, including pulsed field gel electrophoresis, multilocus sequence typing (MLST) and whole genome sequencing (WGS). Using these techniques, four distinct lineages (I–IV) were identified and further subdivided into clonal complexes (CCs) and sequence types (STs), or sublineages (SLs) and core genome MLST types (CTs), respectively [101].

L. monocytogenes that can be frequently isolated from diverse sources are binned into two major lineages (I, II), with lineage I being overrepresented in human clinical isolates and ruminant neurolisteriosis cases as well as being the most genetically homogeneous, while *L. monocytogenes* that are sporadically isolated from animal infections are binned into two minor lineages (III, IV) [63,93,97,102–104].

Several CCs were found to be hypervirulent in experimental models, including CCs from lineage I belonging to serotype 4b (such as CC1, CC2, CC4, and CC6), these were also significantly linked to human clinical cases and well-adapted to host colonization compared to clones overrepresented in food and the environment (such as CC9 and CC121) [91,92,103].

Within clinical isolates and particularly neurolisteriosis isolates from ruminants, lineage I, specifically CC1 and CC4, were found to be significantly overrepresented compared with other clinical listeriosis syndromes in ruminants, such as abortion, mastitis or gastroenteritis. Additional isolates, from diseased animals and diseased animal environments, that are commonly found include isolates from both lineage I (CC2, CC217, CC6, CC191, CC59) and lineage II (CC7, CC11, CC14, CC37, CC204, CC412) [63,87,93,100].

We can conclude that preventing disease in ruminants and its concomitant transmission to humans is a challenging task, requiring efficient surveillance and control measures. As ruminants, humans and the environment are indelibly connected, achieving a more comprehensive understanding of the pathogenesis of listeriosis and its molecular epidemiology within these domains is critical for developing methodologies to meet the challenge in congruence with the "Farm to Fork" strategy and One Health concepts.

2.4. Transmission Scenario 3: Impact of Farm Management and Feeding Practices

Transmission of foodborne pathogens frequently involves complex interactions among the pathogen, the environment and one or multiple host species [105]. *L. monocytogenes* is a ubiquitous pathogen that can be found in moist environments, soil, water and decaying vegetation [106]. However, does *L. monocytogenes* still have its origin and main habitat in the natural environment and wildlife? Or, can we assume that this major pathogen acts as a cultural successor, which has already successfully colonized the farm- and food-production environment, creating new reservoirs there? Interestingly, the prevalence of *L. monocytogenes* in the dairy cattle environment is well documented [5,6]). There are also numerous studies focusing on the occurrence of *L. monocytogenes* in small ruminant farms. However, the knowledge of *Listeria* transmission dynamics and ecology in on-farm dairies is limited (Figure 1).

As far as we know from literature, *L. monocytogenes* prevalence is normally lower during the pasture season than it is during the indoor season [107,108]. Furthermore, the pathogen has been isolated from both clinically infected and clinically symptomless ruminants. In fact, *L. monocytogenes* can be shed by (i) healthy sheep and goats (so called transient "carriers" and asymptomatic carriers); and (ii) by ruminants suffering from a clinical listeriosis (Figure 3).

Faecal shedding of *L. monocytogenes* has several effects on food safety: (i) *L. monocytogenes* accumulation within the immediate environment of the barn increases the probability that more animals will become infected; (ii) contamination of feed and crops with *L. monocytogenes* can occur when the manure of infected animals is used as fertilizer in agriculture, whilst water sources can be contaminated by runoff from farms [109,110]; (iii) raw milk contamination may occur due to poor hygiene standards during the milking of animals in which infection has gone undetected (Figure 3).

Ingestion of contaminated feed, multiplication of the pathogen in animal hosts, and subsequent excretion of the bacterium via faeces, which are in turn used as fertilizers, form a recurring cycle which favours the persistence of *L. monocytogenes* (Figure 3), [5]. It cannot be denied that there is a high contamination pressure of *L. monocytogenes* on dairy farms and we have to admit that the problem is entirely self-generated. Alarmingly, *L. monocytogenes* may be present in 8% up to as much as 50% of faecal samples collected from dairy sheep and goats [10,13,18,78]. The shedding itself is associated with animal stress and is strongly connected to the contamination of silage [111]. While *L. monocytogenes* is rarely detected on growing grasses prior to processing, detection rates in clamp silages range from between 2.5% and 5.9% and reach up to 22.2% in large bales. This further increases to an alarming 44% in mouldy silage samples [112]. Alternatively, use of inadequately fermented silage (pH of 5.0 to 5.5) contaminated by soil and tainted crops can permit subsequent amplification of *L. monocytogenes* numbers to high levels. In this way, field studies consistently highlight silage feeding as the main factor associated with farm animal exposure. However, the pathogen could also be isolated from a number of other sources, including bedding material, feed bunks, and water troughs [113,114].

Once ingested via feed, *L. monocytogenes* transforms its metabolism and colonises the ruminant gastrointestinal tract intracellularly as a cytosol-adapted pathogen, thereby escaping immune defence. According to Zundel and Bernard [90], *L. monocytogenes* multiplied in the rumen of sheep who were asymptomatic carriers due to the favourable environment of the organ (pH 6.5–7.2 and body temperatures from 38.0 to 40.5 °C). Thus, the rumen content serves as an important reservoir for *Listeria*.

However, there is still the widespread opinion that grass and soil are initially contaminated by wildlife such as deer and birds, which means that dairy farm animals are mainly subsequently challenged with *L. monocytogenes*, either during grazing or after consumption of silage: Indeed, asymptomatic carriage of *L. monocytogenes* is thought to be prevalent in up to 36% of wild birds. This includes a variety of species, such as crows, gulls, pheasants, pigeons, rooks, and sparrows [4]. Interestingly, it was suggested that birds may be somewhat responsible for spreading strains of *L. monocytogenes* within the human food chain, as there were often similarities in the pulsotypes isolated from the birds with those found in the food chain [115]. However, the findings do not explain if the birds are infected when feeding on fertilized fields contaminated with *Listeria*, if birds contaminate the environment or if both situations apply. Additionally, a wide range of mammals, such as red fox (3.5%), wild boars (25%), and deer (42%), also harbour *L. monocytogenes* [4]. Silage winter feeding is a common practice for free-living [116,117]) as well as farmed [118] wild ruminants in alpine regions, and it remains to be explored to what extent this practice contributes to *Listeria* infections in wildlife. Again, there is considerable evidence that the high prevalence rate in wild animals is entirely self-generated.

Finally, faecal transmission of *L. monocytogenes* is not exclusively driven by animals, either wild or domestic, as it has been shown to occur regularly in humans also [4]. A number of studies have investigated such transmission within specific occupational groups. Laboratory technicians had a 77% high cumulative prevalence rate of faecal carriage. However, prevalence was also very high (62%) in office workers, who were not occupationally exposed to *L. monocytogenes* [119]. Furthermore, 16% of swab samples from the hands of farmers [120] and 5.7% of swab samples from hands and working clothes of abattoir workers [121] were positive for *L. monocytogenes* [4].

Faecal shedding of *L. monocytogenes* by asymptomatic farm animals increases its presence within the farm environment, which leads to an increased risk of feed and food contamination (Figure 3). Therefore, the ecology of *L. monocytogenes* within the agricultural environment should be thorough analysed and *Listeria* reservoirs should be identified and removed as part pathogen reduction programs [99].

Hence, in order to gain a more comprehensive understanding of the transmission dynamics and ecology of *L. monocytogenes*, a prevalence study was conducted in the dairy-intensive region of Austria, focusing on small ruminant on-farm dairies [13]. The study focused on dairy farms that manufactured cheese from raw caprine and ovine milk, and aimed to identify the routes of transmission of *Listeria* spp. and to investigate the link between *L. monocytogenes* mastitis and the contamination of raw milk. A total of 5799 samples were taken from 53 Austrian dairy farms, and the pathogen was found in 0.9% of them. However, none of the samples taken from the udders of the sheep or goats tested positive, meaning that raw milk contamination was not significantly impacted by listerial mastitis.

The prevalence levels from swab samples of working boots and faecal samples were 15.7% and 13.0%, respectively. The investigators concluded that silage feeding practices correlated significantly with the prevalence of *L. monocytogenes* in the farm and milk processing environments. Again, silage was a main culprit, such that *L. monocytogenes* was between three to seven times more likely to be present in farms that fed silage to animals year-round than in farms that did not use silage [13].

Appraisal of state-of-the-art studies now leads us to conclude that silage and the rumen itself serve as the most important *Listeria* reservoirs. While the pathogen persists in a cyclic infection (from faecal excretion to contamination of feed to multiplication in the gastrointestinal tract) [5], it can enter the food chain either by contaminating raw milk or by being excreted from the udder of an infected animal. In turn, this contamination can spread silently to the milk and cheese processing environment (Figure 3); once contaminated, milk and cheese processing devices and premises can act as a reservoir for *Listeria* and contaminate product batches that were originally free from the pathogen.

3. Risk Factor: Consumer Habits

3.1. Trends in Food Supply and Consumers' Preferences

As the availability and variety of foods in developed countries have increased over the past several decades, consumer perceptions of these essentials are also changing. Perhaps the greatest influence on European eating habits in modern times was the widespread introduction of efficient and affordable domestic refrigeration in the 1960s. For instance, a majority (58%) of British households owned an electric refrigerator by 1970 [122]. Together with social changes at that time, in the context of post-war reconstruction in Europe, where more women entered the workforce, this led to the growth of supermarkets. Shopping for food each day was no longer necessary. Wartime and rationing survivors, as well as baby boomers, began to enjoy ongoing food abundance.

In the meantime, consumers throughout industrialized countries are becoming increasingly alert to the environmental, social and health consequences of mass-produced, refined foods and the globalization of food production and trade. Opinions now abound as to how further environmental damage by mass agriculture can be prevented, how food production can become sustainable without long transportation distances and how to maintain local economies [123].

These everyday messages are motivating a significant number of people to prefer foods that have been produced in a transparent and sustainable way, that are free from pesticides, agrochemicals, processing contaminants, produced without genetic manipulation, and ideally sourced locally. Perceptions that such food tastes better than superstore alternatives and comes without plastic packaging are self-motivating. These sentiments reasonably converge on the local farms, farmers' markets or street stallholders as opposed to the local supermarket.

The trend to shop for locally harvested food is likely to increase due to our growing love of organic produce, renewed enthusiasm for vegetarianism and veganism and simultaneously calls for less meat consumption to slow climate change and improve health. Such calls have been advocated by, among others, the Intergovernmental Panel on Climate Change [124] and the World Scientists' Warning to Humanity [125]. An increasing number of people are vegetarian, vegan, or flexitarian—those who adopt a predominantly plant-based diet with occasional meat consumption.

All of this is intensified by the actions of farmers to keep their businesses viable. It is noteworthy that the number of farms in the EU is in steep decline [37]. There can also be subsidies for farmers to diversify their activities from national and international bodies in the context of development programs for weakened rural communities. Local food systems are purported to promote sustainability, improve local economies, increase access to healthy foods, improve local diets [126] and encourage entrepreneurship and innovation. Direct sales from farm producers to consumers, which include farm gate sales, farmers' markets and internet-direct marketing, are becoming new marketing channels that retain profits in local communities [127]. Indeed, some forms of direct marketing are integrally linked with tourism in local communities. The complement is that farmers who sell products directly to consumers can attract a number of visitors to a community [128].

To demonstrate an alternative market, farmers' markets generally also create a context for closer social ties between farmers and consumers—a human connection—but they remain fundamentally rooted in community relations [129]. The obverse, according to Hinrichs, is a distant and anonymous relation between consumers and a few seemingly unpeopled yet powerful transnational corporations. As for the farmers, the higher costs associated with direct marketing can be compensated for by higher revenues from higher prices and reduced uncertainty, which encourages them to enter into quality food projects without investing excessive labour or capital [130]. Farmers can attract premium prices with minimum costs for handling, transportation, refrigeration, storage and retail premise overheads.

These consumer-led changes are certainly encouraging for consumer wellbeing and the planet. However, the production of industrially produced foods can be regulated legally in the interests of consumer health, while this is not so easy to ensure on the smaller scale.

3.2. On-Farm Dairies and Raw Milk Consumption

The concept of "produce, sell and buy local" has also resulted in an increased interest in the consumption of raw milk [131]. Raw milk advocates argue that it is a complete, natural food containing more amino acids, antimicrobials, vitamins, minerals and fatty acids than pasteurized milk. Furthermore, raw sheep and goat milk is seen to be a better choice for those with lactose intolerance, asthma, and autoimmune and allergic conditions [38–40]. It is estimated that 35–60% of farm families and farm employees consume raw milk on a regular basis, whereas the consumption of raw milk by the urban community is more difficult to estimate [132]. However, raw milk and milk product consumption pose a significant health risk associated with ingestion of *L. monocytogenes*. Surveys from various countries monitoring the presence of this major foodborne pathogen in raw bovine milk (including in-line milk filters), have shown prevalence levels as high as 13% [132]. Studies referring to small ruminant milk revealed a prevalence of up to 17% [133], with the prevalence of pathogens in milk being influenced by numerous factors, including farm size, number of animals on the farm, hygiene, farm management practices, milking facilities, and season [13].

Not surprisingly, numerous foodborne outbreaks caused by milk products contaminated with *L. monocytogenes* are reported [35]. Interestingly, to our knowledge there is no documented case of an outbreak scenario due to cheese manufactured at on-farm dairies. However, Bellemare et al. [134] claimed that the emergence of farmers' markets in the USA increased the number of outbreaks and cases of foodborne illnesses. They detailed a positive relationship between the number of farmers' markets per million individuals and the number per million of reported total outbreaks and cases of foodborne illness in the average state by year.

3.3. Management of Foodborne Hazards in On-Farm Dairies

For its part, the EU has attempted to reduce food safety risks through programs such as "Farm to Fork" food safety legislation [135]. A broad weakness, however, is that farmers' markets, for example, tend to be less rigidly regulated than bricks and mortar shops. Consequentially, this opens up the potential for new routes of food contamination that have until now been neglected. Notwithstanding, several EU countries have developed legal frameworks and incentives to support these so-called "short food (supply) chains" [136]. As direct marketing of food from producers to consumers in Europe grows in popularity, we also must be vigilant about new patterns and scales of food contamination.

One of the largest developments in recent years in nutrition is that consumers are increasingly demanding minimally processed, ready-to-eat (RTE) foods that can be stored in refrigerators for up to several weeks. These foods are challenging hygienists' attempts to ensure microbiological quality and safety [20], not least due to the fact that domestic refrigerators are usually not cold enough [137].

L. monocytogenes is psychrotrophic, which means it is able to multiply even at a few degrees above zero. Nevertheless, in general, optimal storage temperatures of 4 °C will usually slow growth of *L. monocytogenes* and may restrict amounts in food to non-harmful doses. However, in the context of a multinational outbreak, the psychrotrophic growth potential of *L. monocytogenes* can be dramatic [138]. The contamination levels of *L. monocytogenes* in lots of acid curd cheese that caused a listeriosis outbreak, which led to a total of 8 deaths among 34 clinical cases, were determined. Contamination levels varied from $\leq 10^2$ CFU/g to 8.1×10^8 CFU/g. Interestingly, contamination levels of $\leq 10^2$ CFU/g were even found in three of the sixteen lots that had been stored under optimal conditions since the beginning of their shelf-life. Nevertheless, by the end of the shelf life, the contamination levels were found to have increased to the health-endangering levels of 10^5 and 10^6 CFU/g.

4. Conclusions and Future Implications

The long shelf life of our food items, inadequate temperature control, abuse at the household level, combined with the ability of *L. monocytogenes* to grow at refrigeration temperatures and its ability to enter the milk chain at almost every stage, makes *L. monocytogenes* a significant threat to public health. In the context of direct marketing of raw milk and cheese, we can conclude that prudent steps must be taken by the farmers to eliminate major contamination routes, to ensure continuous compliance with the legally prescribed cooling temperature and to offer products with a short shelf life.

European Union legislation requires that food business operators not only comply to basic rules of hygiene (Good Hygiene Practices) [24,27] but, more specifically, in Article 5 of

Regulation (EC) No 852/2004, "shall put in place, implement and maintain a permanent procedure or procedures based on the HACCP principles". More recently, the establishment of "Operational Prerequisite Control Programs" (oPRP) has filled a gap between Good Hygiene Practices (GHP) and HACCP-based procedures [139]. In this context, examination of the udder and determination of the somatic cell count (CMT) of the milk are measures to detect animals with clinical or subclinical mastitis and to discard milk from such infected animals. In fresh cheese making, the addition of appropriate starter cultures can prevent multiplication of *L. monocytogenes*, or even reduce their numbers. Adequate sanitation of the milk processing area is one of the basics in GHP. However, under real-life conditions *L. monocytogenes* is sometimes able to persist in dairy plants, with severe consequences [140].

From the references we retrieved, it is obvious that in some cases, non-compliance to GHP and a lack of HACCP-based procedures were identified as factors creating hazardous situations. We cannot conclude that strict adherence to food safety management programs would render a 100% safe food. Thus, risk management by a shift towards heat-processed products would more likely allow a fully HACCP compliant food safety system for control of *L. monocytogenes* on small on-farm dairies.

Author Contributions: Conceptualization, D.S.; methodology, D.S.; formal analysis, A.P. and P.P.; investigation, D.S. and A.P.; writing—original draft preparation, D.S.; writing—review and editing, D.S., A.P. and P.P.; visualization, A.P.; supervision, D.S. All authors have read and agreed to the published version of the manuscript.

Funding: This research received no external funding.

Data Availability Statement: No new data were created or analyzed in this study. Data sharing is not applicable to this article.

Acknowledgments: Open Access Funding by the University of Veterinary Medicine Vienna.

Conflicts of Interest: The authors declare no conflict of interest.

References

1. Quereda, J.J.; Leclercq, A.; Moura, A.; Vales, G.; Gómez-Martín, Á.; García-Muñoz, Á.; Thouvenot, P.; Tessaud-Rita, N.; Bracq-Dieye, H.; Lecuit, M. *Listeria valentina* sp. nov., isolated from a water trough and the faeces of healthy sheep. *Int. J. Syst. Evol. Microbiol.* **2020**, *70*, 5868–5879. [CrossRef]

2. De Noordhout, C.M.; Devleesschauwer, B.; Angulo, F.J.; Verbeke, G.; Haagsma, J.; Kirk, M.; Havelaar, A.; Speybroeck, N. The global burden of listeriosis: A systematic review and meta-analysis. *Lancet Infect. Dis.* **2014**, *14*, 1073–1082. [CrossRef] [PubMed]

3. Schoder, D. Chapter 425—Listeria: Listeriosis. In *The Encyclopedia of Food and Health*, 1st ed.; Caballero, B., Finglas, P., Toldrá, F., Eds.; Academic Press: Oxford, UK, 2015; pp. 561–566.

4. Schoder, D.; Guldimann, C.; Märtlbauer, E. Asymptomatic Carriage of *Listeria monocytogenes* by Animals and Humans and Its Impact on the Food Chain. *Foods* **2022**, *11*, 3472. [CrossRef] [PubMed]

5. Oliver, S.P.; Jayarao, B.M.; Almeida, R.A. Food-borne pathogens in milk and the dairy farm environment: Food safety and public health implications. *Foodborne Pathog. Dis.* **2005**, *2*, 115–129. [CrossRef] [PubMed]

6. Waak, E.; Tham, E.; Danielsson-Tham, M.-L. Prevalence and Fingerprinting of *Listeria monocytogenes* Strains Isolated from Raw Whole Milk in Farm Bulk Tanks and in Dairy Plant Receiving Tanks. *Appl. Environ. Microbiol.* **2022**, *68*, 3366–3370. [CrossRef]

7. Kulesh, R.; Shinde, S.V.; Khan, W.A.; Chaudhari, S.P.; Patil, A.R.; Kurkure, N.V.; Paliwal, N.; Likhite, A.V.; Zade, N.N.; Barbuddhe, S.B. The occurrence of *Listeria monocytogenes* in goats, farm environment and invertebrates. *Biol. Rhythm. Res.* **2022**, *53*, 831–840. [CrossRef]

8. Zhao, Q.; Hu, P.; Li, Q.; Zhang, S.; Li, H.; Chang, J.; Jiang, Q.; Zheng, Y.; Li, Y.; Liu, Z.; et al. Prevalence and transmission characteristics of *Listeria* species from ruminants in farm and slaughtering environments in China. *Emerg. Microbes Infect.* **2021**, *10*, 356–364. [CrossRef]

9. Condoleo, R.; Giangolini, G.; Chiaverini, A.; Patriarca, D.; Scaramozzino, P.; Mezher, Z. Occurrence of *Listeria monocytogenes* and *Escherichia coli* in Raw Sheep's Milk from Farm Bulk Tanks in Central Italy. *J. Food Prot.* **2020**, *83*, 1929–1933. [CrossRef]

10. Hurtado, A.; Ocejo, M.; Oporto, B. *Salmonella* spp. and *Listeria monocytogenes* shedding in domestic ruminants and characterization of potentially pathogenic strains. *Vet. Microbiol.* **2017**, *210*, 71–76. [CrossRef]

11. Jamali, H.; Radmehr, B.; Thong, K.L. Prevalence, characterisation, and antimicrobial resistance of *Listeria* species and *Listeria monocytogenes* isolates from raw milk in farm bulk tanks. *Food Control* **2013**, *34*, 121–125. [CrossRef]

12. Al-Mariri, A.; Younes, A.A.; Ramadan, L. Prevalence of *Listeria* spp. in Raw Milk in Syria. *Bulg. J. Vet. Med.* **2013**, *16*, 112–122.

13. Schoder, D.; Melzner, D.; Schmalwieser, A.; Zangana, A.; Winter, P.; Wagner, M. Important vectors for *Listeria monocytogenes* transmission at farm dairies manufacturing fresh sheep and goat cheese from raw milk. *J. Food Prot.* **2011**, *74*, 919–924. [CrossRef] [PubMed]

14. Ho, A.J.; Lappi, V.R.; Wiedmann, M. Longitudinal monitoring of *Listeria monocytogenes* contamination patterns in a farmstead dairy processing facility. *J. Dairy Sci.* **2007**, *90*, 2517–2524. [CrossRef]

15. D'Amico, D.J.; Groves, E.; Donnelly, C.W. Low incidence of foodborne pathogens of concern in raw milk utilized for farmstead cheese production. *J. Food Prot.* **2008**, *71*, 1580–1589. [CrossRef]

16. D'Amico, D.J.; Donnelly, C.W. Microbiological quality of raw milk used for small-scale artisan cheese production in Vermont: Effect of farm characteristics and practices. *J. Dairy Sci.* **2010**, *93*, 134–147. [CrossRef]

17. Bagatella, S.; Tavares-Gomes, L.; Oevermann, A. *Listeria monocytogenes* at the interface between ruminants and humans: A comparative pathology and pathogenesis review. *Vet. Pathol.* **2022**, *59*, 186–210. [CrossRef]

18. Nightingale, K.K.; Schukken, Y.H.; Nightingale, C.R.; Fortes, E.D.; Ho, A.J.; Her, Z.; Grohn, Y.T.; McDonough, P.L.; Wiedmann, M. Ecology and transmission of *Listeria monocytogenes* infecting ruminants and in the farm environment. *Appl. Environ. Microbiol.* **2004**, *70*, 4458–4467. [CrossRef] [PubMed]

19. Jordan, K.; McAuliffe, O. *Listeria monocytogenes* in Foods. *Adv. Food Nutr. Res.* **2018**, *86*, 181–213. [CrossRef] [PubMed]

20. EFSA Panel on Biological Hazards (BIOHAZ); Ricci, A.; Allende, A.; Bolton, D.; Chemaly, M.; Davies, R.; Fernández Escámez, P.S.; Girones, R.; Herman, L.; Koutsoumanis, K.; et al. *Listeria monocytogenes* contamination of ready-to-eat foods and the risk for human health in the EU. *EFSA J.* **2018**, *16*, e05134. [PubMed]

21. Rivera, M.; Guarín, A.; Pinto-Correia, T.; Almaas, H.; Mur, L.A.; Burns, V.; Czekaj, M.; Ellis, R.; Galli, F.; Grivins, M.; et al. Assessing the role of small farms in regional food systems in Europe: Evidence from a comparative study. *Glob. Food Sec.* **2020**, *26*, 100417. [CrossRef]

22. Ryffel, S.; Piccinali, P.; Bütikofer, U. Sensory descriptive analysis and consumer acceptability of selected Swiss goat and sheep cheeses. *Small Rumin. Res.* **2008**, *79*, 80–86. [CrossRef]

23. LWKO. Landwirtschaftskammer Österreich: Richtlinien für die Dachmarke „Gutes vom Bauernhof". *Das Qualitätsprogramm für die bäuerliche Direktvermarktung in Österreich.* Available online: https://www.gutesvombauernhof.at/uploads/pics/Oesterreich/GvB_Registerblatt/Richtlinien_GvB_Version_20150630.pdf (accessed on 4 June 2018).

24. Regulation (EC) No 178/2002 of the European Parliament and of the Council of 28 January 2002 Laying Down the General Principles and Requirements of Food Law, Establishing the European Food Safety Authority and Laying Down Procedures in Matters of Food Safety; O.J. L31/1–24. Available online: https://eur-lex.europa.eu/legal-content/EN/ALL/?uri=CELEX%3A32002R0178 (accessed on 22 September 2022).

25. BMG. Leitlinie für Eine Gute Hygienepraxis und Die Anwendung der Grundsätze des HACCP für Bäuerliche Milchverarbeitungsbetriebe. BMGF-75220/0054-IV/B/10/2005 vom 19.12.2005. Available online: https://www.verbrauchergesundheit.gv.at/lebensmittel/buch/hygieneleitlinien/LL_Milchverarbeitungsbetriebe_baeuerliche_.pdf?7vj8co (accessed on 22 September 2022).

26. LFI. Ländliches Fortbildungsinstitut Österreich: Bäuerliche Direktvermarktung von A-Z. Available online: https://www.gutesvombauernhof.at/intranet/allgemeines-recht/direktvermarktung-von-a-bis-z.html (accessed on 4 June 2018).

27. Regulation (EC) No 852/2004 of the European Parliament and of the Council of 29 April 2004 on the Hygiene of Foodstuffs; O.J. L139/1. Available online: https://eur-lex.europa.eu/legal-content/EN/TXT/?uri=celex%3A32004R0852 (accessed on 22 September 2022).

28. Regulation (EC) No 853/2004 of the European Parliament and of the Council of 29 April 2004 on the Laying Down Specific Hygiene Rules for on the Hygiene of Foodstuffs; O.J. L139/55. Available online: https://eur-lex.europa.eu/LexUriServ/LexUriServ.do?uri=OJ:L:2004:139:0055:0205:en:PDF (accessed on 22 September 2022).

29. Commission Regulation (EC) No 2073/2005 of 15 November 2005 on Microbiological Criteria for Foodstuffs; O.J. L338/1–26. Available online: https://eur-lex.europa.eu/legal-content/EN/ALL/?uri=CELEX%3A32005R2073 (accessed on 22 September 2022).

30. BMG. Bundesgesetz über Sicherheitsanforderungen und weitere Anforderungen an Lebensmittel, Gebrauchsgegenstände und kosmetische Mittel zum Schutz der Verbraucherinnen und Verbraucher Lebensmittelsicherheits-und Verbraucherschutzgesetz-LMSVG). BGBl. I Nr. 13/2006 Lebensmittelsicherheits- und Verbraucherschutzgesetz vom 20.1.2006. Available online: https://www.ris.bka.gv.at/eli/bgbl/I/2006/13 (accessed on 22 September 2022).

31. BMG. Verordnung der Bundesministerin für Gesundheit und Frauen BGBl. II Nr. 92/2006 über Lebensmittelhygieneanforderungen an Einzelhandelsunternehmen (Lebensmittel-Einzelhandelsverordnung). BGBl. II Nr. 92/2006 Lebensmittelhygiene-Einzelhandelsverordnung vom 1.1.2006. Available online: https://www.ris.bka.gv.at/eli/bgbl/II/2006/92 (accessed on 22 September 2022).

32. BMG. Verordnung der Bundesministerin für Gesundheit und Frauen über die Eintragung und Zulassung von Betrieben von Lebensmittelunternehmern (Eintragungs- und Zulassungsverordnung). BGBl. II Nr. 93/2006 Eintragungs- und Zulassungsverordnung vom 1.1.2006. Available online: https://www.ris.bka.gv.at/eli/bgbl/II/2006/93 (accessed on 22 September 2022).

33. BMG. Empfehlung: Leitlinien zur Umsetzung der Rückverfolgbarkeit bei Lebensmitteln gemäß Art. 18 und 9 der Verordnung (EG) Nr. 178/2002 zur Festlegung der allgemeinen Grundsätze und Anforderungen des Lebensmittelrechts, zur Errichtung der Europäischen Behörde für Lebensmittelsicherheit und zur Festlegung von Verfahren zur Lebensmittelsicherheit vom 28. Jänner 2002 (V 178/2002). BMGF-30.721/43-IV/B/10/04 vom 5.5.2004. Available online: https://www.verbrauchergesundheit.gv.at/Lebensmittel/rechtsvorschriften/eu/Leitlinie_zur_Umsetzung_Rueckverfolgbarkeit.pdf?8ksxr7 (accessed on 22 September 2022).

34. BMG. Leitlinie Über Mikrobiologische Kriterien für Milch und Milchprodukte für Eine Gute Hygienepraxis und die Anwendung der Grundsätze des HACCP bei der Milchverarbeitung auf Almen. BMG-75220/0054-IV/B/10/2005 vom 19.12.2005. Available online: https://www.verbrauchergesundheit.gv.at/Lebensmittel/buch/hygieneleitlinien/LL_Milchverarbeitung_auf_Almen.pdf?8ksx1i (accessed on 22 September 2022).

35. Anonymous. The European Union One Health 2020 Zoonoses Report. *EFSA J.* **2021**, *19*, e06971.

36. FAO. Dairy Animals. Available online: https://www.fao.org/dairy-production-products/production/dairy-animals/en/ (accessed on 10 November 2022).

37. EUROSTAT: Database on Agriculture Statistics. 2021. Available online: https://ec.europa.eu/eurostat/web/agriculture/data/database (accessed on 10 November 2022).

38. Bawa, S. An update on the beneficial role of conjugated linoleic acid (CLA) in modulating human health: Mechanisms of action—A review. *Pol. J. Food Nutr. Sci.* **2003**, *12*, 3–13.

39. Tricon, S.; Burdge, G.C.; Kew, S.; Banerjee, T.; Russell, J.J.; Jones, E.L.; Grimble, R.F.; Williams, C.M.; Yaqoob, P.; Calder, P.C. Opposing effects of cis-9,trans-11 and trans-10,cis-12 conjugated linoleic acid on blood lipids in healthy humans. *Am. J. Clin. Nutr.* **2004**, *80*, 614–620. [CrossRef]

40. Dario, C.; Carnicella, D.; Dario, M.; Bufano, G. Genetic polymorphism of β-lactoglobulin gene and effect on milk composition in Leccese sheep. *Small Rumin. Res.* **2008**, *74*, 270–273. [CrossRef]

41. Wang, Y.; Jones, P.J. Dietary conjugated linoleic acid and body composition. *Am. J. Clin. Nutr.* **2004**, *79* (Suppl. S6), 1153S–1158S. [CrossRef]

42. Gitter, M.; Bradley, R.; Blampied, P.H. *Listeria monocytogenes* in bovine mastitis. *Vet. Rec.* **1980**, *107*, 390–393. [CrossRef]

43. Sharp, M.W. Bovine mastitis and *Listeria monocytogenes*. *Vet. Rec.* **1989**, *125*, 512–513. [CrossRef]

44. Fedio, W.M.; Schoonderwoerd, M.; Shute, R.H.; Jackson, H. A case of bovine mastitis caused by *Listeria monocytogenes*. *Can. Vet. J.* **1990**, *31*, 773–775. [PubMed]

45. Bourry, A.; Poutrel, B.; Rocourt, J. Bovine mastitis caused by *Listeria monocytogenes*: Characteristics of natural and experimental infections. *J. Med. Microbiol.* **1995**, *43*, 125–132. [CrossRef]

46. Jensen, N.E.; Aarestrup, F.M.; Jensen, J.; Wegener, H.C. *Listeria monocytogenes* in bovine mastitis. Possible implication for human health. *Int. J. Food Microbiol.* **1996**, *32*, 209–216. [CrossRef] [PubMed]

47. Wagner, M.; Podstatzky-Lichtenstein, L.; Lehner, A.; Asperger, H.; Baumgartner, W.; Brandl, E. Prolonged excretion of *Listeria monocytogenes* in a subclinical case of mastitis. *Milchwissenschaft* **2000**, *55*, 3–6.

48. Rawool, D.B.; Malik, S.V.; Shakuntala, I.; Sahare, A.M.; Barbuddhe, S.B. Detection of multiple virulence-associated genes in *Listeria monocytogenes* isolated from bovine mastitis cases. *Int. J. Food Microbiol.* **2007**, *113*, 201–207. [CrossRef] [PubMed]

49. Hunt, K.; Drummond, N.; Murphy, M.; Butler, F.; Buckley, J.; Jordan, K. A case of bovine raw milk contamination with *Listeria monocytogenes*. *Ir. Vet. J.* **2012**, *65*, 13. [CrossRef] [PubMed]

50. Bourry, A.; Cochard, T.; Poutrel, B. Serological diagnosis of bovine, caprine, and ovine mastitis caused by *Listeria monocytogenes* by using an enzyme-linked immunosorbent assay. *J. Clin. Microbiol.* **1997**, *35*, 1606–1608. [CrossRef]

51. Fthenakis, G.C.; Saratsis, P.; Tzora, A.; Linde, K. Naturally occurring subclinical ovine mastitis associated with *Listeria monocytogenes*. *Small Rumin. Res.* **1998**, *31*, 23–27. [CrossRef]

52. Tzora, A.; Fthenakis, G.C.; Linde, K. The effects of inoculation of *Listeria monocytogenes* into the ovine mammary gland. *Vet. Microbiol.* **1998**, *59*, 193–202. [CrossRef]

53. Schoder, D.; Winter, P.; Kareem, A.; Baumgartner, W.; Wagner, M. A case of sporadic ovine mastitis caused by *Listeria monocytogenes* and its effect on contamination of raw milk and raw-milk cheeses produced in the on-farm dairy. *J. Dairy Res.* **2003**, *70*, 395–401. [CrossRef]

54. Winter, P.; Schilcher, F.; Bagò, Z.; Schoder, D.; Egerbacher, M.; Baumgartner, W.; Wagner, M. Clinical and histopathological aspects of naturally occurring mastitis caused by *Listeria monocytogenes* in cattle and ewes. *J. Vet. Med. Ser. B Infect. Dis. Vet. Public Health* **2004**, *51*, 176–179. [CrossRef]

55. Schoder, D.; Zangana, A.; Paulsen, P.; Winter, P.; Baumgartner, W.; Wagner, M. Ovine *Listeria monocytogenes* mastitis and human exposure via fresh cheese from raw milk—The impact of farm management, milking and cheese manufacturing practices. *Milchwissenschaft* **2008**, *63*, 258–262.

56. Pintado, C.M.; Grant, K.A.; Halford-Maw, R.; Hampton, M.D.; Ferreira, M.A.; McLauchlin, J. Association between a case study of asymptomatic ovine listerial mastitis and the contamination of soft cheese and cheese processing environment with *Listeria monocytogenes* in Portugal. *Foodborne Pathog. Dis.* **2009**, *6*, 569–575. [CrossRef] [PubMed]

57. Addis, M.F.; Cubeddu, T.; Pilicchi, Y.; Rocca, S.; Piccinini, R. Chronic intramammary infection by *Listeria monocytogenes* in a clinically healthy goat—A case report. *BMC Vet. Res.* **2019**, *15*, 229. [CrossRef] [PubMed]
58. Samkange, A.; van der Westhuizen, J.; Voigts, A.S.; Chitate, F.; Kaatura, I.; Khaiseb, S.; Hikufe, E.H.; Kabajani, J.; Bishi, A.S.; Mbiri, P.; et al. Investigation of the outbreaks of abortions and orchitis in livestock in Namibia during 2016–2018. *Trop. Anim. Health Prod.* **2022**, *54*, 346. [CrossRef]
59. Kotzamanidis, C.; Papadopoulos, T.; Vafeas, G.; Tsakos, P.; Giantzi, V.; Zdragas, A. Characterization of *Listeria monocytogenes* from encephalitis cases of small ruminants from different geographical regions, in Greece. *J. Appl. Microbiol.* **2019**, *126*, 1373–1382. [CrossRef] [PubMed]
60. Rissi, D.R.; Kommers, G.D.; Marcolongo-Pereira, C.; Schild, A.L.; Barros, C.S.L. Meningoencephalitis in sheep caused by *Listeria monocytogenes* [Meningoencefalite por *Listeria monocytogenes* em ovinos]. *Pesqui. Vet. Bras.* **2010**, *30*, 51–56. [CrossRef]
61. Hamidi, A.; Bisha, B.; Goga, I.; Wang, B.; Robaj, A.; Sylejmani, D.S. A case report of sporadic ovine listerial meningoencephalitis in Kosovo. *Vet. Ital.* **2020**, *56*, 205–211. [CrossRef]
62. Papić, B.; Kušar, D.; Zdovc, I.; Golob, M.; Pate, M. Retrospective investigation of listeriosis outbreaks in small ruminants using different analytical approaches for whole genome sequencing-based typing of *Listeria monocytogenes*. *Infect. Genet. Evol.* **2020**, *77*, 104047. [CrossRef]
63. Dreyer, M.; Thomann, A.; Böttcher, S.; Frey, J.; Oevermann, A. Outbreak investigation identifies a single *Listeria monocytogenes* strain in sheep with different clinical manifestations, soil and water. *Vet. Microbiol.* **2015**, *179*, 69–75. [CrossRef]
64. Wesley, I.V.; Larson, D.J.; Harmon, K.M.; Luchansky, J.B.; Schwartz, A.R. A case report of sporadic ovine listerial menigoen-cephalitis in Iowa with an overview of livestock and human cases. *J. Vet. Diagn. Investig.* **2002**, *14*, 314–321. [CrossRef]
65. Al-Dughaym, A.M.; Elmula, A.F.; Mohamed, G.E.; Hegazy, A.A.; Radwan, Y.A.; Housawi, F.M.; Gameel, A.A. First report of an outbreak of ovine septicaemic listeriosis in Saudi Arabia. *Rev. Sci. Tech.* **2001**, *20*, 777–783. [CrossRef]
66. Wiedmann, M.; Czajka, J.; Bsat, N.; Bodis, M.; Smith, M.C.; Divers, T.J.; Batt, C.A. Diagnosis and epidemiological association of *Listeria monocytogenes* strains in two outbreaks of listerial encephalitis in small ruminants. *J. Clin. Microbiol.* **1994**, *32*, 991–996. [CrossRef]
67. Meredith, C.D.; Schneider, D.J. An outbreak of ovine listeriosis associated with poor flock management practices. *J. S. Afr. Vet. Assoc.* **1984**, *55*, 55–56.
68. Wardrope, D.D.; MacLeod, N.S. Outbreak of listeria meningoencephalitis in young lambs. *Vet. Rec.* **1983**, *113*, 213–214. [CrossRef]
69. Vandegraaff, R.; Borland, N.A.; Browning, J.W. An outbreak of listerial meningo-encephalitis in sheep. *Aust. Vet. J.* **1981**, *57*, 94–96. [CrossRef] [PubMed]
70. Grønstøl, H. Listeriosis in sheep. Isolation of *Listeria monocytogenes* from grass silage. *Acta Vet. Scand.* **1979**, *20*, 492–497. [CrossRef] [PubMed]
71. Du Toit, I.F. An outbreak of caprine listeriosis in the Western Cape. *J. S. Afr. Vet. Assoc.* **1977**, *48*, 39–40. [PubMed]
72. McDonald, J.W. An outbreak of abortion due to *Listeria monocytogenes* in an experimental flock of sheep. *Aust. Vet. J.* **1967**, *43*, 564–567. [CrossRef] [PubMed]
73. Gitter, M.; Terlecki, S.; Turnbull, P.A. An outbreak of visceral and cerebral listeriosis in a flock of sheep in South East England. *Vet. Rec.* **1965**, *77*, 11–15.
74. Schlech, W.F., III. New perspectives on the gastrointestinal mode of transmission in invasive *Listeria monocytogenes* infection. *Clin. Investig. Med.* **1984**, *7*, 321–324.
75. Grønstøl, H. Listeriosis in sheep. *Listeria monocytogenes* excretion and immunological state in healthy sheep. *Acta Vet. Scand.* **1979**, *20*, 168–179. [CrossRef]
76. Nightingale, K.K.; Fortes, E.D.; Ho, A.J.; Schukken, Y.H.; Grohn, Y.T.; Wiedmann, M. Evaluation of farm management practices as risk factors for clinical listeriosis and fecal shedding of *Listeria monocytogenes* in ruminants. *J. Am. Vet. Med. Assoc.* **2005**, *227*, 1808–1814. [CrossRef]
77. Wagner, M.; Melzner, D.; Bagò, Z.; Winter, P.; Egerbacher, M.; Schilcher, F.; Zangana, A.; Schoder, D. Outbreak of clinical listeriosis in sheep: Evaluation from possible contamination routes from feed to raw produce and humans. *J. Vet. Med. Ser. B Infect. Dis. Vet. Public Health* **2005**, *52*, 278–283. [CrossRef]
78. Palacios-Gorba, C.; Moura, A.; Gomis, J.; Leclercq, A.; Gomez-Martin, A.; Bracq-Dieye, H.; Moce, M.L.; Tessaud-Rita, N.; Jimenez-Trigos, E.; Vales, G.; et al. Ruminant-associated *Listeria monocytogenes* isolates belong preferentially to dairy-associated hypervirulent clones: A longitudinal study in 19 farms. *Environ. Microbiol.* **2021**, *23*, 7617–7631. [CrossRef]
79. Chitura, T.; Shai, K.; Ncube, I.; van Heerden, H. Contamination of the environment by pathogenic bacteria in a livestock farm in Limpopo Province, South Africa. *Appl. Ecol. Environ. Res.* **2019**, *17*, 2943–2963. [CrossRef]
80. Cavicchioli, V.Q.; Scatamburlo, T.M.; Yamazi, A.K.; Pieri, F.A.; Nero, L.A. Occurrence of *Salmonella*, *Listeria monocytogenes*, and enterotoxigenic *Staphylococcus* in goat milk from small and medium-sized farms located in Minas Gerais State, Brazil. *J. Dairy Sci.* **2015**, *98*, 8386–8390. [CrossRef]

81. Sarangi, L.N.; Panda, H.K. Isolation, characterization and antibiotic sensitivity test of pathogenic listeria species in livestock, poultry and farm environment of Odisha. *Indian J. Anim. Res.* **2012**, *46*, 242–247.

82. Sharif, J.; Willayat, M.M.; Sheikh, G.N.; Roy, S.S.; Altaf, S. Prevalence of *Listeria monocytogenes* in Organised Sheep Farms of Kashmir. *J. Pure Appl. Microbiol.* **2010**, *4*, 871–873.

83. Soncini, G.; Valnegri, L. Analysis of bulk goats' milk and milk-filters from Valtellina and Valchiavenna (Lombardy Prealps) for the presence of *Listeria* species. *Small Rumin. Res.* **2005**, *58*, 143–147. [CrossRef]

84. Schoder, D.; Zangana, A.; Wagner, M. Sheep and goat raw milk consumption: A hygienic matter of concern? *Arch. Lebensm.* **2010**, *61*, 229–234. [CrossRef]

85. Almeida, G.; Magalhães, R.; Carneiro, L.; Santos, I.; Silva, J.; Ferreira, V.; Hogg, T.; Teixeira, P. Foci of contamination of *Listeria monocytogenes* in different cheese processing plants. *Int. J. Food Microbiol.* **2013**, *167*, 303–309. [CrossRef]

86. Miller, B.A.; Lu, C.D. Current status of global dairy goat production: An overview. *Asian-Australas. J. Anim. Sci.* **2019**, *32*, 1219–1232. [CrossRef] [PubMed]

87. Papić, B.; Golob, M.; Kušar, D.; Pate, M.; Zdovc, I. Source tracking on a dairy farm reveals a high occurrence of subclinical mastitis due to hypervirulent *Listeria monocytogenes* clonal complexes. *J. Appl. Microbiol.* **2019**, *127*, 1349–1361. [CrossRef]

88. Moroni, P.; Pisoni, G.; Ruffo, G.; Boettcher, P.J. Risk factors for intramammary infections and relationship with somatic-cell counts in Italian dairy goats. *Prev. Vet. Med.* **2005**, *69*, 163–173. [CrossRef] [PubMed]

89. Brugère-Picoux, J. Ovine listeriosis. *Small Rumin. Res.* **2008**, *76*, 12–20. [CrossRef]

90. Zundel, E.; Bernard, S. *Listeria monocytogenes* translocates throughout the digestive tract in asymptomatic sheep. *J. Med. Microbiol.* **2006**, *55*, 1717–1723. [CrossRef]

91. Maury, M.M.; Bracq-Dieye, H.; Huang, L.; Vales, G.; Lavina, M.; Thouvenot, P.; Disson, O.; Leclercq, A.; Brisse, S.; Lecuit, M. Hypervirulent *Listeria monocytogenes* clones' adaption to mammalian gut accounts for their association with dairy products. *Nat. Commun.* **2019**, *10*, 2488. [CrossRef]

92. Moura, A.; Criscuolo, A.; Pouseele, H.; Maury, M.M.; Leclercq, A.; Tarr, C.; Björkman, J.T.; Dallman, T.; Reimer, A.; Enouf, V.; et al. Whole genome-based population biology and epidemiological surveillance of *Listeria monocytogenes*. *Nat. Microbiol.* **2016**, *10*, 16185. [CrossRef]

93. Papić, B.; Pate, M.; Félix, B.; Kušar, D. Genetic diversity of *Listeria monocytogenes* strains in ruminant abortion and rhomben-cephalitis cases in comparison with the natural environment. *BMC Microbiol.* **2019**, *19*, 299. [CrossRef]

94. Disson, O.; Moura, A.; Lecuit, M. Making Sense of the Biodiversity and Virulence of *Listeria monocytogenes*. *Trends Microbiol.* **2021**, *29*, 811–822. [CrossRef]

95. Swaminathan, B.; Gerner-Smidt, P. The epidemiology of human listeriosis. *Microbes Infect.* **2007**, *9*, 1236–1243. [CrossRef]

96. Johnson, G.C.; Maddox, C.W.; Fales, W.H.; Wolff, W.A.; Randle, R.F.; Ramos, J.A.; Schwartz, H.; Heise, K.M.; Baetz, A.L.; Wesley, I.V.; et al. Epidemiologic evaluation of encephalitic listeriosis in goats. *J. Am. Vet. Med. Assoc.* **1996**, *208*, 1695–1699. [PubMed]

97. Rocha, P.R.; Lomonaco, S.; Bottero, M.T.; Dalmasso, A.; Dondo, A.; Grattarola, C.; Zuccon, F.; Iulini, B.; Knabel, S.J.; Capucchio, M.T.; et al. Ruminant rhombencephalitis-associated *Listeria monocytogenes* strains constitute a genetically homogeneous group related to human outbreak strains. *Appl. Environ. Microbiol.* **2013**, *79*, 3059–3066. [CrossRef] [PubMed]

98. Borucki, M.K.; Gay, C.C.; Reynolds, J.; McElwain, K.L.; Kim, S.H.; Call, D.R.; Knowles, D.P. Genetic diversity of *Listeria monocytogenes* strains from a high-prevalence dairy farm. *Appl. Environ. Microbiol.* **2005**, *71*, 5893–5899. [CrossRef] [PubMed]

99. Esteban, J.I.; Oporto, B.; Aduriz, G.; Juste, R.A.; Hurtado, A. Faecal shedding and strain diversity of *Listeria monocytogenes* in healthy ruminants and swine in Northern Spain. *BMC Vet. Res.* **2009**, *5*, 2. [CrossRef] [PubMed]

100. Steckler, A.J.; Cardenas-Alvarez, M.X.; Townsend Ramsett, M.K.; Dyer, N.; Bergholz, T.M. Genetic characterization of *Listeria monocytogenes* from ruminant listeriosis from different geographical regions in the U.S. *Vet. Microbiol.* **2018**, *215*, 93–97. [CrossRef]

101. Datta, A.R.; Burall, L.S. Serotype to genotype: The changing landscape of listeriosis outbreak investigations. *Food Microbiol.* **2018**, *75*, 18–27. [CrossRef] [PubMed]

102. Orsi, R.H.; den Bakker, H.C.; Wiedmann, M. *Listeria monocytogenes* lineages: Genomics, evolution, ecology, and phenotypic characteristics. *Int. J. Med. Microbiol.* **2011**, *301*, 79–96. [CrossRef] [PubMed]

103. Maury, M.M.; Tsai, Y.H.; Charlier, C.; Touchon, M.; Chenal-Francisque, V.; Leclercq, A.; Criscuolo, A.; Gaultier, C.; Roussel, S.; Brisabois, A.; et al. Uncovering *Listeria monocytogenes* hypervirulence by harnessing its biodiversity. *Nat. Genet.* **2016**, *48*, 308–313. [CrossRef]

104. Orsi, R.H.; Wiedmann, M. Characteristics and distribution of *Listeria* spp., including *Listeria* species newly described since 2009. *Appl. Microbiol. Biotechnol.* **2016**, *100*, 5273–5287. [CrossRef]

105. Sauders, B.D.; Fortes, E.D.; Morse, D.L.; Dumas, N.; Kiehlbauch, J.A.; Schukken, Y.; Hibbs, J.R.; Wiedmann, M. Molecular subtyping to detect human listeriosis clusters. *Emerg. Infect. Dis.* **2003**, *9*, 672–680. [CrossRef]

106. Haase, J.K.; Didelot, X.; Lecuit, M.; Korkeala, H.; Group, L.m.M.S.; Achtman, M. The ubiquitous nature of *Listeria monocytogenes* clones: A large-scale Multilocus Sequence Typing study. *Environ. Microbiol.* **2014**, *16*, 405–416. [CrossRef]

107. Husu, J.R. Epidemiological studies on the occurrence of *Listeria monocytogenes* in the feces of dairy cattle. *Zent. Vet. Ser. B* **1990**, *37*, 276–282. [CrossRef]
108. Husu, J.R.; Seppanen, J.T.; Sivela, S.K.; Rauramaa, A.L. Contamination of raw milk by *Listeria monocytogenes* on dairy farms. *Zent. Vet. Ser. B* **1990**, *37*, 268–275. [CrossRef]
109. Wilkes, G.; Edge, T.A.; Gannon, V.P.J.; Jokinen, C.; Lyautey, E.; Neumann, N.F.; Ruecker, N.; Scott, A.; Sunohara, M.; Topp, E.; et al. Associations among pathogenic bacteria, parasites, and environmental and land use factors in multiple mixed-use watersheds. *Water Res.* **2011**, *45*, 5807–5825. [CrossRef]
110. Lyautey, E.; Lapen, D.R.; Wilkes, G.; McCleary, K.; Pagotto, F.; Tyler, K.; Hartmann, A.; Piveteau, P.; Rieu, A.; Robertson, W.J.; et al. Distribution and characteristics of *Listeria monocytogenes* isolates from surface waters of the South Nation River watershed, Ontario, Canada. *Appl. Environ. Microbiol.* **2007**, *73*, 5401–5410. [CrossRef]
111. Ivaneka, R.; Grohn, Y.T.; Ho, A.J.J.; Wiedmann, M. Markov chain approach to analyze the dynamics of pathogen fecal shedding—Example of *Listeria monocytogenes* shedding in a herd of dairy cattle. *J. Theor. Biol.* **2007**, *245*, 44–58. [CrossRef]
112. Fenlon, D.R. Wild birds and silage as reservoirs of Listeria in the agricultural environment. *J. Appl. Bacteriol.* **1985**, *59*, 537–543. [CrossRef]
113. Mohammed, H.O.; Stipetic, K.; McDonough, P.L.; Gonzalez, R.N.; Nydam, D.V.; Atwill, E.R. Identification of potential on-farm sources of *Listeria monocytogenes* in herds of dairy cattle. *Am. J. Vet. Res.* **2009**, *70*, 383–388. [CrossRef]
114. Walland, J.; Lauper, J.; Frey, J.; Imhof, R.; Stephan, R.; Seuberlich, T.; Oevermann, A. *Listeria monocytogenes* infection in ruminants: Is there a link to the environment, food and human health? A review. *Schweiz Arch. Tierheilkd.* **2015**, *157*, 319–328. [CrossRef]
115. Hellstrom, S.; Laukkanen, R.; Siekkinen, K.M.; Ranta, J.; Maijala, R.; Korkeala, H. *Listeria monocytogenes* contamination in pork can originate from farms. *J. Food Prot.* **2010**, *73*, 641–648. [CrossRef]
116. Deutz, A.; Gasteiner, J.; Buchgraber, K. *Fütterung von Reh- und Rotwild: Ein Praxisratgeber*; Leopold Stocker Verlag: Graz, Austria, 2009.
117. Resch, R.; Pötsch, E.; Klansek, E.; Gahr, F.; Leitner, A.; Rothmann, G.; Stein, M.; Buchgraber, K. Beste Heu- und Silagequalitäten für Reh- und Rotwild. *Tirol. Jagdaufseher* **2017**, *33*.
118. Bundesverband Österreichischer Wildhalter. Fütterung. Available online: https://wildhaltung.at/haltung/fuetterung/ (accessed on 10 November 2022).
119. Schoder, D.; Wagner, M. Growing awareness of asymptomatic carriage of *Listeria monocytogenes*. *Wien Tierarztl. Monat.* **2012**, *99*, 322–329.
120. Tahoun, A.; Abou Elez, R.M.M.; Abdelfatah, E.N.; Elsohaby, I.; El-Gedawy, A.A.; Elmoslemany, A.M. *Listeria monocytogenes* in raw milk, milking equipment and dairy workers: Molecular characterization and antimicrobial resistance patterns. *J. Glob. Antimicrob. Resist.* **2017**, *10*, 264–270. [CrossRef] [PubMed]
121. Akkaya, L.; Alisarli, M.; Cetinkaya, Z.; Kara, R.; Telli, R. Occurrence of *Escherichia coli* O157: H7/O157, *Listeria monocytogenes* and *Salmonella* spp. in beef slaughterhouse environments, equipment and workers. *J. Muscle Foods* **2008**, *19*, 261–274. [CrossRef]
122. Wilson, B. The Rise of the Fridge. 2011. Available online: https://www.telegraph.co.uk/foodanddrink/8605239/The-rise-of-the-fridge.html (accessed on 30 August 2019).
123. Racine, E.F.; Mumford, E.A.; Laditka, S.B.; Lowe, A.E. Understanding characteristics of families who buy local produce. *J. Nutr. Educ. Behav.* **2013**, *45*, 30–38. [CrossRef]
124. Schiermeier, Q. Eat less meat: UN climate-change report calls for change to human diet. *Nature* **2019**, *572*, 291–292. [CrossRef]
125. Ripple, W.J.; Wolf, C.; Newsome, T.M.; Galetti, M.; Alamgir, M.; Crist, E.; Mahmoud, M.I.; Laurance, W.F.; 15,364 Scientist Signatories from 184 Countries. World Scientists' Warning to Humanity: A Second Notice. *BioScience* **2017**, *67*, 1026–1028. [CrossRef]
126. Sitaker, M.; Kolodinsky, J.; Jilcott Pitts, S.B.; Seguin, R.A. Do entrepreneurial food systems innovations impact rural economies and health? Evidence and gaps. *Am. J. Entrep.* **2014**, *7*, 3–16.
127. Schoder, D.; Strauss, A.; Szakmary-Brändle, K.; Wagner, M. How safe is European Internet cheese? A purchase and microbiological investigation. *Food Control* **2015**, *54*, 225–230. [CrossRef]
128. Leones, J. Tourism Trends and Rural Economic Impacts. In *Direct Farm Marketing and Tourism Handbook*; University of Arizona, Department of Agricultural and Resource Economics, Ed.; University of Arizona: Tucson, AZ, USA, 2005; Available online: https://farmanswers.org/Library/Record/tourism_trends_and_rural_economic_impacts (accessed on 10 November 2022).
129. Hinrichs, C.C. Embeddedness and local food systems: Notes on two types of direct agricultural market. *J. Rural Stud.* **2000**, *16*, 295–303. [CrossRef]
130. Verhaegen, I.; Van Huylenbroeck, G. Costs and Benefits for Farmers Participating in Innovative Marketing Channels for Quality Food Products. *J. Rural Stud.* **2001**, *17*, 443–456. [CrossRef]
131. Pierezan, M.D.; Marchesan Maran, B.; Marchesan Maran, E.; Verruck, S.; Pimentel, T.C.; Gomes da Cruz, A. Relevant safety aspects of raw milk for dairy foods processing. *Adv. Food Nutr. Res.* **2022**, *100*, 211–264. [CrossRef] [PubMed]
132. Oliver, S.P.; Boor, K.J.; Murphy, S.C.; Murinda, S.E. Food safety hazards associated with consumption of raw milk. *Foodborne Pathog. Dis.* **2009**, *6*, 793–806. [CrossRef]

133. Abou-Eleinin, A.A.; Ryser, E.T.; Donnelly, C.W. Incidence and seasonal variation of Listeria species in bulk tank goat's milk. *J. Food Prot.* **2000**, *63*, 1208–1213. [CrossRef] [PubMed]
134. Bellemare, M.F.; Nguyen, N.J. Farmers Markets and Food-Borne Illness. *Am. J. Agric. Econ.* **2018**, *100*, 676–690. [CrossRef]
135. European Commission. Farm to Fork Strategy. Available online: https://food.ec.europa.eu/horizontal-topics/farm-fork-strategy_en (accessed on 10 November 2022).
136. CEEweb. DM-Small-Scale-Farming-Legislative-Problems. Available online: http://www.ceeweb.org/wp-content/uploads/2012/11/DM-small-scale-farming-legislative-problems.pdf (accessed on 26 August 2019).
137. James, C.; Onarinde, B.A.; James, S.J. The Use and Performance of Household Refrigerators: A Review. *Compr. Rev. Food Sci. Food Saf.* **2017**, *16*, 160–179. [CrossRef]
138. Schoder, D.; Rossmanith, P.; Glaser, K.; Wagner, M. Fluctuation in contamination dynamics of *L. monocytogenes* in quargel (acid curd cheese) lots recalled during the multinational listeriosis outbreak 2009/2010. *Int. J. Food Microbiol.* **2012**, *157*, 326–331. [CrossRef]
139. European Commission. Commission Notice on the Implementation of Food Safety Management Systems Covering Good Hygiene Practices and Procedures Based on the HACCP Principles, Including the Facilitation/Flexibility of the Implementation in Certain Food Businesses; 2022/C 355/01. Available online: https://eur-lex.europa.eu/legal-content/EN/TXT/?uri=OJ:C:2022:355:TOC (accessed on 12 December 2022).
140. Nüesch-Inderbinen, M.; Bloemberg, G.V.; Müller, A.; Stevens, M.J.A.; Cernela, N.; Kollöffel, B.; Stephan, R. Listeriosis Caused by Persistence of *Listeria monocytogenes* Serotype 4b Sequence Type 6 in Cheese Production Environment. *Emerg. Infect. Dis.* **2021**, *27*, 284–288. [CrossRef]

Article

The Heat Is On: Modeling the Persistence of ESBL-Producing *E. coli* in Blue Mussels under Meal Preparation

Kyrre Kausrud [1], Taran Skjerdal [1,*], Gro S. Johannessen [1], Hanna K. Ilag [1,2] and Madelaine Norström [1]

[1]   Norwegian Veterinary Institute, 1431 Ås, Norway
[2]   Department of Microbiology, Oslo University Hospital, 0424 Oslo, Norway
*   Correspondence: taran.skjerdal@vetinst.no

Abstract: Pathways for exposure and dissemination of antimicrobial-resistant (AMR) bacteria are major public health issues. Filter-feeding shellfish concentrate bacteria from the environment and thus can also harbor extended-spectrum β-lactamase—producing *Escherichia coli* (ESBL *E. coli*) as an example of a resistant pathogen of concern. Is the short steaming procedure that blue mussels (*Mytilus edulis*) undergo before consumption enough for food safety in regard to such resistant pathogens? In this study, we performed experiments to assess the survival of ESBL *E. coli* in blue mussel. Consequently, a predictive model for the dose of ESBL *E. coli* that consumers would be exposed to, after preparing blue mussels or similar through the common practice of brief steaming until opening of the shells, was performed. The output of the model is the expected number of colony forming units per gram (cfu/g) of ESBL *E. coli* in a meal as a function of the duration and the temperature of steaming and the initial contamination. In these experiments, the heat tolerance of the ESBL-producing *E. coli* strain was indistinguishable from that of non-ESBL *E. coli*, and the heat treatments often practiced are likely to be insufficient to avoid exposure to viable ESBL *E. coli*. Steaming time (>3.5–4.0 min) is a better indicator than shell openness to avoid exposure to these ESBL or indicator *E. coli* strains.

Keywords: heat treatment; *E. coli*; mussels; ESBL; AMR; exposure models

Citation: Kausrud, K.; Skjerdal, T.; Johannessen, G.S.; Ilag, H.K.; Norström, M. The Heat Is On: Modeling the Persistence of ESBL-Producing *E. coli* in Blue Mussels under Meal Preparation. *Foods* **2023**, *12*, 14. https://doi.org/10.3390/foods12010014

Academic Editor: Frans J.M. Smulders

Received: 18 November 2022
Revised: 6 December 2022
Accepted: 9 December 2022
Published: 20 December 2022

1. Introduction

Shellfish such as blue mussels are consumed after a short heat treatment. As shellfish filter seawater, they accumulate microbes and the effectivity of the heat treatment for decontamination is therefore critical for food safety.

Antimicrobial-resistant (AMR) bacteria in seafood represent a potential risk for human beings by two main mechanisms, either clonal transfer of resistant bacteria or by horizontal gene transfer (HGT) of mobile genetic elements (MGEs) to previously susceptible pathogens. The emergence of successful multi-drug-resistant (MDR) variants of *Escherichia coli* and *Klebsiella pneumoniae*, belonging to certain clonal lineages, has contributed to the rapid global spread of extended-spectrum beta lactamase-producing Gram negative bacteria (ESBLs) and carbapenemases [1]. These clones are considered "global high risk clones" and have an excellent ability to colonize human hosts, disseminate and cause infections, with *E. coli* Sequence type (ST) 131 and *K. pneumoniae* ST258 as pertinent examples [2]. To be better equipped for the emerging AMR challenge, a thorough investigation of transmission routes and reservoirs is needed. WHO underlines the knowledge gap of the food chain in transmission of AMR bacteria, and AMR bacteria from seafood have been identified by EFSA as an issue for monitoring [3]. ESBL-producing *E. coli* is one of several emerging AMR microbes that have been detected in blue mussels [4,5]. The origin of such resistances may be both from human or animal sources contaminating seawater [4]. The filtration rate of water in blue mussels is temperature dependent, but at 15 °C it may exceed 120 L of water per day [6]. They therefore constitute potential hot spots for accumulating pathogenic and antimicrobial-resistant bacteria from the marine environment. If ESBL organisms accumulated from the environment survive the light–heat treatment that is traditionally preferred for mussels before consumption, AMR genetic elements can be transferred to the human microbial community. Few prevalence studies have estimated the occurrence of AMR in blue mussels; however, the occurrence of AMR in shellfish will most probably reflect the occurrence in the environment where they have been grown [4,5,7,8].

The potential for blue mussels to be a significant source of ESBL-producing *E. coli* to human beings is unknown. A study performed under the Norwegian monitoring program for antimicrobial resistance in the veterinary and food production sectors (NORM-VET) in 2016 reported that 4.2% of *E. coli* isolates obtained from bivalve molluscs (n = 261) in Norway were resistant to at least one antibiotic, while the prevalence of resistance to three or more antibiotics was 1.4% [9]. By using a selective screening methodology, 3.3% of the 391 samples showed resistance to third-generation cephalosporins and ten of these carried the globally common plasmid-encoded ESBL resistance gene blaCTX-M-15 [9].

Shellfish such as oysters and mussels are often consumed raw or undercooked for culinary reasons, which may pose a risk for the consumer [10]. In addition, consumption of wild-harvested mussels, i.e., uncontrolled mussels, occurs in many coastal areas, particularly during vacation times. In these cases, the heat treatment is the only hurdle for ESBL exposure as the local contamination levels may be unknown or disregarded. Commercially produced blue mussels have a food safety regulation limit of maximum 10 *E. coli*/g by the end of manufacturing process for direct human consumption (Commission regulation (EC) No 2073/2005). Class A shellfish by harvest should not contain more than 230 cfu/g of *E. coli* by harvest. A recommended practice has been to move shellfish with higher *E. coli* concentrations to cleaner water until the concentration falls below this level or they will undergo an industrial heat treatment before being marketed [11]. Whether adhering to the required limit is enough to avoid ESBL *E. coli* exposure may depend on their initial concentration and how well they survive heat treatment.

It is therefore a need for knowledge about the trade-off between safety and preferred sensory quality, i.e., the potential for survival of both *E. coli* and ESBL-producing *E. coli* and minimum heating conditions for elimination of these microbes in shellfish. There exists little knowledge about the persistence of viable ESBL-producing *E. coli* in different food matrixes where only light–heat treatments are performed before consumption, but both the maximum obtained temperature within the mussels as well as the duration of certain temperatures will likely have an impact on bacterial survival rates

The aim of this study was therefore to assess the survival of ESBL-producing *E. coli* in blue mussels following different heat treatment regimes, and to develop a corresponding exposure model and tool for risk assessment and guideline development. To achieve this, we conducted experiments inoculating live shellfish with *E. coli* as an indicator for ESBL-producing *E. coli* to avoid unacceptable contamination risks and used ESBL-producing *E. coli* in controlled heat treatment experiments comparing it to *E. coli* to verify their role as an indicator and accumulate additional data on heat inactivation.

The model for the mussel production chain developed in this work incorporates two sets of experiments. The first set of experiments involves inoculating live mussels with non-ESBL-producing *E. coli* by allowing them to naturally filter contaminated water in an aquarium. This experiment could, however, not utilize ESBL strains due to biohazard procedures, and thus a second set of experiments addressed this by homogenizing a mix of mussel flesh and either non-ESBL or ESBL-producing *E. coli* in a series of heat-resistant plastic bags.

2. Materials and Methods

Blue mussels for the study were purchased at a local supermarket, where they had been stored on ice. The mussels were transported to the lab and within 45 min placed at 5 °C or in the aquarium for acclimatization.

Our approach consisted of two main steps:

Inoculating live blue mussels with *E. coli* in an aquarium and then steaming them for different lengths of time in a kettle, thus closely simulating normal consumer pro-cedure and materials. (See Sections 2.1–2.3 and 2.5).

Inoculating raw de-shelled blue mussel flesh with non-ESBL-producing (indicator) *E. coli* and an ESBL-producing *E. coli*, respectively, in sealed plastic bags and subjecting them to heat treatments at various durations and controlled temperatures in water baths (See Sections 2.1, 2.2, 2.4 and 2.5).

All experiments and analyses were conducted at the Norwegian Veterinary Insti-tute facilities in Oslo.

2.1. Preparation of Inoculum

For the inoculation studies described below, indicator *E. coli* (three isolates, 2016-22-55-1-1-1-1, 2016-01-4162-1-1-1-1 and 2016-01-4220-1-1-1-1) and ESBL-producing *E. coli* (ESBL) (two isolates, 2016-01-4162-1-3-1-1 and 2016-01-4220-1-3-1-1) were used. The five isolates were all isolated from blue mussels analyzed previously [9] and kindly provided by NORM-VET.

One isolate of indicator *E. coli* (2016-22-55-1-1-1-1) originating from a blue mussel purchased from a retail store was used for inoculation of the water in the aquarium experiment. Briefly, the inoculum was prepared from frozen glycerol stocks where a loopful (1 µL) was plated directly from the stock onto a blood agar plate (bovine blood) that was incubated at 37 ± 1 °C overnight. A single colony of *E. coli* from the blood agar was added to 100 mL Buffered Peptone Water (BPW-ISO, OXOID) and incubated over night at 37 ± 1 °C. The overnight broth culture was equally distributed in four 50 mL sterile tubes and washed twice by centrifugation for 10 min at $3800 \times g$ (Beckman GS-15R Centrifuge), removal of the supernatant and resuspension of the bacterial pellet in 10 mL 0.9% saline water. After the second wash, the pellets were resuspended in 10 mL 0.9% saline separately prior to adding to the aquarium.

For the experimental inoculation of blue mussel flesh, samples were spiked with two isolates of indicator *E. coli* (2016-01-4162-1-1-1-1 and 2016-01-4220-1-1-1-1) and two strains of ESBL-producing *E. coli* (2016-01-4162-1-3-1-1 and 2016-01-4220-1-3-1-1), respectively. Both ESBL-producing *E. coli* harbored blaCTX-M-15, where strain 2016-01-4162-1-3-1-1 harbored blaCTX-M-15 alone, while strain 2016-01-4220-1-3-1-1 also had blaCMY-2. These isolates originated from two samples in which both an indicator *E. coli* as well as an ESBL isolate had been detected through selective screening within the NORM-VET program. The inocula were prepared from frozen glycerol stocks by plating a loopful of stock on blood agar plates followed by incubation at 37 ± 1 °C overnight. A single colony from each isolate was transferred to separate tubes of 10 mL BPW-ISO and incubated as described above. After incubation, the two isolates of indicator *E. coli* and the two strains of ESBLs *E. coli* were mixed to an *E. coli* mix and an ESBL -*E. coli* mix, respectively, and used for direct inoculation of samples of blue mussel flesh.

2.2. Aquarium Experiment

Artificial seawater (3%) was made by adding 2100 g Red Sea Salt (© 2020 Red Sea) to approximately 58 L ice and cold 35 L tap water to an aquarium (equipped with an electric pump, clean but not sterile). The aquarium was situated in a room with a constant temperature of 16 °C. Immediately after the preparation of the artificial seawater, a total of 90–100 blue mussels were transferred to the aquarium for acclimation for 24 h before 40 mL inoculum of $\sim 2 \times 10^8$ pr mL *E. coli* (2016-22-55-1-1-1-1) was added. The aquarium experiment was carried out three times.

Water samples were taken after the *E. coli* overnight broth culture was added and before the blue mussels were harvested 17 h after inoculation. After being removed from the aquarium the mussels were brought to the laboratory on ice. Each was marked with a waterproof marker and kept at 5 °C until steaming. Uninoculated blue mussels with temperature loggers (Signatrol SL53T Temperature logger 0/125 °C) were included in the pot during steaming in order to obtain information on the temperature inside the mussel shells flesh during the cooking period.

There was no growth of *E. coli* from the negative controls. A total of 10 inoculated blue mussels were transferred to a pot with approximately 70–80 not-inoculated blue mussels and steamed for pre-determined periods of time (30 s intervals from 60 to 210 s). After steaming, the flesh from the inoculated blue mussels were allowed to cool in room temperature before being distributed in Stomacher bags, one bag per mussel, for quantitative analyses of *E. coli* as described in Section 2.4 below. A total of 10 uninoculated blue mussels and 10 inoculated blue mussels prior to steaming were also distributed in Stomacher bags, one bag per mussel, for quantitative analysis for *E. coli* as negative controls and to check the level of *E. coli* present in the inoculated mussels, respectively.

2.3. Heat Treatment with Inoculated Blue Mussel Flesh

For the experiments with the heat treatment of the inoculated flesh from blue mussels, the flesh was obtained from blue mussels purchased from a local supermarket (see above). Aliquots of 10 g were distributed in separate 400 mL Stomacher bags (Grade Blender Bags, Standard 400) and inoculated separately with 100 µL of ESBL mix (2016-01-4162-1-3-1-1 and 2016-01-4220-1-3-1-1) or *E. coli* (2016-01-4162-1-1-1-1 and 2016-01-4220-1-1-1-1) mix (prior to heat sealing of the bags (for concentration of bacteria in the inocula). Before sealing, as much air as possible was squeezed out of the bags, and the mussel flesh was not flattened beyond this before heat treatment, retaining the effect of slow heat transfer within the mussel during cooking. The inoculated bags were then stored at 5 °C for at least 30 min prior to heat treatment. The heat treatment was carried out by completely submerging the parts of the bags with blue mussel flesh in a water bath (Nüve BM 30) at fixed temperatures (55 °C, 65 °C and 75 °C) for pre-determined periods of time (20–270 s). The bags were allowed to cool at room temperature analogous to the whole mussels prior to quantitative analyses for *E. coli* and ESBL, respectively, as described in Section 2.4.

Temperature loggers were included in similar samples with uninoculated blue mussel flesh.

Inoculated samples were analyzed for *E. coli* and ESBL *E. coli* prior to heat treatment to estimate the initial concentration in the mussel flesh. Uninoculated control samples without heat treatment were analyzed for *E. coli* and ESBL.

2.4. Microbiological Analyses

2.4.1. Quantitative Analyses of Indicator *E. coli* and ESBL-Producing *E. coli*

The blue mussel flesh was weighed and diluted 1:10 by adding BPW-ISO to the Stomacher bag, prior to stomaching or shaking for 30 s to two minutes to homogenize. The samples were further serially diluted in BPW-ISO (aquarium experiment) or Peptone Saline (1 g peptone, 8.5 g NaCl/L) (heat treatment with inoculated blue mussel flesh) and 100 µL of the appropriate dilutions was plated out with a L-rod on TBX (Oxoid) or MacConkey (Becton Dickinson) supplemented with 1 mg/L cefotaxime (Sigma) (MaC-CO) for quantification of *E. coli* or ESBL-producing *E. coli*, respectively. In order to obtain a detection limit of 10 cfu/g, one mL of the initial homogenate was distributed equally on the surface of three plates. The TBX and MaC-CO plates were incubated at 37 ± 1 °C and 41.5 ± 1 °C, respectively. Typical colonies on the different agars were counted and a selection of colonies was further confirmed (Section 2.4.3).

2.4.2. Detection of *E. coli* and ESBL-Producing *E. coli* in Enrichment of Samples

After serial dilutions and plating had been made for quantitative analysis, the rest of the initial homogenate (10 g sample and 90 mL BPW-ISO) from the heat treatment experiment and the aquarium experiment (flesh from one blue mussel diluted 1:10 with BPW-ISO) were incubated at 37 ± 1 °C overnight, followed by plating of a loopful (10 µL) of enrichment on TBX or MaC-CO, depending on which organism analyzed for. All samples were enriched, but plating was only performed if the result from the quantitative analysis was below the detection limit (i.e., <10 cfu/g). The plates were inspected for typical colonies and a selection of colonies was pure-cultured on blood agar and further confirmed (Section 2.4.3).

2.4.3. Confirmation by MALDI-TOF and PCR

A selection of colonies from both the blue mussel inoculation experiment and the heat treatment experiment were confirmed as *E. coli* by MALDI Biotyper MS (MALDI-TOF MS, Bruker Daltronics GmbH). Presumptive ESBL-producing *E. coli* were confirmed by using PCR specific for the genes harbored by the strains [11–13].

2.5. Statistical Analysis and Modeling

Data analysis was conducted using the R 3.5 software [14] with the mgcv package for generalized additive models (GAMs) with smoothness estimators, and application of results was performed using R Shiny [15,16].

We first used a binomial model of steam time vs. probability *P* of mussels opening so that the estimate proportion *EP* of mussels opened for a given time point (Figure 1):

$$EP(\text{Open}|\text{Time}) = \frac{1}{1 + e^{f(T)}} \tag{1}$$

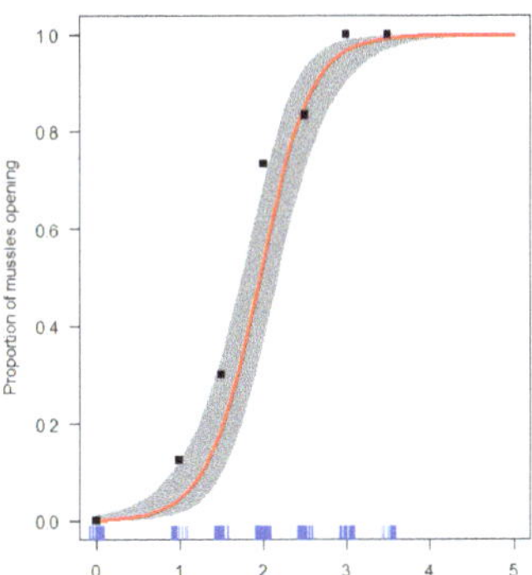

Figure 1. The effect of steaming time on the proportion of mussels opening.

Data were treated to account for "worst case scenarios", so when *E. coli* were not detected by enumeration (i.e quantitative analysis with detection level 10 cfu/g), but only after enrichment, *E. coli* numbers were set to 9 cfu/g.

When modeling how many bacteria that would remain viable after heat treatment, repeated measures on the same mussel were not feasible. Hence, estimates of contamination level had to be made independently on different individual samples. As a measure Pe of proportional survival of bacteria, this means taking the cfu/g for a given sample at time t minutes of the heat treatment and dividing it by the average cfu/g found in inoculated samples at t = 0, i.e., before any heat above room temperature was applied, rounding any number

$$P_{CFU,t} = \min\left\{ \frac{cfu_t}{cfu_0}, 1 \right\} \tag{2}$$

as we are assuming no significant further bacterial growth happening in the few minutes between initial samples being taken and heat treatment being applied, and thus sample variance being the cause for any number over 1.

The proportion of the remaining viable cfus were estimated by the GAM regression models with the non-parametric penalized thin spline model with quasi-binomial error distribution to allow for overdispersion and allow the estimation of survival curves to be data-driven and flexible rather than being bound to specific a priori survival functions such as the Weibull [17] or exponential. Logistic regressions on the proportion as given in the results and discussion all follow the general format of

$$Y = \ln\left(\frac{1}{1 - P_{CFU}}\right) \tag{3}$$

$$Y = b + B(X) + f(X,k) + \varepsilon \tag{4}$$

where b is a constant (intercept), B a vector of constants and X a matrix of explanatory variables. f(X) denotes a set of zero or more penalized regression splines [18] with smoothing parameters selected by the GCV criterion limited upwards to a maximum number of degrees of freedom k, and the conditional distribution of the response a quasi-binomial distribution

$$P(Y = k) = \binom{n}{k} p(p + k\phi)^{k-1}(1 - p - k\phi)^{n-k} \tag{5}$$

which is similar to the binomial distribution except for the parameter ϕ, which captures excess variance. Some models incorporate only linear predictors B(X), others non-linear effects (f(X,k)), and this is made clear in the text for each relevant model.

When modeling the number of cfu/g directly, not as a proportion remaining of the initial concentration after inoculation, the same model framework was used. Except for the response variable

$$Y = \ln(cfu) \tag{6}$$

And when the conditional distribution of the response is a quasi-Poisson distribution [19], i.e., where if

$$E(Y) = \mu \tag{7}$$

$$Var(Y) = \theta\mu \tag{8}$$

$$P(Y = k) = \frac{\lambda^k e^{-\lambda}}{k!} \tag{9}$$

making it a Poisson distribution with an overdispersion parameter θ regulates the variance/mean relationship.

When estimating the exposure, we assumed that the observed concentrations of bacteria were representative of an underlying probability density. We then estimated smoothed empirical probability densities on the observed concentrations and simulated expected exposures by drawing 20 hypothetical shells as a typical meal from these distributions, taking the average bacterial concentration (cfu/g) and multiplying by 250, as 250 g is assumed to be a typical portion of blue mussels.

For temperature logger data, a set of algorithms was developed to identify and homogenize logger time series, but an element of manual delineating was most efficient as some treatment times had not left peaks identifiable beyond noise and temperature fluctuations between refrigerators, water and air.

3. Results

3.1. Experiments

In traditional preparation, where a culinary value is placed on minimizing heat exposure, looking for the mussel shells opening under steaming is a common indicator for when they can be taken off the heat. The opening rate is well described by a logistic model of opening as a function of the steaming time T, which explains about 70% of the variance in openness status for mussels (Methods Section 2.2 and Equation (1), Figure 1). The opening rate as a functioning of heat (i.e., steam) exposure was consistent between experiments, and showed no statistically significant differences between steaming batches.

Black squares are averages. y-axis values represent the fraction of mussels open (i.e., 0 = all closed; 1 = all open). The blue rug lines indicate datapoints along the time axis.

Model fit is shown in red, with a 2SE (SE = Standard error) confidence interval for model fit shaded gray.

Whether or not a mussel is fully open is highly correlated with the proportion P_{CFU} (see Section 2.5 of *E. coli* being cultivable from that mussel, but alone it explains only about 40% of the variance in the proportion of bacteria being viable after steaming. Including the starting concentration (mean cfu/g in samples from the same batch taken before steaming) as an explanatory variable was not significant, suggesting that at these concentrations the survival rate of bacteria was independent of concentration. See Figure 2. We see a strong reduction in average bacterial concentrations as shells open, but also that even some opened shells retain fairly high bacterial concentrations.

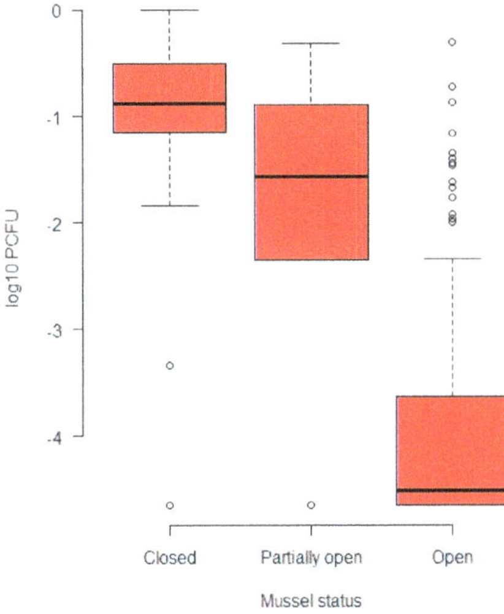

Figure 2. The effect of steam-exposed shell openness status on bacterial concentrations (cfu/g) relative to raw (unopened) mussels (see Equation (2)).

Simulating meal exposures from eating 20 mussels (assumed to be a typical meal) suggests a 10% risk of ingesting a dose over 10% of the original (pre-steaming) concentration of bacteria. If partially opened mussels are included in the meal, the risk increases significantly, as 10% of meals will contain bacteria corresponding to 20% or more of the original concentration in the meal as a whole. See Figure 3.

3.1.1. Steaming Time Effects on Bacterial Concentration

Modeling the cfu/g directly in an overdispersed Poisson regression model (see Section 2.5) using the steaming time and the mean initial bacterial concentration in the unsteamed mussels (cfug0) as the variables explains about 60% of the variance. The remaining variation is likely to be due to the cooling period after the steaming and until the analysis of the sample, and to the random variation in the sampling and culturing. The open status loses all significant explanatory power when the steaming time is allowed to enter as a non-linear effect (see Section 2.5). After the mussels had been steamed for >3.5 min, *E. coli* was not detected in any of the samples. We thus obtained a range of steam times predicted to bring exposure down to regulation levels depending on meal size and contamination level estimated from this. See Figure 4.

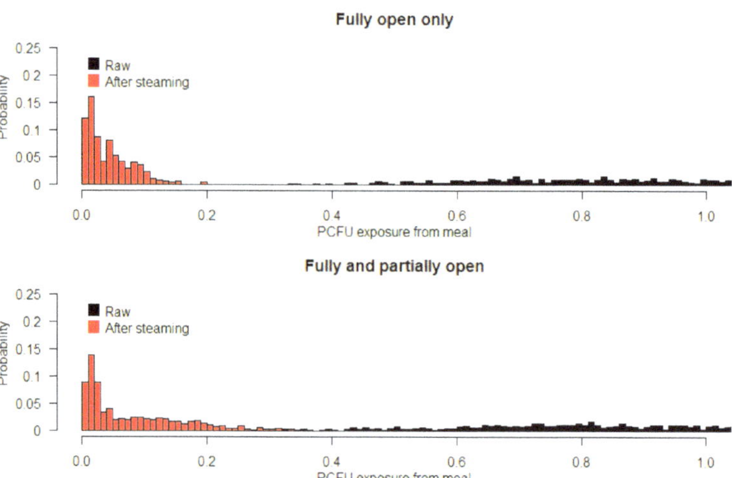

Figure 3. Simulated doses of *E. coli* in a meal size consisting of 20 randomly drawn open mussels (**top** panel) as a proportion of original (un-steamed) contamination (Equation (2)). The red bars are a histogram of such meals consisting of mussels steamed to opening, compared to the load from a similar number of raw shells (black bars). The risk is markedly higher if partially opened mussels are included in the meal at the same frequency they were found in this experiment (**bottom** panel).

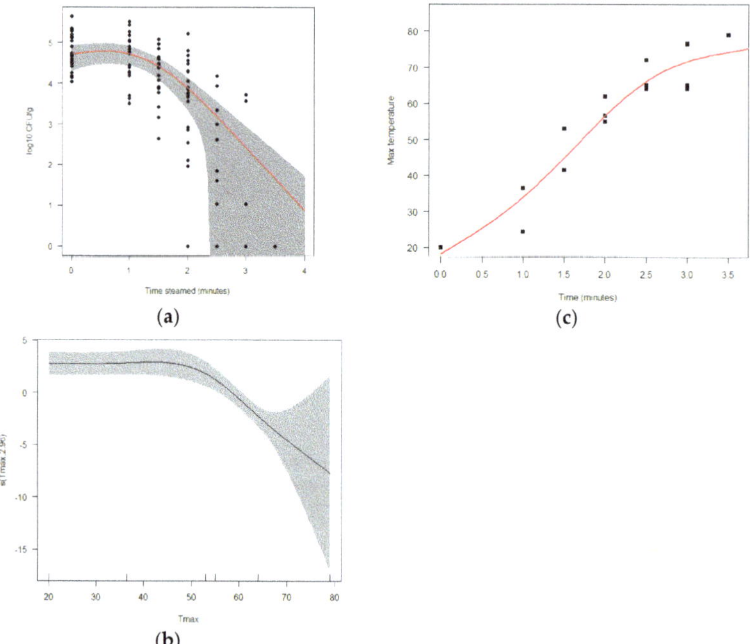

Figure 4. (**a**) Log_{10} bacterial concentrations of *E. coli* as a function of steaming times. The red line gives the best model for inactivation by steaming, with the 2SE confidence interval in gray. (**b**) Maximum temperature registered on loggers glued inside mussel shells as a function of steaming time. (**c**) The effect of maximum temperature on viable bacterial concentration (see equations in Section 2.5). We see that it suggests inactivation from a threshold value a little over 55 °C.

These steam times map closely to the maximum temperature found by temperature loggers to be attained within steamed mussels and exchanging the max temperature for the time had the same explanatory power. The effect of the steaming time is thus suggested to be largely mediated through the maximum temperature and the time over which it is applied. However, using maximum temperature as an explanatory variable does suggest that the effect of temperature is strongly non-linear and that the *E. coli* strains used in the present study start to be inactivated at temperatures exceeding 55°C. See Figure 4.

3.1.2. Heat Treatment

For comparisons of indicator strains of *E. coli* and ESBL producing *E. coli*, experiments spanning the temperatures relevant for steamed shells were chosen. Mussel flesh was therefore taken out of the shell prior to heating, and the mussel flesh was inoculated with the bacteria.

Blue mussel flesh was then inoculated directly and treated at constant temperatures for pre-determined periods.

At 55 °C, both ESBL-producing *E. coli* and indicator *E. coli* remained in high concentrations even after 270 seconds of heat treatment. Surprisingly, the germination rate seemed possibly even higher after warming.

At 65°C, there was more variation, for both indicator *E. coli* and ESBL-producing *E. coli* being detected in some of the samples that were treated for 90 to 240 s, but after 270 s at 65 °C, neither indicator *E. coli* nor ESBL-producing *E. coli* could be detected.

At 75 °C, the mussel flesh samples were generally negative after 40–60 s treatment, but some *E. coli* and ESBL-producing *E. coli* could be detected in some of the samples up to 100 s treatment. This could be explained by the uneven distribution of the inoculum in the samples. After 110 s at 75 °C, neither indicator *E. coli* nor ESBL-producing *E. coli* could be detected.

None of the control samples contained indicator *E. coli*, except for one sample from which indicator *E. coli* was detected after enrichment of the sample (<10 cfu/g).

When modeling the effects of heat treatment in water baths at constant temperature (see Section 2.5), we saw no robust effect from ESBL status on the proportion of bacteria remaining viable after heat treatment. We saw only a weak and not robustly significant trend towards *lower* survival for ESBL *E. coli* at the very lowest (55 °C) treatment, where bacterial survival rates were nevertheless very high. In general, survival seemed indistinguishable between our genotypes of ESBL *E. coli* and the indicator strains used here. We also saw that the survival time at the higher temperature treatments was very short, and that models using exposure time, temperature and ESBL status explain approximately 90% of the variance, leaving little noise. See Figure 5.

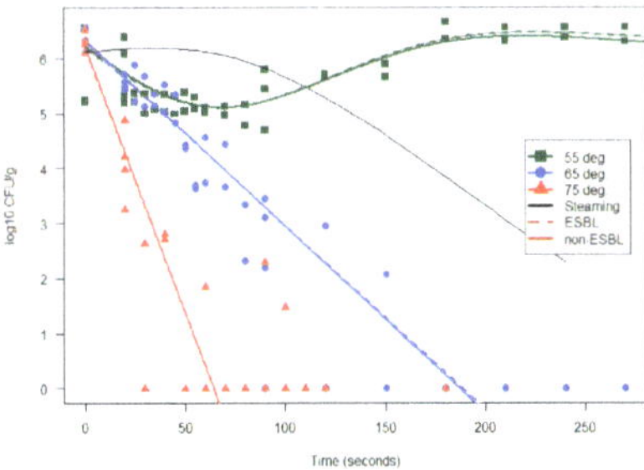

Figure 5. Concentrations (log(10) cfu/g) of *E. coli* and ESBL-producing *E. coli* during heat treatment in inoculated mussel flesh in constant-temperature water baths at 55 °C (green), 65 °C (blue) and 75 °C (red). ESBL and non-ESBL-producing *E.coli* inoculates are shown as solid and broken lines respectively. Only one line indicates overlap.

4. Discussion

It is mostly unknown how ESBL-producing *E. coli* and other AMR bacteria behave in raw and lightly cooked conditions. Conditions of stress, such as heat, may trigger several mechanisms in bacterial cells, e.g., adaptation, cellular repair, application of response mechanisms and enhanced virulence [19]. Several studies [20,21] have shown that sublethal food preservation stresses, such as heat and salt, can significantly alter phenotypic AMR in food-related pathogens such as *E. coli, Salmonella typhimurium, Staphylococcus aureus* and *Cronobacter sakazakii*.

For this study we successfully inoculated live blue mussels with *E. coli* by allowing the blue mussels to acclimatize in artificial seawater in the aquarium prior to adding *E. coli*. This is, to our knowledge, the first time an experimental study has been carried out to estimate the survival of *E. coli* as well as ESBLs during the steaming of blue mussels mimicking the cooking procedure in the consumer's home.

A notable finding is that the study showed that about 1.5% of the *E. coli* present before steaming is expected to survive if the mussels are only steamed until they are opened and not longer, and that the variation between mussels gives an overall likelihood of about 10% to ingest meals with an exposure corresponding to 10% of the pre-cooking bacterial concentration. It is more if half-opened mussels are included.

As the ESBL *E. coli* survived as well as the indicator strains in our other experiments, this means that the traditional preparation method cannot be trusted to inactivate AMR *E. coli* or other bacteria with similar heat inactivation profiles if they are present in the raw mussels. Considering the mussels ability to concentrate bacteria from the surrounding seawater, and that the time point when >95% are open probably represents an optimistic estimate for how long a consumer would keep heat treating, this suggests that they can be a significant source of human exposure to environmental AMR genotypes present in coastal waters or mussel farms, unless effective monitoring and/or heat treatment practices are in place.

The inactivation curves found in our water bath experiments seem consistent with previous reports on *E. coli* heat tolerance [22], where the "shoulder" before inactivation starts is too small to be measured for higher temperatures and probably reflects a combination of heat tolerance and a delay of heat penetrating into the mussel flesh for lower temperatures. No particularly robust "tails" show up in our data, except possibly for the 65 °C treatment where a weak tail effect may be present. For the 55 °C treatment, the inactivation never entered a tail phase, and for 75 °C and steam treatments the inactivation could not be distinguished from log-linear as time progressed. The steaming treatment on the other hand shows a significant "shoulder", or delay between putting the mussels in the pot and inactivation starting. As shown by the temperature loggers, this likely is due to the time it takes for the inside of mussels to attain a critical temperature seeming to be between 55 °C and 65 °C.

The survival of indicator and ESBL *E. coli* followed indistinguishable trajectories under heat treatment, indicating that thermal inactivation curves for *E. coli* can be used for risk exposure models of resistant isolates, as have been conducted in a recent risk exposure study of ground beef [23], which used the thermal inactivation curve in *E. coli* O157:H7 for hamburgers [23]. Nevertheless, the use of available data needs to be carefully assessed as the different food matrixes and food preparations will have impact of the survival of the specific agent under study [24].

A factor we did not have the opportunity to explore is the differences in heat tolerance between different genotypes of *E. coli*. While *E. coli* is often seen as a heat-sensitive organism, some strains are among the most heat-resistant of foodborne pathogens with D_{60} values >6 min [22,25,26], which suggests inactivation curves with considerably less steep slopes than we observe here are possible. Thus, our model should be treated as a guideline, keeping in mind that judging from the inactivation curves reflecting D_{55} and D_{65} in our experiment, we note that our strain seems to be representative of the most commonly tested of *E. coli* strains [22], but not the most heat tolerant. Further work should take this into account and base risk models for recommendations on inactivation on a wider range of strains found in the relevant environments.

Another factor of concern that needs addressing in this context is the possible transfer of MGEs remaining after thermal inactivation, which cannot be ruled out and needs further studies. Work on post-inactivation MGE transfer and inoculation studies addressing differences in heat tolerance between strains and any possible links between heat resistance and AMR phenotype require further study.

5. Conclusions

The present study has indicated that

- Shellfish prepared traditionally is a potential pathway for exposure to viable AMR bacteria concentrated from the environment.
- Inoculation studies mimicking natural bacterial accumulation and realistic preparation have been shown to be feasible and a useful model system.
- Consuming blue mussels only steamed to opening carries a significant risk of viable bacteria being present in concentrations just one order of magnitude reduced from the raw state.
- Steaming time (>3.5–4.0 min) is a better indicator than shell openness to avoid exposure to these ESBL or indicator *E. coli* strains.
- Further studies including more genotypes and relating them to what is found in the environment are needed.
- No difference in heat tolerance was found between ESBL *E. coli* and an indicator *E. coli* strain in the studied food matrix.

Author Contributions: Conceptualization, M.N., K.K., T.S. and G.S.J.; methodology, all authors; software, K.K.; formal analysis, K.K.; investigation, H.K.I.; resources, M.N.; data curation, K.K., M.N. and T.S.; writing—original draft preparation, all authors; writing—review and editing, K.K., T.S., G.S.J. and M.N.; visualization, K.K.; project administration, M.N.; funding acquisition, M.N. All authors have read and agreed to the published version of the manuscript.

Funding: This work was carried out in the RaDAR project as a part of the One Health European Joint Programme funded by Horizon 2020 (Grant agreement No 773830).

Data Availability Statement: Models and research data are available upon request to kyrre.kausrud@vetinst.no.

Conflicts of Interest: The authors declare no conflict of interest.

References

1. Dunn, S.J.; Connor, C.; McNally, A. The evolution and transmission of multi-drug resistant *Escherichia coli* and *Klebsiella pneumoniae*: The complexity of clones and plasmids. *Curr. Opin. Microbiol.* **2019**, *51*, 51–56. [CrossRef] [PubMed]
2. Nascimento, T.; Cantamessa, R.; Melo, L.; Fernandes, M.R.; Fraga, E.; Dropa, M.; Sato, M.I.; Cerdeira, L.; Lincopan, N. International high-risk clones of *Klebsiella pneumoniae* KPC-2/CC258 and *Escherichia coli* CTX-M-15/CC10 in urban lake waters. *Sci. Total Environ.* **2017**, *598*, 910–915. [CrossRef]
3. European Food Safety Authority (EFSA); Aerts, M.; Battisti, A.; Hendriksen, R.; Kempf, I.; Teale, C.; Tenhagen, B.-A.; Veldman, K.; Wasyl, D.; Guerra, B.; et al. Scientific report on the technical specifications on harmonised monitoring of antimicrobial resistance in zoonotic and indicator bacteria from food-producing animals and food. *EFSA J.* **2019**, *17*, 122. [CrossRef]
4. Rees, E.E.; Davidson, J.; Fairbrother, J.M.; St. Hilaire, S.; Saab, M.; McClure, J.T. Occurrence and antimicrobial resistance of *Escherichia coli* in oysters and mussels from Atlantic Canada. *Foodborne Pathog. Dis.* **2015**, *12*, 164–169. [CrossRef] [PubMed]
5. Singh, A.S.; Nayak, B.B.; Kumar, S.H. High Prevalence of Multiple Antibiotic-Resistant, Extended-Spectrum β-Lactamase (ESBL)-Producing *Escherichia coli* in Fresh Seafood Sold in Retail Markets of Mumbai, India. *Vet. Sci.* **2020**, *7*, 46. [CrossRef] [PubMed]
6. Kittner, C.; Riisgård, H.U. Effect of temperature on filtration rate in the mussel *Mytilus edulis*: No evidence for temperature compensation. *Mar. Ecol. Prog. Ser.* **2005**, *305*, 147–152. [CrossRef]
7. Kumar, S.; Lekshmi, M.; Parvathi, A.; Nayak, B.B.; Varela, M.F. Antibiotic Resistance in Seafood-Borne Pathogens. *Food Borne Pathog. Antibiot. Resist.* **2017**, *17*, 397–415. [CrossRef]
8. Van, T.T.H.; Chin, J.; Chapman, T.; Tran, L.T.; Coloe, P.J. Safety of raw meat and shellfish in Vietnam: An analysis of *Escherichia coli* isolations for antibiotic resistance and virulence genes. *Int. J. Food Microbiol.* **2008**, *124*, 217–223. [CrossRef] [PubMed]

9. Simonsen, G.S.; Urdahl, A.M. *NORM/NORM-VET 2016-Usage of Antimicrobial Agents and Occurrence of Antimicrobial Resistance in Norway*; Norwegian Veterinary Institute: Oslo, Norway, 2017.
10. Ottaviani, D.; Santarelli, S.; Bacchiocchi, S.; Masini, L.; Ghittino, C.; Bacchiocchi, I. Occurrence and characterization of Aeromonas spp. in mussels from the Adriatic Sea. *Food Microbiol.* **2006**, *23*, 418–422. [CrossRef] [PubMed]
11. World Health Organization. *Proposed Draft Code of Practice for Fish and Fishery Products*; WHO: Geneva, Switzerland, 1998.
12. Schmidt, G.V.; Mellerup, A.; Christiansen, L.E.; Ståhl, M.; Olsen, J.E.; Angen, Ø. Sampling and pooling methods for capturing herd level antibiotic resistance in swine feces using qPCR and CFU approaches. *PLoS ONE* **2015**, *10*, e0131672. [CrossRef] [PubMed]
13. Hasman, H.; Mevius, D.; Veldman, K.; Olesen, I.; Aarestrup, F.M. β-Lactamases among extended-spectrum β-lactamase (ESBL)-resistant Salmonella from poultry, poultry products and human patients in The Netherlands. *J. Antimicrob. Chemother.* **2005**, *56*, 115–121. [CrossRef] [PubMed]
14. R Development Core Team. *A Language and Environment for Statistical Computing, Version 3.5.3*; R Foundation for Statistical Computing: Vienna, Austria, 2019.
15. Chang, W.; Cheng, J.; Allaire, J.; Xie, Y.; McPherson, J. Package 'Shiny'. Available online: https://cran.r-project.org/web/packages/shiny/index.html (accessed on 12 August 2020).
16. Wood, S. mgcv: Mixed GAM Computation Vehicle with GCV/AIC/REML Smoothness Estimation. Available online: https://researchportal.bath.ac.uk/en/publications/mgcv-mixed-gam-computation-vehicle-with-gcvaicreml-smoothness-est (accessed on 5 June 2019).
17. Lorentzen, G.; Ytterstad, E.; Olsen, R.L.; Skjerdal, T. Thermal inactivation and growth potential of Listeria innocua in rehydrated salt-cured cod prepared for ready-to-eat products. *Food Control* **2010**, *21*, 1121–1126. [CrossRef]
18. Wood, S. Mixed GAM computation vehicle with GCV/AIC/REML Smoothness Estimation and GAMMs by REML/PQL. Available online: https://stat.ethz.ch/R-manual/R-devel/library/mgcv/html/mgcv-package.html (accessed on 5 May 2019).
19. Wesche, A.M.; Gurtler, J.B.; Marks, B.P.; Ryser, E.T. Stress, sublethal injury, resuscitation, and virulence of bacterial foodborne pathogens. *J. Food Prot.* **2009**, *72*, 1121–1138. [CrossRef] [PubMed]
20. McMahon, M.A.S.; Xu, J.; Moore, J.E.; Blair, I.S.; McDowell, D.A. Environmental stress and antibiotic resistance in food-related pathogens. *Appl. Environ. Microbiol* **2007**, *73*, 211–217. [CrossRef] [PubMed]
21. Al-Nabulsi, A.A.; Osaili, T.M.; Elabedeen, N.A.Z.; Jaradat, Z.W.; Shaker, R.R.; Kheirallah, K.A.; Tarazi, Y.H.; Holley, R.A. Impact of environmental stress desiccation, acidity, alkalinity, heat or cold on antibiotic susceptibility of Cronobacter sakazakii. *Int. J. Food Microbiol.* **2011**, *146*, 137–143. [CrossRef] [PubMed]
22. Li, H.; Gänzle, M. Some like it hot: Heat resistance of *Escherichia coli* in food. *Front. Microbiol.* **2016**, *7*, 1763. [CrossRef] [PubMed]
23. Juneja, V.K.; Snyder, O.P.; Marmer, B.S. Thermal destruction of *Escherichia coli* O157:H7 in beef and chicken: Determination of D- and Z-values. *Int. J. Food Microbiol.* **1997**, *35*, 231–237. [CrossRef] [PubMed]
24. Evers, E.G.; Pielaat, A.; Smid, J.H.; van Duijkeren, E.; Vennemann, F.B.C.; Wijnands, L.M.; Chardon, J.E. Comparative Exposure Assessment of ESBL-Producing *Escherichia coli* through Meat Consumption. *PLoS ONE* **2017**, *12*, e0169589. [CrossRef] [PubMed]
25. Duffy, G.; Walsh, C.; Blair, I.; McDowell, D. Survival of antibiotic resistant and antibiotic sensitive strains of E. coli O157 and E. coli O26 in food matrices. *Int. J. Food Microbiol.* **2006**, *109*, 179–186. [CrossRef] [PubMed]
26. Mercer, R.; Zheng, J.; Garcia-Hernandez, R.; Ruan, L.; Gänzle, M.; McMullen, L. Genetic determinants of heat resistance in *Escherichia coli*. *Front. Microbiol.* **2015**, *6*, 932. [CrossRef] [PubMed]

Article

Bacterial Skin Microbiota of Seabass from Aegean Fish Farms and Antibiotic Susceptibility of Psychrotrophic *Pseudomonas*

Ali Aydin [1,*], Mert Sudagidan [2], Zhanylbubu Mamatova [1], Mediha Nur Zafer Yurt [2], Veli Cengiz Ozalp [3], Jacob Zornu [4], Saraya Tavornpanich [4] and Edgar Brun [4]

[1] Department of Food Hygiene and Technology, Faculty of Veterinary Medicine, Istanbul University-Cerrahpasa, Avcilar, Istanbul 34320, Turkey
[2] KIT-ARGEM R&D Center, Konya Food and Agriculture University, Meram, Konya 42080, Turkey
[3] Department of Medical Biology, Medical School, Atilim University, Golbasi, Ankara 06830, Turkey
[4] Norwegian Veterinary Institute, 1433 Ås, Norway; jacob.zornu@vetinst.no (J.Z.)
* Correspondence: aliaydin@istanbul.edu.tr; Tel.: +90-532-473-77-79

Abstract: Farming seabass (*Dicentrarchus labrax*) is an essential activity in the Mediterranean basin including the Aegean Sea. The main seabass producer is Turkey accounting for 155,151 tons of production in 2021. In this study, skin swabs of seabass farmed in the Aegean Sea were analysed with regard to the isolation and identification of *Pseudomonas*. Bacterial microbiota of skin samples ($n = 96$) from 12 fish farms were investigated using next-generation sequencing (NGS) and metabarcoding analysis. The results demonstrated that Proteobacteria was the dominant bacterial phylum in all samples. At the species level, *Pseudomonas lundensis* was identified in all samples. *Pseudomonas, Shewanella,* and *Flavobacterium* were identified using conventional methods and a total of 46 viable (48% of all NGS+) *Pseudomonas* were isolated in seabass swab samples. Additionally, antibiotic susceptibility was determined according to standards of the European Committee on Antimicrobial Susceptibility Testing (EUCAST) and Clinical and Laboratory Standards Institute (CLSI) in psychrotrophic *Pseudomonas*. *Pseudomonas* strains were tested for susceptibility to 11 antibiotics (piperacillin-tazobactam, gentamicin, tobramycin, amikacin, doripenem, meropenem, imipenem, levofloxacin, ciprofloxacin, norfloxacin, and tetracycline) from five different groups of antibiotics (penicillins, aminoglycosides, carbapenems, fluoroquinolones, and tetracyclines). The antibiotics chosen were not specifically linked to usage by the aquaculture industry. According to the EUCAST and CLSI, three and two *Pseudomonas* strains were found to be resistant to doripenem and imipenem (E-test), respectively. All strains were susceptible to piperacillin-tazobactam, amikacin, levofloxacin, and tetracycline. Our data provide insight into different bacteria that are prevalent in the skin microbiota of seabass sampled from the Aegean Sea in Turkey, and into the antibiotic resistance of psychrotrophic *Pseudomonas* spp.

Keywords: seabass; microbiota; fish farms; *Pseudomonas*; antibiotic resistance

Citation: Aydin, A.; Sudagidan, M.; Mamatova, Z.; Yurt, M.N.Z.; Ozalp, V.C.; Zornu, J.; Tavornpanich, S.; Brun, E. Bacterial Skin Microbiota of Seabass from Aegean Fish Farms and Antibiotic Susceptibility of Psychrotrophic *Pseudomonas*. *Foods* **2023**, *12*, 1956. https://doi.org/10.3390/foods12101956

Academic Editor: Frans J.M. Smulders

Received: 15 December 2022
Revised: 9 April 2023
Accepted: 17 April 2023
Published: 11 May 2023

1. Introduction

Seafood, especially fish, is an increasingly important component of human diets. Thus, aquaculture is an important source of food suitable for human consumption [1], and could provide a sustainable supply of affordable seafood to an increasing global population. Mediterranean marine aquaculture grew exponentially during the last decades of the 20th century, though at a slower pace over the past 20 years or so [2]. European seabass (*Dicentrarchus labrax*) is the 31st most-reared fish in worldwide aquaculture [3]. Seabass production increased by 2.9% in 2020 and reached 243,900 tons globally [4]. More than 95% of the world's seabass and sea bream (*Sparus aurata*) production comes from aquaculture, of which, 97% accounts for the production in Mediterranean countries. Turkey and Greece are the primary producers, while Spain, France, Italy, Greece, and Turkey are the primary consumers [5].

Skin microbiota of fish species such as seabass have, however, hardly been investigated. To fill this knowledge gap, sampled seabass could be analyzed e.g., using next generation sequencing (NGS) whole genome sequencing and metabarcoding analysis. Such an approach would generate essential information on the profiles of both culturable and non-culturable microbial communities [6]. Furthermore, determining dominant microorganisms by NGS could contribute to the identification of pathogenic and/or potentially pathogenic bacteria in the aquaculture industry.

Although *Pseudomonas* species (including *P. aeruginosa, P. fluorescens, P. baetica, P. putida,* and *P. lundensis*) have been described as opportunistic human pathogens, many *Pseudomonas* species have also been associated with several diseases in farmed fish [7,8]. Additionally, psychrophilic *Pseudomonas* spp. cause spoilage of fishery products.

Apart from considerably limiting the success of aquaculture, the prevalence of fish diseases of microbial origin also necessitates the use of antibiotic treatments. Such treatments, particularly when applied without prudent justification, are known to cause the emergence of antibiotic-resistant bacteria [9]. Consequently, there is a continuous risk of the emergence of antibiotic resistance (AR) or multidrug resistance (MDR), i.e., the ability of a microorganism to withstand the action of one or more antimicrobial compounds [10]. Research has demonstrated the predominance and persistence of *Pseudomonas* spp. in, and on the surface of, seafood and in food processing plants, which reflects the ability of these microorganisms to withstand adverse conditions, including several antimicrobial treatments [11]. In addition, antibiotics are frequently used in the treatment of diseases in fish farming. Microbial communities on fish skin are highly variable, may be responsible for causing fish diseases, and may threaten the health of consumers [12]. Commonly, standard s of the European Committee on Antimicrobial Susceptibility Testing (EUCAST) [13] and Clinical and Laboratory Standards Institute (CLSI) [14] are used to determine the antibiotic susceptibility of bacteria in food intended for human consumption.

This study aimed to use NGS and metabarcoding analysis to determine the bacterial microbiota of seabass skin samples collected from fish farms in different parts of the Aegean Sea of Turkey. In addition, agar diffusion assays were performed to evaluate the antibiotic susceptibility against 11 antibiotics (piperacillin-tazobactam, gentamicin, tobramycin, amikacin, doripenem, meropenem, imipenem, levofloxacin, ciprofloxacin, norfloxacin, and tetracycline) from five antibiotics groups (penicillins, aminoglycosides, carbapenems, fluoroquinolones, and tetracyclines). Based on results from agar-disc diffusion assays and the E-test, Minimum Inhibitory Concentration (MIC) values were utilized to evaluate resistant psychrotrophic *Pseudomonas* strains in accordance with EUCAST and CLSI criteria [13,14].

2. Materials and Methods

2.1. Sampling

During June 2022, 96 seabass with an average weight of 300 g and average length of 220 mm were obtained from fish farms in 12 locations (8 samples per farm) in the Aegean Sea. These fish farms belonged to five different aquaculture companies and were labeled using capital letters with a numerical subscript (i.e., A1, A2, A3, B1, B2, C1, C2, D1, D2, E1, E2, and E3) (Figure 1). The collected fish were stored in styrofoam boxes containing aseptic ice and transported within 4–6 h in refrigerated vehicles (+4 °C) to the international market chain in Istanbul. The styrofoam boxes were opened immediately on arrival under aseptic conditions. The central temperature in the boxes was ≤+4 °C measured with a thermometer (Testo, Lenzkirch, Germany). Under the same conditions, the samples were taken by rubbing off the skin of the seabass with sterile swabs containing a transport liquid medium (Becton Dickinson, NJ, USA). The swabs were transported under refrigeration temperatures in thermal boxes (≤+4 °C) to the laboratory (Department of Food Science and Technology, Istanbul University-Cerrahpasa) for immediate analyses.

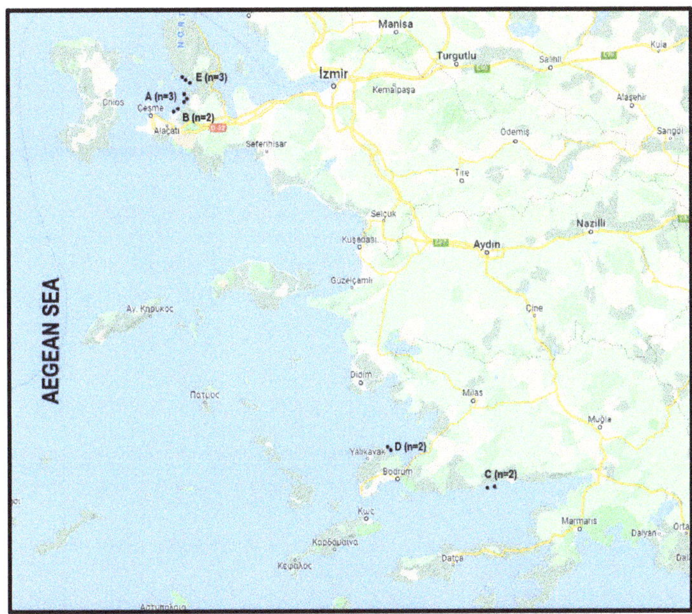

Figure 1. Seabass aquaculture companies and fish farms locations in the Aegean Sea. (Aquaculture Company A: three fish farms in Izmir; Aquaculture Company B: two fish farms in Izmir; Aquaculture Company C: two fish farms in Mugla; Aquaculture Company D: two fish farms in Mugla; and Aquaculture Company E: three fish farms in Izmir).

2.2. Next Generation Sequencing (NGS) and Metabarcoding Analysis

2.2.1. Total DNA Extraction

Total DNA extraction was carried out directly from the swab samples by applying the phenol/chloroform/isoamyl alcohol method [15]. For this purpose, 2 mL swab samples were centrifuged at 14,000 rpm for 5 min at room temperature. The pellet was resuspended in 500 µL 1×TE buffer (10 mM Tris-HCl, 1 mM EDTA, pH 8.0) containing 5 mg/mL lysozyme (Applichem, Darmstadt, Germany) and the phenol/chloroform/isoamyl alcohol was applied. Finally, the extracted DNA samples were resuspended in 30 µL sterile deionized water and stored at −20 °C for amplicon PCR experiments in NGS studies.

2.2.2. Next-Generation Sequencing

16S rRNA amplicon sequencing and DNA library preparation were carried out according to the 16S metabarcoding sequencing library preparation guide [16]. The primers for the amplicon PCR were F-primer: 5′-TCGTCGGCAGCGTCAGATGTGTATAAGAGAC AGCCTACGGGNGGCWGCAG-3′ and R-primer: 5′-GTCTCGTGGGCTCGGAGATGTGTATAA GAGACAGGACTACHVGGGTATCTAACC-3′. Bacterial 16S rRNA V3-V4 gene regions were amplified using a KAPA HiFi HS kit (Roche, Mannheim, Germany). PCR products from each seabass sample were indexed with dual indexes using a Nextera® XT Index Kit v2 Set-A (Illumina, San Diego, CA, USA). All the amplicon PCR products and indexed amplicons were purified using AMPure XP magnetic beads (Beckman Coulter, Indianapolis, IN, USA). The prepared equimolar proportions (10 nM) of the samples were pooled, and diluted to a 35 pM library containing 5% (*v*/*v*) PhiX control DNA (Illumina). Subsequently, a 20 µL library was loaded into an iSeq100 v1 cartridge. The sequencing was carried out using the iSeq100 system (Illumina) pair end read type and two reads of 151 bp read length.

2.2.3. Metabarcoding Analysis

The sequencing reads from the 16S rRNA gene were analyzed using Silva NGS software version 138.1, VSEARCH 2.17.0, SINA v1.2.10 for ARB SVN (revision 21008), and BLASTn version 2.11.0+. Trimming of adapter sequences from short NGS read data was performed using Genious Prime software. The amplicons were clustered based on the sequence identity operational taxonomic unit (OTU) approach. Clustering Ward's analysis was applied using the PAleontological STatistics (PAST) Software version 4.11 package (2022) at the genus levels in the seabass samples [17].

2.3. Isolation and Identification of Pseudomonas

Pseudomonas spp. isolation and identification were performed using the modified conventional TS EN ISO 13720 standard [18]. First, 250 μL of the swab sample containing each liquid medium was taken and placed in 2 mL of Pseudomonas Broth (Z699101 Merck, Darmstadt, Germany) and incubated at 22 ± 2 °C for 44 ± 4 h (Pre-enrichment). Subsequently, 0.1 mL of the suspension in Pseudomonas Broth was taken and spread onto Pseudomonas Agar (CM 559 Oxoid, Basingstoke, UK) containing Pseudomonas CFC Selective Supplement (SR103 Oxoid). The plates were incubated at 22 ± 2 °C for 44 ± 4 h. After incubation, suspected *Pseudomonas* spp. were transferred to Tryptic Soy Agar (CM 131, Oxoid, Basingstoke, UK) for purification. Biochemical tests such as Gram staining, oxidase test, catalase test, and fluorescence properties with UV light (365 nm) were applied to confirm *Pseudomonas* strains [9,18].

2.4. Determination of Antibiotic Susceptibility in Psychrotrophic Pseudomonas Strains

Pseudomonas strains were tested for antibiotic susceptibility using the agar disk diffusion method on Mueller–Hinton agar (CM 337 Oxoid) [19]. The plates were incubated at 22 ± 2 °C for 24 h. Eleven (11) different antibiotics were used: Piperacillin-tazobactam (Oxoid-CT1616, 30–6 μg), gentamicin (Oxoid-CT0024, 10 μg), tobramycin (Oxoid-CT0056, 10 μg), amikacin (Oxoid-CT0107, 30 μg), doripenem (Oxoid-CT1880, 10 μg), meropenem (Oxoid-CT0774, 10 μg), imipenem (Oxoid-CT0455, 10 μg), levofloxacin (Oxoid-CT1587, 5 μg), ciprofloxacin (Oxoid-CT0425, 5 μg), norfloxacin (Oxoid-CT0434, 10 μg) and tetracycline (Oxoid-CT0054, 30 μg) according to the CLSI [14] from five preferred antibiotic groups (penicillins, aminoglycosides, carbapenems, fluoroquinolones, and tetracyclines).

The E-test (Bioanalyse, Turkey) was applied to determine the Minimum Inhibitory Concentration (MIC) of *Pseudomonas* strains that were found to be resistant to antibiotics in the disc diffusion test. Results were evaluated according to the EUCAST [13] and CLSI [14] breakpoint tables.

3. Results and Discussion

3.1. NGS and Metabarcoding Analysis Results

Modern high-throughput methods have substituted conventional culture-based microbiological techniques, increasing our understanding of fish microbial communities throughout the production chain, from harvesting through storage distribution, until the end of shelf life [20]. In this study, the alpha diversity of bacteria was estimated to determine the diversity within samples, and the Shannon species diversity index values were determined using Silva NGS software (Table 1). This diversity index is a quantitative measure for estimating the number of different species in a given environment and their relative abundance [21]. This can be relevant for identifying the bacterial diversity in skin seabass samples because skin mucus harbors a complex bacterial community [22].

Table 1. Shannon species diversity index values * of seabass skin samples.

Company Code	Sample Name	Shannon Index	Company Code	Sample Name	Shannon Index	Company Code	Sample Name	Shannon Index
A1	Fish_S1	7.02	E1	Fish_S33	7.03	E2	Fish_S65	6.73
	Fish_S2	6.84		Fish_S34	6.83		Fish_S66	6.51
	Fish_S3	7.25		Fish_S35	7.14		Fish_S67	6.88
	Fish_S4	7.11		Fish_S36	6.93		Fish_S68	6.61
	Fish_S5	7.10		Fish_S37	6.95		Fish_S69	6.78
	Fish_S6	7.21		Fish_S38	6.58		Fish_S70	6.94
	Fish_S7	6.99		Fish_S39	6.82		Fish_S71	6.66
	Fish_S8	6.88		Fish_S40	6.58		Fish_S72	6.82
B1	Fish_S9	6.89	D2	Fish_S41	6.76	B2	Fish_S73	6.56
	Fish_S10	7.14		Fish_S42	6.68		Fish_S74	6.29
	Fish_S11	7.14		Fish_S43	6.88		Fish_S75	6.62
	Fish_S12	7.00		Fish_S44	6.84		Fish_S76	6.44
	Fish_S13	7.25		Fish_S45	6.92		Fish_S77	6.38
	Fish_S14	6.66		Fish_S46	7.02		Fish_S78	6.52
	Fish_S15	7.12		Fish_S47	6.65		Fish_S79	6.78
	Fish_S16	6.91		Fish_S48	6.55		Fish_S80	6.52
C1	Fish_S17	6.86	C2	Fish_S49	6.82	E3	Fish_S81	6.83
	Fish_S18	6.88		Fish_S50	6.33		Fish_S82	6.73
	Fish_S19	6.87		Fish_S51	7.06		Fish_S83	6.80
	Fish_S20	6.53		Fish_S52	6.30		Fish_S84	6.75
	Fish_S21	6.92		Fish_S53	7.01		Fish_S85	6.69
	Fish_S22	6.85		Fish_S54	6.79		Fish_S86	6.61
	Fish_S23	7.13		Fish_S55	6.70		Fish_S87	7.00
	Fish_S24	7.02		Fish_S56	6.84		Fish_S88	6.90
D1	Fish_S25	7.16	A2	Fish_S57	6.99	A3	Fish_S89	6.87
	Fish_S26	6.82		Fish_S58	7.00		Fish_S90	7.08
	Fish_S27	6.68		Fish_S59	6.98		Fish_S91	6.94
	Fish_S28	6.16		Fish_S60	6.81		Fish_S92	6.69
	Fish_S29	6.75		Fish_S61	6.65		Fish_S93	6.66
	Fish_S30	6.95		Fish_S62	6.34		Fish_S94	6.61
	Fish_S31	6.87		Fish_S63	6.87		Fish_S95	6.96
	Fish_S32	6.52		Fish_S64	6.74		Fish_S96	6.89

* The higher the index values, the more diverse the species in the habitat.

Metabarcoding analysis of 189,207 sequences from 96 seabass skin samples led to 123,391 OTUs, 39,737 clustered sequences, and 164,870 classified sequences. The results indicated that the phylum Proteobacteria was dominant in all seabass skin samples. The skin microbiota samples also contained bacteria belonging to the phyla Firmicutes and Bacteroidota (Figure 2). At the genus level, *Pseudomonas* was the dominant genus among the 96 seabass swab samples. (Figure 3). Additionally, *Shewanella*, *Acinetobacter*, and *Flavobacterium* were also among the most prevalent genera (Figure 3). Similar results were reported from the Bodrum coast in seawater, Mugla [23]. The genus *Pseudomonas* is considered to be an important fish pathogen as it comprises some (sub) species which are opportunistic pathogens to humans [23]. Another study dedicated to examining the microbiota of whole and filleted seabass [20] presented results similar to those we obtained. *Pseudomonas* was dominant in seabass samples, based on the 16S rRNA metabarcoding analysis, followed by the presence of *Shewanella*. Among animal food products, fish are the most vulnerable to bacterial spoilage and *Shewanella* has previously been reported as a main contributor in the microbiota of spoiled seafood, such as hake fillets [24]. Additionally, *Shewanella* was the dominant genus in MAP-stored seabass fillets, but its relative abundance declined dramatically towards the end of the products' shelf life [19]. *Acinetobacter* are abundant in aquatic environments and frequently isolated from the skin and gills of fresh fish [25]. In a previous study, *Acinetobacter* were the dominant bacteria in seabass fillets [20] and rainbow trout samples [26]. However, *Acinetobacter* are not recognized as important spoilage bacteria [27] as they cannot hydrolyze fish proteins and are thus, a weak producer of biogenic amines, as well as a weak degrader of ATP-related compounds [28].

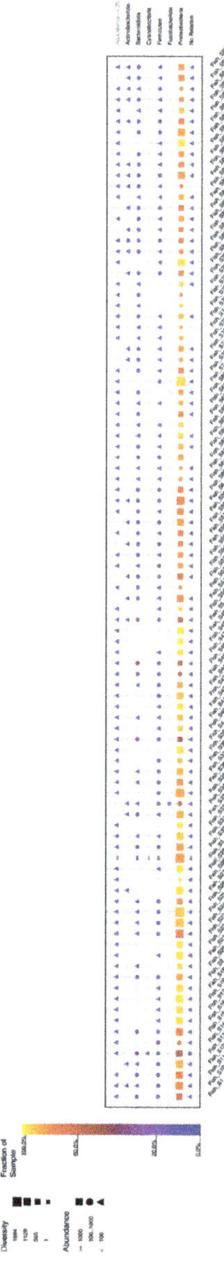

Figure 2. Distribution of bacterial communities in seabass swab samples at the phylum level.

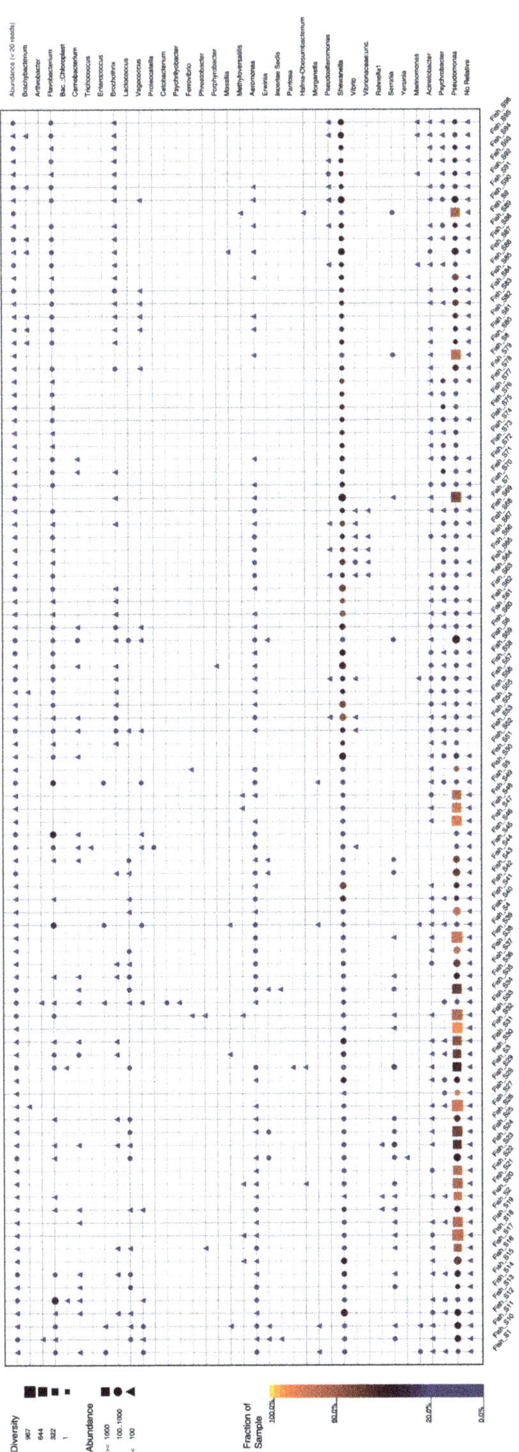

Figure 3. Distribution of bacterial communities in seabass swab samples at the genus level.

P. lundensis was identified by NGS analysis of all seabass samples. Similar to our results, Elbehiry et al. [29] reported that, in red meat samples, *P. lundensis* was the dominant species. Pseudomonads are highly opportunistic and may become a highly threatening fish pathogen causing serious illness including ulcerative syndrome and hemorrhagic septicemia [30]. *Enterococcus* were found in 15 seabass samples, of which samples S1, S4, and S5 were sampled from the same fish farm. The other *Enterococcus*-containing samples were S11, S34, S45, S53, S54, S57, S64, S65, S66, S67, S68, and S69 identified from four different fish farms (D2 and C2 located in Mugla, A2 and E2 located in Izmir). Detection of *Enterococcus* spp. in sea bass skin samples may indicate fecal contamination in seawater.

The highest Shannon diversity index in this study (7.25) was obtained for samples S3 and S13, indicating that these samples had the highest diversity of skin microbiota. The S28 sample contained the lowest species diversity with a value of 6.16 (Table 1). Ward's analysis demonstrated that two main clusters were present at the genus level (Figure 4). The composition of the microbiota, however, did not cluster at the genus level. This might be attributable to differences in the composition of the fish skin microbiomes between individual fish from the same population and differences between the skin microbiome and the surrounding water [6].

Foodborne pathogens such as *Salmonella*, *Escherichia*, and *Mycobacterium* genera were not found in the samples. On the other hand, *Vibrio* (*V.*) *ordalii* was detected in three seabass swab samples (numbers 65, 66, and 67) originating from E-2 fish farms in Izmir. Similarly, many researchers have reported *V. ordalii* from seabass in the Aegean Sea [31,32], including Izmir [33]. Bacterial infections most frequently detected in cultured seabass and gilthead sea bream are caused by bacteria belonging to the family *Vibrionaceae*. Associated losses have been reported with *Vibrionaceae* in many fish species, including seabass, sea bream, and salmonid species etc. [34].

3.2. Temperature Measurement of Seabass Samples in Styrofoam Boxes Containing Ice

The lowest average temperature was 1.7 °C in the samples from the fish farm B2 located in Izmir, and the highest temperature was 3.4 °C in the samples from the fish farms A3 (Izmir) and E2 (Izmir) (Table 2). The average and standard deviation of the inner temperature of seabass samples were 2.58 ± 0.53 °C. In addition, the internal temperature values measured in all fish samples were below +4 °C. Similarly, a study reported the internal temperature of iced styrofoam-packaged seabass from the Aegean Sea to be 4.15 ± 1.12 °C [35]. The extension of shelf life by chilling is essentially due to the reduction in the growth rate and metabolic activity of spoilage microorganisms such as *Pseudomonas* spp. [35] and *Acinetobacter* spp. *Acinetobacter* species have been found in great abundance in fresh seabass at 12 °C [19] and fish fillets at 10 °C [36], and were the dominant species at the end of the shelf life of rainbow trout stored aerobically at 4 °C [25]. Indeed, upon storage the psychrophilic bacteria proliferated slowly and dominated the mesophilic load, as the low temperature favored their growth [37]. Similar to our study, Syropoulou et al. [38] reported that *Pseudomonas* spp. were found from the beginning of shelf life, whilst in seabass products from Greece, *Shewanella* were detected at later storage stages.

3.3. Isolation of Psychrotrophic Pseudomonas spp. in Seabass Swab Samples using Conventional Methods

In total, 46 seabass swab samples (48%) were positive for psychrotrophic *Pseudomonas* strains isolated with the conventional ISO method [18] (Table 2). *Pseudomonas* strains were isolated from four fish farms in Izmir, i.e., A2 ($n = 6$), E2 ($n = 6$), E3 ($n = 6$), and A3 ($n = 5$), and farm C1 ($n = 5$) in Mugla. The cultivation-based method will detect live *Pseudomonas* strains, which is an important characteristic when compared to NGS and metabarcoding methods that are used in the detection of DNA fragments and DNA structures, as these do not necessarily indicate the presence of living bacteria [39].

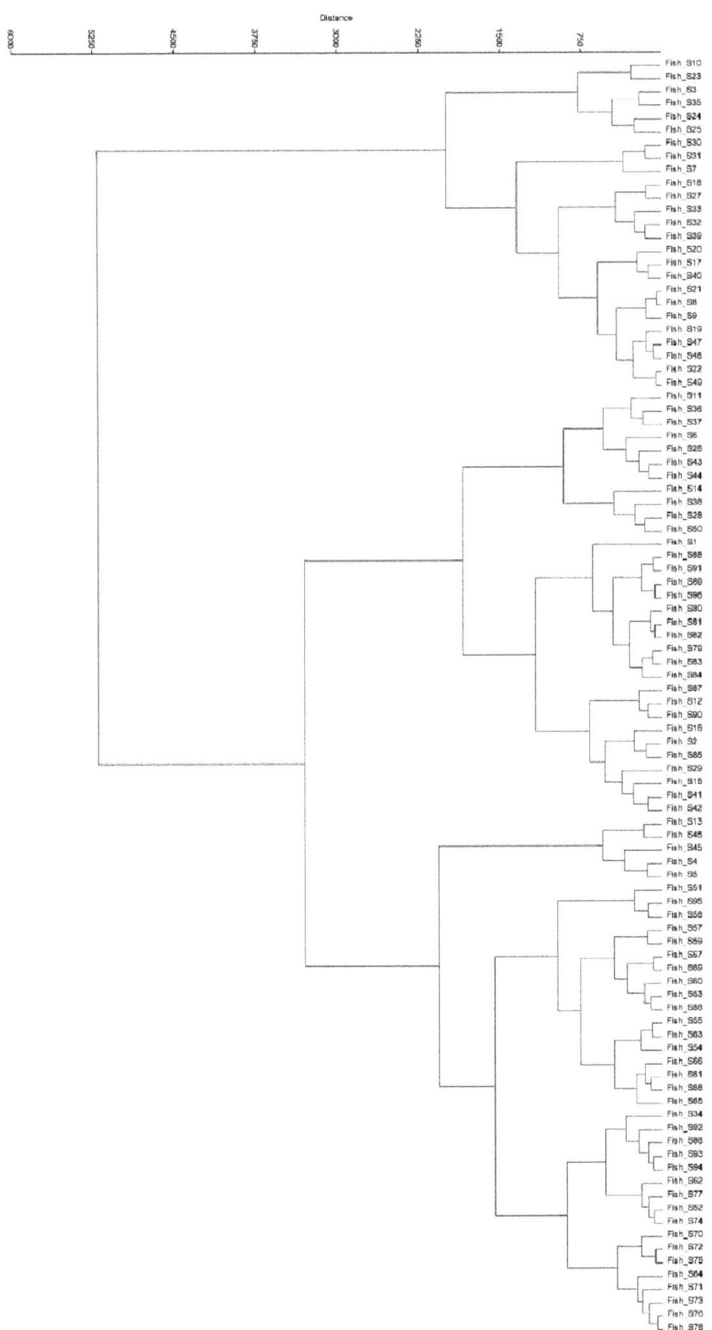

Figure 4. Dendrogram based on Ward's method of clustering.

Table 2. Temperature of seabass samples and verification of viable psychrotrophic *Pseudomonas* strains after Next Generation Sequencing analysis (NGS) using conventional methods [18].

Company Code and Fish Farm Number	Temperature Measurement of Seabass Samples (°C)	Samples with DNA Fragments from *Pseudomonas*	Number Samples with Viable *Pseudomonas* Strains (out of NGS Positive Samples)
1A1	2.3	8	3
A2	2.7	8	6
A3	3.4	8	4
2B1	2.7	8	1
B2	1.7	8	2
3C1	2.1	8	5
C2	2.3	8	4
4D1	3.2	8	4
D2	2.5	8	4
5E1	1.9	8	2
E2	3.4	8	6
E3	2.8	8	5
Totally	Total x-Sx 2.58 ± 0.53	96	46

[1] Fish Company A: A1–A3, three different fish farms of fish company A in Izmir Province; [2] Fish Company B: B1–B2, two different fish farms of fish company B in Izmir Province; [3] Fish Company C: C1–C2, two different fish farms of fish company C in Mugla Province; [4] Fish Company D: D1–D2, two different fish farms of fish company D in Mugla Province; [5] Fish Company E: E1–E3, three different fish farms of fish company E in Izmir Province.

3.4. Antibiotic Susceptibility of Pseudomonas spp. Using Disc Diffusion

Susceptibility to 11 antibiotics was tested among 46 viable *Pseudomonas* spp. isolates. Some of the strains (13/46; 28.3%) were found to be resistant to doripenem, according to EUCAST [13] and CLSI [14] (Table 3).

Table 3. Clinical and Laboratory Standards Institute (CLSI) and the European Committee on Antimicrobial Susceptibility Testing (EUCAST) as assessed using the disc diffusion method of psychrotrophic *Pseudomonas* strains (*n* = 46) [Resistant ("R"); Intermediate susceptibility ("I") or Susceptible ("S")].

Antibiotic Groups	Name of Antibiotics	Distribution of *Pseudomonas* Strains according to CLSI			Distribution of *Pseudomonas* Strains According to EUCAST	
		R (%)	I (%)	S (%)	R (%)	S (%)
Penicillins	Piperacillin-tazobactam 30 µg	-	-	46 (100)	-	46 (100)
Aminoglycosides	Gentamicin 10 µg	-	-	46 (100)	*n* *	*n*
	Tobramycin 10 µg	2 (4.3)	-	44 (95.7)	*n*	*n*
	Amikacin 30 µg	-	-	46 (100)	-	46 (100)
Carbapenems	Doripenem 10 µg	13 (28.3)	-	33 (71.7)	13 (28.3)	33 (71.7)
	Meropenem 10 µg	1 (2.3)	4 (8.6)	41 (89.1)	5 (10.9)	41 (89.1)
	Imipenem 10 µg	6 (13)	2 (4.4)	38 (82.6)	6 (13)	40 (87)
Fluoroquinolones	Levofloxacin 5 µg	-	-	46 (100)	-	46 (100)
	Ciprofloxacin 5 µg	-	1 (2.2)	45 (97.8)	1 (2.2)	45 (97.8)
	Norfloxacin 10 µg	-	-	46 (100)	*n*	*n*
Tetracyclines	Tetracycline 30 µg	-	-	46 (100)	*n*	*n*

* *n*: A breakpoint value of this antibiotic is not available in the CLSI standard.

Thirty (65.2%) *Pseudomonas* strains were susceptible to all antibiotics according to the CLSI [14]. On the other hand, thirty-three (71.7%) *Pseudomonas* strains were susceptible to all antibiotics according to the EUCAST [13]. All *Pseudomonas* strains from A1 (Izmir), B1 (Izmir), and E1 (Izmir) fish farms were susceptible to all antibiotics (Table 4).

Table 4. Distribution of susceptible, resistant, and multidrug resistant *Pseudomonas* spp. according to the Clinical and Laboratory Standards Institute (CLSI) [14] and the European Committee on Antimicrobial Susceptibility Testing (EUCAST) [13].

Company Name (Number of Isolated Strains per Farm)	CLSI				EUCAST			
	Number of Susceptible Strains	Number of *Pseudomonas* Strains Resistant to One Antibiotic	Number of *Pseudomonas* Strains Resistant to Two Antibiotics	* Number of MDR Strains	Number of Susceptible Strains	Number of *Pseudomonas* Strains Resistant to One Antibiotic	Number of *Pseudomonas* Strains Resistant to Two Antibiotics	Number of MDR Strains
A1 ($n = 3$)	3	-	-	-	3	-	-	-
A2 ($n = 6$)	5	-	1 (DOR **,IPM)	-	5	-	-	1 (DOR,IPM,MEM)
A3 ($n = 4$)	2	1 (DOR)	1 (DOR,IPM)	-	2	1 (DOR)	1 (DOR, IPM)	-
B1 ($n = 1$)	1	-	-	-	1	-	-	-
B2 ($n = 2$)	0	1 (DOR), 1 (TOB)	-	-	-	1 (TOB)	1 (CIP,DOR)	-
C1 ($n = 5$)	3	1 (TOB)	1 (DOR,IPM)	-	3	1 (TOB)	-	1 (DOR,IPM,MEM)
C2 ($n = 4$)	3	-	1 (DOR,IPM)	-	3	-	-	1 (DOR,IPM,MEM)
D1 ($n = 4$)	2	1 (DOR)	1 (DOR,IPM)	-	3	-	-	1 (DOR,IPM,MEM)
D2 ($n = 4$)	1	-	-	3 (DOR, IPM, MEM)	3	-	-	1 (DOR,IPM,MEM)
E1 ($n = 2$)	2	-	-	-	2	-	-	-
E2 ($n = 6$)	5	1 (DOR)	-	-	5	1 (DOR)	-	-
E3 ($n = 5$)	3	2 (DOR)	-	-	3	2 (DOR)	-	-

* Number of multidrug (3 or more) resistant *Pseudomonas* spp. ** Abbreviated Name of antibiotics (DOR: Doripenem; IPM: Imipenem; TOB: Tobramycin; MEM: Meropenem, and CIP: Ciprofloxacin).

Sixteen (34.8%) *Pseudomonas* strains were resistant to more than one antibiotic based on the CLSI [14]. Eight (17.4%) *Pseudomonas* strains were resistant to one antibiotic only, including carbapenem (doripenem) and aminoglycoside group (tobramycin). Six *Pseudomonas* strains were resistant to doripenem, and two strains were resistant to tobramycin based on the CLSI [14]. However, only five (10.9%) *Pseudomonas* strains were resistant to two antibiotics, according to the CLSI [14]. All *Pseudomonas* strains from fish farms in Izmir [(A2; $n = 1$) and (A3; $n = 1$)] and Mugla [(C1; $n = 1$) and (D1; $n = 1$)] were resistant to doripenem and imipenem (carbapenem group). In addition, one strain originating from D2 fish farms (Mugla) was found to be MDR to doripenem, imipenem, and meropenem, all included in the carbapenem group, based on the CLSI (Table 4).

Thirteen (28.3%) *Pseudomonas* strains were found to be resistant to several antibiotics according to EUCAST [13], seven (13.4%) to only one antibiotic, including carbapenem (doripenem) and fluoroquinolones (norfloxacin) group. Four *Pseudomonas* strains were resistant to doripenem, and two strains to tobramycin according to the EUCAST standard [13]. Additionally, only two (4.3%) *Pseudomonas* strains isolated from Izmir (A3 and B2) were resistant to two antibiotics, according to the EUCAST standard [13]. Moreover, five *Pseudomonas* strains originating from Izmir (A2), and Mugla (C1, C2, D1, and D2) fish farms were found to be MDR to doripenem, imipenem, and meropenem including carbapenem group based on the EUCAST [13] (Table 4).

Pseudomonas spp. have been identified as primarily invasive or opportunistic pathogens for many organisms and this genus has also grown in importance in terms of antimicrobial resistance [9]. Many researchers have evaluated the antimicrobial sensitivity of *Pseudomonas* species isolated from fish, and have reported them as MDR, based on their resistance to ampicillin, cefotaxime, aztreonam, trimethoprim-sulfamethoxazole, nitrofurantoin and other groups of antimicrobials [9,40]. Recently, Rezgui et al. [41] showed an abundance of antibiotic-resistant bacteria isolated from the gills and intestinal tract of seabass and sea bream. The antibiotic-resistant bacteria belong to several species of the genera *Pseudomonas*, *Vibrio*, *Aeromonas*, and *Enterobacterales*. They were resistant to tetracycline and penicillin, which are commonly used in treating infections in animals and humans. In another study, almost all *Pseudomonas* strains were resistant to penicillins (ampicillin), macrolides (erythromycin, clindamycin), sulfonamides (trimethoprim-sulphamethoxazole-), and chloramphenicol [9]. We report similar results, i.e., that the *Pseudomonas* strains were susceptible to penicillins (piperacillin-tazobactam), aminoglycosides (amikacin and gentamycin), fluoroquinolones (levofloxacin, norfloxacin), and tetracyclines (tetracycline, ciprofloxacin) based on the CLSI [14]. Likewise, a study reported that enrofloxacin, oxytetracycline, and ciprofloxacin were found to be effective antibiotics against fish disease agents such as *Pseudomonas* spp., *Vibrio* spp. and *Staphylococcus* spp. in Turkey [42]. On the other hand, all *P. fluorescence* strains isolated from fish were resistant to piperacillin, ceftazidime, and cefepime in Egypt [43]. In the present study, psychrotrophic *Pseudomonas* strains were partially resistant (based on the EUCAST and CLSI) to antibiotics commonly used in fish farms. This fact should be carefully addressed in the context of the environmental spread of antibiotic resistance.

According to the CLSI, psychrotrophic *Pseudomonas* strains showed different resistance patterns to doripenem (28.3%), imipenem (13%), tobramycin (4.3%), and meropenem (2.3%). Similarly, *Pseudomonas* were resistant to doripenem (28.3%), imipenem (13%), meropenem (10.9%) and ciprofloxacin (2.2%) based on the EUCAST. In total, *Pseudomonas* strains resistant to nine antibiotics were isolated from nine different fish farms [A2 ($n = 1$), A3 ($n = 2$), B2 ($n = 2$), C1 ($n = 1$), C2 ($n = 1$), D1 ($n = 1$), D2 ($n = 1$), E2 ($n = 1$), and E3 ($n = 2$)]. *Pseudomonas* strains were resistant to the same antibiotics (imipenem, meropenem, and doripenem) (Table 4). Additionally, one *Pseudomonas* strain belonging to B2 ($n = 1$) fish farm showed resistance to ciprofloxacin and doripenem based on the EUCAST [13]. Finally, five *Pseudomonas* strains resistant to three antibiotics were identified according to the EUCAST [13]. These strains originated from five different fish farms: A2 ($n = 1$, Izmir), C1($n = 1$, Mugla), C2 ($n = 1$, Mugla), D1 ($n = 1$, Mugla), and D2 ($n = 1$, Mugla). Fish diseases are limiting factors in fish production, causing high mortality, especially in hatcheries, which affects profit negatively [29]. Antibacterial therapy is often chosen as the way to control bacterial disease outbreaks that pose economic challenges [43]. Additionally, antibiotic resistance is one of the most significant challenges to human health and food security [28]. Some studies are available on antibiotic susceptibility in human pathogenic bacteria, including *Pseudomonas* spp. [44].

3.5. MICs of Psychrotrophic Pseudomonas spp.

Pseudomonas strains that had shown resistance to antibiotics in the disc diffusion assay were selected for examination using the E-Test (gradient diffusion method) to determine their MIC (Table 5). From the 13 strains that showed resistance to doripenem in the disc diffusion test, two had an MIC exceeding the threshold $\geq$8 g/mL for antibiotic resistance (12 and 125 g/mL; the latter isolate originated from farm A3 in İzmir). For imipenem, three out of six isolates resistant according to disc-diffusion assay were confirmed as resistant by E-test. The MIC of these three resistant strains was >32 µg/mL. All these isolates originated from farms A3 (in Izmir), C1, and D1 (both in Mugla). Similarly, isolates resistant to tobramycin, meropenem, or ciprofloxacin according to the disc diffusion assay, were classified as susceptible based on the E-test MIC [13,14]. Only one *Pseudomonas* strain from C1 fish farms (Sample no. 24) was resistant to doripenem and imipenem, as assessed by MIC determination.

Table 5. The Minimum Inhibitory Concentrations (MIC), as assessed by E-Test, for four antimicrobial agents against *Pseudomonas* strains isolated from sea bass samples.

Group	Antimicrobial	Tested $n = 22$	MIC (µg/mL), $n = 22$										Resistant Isolates, $n = 5$
			0.012–0.025	0.026–0.50	0.051–0.999	1–1.5	3	4	6	12	>32	125	
Carbapenems	Doripenem [1]	13		5	1			2	3	1		1	2
	Meropenem [1]	1				1							0
	Imipenem [1]	6	1		1				1		3		3
Aminoglycosides	Tobramycin [2]	2						1	1				0
Fluoroquinolones	Ciprofloxacin [3]	1	1										0

n = number of isolates; [1] = MIC $\geq$ 8 µg/mL indicates antimicrobial resistance according to CLSI and EUCAST; [2] = MIC $\geq$ 16 µg/mL indicates antimicrobial resistance according to CLSI; [3] = MIC $\geq$ 2 µg/mL indicates antimicrobial resistance according to EUCAST.

The different results obtained by the gradient diffusion (E-test) and the disc diffusion methods for *Enterobacterales* and *Pseudomonas aeruginosa* strains are not unexpected since the E-test generally performs better [45]. Despite the different outcomes from different methods, our results are in line with reports on antimicrobial resistance in *Pseudomonas* and *Escherichia coli* in general. The European Antimicrobial Resistance Surveillance Network reported on samples from human patients in 2017, of which, 30.8% of the *Pseudomonas aeruginosa* strains isolated were resistant to at least one of the antimicrobial groups under regular surveillance (fluoroquinolones, aminoglycosides, and carbapenems) [46]. Moreover, the European Centre for Disease Prevention and Control has shown significant increments in the percentage of antibiotic-resistance among pathogenic bacteria, such as carbapenem-resistance in *Pseudomonas aeruginosa* and *Acinetobacter* spp. in several countries in the European region of concern [47].

With respect to fish, a study from Egypt reported that *Pseudomonas aeruginosa* and *E. coli* strains were resistant to third-generation cephalosporin and last-resort carbapenems isolated from Nile tilapia [41]. Interestingly, 29.7% of *P. fluorescens* strains isolated showed MDR, especially to penicillin and cephalosporin groups [41].

4. Conclusions

Results from this study show that psychrotrophic *Pseudomonas* were the dominant bacterial species in seabass skin samples from 12 selected fish farms in the Aegean Sea.

Ninety-six fish were sampled by skin swab, and in all samples, NGS analysis indicated the presence of *Pseudomonas*. Viable isolates were cultured from 46 of these samples. Testing the isolates against 11 different antibiotics (five main groups), showed that all samples were susceptible to piperacillin-tazobactam, gentamicin, amikacin, levofloxacin, norfloxacin, and tetracycline. Based on the CLSI, the isolates from across the farms showed various resistance patterns to the carbapenem group [doripenem (28.3%), imipenem (13%), and meropenem (2.3%)] and aminoglycosides [tobramycin (4.3%)]. Using the EUCAST standard, there was additional resistance to doripenem (28.3%), imipenem (13%), meropenem (2.3%), and ciprofloxacin (2.2%). MDR was found among three *Pseudomonas* strains from Mugla (D = 2) based on the CLSI and five *Pseudomonas* strains based on the EUCAST criteria (disc diffusion method). Three farms with six isolates showed no antibiotic resistance based on EUCAST and CLSI criteria.

This study has shown that resistance to a broad range of antibiotics prevails in *Pseudomonas* from the selected farms. As the farms were chosen without looking at their histories of disease and antibiotic use, our results may indicate a representative situation for the industry in the region. This should, however, be confirmed in a broader study, including records of antibiotic use at the farm level.

The use of antibiotics is generally regarded as the main driver for developing resistance. Exposure to antibiotics may be due to own use or external exposure. The industry uses antibiotics for prophylactic and therapeutic treatments to keep farmed fish free of diseases. Prudent use of antibiotics is therefore essential also for the aquaculture industry to minimize antibiotic resistance and the spread of resistant bacteria or genes to the environment. Ultimately, this will serve consumer protection and lead to a more efficient application of antibiotics in human therapy.

Author Contributions: Conceptualization, A.A., M.S. and E.B.; writing, A.A., Z.M., J.Z., M.S. and S.T.; software, A.A., M.S. and Z.M.; formal analysis, Z.M., M.S., M.N.Z.Y. and A.A.; investigation, Z.M., M.N.Z.Y., M.S., S.T. and A.A.; resources, A.A., M.S. and Z.M.; data curation, A.A., J.Z., M.S., Z.M., M.N.Z.Y. and S.T.; writing—original draft preparation, A.A., J.Z., M.S., Z.M., S.T. and V.C.O. writing—review and editing, E.B., A.A., J.Z. and S.T.; visualization, A.A. and M.S.; supervision, E.B.; project administration, A.A. and S.T. funding acquisition, none. All authors have read and agreed to the published version of the manuscript.

Funding: This research received no external funding.

Data Availability Statement: The data presented in this study are available on request from the corresponding author.

Acknowledgments: The authors would like to thank Peter PAULSEN for assistance with English editing, and Alper YANKIN for assistance with sampling. The authors would like to acknowledge Sezer OZKAN (Department of Food Hygiene and Technology, Faculty of Veterinary Medicine, Istanbul University-Cerrahpasa) and Berke TUMBAL (Department of Molecular Biology and Genetics, Istanbul Kultur University) for their technical assistance. Open Access Funding by the Norwegian Veterinary Institute.

Conflicts of Interest: The authors declare no conflict of interest.

References

1. Gozutok, E.; Aydin, A. Presence and virulence characterization of *Listeria monocytogenes* from fish samples in the Black Sea, Türkiye. *Ankara Univ. Vet. Fak. Derg.* **2022**, *69*, 387–394.
2. Muniesa, A.; Basurco, B.; Aguilera, C.; Furones, D.; Reverté, C.; Sanjuan-Vilaplana, A.; Jansen, M.D.; Brun, E.; Tavornpanich, S. Mapping the knowledge of the main diseases affecting sea bass and sea bream in the Mediterranean. *Transbound. Emerg. Dis.* **2020**, *67*, 1089–1100. [CrossRef]
3. FAO. Food and Agriculture Organization: Fishery and Aquaculture Statistics. 1976–2019. Available online: https://www.fao.org/fishery/statistics/software/fishstatj/en (accessed on 1 November 2022).
4. FAO. Food and Agriculture Organization: The State of World Fisheries and Aquaculture. Available online: https://www.fao.org/3/cc0461en/cc0461en.pdf (accessed on 1 November 2022).

5.   Carvalho, N.; Guillen, J. Aquaculture in the Mediterranean. Available online: https://www.iemed.org/publication/aquaculture-in-the-mediterranean/ (accessed on 1 November 2022).

6.   Ucak, S.; Zafer Yurt, M.N.; Tasbasi, B.B.; Acar, E.E.; Altunbas, O.; Soyucok, A.; Aydin, A.; Ozalp, V.C.; Sudagidan, M. Identification of bacterial communities of fermented cereal beverage Boza by metagenomic analysis. *LWT* **2022**, *153*, 112465. [CrossRef]

7.   Austin, B.; Austin, D.A. Pseudomonads. In *Bacterial Fish Pathogens*; Springer International Publishing: Berlin/Heidelberg, Germany, 2016; pp. 475–498.

8.   Duman, M.; Mulet, M.; Altun, S.; Saticioğlu, I.B.; Ozdemir, B.; Ajmi, N.; Lalucat, J.; Garcia-Valdes, E. The diversity of *Pseudomonas* species isolated from fish farms in Turkey. *Aquaculture* **2021**, *535*, 736–739. [CrossRef]

9.   Aydin, A.; Sudagidan, M. Increasing global threat: Antibiotic Resistance. In *Current Factors Risking the Future of Food One Health One Medicine Concept Approach*; Buyukunal, S., Ed.; Turkiye Klinikleri Publishing: Ankara, Turkiye, 2022; pp. 8–17.

10. Vaishali, G.M.; Geetha, R.V. The superbug threat. *Res. J. Pharm. Technol.* **2015**, *8*, 343. [CrossRef]

11. Møretrø, T.; Langsrud, S. Residential bacteria on surfaces in the food industry and their implications for food safety and quality. *Compr. Rev. Food Sci. Food Saf.* **2017**, *16*, 1022–1041. [CrossRef]

12. Berggren, H.; Tibblin, P.; Yıldırım, Y.; Broman, E.; Larsson, P.; Lundin, D.; Forsman, A. Fish skin microbiomes are highly variable among individuals and populations but not within individuals. *Front. Microbiol.* **2022**, *12*, 767770. [CrossRef]

13. EUCAST, European Committee on Antimicrobial Susceptibility Testing. Breakpoints Tables for Interpretation of MICs and Zone Diameters, Version 12.0. 2022. Available online: http://www.eucast.org (accessed on 26 September 2022).

14. CLSI, Clinical and Laboratory Standards Institute. *M100-Ed31 Performance Standards for Antimicrobial Susceptibility Testing*, 31st ed; Clinical and Laboratory Standards Institute Publishing Malvern: Malvern, PA, USA, 2021; ISBN 978-1-68440-105-5.

15. Liu, D.; Ainsworth, A.J.; Austin, F.W.; Lawrence, M.L. Use of PCR primers derived from a putative transcriptional regulator gene for species-specific determination of *Listeria monocytogenes*. *Int. J. Food Microbiol.* **2004**, *91*, 297–304. [CrossRef]

16. Anonymous. 16S Metagenomic Sequencing Library Preparation. Part # 15044223 Rev. B. Available online: https://support.illumina.com/documents/documentation/chemistry_documentation/16s/16smetagenomic-library-prep-guide-15044223-b.pdf) (accessed on 26 September 2022).

17. Hammer, Ø.; Harper, D.A.T.; Ryan, P.D. Past: Paleontological statistics software package for education and data analysis. *Palaeontol. Electron.* **2001**, *4*, 1–9.

18. *TS EN ISO 1370*; Meat and Meat Products—Enumeration of Presumptive *Pseudomonas* spp. ISO. International Standardization Organization: Geneva, Switzerland, 2010.

19. EUCAST, European Committee on Antimicrobial Susceptibility Testing. Antimicrobial Susceptibility Testing EUCAST Disk Diffusion Method. Version 10.0. 2022. Available online: Manual_v_10.0_EUCAST_Disk_Test_2022.pdf (accessed on 26 September 2022).

20. Syropoulou, F.; Anagnostopoulos, D.A.; Parlapani, F.F.; Karamani, E.; Stamatiou, A.; Tzokas, K.; Nychas, G.-J.E.; Boziaris, I.S. Microbiota succession of whole and filleted European Sea Bass (*Dicentrarchus labrax*) during storage under aerobic and MAP conditions via 16S rRNA gene high throughput sequencing approach. *Microorganisms* **2022**, *10*, 1870. [CrossRef]

21. Konopiński, M.K. Shannon diversity index: A call to replace the original Shannon's formula with unbiased estimator in the population genetics studies. *Peer J.* **2020**, *8*, e9391. [CrossRef]

22. Ruiz-Rodríguez, M.; Scheifler, M.; Sanchez-Brosseau, S.; Magnanou, E.; West, N.; Suzuki, M.; Duperron, S.; Desdevises, Y. Host species and body site explain the variation in the microbiota associated to wild sympatric Mediterranean teleost fishes. *Microb. Ecol.* **2020**, *80*, 212–222. [CrossRef]

23. Ugur, A.; Ceylan, O.; Aslım, B. Characterization of *Pseudomonas* spp. from seawater of the southwest coast of Turkey. *J. Biol. Environ. Sci.* **2012**, *6*, 15–23.

24. Antunes-Rohling, A.; Calero, S.; Halaihel, N.; Marquina, P.; Raso, J.; Calanche, J.; Beltr, A.; Ignacio, Á.; Cebri, G. Characterization of the spoilage microbiota of hake fillets packaged under a modified atmosphere (MAP) rich in CO_2 (50% CO_2/50% N_2) and stored at different temperatures. *Foods* **2019**, *2*, 489. [CrossRef]

25. Kozinska, A.; Pazdzior, E.; Pekala, A.; Niemczuk, W. *Acinetobacter johnsonii* and *Acinetobacter lwoffii*—The emerging fish pathogens. *Bull. Vet. Inst. Pulawy* **2014**, *58*, 193–199. [CrossRef]

26. Du, G.; Gai, Y.; Zhou, H.; Fu, S.; Zhang, D. Assessment of spoilage microbiota of rainbow trout (*Oncorhynchus mykiss*) during storage by 16S rDNA sequencing. *J. Food Qual.* **2022**, *5367984*. [CrossRef]

27. Zhu, S.; Wu, H.; Zhang, C.; Jie, J.; Liu, Z.; Zeng, M.; Wang, C. Spoilage of refrigerated *Litopenaeus vannamei*: Eavesdropping on *Acinetobacter* acyl-homoserine lactones promotes the spoilage potential of *Shewanella baltica*. *J. Food Sci. Technol.* **2018**, *55*, 1903–1912. [CrossRef]

28. Liu, X.; Huang, Z.; Jia, S.; Zhang, J.; Li, K.; Luo, Y. The roles of bacteria in the biochemical changes of chill-stored bighead carp (*Aristichthys nobilis*): Proteins degradation, biogenic amines accumulation, volatiles production, and nucleotides catabolism. *Food Chem.* **2018**, *255*, 174–181. [CrossRef]

29. Elbehiry, A.; Marzouk, E.; Aldulbaib, M.; Moussa, I.; Abalkhail, A.; Ibrahem, M.; Hamada, M.; Sindi, W.; Alzaben, F.; Almuzaini, A.M.; et al. *Pseudomonas* species prevalence, protein analysis, and antibiotic resistance: An evolving public health challenge. *AMB Express* **2022**, *12*, 53. [CrossRef]

30. Algammal, A.M.; Mabrok, M.; Sivaramasamy, E.; Youssef, F.M.; Atwa, M.H.; El-Kholy, A.W.; Hetta, H.F.; Hozzein, W.N. Emerging MDR-*Pseudomonas aeruginosa* in fish commonly harbor *opr*L and *tox*A virulence genes and *bla*$_{TEM}$, *bla*$_{CTX-M}$, and *tet*A antibiotic-resistance genes. *Sci. Rep.* **2020**, *10*, 15961. [CrossRef]

31. Turk, N. Çipura (*Sparus aurata*) ve levrek (*Dicentrarchus labrax*) yavrularından bakteri izolasyonu ve antibiyotiklere duyarlılıklarının belirlenmesi. *Bornova Vet. Kontr. Araş. Enst. Derg.* **2002**, *27*, 7–14.

32. Korun, J.; Timur, G. Marine vibrios associated with diseased sea bass (*Dicentrarchus labrax*) in Turkey. *J. Fish. Sci.* **2008**, *2*, 66–76. [CrossRef]

33. Korun, J. A study on diagnosis of vibriosis and pasteurellosis some diagnostic kits and laboratory methods in cultured Sea bass (*Dicentrarchus labrax*). Ph.D. Thesis, Istanbul University, Istanbul, Türkiye, 2004.

34. Ozturk, R.C.; Altınok, I. Bacterial and viral fish diseases in Turkey. *Turkish J. Fish. Aquat. Sci.* **2014**, *14*, 275–297.

35. Cakli, S.; Kilinc, B.; Cadun, A.; Dincer, T.; Tolasa, S. Quality differences of whole ungutted sea bream (*Sparatus aurata*) and sea bass (*Dicentrarchus labrax*) while stored in ice. *Food Control* **2007**, *18*, 391–397. [CrossRef]

36. Zotta, T.; Parente, E.; Ianniello, R.G.; De Filippis, F.; Ricciardi, A. Dynamics of bacterial communities and interaction networks in thawed fish fillets during chilled storage in air. *Int. J. Food Microbiol.* **2019**, *293*, 102–113. [CrossRef]

37. Lakshmanan, R.; Jeya Shakıla, R.; Jeyasekaran, G. Survival of amine forming bacteria during the ice storage of fish and schrimp. *Food Microbiol.* **2002**, *19*, 617–625. [CrossRef]

38. Syropoulu, F.; Parlapani, F.F.; Kakasis, S.; Nychas, G.-J.E.; Boziaris, I.S. Primary processing and storage affect the dominant microbiota of fresh and chill-stored sea bass products. *Foods* **2021**, *10*, 671. [CrossRef]

39. Chiu, C.Y.; Miller, S.A. Clinical metagenomics. *Microb. Genom.* **2019**, *20*, 341–355. [CrossRef]

40. Matyar, F.; Kaya, A.; Dinçer, S. Antibacterial agents and heavy metal resistance in gram-negative bacteria isolated from seawater, shrimp and sediment in Iskenderun Bay, Turkey. *Sci. Total Environ.* **2008**, *407*, 279–285. [CrossRef]

41. Rezgui, I. Étude de l'antibioresistance chez deux espèces de poisons aquacoles en Tunisie: Le loup (*Dicentrarchus labrax*) et la daurade (*Sparus aurata*). Master's Thesis, University of Sciences of Tunis, Tunis, Tunisia, Institut des Sciences et Technologies de la Mer, Tunis, Tunisia, 2010.

42. Canak, O.; Akayli, T. Bacteria recovered from cultured gilt-head seabream (*Sparus aurata*) and their antibiotic susceptibilities. *Eur. J. Biol.* **2018**, *77*, 11–17. [CrossRef]

43. Shabana, B.M.; Elkenany, R.M.; Younis, G. Sequencing and multiple antimicrobial resistance of *Pseudomonas* fluorescens isolated from Nile tipalipa fish in Egypt. *Brazil J. Biol.* **2022**, *84*, e257144. [CrossRef]

44. Rigos, G.; Kogiannou, D.; Padros, F.; Cristofol, C.; Florio, D.; Fioravanti, M.; Zarga, C. Best European seabass (*Dicentrarchus labrax*) and gilthead seabream (*Sparus aurata*) in Mediterranean aquaculture. *Rev. Aquac.* **2021**, *13*, 1285–1323. [CrossRef]

45. Zhang, J.; Li, G.; Zhang, G.; Kang, W.; Duan, S.; Wang, T.; Li, J.; Huangfu, Z.; Yang, Q.; Xu, Y.; et al. Performance evaluation of the gradient diffusion strip method and disk diffusion method for ceftazidime-avibactam against *Enterobacterales* and *Pseudomonas aeruginosa*: A dual-center study. *Front. Microbiol* **2021**, *12*, 710526. [CrossRef]

46. ECDC, European Centre for Disease Prevention and Control. Surveillance of Antimicrobial Resistance in Europe-Annual Report of the European Antimicrobial Resistance Surveillance Network (EARS-Net) 2017. ECDC: Stockholm, Sweden, 2018. Available online: https://www.ecdc.europa.eu/sites/default/files/documents/EARS-Net-report-2017-update-jan-2019.pdf (accessed on 26 September 2022).

47. ECDC, European Centre for Disease Prevention and Control. Antimicrobial Resistance Surveillance in Europe 2022—2020 Data. ECDC: Solna, Sweden, 2022. Available online: https://www.ecdc.europa.eu/sites/default/files/documents/Joint-WHO-ECDC-AMR-report-2022.pdf (accessed on 26 November 2022).

 *foods*

Review

The Zoonotic Potential of Chronic Wasting Disease—A Review

Michael A. Tranulis [1,*] and Morten Tryland [2]

1. Department of Preclinical Sciences and Pathology, Faculty of Veterinary Medicine, Norwegian University of Life Sciences, 5003 Ås, Norway
2. Department of Forestry and Wildlife Management, Faculty of Applied Ecology, Agricultural Sciences and Biotechnology, Inland Norway University of Applied Sciences, 2480 Koppang, Norway
* Correspondence: michael.tranulis@nmbu.no; Tel.: +47-67232040

Abstract: Prion diseases are transmissible neurodegenerative disorders that affect humans and ruminant species consumed by humans. Ruminant prion diseases include bovine spongiform encephalopathy (BSE) in cattle, scrapie in sheep and goats and chronic wasting disease (CWD) in cervids. In 1996, prions causing BSE were identified as the cause of a new prion disease in humans; variant Creutzfeldt-Jakob disease (vCJD). This sparked a food safety crisis and unprecedented protective measures to reduce human exposure to livestock prions. CWD continues to spread in North America, and now affects free-ranging and/or farmed cervids in 30 US states and four Canadian provinces. The recent discovery in Europe of previously unrecognized CWD strains has further heightened concerns about CWD as a food pathogen. The escalating CWD prevalence in enzootic areas and its appearance in a new species (reindeer) and new geographical locations, increase human exposure and the risk of CWD strain adaptation to humans. No cases of human prion disease caused by CWD have been recorded, and most experimental data suggest that the zoonotic risk of CWD is very low. However, the understanding of these diseases is still incomplete (e.g., origin, transmission properties and ecology), suggesting that precautionary measures should be implemented to minimize human exposure.

Keywords: cervids; CWD; wildlife; zoonosis

Citation: Tranulis, M.A.; Tryland, M. The Zoonotic Potential of Chronic Wasting Disease—A Review. *Foods* **2023**, *12*, 824. https://doi.org/10.3390/foods12040824

Academic Editors: Frans J.M. Smulders and Arun K. Bhunia

Received: 11 November 2022
Revised: 1 February 2023
Accepted: 6 February 2023
Published: 15 February 2023

1. Introduction

Zoonoses are human diseases caused by pathogens derived from natural vertebrate animal reservoirs either directly or via intermediate animal hosts. It is estimated that of the emerging infectious diseases in humans after 1940, at least 60% are zoonotic and that the majority of these (>70%) are caused by pathogens originating in wildlife [1].

Prions are unique pathogens consisting of protein aggregates that cause incurable transmissible neurodegenerative diseases in humans and some other mammalian species [2]. These diseases (Tables 1 and 2) are, with three notable exceptions, very rare and, although transmissible, not normally contagious. Rather, they occur naturally as sporadic and/or genetic diseases, although outbreaks can occur under conditions created by humans (e.g., recycling of prion infected feedstuff or iatrogenic) [3]. The exceptions are classical scrapie in sheep, chronic wasting disease (CWD) in deer, and camelid prion disease in dromedary camels [4]. For these diseases, the infectious prions are present at high titers in lymphoid organs [5–7] and detectable in bodily excretions, allowing horizontal (nose-to-nose) or indirect transmission via contaminated environs [8]. These prion diseases, therefore, pose particular problems, not only because infectious prions are abundantly present in musculature and other edible tissues, thus entering the human food chain, but also because the release of prions to the environment is building a transmission potential over time, contributing to increased infection pressure for animals sharing these habitats [9–11]. The latter problem is compounded by the extraordinary physiochemical stability of prions, making prion-contaminated environs a long-term challenge [9,12,13].

Table 1. Human prion diseases and their epidemiological profile.

Disease	Mode of Occurrence	References
Creutzfeldt-Jakob disease		
Sporadic, sCJD	Sporadic	[14]
Sporadic fatal insomnia	Sporadic	[15]
Genetic CJD, gCJD	Familial, *PRNP* mutations	[16]
Iatrogenic CJD, iCJD	Acquired, medical or surgical treatment	[17]
Variant CJD, vCJD	Acquired, foodborne zoonosis	[18]
Kuru	Acquired, cannibalism (disease eradicated)	[19]
Gerstmann-Sträussler-Scheinker disease, GSS	Familial, *PRNP* mutations	[20]
Fatal familiar insomnia, FFI	Familial, *PRNP* mutations	[20]
Variable proteinase sensitive prionopathy VPSPr	Sporadic	[21]

Table 2. Animal prion diseases and their epidemiological profile.

Disease and Species of Occurrence	Mode of Occurrence	References
Scrapie in sheep and goats		
Classical	Contagious	[22]
Atypical/Nor98	Sporadic	[23]
Bovine spongiform encephalopathy in cattle, BSE		
Classical C-BSE	Foodborne	[24]
Atypical L-BSE	Sporadic	[25,26]
Atypical H-BSE	Sporadic	[25,27]
Chronic wasting disease in deer, CWD		
Classical C-CWD	Contagious	[28]
Moose sporadic CWD, Mo-sCWD	Sporadic	[29]
Red deer sporadic CWD, Rd-sCWD	Sporadic	[30]
Camelid prion disease	Contagious	[4]
Transmissible mink encephalopathy TME	Foodborne (BSE L-form)	[31]
Transmissible feline encephalopathy FSE	Foodborne (C-BSE)	[32]

Natural transmission of CWD occurs most frequently between genetically susceptible individuals of the same or a closely related species [33]. This is explained by the molecular composition of prions and their peculiar way of propagation [34]. The normal cellular prion protein (PrP^C), encoded by the *PRNP* gene [35,36], is a cell surface protein expressed in most tissues and at high levels in the central and peripheral nervous systems [37]. Its physiological functions are not fully understood [38–40]. Prions are multi-molecular aggregates of a misfolded conformer (termed PrP^{Sc}) of PrP^C [41,42]. In prion propagation, PrP^{Sc} binds to PrP^C and templates the misfolding of PrP^C into the PrP^{Sc} conformational state i.e., adding building blocks to the PrP^{Sc} aggregate. This process is most efficient when the primary structures (amino acid sequence) of the interacting PrP molecules are identical [43]. Even a single amino acid difference can impose a significant energy barrier on the misfolding process, thus slowing or even blocking the molecular event that drives prion disease pathogenesis and transmission dynamics [44]. This explains, for a large part, the sometimes-potent genetic modulation of prion disease susceptibility observed in scrapie [45–48] and CWD [49–52], which is governed by alteration of the *PRNP* gene causing amino acid substitutions in the PrP^C structure.

Conversion of PrP^C to PrP^{Sc} was demonstrated in cell-free, in vitro assays almost 30 years ago [53]. Today, ultrasensitive methods are available for detection of PrP amyloid seeding activity, which correlates strongly with prion infectivity [54–57]. The barrier to transmission of prion disease between different species has been demonstrated in many experimental studies and has also been observed in practical husbandry.

For instance, classical scrapie in sheep has been a problem in European sheep production for about 250 years [58]. Scrapie-infected sheep were often co-housed with other production animals, horses, and pets. Still, spillover to these species was never recorded, except for goats, which are susceptible [59,60]. Human exposure must have been common, since scrapie was widely distributed, and no tests were available to remove infected animals from consumption. In most regions, wildlife, such as red deer (*Cervus elaphus*), roe deer (*Capreolus capreolus*) and other cervids, were probably also exposed by sharing grassland with scrapie-infected sheep over the centuries. It thus seems likely that a spillover of scrapie to cervids, resulting in CWD or a CWD-like disease (i.e., with subsequent horizontal transmission), would have resulted in disease outbreaks that would not have gone unnoticed. However, no such outbreaks have been recorded among European cervids, indicating a barrier for transmission of prions between sheep and cervids.

In addition, transmission properties of prions can be modulated by structural arrangements of the prion particle, implying that a PrP molecule with a given primary structure can build up PrP^{Sc} aggregates with distinct features, such as altered transmission properties [61,62].

Another important aspect of prion biology is that the above-mentioned model for prion propagation may allow a spectrum of conformational states to be propagated in parallel. This, "cloud of conformations" model, is one way of understanding prion adaptability and plasticity [63]. Different prion structures in an isolate may compete in a structure-selection process, i.e., those that most effectively misfold the available PrP substrate will dominate. This may therefore vary between host tissues and between individuals and/or species encoding different PrPs. In this way, the transmission of a prion to a new host species may elicit adaptations that alter the characteristics of the original prion structure and thereby also its characteristics, for instance concerning clinical symptoms (or lack thereof), prion tissue distribution, and transmission capacity to other species [64–67].

Thus, a prion that appears harmless to humans in its original host may, via one or more intermediate hosts, be altered so that its zoonotic potential is increased. Such alterations in transmission properties and hence zoonotic potential of prion agents are difficult to predict. Thus, the occurrence of prion diseases in humans and animals must be closely monitored and measures that minimize the entry of prions into the human food web should be continued.

2. Chronic Wasting Disease

2.1. Historical Background North America

During 1967–1979, a syndrome called chronic wasting disease was observed in 53 mule deer (*Odocoileus hemionus hemionus*) and in one black-tailed deer (*Odocoileus hemionus columbianus*) in captivity in Colorado, USA. The clinical signs appeared in adult animals and consisted of altered behavior, progressive weight loss and death within two weeks to eight months after onset of clinical signs. Diseased animals had specific CNS pathology suggesting a spontaneously occurring form of transmissible spongiform encephalopathy (TSE), not previously reported in deer species, and with an unknown origin [68].

2.2. Geographical Expansion, Increasing Exposure and Prevalence

A typical feature of CWD is that infected animals shed prions via saliva, feces, urine and blood, and possibly also through nasal secretions, milk and semen, and oral exposure is regarded as the main route of natural infection [69–71]. Susceptible hosts may be exposed to CWD prions through physical contact with an infected animal, or indirectly via contaminated food, water, and other environmental factors. In contrast to many infectious diseases in wildlife, field and modeling data from North America have indicated that CWD epizootics develop relatively slowly and that the disease remains at a low prevalence and spatially localized for a decade or more after introduction [72]. Depending on management strategy and test regimes, this may explain why the disease is often identified 10–20 years after its introduction to a cervid population [73]. However, prevalence is increasing with time after disease introduction, presumingly due to indirect transmission through contaminated food, water and the environment [74].

After the recognition of CWD in free-ranging mule deer and wapiti in 1981, a contiguous area in north-eastern Colorado, south-eastern Wyoming and western Nebraska was regarded as an enzootic region, in which CWD probably had been present for several decades prior to its recognition [72]. The introduction of CWD to Toronto Zoo probably took place via the import of infected animals from Denver Zoo, USA, and further spread of the disease from Toronto Zoo remains a possibility, but no evidence for such spread could be documented in a retrospective investigation of available material [75]. CWD was also imported to South Korea via infected live cervids [76]. More recently, CWD has been diagnosed in captive and free-ranging moose (*A. a. shirasi*) in the USA [77,78]. Since 2000, CWD has continued to spread and has been detected in many other foci in Northern America. The disease now affects 30 states in the USA and four Canadian provinces, for a detailed overview of CWD occurrence in North American wild and captive deer see [79].

2.3. CWD in Northern Europe

In North America, CWD has been observed in several deer species [80], including a recent case in captive reindeer (Chronic Wasting Disease Alliance, 2018), but hitherto not in free-ranging reindeer or caribou (*Rangifer tarandus*), despite overlapping habitats with other cervid species known to be affected with CWD. Inoculation studies, however, have indicated that two of three reindeer that were orally inoculated with brain homogenates from white-tailed deer (WTD) with CWD were susceptible, developing clinical signs 17–18 months post inoculation (p.i.) and died within weeks of developing clinical signs. In contrast, three reindeer inoculated in the same manner with brain homogenates from elk did not develop clinical signs and were euthanized 22–61 months p.i. [81]. Although the results could indicate that reindeer are less susceptible to elk derived CWD, the authors argue that host *PRNP* genetics are the most likely explanation. The reindeer inoculated with the elk isolate were heterozygous at codon 138 (S/N) whereas the two clinically affected reindeer inoculated with the WTD isolate were homozygous 138SS. The one that remained healthy after inoculation with the WTD isolate carried the 138S/N genotype, suggesting that this polymorphism may be protective [81]. The 138S/N polymorphism appears to be absent among Norwegian wild and semi-domesticated reindeer [49,82].

Norway hosts about 25,000 wild reindeer, distributed between 24 more or less separated populations. In March 2016, a wild European tundra reindeer (*R. t. tarandus*) was found moribund during a research field study in Nordfjella, Norway, when a reindeer flock was approached by helicopter. The animal died and was necropsied. Except for muscle hemorrhages, no other gross pathological findings were observed, but analysis of brain tissue indicated CWD [83]. This represented the first naturally occurring CWD case outside North America and the first case in a *Rangifer* subspecies. During a stamping out procedure of the Nordfjella reindeer population, 19 animals tested positive for CWD. As a result of increased surveillance of other wild reindeer populations, two cases have been diagnosed, both in the Hardangervidda population. Hardangervidda is the largest national park in Norway, hosting the largest remaining wild reindeer population in Western Europe, about 6000 to 9000 animals.

In addition to the wild reindeer, Norway hosts (2020) about 215,000 semi-domesticated reindeer of the same sub-species, the Eurasian tundra reindeer [84]. Most of the semi-domesticated reindeer in Norway is comprising a traditional cornerstone of the Sami people and culture in Fennoscandia, whereas a non-Sami reindeer herding is conducted north of the Nordfjella mountain region where CWD was first recognized. Although an exchange of animals between the wild reindeer in Nordfjella and the adjacent semi-domesticated reindeer has been observed, in particular bulls drifting north during the rut season, no CWD-positive animals have been found in this or other herds of semi-domesticated reindeer in Norway (about 57,000 animals tested, 2016–Jan. 2023). Semi-domesticated reindeer are tagged by each owner and are typically gathered twice a year, for transition to the calving ground and summer pasture regions in early spring, and again during late summer and fall for other purposes, such as tagging calves, separation of herds, selecting animals for slaughter, and parasite treatment. During the gathering and handling, reindeer are in close contact with their owners and family members, comprising the herding unit, the siida. Animals for slaughter are driven by foot if feasible, or more commonly transported on trucks to the slaughterhouse. The reindeer are subjected to veterinary inspection before and after slaughter (i.e., ante mortem control and meat control). For 2020, 52,642 reindeer were slaughtered, comprising 1,253 tons of meat, representing an economical value of about 100 million NOK [84].

2.4. CWD with Unusual Features in Moose and Red Deer in Northern Europe

In May 2016, two moose (*Alces alces*) were diagnosed with CWD in Selbu, not far from Trondheim, and approximately 300 km north of Nordfjella where the first reindeer case was located. Following increased surveillance of cervid populations and species in Norway, CWD has been diagnosed in 11 moose in Norway, four in Sweden [85] and three in Finland, in addition to three red deer (*Cervus elaphus*) in Norway (November 2022). Data from the investigations of moose and red deer showed that, whereas reindeer with CWD were 2.5–8 years old, CWD affected moose and red deer were 12–15 years old. In reindeer, all CWD cases tested positive for PrP^{Sc} in lymphoid tissues, whereas in moose and red deer, PrP^{Sc} deposits appeared to be confined to the CNS, and lymphoid tissues were negative [29,30,85]. Further investigations have confirmed that North American CWD strains differ from those observed in Europe, and that the European strains causing CWD in reindeer, moose and red deer are all separate strains [79,86].

The CWD cases in moose and red deer were strikingly different from CWD as observed in North America and from the outbreak in wild reindeer; in terms of age-category, organ distribution of PrP^{Sc}, histopathology and epidemiology, with a seemingly sporadic appearance. By analogy to the well-established dichotomy of "classical" vs. "atypical" scrapie and BSE [87], scientists and governmental bodies in Northern Europe have arbitrarily adopted the term "atypical" CWD to distinguish the newly discovered variants in European moose and red deer, from the well-described contagious forms of CWD, reviewed in [88]. In Table 2, we use the descriptive epidemiological terms "moose sporadic CWD" and "red deer sporadic CWD".

The expansion of CWD in North America and its appearance in Northern Europe will inevitably increase human exposure. Further, CWD prions are more diverse and adaptable than previously recognized. This diversity and adaptability are seen in both North America and Europe [86,89–95], suggesting that inter-species transmission properties and zoonotic potential may also be altered. The emerging dynamic character identifies CWD as a worrisome animal prion disease deserving our close attention.

In the following paragraphs we will recapitulate epidemiological, in vitro, and bioassay data addressing the zoonotic potential of CWD.

3. Zoonotic Potential

3.1. Case Reports, Epidemiological Observations, and Active Surveillance

Prion diseases have long incubation periods; in humans reaching up to fifty years [19]. The long time from potential exposure to disease manifestation makes epidemiological investigation of the zoonotic potential of animal prion disease difficult and retrospective. In addition, disease phenotypes may deviate. Although recognized as a problem, phenotypic diversity played an important role when establishing an association between variant Creutzfeldt-Jakob disease (vCJD) and exposure to BSE infected meat. The vCJD cases were unusually young (mean age around 30) as opposed to sporadic Creutzfeldt-Jakob disease (sCJD), which has a mean age of onset around 60. The clinical symptoms and disease duration also differed, and based on analyses of the *PRNP* gene, genetic prion disease could be ruled out, rendering the newly discovered disease a "new variant" of CJD [18]. The epidemic of vCJD peaked in 2000, affecting mainly UK citizens, but also appearing in many other countries [96]. Molecular analysis of proteinase K resistant PrP fragments from vCJD cases revealed a band pattern identical to that seen in cattle and rodents inoculated with material from BSE infected cattle [97]. Cases also presented with unique neuropathological features, most notably the presence of multiple kuru-like plaques, surrounded by vacuolization, clearly distinguishing the condition from sCJD [18]. In addition, the vCJD cases appeared in geographical areas that had been heavily affected with BSE 10 years earlier.

What would have been the situation if vCJD had presented disease characteristics similar or indistinguishable to sCJD; would it still have been recognized as a distinct disease and linked to BSE? The answer is "probably not", illustrating the importance of diagnostic accuracy i.e., the ability to discriminate between similar disease pathologies and varieties of prion agents. This has been explored for sCJD [14,98–100], genetic Creutzfeldt-Jacob disease (gCJD) [16] and some animal prion diseases [101–103] and has resulted in a growing catalogue of disease sub-types and agent varieties. Thus, criteria for detailed active surveillance and diagnostics are to some extent available. Implementation of these tools in routine diagnostics and surveillance is however technically demanding and costly.

For an extensive review of the global incidence of CJD and inherent challenges related to diagnosis and surveillance see [104].

In 2006, Mawhinney and collaborators investigated the relative risk of contracting CJD for residents in CWD-endemic areas in Colorado with those living in non-CWD endemic areas [105]. The assumption was that people living in CWD-endemic areas were more exposed to CWD since most of the venison was consumed locally. They investigated a total of 65 CJD cases from 1979 through 2001 (of 506,335 deaths) and found no significant difference in CJD occurrence between the groups. Nor did they observe any increase in CJD rate in CWD affected areas, or in Colorado as a whole, concluding that CWD related human prion disease must be rare or nonexistent in Colorado.

The scientific literature contains a few case reports of rapid neurodegenerative disease in subjects with known exposure to CWD. Some of the cases have presented with unusual clinicopathological features, such as young age, but detailed analysis has failed to associate any of the cases to CWD [106,107]. Further, a cohort analysis (six years follow up) of 81 individuals attending a barbeque where CWD infected venison was unknowingly served, did not observe any neurodegenerative disease that could be linked to the exposure [108]. In conclusion, there is currently no epidemiological evidence of human prion disease caused by CWD. The datasets are however limited, for instance concerning the clinicopathological spectrum of potential human conditions caused by CWD, and the time of observations, which needs to span many decades.

3.2. In Vitro Amplification Methods for Assessment of Transmission Barriers

Conformational conversion of PrP^C, seeded by the presence of preexisting PrP^{Sc} molecules, was demonstrated in cell-free in vitro systems, using purified components already in 1994 [53] and soon the barrier to transmission of prions between species was elegantly explored and demonstrated by this method [109]. In the protocol, PrP^C and PrP^{Sc} were mixed under denaturing conditions, with an excess of PrP^{Sc} roughly 50-fold over PrP^C. Prior to incubation at 37 °C for two days, samples were sonicated [109]. Soto and collaborators developed this further by using fresh brain homogenates as a PrP^C source and by including repeated short bursts of intense sonication during the incubation, which dramatically sped up the conversion process [57]. The new method, designated protein misfolding cyclic amplification (PMCA) was highly sensitive and mimicked in vivo prion propagation, with de novo generation of infective prions, inter-species transmission potential and prion strain features [56]. This method has been used to detect and quantify prions in bodily fluids of infected animals with extreme sensitivity [110].

As an alternative to sonication, mechanical disruption of PrP^{Sc} aggregates can successfully be achieved by vigorous shaking, so-called quaking, used in quaking-induced conversion (QuIC) assays [111], which use recombinant PrP (recPrP) as substrate for the conversion reaction. QuIC assays were shown to have a sensitivity matching that of mouse bioassays (see below) [112]. Both PMCA and QuIC assays depend on handling of individual test-tubes for analysis of reaction products with western blot (WB) and are therefore less suited for high-throughput screening.

Another method known as amyloid seeding assay (ASA) also involved shaking and recPrP, but with the addition of Thioflavin T (ThT) that intercalates with misfolded PrP, and allows high-throughput multi-well readouts of fluorescence [113]. A modified, real-time version of the QuIC assay (RT-QuIC), using ThT as with ASA, but less prone to false positive signals, has been developed [114] and is today the most widely used method for ultrasensitive detection of PrP seeding activity, together with, and/or combined with the original PMCA method.

RT-QuIC and PMCA have been used to detect trace levels of amyloid seeding activity in tissues and body fluids of deer with pre-clinical or clinical CWD, such as saliva [115–117], urine [118], feces [119,120] and blood [121]. For a detailed comparative analysis of CWD prion detection by conventional, bioassay and amplification methods see [122]. The main advantage of the RT-QuIC method is that a standardized "universal" recombinant PrP substrate, for instance recombinant bank vole (*Myodes glareolus*) PrP, can be used to test amyloid seeding activity in tissues from a variety of different species, which makes the method well-suited for screening purposes [123]. It is also a benefit that the generated product contains no prion infectivity, which constitutes a laboratory health and safety issue. Conversely, the product generated with PMCA is infectious and the reaction depends on species and sequence specific PrP^C brain homogenate as substrate, which matches the incoming prion seed. This makes the PMCA method less suited for screening of samples of unknown origin but more feasible for the analysis of prion strain features and for estimating within- and inter-species transmission potential of prions [69].

Early in vitro evidence of a strong molecular barrier for transmission of CWD to humans came from a study using cell-free conversion. It was demonstrated that CWD isolates from elk, white-tailed deer and mule deer could convert human and bovine PrP, but were more than 10-fold less efficient than cervid PrP substrates, while conversion of sheep PrP was intermediate [124]. Furthermore, PMCA experiments with brain homogenates from Tg1536 mice overexpressing human PrP (MM129 genotype), gave no conversion when seeded with mule deer CWD or material from infected Tg1536 mice. Conversion of human PrP required several rounds of strain adaptation in PMCA or serial passage in transgenic mice [125], demonstrating that in vitro or in vivo adaptation of a prion strain can alter its transmission properties independent of the PrP primary structure.

To identify structural differences between human and deer PrP that impede conversion and cross-species transmission, Kurt and co-workers [126] cloned and expressed chimeric human and elk PrP, in which specific amino acids in the human PrP were substituted with those of the elk structure. They used cell lysates of transfected cells as substrate for PMCA. They did not observe any conversion of huPrP but achieved very efficient conversion with some of the hu-elk chimeric PrP substrates, results which fitted well with inoculation experiments of Tg-Hu mice and Tg-Hu-Elk chimeric mice (see below).

Further experiments with PMCA [127] have showed that CWD isolates from WTD, elk and reindeer experimentally inoculated with WTD isolate were capable of converting huPrP substrates covering the 129MM, MV and VV genotypes, although with varying efficiency. Recently, CWD isolates from six cervid species; WTD, mule deer, and elk from North America, and reindeer, red deer, and moose from Norway, were compared with the PMCA method for their inter-species transmission potential [79]. Some conversion of huPrP 129M and 129V was observed with North American CWD isolates, but no conversion was observed with any of the Norwegian isolates, suggesting that the Norwegian isolates might have a somewhat lower zoonotic potential. Conversely, the Norwegian reindeer isolate effectively converted sheep, bovine and hamster PrP, thus displaying a potential capacity to cross species barriers, comparable to that of CWD isolates from WTD. Interestingly, the Norwegian reindeer isolate had previously been shown to transmit poorly to bank vole, compared with North American CWD isolates [86].

3.3. Transmission of CWD to Transgenic Mice Expressing Human PrP

Natural occurrence of CWD has been recorded in several cervid species including white-tailed deer, mule deer, Rocky Mountain elk, moose, and reindeer. In addition, CWD has been experimentally transmitted to laboratory rodents and either intracerebrally and/or orally to sheep [128], cattle [129], pigs [130], cats [131], ferrets [132] and to squirrel monkeys [133]. Although this species spectrum may indicate a cause for concern, transmission of CWD between cervids is facilitated by cervid specific structural features of the prion protein [134–136], lowering the transmission barrier. Thus, transmission of CWD to non-cervid species, has been relatively inefficient, for instance compared with BSE.

Transgenic mice, engineered to express human PrP (huPrP, "humanized mice") have been used to assess the human barrier for transmission of CWD (Table 3). To optimize transmission success, mouse lines that overexpress huPrP are often used. Moreover, mouse lines known to be sensitive to human prion isolates or the zoonotic BSE agent are used and infectivity of CWD isolates is demonstrated by inoculation in mice expressing cervid PrP ("cervidized" mice) or bank voles. In an elegant study, mice were engineered to express a human-elk chimeric PrP, in which four amino acids were substituted in huPrP, creating a loop sequence (aa165–177) identical to the elk PrP sequence. In contrast to huPrP mice, the chimeric (huPrP$^{elk165-177}$) mice proved susceptible to CWD isolates, but they were concurrently less sensitive toward human CJD prions than their huPrP counterparts [134]. This study pinpointed important structural elements contributing to the barrier for CWD transmission to humans.

Table 3. Chronic wasting disease transmission experiments with transgenic mice expressing Human PrP ("humanized mice").

CWD Source		huPrP, 129MV	Readouts							Reference
North America	Europe		Clinical signs	Brain pathology, IHC, PrP^{Sc}	WB PrP^{Res}	Other	RT-QuIC	PMCA	Serial passage	
Elk	NA	Tg40,1X,129M Tg1, 2X, 129M	Neg. (0/29) Neg. (0/22)	NA	Neg.	PTA Neg.	NA	NA	NA	[137]
Elk, MD [1], WTD [2]	NA	Tg440, 2X	Neg. (0/67)	Neg. (selected mice tested)	NA	NA	NA	NA	NA	[138]
MD	NA	Tg152, 2X 129VV Tg45, 4X 129MM Tg35, 6X, 129MM	Neg. (0/41)	Neg.	Neg.	PTA Neg.	NA	NA	NA	[139]
WTD	NA	HuMM129, 1X HuMV129, 1X HuVV129, 1X	Neg. (0/72)	NA	NA	IDEXX Spleen, Neg.	NA	NA	NA	[140]
Elk and MD	NA	Tg(huPrP) 1-2X Tg(huPrP$^{elk166-174}$)	Neg. (0/12) Pos. (7/8 Elk CWD), (3/4 MD CWD)	Neg. Pos.	Neg. Pos.	PTA Neg. Pos.	NA	NA	NA	[126]
Elk, WTD, MD	NA	Tg66, 8-16X 129M TgRM, 2-4X 129M	4/52 suspicious 0/45	Neg.	Neg.	PTA Neg.	Inconclusive	NA	NA	[141]
	One reindeer, two moose	Tg35 2X, 129VV Tg152c 6X 129MM	0/19 RD CWD 0/39 MO CWD	Neg.	Neg.	NA	NA	NA	NA	[142]
WTD, Wisc-1, 116AG isolates	NA	Tg650, 6X, 129MM	Myoclonus, variable CNS signs in 93.8%	1/5, remaining animals NA	1/20	NA	7/18 Pos. Brain 8/18 Neg. brain 3/18 Inconclusive	NA	2nd passage to Tg650 mice 5/10 Pos. Bank vole 4/9 Pos.	[143]

[1] Mule deer, [2] White-tailed deer.

In prion bioassays, the primary clinical readout is progressive neurological disease. The prion disease diagnosis is then according to conventional methods confirmed by brain pathology and immunohistochemistry (IHC) detection of PrP^{Sc}, often combined with WB analysis.

A challenge inherent to mouse bioassays is the short lifespan of mice (around 2.5 years) compared with the extended incubation periods frequently seen in primary transmissions of prion isolates. When primary diagnostic results are inconclusive and/or negative, other, more sensitive methods are available to test for subclinical transmission and/or asymptomatic carrier status. This is important not only to detect minute levels of PrP^{Sc}, but also because prion infectivity titers do not always correlate with conventional diagnostic markers i.e., prion replication can occur without recognizable pathology and without proteinase resistant PrP^{Sc} accumulations [144]. As evident from Table 3, only two of the CWD transmission studies using huPrP mice have reported data with the aforementioned highly sensitive in vitro conversion methods or from serial passage experiments.

Race and co-workers [141] found that Tg66 and TgRM mice, overexpressing huPrP 8–16-fold and four-fold, respectively, did not develop typical or terminal prion disease after more than 700 days post inoculation with three different CWD isolates. They did not observe PrP^{Sc} deposits in IHC or PrP^{Res} fragments in WB, hallmarks of prion disease. They did, however, observe 18 clinically suspect mice of the 108 inoculated. All mice were analyzed with RT-QuIC for detection of PrP amyloid seeding activity. In four mice from the Tg66 group, results were inconclusive, reaching slightly above detection limit of the assay, suggesting that the observed clinical abnormalities could be early signs of prion disease. Race and co-workers discuss whether the RT-QuIC data could be false positive caused by residual inoculum or by the abnormally high PrP expression levels in the Tg66 mouse line, potentially causing spontaneous PrP amyloids/aggregates, detectable with the RT-QuIC method. The low number of uninoculated control mice tested was insufficient to rule out the latter possibility.

Another method for increasing prion detection sensitivity is by precipitating misfolded PrP with sodium phosphotungstic acid (PTA) prior to analysis by WB. PTA-enhancement has been shown to increase detection sensitivity for CWD approximately 100-fold compared with crude extracts [145]. In experiments with CWD inoculated huPrP mice, PTA-enhancement has not resulted in PrP^{Res} detection.

In a recently published report, Tg650 mice, overexpressing huPrP (129MM) approximately six-fold, developed unusual clinical signs with progressive myoclonus (involuntary twitching of a muscle or group of muscles) after inoculation with two CWD isolates (Wisc-1, 116AG) from white-tailed deer [143]. Despite alarming neurological signs in many inoculated mice, histopathological analysis of the brain did not indicate TSE-pathology, whereas IHC analysis was performed in six animals, of which one (#328), displayed pericellular, granular PrP deposits, in the brain. Western blot analysis of brain material from this animal was negative for PrP^{Res}. Only one of the nine mice analyzed with WB was weakly positive, with an unusual two-band PrP^{Res} profile at 12 kDa and 7–8 kDa. Brain material from all mice was analyzed with a modified RT-QuIC protocol with enhanced sensitivity. With this protocol, all mice inoculated with the 116AG isolate were negative. The apparent disconnection between clinical signs and highly sensitive prion diagnostic markers suggests that the clinical signs could stem from a hard-to-detect prion agent. Unfortunately, secondary transmissions, which would provide a test for prion infectivity in these mice, were not reported. Among the Wisc-1 inoculated mice, a majority tested positive with RT-QuIC, although results also showed some inconsistencies, which was attributed to very low seeding activity. One such case was mouse #327 which had terminal illness but very low/inconsistent seeding activity in the brain. Interestingly, this mouse showed high seeding activity in feces, which was also detectable in 50% of the inoculated mice, suggesting that prion infectivity could be shed from some of the inoculated animals.

Transmission of sonicated fecal homogenate from mouse #327 to Tg650 mice and bank vole produced different results. In clinically ill Tg650 mice, no PrP^{Res} could be detected in brain homogenates and RT-QuIC analysis of the animals was not reported. In bank voles, six out of nine developed clinical symptoms. Three voles were tested for RT-QuIC seeding activity in brain and they were all positive and one (#3430) was also positive in spinal cord. Western blot analysis of brain and spinal cord homogenates from this animal revealed a typical three banding PrP^{Res} profile, dramatically different from that observed in the Wisc-1 inoculated Tg650 mice. Interestingly, the PrP^{Res} profile in the bank vole #3430 resembled that of the original WTD Wisc-1 isolate, but not the PrP^{Res} signature seen in bank voles inoculated with the WTD Wisc-1 isolate (first or second transmission).

This study [143] stands out in several ways from other investigations of CWD in humanized mice. Most strikingly, the high incidence of profound, albeit unusual, clinical signs among inoculated mice. Next, the lack of coherence between clinical signs and conventional and ultrasensitive diagnostic markers of prion disease, suggestive of toxicity driven by an easily misdiagnosed "stealth prion" evading most diagnostic modalities. The observation of seeding activity and prion infectivity in feces is also remarkable. Whether this is a phenomenon specific to the Tg650 mouse line or CWD strain, or a more widespread and previously overlooked feature of huPrP mice inoculated with CWD prions must be investigated. If the latter is shown to be the case, it will impact our understanding of the zoonotic potential of CWD, as interpreted from mouse bioassays.

Still, it can be argued from an epidemiological perspective that the traditional readout from primary prion bioassays, namely clinical neurological signs, and diagnosis of bona fide prion disease by conventional methods, provides the most relevant and informative analysis of the cross-species transmission potential for a prion. It is evident from Table 3, that primary transmission of a variety of CWD isolates to several huPrP mouse lines, overexpressing huPrP, has been uniformly negative. Although sub-passage and further use of ultrasensitive diagnostic tools, involving extra neural tissues, may identify aspects that can have been missed in earlier studies, the overall conclusion from mouse bioassays is that the human barrier for CWD transmission is very strong.

Finally, as seen in Table 3 only one of the published reports has used prion isolates from Europe [142]. Material from one reindeer and two moose CWD cases, all from Norway, were inoculated in huPrP Tg35 and Tg152c mice, covering 129 genotypes MM and VV. All inoculated mice remained healthy, and no signs of prion disease could be detected, suggesting that the human transmission barrier for these novel CWD strains is robust.

A potential weakness of the traditional Tg mouse lines, overexpressing huPrP, is that these do not precisely recapitulate tissue and organ-specific expression profiles of *Prnp* [146]. Many of the models have relatively low expression of *Prnp* in peripheral tissues, which may be important for studies of inter-species transmission potential, lymphotrophism and pathogenesis of experimental prion disease, arguing that further refinement of mouse models, for instance with gene-targeting could be beneficial [146].

3.4. Transmission of CWD to Non-Human Primates

The history of using nonhuman primates as models for human prion disease and in risk assessments has recently been comprehensively reviewed [147] and will therefore not be recapitulated in detail here. It is well established that the Squirrel monkey (*Saimiri sciureus*) is susceptible to both oral and intracerebral inoculation with different CWD isolates [133,148,149]. Indeed, the squirrel monkey is a permissive host to many prion agents such as kuru, vCJD, sCJD, Gerstmann-Sträussler-Scheinker disease (GSS), BSE, transmissible mink encephalopathy (TME) and sheep scrapie with relatively short incubation periods from 20 months to 46 months after intracerebral inoculation [150–152]. In contrast to the efficient transmission in squirrel monkeys, transmission experiments with cynomolgus macaques (*Macaca fascicularis*) have shown these to be less susceptible to animal prions, including CWD [153]. Macaques are evolutionary closer to humans [154] and considered a more precise animal model for human prion disease, although Macaque and Squirrel monkey PrPs are equally distant from human PrP [155].

Macaques have been shown to be susceptible to intracerebral inoculation of vCJD, atypical L-type BSE (L-BSE) and classical BSE (C-BSE) with incubation periods of two to three years and to sCJD with incubation period of around five years [156–158]. Classical scrapie was evident in a macaque after a 10-year incubation period, following a high-dose intracerebral inoculation of a classical scrapie isolate [159], illustrating the importance of very long and costly observation periods in this type of study.

In 2018, Race and co-workers [153], summarized a large study with oral and intracerebral inoculation of macaques with CWD prions. Some animals had been observed for up to 13 years after inoculation, without evidence of prion disease. The RT-QuIC assay was used to test brain, brain stem and spinal cord tissue for amyloid seeding activity, but results were similar between CWD inoculated and uninoculated animals. They observed some irregularities in the brain PrP-staining pattern of both inoculated and uninoculated animals and in two of the inoculated macaques PrP deposits that could potentially be disease-associated were observed. However, no histopathological or WB evidence of prion disease could be detected in these animals and tests with RT-QuIC were negative. Thus, the authors found no evidence of transmission of CWD to macaque.

In another, ongoing and unpublished study of CWD transmission to macaque that included oral infection with muscle tissue from cervids, preliminary congress interim reports and presentations have suggested that CWD has been transmitted to some macaques, albeit with atypical and subclinical disease manifestations. In tissues from some animals, a low level of PrP converting activity was observed with RT-QuIC and PMCA assays and sub-passage in mice overexpressing elk PrP (TgElk) or deer PrP (Tg1536) gave low attack rates, but subsequent passage from 2nd passage in mice, to bank voles resulted in 100% attack rates and appearance of typical prion disease pathology. Interestingly, infectivity was found also in the gastrointestinal tract [160]. These findings indicate that the species barrier to humans is not absolute, and it is likely that it can be crossed (Schätzl, personal communication).

Full comparative analysis of the two apparently contradicting macaque investigations must await publication of the latter, still ongoing investigation. However, both studies clearly demonstrate that the barrier for transmission of CWD to macaque is very strong, but probably not absolute, which is in accordance with data from transgenic mice and in vitro experiments. Differences between studies could be related to differences in CWD inocula i.e., strain differences and infective doses as well as differences among the recipient macaques.

4. Discussion

We have summarized available epidemiological, in vitro and bioassay derived data concerning the zoonotic potential of CWD. We have identified only one report in which CWD strains recently identified in Northern Europe have been analyzed for their zoonotic potential, by inoculation in huPrP mice [142] and one study exploring this by in vitro methods [79]. Since CWD strains identified in Northern Europe clearly are different from strains from North America, further experiments are needed (and ongoing) to map this out in further detail.

Data from recent bioassays in mice and macaques suggest that conventional readouts for prion disease should be strengthened by ultra-sensitive RT-QuIC and PMCA assays in combination with serial passage to analyze for prion infectivity. The phenomenon of unusual/atypical clinicopathological disease presentation and proteinase sensitive prion strains, evading traditional PrP^{Sc}/PrP^{Res} detection, is still incompletely understood, including its real-life epidemiological relevance. For instance, are the rare observations of abnormal PrP deposits in peripheral tissues in healthy individuals merely rarities reflecting the ultrasensitive methods used, or representations of phenomena directly relevant to surveillance programs and risk assessments? We know that prion agents can adapt and change characteristics when propagated within the host or in a new host according to mechanisms that are poorly understood.

Controlling a transmissible and potentially zoonotic disease in wild cervid populations is complicated, and many disease characteristics, such as long incubation time, no antibody production (i.e., no immunity), pathogen robustness in the environment and many other factors are further challenges to surveillance strategies. Furthermore, most of the affected cervid populations are in remote areas with restricted availability and infrastructure. Today, the management of these cervid populations in Fennoscandia is based on hunting, with a private motivation for preparing and consuming the game. During the culling of the affected reindeer population in Nordfjella, where CWD was first diagnosed, hunted carcasses were held in arrest until CWD test results were available. This practice, however, is very time consuming and costly, and may be evaluated against the precautionary principles. Thus, the appearance of CWD in wild cervids in Fennoscandia necessitates new management practices, for Norway and for the European Union.

A major goal for the management of CWD in Norway has been to prevent the disease from entering the semi-domesticated reindeer herds [161]. The non-Sami reindeer herding is conducted north of and in close proximity to Nordfjella, and exchange of animals between wild and semi-domesticated herds have been observed, opening for the possibility that infected wild reindeer may already have had contact with reindeer herding. However, about 14,000 semi-domesticated reindeer from these herds (Jan. 2023) have been tested with no CWD-positive animals detected [162].

It is important to keep in mind that also semi-domesticated reindeer are free-ranging year around just as much as the wild reindeer, and only gathered and handled a couple of times during the year. Despite being routinely inspected and herded, it is challenging to address disease among free-ranging animals in remote high mountain pastures, and fallen stock is quickly scavenged and decomposed making it difficult to address cause of death.

Exposure of people through consumption is very similar whether it is a wild, hunted reindeer or a semi-domesticated, slaughtered reindeer. Reindeer herders are probably consuming more reindeer meat than the general consumer. In addition, their work involves close contact with reindeer during gathering and handling of animals, but also through periods of supplementary feeding which is becoming increasingly common. Although the chance of CWD eradication may be greater in a semi-domesticated reindeer herd than in the wild populations, an appearance of CWD in reindeer herding will necessitate dramatic measures which may have a major impact on the herd size and structure, the use of pastures, collaboration between herders, the economy, as well as the social, traditional, and cultural aspects associated with reindeer herding.

5. Conclusions

No cases of human prion disease caused by CWD have been reported and most experimental data suggest that the zoonotic potential of CWD is very low. Based on the current knowledge and identified knowledge gaps regarding the zoonotic potential of the new CWD strains in Fennoscandia, it is good advice to keep human and animal exposure to prions to an absolute minimum and closely monitor and restrict CWD and other animal prion diseases to prevent these agents from entering the human food chain.

Author Contributions: Conceptualization, M.A.T., writing—original draft preparation, M.A.T. and M.T.; writing—review and editing, M.A.T. and M.T. All authors have read and agreed to the published version of the manuscript.

Funding: This research received no external funding.

Data Availability Statement: Not applicable.

Acknowledgments: We are thankful to Charles McLean Press and Gjermund Gunnes for critical reading of the manuscript.

Conflicts of Interest: The authors declare no conflict of interest.

Abbreviations

ASA	Amyloid seeding assay is a method by which recombinant prion protein is polymerized into amyloid fibrils in the presence of partly purified prion preparations. The newly generated fibrils that can be detected with dyes like Thioflavin T.
BSE	Bovine spongiform encephalopathy is a prion disease of cattle caused by prion-contaminated meat and bone meal.
CJD	General abbreviation of Creutzfeldt-Jakob disease in humans
sCJD	Sporadic Creutzfeldt-Jakob disease, caused by spontaneous conversion of PrP^C into PrP^{Sc} or by somatic mutation in PRNP which is the gene encoding PrP.
sFI	Sporadic form of fatal insomnia. Extremely rare sporadic form of the inherited familial fatal insomnia
FFI	Inherited prion disease caused by germ-line mutation in PRNP.
gCJD	Inherited form of Creutzfeldt-Jakob disease, caused by germ-line mutation in PRNP.
iCJD	Creutzfeldt-Jakob disease, caused by infection with prion-contaminated tissue grafts or medical preparations.
vCJD	Variant Creutzfeldt-Jakob disease, caused by BSE-contaminated feedstuff
CM	Cynomolgus macaques, Macaca fascicularis, Old World monkey used in experimental transmission studies of prion diseases, to test for zoonotic potential
CWD	Chronic Wasting disease, is an infectious prion disease affecting cervid species
FSE	Feline spongiform encephalopathy is prion disease of fields caused by intake of BSE-contaminated feedstuff
GSS	Gerstmann-Straussler-Sheinker syndrome is a human prion disease caused by germ-line mutations in PRNP
IHC	Immunohistochemistry is a commonly used method for selective identification of proteins in biological tissues by use of antibodies that binds specifically to the proteins of interest.
Mo-sCWD	Moose sporadic CWD is a prion disease recently identified in Fennoscandia (Norway, Sweden, and Finland). The disease has an apparently sporadic occurrence, affecting old animals, and prions appear confined to the central nervoussystem i.e., not detectable in peripheral lymphoid tissues. Our understanding of this disease, including its epidemiology and potential to infect other species is still incomplete and an area of intense investigation.
PMCA	Protein misfolding cyclic amplification is a method whereby in vitro nucleation-dependent conversion of PrP^C into PrP^{Sc} is accelerated by use of periodic fragmentation of PrP^{Sc} fibrils by intense bursts of ultrasonic waves, followed by incubation, allowing new PrP^{Sc} fibrils to form, amplifying the original signal. The PMCA method is used for ultrasensitive detection of prions in tissue samples of environmental samples, and for investigation of many aspects of prions.
PRNP	The gene encoding the prion protein
PrP	General abbreviation of the prion protein
PrP^C	The physiological cellular prion protein
PrP^{Res}	A misfolded and proteinase K resistant protein core of the prion protein detected in gel-electrophoresis and protein immunoblots (western blots)
PrP^{Sc}	An abnormal, pathogenic, and infectious conformer of the prion protein, isolated from patients with prion disease
recPrP	Recombinant prion protein, produced in bacteria
huPrP	Human prion protein
PTA	Phosphotungstic acid is used to precipitate and thus concentrate prions from tissue preparations to enhance detection sensitivity
QuIC	Quake induced conversion is a method for sensitive detection of misfolded PrP by in vitro conversion of an excess of recombinant PrP in the presence of a tissue derived seed, for instance from an animal suspected to be prion infected. While the PMCA method uses ultrasound to break apart PrP fibrils, the QuIC method achieves this by vigorous shaking (quaking).
Rd-sCWD	Red deer sporadic CWD is a prion disease observed in three red deer in Norway with what appears to be sporadic occurrence. Prions appear confined to the central nervous system i.e., not detectable in peripheral lymphoid tissues. Our understanding of this disease, including its epidemiology and potential to infect other species is still incomplete and an area of intense investigation.
RT-QuIC	Real-time quake induced conversion is a modified and improved variant of the QuIC method, allowing real-time detection of newly formed PrP aggregates with fluorescence detection of thioflavin T. The RT-QuIC method allows ultrasensitive detection of misfolded PrP in tissue samples, lymph, and environmental samples.
SM	Squirrel monkey, Saimiri sciureus, New World monkey, used in experimental transmission studies of prion diseases, to test for zoonotic potential.
ThT	Thioflavin T is a fluorescent dye which binds to proteins rich in beta-sheet structures, such as amyloid. Upon binding, the dye displays an enhanced fluorescence and emits at about 480 nm after excitation at 450 nm. ThT is widely used for detection of amyloid protein aggregates in tissues and in vitro, for instance with the RT-QuIC method.
TSE	Transmissible spongiform encephalopathy is a previously used term for the group of neurodegenerative diseases today known as prion diseases

WB	Western blot, a commonly used method for analysis of proteins, separated with electrophoresis and transferred to membranes for specific detection with antibodies raised against the protein(s) of interest. The term Western stems from a lab-jargon following a method for detection DNA, named after its inventor Edwin Southern. Similar detection of RNA is called Northern blot.
WTD	White-tailed deer, Odocoileus virginianus.
Zoonosis	An infectious disease that can be transmitted between animals and humans

References

1. Jones, K.E.; Patel, N.G.; Levy, M.A.; Storeygard, A.; Balk, D.; Gittleman, J.L.; Daszak, P. Global trends in emerging infectious diseases. *Nature* **2008**, *451*, 990–993. [CrossRef]
2. Prusiner, S.B. Prions. *Proc. Natl. Acad. Sci. USA* **1998**, *95*, 13363–13383. [CrossRef]
3. Ironside, J.W.; Ritchie, D.L.; Head, M.W. Prion diseases. *Handb. Clin. Neurol.* **2017**, *145*, 393–403. [CrossRef]
4. Babelhadj, B.; Di Bari, M.A.; Pirisinu, L.; Chiappini, B.; Gaouar, S.B.S.; Riccardi, G.; Marcon, S.; Agrimi, U.; Nonno, R.; Vaccari, G. Prion Disease in Dromedary Camels, Algeria. *Emerg. Infect. Dis.* **2018**, *24*, 1029–1036. [CrossRef]
5. Heggebo, R.; Press, C.M.; Gunnes, G.; Gonzalez, L.; Jeffrey, M. Distribution and accumulation of PrP in gut-associated and peripheral lymphoid tissue of scrapie-affected Suffolk sheep. *J. Gen. Virol.* **2002**, *83*, 479–489. [CrossRef]
6. Sigurdson, C.J.; Williams, E.S.; Miller, M.W.; Spraker, T.R.; O'Rourke, K.I.; Hoover, E.A. Oral transmission and early lymphoid tropism of chronic wasting disease PrPres in mule deer fawns (*Odocoileus hemionus*). *J. Gen. Virol.* **1999**, *80 Pt 10*, 2757–2764. [CrossRef]
7. van Keulen, L.J.; Schreuder, B.E.; Meloen, R.H.; Mooij-Harkes, G.; Vromans, M.E.; Langeveld, J.P. Immunohistochemical detection of prion protein in lymphoid tissues of sheep with natural scrapie. *J. Clin. Microbiol.* **1996**, *34*, 1228–1231. [CrossRef]
8. Gough, K.C.; Maddison, B.C. Prion transmission: Prion excretion and occurrence in the environment. *Prion* **2010**, *4*, 275–282. [CrossRef]
9. Marin-Moreno, A.; Espinosa, J.C.; Fernandez-Borges, N.; Piquer, J.; Girones, R.; Andreoletti, O.; Torres, J.M. An assessment of the long-term persistence of prion infectivity in aquatic environments. *Environ. Res.* **2016**, *151*, 587–594. [CrossRef]
10. Maddison, B.C.; Baker, C.A.; Terry, L.A.; Bellworthy, S.J.; Thorne, L.; Rees, H.C.; Gough, K.C. Environmental sources of scrapie prions. *J. Virol.* **2010**, *84*, 11560–11562. [CrossRef]
11. Miller, M.W.; Williams, E.S.; Hobbs, N.T.; Wolfe, L.L. Environmental sources of prion transmission in mule deer. *Emerg. Infect. Dis.* **2004**, *10*, 1003–1006. [CrossRef]
12. Hawkins, S.A.; Simmons, H.A.; Gough, K.C.; Maddison, B.C. Persistence of ovine scrapie infectivity in a farm environment following cleaning and decontamination. *Vet. Rec.* **2015**, *176*, 99. [CrossRef]
13. Georgsson, G.; Sigurdarson, S.; Brown, P. Infectious agent of sheep scrapie may persist in the environment for at least 16 years. *J. Gen. Virol.* **2006**, *87*, 3737–3740. [CrossRef]
14. Head, M.W.; Ironside, J.W. Review: Creutzfeldt-Jakob disease: Prion protein type, disease phenotype and agent strain. *Neuropathol. Appl. Neurobiol.* **2012**, *38*, 296–310. [CrossRef]
15. Parchi, P.; Capellari, S.; Chin, S.; Schwarz, H.B.; Schecter, N.P.; Butts, J.D.; Hudkins, P.; Burns, D.K.; Powers, J.M.; Gambetti, P. A subtype of sporadic prion disease mimicking fatal familial insomnia. *Neurology* **1999**, *52*, 1757–1763. [CrossRef]
16. Baiardi, S.; Rossi, M.; Mammana, A.; Appleby, B.S.; Barria, M.A.; Cali, I.; Gambetti, P.; Gelpi, E.; Giese, A.; Ghetti, B.; et al. Phenotypic diversity of genetic Creutzfeldt-Jakob disease: A histo-molecular-based classification. *Acta Neuropathol.* **2021**, *142*, 707–728. [CrossRef]
17. Douet, J.Y.; Huor, A.; Cassard, H.; Lugan, S.; Aron, N.; Mesic, C.; Vilette, D.; Barrio, T.; Streichenberger, N.; Perret-Liaudet, A.; et al. Prion strains associated with iatrogenic CJD in French and UK human growth hormone recipients. *Acta Neuropathol. Commun.* **2021**, *9*, 145. [CrossRef]
18. Will, R.G.; Ironside, J.W.; Zeidler, M.; Cousens, S.N.; Estibeiro, K.; Alperovitch, A.; Poser, S.; Pocchiari, M.; Hofman, A.; Smith, P.G. A new variant of Creutzfeldt-Jakob disease in the UK. *Lancet* **1996**, *347*, 921–925. [CrossRef]
19. Collinge, J.; Whitfield, J.; McKintosh, E.; Beck, J.; Mead, S.; Thomas, D.J.; Alpers, M.P. Kuru in the 21st century—An acquired human prion disease with very long incubation periods. *Lancet* **2006**, *367*, 2068–2074. [CrossRef]
20. Collins, S.; McLean, C.A.; Masters, C.L. Gerstmann-Straussler-Scheinker syndrome, fatal familial insomnia, and kuru: A review of these less common human transmissible spongiform encephalopathies. *J. Clin. Neurosci.* **2001**, *8*, 387–397. [CrossRef]
21. Zou, W.Q.; Puoti, G.; Xiao, X.; Yuan, J.; Qing, L.; Cali, I.; Shimoji, M.; Langeveld, J.P.; Castellani, R.; Notari, S.; et al. Variably protease-sensitive prionopathy: A new sporadic disease of the prion protein. *Ann. Neurol.* **2010**, *68*, 162–172. [CrossRef]
22. Acin, C.; Bolea, R.; Monzon, M.; Monleon, E.; Moreno, B.; Filali, H.; Marin, B.; Sola, D.; Betancor, M.; Guijarro, I.M.; et al. Classical and Atypical Scrapie in Sheep and Goats. Review on the Etiology, Genetic Factors, Pathogenesis, Diagnosis, and Control Measures of Both Diseases. *Animals* **2021**, *11*, 691. [CrossRef]
23. Benestad, S.L.; Sarradin, P.; Thu, B.; Schonheit, J.; Tranulis, M.A.; Bratberg, B. Cases of scrapie with unusual features in Norway and designation of a new type, Nor98. *Vet. Rec.* **2003**, *153*, 202–208. [CrossRef]
24. Ducrot, C.; Arnold, M.; de Koeijer, A.; Heim, D.; Calavas, D. Review on the epidemiology and dynamics of BSE epidemics. *Vet. Res.* **2008**, *39*, 15. [CrossRef]
25. Dudas, S.; Czub, S. Atypical BSE: Current Knowledge and Knowledge Gaps. *Food Saf.* **2017**, *5*, 10–13. [CrossRef]

26. Casalone, C.; Zanusso, G.; Acutis, P.; Ferrari, S.; Capucci, L.; Tagliavini, F.; Monaco, S.; Caramelli, M. Identification of a second bovine amyloidotic spongiform encephalopathy: Molecular similarities with sporadic Creutzfeldt-Jakob disease. *Proc. Natl. Acad. Sci. USA* **2004**, *101*, 3065–3070. [CrossRef]

27. Comoy, E.E.; Casalone, C.; Lescoutra-Etchegaray, N.; Zanusso, G.; Freire, S.; Marce, D.; Auvre, F.; Ruchoux, M.M.; Ferrari, S.; Monaco, S.; et al. Atypical BSE (BASE) transmitted from asymptomatic aging cattle to a primate. *PLoS ONE* **2008**, *3*, e3017. [CrossRef]

28. Haley, N.J.; Hoover, E.A. Chronic wasting disease of cervids: Current knowledge and future perspectives. *Annu. Rev. Anim. Biosci.* **2015**, *3*, 305–325. [CrossRef]

29. Pirisinu, L.; Tran, L.; Chiappini, B.; Vanni, I.; Di Bari, M.A.; Vaccari, G.; Vikoren, T.; Madslien, K.I.; Vage, J.; Spraker, T.; et al. Novel Type of Chronic Wasting Disease Detected in Moose (*Alces alces*), Norway. *Emerg. Infect. Dis.* **2018**, *24*, 2210–2218. [CrossRef]

30. Vikoren, T.; Vage, J.; Madslien, K.I.; Roed, K.H.; Rolandsen, C.M.; Tran, L.; Hopp, P.; Veiberg, V.; Heum, M.; Moldal, T.; et al. First Detection of Chronic Wasting Disease in a Wild Red Deer (*Cervus elaphus*) in Europe. *J. Wildl. Dis.* **2019**, *55*, 970–972. [CrossRef]

31. Baron, T.; Bencsik, A.; Biacabe, A.G.; Morignat, E.; Bessen, R.A. Phenotypic similarity of transmissible mink encephalopathy in cattle and L-type bovine spongiform encephalopathy in a mouse model. *Emerg. Infect. Dis.* **2007**, *13*, 1887–1894. [CrossRef]

32. Sigurdson, C.J.; Miller, M.W. Other animal prion diseases. *Br. Med. Bull.* **2003**, *66*, 199–212. [CrossRef]

33. Kurt, T.D.; Sigurdson, C.J. Cross-species transmission of CWD prions. *Prion* **2016**, *10*, 83–91. [CrossRef]

34. Stahl, N.; Prusiner, S.B. Prions and prion proteins. *FASEB J.* **1991**, *5*, 2799–2807. [CrossRef]

35. Basler, K.; Oesch, B.; Scott, M.; Westaway, D.; Walchli, M.; Groth, D.F.; McKinley, M.P.; Prusiner, S.B.; Weissmann, C. Scrapie and cellular PrP isoforms are encoded by the same chromosomal gene. *Cell* **1986**, *46*, 417–428. [CrossRef]

36. Oesch, B.; Westaway, D.; Walchli, M.; McKinley, M.P.; Kent, S.B.; Aebersold, R.; Barry, R.A.; Tempst, P.; Teplow, D.B.; Hood, L.E.; et al. A cellular gene encodes scrapie PrP 27-30 protein. *Cell* **1985**, *40*, 735–746. [CrossRef]

37. Brown, H.R.; Goller, N.L.; Rudelli, R.D.; Merz, G.S.; Wolfe, G.C.; Wisniewski, H.M.; Robakis, N.K. The mRNA encoding the scrapie agent protein is present in a variety of non-neuronal cells. *Acta Neuropathol.* **1990**, *80*, 1–6. [CrossRef]

38. Salvesen, O.; Tatzelt, J.; Tranulis, M.A. The prion protein in neuroimmune crosstalk. *Neurochem. Int.* **2019**, *130*, 104335. [CrossRef]

39. Wulf, M.A.; Senatore, A.; Aguzzi, A. The biological function of the cellular prion protein: An update. *BMC Biol.* **2017**, *15*, 34. [CrossRef]

40. Linden, R. The Biological Function of the Prion Protein: A Cell Surface Scaffold of Signaling Modules. *Front. Mol. Neurosci.* **2017**, *10*, 77. [CrossRef]

41. Wang, L.Q.; Zhao, K.; Yuan, H.Y.; Wang, Q.; Guan, Z.; Tao, J.; Li, X.N.; Sun, Y.; Yi, C.W.; Chen, J.; et al. Cryo-EM structure of an amyloid fibril formed by full-length human prion protein. *Nat. Struct. Mol. Biol.* **2020**, *27*, 598–602. [CrossRef]

42. Glynn, C.; Sawaya, M.R.; Ge, P.; Gallagher-Jones, M.; Short, C.W.; Bowman, R.; Apostol, M.; Zhou, Z.H.; Eisenberg, D.S.; Rodriguez, J.A. Cryo-EM structure of a human prion fibril with a hydrophobic, protease-resistant core. *Nat. Struct. Mol. Biol.* **2020**, *27*, 417–423. [CrossRef]

43. Prusiner, S.B.; Scott, M.; Foster, D.; Pan, K.M.; Groth, D.; Mirenda, C.; Torchia, M.; Yang, S.L.; Serban, D.; Carlson, G.A.; et al. Transgenetic studies implicate interactions between homologous PrP isoforms in scrapie prion replication. *Cell* **1990**, *63*, 673–686. [CrossRef]

44. Priola, S.A. Prion protein and species barriers in the transmissible spongiform encephalopathies. *Biomed. Pharmacother.* **1999**, *53*, 27–33. [CrossRef]

45. Goldmann, W. PrP genetics in ruminant transmissible spongiform encephalopathies. *Vet. Res.* **2008**, *39*, 30. [CrossRef]

46. Tranulis, M.A. Influence of the prion protein gene, *Prnp*, on scrapie susceptibility in sheep. *APMIS* **2002**, *110*, 33–43. [CrossRef]

47. Bossers, A.; Schreuder, B.E.; Muileman, I.H.; Belt, P.B.; Smits, M.A. PrP genotype contributes to determining survival times of sheep with natural scrapie. *J. Gen. Virol.* **1996**, *77 Pt 10*, 2669–2673. [CrossRef]

48. Goldmann, W.; Hunter, N.; Foster, J.D.; Salbaum, J.M.; Beyreuther, K.; Hope, J. Two alleles of a neural protein gene linked to scrapie in sheep. *Proc. Natl. Acad. Sci. USA* **1990**, *87*, 2476–2480. [CrossRef]

49. Guere, M.E.; Vage, J.; Tharaldsen, H.; Benestad, S.L.; Vikoren, T.; Madslien, K.; Hopp, P.; Rolandsen, C.M.; Roed, K.H.; Tranulis, M.A. Chronic wasting disease associated with prion protein gene (*PRNP*) variation in Norwegian wild reindeer (*Rangifer tarandus*). *Prion* **2020**, *14*, 1–10. [CrossRef]

50. Brandt, A.L.; Kelly, A.C.; Green, M.L.; Shelton, P.; Novakofski, J.; Mateus-Pinilla, N.E. Prion protein gene sequence and chronic wasting disease susceptibility in white-tailed deer (*Odocoileus virginianus*). *Prion* **2015**, *9*, 449–462. [CrossRef]

51. Robinson, S.J.; Samuel, M.D.; O'Rourke, K.I.; Johnson, C.J. The role of genetics in chronic wasting disease of North American cervids. *Prion* **2012**, *6*, 153–162. [CrossRef]

52. Wilson, G.A.; Nakada, S.M.; Bollinger, T.K.; Pybus, M.J.; Merrill, E.H.; Coltman, D.W. Polymorphisms at the *PRNP* gene influence susceptibility to chronic wasting disease in two species of deer (*Odocoileus* Spp.) in western Canada. *J. Toxicol. Environ. Health A* **2009**, *72*, 1025–1029. [CrossRef]

53. Kocisko, D.A.; Come, J.H.; Priola, S.A.; Chesebro, B.; Raymond, G.J.; Lansbury, P.T.; Caughey, B. Cell-free formation of protease-resistant prion protein. *Nature* **1994**, *370*, 471–474. [CrossRef]

54. Haley, N.J.; Donner, R.; Henderson, D.M.; Tennant, J.; Hoover, E.A.; Manca, M.; Caughey, B.; Kondru, N.; Manne, S.; Kanthasamay, A.; et al. Cross-validation of the RT-QuIC assay for the antemortem detection of chronic wasting disease in elk. *Prion* **2020**, *14*, 47–55. [CrossRef]

55. Haley, N.J.; Siepker, C.; Hoon-Hanks, L.L.; Mitchell, G.; Walter, W.D.; Manca, M.; Monello, R.J.; Powers, J.G.; Wild, M.A.; Hoover, E.A.; et al. Seeded Amplification of Chronic Wasting Disease Prions in Nasal Brushings and Recto-anal Mucosa-Associated Lymphoid Tissues from Elk by Real-Time Quaking-Induced Conversion. *J. Clin. Microbiol.* **2016**, *54*, 1117–1126. [CrossRef]

56. Castilla, J.; Morales, R.; Saa, P.; Barria, M.; Gambetti, P.; Soto, C. Cell-free propagation of prion strains. *EMBO J.* **2008**, *27*, 2557–2566. [CrossRef]

57. Saborio, G.P.; Permanne, B.; Soto, C. Sensitive detection of pathological prion protein by cyclic amplification of protein misfolding. *Nature* **2001**, *411*, 810–813. [CrossRef]

58. Detwiler, L.A. Scrapie. *Rev. Sci. Tech.* **1992**, *11*, 491–537. [CrossRef]

59. Gonzalez, L.; Martin, S.; Siso, S.; Konold, T.; Ortiz-Pelaez, A.; Phelan, L.; Goldmann, W.; Stewart, P.; Saunders, G.; Windl, O.; et al. High prevalence of scrapie in a dairy goat herd: Tissue distribution of disease-associated PrP and effect of PRNP genotype and age. *Vet. Res.* **2009**, *40*, 65. [CrossRef]

60. Pattison, I.H.; Smith, K. Experimental Scrapie in Goats: A Modification of Incubation Period and Clinical Response Following Pre-Treatment with Normal Goat Brain. *Nature* **1963**, *200*, 1342–1343. [CrossRef]

61. Collinge, J.; Clarke, A.R. A general model of prion strains and their pathogenicity. *Science* **2007**, *318*, 930–936. [CrossRef]

62. Hill, A.F.; Desbruslais, M.; Joiner, S.; Sidle, K.C.; Gowland, I.; Collinge, J.; Doey, L.J.; Lantos, P. The same prion strain causes vCJD and BSE. *Nature* **1997**, *389*, 448–450. [CrossRef]

63. Collinge, J. Medicine. Prion strain mutation and selection. *Science* **2010**, *328*, 1111–1112. [CrossRef]

64. Marin-Moreno, A.; Huor, A.; Espinosa, J.C.; Douet, J.Y.; Aguilar-Calvo, P.; Aron, N.; Piquer, J.; Lugan, S.; Lorenzo, P.; Tillier, C.; et al. Radical Change in Zoonotic Abilities of Atypical BSE Prion Strains as Evidenced by Crossing of Sheep Species Barrier in Transgenic Mice. *Emerg. Infect. Dis.* **2020**, *26*, 1130–1139. [CrossRef]

65. Foster, J.D.; Parnham, D.W.; Hunter, N.; Bruce, M. Distribution of the prion protein in sheep terminally affected with BSE following experimental oral transmission. *J. Gen. Virol.* **2001**, *82*, 2319–2326. [CrossRef]

66. Foster, J.D.; Parnham, D.; Chong, A.; Goldmann, W.; Hunter, N. Clinical signs, histopathology and genetics of experimental transmission of BSE and natural scrapie to sheep and goats. *Vet. Rec.* **2001**, *148*, 165–171. [CrossRef]

67. Bartz, J.C.; Marsh, R.F.; McKenzie, D.I.; Aiken, J.M. The host range of chronic wasting disease is altered on passage in ferrets. *Virology* **1998**, *251*, 297–301. [CrossRef]

68. Williams, E.S.; Young, S. Chronic wasting disease of captive mule deer: A spongiform encephalopathy. *J. Wildl. Dis.* **1980**, *16*, 89–98. [CrossRef]

69. Escobar, L.E.; Pritzkow, S.; Winter, S.N.; Grear, D.A.; Kirchgessner, M.S.; Dominguez-Villegas, E.; Machado, G.; Townsend Peterson, A.; Soto, C. The ecology of chronic wasting disease in wildlife. *Biol. Rev. Camb. Philos. Soc.* **2020**, *95*, 393–408. [CrossRef]

70. Hazards, E.P.o.B.; Ricci, A.; Allende, A.; Bolton, D.; Chemaly, M.; Davies, R.; Fernandez Escamez, P.S.; Girones, R.; Herman, L.; Koutsoumanis, K.; et al. Chronic wasting disease (CWD) in cervids. *EFSA J.* **2017**, *15*, e04667. [CrossRef]

71. Williams, E.S.; Miller, M.W. Chronic wasting disease in deer and elk in North America. *Rev. Sci. Tech.* **2002**, *21*, 305–316. [CrossRef] [PubMed]

72. Miller, M.W.; Williams, E.S.; McCarty, C.W.; Spraker, T.R.; Kreeger, T.J.; Larsen, C.T.; Thorne, E.T. Epizootiology of chronic wasting disease in free-ranging cervids in Colorado and Wyoming. *J. Wildl. Dis.* **2000**, *36*, 676–690. [CrossRef]

73. Miller, M.W.; Fischer, J.R. The first five (or more) decades of Chronic Wasting Disease: Lessons for the five decades to come. In Proceedings of the Transactions of the North American Wildlife and Natural Resources Conference, Pittsburgh, PA, USA, 13–18 March 2016; Volume 81.

74. Mathiason, C.K.; Hays, S.A.; Powers, J.; Hayes-Klug, J.; Langenberg, J.; Dahmes, S.J.; Osborn, D.A.; Miller, K.V.; Warren, R.J.; Mason, G.L.; et al. Infectious prions in pre-clinical deer and transmission of chronic wasting disease solely by environmental exposure. *PLoS ONE* **2009**, *4*, e5916. [CrossRef]

75. Dube, C.; Mehren, K.G.; Barker, I.K.; Peart, B.L.; Balachandran, A. Retrospective investigation of chronic wasting disease of cervids at the Toronto Zoo, 1973–2003. *Can. Vet. J.* **2006**, *47*, 1185–1193. [PubMed]

76. Kim, T.Y.; Shon, H.J.; Joo, Y.S.; Mun, U.K.; Kang, K.S.; Lee, Y.S. Additional cases of Chronic Wasting Disease in imported deer in Korea. *J. Vet. Med. Sci.* **2005**, *67*, 753–759. [CrossRef]

77. Baeten, L.A.; Powers, B.E.; Jewell, J.E.; Spraker, T.R.; Miller, M.W. A natural case of chronic wasting disease in a free-ranging moose (*Alces alces shirasi*). *J. Wildl. Dis.* **2007**, *43*, 309–314. [CrossRef] [PubMed]

78. Kreeger, T.J.; Montgomery, D.L.; Jewell, J.E.; Schultz, W.; Williams, E.S. Oral transmission of chronic wasting disease in captive Shira's moose. *J. Wildl. Dis.* **2006**, *42*, 640–645. [CrossRef]

79. Expanding Distribution of Chronic Wasting Disease. Available online: https://www.usgs.gov/centers/nwhc/science/expanding-distribution-chronic-wasting-disease (accessed on 26 January 2023).

80. Williams, E.S. Chronic wasting disease. *Vet. Pathol.* **2005**, *42*, 530–549. [CrossRef]

81. Mitchell, G.B.; Sigurdson, C.J.; O'Rourke, K.I.; Algire, J.; Harrington, N.P.; Walther, I.; Spraker, T.R.; Balachandran, A. Experimental oral transmission of chronic wasting disease to reindeer (*Rangifer tarandus tarandus*). *PLoS ONE* **2012**, *7*, e39055. [CrossRef]

82. Guere, M.E.; Vage, J.; Tharaldsen, H.; Kvie, K.S.; Bardsen, B.J.; Benestad, S.L.; Vikoren, T.; Madslien, K.; Rolandsen, C.M.; Tranulis, M.A.; et al. Chronic wasting disease in Norway-A survey of prion protein gene variation among cervids. *Transbound. Emerg. Dis.* **2022**, *69*, e20–e31. [CrossRef]

83. Benestad, S.L.; Mitchell, G.; Simmons, M.; Ytrehus, B.; Vikoren, T. First case of chronic wasting disease in Europe in a Norwegian free-ranging reindeer. *Vet. Res.* **2016**, *47*, 88. [CrossRef]

84. Totalregnskap for Reindriftsnæringen. Available online: https://www.landbruksdirektoratet.no/nb/nyhetsrom/rapporter/totalregnskap-for-reindriftsnaeringen (accessed on 10 January 2023).

85. Agren, E.O.; Soren, K.; Gavier-Widen, D.; Benestad, S.L.; Tran, L.; Wall, K.; Averhed, G.; Doose, N.; Vage, J.; Noremark, M. First Detection of Chronic Wasting Disease in Moose (*Alces alces*) in Sweden. *J. Wildl. Dis.* **2021**, *57*, 461–463. [CrossRef]

86. Nonno, R.; Di Bari, M.A.; Pirisinu, L.; D'Agostino, C.; Vanni, I.; Chiappini, B.; Marcon, S.; Riccardi, G.; Tran, L.; Vikoren, T.; et al. Studies in bank voles reveal strain differences between chronic wasting disease prions from Norway and North America. *Proc. Natl. Acad. Sci. USA* **2020**, *117*, 31417–31426. [CrossRef]

87. Tranulis, M.A.; Benestad, S.L.; Baron, T.; Kretzschmar, H. Atypical prion diseases in humans and animals. *Prion Proteins* **2011**, *305*, 23–50. [CrossRef]

88. Tranulis, M.A.; Gavier-Widen, D.; Vage, J.; Noremark, M.; Korpenfelt, S.L.; Hautaniemi, M.; Pirisinu, L.; Nonno, R.; Benestad, S.L. Chronic wasting disease in Europe: New strains on the horizon. *Acta Vet. Scand.* **2021**, *63*, 48. [CrossRef]

89. Moore, J.; Tatum, T.; Hwang, S.; Vrentas, C.; West Greenlee, M.H.; Kong, Q.; Nicholson, E.; Greenlee, J. Novel Strain of the Chronic Wasting Disease Agent Isolated From Experimentally Inoculated Elk with LL132 Prion Protein. *Sci. Rep.* **2020**, *10*, 3148. [CrossRef]

90. Duque Velasquez, C.; Kim, C.; Haldiman, T.; Kim, C.; Herbst, A.; Aiken, J.; Safar, J.G.; McKenzie, D. Chronic wasting disease (CWD) prion strains evolve via adaptive diversification of conformers in hosts expressing prion protein polymorphisms. *J. Biol. Chem.* **2020**, *295*, 4985–5001. [CrossRef]

91. Herbst, A.; Velasquez, C.D.; Triscott, E.; Aiken, J.M.; McKenzie, D. Chronic Wasting Disease Prion Strain Emergence and Host Range Expansion. *Emerg. Infect. Dis.* **2017**, *23*, 1598–1600. [CrossRef]

92. Duque Velasquez, C.; Kim, C.; Herbst, A.; Daude, N.; Garza, M.C.; Wille, H.; Aiken, J.; McKenzie, D. Deer Prion Proteins Modulate the Emergence and Adaptation of Chronic Wasting Disease Strains. *J. Virol.* **2015**, *89*, 12362–12373. [CrossRef]

93. Perrott, M.R.; Sigurdson, C.J.; Mason, G.L.; Hoover, E.A. Evidence for distinct chronic wasting disease (CWD) strains in experimental CWD in ferrets. *J. Gen. Virol.* **2012**, *93*, 212–221. [CrossRef]

94. Bessen, R.A.; Robinson, C.J.; Seelig, D.M.; Watschke, C.P.; Lowe, D.; Shearin, H.; Martinka, S.; Babcock, A.M. Transmission of chronic wasting disease identifies a prion strain causing cachexia and heart infection in hamsters. *PLoS ONE* **2011**, *6*, e28026. [CrossRef]

95. Raymond, G.J.; Raymond, L.D.; Meade-White, K.D.; Hughson, A.G.; Favara, C.; Gardner, D.; Williams, E.S.; Miller, M.W.; Race, R.E.; Caughey, B. Transmission and adaptation of chronic wasting disease to hamsters and transgenic mice: Evidence for strains. *J. Virol.* **2007**, *81*, 4305–4314. [CrossRef]

96. Watson, N.; Brandel, J.P.; Green, A.; Hermann, P.; Ladogana, A.; Lindsay, T.; Mackenzie, J.; Pocchiari, M.; Smith, C.; Zerr, I.; et al. The importance of ongoing international surveillance for Creutzfeldt-Jakob disease. *Nat. Rev. Neurol.* **2021**, *17*, 362–379. [CrossRef]

97. Bruce, M.E.; Will, R.G.; Ironside, J.W.; McConnell, I.; Drummond, D.; Suttie, A.; McCardle, L.; Chree, A.; Hope, J.; Birkett, C.; et al. Transmissions to mice indicate that 'new variant' CJD is caused by the BSE agent. *Nature* **1997**, *389*, 498–501. [CrossRef]

98. Gelpi, E.; Baiardi, S.; Nos, C.; Dellavalle, S.; Aldecoa, I.; Ruiz-Garcia, R.; Ispierto, L.; Escudero, D.; Casado, V.; Barranco, E.; et al. Sporadic Creutzfeldt-Jakob disease VM1: Phenotypic and molecular characterization of a novel subtype of human prion disease. *Acta Neuropathol. Commun.* **2022**, *10*, 114. [CrossRef]

99. Galeno, R.; Di Bari, M.A.; Nonno, R.; Cardone, F.; Sbriccoli, M.; Graziano, S.; Ingrosso, L.; Fiorini, M.; Valanzano, A.; Pasini, G.; et al. Prion Strain Characterization of a Novel Subtype of Creutzfeldt-Jakob Disease. *J. Virol.* **2017**, *91*, e02390-16. [CrossRef]

100. Petrovic, I.N.; Martin-Bastida, A.; Massey, L.; Ling, H.; O'Sullivan, S.S.; Williams, D.R.; Holton, J.L.; Revesz, T.; Ironside, J.W.; Lees, A.J.; et al. MM2 subtype of sporadic Creutzfeldt-Jakob disease may underlie the clinical presentation of progressive supranuclear palsy. *J. Neurol.* **2013**, *260*, 1031–1036. [CrossRef]

101. Fatola, O.I.; Keller, M.; Balkema-Buschmann, A.; Olopade, J.; Groschup, M.H.; Fast, C. Strain Typing of Classical Scrapie and Bovine Spongiform Encephalopathy (BSE) by Using Ovine PrP (ARQ/ARQ) Overexpressing Transgenic Mice. *Int. J. Mol. Sci.* **2022**, *23*, 6744. [CrossRef]

102. Bruce, M.E. Scrapie strain variation and mutation. *Br. Med. Bull.* **1993**, *49*, 822–838. [CrossRef]

103. Bruce, M.E.; Fraser, H. Scrapie strain variation and its implications. *Curr. Top. Microbiol. Immunol.* **1991**, *172*, 125–138. [CrossRef]

104. Uttley, L.; Carroll, C.; Wong, R.; Hilton, D.A.; Stevenson, M. Creutzfeldt-Jakob disease: A systematic review of global incidence, prevalence, infectivity, and incubation. *Lancet Infect. Dis.* **2020**, *20*, e2–e10. [CrossRef]

105. Mawhinney, S.; Pape, W.J.; Forster, J.E.; Anderson, C.A.; Bosque, P.; Miller, M.W. Human prion disease and relative risk associated with chronic wasting disease. *Emerg. Infect. Dis.* **2006**, *12*, 1527–1535. [CrossRef]

106. Anderson, C.A.; Bosque, P.; Filley, C.M.; Arciniegas, D.B.; Kleinschmidt-Demasters, B.K.; Pape, W.J.; Tyler, K.L. Colorado surveillance program for chronic wasting disease transmission to humans: Lessons from 2 highly suspicious but negative cases. *Arch. Neurol.* **2007**, *64*, 439–441. [CrossRef]

107. Belay, E.D.; Gambetti, P.; Schonberger, L.B.; Parchi, P.; Lyon, D.R.; Capellari, S.; McQuiston, J.H.; Bradley, K.; Dowdle, G.; Crutcher, J.M.; et al. Creutzfeldt-Jakob disease in unusually young patients who consumed venison. *Arch. Neurol.* **2001**, *58*, 1673–1678. [CrossRef]

108. Olszowy, K.M.; Lavelle, J.; Rachfal, K.; Hempstead, S.; Drouin, K.; Darcy, J.M., 2nd; Reiber, C.; Garruto, R.M. Six-year follow-up of a point-source exposure to CWD contaminated venison in an Upstate New York community: Risk behaviours and health outcomes 2005–2011. *Public Health* **2014**, *128*, 860–868. [CrossRef]

109. Kocisko, D.A.; Priola, S.A.; Raymond, G.J.; Chesebro, B.; Lansbury, P.T., Jr.; Caughey, B. Species specificity in the cell-free conversion of prion protein to protease-resistant forms: A model for the scrapie species barrier. *Proc. Natl. Acad. Sci. USA* **1995**, *92*, 3923–3927. [CrossRef]

110. Chen, B.; Morales, R.; Barria, M.A.; Soto, C. Estimating prion concentration in fluids and tissues by quantitative PMCA. *Nat. Methods* **2010**, *7*, 519–520. [CrossRef]

111. Atarashi, R.; Wilham, J.M.; Christensen, L.; Hughson, A.G.; Moore, R.A.; Johnson, L.M.; Onwubiko, H.A.; Priola, S.A.; Caughey, B. Simplified ultrasensitive prion detection by recombinant PrP conversion with shaking. *Nat. Methods* **2008**, *5*, 211–212. [CrossRef]

112. Wilham, J.M.; Orru, C.D.; Bessen, R.A.; Atarashi, R.; Sano, K.; Race, B.; Meade-White, K.D.; Taubner, L.M.; Timmes, A.; Caughey, B. Rapid end-point quantitation of prion seeding activity with sensitivity comparable to bioassays. *PLoS Pathog.* **2010**, *6*, e1001217. [CrossRef]

113. Colby, D.W.; Zhang, Q.; Wang, S.; Groth, D.; Legname, G.; Riesner, D.; Prusiner, S.B. Prion detection by an amyloid seeding assay. *Proc. Natl. Acad. Sci. USA* **2007**, *104*, 20914–20919. [CrossRef]

114. Atarashi, R.; Sano, K.; Satoh, K.; Nishida, N. Real-time quaking-induced conversion: A highly sensitive assay for prion detection. *Prion* **2011**, *5*, 150–153. [CrossRef]

115. Davenport, K.A.; Mosher, B.A.; Brost, B.M.; Henderson, D.M.; Denkers, N.D.; Nalls, A.V.; McNulty, E.; Mathiason, C.K.; Hoover, E.A. Assessment of Chronic Wasting Disease Prion Shedding in Deer Saliva with Occupancy Modeling. *J. Clin. Microbiol.* **2018**, *56*, e01243-17. [CrossRef]

116. Henderson, D.M.; Davenport, K.A.; Haley, N.J.; Denkers, N.D.; Mathiason, C.K.; Hoover, E.A. Quantitative assessment of prion infectivity in tissues and body fluids by real-time quaking-induced conversion. *J. Gen. Virol.* **2015**, *96*, 210–219. [CrossRef]

117. Henderson, D.M.; Manca, M.; Haley, N.J.; Denkers, N.D.; Nalls, A.V.; Mathiason, C.K.; Caughey, B.; Hoover, E.A. Rapid antemortem detection of CWD prions in deer saliva. *PLoS ONE* **2013**, *8*, e74377. [CrossRef]

118. John, T.R.; Schatzl, H.M.; Gilch, S. Early detection of chronic wasting disease prions in urine of pre-symptomatic deer by real-time quaking-induced conversion assay. *Prion* **2013**, *7*, 253–258. [CrossRef]

119. Cheng, Y.C.; Hannaoui, S.; John, T.R.; Dudas, S.; Czub, S.; Gilch, S. Early and Non-Invasive Detection of Chronic Wasting Disease Prions in Elk Feces by Real-Time Quaking Induced Conversion. *PLoS ONE* **2016**, *11*, e0166187. [CrossRef]

120. Pulford, B.; Spraker, T.R.; Wyckoff, A.C.; Meyerett, C.; Bender, H.; Ferguson, A.; Wyatt, B.; Lockwood, K.; Powers, J.; Telling, G.C.; et al. Detection of PrPCWD in feces from naturally exposed Rocky Mountain elk (*Cervus elaphus nelsoni*) using protein misfolding cyclic amplification. *J. Wildl. Dis.* **2012**, *48*, 425–434. [CrossRef]

121. Kramm, C.; Pritzkow, S.; Lyon, A.; Nichols, T.; Morales, R.; Soto, C. Detection of Prions in Blood of Cervids at the Asymptomatic Stage of Chronic Wasting Disease. *Sci. Rep.* **2017**, *7*, 17241. [CrossRef]

122. McNulty, E.; Nalls, A.V.; Mellentine, S.; Hughes, E.; Pulscher, L.; Hoover, E.A.; Mathiason, C.K. Comparison of conventional, amplification and bio-assay detection methods for a chronic wasting disease inoculum pool. *PLoS ONE* **2019**, *14*, e0216621. [CrossRef]

123. Bistaffa, E.; Vuong, T.T.; Cazzaniga, F.A.; Tran, L.; Salzano, G.; Legname, G.; Giaccone, G.; Benestad, S.L.; Moda, F. Use of different RT-QuIC substrates for detecting CWD prions in the brain of Norwegian cervids. *Sci. Rep.* **2019**, *9*, 18595. [CrossRef]

124. Raymond, G.J.; Bossers, A.; Raymond, L.D.; O'Rourke, K.I.; McHolland, L.E.; Bryant, P.K., 3rd; Miller, M.W.; Williams, E.S.; Smits, M.; Caughey, B. Evidence of a molecular barrier limiting susceptibility of humans, cattle and sheep to chronic wasting disease. *EMBO J.* **2000**, *19*, 4425–4430. [CrossRef]

125. Barria, M.A.; Telling, G.C.; Gambetti, P.; Mastrianni, J.A.; Soto, C. Generation of a new form of human PrP(Sc) in vitro by interspecies transmission from cervid prions. *J Biol. Chem.* **2011**, *286*, 7490–7495. [CrossRef]

126. Kurt, T.D.; Jiang, L.; Fernandez-Borges, N.; Bett, C.; Liu, J.; Yang, T.; Spraker, T.R.; Castilla, J.; Eisenberg, D.; Kong, Q.; et al. Human prion protein sequence elements impede cross-species chronic wasting disease transmission. *J. Clin. Investig.* **2015**, *125*, 2548. [CrossRef]

127. Barria, M.A.; Libori, A.; Mitchell, G.; Head, M.W. Susceptibility of Human Prion Protein to Conversion by Chronic Wasting Disease Prions. *Emerg. Infect. Dis.* **2018**, *24*, 1482–1489. [CrossRef]

128. Hamir, A.N.; Kunkle, R.A.; Cutlip, R.C.; Miller, J.M.; Williams, E.S.; Richt, J.A. Transmission of chronic wasting disease of mule deer to Suffolk sheep following intracerebral inoculation. *J. Vet Diagn. Investig.* **2006**, *18*, 558–565. [CrossRef]

129. Hamir, A.N.; Kunkle, R.A.; Miller, J.M.; Greenlee, J.J.; Richt, J.A. Experimental second passage of chronic wasting disease (CWD(mule deer)) agent to cattle. *J. Comp. Pathol.* **2006**, *134*, 63–69. [CrossRef]

130. Moore, S.J.; West Greenlee, M.H.; Kondru, N.; Manne, S.; Smith, J.D.; Kunkle, R.A.; Kanthasamy, A.; Greenlee, J.J. Experimental Transmission of the Chronic Wasting Disease Agent to Swine after Oral or Intracranial Inoculation. *J. Virol.* **2017**, *91*, e00926-17. [CrossRef]

131. Mathiason, C.K.; Nalls, A.V.; Seelig, D.M.; Kraft, S.L.; Carnes, K.; Anderson, K.R.; Hayes-Klug, J.; Hoover, E.A. Susceptibility of domestic cats to chronic wasting disease. *J. Virol.* **2013**, *87*, 1947–1956. [CrossRef]

132. Sigurdson, C.J.; Mathiason, C.K.; Perrott, M.R.; Eliason, G.A.; Spraker, T.R.; Glatzel, M.; Manco, G.; Bartz, J.C.; Miller, M.W.; Hoover, E.A. Experimental chronic wasting disease (CWD) in the ferret. *J. Comp. Pathol.* **2008**, *138*, 189–196. [CrossRef]

133. Marsh, R.F.; Kincaid, A.E.; Bessen, R.A.; Bartz, J.C. Interspecies transmission of chronic wasting disease prions to squirrel monkeys (*Saimiri sciureus*). *J. Virol.* **2005**, *79*, 13794–13796. [CrossRef]

134. Kurt, T.D.; Bett, C.; Fernandez-Borges, N.; Joshi-Barr, S.; Hornemann, S.; Rulicke, T.; Castilla, J.; Wuthrich, K.; Aguzzi, A.; Sigurdson, C.J. Prion transmission prevented by modifying the beta2-alpha2 loop structure of host PrPC. *J. Neurosci.* **2014**, *34*, 1022–1027. [CrossRef]

135. Kyle, L.M.; John, T.R.; Schatzl, H.M.; Lewis, R.V. Introducing a rigid loop structure from deer into mouse prion protein increases its propensity for misfolding in vitro. *PLoS ONE* **2013**, *8*, e66715. [CrossRef]

136. Kurt, T.D.; Telling, G.C.; Zabel, M.D.; Hoover, E.A. Trans-species amplification of PrP(CWD) and correlation with rigid loop 170N. *Virology* **2009**, *387*, 235–243. [CrossRef]

137. Kong, Q.; Huang, S.; Zou, W.; Vanegas, D.; Wang, M.; Wu, D.; Yuan, J.; Zheng, M.; Bai, H.; Deng, H.; et al. Chronic wasting disease of elk: Transmissibility to humans examined by transgenic mouse models. *J. Neurosci.* **2005**, *25*, 7944–7949. [CrossRef]

138. Tamguney, G.; Giles, K.; Bouzamondo-Bernstein, E.; Bosque, P.J.; Miller, M.W.; Safar, J.; DeArmond, S.J.; Prusiner, S.B. Transmission of elk and deer prions to transgenic mice. *J. Virol.* **2006**, *80*, 9104–9114. [CrossRef]

139. Sandberg, M.K.; Al-Doujaily, H.; Sigurdson, C.J.; Glatzel, M.; O'Malley, C.; Powell, C.; Asante, E.A.; Linehan, J.M.; Brandner, S.; Wadsworth, J.D.; et al. Chronic wasting disease prions are not transmissible to transgenic mice overexpressing human prion protein. *J. Gen. Virol.* **2010**, *91*, 2651–2657. [CrossRef]

140. Wilson, R.; Plinston, C.; Hunter, N.; Casalone, C.; Corona, C.; Tagliavini, F.; Suardi, S.; Ruggerone, M.; Moda, F.; Graziano, S.; et al. Chronic wasting disease and atypical forms of bovine spongiform encephalopathy and scrapie are not transmissible to mice expressing wild-type levels of human prion protein. *J. Gen. Virol.* **2012**, *93*, 1624–1629. [CrossRef]

141. Race, B.; Williams, K.; Chesebro, B. Transmission studies of chronic wasting disease to transgenic mice overexpressing human prion protein using the RT-QuIC assay. *Vet. Res.* **2019**, *50*, 6. [CrossRef]

142. Wadsworth, J.D.F.; Joiner, S.; Linehan, J.M.; Jack, K.; Al-Doujaily, H.; Costa, H.; Ingold, T.; Taema, M.; Zhang, F.; Sandberg, M.K.; et al. Humanised transgenic mice are resistant to chronic wasting disease prions from Norwegian reindeer and moose. *J. Infect. Dis.* **2021**, *226*, 933–937. [CrossRef]

143. Hannaoui, S.; Zemlyankina, I.; Chang, S.C.; Arifin, M.I.; Beringue, V.; McKenzie, D.; Schatzl, H.M.; Gilch, S. Transmission of cervid prions to humanized mice demonstrates the zoonotic potential of CWD. *Acta Neuropathol.* **2022**, *144*, 767–784. [CrossRef]

144. Hill, A.F.; Collinge, J. Subclinical prion infection. *Trends Microbiol.* **2003**, *11*, 578–584. [CrossRef] [PubMed]

145. Huang, H.; Rendulich, J.; Stevenson, D.; O'Rourke, K.; Balachandran, A. Evaluation of Western blotting methods using samples with or without sodium phosphotungstic acid precipitation for diagnosis of scrapie and chronic wasting disease. *Can. J. Vet. Res.* **2005**, *69*, 193–199. [PubMed]

146. Bian, J.; Kim, S.; Kane, S.J.; Crowell, J.; Sun, J.L.; Christiansen, J.; Saijo, E.; Moreno, J.A.; DiLisio, J.; Burnett, E.; et al. Adaptive selection of a prion strain conformer corresponding to established North American CWD during propagation of novel emergent Norwegian strains in mice expressing elk or deer prion protein. *PLoS Pathog.* **2021**, *17*, e1009748. [CrossRef]

147. Comoy, E.E.; Mikol, J.; Deslys, J.P. Non-human primates in prion diseases. *Cell Tissue Res.* **2022**. [CrossRef]

148. Race, B.; Meade-White, K.D.; Phillips, K.; Striebel, J.; Race, R.; Chesebro, B. Chronic wasting disease agents in nonhuman primates. *Emerg. Infect. Dis.* **2014**, *20*, 833–837. [CrossRef] [PubMed]

149. Race, B.; Meade-White, K.D.; Miller, M.W.; Barbian, K.D.; Rubenstein, R.; LaFauci, G.; Cervenakova, L.; Favara, C.; Gardner, D.; Long, D.; et al. Susceptibilities of nonhuman primates to chronic wasting disease. *Emerg. Infect. Dis.* **2009**, *15*, 1366–1376. [CrossRef]

150. Piccardo, P.; Cervenak, J.; Yakovleva, O.; Gregori, L.; Pomeroy, K.; Cook, A.; Muhammad, F.S.; Seuberlich, T.; Cervenakova, L.; Asher, D.M. Squirrel monkeys (*Saimiri sciureus*) infected with the agent of bovine spongiform encephalopathy develop tau pathology. *J. Comp. Pathol.* **2012**, *147*, 84–93. [CrossRef]

151. Williams, L.; Brown, P.; Ironside, J.; Gibson, S.; Will, R.; Ritchie, D.; Kreil, T.R.; Abee, C. Clinical, neuropathological and immunohistochemical features of sporadic and variant forms of Creutzfeldt-Jakob disease in the squirrel monkey (*Saimiri sciureus*). *J. Gen. Virol.* **2007**, *88*, 688–695. [CrossRef]

152. Gibbs, C.J., Jr.; Amyx, H.L.; Bacote, A.; Masters, C.L.; Gajdusek, D.C. Oral transmission of kuru, Creutzfeldt-Jakob disease, and scrapie to nonhuman primates. *J. Infect. Dis.* **1980**, *142*, 205–208. [CrossRef]

153. Race, B.; Williams, K.; Orru, C.D.; Hughson, A.G.; Lubke, L.; Chesebro, B. Lack of Transmission of Chronic Wasting Disease to Cynomolgus Macaques. *J. Virol.* **2018**, *92*, e00550-18. [CrossRef]

154. Hayasaka, K.; Gojobori, T.; Horai, S. Molecular phylogeny and evolution of primate mitochondrial DNA. *Mol. Biol. Evol.* **1988**, *5*, 626–644. [CrossRef]

155. Schatzl, H.M.; Da Costa, M.; Taylor, L.; Cohen, F.E.; Prusiner, S.B. Prion protein gene variation among primates. *J. Mol. Biol.* **1995**, *245*, 362–374. [CrossRef]

156. Holznagel, E.; Yutzy, B.; Schulz-Schaeffer, W.; Kruip, C.; Hahmann, U.; Bierke, P.; Torres, J.M.; Kim, Y.S.; Thomzig, A.; Beekes, M.; et al. Foodborne transmission of bovine spongiform encephalopathy to nonhuman primates. *Emerg. Infect. Dis.* **2013**, *19*, 712–720. [CrossRef]

157. Ono, F.; Tase, N.; Kurosawa, A.; Hiyaoka, A.; Ohyama, A.; Tezuka, Y.; Wada, N.; Sato, Y.; Tobiume, M.; Hagiwara, K.; et al. Atypical L-type bovine spongiform encephalopathy (L-BSE) transmission to cynomolgus macaques, a non-human primate. *Jpn. J. Infect. Dis.* **2011**, *64*, 81–84. [CrossRef]

158. Lasmezas, C.I.; Comoy, E.; Hawkins, S.; Herzog, C.; Mouthon, F.; Konold, T.; Auvre, F.; Correia, E.; Lescoutra-Etchegaray, N.; Sales, N.; et al. Risk of oral infection with bovine spongiform encephalopathy agent in primates. *Lancet* **2005**, *365*, 781–783. [CrossRef]

159. Comoy, E.E.; Mikol, J.; Luccantoni-Freire, S.; Correia, E.; Lescoutra-Etchegaray, N.; Durand, V.; Dehen, C.; Andreoletti, O.; Casalone, C.; Richt, J.A.; et al. Transmission of scrapie prions to primate after an extended silent incubation period. *Sci. Rep.* **2015**, *5*, 11573. [CrossRef]

160. Hannaoui, S.; Cheng, G.; Wemheuer, W.; Schulz-Schaeffer, W.J.; Gilch, S.; Schatzl, M. Prion 2022 Conference abstract: Transmission of prion infectivity from CWD-infected macaque tissues to rodent models demonstrates the zoonotic potential of chronic wasting disease. *Prion* **2022**, *16*, 95–253. [CrossRef]

161. VKM. *Factors That Can Contribute to Spread of CWD—An Update on the Situation in Nordfjella. Opinion of the Panel of Biological Hazards*; Norwegian Scientific Committee for Food and Environmenmt: Oslo, Norway, 2018; pp. 1–92; ISBN 978-82-8259-316-8.

162. NVI. Norwegian CWD Surveillance—Live Updates. Web-Page in Norwegian. Available online: http://apps.vetinst.no/skrantesykestatistikk/NO/#omrade (accessed on 10 January 2023).

Article

Characterisation of *Listeria monocytogenes* Isolates from Hunted Game and Game Meat from Finland

Maria Fredriksson-Ahomaa [1,*], Mikaela Sauvala [1], Paula Kurittu [1], Viivi Heljanko [1], Annamari Heikinheimo [1,2] and Peter Paulsen [3]

1   Department of Food Hygiene and Environmental Health, Faculty of Veterinary Medicine, University of Helsinki, 00014 Helsinki, Finland
2   Microbiology Unit, Finnish Food Authority, 60100 Seinäjoki, Finland
3   Unit of Food Hygiene and Technology, Institute of Food Safety, Food Technology and Veterinary Public Health, University of Veterinary Medicine, 1210 Vienna, Austria
*   Correspondence: maria.fredriksson-ahomaa@helsinki.fi

Abstract: *Listeria monocytogenes* is an important foodborne zoonotic bacterium. It is a heterogeneous species that can be classified into lineages, serogroups, clonal complexes, and sequence types. Only scarce information exists on the properties of *L. monocytogenes* from game and game meat. We characterised 75 *L. monocytogenes* isolates from various game sources found in Finland between 2012 and 2020. The genetic diversity, presence of virulence and antimicrobial genes were studied with whole genome sequencing. Most (89%) of the isolates belonged to phylogenetic lineage (Lin) II and serogroup (SG) IIa. SGs IVb (8%) and IIb (3%) of Lin I were sporadically identified. In total, 18 clonal complexes and 21 sequence types (STs) were obtained. The most frequent STs were ST451 (21%), ST585 (12%) and ST37 (11%) found in different sample types between 2012 and 2020. We observed 10 clusters, formed by closely related isolates with 0–10 allelic differences. Most (79%) of the virulence genes were found in all of the *L. monocytogenes* isolates. Only *fosX* and *lin* were found out of 46 antimicrobial resistance genes. Our results demonstrate that potentially virulent and antimicrobial-sensitive *L. monocytogenes* isolates associated with human listeriosis are commonly found in hunted game and game meat in Finland.

Keywords: *Listeria monocytogenes*; game; whole genome sequencing; sequence type; virulence; antimicrobial resistance

Citation: Fredriksson-Ahomaa, M.; Sauvala, M.; Kurittu, P.; Heljanko, V.; Heikinheimo, A.; Paulsen, P. Characterisation of *Listeria monocytogenes* Isolates from Hunted Game and Game Meat from Finland. *Foods* **2022**, *11*, 3679. https://doi.org/10.3390/foods11223679

Academic Editor: Arun K. Bhunia

Received: 11 October 2022
Accepted: 14 November 2022
Published: 17 November 2022

Publisher's Note: MDPI stays neutral with regard to jurisdictional claims in published maps and institutional affiliations.

1. Introduction

Listeria monocytogenes has emerged over recent decades as an important foodborne pathogen responsible for numerous outbreaks [1]. *L. monocytogenes* is responsible for listeriosis, a disease affecting both humans and animals. Foodborne listeriosis typically causes a self-limited gastroenteritis among healthy people [2]. However, invasive infection leading to hospitalisation and even death may occur, especially among immunocompromised people [1]. Invasive listeriosis may also lead to abortion in pregnant women. The severity of listeriosis depends, inter alia, on the virulence of the bacterial strain [2]. Invasive listeriosis, in particular, requires antimicrobial treatment. Listeriosis had the highest proportion of hospitalised cases of all zoonoses in 2020 in the EU [3].

L. monocytogenes is a ubiquitous bacterium that can survive in a variety of environments and grow at low temperatures, e.g., in cold-stored foods [4]. Soil and decaying organic material are important sources of *L. monocytogenes*, and mammals and birds can spread this pathogen through faecal shedding [5]. *L. monocytogenes*-contaminated food is an important source attributed to human infections [6]. The consumption of contaminated food has been linked to both epidemic and sporadic listeriosis. Poor hygiene practices and inadequate sanitation procedures in the food processing industry can lead to listeriosis outbreaks [4,7].

L. monocytogenes has sporadically been found in game animals and on game carcasses [8–11]. Detection rates of *L. monocytogenes* in deer and wild boar faeces have been low, varying between 0 and 6% [12]. However, this pathogen is more frequently present in the tonsils than in faeces [12–15]. In Spain, *L. monocytogenes* was detected in 44% and 41% of deer and wild boar tonsils, respectively [12]. Recently, *L. monocytogenes* was detected in 5% of deer carcasses in Austria using an antigen test [16] and in 12% of deer carcasses in Finland using a polymerase chain reaction (PCR) [17]. This relatively high prevalence of *L. monocytogenes* on deer carcasses shows the importance of observing hygiene practices during hunting and slaughtering.

L. monocytogenes is a very heterogeneous species, which can be divided into at least 14 serotypes based on variation in the somatic (O) and flagellar (H) antigens [18,19]. Over 95% of the human and food strains are linked to four (1/2a, 1/2b, 1/2c, and 4b) serotypes. Genetically, *L. monocytogenes* can be divided into four phylogenetic lineages (Lin), six serogroups (SGs), multiple clonal complexes (CCs), and sequence types (STs) [2,5]. Most clinical strains found in humans belong to Lin I (SGs IIb and IVb) and II (SG IIa) [20], whereas food strains more frequently belong to SG IIa [21,22].

Whole genome sequencing (WGS)—an accurate method with a high resolution—is currently becoming the method of choice for characterising *L. monocytogenes* isolates [20]. It has emerged as a powerful tool for outbreak investigations and is increasingly also used for the surveillance and monitoring of listeriosis [23]. Investigating the diversity of *L. monocytogenes* isolates from game and game meat will provide valuable information on the significance of game in the meat production chain and in human infections.

Studies on the genetic relationships between *L. monocytogenes* isolates from game sources remain scarce. Game and game meat may play an important role in the *L. monocytogenes* infection cycle. The aim of our study was to use WGS to investigate the diversity and genetic relationships between *L. monocytogenes* isolates from hunted game and game meat in Finland. Furthermore, we studied the presence of important virulence and resistance genes obtained from the WGS data.

2. Materials and Methods

2.1. Listeria Monocytogenes Isolates

L. monocytogenes has been detected in hunted game and game meat in Finland between 2012 and 2020, especially in deer and mallard meat (Table 1). We characterised a total of 75 *L. monocytogenes* isolates from various game sources in this study (Table 1). One isolate per positive sample from the earlier studies was characterised. The sampling plan and time frame differed between the earlier studies (Table 1). Moose, deer, and wild boar samples were collected from wild hunted animals and game bird samples from game birds that were farmed for hunting. Deer and mallard meat samples were collected from a small meat processing plant. *L. monocytogenes* isolates were found after PCR screening in our microbiological laboratory at the Department of Food Hygiene and Environmental Health of the Faculty of Veterinary Medicine, University of Helsinki, in Helsinki, Finland. PCR screening and isolation of the isolates have been described in earlier studies [17,24].

Table 1. Isolation rates of *Listeria monocytogenes* from hunted game and game meat in Finland between 2012 and 2020.

Scheme 2012	Sampling Year	Number of Samples	Positive Samples		Reference
Moose carcass	2012–2014	100	5	(5%)	[17]
Deer carcass	2013–2015	100	5	(5%)	[17]
Deer meat	2019–2020	50	9	(18%)	Not published
Wild boar organ	2016	130	40	(31%)	[25]
Pheasant faeces	2013–2014	101	9	(9%)	[24]
Teal faeces	2013	30	1	(3%)	[24]
Mallard faeces	2013–2014	110	15	(14%)	[24]
Mallard meat	2016	100	13	(13%)	[24]

2.2. Whole Genome Sequencing (WGS)

DNA of *L. monocytogenes* isolates was purified from overnight enrichment at 37 °C in tryptic soya broth using PureLink Genomic DNA Mini Kit (Invitrogen, Carlsbaden, CA, USA) according to the manufacturer's protocol. DNA quality was measured with a NanoDrop™ spectrophotometer (ThermoFisher Scientific, Waltham, MA, USA) and DNA quantity with a Qubit fluorometer (ThermoFisher Scientific). WGS was performed on the Illumina platform by CeGaT (Center for Genomics and Transcriptomics, Tübingen, Germany). Illumina DNA Prep library preparation kit and NovaSeq6000 were used to generate 100 bp paired end reads. The short raw reads were assembled de novo using a Unicycler v0.4.8 assembler available on the PATRIC 3.6.12 platform (https://www.patricbrc.org/app/Assembly, accessed on 11 November 2022).

2.3. Characterisation of Listeria Monocytogenes Isolates

Species identification was confirmed in silico from the assemblies with KmerFinder v3.2 and SpeciesFinder v2.0 [26] available on the CGE (Center for Genomic Epidemiology) platform (http://www.genomicepidemiology.org/services/, accessed on 11 November 2022). In silico typing using 7-gene multi-locus sequence typing (MLST) [27] was performed on the CGE and BIGSdb-Lm (https://bigsdb.pasteur.fr/listeria/, accessed on 11 November 2022) platforms. STs obtained through the 7-gene MLST were grouped into CCs and phylogenetic Lin [27].

Assembled sequence data of 55 *L. monocytogenes* isolates were genotyped with core genome MLST (cgMLST) based on 1748 genes [28] using the open-source tool available on the BIGSdb-Lm platform. The nearest cgMLST profile (CT) from the database was recorded. Additionally, cgMLST targeting 1701 genes was performed using Ridom SeqSphere+ software v7.7.5 (Ridom GmbH, Muenster, Germany) [29] and the results were visualised with a minimum spanning tree (MST). Isolates forming a cluster (CL) displayed a maximum of 10 allelic differences from each other. The CLs were shaded in grey, and the number of allelic differences between the isolates was indicated on the connecting lines. Using the default parameters in the Ridom software, (1) STs, (2) PCR serogroups (SGs), (3) virulence genes and (4) antimicrobial resistance genes were also determined. Presence of the virulence genes was additionally studied with the VirulenceFinder 2.0 available on the CGE platform and on the Virulence Factor Database (VFDB) [30] (http://www.mgc.ac.cn/VFs/, accessed on 11 November 2022). In total, the presence of 33 virulence genes and 46 AMR genes among the 55 *L. monocytogenes* isolates was recorded.

3. Results

In total, 75 *L. monocytogenes* isolates from 75 hunted game and game meat samples—isolated in Finland between 2012 and 2020—were serotyped and characterised by seven-gene MLST (Table 2). Most (89%) of the isolates belonged to serotype 1/2a and were found in all sample types. Serotypes 4b and 2b were identified among 8% and 3% of the isolates, respectively. *L. monocytogenes* 4b was found on deer carcasses (n = 3), wild boar organs (n = 2) and in pheasant faeces (n = 1) (Table 2). *L. monocytogenes* 2b was only found in mallard faeces.

Based on the seven-gene MLST, 75 *L. monocytogenes* isolates from hunted game and game meat samples (n = 75) were classified into Lin I and II, 18 CCs and 21 STs (Table 3). Most of the isolates (89%) belonged to Lin II, including all serotype 1/2a isolates. Lin I included isolates of serotypes 1/2b and 4b. ST451 (16/75) was the most frequent ST followed by ST585 (9/75) and ST37 (8/75). These STs were found from different sample types between 2012 and 2020 (Table 3). ST451 and ST37 have frequently been identified in human listeriosis during recent years in Finland (Table 3). In total, 8 out of 21 STs found in game have been identified in cases of human listeriosis in Finland between 2016 and 2021. Most (17/75) of the isolates from wild boars hunted in 2016 belonged to several (11/21) STs (Figure 1). *L. monocytogenes* isolates were also frequently found from mallard faeces (15/75) and mallard meat (13/75) (Table 2). These samples were contaminated with less (5/75) STs compared with wild boar organs.

Table 2. Serotypes and sequence type (STs) using 7-gene multi-locus sequence typing (MLST) of 75 *Listeria monocytogenes* isolates obtained from hunted game and game meat in Finland between 2012 and 2022.

Source	Isolation Year	Number of Isolates	Serotype	MLST
Moose carcass	2012–2013	5	1/2a	ST7, 29, 37, 451
Deer carcass	2013–2014	5	1/2a (2), 4b (3)	ST4, 18, 315, 412
Deer meat	2019	9	1/2a	ST8, 11, 155
Mallard faeces	2013–2014	15	1/2a (13), 1/2b (2)	ST11, 37, 224, 391, 585
Mallard meat	2016	13	1/2a	ST8, 18, 412, 451, 585
Teal faeces	2013	1	1/2a	ST37
Pheasant faeces	2013–2014	9	1/2a (8), 4b (1)	ST1, 7, 20, 37, 585
Wild boar faeces	2020	1	1/2a	ST451
Wild boar organ	2016	17	1/2a (15), 4b (2)	ST1, 7, 8, 18, 20, 21, 37, 91, 399, 451, 573

Table 3. Characteristics of 75 *Listeria monocytogenes* isolates from hunted game and game meat from Finland between 2012 and 2020.

MLST	Finland [b]	Clonal Complex	Lineage	Serotype	Number of Isolates	Source (Isolation Year)
ST1 [a]	2016, 2017	CC1	I	4b	3	Pheasant faeces (2013), Wild boar organ (2016)
ST4		CC4	I	4b	1	Deer carcass (2014)
ST7	2018–2021	CC7	II	1/2a	4	Moose carcass (2012), Pheasant faeces (2013), Wild boar organ (2016)
ST8	2017–2018, 2020–2021	CC8	II	1/2a	3	Mallard meat (2016), Wild boar organ (2016), Deer meat (2019)
ST11		CC11	II	1/2a	1	Deer meat (2019)
ST18	2016	CC18	II	1/2a	5	Deer carcass (2013), Wild boar organ (2016), Mallard meat (2019)
ST20		CC20	II	1/2a	3	Pheasant faeces (2013), Wild boar organ (2016)
ST21		CC21	II	1/2a	2	Wild boar organ (2016)
ST29		CC29	II	1/2a	1	Moose carcass (2013)
ST37	2016, 2018–2020	CC37	II	1/2a	8	Moose carcass (2013), Teal faeces (2013), Mallard faeces (2013–2014), Pheasant faeces (2013–2014) Wild boar organ (2016)
ST91	2021	CC14	II	1/2a	1	Wild boar organ (2019)
ST155	2020	CC155	II	1/2a	3	Deer meat (2019)
ST224		CC224	I	1/2b	2	Mallard faeces (2013)
ST249		CC315	I	4b	2	Deer carcass (2013)
ST391		CC89	II	1/2a	2	Mallard faeces (2013)
ST399		CC14	II	1/2a	1	Wild boar organ (2016)
ST400		CC11	II	1/2a	2	Mallard faeces (2013)
ST412		CC412	II	1/2a	5	Deer carcass (2013), Mallard meat (2019)
ST451	2017–2021	CC11	II	1/2a	16	Moose carcass (2012), Mallard faeces (2013–2014), Mallard meat (2016), Wild boar organ (2016), Deer meat (2019), Wild boar faeces (2020)
ST573		CC573	II	1/2a	1	Wild boar organ (2016)
ST585		ST585	II	1/2a	9	Pheasant faeces (2013–2014), Mallard faeces (2013–2014), Mallard meat (2016)

[a] Sequence types in bold have recently been published in other European countries [31–33]. [b] Reporting year of the most common sequence types found in human listeriosis in Finland during 2016–2021.

A subset of 55 out of 75 *L. monocytogenes* isolates were characterised with cgMLST based on 1748 genes. In total, 35 CTs among 21 STs were obtained using the BIGSdb-Lm platform (Table 4). Overall, 10 CTs (CT5208, 11797, 20896, 20939, 25365, 26674, 26763,

28125, 28250, and 28251) included more than one *L. monocytogenes* isolate. In total, 10 CLs (CL1, 8, 18, 155, 412, 451a, 451b, 451c, 451d, and 585) formed by closely related genotypes were obtained with the cgMLST based on 1701 genes using the Ridom software (Table 4). Isolates belonging to the same CL showed an allelic difference between 0 and 10 (Figure 2). All CLs obtained by Ridom software had their own CT obtained from the BIGSSdb-Lm platform (Table 4). Most (9/10) of the CLs included isolates of SG IIa. CL1 included two undistinguishable isolates of SG IVb, both found from wild boar organs. Five CLs (CL1, 18, 155, 412 and 451c) included very closely related isolates, each with 0 to 5 allelic differences (Figure 2). Each of these CLs included *L. monocytogenes* isolates found from one source during the same year (Table 4). CL18 and CL412 were formed by isolates from mallard meat and CL155 and CL451c from deer meat. The other CLs (CL8, 451a, 451b, 451d and 585) were formed by closely related isolates (0 to 10 allele differences) found from different sources between 2012 and 2019 (Table 4).

Table 4. Distribution of different serogroups (SGs), MLST (STs) and cgMLST (CTs) profiles, clusters (CLs) and virulence profiles (VPs) among 55 *Listeria monocytogenes* isolates from hunted game and game meat between 2012 and 2020 in Finland.

ST [a]	SG	CT [b]	CL [c]	VP [d]	No.	Source		Year
1	IVb	1430		2c	1	Pheasant	Faeces	2013
1	IVb	25365	1	2c	2	Wild boar	Organ	2016
4	IVb	27292		1d	1	Deer	Carcass	2014
7	IIa	21218		2a	1	Wild boar	Organ	2016
7	IIa	22874		1e	1	Moose	Carcass	2012
8	IIa	19030		2a	1	Wild boar	Organ	2016
8	IIa	28125	8	1e	2	Mallard, Deer	Meat	2016, 2019
11	IIa	30149		1b	1	Deer	Meat	2019
18	IIa	26674	18	1a	3	Mallard	Meat	2016
18	IIa	28197		1a	1	Wild boar	Organ	2016
20	IIa	21358		1a	1	Wild boar	Organ	2016
20	IIa	26681		0	1	Pheasant	Faeces	2013
21	IIa	9841		2a	1	Wild boar	Organ	2016
21	IIa	25363		2a	1	Wild boar	Organ	2016
29	IIa	23920		1e	1	Moose	Carcass	2013
37	IIa	22893		1e	1	Teal	Faeces	2013
37	IIa	28230		1e	1	Pheasant	Faeces	2014
37	IIa	30787		2a	1	Wild boar	Organ	2016
91	IIa	20232		1a	1	Wild boar	Organ	2016
155	IIa	26763	155	0	3	Deer	Meat	2019
224	IIb	8887		1c	1	Mallard	Faeces	2013
249	IVb	20958		1d	1	Deer	Carcass	2013
391	IIa	29935		0	1	Mallard	Meat	2013
399	IIa	22031		2b	1	Wild boar	Organ	2016
400	IIa	28173		1b	1	Mallard	Meat	2013
412	IIa	8287		3b	1	Deer	Carcass	2013
412	IIa	20896	412	2d	4	Mallard	Meat	2016
451	IIa	24184		0	1	Deer	Meat	2019
451	IIa	11793		0	1	Wild boar	Faeces	2020
451	IIa	5208	451a	1a	3	Mallard, wild boar	Meat, organ	2016
451	IIa	11797	451b	0,1a	2	Moose, wild boar	Carcass, organ	2012, 2016
451	IIa	20939	451c	0	3	Deer	Meat	2019
451	IIa	28250	451d	0,1a	4	Mallard, wild boar	Faeces, organ	2014, 2016
573	IIa	1569		2a	1	Wild boar	Faeces	2016
585	IIa	28251	585	1e,2a,3a	4	Mallard	Faeces, meat	2013, 2016

[a] ST based on 7-gene MLST using the BIGSdb-Lm platform and Ridom software. [b] Nearest CT based on cgMLST (1748 target genes) using the BIGSdb-Lm platform. [c] CL based cgMLST (1708 target genes) using the Ridom software. [d] VPs using the Ridom software and CGE platform.

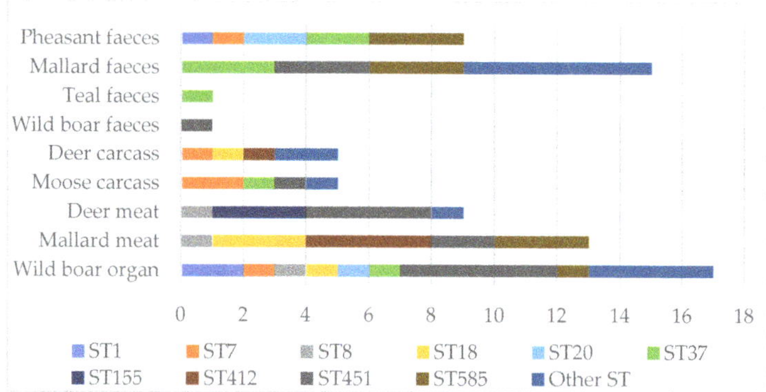

Figure 1. Sequence types (ST), which include at least three *Listeria monocytogenes* isolates, found in hunted game and game meat in Finland between 2012 and 2020.

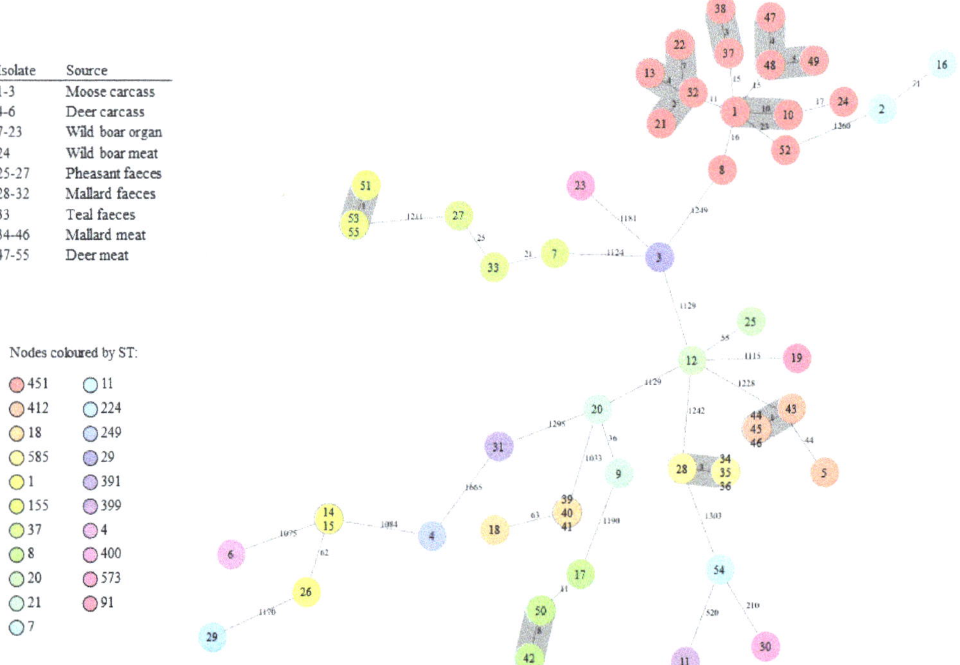

Figure 2. Minimum spanning tree of 55 *Listeria monocytogenes* isolates from hunted game and game meat in Finland during 2012–2020. The tree was calculated in Ridom SeqSphere+ with 1708 core genome multi-locus sequence typing (MLST) targets and 7-gene MLST targets (pairwise ignoring missing values, logarithmic scale). Nodes are coloured according to sequence type. Number of allelic differences between the isolates are indicated on the connecting lines. Clusters are shaded in grey and a cluster distance threshold of maximum 10 was used according to Ruppitsch et al. [29].

We studied the presence of 33 virulence genes available in the Ridom software [34] among the 55 *L. monocytogenes* isolates. Most (26/33) of the genes were detected in all isolates. Seven virulence genes (*act, ami, aut, inl*F, *inl*J, *lap*B and *vip*) were not present in all isolates. We designed 12 virulence profiles (VPs) based on these genes (Table 5). All virulence genes (VP0) were detected in 12 (22%) *L. monocytogenes* isolates, all belonging to SG IIa. The most frequently missing genes were *ami* and *vip*, which were missing in 42% and 38% of *L. monocytogenes* isolates, respectively. The VP did not correlate with ST, but isolates belonging to the same cluster mostly (66%) had the same VP (Table 4).

Table 5. Virulence profiles detected among 55 *Listeria monocytogenes* isolates.

Virulence Profile	Number of Isolates	Sequence Types	Missing Virulence Genes (=1)						
			act	*ami*	*aut*	*inl*F	*inl*J	*lap*B	*vip*
VP0	12	ST20,155,391,451	0	0	0	0	0	0	0
VP1a	13	ST18,20,91,451	0	1	0	0	0	0	0
VP1b	2	ST11,400	0	0	0	1	0	0	0
VP1c	1	ST224	0	0	0	0	1	0	0
VP1d	2	**ST4,249** [a]	0	0	1	0	0	0	0
VP1e	7	ST7,8,29,37, 585	0	0	0	0	0	0	1
VP2a	7	ST7,8,21,37,573,585	0	1	0	0	0	0	1
VP2b	1	ST399	0	1	0	1	0	0	0
VP2c	3	**ST1**	0	0	1	0	1	0	0
VP2d	4	ST412	0	0	0	0	0	1	1
VP3a	2	ST585	0	1	0	0	1	0	1
VP3b	1	ST412	1	0	0	0	0	1	1

[a] Isolates with sequence types in bold belong to serogroup IVb.

We studied the presence of 46 AMR genes available in the Ridom software. Only the *fos*X and *lin* genes were detected in all 55 *L. monocytogenes* isolates.

4. Discussion

L. monocytogenes is a common finding in hunted game and game meat in Finland. Most of the *L. monocytogenes* isolates originating from game belonged to serotype 1/2a (SG IIa, Lin II) but serotype 4b (SG IVb, Lin I) was also found. *L. monocytogenes* strains belonging to SG IIa/Lin II and SG IVb/Lin I are responsible for the largest share of listeriosis [20,35,36]. However, SG IVb is more frequently associated with human diseases and outbreaks than SG IIa, which is more often identified among isolates found in animal, environmental and food samples [6,18,21]. Recently, *L. monocytogenes* IIa and IVb were found in deer and wild boar tonsils in Spain [12]. Serotyping and serogrouping provide useful information about *L. monocytogenes* isolates found in epidemiological studies, surveys and during monitoring.

Very little is known about the genetic diversity of *L. monocytogenes* isolates from game origin [12]. We found several CCs and STs in hunted game and game meat from Finland showing a large genetic diversity among the *L. monocytogenes* isolates studied. This was expected because *L. monocytogenes* isolates were found from various sources and locations during a ten-year period [17,24,25]. All CCs identified among our hunted game and game meat isolates from Finland have recently been identified among various environmental and animal sources in Europe [5]. In our data, the most common CC was CC11 (25%), which included three STs: ST11, ST400 and ST451. CC11 is also a prevalent clonal type found in Europe [23,31]. Most (67%) of the STs found in our study have also been found in Europe from various sources [23,31,32]. Several (7/21) STs found in game in our study have been associated with human listeriosis in Finland (https://thl.fi/en/web/infectious-diseases-and-vaccinations accessed on 11 November 2022). ST451 (21%) was the most frequently found ST in our study, as it was found in different sample types between 2012 and 2020. This type has also been reported in human listeriosis in Finland yearly between 2017 and 2021. ST451 is a common universal ST found in humans, animals, foods, and the environment in Europe [23,31,32,37]. To obtain more accurate information about the link between human and game isolates, STs based on the core or whole genome (cgMLST or wgMLST) should be used instead of seven-gene MLST.

Wild boar organs were contaminated with several *L. monocytogenes* isolates of different STs. This is very understandable because the isolates originated from wild boars hunted in various geographical locations in Finland [25]. Fewer STs were found in isolates from deer and mallard meat than from wild boar. Deer and mallard meat were processed in one meat processing plant each, which may explain the limited genetic diversity among the meat isolates. Interestingly, only 4 STs were identified among 13 *L. monocytogenes* SG IIa isolates from mallard faeces. The hunted mallards were reared and fed in a natural pond before being hunted, which could be a common contamination source for the mallards [24]. *L. monocytogenes* is relatively commonly found in various environments, and *L. monocytogenes*-contaminated soil and water are therefore important *L. monocytogenes* sources [5,38]. ST18, ST20, ST37, ST91 and ST451, identified among our game isolates, are reportedly common STs among isolates from environmental samples in Finland [37] and Latvia [31].

We identified some CLs of *L. monocytogenes* isolates with 0 to 10 allelic differences among the hunted game and game meat isolates using cgMLST, which is the method capable of differentiating related strains from unrelated ones [39]. Very closely related isolates, with a maximum of five allelic differences, were found in five CLs, and they originated from the same source and year, which could explain the high genetic similarity and may indicate a common source of contamination. Three very closely related *L. monocytogenes* isolates—forming CL18—were isolated from mallard meat originated from various mallards sampled on the same day in the same game meat processing plant, indicating a cross-contamination during processing. In CL412, four very closely related isolates from mallards were sampled on two different days in the same plant, indicating a plant contamination possibly due to inadequate cleaning. Deer meat isolates also formed two clusters—CL155 and CL451c—both with three very closely related isolates. The isolates in CL155 and CL451c were from deer meat samples cut on different days in the same plant. Cross-contamination during meat cutting occurs easily if working hygiene is poor. *L. monocytogenes* can easily persist in the plant, and thorough cleaning of the meat processing plant after each working day is therefore very important.

L. monocytogenes has shown heterogeneity in its virulence [35,40]. Virulence factors are essential for adapting *L. monocytogenes* to spread optimally within the environment [35]. The virulence of *L. monocytogenes* is encoded by a wide range of virulence genes [2]. In our study, most (79%) of the 33 studied virulence genes were present in all 55 *L. monocytogenes* isolates of game origin in Finland. The *act*A gene located on the *Listeria* pathogenicity island (LIPI-1) was missing in only one isolate (a deer carcass isolate). LIP-I is composed of important virulence genes (including *act*A, *hly*, *mpl*, *plc*A, *plc*B, *prf*A and *orf*X) and is necessary for intracellular survival and spread from cell to cell [35]. LPI-1 is typically present in all *L. monocytogenes* strains [2,41]. This *act*A-negative deer carcass isolate (with VP3b, SG IIa and ST412) also missed the *lap*B (coding for an adhesion protein) and *vip* (coding for an invasion protein) genes, indicating a reduced virulence in this isolate. The most frequently missing virulence gene *ami*, which is coding an autolysin protein for adherence, was not found in 42% of the isolates. However, the meaning of *ami* in the virulence remains unclear. The invasion gene *aut* was missing only in the isolates belonging to SG IVb. All SG IVb isolates (with ST1, ST4 and ST249) were *aut*-negative. The three SG IVb isolates belonging to ST1 were also *inl*J-negative. The *aut* gene codes for an autolysin protein needed for invasion and the *inl*J for an internalin protein needed for adherence [35]. How the absence of these two genes affects the virulence of *L. monocytogenes* IVb isolates needs to be further studied. Typically, ST1 (CC1) and ST4 (CC4) have been associated with clinical cases more often than other STs [35,41].

AMR is a serious public health issue due to increasing resistance. There is also a trend of increasing AMR among *L. monocytogenes* strains of animal and food origin. Resistance, e.g., to penicillin, ampicillin, gentamycin, streptomycin, tetracycline, and trimethoprim-sulfamethoxazole has been reported [42,43]. However, in our study, only the *fos*X and *lin* genes were detected. These two genes were present in all 55 *L. monocytogenes* isolates. Earlier studies have shown *fos*X and *lin* to be present in nearly all *L. monocytogenes* isolates [42]. This can be explained by native resistance to fosfomycin and lincosamides reported in *L. monocytogenes* strains [34,43]. Our results indicate that *L. monocytogenes* of game and game meat origin found in Finland are so far sensitive to antimicrobials. One explanation may be that hunted game in Finland have no access to feed contaminated with resistant *L. monocytogenes* strains.

5. Conclusions

In this study, we analysed the sequence data of *L. monocytogenes* isolates of game origin using tools available on open-source platforms and Ridom software. Our study demonstrates that game meat is contaminated with various STs associated with human listeriosis. All *L. monocytogenes* isolates were potentially pathogenic, carrying most important virulence genes. No acquired AMR genes were found, indicating that all isolates were sensitive to most of the important antimicrobials used to treat listeriosis. Some of the isolates from mallard and deer meat belonged to CLs that were formed by very closely related isolates, indicating common contamination sources. Contaminated game meat may pose a public health problem, and game meat should therefore be handled and stored correctly.

Author Contributions: Conceptualisation: all authors; Investigation: M.S., P.K. and V.H.; Methodology: M.S. and A.H.; Software: P.K. and V.H.; Visualisation: P.K.; Validation: all authors; Analysis: M.F.-A.; Writing—original draft preparation: M.F.-A.; Writing—review and editing: P.P.; Resources: M.F.-A., A.H. and P.P.; Supervision, project administration and funding acquisition: M.F.-A. All authors have read and agreed to the published version of the manuscript.

Funding: This study was partly funded by the "Kyllikki ja Uolevi Lehikoinen" (IQUJA:10019) and "Walter Ehrström" Foundations. Proofreading and APC were funded by the University of Helsinki, Finland.

Data Availability Statement: Data are contained within the article.

Acknowledgments: Open access funding provided by University of Helsinki. Maria Stark is gratefully acknowledged for her technical assistance.

Conflicts of Interest: The authors declare no conflict of interest.

References

1. Lecuit, M. Listeria Monocytogenes, a Model in Infection Biology. *Cell. Microbiol.* **2020**, *22*, e13186. [CrossRef] [PubMed]
2. Quereda, J.J.; Morón-García, A.; Palacios-Gorba, C.; Dessaux, C.; García-del Portillo, F.; Pucciarelli, M.G.; Ortega, A.D. Pathogenicity and Virulence of Listeria Monocytogenes: A Trip from Environmental to Medical Microbiology. *Virulence* **2021**, *12*, 2509–2545. [CrossRef]
3. EFSA; ECDC. The European Union One Health 2020 Zoonoses Report. *EFSA J.* **2021**, *19*, e06971.
4. Matereke, L.T.; Okoh, A.I. Listeria Monocytogenes Virulence, Antimicrobial Resistance and Environmental Persistence: A Review. *Pathogens* **2020**, *9*, 528. [CrossRef] [PubMed]
5. Félix, B.; Sevellec, Y.; Palma, F.; Douarre, P.E.; Felten, A.; Radomski, N.; Mallet, L.; Blanchard, Y.; Leroux, A.; Soumet, C. A European-Wide Dataset to Uncover Adaptive Traits of Listeria Monocytogenes to Diverse Ecological Niches. *Sci. Data* **2022**, *9*, 1–12. [CrossRef] [PubMed]
6. Acciari, V.A.; Ruolo, A.; Torresi, M.; Ricci, L.; Pompei, A.; Marfoglia, C.; Valente, F.M.; Centorotola, G.; Conte, A.; Salini, R. Genetic Diversity of Listeria Monocytogenes Strains Contaminating Food and Food Producing Environment as Single Based Sample in Italy (Retrospective Study). *Int. J. Food Microbiol.* **2022**, *366*, 109562. [CrossRef]
7. Osek, J.; Lachtara, B.; Wieczorek, K. Listeria Monocytogenes in Foods—From Culture Identification to Whole-genome Characteristics. *Food Sci. Nutr.* **2022**, *10*, 2825–2854. [CrossRef]
8. Gomes-Neves, E.; Abrantes, A.C.; Vieira-Pinto, M.; Müller, A. Wild Game Meat—A Microbiological Safety and Hygiene Challenge? *Curr. Clin. Microbiol. Rep.* **2021**, *8*, 31–39. [CrossRef]
9. Obwegeser, T.; Stephan, R.; Hofer, E.; Zweifel, C. Shedding of Foodborne Pathogens and Microbial Carcass Contamination of Hunted Wild Ruminants. *Vet. Microbiol.* **2012**, *159*, 149–154. [CrossRef]
10. Paulsen, P.; Winkelmayer, R. Seasonal Variation in the Microbial Contamination of Game Carcasses in an Austrian Hunting Area. *Eur. J. Wildl. Res.* **2004**, *50*, 157–159. [CrossRef]
11. Sasaki, Y.; Goshima, T.; Mori, T.; Murakami, M.; Haruna, M.; Ito, K.; Yamada, Y. Prevalence and Antimicrobial Susceptibility of Foodborne Bacteria in Wild Boars (Sus Scrofa) and Wild Deer (Cervus Nippon) in Japan. *Foodborne Pathog. Dis.* **2013**, *10*, 985–991. [CrossRef]

12. Palacios-Gorba, C.; Moura, A.; Leclercq, A.; Gómez-Martín, Á.; Gomis, J.; Jiménez-Trigos, E.; Mocé, M.L.; Lecuit, M.; Quereda, J.J. *Listeria* spp. Isolated from Tonsils of Wild Deer and Boars: Genomic Characterization. *Appl. Environ. Microbiol.* **2021**, *87*, e02651-20. [CrossRef] [PubMed]

13. Stella, S.; Tirloni, E.; Castelli, E.; Colombo, F.; Bernardi, C. Microbiological Evaluation of Carcasses of Wild Boar Hunted in a Hill Area of Northern Italy. *J. Food Prot.* **2018**, *81*, 1519–1525. [CrossRef]

14. Wacheck, S.; Fredriksson-Ahomaa, M.; König, M.; Stolle, A.; Stephan, R. Wild Boars as an Important Reservoir for Foodborne Pathogens. *Foodborne Pathog. Dis.* **2010**, *7*, 307–312. [CrossRef] [PubMed]

15. Weindl, L.; Frank, E.; Ullrich, U.; Heurich, M.; Kleta, S.; Ellerbroek, L.; Gareis, M. Listeria Monocytogenes in Different Specimens from Healthy Red Deer and Wild Boars. *Foodborne Pathog. Dis.* **2016**, *13*, 391–397. [CrossRef] [PubMed]

16. Paulsen, P.; Smulders, F.; Kukleci, E.; Tichy, A. Bacterial Surface Counts and Visually Assessed Cleanliness of Carcasses from Hunted Roe Deer (Capreolus Capreolus). *Wien. Tierarztl. Mon.* **2022**, *109*, doc3.

17. Sauvala, M.; Laaksonen, S.; Laukkanen-Ninios, R.; Jalava, K.; Stephan, R.; Fredriksson-Ahomaa, M. Microbial Contamination of Moose (Alces Alces) and White-Tailed Deer (Odocoileus Virginianus) Carcasses Harvested by Hunters. *Food Microbiol.* **2019**, *78*, 82–88. [CrossRef]

18. Alía, A.; Andrade, M.J.; Córdoba, J.J.; Martín, I.; Rodríguez, A. Development of a Multiplex Real-Time PCR to Differentiate the Four Major Listeria Monocytogenes Serotypes in Isolates from Meat Processing Plants. *Food Microbiol.* **2020**, *87*, 103367. [CrossRef]

19. Li, F.; Ye, Q.; Chen, M.; Zhou, B.; Xiang, X.; Wang, C.; Shang, Y.; Zhang, J.; Pang, R.; Wang, J. Mining of Novel Target Genes through Pan-Genome Analysis for Multiplex PCR Differentiation of the Major Listeria Monocytogenes Serotypes. *Int. J. Food Microbiol.* **2021**, *339*, 109026. [CrossRef] [PubMed]

20. Hyden, P.; Pietzka, A.; Lennkh, A.; Murer, A.; Springer, B.; Blaschitz, M.; Indra, A.; Huhulescu, S.; Allerberger, F.; Ruppitsch, W. Whole Genome Sequence-Based Serogrouping of Listeria Monocytogenes Isolates. *J. Biotechnol.* **2016**, *235*, 181–186. [CrossRef]

21. Lachtara, B.; Wieczorek, K.; Osek, J. Genetic Diversity and Relationships of Listeria Monocytogenes Serogroup IIa Isolated in Poland. *Microorganisms* **2022**, *10*, 532. [CrossRef] [PubMed]

22. Zhang, H.; Chen, W.; Wang, J.; Xu, B.; Liu, H.; Dong, Q.; Zhang, X. 10-Year Molecular Surveillance of Listeria Monocytogenes Using Whole-Genome Sequencing in Shanghai, China, 2009–2019. *Front. Microbiol.* **2020**, *11*, 551020. [CrossRef] [PubMed]

23. Fagerlund, A.; Idland, L.; Heir, E.; Møretrø, T.; Aspholm, M.; Lindbäck, T.; Langsrud, S. Whole-Genome Sequencing Analysis of Listeria Monocytogenes from Rural, Urban, and Farm Environments in Norway: Genetic Diversity, Persistence, and Relation to Clinical and Food Isolates. *Appl. Environ. Microbiol.* **2022**, *88*, e02136-e21. [CrossRef] [PubMed]

24. Sauvala, M.; Woivalin, E.; Kivistö, R.; Laukkanen-Ninios, R.; Laaksonen, S.; Stephan, R.; Fredriksson-Ahomaa, M. Hunted Game Birds–Carriers of Foodborne Pathogens. *Food Microbiol.* **2021**, *98*, 103768. [CrossRef] [PubMed]

25. Fredriksson-Ahomaa, M.; London, L.; Skrzypczak, T.; Kantala, T.; Laamanen, I.; Biström, M.; Maunula, L.; Gadd, T. Foodborne Zoonoses Common in Hunted Wild Boars. *EcoHealth* **2020**, *17*, 512–522. [CrossRef]

26. Larsen, M.V.; Cosentino, S.; Lukjancenko, O.; Saputra, D.; Rasmussen, S.; Hasman, H.; Sicheritz-Pontén, T.; Aarestrup, F.M.; Ussery, D.W.; Lund, O. Benchmarking of Methods for Genomic Taxonomy. *J. Clin. Microbiol.* **2014**, *52*, 1529–1539. [CrossRef]

27. Ragon, M.; Wirth, T.; Hollandt, F.; Lavenir, R.; Lecuit, M.; Le Monnier, A.; Brisse, S. A New Perspective on Listeria Monocytogenes Evolution. *PLoS Pathog.* **2008**, *4*, e1000146. [CrossRef]

28. Moura, A.; Criscuolo, A.; Pouseele, H.; Maury, M.M.; Leclercq, A.; Tarr, C.; Björkman, J.T.; Dallman, T.; Reimer, A.; Enouf, V. Whole Genome-Based Population Biology and Epidemiological Surveillance of Listeria Monocytogenes. *Nat. Microbiol.* **2016**, *2*, 16185. [CrossRef]

29. Ruppitsch, W.; Pietzka, A.; Prior, K.; Bletz, S.; Fernandez, H.L.; Allerberger, F.; Harmsen, D.; Mellmann, A. Defining and Evaluating a Core Genome Multilocus Sequence Typing Scheme for Whole-Genome Sequence-Based Typing of Listeria Monocytogenes. *J. Clin. Microbiol.* **2015**, *53*, 2869–2876. [CrossRef]

30. Liu, B.; Zheng, D.; Zhou, S.; Chen, L.; Yang, J. VFDB 2022: A General Classification Scheme for Bacterial Virulence Factors. *Nucleic Acids Res.* **2022**, *50*, D912–D917. [CrossRef]

31. Terentjeva, M.; Šteingolde, Ž.; Meistere, I.; Elferts, D.; Avsejenko, J.; Streikiša, M.; Gradovska, S.; Alksne, L.; Ķibilds, J.; Bērziņš, A. Prevalence, Genetic Diversity and Factors Associated with Distribution of Listeria Monocytogenes and Other Listeria spp. in Cattle Farms in Latvia. *Pathogens* **2021**, *10*, 851. [CrossRef] [PubMed]

32. Filipello, V.; Mughini-Gras, L.; Gallina, S.; Vitale, N.; Mannelli, A.; Pontello, M.; Decastelli, L.; Allard, M.W.; Brown, E.W.; Lomonaco, S. Attribution of Listeria Monocytogenes Human Infections to Food and Animal Sources in Northern Italy. *Food Microbiol.* **2020**, *89*, 103433. [CrossRef] [PubMed]

33. Fagerlund, A.; Langsrud, S.; Møretrø, T. Microbial Diversity and Ecology of Biofilms in Food Industry Environments Associated with Listeria Monocytogenes Persistence. *Curr. Opin. Food Sci.* **2021**, *37*, 171–178. [CrossRef]

34. Parra-Flores, J.; Holý, O.; Bustamante, F.; Lepuschitz, S.; Pietzka, A.; Contreras-Fernández, A.; Castillo, C.; Ovalle, C.; Alarcón-Lavín, M.P.; Cruz-Córdova, A. Virulence and Antibiotic Resistance Genes in Listeria Monocytogenes Strains Isolated from Ready-to-Eat Foods in Chile. *Front. Microbiol.* **2021**, *12*, 796040. [CrossRef]

35. Disson, O.; Moura, A.; Lecuit, M. Making Sense of the Biodiversity and Virulence of Listeria Monocytogenes. *Trends Microbiol.* **2021**, *29*, 811–822. [CrossRef]

36. Gorski, L.; Cooley, M.B.; Oryang, D.; Carychao, D.; Nguyen, K.; Luo, Y.; Weinstein, L.; Brown, E.; Allard, M.; Mandrell, R.E. Prevalence and Clonal Diversity of over 1,200 Listeria Monocytogenes Isolates Collected from Public Access Waters near Produce Production Areas on the Central California Coast during 2011 to 2016. *Appl. Environ. Microbiol.* **2022**, *88*, e00357-22. [CrossRef]

37. Castro, H.; Douillard, F.P.; Korkeala, H.; Lindström, M. Mobile Elements Harboring Heavy Metal and Bacitracin Resistance Genes Are Common among Listeria Monocytogenes Strains Persisting on Dairy Farms. *Msphere* **2021**, *6*, e00383-21. [CrossRef]

38. Farber, J.M.; Zwietering, M.; Wiedmann, M.; Schaffner, D.; Hedberg, C.W.; Harrison, M.A.; Hartnett, E.; Chapman, B.; Donnelly, C.W.; Goodburn, K.E. Alternative Approaches to the Risk Management of Listeria Monocytogenes in Low Risk Foods. *Food Control* **2021**, *123*, 107601. [CrossRef]

39. Chen, Y.; Gonzalez-Escalona, N.; Hammack, T.S.; Allard, M.W.; Strain, E.A.; Brown, E.W. Core Genome Multilocus Sequence Typing for Identification of Globally Distributed Clonal Groups and Differentiation of Outbreak Strains of Listeria Monocytogenes. *Appl. Environ. Microbiol.* **2016**, *82*, 6258–6272. [CrossRef]

40. Vázquez-Boland, J.A.; Kuhn, M.; Berche, P.; Chakraborty, T.; Domínguez-Bernal, G.; Goebel, W.; González-Zorn, B.; Wehland, J.; Kreft, J. Listeria Pathogenesis and Molecular Virulence Determinants. *Clin. Microbiol. Rev.* **2001**, *14*, 584–640. [CrossRef]

41. Schiavano, G.F.; Ateba, C.N.; Petruzzelli, A.; Mele, V.; Amagliani, G.; Guidi, F.; De Santi, M.; Pomilio, F.; Blasi, G.; Gattuso, A. Whole-Genome Sequencing Characterization of Virulence Profiles of Listeria Monocytogenes Food and Human Isolates and in Vitro Adhesion/Invasion Assessment. *Microorganisms* **2021**, *10*, 62. [CrossRef] [PubMed]

42. Hanes, R.M.; Huang, Z. Investigation of Antimicrobial Resistance Genes in Listeria Monocytogenes from 2010 through to 2021. *Int. J. Environ. Res. Public Health* **2022**, *19*, 5506. [CrossRef] [PubMed]

43. Olaimat, A.N.; Al-Holy, M.A.; Shahbaz, H.M.; Al-Nabulsi, A.A.; Abu Ghoush, M.H.; Osaili, T.M.; Ayyash, M.M.; Holley, R.A. Emergence of Antibiotic Resistance in Listeria Monocytogenes Isolated from Food Products: A Comprehensive Review. *Compr. Rev. Food Sci. Food Saf.* **2018**, *17*, 1277–1292. [CrossRef] [PubMed]

Review

Presence of Foodborne Bacteria in Wild Boar and Wild Boar Meat—A Literature Survey for the Period 2012–2022

Caterina Altissimi [1], Clara Noé-Nordberg [2], David Ranucci [1] and Peter Paulsen [3,*]

[1] Department of Veterinary Medicine, University of Perugia, Via San Costanzo 4, 06121 Perugia, Italy; caterina.altissimi@studenti.unipg.it (C.A.); david.ranucci@unipg.it (D.R.)
[2] Esterhazy Betriebe GmbH, Esterházyplatz 5, 7000 Eisenstadt, Austria; c.noe-nordberg@esterhazy.at
[3] Unit of Food Hygiene and Technology, Institute of Food Safety, Food Technology and Veterinary Public Health, University of Veterinary Medicine Vienna, Veterinärplatz 1, 1210 Vienna, Austria
* Correspondence: peter.paulsen@vetmeduni.ac.at; Tel.: +43-12-5077-3318

Abstract: The wild boar is an abundant game species with high reproduction rates. The management of the wild boar population by hunting contributes to the meat supply and can help to avoid a spillover of transmissible animal diseases to domestic pigs, thus compromising food security. By the same token, wild boar can carry foodborne zoonotic pathogens, impacting food safety. We reviewed literature from 2012–2022 on biological hazards, which are considered in European Union legislation and in international standards on animal health. We identified 15 viral, 10 bacterial, and 5 parasitic agents and selected those nine bacteria that are zoonotic and can be transmitted to humans via food. The prevalence of *Campylobacter*, *Listeria monocytogenes*, *Salmonella*, Shiga toxin-producing *E. coli*, and *Yersinia enterocolitica* on muscle surfaces or in muscle tissues of wild boar varied from 0 to ca. 70%. One experimental study reported the transmission and survival of *Mycobacterium* on wild boar meat. *Brucella*, *Coxiella burnetii*, *Listeria monocytogenes*, and Mycobacteria have been isolated from the liver and spleen. For *Brucella*, studies stressed the occupational exposure risk, but no indication of meat-borne transmission was evident. Furthermore, the transmission of *C. burnetii* is most likely via vectors (i.e., ticks). In the absence of more detailed data for the European Union, it is advisable to focus on the efficacy of current game meat inspection and food safety management systems.

Keywords: wildlife; game meat; *Salmonella*; *Listeria*; *Campylobacter*; *Yersinia*; mycobacteria; verotoxinogenic *E. coli*; *Brucella*; *Staphylococcus aureus*

Citation: Altissimi, C.; Noé-Nordberg, C.; Ranucci, D.; Paulsen, P. Presence of Foodborne Bacteria in Wild Boar and Wild Boar Meat—A Literature Survey for the Period 2012–2022. *Foods* **2023**, *12*, 1689. https://doi.org/10.3390/foods12081689

Academic Editor: Evandro Leite de Souza

Received: 22 March 2023
Revised: 12 April 2023
Accepted: 15 April 2023
Published: 18 April 2023

1. Introduction

During the last decade, numbers of wild ungulates, in particular wild boars, have been rising significantly worldwide, generating environmental, economic, public health, and social concerns. Wild boar is the most widespread species due to its high adaptability and fertility rate, and its spread has been facilitated by climate change, the abandonment of rural areas, reforestation, a lack of predators, animal introductions, and supplementary feeding for hunting purposes [1–4]. The high density of this expanding species is causing, in particular, in Europe, not only relevant damages to agriculture and ecosystems and an increase in road accidents but also increases the risk of transmission of pathogens from wild boar to humans, livestock, and domestic animals [5,6]. The synanthropic behavior of wild boars is an important co-factor in creating disease-transmission scenarios [7]. Furthermore, the attention being paid to wild boar population control is leading to an increase in the availability of game meat. Additionally, the market has to face different harvesting practices, the wider distribution of this product, and, simultaneously, guarantee its safety aspects. In this context, it is of the utmost importance to understand the epidemiological situation and the major hazards due to the consumption of such meat.

Indeed, it has been highlighted by several authors how wild boar could act as a reservoir, playing an important role in the maintenance, circulation, and diffusion of certain pathogens for humans and animals [8–12]. In particular, the same authors focused their attention on the most relevant bacterial food hazards that: cause disease to wild boar and can be present in the meat (e.g., *Brucella* spp., *Mycobacterium tuberculosis* complex); are harbored in the gut or other tissues and then transferred to the meat during processing (e.g., *Salmonella* spp., *Campylobacter* spp., *Escherichia coli*, *Yersinia enterocolitica*); contaminate the carcass due to their presence on animal skin and in the environment (e.g., Listeria spp., *Staphylococcus aureus*).

In a framework of global health, it is essential to consider not only zoonotic diseases but also animal diseases with an impact on food security. The aim of this review is to give an overview of publications from the period 2012–2022 on the presence of biological hazards in the wild boar population. In particular, foodborne zoonotic bacteria commonly reported in meat from domestic animals will be the focus, and their presence in wild boars will be reviewed.

2. Materials and Methods

A list of infectious agents was compiled, combining zoonotic agents included in compulsory monitoring in the European Union (Directive 2003/99/EC List A) [13], zoonotic agents monitored according to the epidemiological situation (Directive 2003/99/EC List B) [13], swine and multiple species diseases, infections, and infestations listed by the World Organisation for Animal Health (OIE), and the most common agents responsible for foodborne outbreaks reported from the European Food Safety Authority (EFSA) during the period 2015–2020 and in the EU Rapid Alert System for Food and Feed (RASFF).

For each agent on the list, a literature search was conducted on SCOPUS using the name of the selected pathogen or the related disease combined with the search string: "wild" AND "boar" OR "feral AND pig" OR "warthog". During the literature search, biological hazards that do not concern wild boars were excluded. The search was then adjusted for (i) the time period 2012–2022, (ii) document type as article or review, and (iii) English as the selected language. Papers about the prevalence and control strategy of selected diseases were considered, whereas articles reporting solely detection methods were included only if relevant for the interpretation of results. Although our work focuses on the relevance of wild boar (meat) in the European Union, we included references from other countries in view of imports of wild boar meat from third countries in the EU; similarly, studies on feral pigs and warthogs were included.

We also report the number of publications per agent and year as a proxy for the relevance of the agent and the interest and effort of the scientific community in this topic [14]. From this long list of biological hazards specifically addressed in national legislation or by international organizations, we selected those with evidence that they are actually transmitted via the handling, processing, and consumption of porcine meat and meat products.

3. Results

3.1. Overview of Biological Hazards in Wild Boar and Their Impact on Food Safety and Security

The array of biological agents addressed in EU legislation and international organizations such as the OIE is displayed in Table 1. Information on zoonotic potential and mode of transmission was taken from OIE, EFSA, and ECDC documentation. Notably, not all agents are zoonotic, and not all zoonotic agents are transmitted by meat. Among the pre-selected (i.e., taken from EU and OIE documents) infectious agents, no scientific literature was retrieved for two viruses and one bacterial genus. A clear increase (i.e., more than one doubling) in the average number of publications per year in the period 2017–2022 compared with that from 2012–2017 was noted for the viral diseases African swine fever, West Nile fever, and Japan encephalitis; the bacterium *Listeria*; and the parasite genera *Cryptosporidium*, *Cysticercus*, and *Echinococcus*.

Table 1. Agents or diseases of wild boar covered in the literature survey (2012–2022), their coverage in legislation, and the number of pertinent publications.

Agent/Disease	Type	Zoonotic	EU Zoonoses Directive	OIE Listed	n, Period 2012–2022	n, period 2012–2016	n, period 2017–2022	Average /Year, Period 2012–2016	Average/Year, Period 2017–2022	Ratio of Averages
African Swine Fever	V	n		y	499	58	441	11.6	73.5	6.3
Aujeszky's Disease	V	n		y	108	43	65	8.6	10.8	1.3
CSF	V	n		y	158	54	104	10.8	17.3	1.6
Foot and Mouth Disease	V	n		y	35	13	22	2.6	3.7	1.4
Porcine Respiratory and Reproductive Syndrome	V	n		y	62	27	35	5.4	5.8	1.1
West Nile Fever	V	n		y	17	4	13	0.8	2.2	2.7
Hepatitis A	V	y	f		0	0	0			
Influenza	V	y	f		0	0	0			
Japan Encephalitis	V	y		y	21	6	15	1.2	2.5	2.1
Rabies	V	y	f	y	19	6	13	1.2	2.2	1.8
Paratuberculosis	B	n		y	9	7	2	1.4	0.3	0.2
Bacillus anthracis	B	y		y	3	2	1	0.4	0.2	0.4
Borrelia	B	y	f		30	9	21	1.8	3.5	1.9
Brucella	B	y	m	y	95	36	59	7.2	9.8	1.4
Campylobacter	B	y	m		22	7	15	1.4	2.5	1.8
Clostridium	B	y	f (*C. botulinum*)		0	0	0			
Francisella	B	y		y	12	6	6	1.2	1.0	0.8
Leptospira	B	y	f		55	17	36	3.4	6.0	1.8
Listeria	B	y	m		12	3	9	0.6	1.5	2.5
Q-Fever	B	y		y	23	7	16	1.4	2.7	1.9
Salmonella	B	y	m		80	25	55	5.0	9.2	1.8
St. aureus	B	y	*		27	10	17	2.0	2.8	1.4
Tuberculosis	B	y	m (*M. bovis*), f (others)		214	97	117	19.4	19.5	1.0
Verotoxinogenic *E. coli*	B	y	m		27	10	17	2.0	2.8	1.4
Yersinia	B	y	f		40	13	27	2.6	4.5	1.7
Cryptosporidium	P	y	f		18	5	13	1.0	2.2	2.2
Cysticercus	P	y	f	y	9	2	7	0.4	1.2	2.9
Echinococcus	P	y	m	y	47	12	35	2.4	5.8	2.4
Toxoplasma	P	y	f		90	35	55	7.0	9.2	1.3
Trichinella	P	y	m	y	167	67	100	13.4	16.7	1.2

V = virus; B = bacterium; P = parasite; f = facultative, according to the epidemiological situation; m = mandatory; * = multi-resistant *St. aureus*.

For a detailed review of the occurrence and significance of biological hazards, we focused on bacteria since these are the main causative agents for foodborne diseases reported in the EU [15].

3.2. Occurrence and Prevalence of Selected Zoonotic Bacteria in Wild Boar

3.2.1. *Brucella*

Brucella (B.) are gram-negative, nonsporeforming, aerobic, short-rod bacteria that include several pathogenic species. In the EU, monitoring of brucellosis is compulsory (Directive 2003/99/EC List A) [13]. In ruminants, swine, and dogs, infection with the agents causes diseases of the reproductive system, e.g., abortion or epididymitis. Symptomless carriers can excrete the pathogen, e.g., via milk. Small ruminants with mastitis caused by Brucella-melitensis can excrete the pathogen via milk. Ingestion of raw milk, inhalation, or close contact with infected animals or parts thereof (e.g., when dressing hunted wild game) can lead to human infections. These may resemble a feverish flu, whereas more severe courses involve splenomegaly and splenic or hepatic abscesses. In 2021, cattle livestock in 21 EU member states was officially free from brucellosis (*B. abortus*, *B. melitensis*, and *B. suis*), and as regards small ruminant livestock, 20 member states were officially free from the pathogen. In 2021, 162 human cases were reported, two of them foodborne. In 2020, there were also 2 cases linked to the consumption of sheep meat products, with *B. melitensis* being the causative species [15]. In the EU rapid alert system for food and feed (RASFF), no notification of the presence of Brucella in food was found.

As regards wild boar and Brucella, 96 documents were retrieved. Those reporting prevalence data were included in Table 2 (seropositivity) and Table 3 (DNA or viable bacteria). With respect to serological testing, the cross-reactivity with the Yersinia enterocolitica O9 antigen is a well-known issue. More recent methods may overcome this problem [16]. Some authors present seroprevalences corrected for cross-reactivity [17]. When tissues/organs of the animal were tested by bacteriological culturing, or PCR, blood, lymphatic organs, genital organs, and fetuses were examined. There was no study on Brucella in muscle tissue or commonly consumed organs, e.g., liver, from wild boar. When Brucella species and biovars are explicitly reported, it is mainly B. suis biovar 2.

While no documented cases of meat-borne brucellosis could be retrieved, several cases of brucellosis in humans hunting wild boar and dressing wild boar carcasses have been published; most reports are from the USA [18–21], but also from France [22] and Australia [23]. In two cases, neurological disorders [18,23] were reported, and in one case, arterial and venous thromboses were reported [20], which are otherwise rarely observed [24]. Similarly, dogs frequently in contact with wild boar are at risk of seropositivity to Brucella [25–27].

Table 2. Prevalence of *Brucella* spp. antibodies in wild boars (2012–2022), by country and continent.

Prevalence/Frequency	Species	Matrix	Country	Comment	Ref.
15.6% (15/96)	*B.* spp.	Sera	Italy (Tuscany)	serology	[28]
5.74% (16/287)	*B.* spp.	Sera	Italy (Tuscany)	RBT, CFT	[29]
5.1% (22/434) 13.5% (58/434)	*B.* spp.	Sera	Italy (Campania)	RBT ELISA	[30]
0.53% (2/374)	*B.* spp.	Sera	Italy (Tuscany)	RBT, CFT	[31]
6.2% (35/570)	*B.* spp.	Sera	Italy (Sardinina)	ELISA	[32]
15% (19/126)	*B. suis*	Sera	Italy (Central)	serology	[33]
59.3% (121/204)	*B.* spp.	Sera	Spain (Extremadura)	ELISA	[34]
9.4% (45/480)	*B. suis* biovar 2	Sera	Serbia	RBT, ELISA	[35]
1.3% (42/3230)	*B.* spp.	Sera	Croatia	RBT; CFT; ELISA	[36]
6.4% (131/2057)	*B.* spp.	Sera	Netherlands	ELISA	[37]
0% (0/286)	*B. suis*	Blood	Sweden	ELISA	[38]
9% (8/87)	*B.* spp.	Blood	Finland	RBT, ELISA; visceral organs from 5 seropos. animals available, in 4 of which *B. suis* biovar 2 was detected	[39]
13.3% (139/1044)	*B. suis*	Sera	Latvia	RBT, CFT, ELISA, data corrected for O9-cross-reactivity	[17]

Table 2. *Cont.*

Prevalence/Frequency	Species	Matrix	Country	Comment	Ref.
0% (0/100)	*B.* spp.	Sera	South Africa	Warthog	[40]
12.5% (1/8)	*B.* spp.	Sera	Kenya	Warthog; Antibody-ELISA	[41]
0% (0/86)	*B.* spp.	Sera	Brazil	Agglutination, 2MET	[42]
0% (0/61)	*B.* spp.	Sera	Brazil (Santa Catarina)		[43]
0.49% (1/205)	*B.* spp.	Blood	Brazil	Feral pigs; serology (BAPA, FPT)	[44]
0% (0/15)	*B.* spp.	Blood	Colombia	Feral pigs	[45]
2.2% (1/46)	*B.* spp.	Blood	Guam	Feral pigs; FPT	[46]
0.7% (2/282)	*B. abortus*	Sera	USA (Oklahoma)	BAPA, RIV, FPT	[47]
2.95% (7/238)	*B. suis*	Sera	Australia (NSW)	RBT, CFT	[48]
9.6% (8/83)	*B. suis*	Blood	Australia (Queensland)	RBT, CFT	[49]
0% (0/303)	*B.* spp.	Sera	Finland	RBT	[50]
54.9% (641/1168)	*B.* spp.	Sera	Belgium	ELISA	[51]

BAPA = Buffered Acidified Plate Antigen, CFT = Complement Fixation Test, RBT = Rose-Bengal-Test, RIV = Rivanol Agglutination, 2MET = 2-Mercapto-Ethanol.

Table 3. Prevalence of *Brucella* spp. (viable bacteria or DNA) in wild boar (2012–2022), by country and continent.

Prevalence/Frequency	Species	Matrix	Country	Comment	Ref.
12.5% (1/8)	*B.* spp.	Sera	Kenya	Warthog; PCR	[41]
1.4% (4/287) 1.7% (5/287) 2.2% 0% (0/287)	*B. suis* biovar 2	Lymph nodes epididymides fetuses pooled livers, spleens	Italy (Tuscany)	DNA	[29]
0.83% (2/240)	*B.* spp.	Inner organs	Denmark	culture	[52]
3.8% (7/180) 10.5% (19/180)	*B.* spp.	Tonsils	Netherlands	culture PCR; confirmed as *B. suis* biovar 2	[37]
22% (19/87)	*B. suis*	Feces	USA (Georgia)	Feral pigs, PCR	[53]
1.3% (5/389)	*B. suis* biovar 2	Retropharyngeal lymph nodes	Italy	culture	[54]
3.7% (7/188)	*B. suis* biovar 2	Reproductive organs	Spain (Extremadura)	culture, PCR	[34]
0% (0/238)	*B.* spp.	Blood	Australia (NSW)	culture	[48]

3.2.2. *Campylobacter*

Campylobacter is a genus of gram-negative, nonsporeforming, microaerophilic, motile spiral-shaped bacteria, with *C. jejuni* and *C. coli* as the main species involved in Campylobacteriosis. The principal symptoms of *Campylobacter* infections are diarrhea, abdominal pain, fever, headache, nausea, and vomiting. The disease is usually self-limiting, and death is rare except in severe cases in elderly people, very young children, or immunocompromised patients [55]. In 2021, campylobacteriosis was the zoonosis with the highest number of human cases reported in the EU, with 127,840 cases of illness and 10,469 hospitalizations. With respect to foodborne outbreaks, it was the fourth most frequently reported agent with

249 outbreaks, 1051 cases, and 134 hospitalizations [15]. *Campylobacter* is common in food animals such as poultry, pigs, and cattle, and the main transmission route is via meat and meat products, as well as raw milk and milk products.

Twenty-two articles have been published from 2012 to 2022 regarding the prevalence of *Campylobacter* in wild boars, five of which were excluded as not relevant. The main matrix considered for the isolation of *Campylobacter* is feces, as reported in Table 4. The references highlighted the role of wild boars as a possible source of *Campylobacter* infection due to the prevalence of *Campylobacter* spp. in feces samples, albeit in a variable range from 12.5% [56] to 66% [57]. Several species have been isolated from fecal samples in varying prevalence ranges, e.g., *C. lanienae* from 1.2% [56] to 69% [58], *C. hyointestinalis* from 0.8% [59] to 22.1% [60], *C. coli* from 0.8% [56] to 16.3% [58], and *C. jejuni* from 0% [61] to 4.1% [58] of samples. As suggested by [59], the degree of urbanization of some areas populated by wild boars could have a relationship with the detection frequency of some *Campylobacter* species; in particular, *C. lanienae* was more frequently isolated in low urbanizations areas, suggesting that this pathogen could be interconnected with the kind of diet available.

During the period considered, only two studies were conducted on carcasses, and they presented similar results, with a prevalence of *Campylobacter* spp. of 11.1% [62] and 16.7% [63]. Peruzy et al. [64] investigated the presence of *Campylobacter* in wild boar meat samples, but the pathogen was not detected.

To date, the EU has set food processing hygiene criteria for *Campylobacter* only for poultry [65].

Table 4. Prevalence of *Campylobacter* spp. in wild boar (2012–2022) feces or on carcasses or meat.

Prevalence/Frequency	Species	Matrix	Country	Comment	Ref.
51.8% (29/56)	*Campylobacter* spp.	Feces	Italy		[63]
50% (38/76) 40.8% (31/76)	*Campylobacter* spp. *C. lanienae*	Feces	Italy	Campylobacter spp. with levels up to 10^3 CFU/g was detected in 39.5% animals	[66]
66% (188/287)	*Campylobacter* spp.	Feces	Spain	One isolate was identified as *C. jejuni*	[57]
60.8% (79/130) 46.2% (60/130) 16.9% (22/130) 0.8% (1/130) 0% (0/130)	*Campylobacter* spp. *C. lanienae* *C. coli* *C. hyointestinalis* *C. jejuni*	Feces	Spain	4% WB had both *C. lanienae* and *C. coli*, and 1% had both *C. lanienae* and *C. hyointestinalis*. All the isolates were resistant to at least one antimicrobial agent considered	[59]
38.9% (49/126) 69.4% (34/49) 16.3% (8/49) 4.1% (2/49)	*Campylobacter* spp. *C. lanienae* *C. coli* *C. jejuni*	Feces	Spain		[58]
19.51% (8/41) 4.88% (2/41) 0% (0/41)	*Campylobacter* spp. *C. coli* *C. jejuni*	Feces	Spain		[61]
43.8% (53/121) 25.6% (31/121) 17.4% (21/121) 0.8% (1/121)	*Campylobacter* spp. *C. lanienae* *C. hyointestinalis* *C. jejuni*	Feces	Japan	Five (16%) and 6 (29%) isolates of *C. lanienae* and *C. hyointestinalis*, respectively, were resistant to enrofloxacin	[67]
22.1% (71/321)	*C. hyointestinalis*	Feces	Japan		[60]
12.5% (31/248) 9.7% (25/248) 1.2% (3/248) 0.8% (2/248)	*Campylobacter* spp. *C. hyointestinalis* *C. lanienae* *C. coli*	Feces	Japan		[56]

Table 4. *Cont.*

Prevalence/Frequency	Species	Matrix	Country	Comment	Ref.
3.5% (13/370) 1.6% (6/370)	*C. coli* *C. jejuni*	Feces	USA	*C. coli* was significantly more frequent in female feral pigs	[68]
0% (0/87)	*C. jejuni*	Feces	USA		[53]
16.7% (5/30)	*Campylobacter* spp.	Carcass	Italy		[63]
11.1% (4/36)	*Campylobacter* spp.	Carcass	Italy		[62]
0% (0/28)	*Campylobacter* spp.	Meat	Italy		[64]

WB = wild boars.

3.2.3. *Coxiella burnetii*—Q-Fever

Coxiella burnetii is a gram-positive short-rod bacterium that grows aerobically within but also outside of host cells. It can form spores and persist under dry and acidic conditions. The bacterium is not only excreted via effluents, but several tick species can act as vectors for the pathogen. Infection of humans can occur via contact with effluents, ingestion of contaminated food, and inhalation of aerosolized pathogens, but also by tick bites. Infection causes a feverish disease (Q-fever) with pneumonia, followed by affections of the heart, liver, and spleen. In the EU, human cases are notifiable. Data indicate that the number of human cases as well as prevalence in animals is declining. However, monitoring of farm and wild animals is not harmonized in the EU [15]. At least 347 of the 460 confirmed human cases of Q-fever in 2021 were acquired within the EU, and the pathogen was prevalent in 5.2%, 5.9%, and 16.5% of samples from cattle, goats, and sheep, respectively. Since not all member states submitted data, the reported percentages are not necessarily representative of the EU [15]. Studies conducted on *C. burnetii* and wild boar can be grouped into three categories: (i) those on ticks collected from wild boars or from hunters or dogs in frequent contact with wild boars; (ii) those on serum or spleen samples from wild boar; and (iii) studies on the genetic diversity of *C. burnetii*.

Within Europe, studies originated in Spain and Italy (Table 5). DNA from *C. burnetii* was detected in 1.9% of spleen samples [69], and antibodies were found in 5.5% of serum samples [70] from wild boar in Spain. In studies from Italy, the pathogen was not recovered from wild boar samples but from ticks feeding on wild boars (0.5%; [71]) and from dogs in contact with wild boars (5.1%; [72]). Wild boar is not a specific or primary host for the pathogen [73], but since the agent is occasionally detected in tissues from wild boar, hunters and consumers handling and processing wild boar (meat) are both occupationally and dietary exposed. Similarly, hunters and dogs often in contact with wild boars are at risk of exposure to tick-borne pathogens, among them *C. burnetii* [71].

Table 5. Presence of *Coxiella burnetii* or antibodies in wild boar or in vectors associated with wild boar, according to country and continent, 2012–2022.

Prevalence/Frequency	Matrix	Country	Comment	Ref.
0% (0/100)	Spleen	Italy (Central)	PCR	[73]
0% (0/93) 0% (0/176)	Spleen Ticks	Italy	PCR	[74]
5.1% (6/117)	Blood of dogs	Italy (Central)	PCR	[72]
0.48% (2/411)	Ticks	Italy (South)	Ticks collected from hunters and dogs	[71]
0% (0/40) feeding ticks 0% (0/489) questing ticks	Ticks	Spain (Northwest)	PCR	[75]

Table 5. *Cont.*

Prevalence/Frequency	Matrix	Country	Comment	Ref.
5.5% (4/73)	Serum	Spain (Northwest)	antibodies	[70]
1.9% (9/484)	Spleen	Spain (North)	PCR	[69]
0% (0/2256) 0% (0/167)	Ticks Spleen	Spain	Near to Barcelona, a highly populated area	[76]
0% (0/8)	Serum	Kenya	antibodies Serology (ELISA)	[41]
0% (0/67)	Blood	Brazil		[77]
5% (4/79)	Ticks	Thailand	PCR	[78]
18.3% (19/104)	Serum of dogs	Australia	Queensland	[79]

No notifications regarding the presence of *C. burnetii* in foods were listed in the EU rapid alarm system (RASFF).

3.2.4. *Listeria monocytogenes*

Listeriosis is a zoonotic disease caused by *Listeria monocytogenes*, a gram-positive, nonsporeforming, facultatively anaerobic bacterium. Foodborne listeriosis is one of the most severe diseases, causing septicemia, neurologic disorders, and reproductive disorders. Pregnant women, elderly people, and individuals with weakened immune systems are at risk for severe courses of the disease. *Listeria* is a ubiquitous microorganism that thrives in soil, water, vegetables, and the digestive tracts of animals. It can survive and proliferate in different environmental conditions since it is tolerating a wide range of pH and temperatures [80]. The main transmission route of *Listeria* is through the ingestion of contaminated food [15].

Twelve studies have been found from 2012 to 2022 regarding the presence of *Listeria* spp. in wild boar carcasses, meat, and related products, two of which were excluded as not relevant (Table 6). *Listeria monocytogenes* was detected by many authors in tonsil samples, highlighting this organ as the preferred matrix for the presence and detection of *Listeria* [63,81,82]. Fredriksson-Ahomaa et al. [39] found *L. monocytogenes* in 48% of spleen and kidney samples from wild boars. Almost all isolates belonged to serotype 2a, except for two isolates identified as serotype 4b. The presence of *Listeria* in tonsils and in visceral organs underlines the necessity of particular attention during handling and evisceration of wild boar carcasses.

Regarding the presence of *Listeria* in wild boar meat products, Roila et al. [83] did not detect the pathogen in wild boar salami, whereas Lucchini et al. [84] isolated *Listeria* spp. in 65% of cured game meat sausages. Three species were identified: *L. monocytogenes*, 24%; *L. innocua*, 32% and *L. welshimeri*, 8%. Counts of *L. monocytogenes* were, however, always below the legal limit of 100 cfu/g set by Regulation (EC) 2073/2005 [65].

In the years 2020–2022, 340 notifications regarding the presence of *L. monocytogenes* in foods were listed in the EU rapid alarm system RASFF, of which 82 implicated meat and meat products; there was no explicit mention of game meat or wild boar meat in particular.

Table 6. Presence of *Listeria* sp. in wild boar, 2012–2022.

Prevalence/Frequency	Species	Matrix	Country	Comment	Ref.
0. 35% (1/287)	*L. monocytogenes*	Rectal swabs	Italy	*L.m.* serogroup IVb, serovar 4b; resistant to cefoxitin, cefotaxime and nalidixic acid	[85]
68.5% (37/54) 35.3% (18/51) 26.7% (8/30) 0% (0/30)	*Listeria* spp. *L. monocytogenes* *Listeria* spp. *L. monocytogenes*	tonsils tonsils Carcass Carcass	Italy	prevalence influenced by animal age and environmental temperature	[63]
48% (63/130)	*L. monocytogenes*	Spleen and kidneys	Finland		[39]
24.5% (12/49)	*L. monocytogenes*	Liver or tonsils or feces or intestinal lymph nodes, caecum content	Germany	Positive in at least one of the different matrices studied	[81]
14.3% (7/49)	*L. monocytogenes*	Tonsils	Germany		[81]
2% (1/49)	*L. monocytogenes*	Liver and intestinal lymph nodes and caecum content and feces	Germany	The same animal resulted positive for *L.m.* in all the matrices analyzed	[81]
51.8% (14/27) 40.7% (11/27) 0% (0/27)	*Listeria* spp. *L. monocytogenes* *L. monocytogenes*	Tonsils Tonsils Feces	Spain		[82]
37.3% (28/75) 0% (0/75)	*Listeria* spp. *L. monocytogenes*	Feces	Japan		[67]
0% (0/72)	*L. monocytogenes*	Carcass	Italy		[86]
65% (24/37) 24% (9/37) 32% (12/37) 8% (3/37)	*Listeria* spp. *L. monocytogenes* *L. innocua* *L. welshimeri*	Game meat cured sausages	Italy	*L.m.* < 10 cfu/g	[84]
0% (0/40)	*L. monocytogenes*	Wild boar salami	Italy		[83]

3.2.5. *Mycobacterium tuberculosis* Complex

Mycobacterium tuberculosis complex is a group of *mycobacteria* that include *M. tuberculosis*, the major cause of human tuberculosis (TB), and other genetically related species that affect livestock and wild animals but are also implicated in human disease [87,88]. Among these species, in the last decade, *M. bovis* [89–115], *M. caprae* [89,104,111,116,117], and *M. microti* [118–124] have been frequently reported from wild boar, feral pigs, and warthogs in different countries.

The MTC bacteria can cause localized granulomas (primary complex) after entering the host through the respiratory or digestive tract, and when the organism´s immune system cannot contain it (which can be the case in the elderly, children, and in people with compromised immune systems), it may be followed by primary or secondary-reactivated TB. Meningitis, extrapulmonary granulomas, miliary tuberculosis, and other disseminated/generalized forms are only a few examples of the various manifestations, along with a variety of clinical symptoms [125]. *M. bovis* is usually transmitted through oral ingestion, and therefore the extrapulmonary lesions in humans are more frequent than for *M. tuberculosis* [126]. In wild boar, the main primary complex is usually located in the submandibular and retropharyngeal lymph nodes, where the MTC is most frequently isolated [89,98,105,117,122,127,128]. Lesions were also reported in the tonsils, lung, mediastinal lymph nodes, spleen, liver, and kidney [106,117,127,128]. The lesion in the lymph nodes is characterized by caseous or necrotic-calcified tubercles that are defined as tuberculosis-like lesions (TBLL), as other *mycobacteria* different from MTC (e.g., *M. avium subsp. hominissuis*) could cause the same lesion [119,129–131]. *M. bovis* and *M. caprae* could also be detected (isolated/PCR) in lymph nodes without visible lesions [94,105,128,131]. Wild boar is reported for MTC shedding through the oral, nasal, and fecal routes [132], and therefore animal aggregation areas could result in contaminated water and soil and the maintenance of the infection in wildlife and livestock [118,133,134].

In addition, 214 studies regarding MTC and non-MTC in wild *Suidae* species have been found in the literature over the considered period, but only 35 were related to prevalence studies of MTC and were therefore considered. These studies were performed both by serology (Table 7) and by isolation or direct identification of *mycobacteria* in organs and tissues (Table 8). The prevalence of MTC varies between countries and between regions/counties inside each nation (e.g., Spain), but also due to the investigated matrix and the diagnostic methods adopted [94,98,135]. In this context, some studies were performed to define the sensitivity of different diagnostic tools on sera and on organs and tissues [94,96,119,136]. The serological prevalence of MTC in wild boar is generally conducted over multi-year studies and ranged from 87.7% in Montes de Toledo and Doñana National Park (Spain) [132] to near 0% in the USA [137]. The prevalence of MTC isolation in tissue and organs, considering studies conducted on more than 100 subjects, ranges from 64.2% for *M. microti* in the Lombardia region (Italy) [123] to 1.1% for *M. bovis* in the Basque Country (Spain) [89].

The presence of MTC in wild boar is still recognized as one of the main barriers to the eradication of the disease in livestock and, subsequently, in humans, particularly when extensive pastoral systems are implemented and there is an interface between farmed and wild animals [93,100,101,104,111,133,138,139]. Although the disease is notifiable in many countries (such as Europe and the United States), its control in wild boar is primarily restricted to standard visual game meat inspection, which is thought to be insufficient to find primary complex and small lesions [117], especially as post-mortem inspection could be carried out also by trained hunters [EC Regulation 853/2004 [140]]. Even the cultural method for bacterial isolation is less effective than other diagnostic tools (e.g., screening PCR directly performed on target tissues, such as head lymph nodes, even when no TBLL are detected) [94,136]. Another topic to be considered is the free movement of wildlife that could spread the disease in different geographic areas. The identification and long-term monitoring of the genotype/spoligotype existing in a territory may aid in specific surveillance plans and control actions [100,141].

Despite the role of wild boar as a reservoir for MTC and the possible transmission through food [11], wild boar meat and meat products as a source for human infection are reported only by Clausi et al. [142]. In this study, PCR tests revealed the presence of MTC DNA on the carcass surface of wild boar without TBLL, but no *Mycobacterium* spp. could be isolated. Clausi et al. [142] added lymph nodes with active TBLL (*M. bovis*) to meat batter during sausage processing. Although live bacteria could be isolated only at day 23 after the contamination of the sausages (neither before nor after), bacterial DNA was detected (PCR) throughout the entire study period (end of sampling at day 41). When *M. bovis* (10^5 CFU/g) was directly added during sausage manufacturing, it was isolated for up to 22 days of ripening. When meat surfaces were experimentally contaminated with *M. bovis*, the bacterium could be recovered after frozen storage for over 5 months [142]. The role of wild boar meat and derived raw meat products could therefore be further investigated, even if other authors consider meat a negligible source of human infection [117].

Table 7. Seroprevalence of MTC in wild boar, feral pigs, and warthogs, 2012–2022.

Prevalence/Frequency	Species	Country	Area	Comment	Ref.
16.7% (5/30)	*MTC*	Malaysia	Selangor	Sampling in 2019–2020 Test used: bovine purified protein derivative (bPPD)-based indirect in-house ELISA	[127]
17% (326/1902)	*MTC*	Spain	Basque Country	Sampling in 2010–2016 Test used: in house validated enzyme-linked immunosorbent assay (ELISA)	[143]
10.6% (46/434)	*MTC*	Italy	Campania Region	Sampling in 2012–2017 Test Used: Indirect ELISA INgezim Tuberculosis DR kit based on recombinant M. bovis protein (MPB83)	[92]
2.4% (16/278)	*MTC*	Portugal	Several Counties	Sampling in 2006–2013 Test used: bPPD-based indirect in-house ELISA	[95]
49.0% (49/100)	*M. bovis*	South Africa	uMhkuze Nature Reserve in Kwa-Zulu Natal, Marloth Park on the southern border of Kruger National Park in Mpumalanga	Sampling in 2013–2015 Test used: Indirect PPD ELISA and TB ELISA-VK®	[96]
87.7% (36/41)	*MTC*	Spain	Montes de Toledo and Doñana National Park	Sampling in 2011–2013 Test used: bPPD-based indirect in-house ELISA Prevalence was obtained adding the number of animals with lesions at necroscopy to the number of positive serological samples	[132]
0.0003% (1/2735)	*MTC*	USA	National survey	Sampling in 2007–2015 Test used: bPPD-based indirect ELISA	[137]
2.4% (18/743)	*MTC*	Switzerland	Geneva, Mittelland, Jura, Thurgau, Tessin	Sampling in 2008–2013 Test used: bPPD-based indirect in-house ELISA	[109]
5.9% (123/2080)	*MTC*	France	58 Departments	Sampling in 2000–2004/2009–2010 Test used: bPPD-based indirect ELISA	[144]
2.1% (22/1057)	*MTC*	Spain	Asturias and Galicia	Sampling in 2010–2012 Test used: bPPD-based indirect ELISA	[111]
67.7% (87/130)	*MTC*	Spain	Andalusia	Sampling in 2006–2010 Test used: MPB83-ELISA	[115]

Table 8. Prevalence of *Mycobacterium* spp. in wild boar, feral pigs and warthog organs and tissues, 2012–2022.

Prevalence/Frequency	Species	Country	Area	Comment	Ref.
37.7% (29/77)	*M. bovis*	Brasil	Rio Grande do Sul	Sampling in 2013–2019 Test used: DNA extraction from lungs, lymph nodes, liver, spleen and kidney followed by PCR	[91]
1.1% (10/894)	*MTC*	Spain	Basque County	Sampling in 2010–2019 Test used: isolation from lymph nodes followed by real time PCR and spoligotyping of the isolates Positive cultures were detected only form head lymph nodes	[89]
2.8% (5/176)	*MTC* (mainly *M. microti*)	Switzerland	Canton of Ticino	Sampling in 2017–2018 Test used: isolation from lymph nodes + direct PCR followed by MALDI-TOF MS identification High prevalence of N-MTC identification (57.4%)	[119]
38.2% (21/55)	*M. caprae*	Poland	Bieszczady Mountains region	Sampling in 2011–2017 Test used: isolation form lymph nodes followed by PCR and spoligotyping of the isolates	[116]
76.7% (946/1235)	*Mycobacterium* spp.	Spain	Doñana National Park	Sampling in 2006–2018 Test used: Visual inspection for TBLL	[133]
1.6% (8/495) Culture 4.4% (17/386) PCR	*M. bovis*	France	Aquitaine, Côte d'Or and Corsica	Sampling 2014–2016 Test used: isolation or direct PCR form lymph nodes followed by spoligotyping of the isolates	[94]
47.1% (16/34)	*M. bovis*	South Africa	Greater Kruger National Park	Sampling in 2015 Test used: Intradermal Tuberculin Test (ITT) on captured warthog. Lymph nodes bacterial culture followed by PCR identification	[97]
2.4% (180/7634)	*M. bovis*	France	National scale (11 at-risk areas)	Sampling in 2011–2017 Test used: Lymph nodes bacterial culture followed by PCR identification Detected in 7 of the 11 at-risk areas	[98]
37.0% (25/67)	*M. bovis*	South Africa	uMhkuze Nature Reserve in Kwa-Zulu Natal, Marloth Park on the southern border of Kruger National Park in Mpumalanga	Sampling in 2013–2015 Test used: Lymph nodes bacterial culture followed by PCR identification	[96]

Table 8. *Cont.*

Prevalence/Frequency	Species	Country	Area	Comment	Ref.
6.8% (19/280)	*Mycobacterium* spp.	Italy	Sicily	Sampling in 2004–2014 Test used: Visual inspection for TBLL. Tissue samples with TBLs were cultures followed by PCR identification. *M. bovis* was isolated from one sample	[100]
16.2% (647/3963)	*Mycobacterium* spp.	Portugal	Idanha-a-Nova	Sampling in 2006–2016 Test used: Visual inspection for tuberculosis-like lesions (TBLL). Considered positive when at least in one organ or lymph node showed TBLs	[129]
4.3% (329/7729)	*MTC*	Spain	Castilla y León	Sampling in 2011–2015 Test used: Lymph nodes bacterial culture followed by PCR identification	[134]
2.5% (3/118)	*M. bovis*	South Korea	Gyeonggi Province	Sampling in 2011–2015 Test used: Lymph nodes and lung bacterial culture followed by PCR identification	[102]
38.3% (16/41)	*M. bovis*	Portugal	Castelo Branco	Sampling in 2009–2013 Test used: first screening by visual inspection for TBLL (41/192 had lesions). Tissue samples with TBLs were cultures followed by PCR identification.	[105]
18.2% (8/44)	*Mycobacterium* spp.	Slovenia	Different areas	Sampling in 2010–2013 Test used: Lymph nodes and liver bacterial culture followed by PCR identification. No MTC were isolated	[130]
13.5% (36/267)	*M. caprae*	Hungary	South-Western Hungary	Sampling in 2008–2013 Test used: bacterial culture followed by PCR identification.	[117]
33.9% (18/58)	*M. bovis*	Spain	Sevilla province	Sampling in 2012–2013 Test used: Lymph nodes bacterial culture followed by PCR identification and spoligotyping. The study was performed on wild boar piglets	[108]
0% (0/9)	*M. bovis*	Brasil	Pantanal area	Test used: bacterial culture of unspecified feral pigs´ tissues followed by PCR identification	[145]
25.2% (61/242) PCR 21.5% (52/242) RPFL	*MTC*	Italy	Lombardia Region	Sampling in 2002–2003 Test used: Lymph nodes histology, bacterial culture, PCR, RFLP *M. microti* in 52 samples and *M. bovis* in 2 samples by RFLS	[123]

Table 8. *Cont.*

Prevalence/Frequency	Species	Country	Area	Comment	Ref.
8.5% (51/602) PCR 5.8% (35/602) RFPL	*M. microti*	Italy	Lombardia Region	Sampling in 2006 Test used: Lymph nodes histology, bacterial culture, direct PCR, direct RFLP	[123]
7.5% (23/307) Culture 64.2% (197/307) PCR 55.0% (169/307) RFPL	*M. microti*	Italy	Lombardia Region	Sampling in 2007–2011 (only wild boar with TBLL) Test used: Lymph nodes histology, bacterial culture, direct PCR, direct RFLP	[123]
59% (1512/2562)	*Mycobacterium* spp.	Spain	Ciudad Real province	Sampling in 2008–2012 Test used: Visual inspection for TBLL in lymph nodes and organs. Generalised TBLs were detected in 51% of the subjects	[146]
2.59% (33/1275)	*MTC*	Spain	Asturias and Galicia	Sampling in 2008–2012 Test used: lymph nodes and organs culture followed by PCR identification and spoligotyping of the isolates Number of *M. bovis* isolates = 19 and *M. caprae* isolates = 14	[111]
3.64% (6/165)	*MTC*	Switzerland and Liechtenstein	Geneva, Thurgovia, Saint Gall, Grisons, Tessin, Liechtenstein	Sampling in 2009–2011 Test used: lymph nodes and tonsil culture followed by PCR identification and spoligotyping of the isolates	[124]
37.3% (293/785)	*M. bovis*	New Zealand	Different areas	Sampling in 1997–2007 Test used: Lymph nodes culture followed by PCR identification	[114]
88.9% (16/18)	*M. bovis*	Spain	Andalusia	Sampling in 2006–2010 Test used: Culture of pool homogenate of lymph nodes and lungs followed by PCR and spoligotyping of the isolates	[115]
13.3% (2/15)	*M. bovis*	Italy		Test used: Culture and PCR of swab samples on muscle surface of wild boar without TBLL	[142]
8.7 R_0	*Mycobacterium* spp.	Spain and Portugal	29 sites	Metadata analyses from 2010–2019. Test used: gross pathology and culture Reproduction number (R_0) defined considering prevalence in the host species, MTC excretion in infected host species, abundance of the host species, transmission rate to host species	[138]

3.2.6. *Salmonella*

Salmonellosis is an enteric infection caused by species of the *Salmonella* genus other than *Salmonella* Typhi and *Salmonella* Paratyphi. Salmonellae are gram-negative bacteria belonging to the Enterobacteriaceae family. They are motile, nonsporeforming, aerobic, or facultatively anaerobic. The transmission of this infection occurs principally by the fecal-oral route: the ingestion of contaminated food or water, contact with infected animals, feces or contaminated environments. The main symptoms of salmonellosis are diarrhea, abdominal cramps, vomiting, and fever. The severity and course of the disease are related to the serotype, the number of microorganisms ingested, and the individual's immune system [147]. *Salmonella* spp. is widely spread for its ability to infect several animal species and survive in different environmental conditions with a wide range of temperatures (2–54 °C) and pH values (3.7–9.4) [148].

Salmonellosis is a public health issue, and it was the second zoonosis reported in the EU in 2021, with 60,050 confirmed human cases, 11,785 hospitalisations, and 71 fatalities [15]. The *Salmonella* genus consists of two species: *Salmonella bongori* and *Salmonella enterica*, the latter divided into six subspecies and several serotypes [149]. The main *Salmonella* serovars implicated in human infections in 2020 and 2021 were *S.* Enteritidis, *S.* Typhimurium, monophasic *S.* Typhimurium (1,4, [5],12:i:-), *S.* Infantis, and *S.* Derby [15,150].

Overall, 80 articles regarding *Salmonella* in wild boars have been found in the literature from 2012 to 2022, seven of which are reviews [10,11,150–155], and 28 articles were not considered relevant for this study. The prevalence of *Salmonella* in the wild boar population has been studied through the analysis of different matrices. Some authors investigated the seroprevalence from blood serum, diaphragm, or muscle samples, achieving different percentages: 1.27% (141/1103) [156], 3.6% (14/393) [157], 4.3% (4/94) [158], 5% (1/20) [159], 17% (21/126) [160], 19.3% (52/269) [161], 38% (69/181) [39], and 66.5% (255/383) [162]. Testing of serum samples can reveal the presence of antibodies against *Salmonella* spp. in wild boars but not the presence of the microorganism on carcass surfaces or meat. The prevalence of *Salmonella* spp. in other matrices such as feces, spleen, kidney, submandibular lymph nodes, ileocecal lymph nodes, mesenteric lymph nodes, and tonsils is reported in Table 9, which shows that feces are the main investigated samples with a prevalence range of 0% to 43%. As shown in Table 10, the prevalence of *Salmonella* spp. in wild boar carcasses is between 0% and 2.5%, while in meat samples it ranges from 0% to 35.7%. This wide variability could be due to different geographic sampling areas, sampling methods, and the hygienic level of process procedures and the environment. The presence of *Salmonella* in wild boar cured meat products was investigated only by Roila et al. [83] in wild boar salami. *Salmonella enterica* serovar typhimurium and *Salmonella enterica* serovar Rissen were found in different batches of meat batter and salami after 7 days of curing, but in the final product after 60 days of aging, *Salmonella* spp. were not detected. However, it was not possible to specify if wild boar had been the source of *Salmonella* since the salami were made with 50% wild boar meat and 50% pork meat.

Table 9. Prevalence of *Salmonella* spp. in wild boar, feces, lymphatic tissues, and inner organs, 2012–2022.

Prevalence/Frequency	Species	Matrix	Country	Comment	Ref.
3.1% (13/425) 0.2% (1/425)	*Salmonella* spp. *Salmonella* spp.	Feces Mesenteric lymph nodes	Serbia	*S.* Enteritidis was the main serotype identified	[163]
3.1% (4/130)	*S. enterica*	Feces	Spain	Serotype identified were monophasic *S.* Typhimurium, *S.* Bardo, *S.* Enteritidis	[59]
35.6% (32/90) 17.8% (16/90)	*Salmonella* spp. *Salmonella* spp.	Feces Lymph nodes	Italy	46.7% (42/90) animals were positive in feces or lymph nodes, of which 11.9% (5/42) were positive at the same time in both matrices. *S.* Abony, *S.* Newport, *S.* Agona, *S.* Derby, *S.* Hermannswerder, *S.* Saintpaul, *S.* Elomrane, *S. salamae* were identified	[164]
7.8% (5/64) 4.7% (3/64)	*Salmonella* spp. *Salmonella* spp.	Mesenteric lymph nodes Carcass	Italy	Sampling from game-handling establishment, game collection point and slaughterhouse	[165]
6% (260/4335)	*Salmonella* spp.	Liver	Italy	Sampling in 2013–2017. Isolated strains belonged to all six *Salmonella enterica* subspecies and the main serotype was *S.* Enteritidis	[166]
4.18% (12/287)	*Salmonella* spp.	Liver or spleen or rectal swab	Italy	*S. diarizonae, S. houtenae, S.* Newport, *S.* Kottbus, *S.* London, *S.* Infantis, *S.* Rubislaw were identified.	[85]
2.4% (13/552)	*Salmonella* spp.	Feces	Germany	*S.* Enteritidis, *S.* Typhimurium, *S.* Stanleyville, were identified	[167]
5% (6/130)	*Salmonella* spp.	Spleen and kidney	Finland		[39]
0% (0/115)	*Salmonella* spp.	Feces	Denmark		[52]
15.9% (30/189) 3.2% (6/189)	*Salmonella* spp. *Salmonella* spp.	Mesenteric lymph nodes Feces	Italy	Three animals were positive in both samples	[168]
18.69% (40/214) 5.06% (21/415) 2.98% (25/838)	*Salmonella* spp. *Salmonella* spp. *Salmonella* spp.	Tonsils Submandibular lymph nodes Feces	Spain	Sampling in 2010–2015 From 148 wild boars the 3 matrices were collected in the same animals and 27.02% (40/148) of them were positive to *Salmonella* spp. (31/148 tonsils, 12/148 lymph nodes, 2/148 feces) but none of them were positive in the three samples simultaneously	[169]
7% (4/57) 3.5% (2/57)	*S. enterica* *S. enterica*	Feces Mesenteric lymph glands	Italy	*S.* Thompson and *S.* Braenderup were identified	[63]
43.9% (194/442)	*Salmonella* spp.	Feces	USA	Sampling from 2013 to 2015. Main serovars identified were *S. Montevideo, S. Newport* and *S. Give*	[170]

Table 9. *Cont.*

Prevalence/Frequency	Species	Matrix	Country	Comment	Ref.
5% (1/21)	*Salmonella* spp.	Feces	Portugal		[171]
5.1% (9/175)	*Salmonella* spp.	Tonsils			
1.8% (1/56)	*Salmonella* spp.	Ileocaecal lymph nodes	Sweden	*S. enterica* and *S. diarizonae* were identified	[172]
1.1% (1/88)	*Salmonella* spp.	Feces			
33.3% (1/3)	*Salmonella* spp.	Tonsils	Argentina	Tonsils carried both *S.* Gaminara and *S.* Newport, while only *S.* Gaminara were isolated from tongue	[173]
33.3% (1/3)	*Salmonella* spp.	Tongue			
5% 2/40	*S. enterica*	Feces	Spain	*Salmonella enterica* serotype Anatum and Corvallis were isolated	[61]
7.4% (9/121)	*Salmonella* spp.	Feces	Japan	*S. enterica* subsp. *enterica* serovar Agona (3), *S.* Narashino (2), *S.* Enteritidis (1), *S.* Havana (1), *S.* Infantis (1), and *S.* Thompson (1) were obtained	[67]
0.3% (1/333)	*Salmonella* spp.	Feces	Spain	One animal was positive in both carcass and feces samples. *S.* Bardo, *S.* Montevideo, *S. arizonae* III (16:i,v:1,5,7) and *S.* Typhimurium were identified	[57]
10.8% (54/499)	*Salmonella* spp.	Feces	Italy	*S. enterica* subsp. *salamae* II, *S. enterica* subsp. *diarizonae* III b, *S. enterica* subsp. *houtenae* IV and *S.* Fischerhuette were the most common isolated	[162]
24.82% (326/1313)	*Salmonella* spp.	Feces	Italy	Sampling from 2007 to 2010 *S. enterica* subsp. *enterica* was the main serovar isolated (79.5%)	[174]
15.4% (33/214)	*Salmonella* spp.	Feces	Spain		[175]

Table 10. Prevalence of *Salmonella* spp. in wild boar meat and carcasses, 2012–2022.

Prevalence/Frequency	Species	Matrix	Country	Comment	Ref.
2.7% (1/36)	*Salmonella* spp.	Meat	Italy		[62]
0% (0/36)	*Salmonella* spp.	Carcass			
35.7% (10/28)	*Salmonella* spp.	Meat	Italy	*S.* Veneziana, *S.* Kasenyi, *S.* Coeln, *S.* Manhattan, *S.* Thompson and *S.* Stanleyville were identified	[64]
2.5% (3/121)	*Salmonella* spp.	Carcass	Italy	Two *S.* Stanleyville and one *S.* Typhimurium were identified	[176]
1.1% (1/90)	*Salmonella* spp.	Carcass	Italy		[164]
0% (0/37)	*Salmonella* spp.	Meat	Italy	Meat cut sampled were fillet and legquarter	[177]

Table 10. *Cont.*

Prevalence/Frequency	Species	Matrix	Country	Comment	Ref.
31.82% (7/22)	*Salmonella* spp.	Meat	Italy	*S.* Stanleyville, *monophasic S.* Typhimurium, and *S.* Kasenyi were identified	[178]
0% (0/30)	*S. enterica*	Carcass	Italy		[63]
0% (0/128)	*Salmonella* spp.	Meat	Japan		[179]
1.4% (3/210) 1.9% (4/210)	*Salmonella* spp. *Salmonella* spp.	Skin Carcass	Serbia		[180]
4.55% (1/22)	*Salmonella* spp.	Meat	Italy	Meat cut sampled was *Longissimus dorsi* muscle	[181]
1.2% (4/333)	*Salmonella* spp.	Carcass	Spain	One animal was positive in both carcass and feces samples	[57]
0% (0/72)	*Salmonella* spp.	Carcass	Italy		[86]

In order to reduce the risk of infection, it is recommended to pay particular attention to the skinning and evisceration processes, maintain the cool chain, have a good hygienic level during meat cutting, and to cook the final product.

3.2.7. *Staphylococcus aureus*

Staphylococcus aureus is a gram-positive, spherical, nonsporeforming, coagulase-positive, aerobic or anaerobic, facultative, halophilic bacterium with the tendency to aggregate in "grape-like" clusters. The usual habitat of this commensal microorganism is the skin and nose of healthy humans and animals, but in some cases, it could lead to a wide range of clinical infections such as bacteremia, endocarditis, pneumonia, infections of the skin and soft tissues, mastitis, and bone and joint infections [182,183]. Some *S. aureus* strains may develop resistance to beta-lactam antibiotics, which are widely used to treat infections, and these strains are termed methicillin-resistant *Staphylococcus aureus* (MRSA). MRSA used to be associated mainly with hospital-related infections, but recently this strain has been found also in people without any contact with hospitals and, in companion animals, livestock, and wild animals [184]. There is an increasing interest in understanding the role of wild boars as possible reservoirs of *S. aureus* and MRSA in particular. About this topic, it has been found in 27 articles from 2012 to 2022, 14 of which were relevant for this study. The majority of studies performed nasal swabs for the detection of *S. aureus*, with a variable prevalence as shown in Table 11. Sousa et al. [185] considered both oral and nasal swabs, with a prevalence of *S. aureus* of 33%. Both studies from Porrero et al. [186,187] considered skin and nasal swabs; in the first study, they found 0.86% of animals positive for MRSA, of which 62.5% were detected from skin swabs and 37.5% from nasal swabs, and only one wild boar was positive in both the skin and nasal samples. Instead, Porrero et al. [187] noticed a higher percentage of positives for *S. aureus* in the nasal sample rather than in skin swabs, but without skin swabs, 18.25% of positives for wild boars would not have been detected. Only Traversa et al. [188] considered lymph nodes for the detection of *S. aureus* in wild boar and revealed a prevalence of 3.2%. No studies on the presence of *S. aureus* in carcasses, raw meat, or processed meat were retrieved in our literature survey.

Table 11. Prevalence of MRSA on wild boar mucosal membranes and in lymphatic organs, 2012–2022.

Prevalence/Frequency	Species	Matrix	Country	Comment	Ref.
36.9% (41/111)	*S. aureus*	Nasal swab	Germany	MRSA were not detected	[189]
33% (30/90)	*S. aureus*	Oral and nasal swab	Portugal	7 isolates showed resistance to at least one of the antibiotics tested; 1 MRSA CC398 (spa-type t899) was identified	[185]
32.2% (57/177)	*S. aureus*	Nasal swab	Portugal	Isolates were resistant to all antimicrobials tested, except of trimethoprim-sulfamethoxazole and vancomycin	[190]
17.8% (66/371) 13.7% (51/371) 1.96% (1/51)	CoPS *S. aureus* MRSA	Nasal swab	Spain	74.5% isolates were susceptible to all the antimicrobials analyzed, 19.6% were resistant to penicillin and 9.8% were resistant to streptomycin	[191]
17.67% (126/713)	MSSA	Skin and/or nasal swabs	Spain		[187]
6.8% (8/117)	*S. aureus*	Nasal swabs	Germany	No antibiotic resistance was detected	[192]
3.2% (23/697)	*S. aureus*	Lymph nodes	Italy	MRSA were not detected	[188]
0.87% (5/577)	MRSA	Nasal swab	Germany		[167]
0.86% (7/817)	MRSA	Skin and nasal swabs	Spain	8 isolates were identified from 7 positive animals: 3 from nasal swabs and 5 from skin swabs. One animal was MRSA positive for both skin and nasal swabs	[186]
0% (0/90)	MRSA	Nasal swab	Spain		[193]
0% (0/439)	MRSA	Nasal swab	Germany		[194]
0% (0/244)	MRSA	Nasal swab	Denmark		[52]

MRSA: methicillin-resistant *Staphylococcus aureus*; MSSA: methicillin-susceptible *Staphylococcus aureus* (MSSA); CoPS: coagulase positive *Staphylococcus*.

3.2.8. Verotoxinogenic/Shigatoxinogenic *E. coli*

Verotoxinogenic/Shigatoxinogenic *E. coli* (VTEC/STEC) form a group of pathogenic *E. coli* (gram-positive short-rods) that elaborate Shiga-like toxins together with other virulence factors. Infections in humans can range from bloody diarrhea to life threatening coagulopathy and renal failure/hemolytic-uremic syndrome. Originally associated with the presence of the O157 antigen, a number of strains with other O-serotypes have been identified as STEC. It has been proposed to use *stx*-gene typing to assess the pathogenicity of STEC (EFSA 2020). In particular, *E. coli* with genes encoding for the stx-2 gene and the virulence factor intimin (*eae*) are associated with severe courses of the disease [15]. In 2021, 6084 confirmed cases were reported in the EU, with 901 hospitalizations and 18 fatalities. From the 5 strong evidence outbreaks, 3 were attributable to meat or meat products [15]. In many animal species, asymptomatic STEC carriers are the rule. In particular, ruminants do not show symptoms since they lack vascular receptors for the Shiga-toxins [195]. A survey of notifications in the RASFF revealed no cases of wild boar meat contamination with STEC.

As regards wild boar, the literature search retrieved 27 documents. The definitions for pathogenic *E. coli* were not consistent between the studies. In 12 studies, the prevalence of STEC was reported, ranging from 0 to 28.3% (Table 12). Data on meat were reported in merely four studies, with a prevalence ranging from 0 to nearly 43% (Table 13). A more detailed view of other isolates with pathogenic potential and antimicrobial resistance described in the studies is outside the scope of our review. E.g., one study reported the isolation of STEC from wild boars with the additional feature of producing enterotoxins (*sta1* and *stb* genes), causing oedema disease [196].

Three studies reported the transmission of STEC from the feces of wild boar to fresh produce [197,198] or to recreational waters [199]. Although not the primary focus of this review, the studies highlight indirect transmission routes of pathogenic bacteria to humans.

Table 12. Prevalence of Shiga toxin-forming *E. coli* in wild boar, fecal samples, lymphatic organs, 2012–2022.

Prevalence/Frequency	Species	Matrix	Country	Comment	Ref.
14% (8/56)	STEC (stx2)	Feces	Portugal	Culture and PCR, WGS	[200]
6.9% (37/536)	STEC	Feces	Germany	Culture, PCR	[167]
1.9% (9/474)	STEC O157	Feces	Japan	Culture, PCR	[201]
6.5% (13/200)	STEC	Feces	Italy (Tuscany)	Culture, PCR	[202]
1.2% (3/248)	STEC	Feces	Japan	Culture, PCR	[56]
28.3% (43/152)	STEC	Feces	Poland	Culture, PCR; includes STEC and AE-STEC	[203]
4.8% (1/21)	STEC	Feces	Portugal	Culture, PCR	[204]
3.33% (3/90)	STEC	Feces	Spain	Culture, PCR	[205]
3.4% (4/117)	E. coli O157	Feces	Spain	Culture	[206]
0% (0/88)	E. coli O157:H7	Tonsils, lymph nodes, feces	Finland	Culture, PCR	[172]
0% (0/121)	STEC O157, O26	Feces	Japan	Culture, PCR	[67]
0% (0/301)	STEC O157	Feces	Spain	Culture, PCR	[57]

Table 13. Prevalence of Shiga toxin-forming *E. coli* in wild boar meat and carcasses.

Prevalence/Frequency	Species	Matrix	Country	Comment	Ref.
42.9% (12/28)	STEC (stx1+ stx2+eae)	Meat (foreleg)	Italy (Campania)	Culture, PCR (27/28 *eae* positive)	[64]
0% (0/128)	STEC	Meat	Japan	Culture	[179]
0% (0/310)	STEC O157	Meat	Spain	Culture, PCR	[57]
5.3% (3/57)	STEC	Meat and meat products	Spain	Culture, PCR	[207]

3.2.9. Yersinia

The Enterobacteriaceae family includes the food-borne pathogen *Yersinia enterocolitica*, responsible for yersiniosis in humans, a gastrointestinal disease that could simulate appendicitis and can cause mesenteric lymphadenitis, reactive arthritis, erythema nodosum, and conjunctivitis [208,209]. The disease appears to be widespread, with ca. 6800 cases in Europe in 2020 and 100,000 illnesses every year in the USA [EFSA, 2022; CDC, 2016] [15,210].

The epidemiological situation could be even more severe, as the role of biotype 1A in human infection and disease symptoms (considered non-pathogenic compared to biotypes 1B, 2, 3, 4 and 5) is still debated and therefore underestimated [211].

Ready-to-eat foods are the major sources of human infection, especially as *Y. enterocolitica* can resist cold environments and even replicate at refrigeration temperatures [211]. Animals, especially pigs, are considered the main reservoir of the bacteria, which could be found mainly in the intestine and tonsils [212]. Nevertheless, the outbreaks reported in 2021 are related to prepared dishes and ready-to-eat vegetables [15], and no reports are available on wild boar meat as an outbreak source.

The database research retrieved 39 studies regarding *Y. enterocolitica* in wild boars and feral pigs between 2012–2022. The articles that reported studies on the prevalence of the microorganism in animal tissue, feces, or carcasses/muscles of wild boars were 21. Only two articles describe the prevalence of antibodies against *Y. enterocolitica* in animal blood samples. Papers on *Yersinia pseudotuberculosis* were not considered. Most of the studies were conducted in Europe (19 out of 21), especially in Italy (10 articles). Samples of different matrices were considered: eight studies on fecal samples, nine on organs different from muscles, four on carcass surfaces (external or internal), and four in muscles (Table 14).

The seroprevalence in wild boar was above 50% (in Finland and the Czech Republic), proving that the microorganism is widespread in this species. Fecal material is considered the main source of contamination of the carcass and, ultimately, of the meat. This contamination could happen during hunting (the precision of the shot), evisceration, or carcass processing and cutting [176,180]. Fecal sample positivity for *Y. enterocolitica* ranges from 0% (different Italian regions) to 74% (Japan). Thus, as for other genus belonging to the Enterobacteriaceae family, the fecal shedding could be intermittent [213]. Regarding organs and tissues that could harbour the microorganism in *Suidae*, the prevalence of the microorganism in the tonsils of wild boar ranges from 14% (Sweden) to 64% (Campania Region, Italy), with a higher percentage than in lymph nodes (ranging from 0% to 4.4%). The presence of the pathogen in such tissues could be considered during carcass processing to avoid the spread of the microorganism to the meat. Nonetheless, in wild boar, in contrast to the domestic pig, the head is removed during carcass dressing at cervical vertebrae level, thus the laryngeal and pharyngeal area is removed from the carcass at an early stage of the processing chain.

The presence of *Y. enterocolitica* on carcass surfaces ranges from 0% to 85.7%. Such a wide range could be due to different sampling methods and areas, but also to differences in the hygienic level of the process. The same might hold true for muscles, where the prevalence ranges from 0% to 71%. The wide range of prevalence denotes that, although wild boar can harbour microorganisms in the intestines and tonsils, the procedures to obtain the meat are relevant to prevent contamination of muscles. In this perspective, the training of the personnel, the presence of suitable structure and equipment, the correct hygienic procedure implementation, and standard sanitation operating procedures are of paramount importance.

Another important aspect that emerged from the literature survey is that the biotype most frequently observed in wild boar is 1A, the least pathogenic but also the most underrated of the *Y. enterocolitica* biotypes.

Table 14. Prevalence of *Yersinia enterocolitica* in wild boar, feral pigs and warthog.

Prevalence/Frequency	Country	Area	Matrix	Comment	Ref.
0% (0/107)	Italy	Valle d'Aosta Region	Feces	Sampling in 2015–2018 Test used: PCR	[214]
85.7% (12/36)	Italy	Campania Region	Carcass	Sampling in 2019 Test used: bacterial isolation and SYBR green PCR-assay for ystA and ystB genes. 12 animals carried ystB gene, and 3 animals both ystA and ystB genes	[62]
64.3% (9/36)			Tonsils		
71.4% (10//36)			Muscle		
0.01% (1/110)	Tunisia	Ariana, Bizerte, Manouba, Nabeul and Siliana	Feces	Sampling in 2018–2020 Test used: bacterial isolation and biochemical identification	[215]
0% (0/64)	Italy	Parma and Bologna province	Carcass and Mesenteric lymph nodes	Sampling in 2020 Test used: bacterial isolation and biochemical identification	[165]
2.6% (126/4890)	Italy	Liguria Region	Liver	Sampling in 2013–2018 Test used: bacterial isolation, Serotyping and Real Time PCR for virulence genes. Biotype 1A was the most isolated (92.9%), then biotype 1B (6.3%) and 2 (0.8%)	[216]
18.8% (54/287)	Italy	Tuscany Region	Rectal swab	Sampling in 2018–2020 Test used: bacterial isolation, biochemical identification. and Real Time PCR for virulence Genes. Identification of gene *ystA* in 14 out of 54 isolates, *inv* in 13, *ail* in 12, *ystB* in 10 and *virF* in 8	[85]
56.4% (102/181)	Finland	12 out of 19 regions	Blood	Sampling in 2016 Test used: seroprevalence ELISA test. Test used: Organs: real-time PCR based on SYBRGreen for *ail* gene	[39]
16.9% (22/130)			Spleen and kidneys		
6.2% (19/305)	Italy	Parma and Piacenza provinces	Feces	Sampling in 2017–2019 Test used: bacterial isolation, biochemical identification, and Real Time PCR for virulence Genes. All isolates belonged to biotype 1A	[217]
3.3% (10/305)			Mesenteric lymph nodes		
74.1% (40/54)	Japan	Not specified	Feces	Sampling in 2014–2016 Test used: bacterial isolation, biochemical identification. Prevalence is reported for *Yersinia* spp. 97.4% of the *Y. enterocolitica* isolates belonged to biotype 1A	[218]
13.6% (3/22)	Italy	Campania region	Muscle	Sampling in 2017 Test used: bacterial isolation, biochemical identification, and Real Time PCR for virulence Genes. All isolates present only *ystB* gene	[178]

Table 14. *Cont.*

Prevalence/Frequency	Country	Area	Matrix	Comment	Ref.
6.7% (6/90)	Sweden	13 counties in southern Sweden	Feces	Sampling in 2014–2016 Test used: bacterial isolation, and Real Time PCR for ail gene	[219]
14.0% (19/136)			Tonsils		
4.4% (4/90)			Mesenteric lymph nodes		
25.3% (110/434)	Poland	12 out of 16 Polish regions	Rectal swab	Sampling in 2013–2014 Test used: bacterial isolation, and multiplex PCR for ail, ystA and ystB genes. 92.5% of the isolates belong to biotype 1A	[220]
0% (0/42)	Italy	Tuscany Region	Muscle	Sampling in 2013–2014 Test used: bacterial isolation, and biochemical identification	[181]
65.9% (89/135)	Czech Republic	Moravian Regions	Blood	Sampling in 2013–2014 Test used: ELISA	[221]
55.5% (11/20)	Poland	North-East Poland	Swab samples from tonsils area, peritoneum and perineum	Sampling in 2013 Test used: bacterial isolation, and biochemical identification biotyping, serotyping and molecular characterisation. 90.5% of the isolates belong to biotype 1A	[222]
33.3% (24/72)	Spain	Basque Country	Tonsils	Sampling in 2001–2012 Test used: bacterial isolation, biochemical identification, and molecular characterization	[223]
15.3% (17/111)	Germany	Lower saxony	Tonsils	Sampling in 2013–2014 Test used: bacterial isolation, MALDI-TOF identification, Real Time PCR for virulence Genes. 89.55% of the isolates belong to biotype 1A	[224]
20.5% (18/88)	Sweden	Central Sweden	Feces and Ileocecal lymph nodes and tonsils	Sampling in 2010–2011 Test used: bacterial isolation, and multiplex PCR for *ail* gene	[219]
27.3% (18/66)	Spain	Basque Country	Tonsils	Sampling in 2010–2012 Test used: bacterial isolation, and biochemical identification and direct real time PCR with new enrichment protocol	[225]
0% (0/3)	Argentina	San Luis city	Tonsils and tongue	Sampling in 2008–2012 Test used: bacterial isolation and biochemical identification	[173]
14.8% (34/230)	Italy	Viterbo Province	Muscle	Sampling in 2012–2013 Test used: bacterial isolation, and multiplex PCR for *ail* gene	[157]

Table 14. *Cont.*

Prevalence/Frequency	Country	Area	Matrix	Comment	Ref.
4.2% (3/72)	Italy	Upper Susa valley Piedmont Region	Carcass	Sampling in Test used: bacterial isolation, biochemical identification and molecular characterisation for *inv, ail* and *yst* genes. *ail* and *yst* genes were not detected	[86]

4. Conclusions

The increasing popularity of meat from wild game is observed in many countries. Diseases in wildlife have often been seen as an issue or spill-over or spill-back of infection agents from farm animals, and exposure of humans and animals in frequent and close contact with wild animals has been studied to some extent. Additionally, while the presence of antibodies against a specific pathogen may be useful for epidemiological purposes, its value for the assessment of meat safety is primarily that the given pathogen must be considered a potential hazard. Similarly, the presence of pathogens in the feces and even in the lymph nodes of the digestive tract mainly indicates that the host organism can keep the pathogen under control. Similar to farm animals, it can be expected that stress, but also the dressing procedures after killing, can cause the spread of the pathogen on/in edible organs. Since these scenarios do not result in any typical lesion, the routine ante- and post-mortem examinations [226] will not give an indication of the presence of a certain pathogen, and minimizing the spread of the agent is a matter of good hygienic practice. However, if serological or other testing has demonstrated the presence of a certain pathogen in wildlife in a certain region, it would be wise to adopt hygienic precautions (i.e., no admittance of carcasses with "gut shots" in the food chain; or disinfecting knives after cutting in the tonsillar area).

For five (*Campylobacter, Listeria monocytogenes, Salmonella*, Shiga toxin-forming *E. coli*, and *Yersinia enterocolitica*) of the nine agents we reviewed, one or more studies dealt with the presence of the pathogen on muscle surfaces or muscle tissues of wild boar, with prevalences ranging from 0 to ca. 70%. One experimental study was retrieved on the transmission and survival of *Mycobacterium* on wild boar meat. As regards edible inner organs, the liver and spleen have been examined for the presence of *Brucella, Coxiella burnetii, Listeria monocytogenes,* and *Mycobacteria,* and the latter four agents have actually been recovered, albeit with varying percentages. For *Brucella,* human case reports and epidemiological studies in (hunting) dogs stressed the occupational exposure risk, but no indication of meat-borne transmission to humans was evidenced. Similarly, the mode of transmission of *C. burnetii* is more likely via vectors (i.e., ticks). In most studies, animals without specific histories or pathologies had been examined.

In essence, the literature we reviewed confirmed that food-borne pathogenic bacteria present in meat from domestic animals [15] and implicated in food-borne disease can also be found in wild boars, with varying prevalence and regional differences. It is unclear to what extent such differences are biased by sampling and analytical procedures. In the absence of more detailed data for the European Union, it might be advisable to focus on the efficacy of current game meat inspection [226] and handling practices [140] to minimize introduction in the game meat chain. Similarly, the implementation of HACCP-based food safety management systems [227] needs to be stressed.

With respect to the placing on the market of meat from wild hunted game, European Union legislation distinguishes an "approved" chain (i.e., the hunted game specimens are collected, post-mortem inspected, and processed in approved establishments) from an unapproved chain, which is largely subject to national regulation (for primary products, i.e., the eviscerated carcass, see Recital 10 and Article 1 of EC Regulation 852/2004 [228]; for processed or unprocessed products, see Recital 11 and Article 1 of EC Regulation

853/2004 [140]). This unapproved chain represents the supply of small quantities of wild game or wild game meat directly from the hunter to the final consumer or to local retail establishments directly supplying the final consumer [140].

Currently, there is no uniform way in which this unapproved sector is regulated in the member states; there is even no consistent definition of "small quantities of wild game or wild game meat" [140]. Admittedly, all national legislation has a common baseline represented by EC Regulation 178/2002 (in particular, Articles 14, 16–19; "safe food", traceability, identification of hazards, and management of risks) [229,230]. An in-depth and comprehensive consideration of said regulation should, in fact, be sufficient to warrant food safety. European Union member states have chosen different approaches [231,232], but there are no real metrics to assess how the systems actually perform in managing the consumers´ risk posed by the presence of foodborne pathogens in game meat.

Author Contributions: Conceptualization, C.A., P.P. and D.R.; methodology, C.A. and P.P.; validation, C.A., C.N.-N. and P.P.; formal analysis, C.A. and D.R.; investigation, C.A., P.P. and D.R.; data curation, C.A.; writing—original draft preparation, C.A., P.P., C.N.-N. and D.R.; writing—review and editing, C.A., P.P. and D.R.; supervision, P.P. and D.R.; funding acquisition, P.P. All authors have read and agreed to the published version of the manuscript.

Funding: Open access funding by the University of Veterinary Medicine Vienna.

Institutional Review Board Statement: Not applicable.

Informed Consent Statement: Not applicable.

Data Availability Statement: Not applicable.

Acknowledgments: Italian Ministry of University and Research—P.O.N. Research and Innovation 2014–2020 (CCI 2014IT16M2OP005), Action IV.5. Project title: Game meat green safety.

Conflicts of Interest: The author, Clara Noé-Nordberg, was employed by the company Esterhazy Betriebe GmbH. The remaining authors declare that the research was conducted in the absence of any commercial or financial relationships that could be construed as a potential conflict of interest.

References

1. Bieber, C.; Ruf, T. Population Dynamics in Wild Boar *Sus scrofa*: Ecology, Elasticity of Growth Rate and Implications for the Management of Pulsed Resource Consumers: Population Dynamics in Wild Boar. *J. Appl. Ecol.* **2005**, *42*, 1203–1213. [CrossRef]
2. Keuling, O.; Baubet, E.; Duscher, A.; Ebert, C.; Fischer, C.; Monaco, A.; Podgórski, T.; Prevot, C.; Ronnenberg, K.; Sodeikat, G.; et al. Mortality Rates of Wild Boar *Sus scrofa* L. in Central Europe. *Eur. J. Wildl. Res.* **2013**, *59*, 805–814. [CrossRef]
3. Massei, G.; Kindberg, J.; Licoppe, A.; Gačić, D.; Šprem, N.; Kamler, J.; Baubet, E.; Hohmann, U.; Monaco, A.; Ozoliņš, J.; et al. Wild Boar Populations up, Numbers of Hunters down? A Review of Trends and Implications for Europe: Wild Boar and Hunter Trends in Europe. *Pest. Manag. Sci.* **2015**, *71*, 492–500. [CrossRef] [PubMed]
4. Johann, F.; Handschuh, M.; Linderoth, P.; Dormann, C.F.; Arnold, J. Adaptation of Wild Boar (*Sus scrofa*) Activity in a Human-Dominated Landscape. *BMC Ecol.* **2020**, *20*, 4. [CrossRef]
5. Sáenz-de-Santa-María, A.; Tellería, J.L. Wildlife-Vehicle Collisions in Spain. *Eur. J. Wildl. Res.* **2015**, *61*, 399–406. [CrossRef]
6. Jori, F.; Massei, G.; Licoppe, A.; Ruiz-Fons, F.; Linden, A.; Václavek, P.; Chenais, E.; Rosell, C. Management of Wild Boar Populations in the European Union before and during the ASF Crisis. In *Understanding and Combatting African Swine Fever*; Iacolina, L., Penrith, M.-L., Bellini, S., Chenais, E., Jori, F., Montoya, M., Ståhl, K., Gavier-Widén, D., Eds.; Wageningen Academic Publishers: Wageningen, The Netherlands, 2021; pp. 197–228.
7. Bacigalupo, S.A.; Dixon, L.K.; Gubbins, S.; Kucharski, A.J.; Drewe, J.A. Wild Boar Visits to Commercial Pig Farms in Southwest England: Implications for Disease Transmission. *Eur. J. Wildl. Res.* **2022**, *68*, 69. [CrossRef]
8. Ruiz-Fons, F.; Segalés, J.; Gortázar, C. A Review of Viral Diseases of the European Wild Boar: Effects of Population Dynamics and Reservoir Rôle. *Vet. J.* **2008**, *176*, 158–169. [CrossRef]
9. Meng, X.J.; Lindsay, D.S.; Sriranganathan, N. Wild Boars as Sources for Infectious Diseases in Livestock and Humans. *Philos. Trans. R. Soc. B.* **2009**, *364*, 2697–2707. [CrossRef]
10. Brown, V.R.; Bowen, R.A.; Bosco-Lauth, A.M. Zoonotic Pathogens from Feral Swine That Pose a Significant Threat to Public Health. *Transbound. Emerg. Dis.* **2018**, *65*, 649–659. [CrossRef]
11. Fredriksson-Ahomaa, M. Wild Boar: A Reservoir of Foodborne Zoonoses. *Foodborne Pathog. Dis.* **2019**, *16*, 153–165. [CrossRef]
12. Gomes-Neves, E.; Abrantes, A.C.; Vieira-Pinto, M.; Müller, A. Wild Game Meat—A Microbiological Safety and Hygiene Challenge? *Curr. Clin. Microbiol. Rep.* **2021**, *8*, 31–39. [CrossRef]

13. Directive 2003/99/EC of the European Parliament and of the Council of 17 November 2003 on the monitoring of zoonoses and zoonotic agents, amending Council Decision 90/424/EEC and repealing Council Directive 92/117/EEC. *Off. J. Eur. Union* **2003**, *L325*, 31–40.

14. Trimmel, N.E.; Walzer, C. Infectious Wildlife Diseases in Austria-A Literature Review From 1980 Until 2017. *Front. Vet. Sci.* **2020**, *7*, 3. [CrossRef] [PubMed]

15. European Food Safety Authority (EFSA) and European Centre for Disease Prevention and Control (ECDC). The European Union One Health 2021 Zoonoses Report. *EFSA J.* **2022**, *20*, 7666. [CrossRef]

16. Touloudi, A.; McGiven, J.; Cawthraw, S.; Valiakos, G.; Kostoulas, P.; Duncombe, L.; Gortázar, C.; Boadella, M.; Sofia, M.; Athanasakopoulou, Z.; et al. Development of a Multiplex Bead Assay to Detect Serological Responses to *Brucella* Species in Domestic Pigs and Wild Boar with the Potential to Overcome Cross-Reactivity with *Yersinia enterocolitica* O:9. *Microorganisms* **2022**, *10*, 1362. [CrossRef]

17. Grantina-Ievina, L.; Avsejenko, J.; Cvetkova, S.; Krastina, D.; Streikisa, M.; Steingolde, Z.; Vevere, I.; Rodze, I. Seroprevalence of *Brucella suis* in eastern Latvian wild boars (*Sus scrofa*). *Acta Vet. Scand.* **2018**, *60*, 19. [CrossRef]

18. Powers, H.R.; Nelson, J.R.; Alvarez, S.; Mendez, J.C. Neurobrucellosis associated with feral swine hunting in the southern United States. *BMJ Case Rep.* **2020**, *13*, 238216. [CrossRef]

19. Landis, M.; Rogovskyy, A.S. The Brief Case: *Brucella suis* Infection in a Household of Dogs. *J. Clin. Microbiol.* **2022**, *60*, e00984-21. [CrossRef]

20. Gowe, I.; Parsons, C.; Vickery, S.; Best, M.; Prechter, S.; Haskell, M.G.; Parsons, E. Venous thrombosis, peripheral aneurysm formation, and fever in a feral pig hunter with Brucellosis. *IDCases* **2022**, *27*, e01449. [CrossRef]

21. Franco-Paredes, C.; Chastain, D.; Taylor, P.; Stocking, S.; Sellers, B. Boar hunting and brucellosis caused by *Brucella suis*. *Travel Med. Infect. Dis.* **2017**, *16*, 18–22. [CrossRef]

22. Mailles, A.; Ogielska, M.; Kemiche, F.; Garin-Bastuji, B.; Brieu, N.; Burnusus, Z.; Creuwels, A.; Danjean, M.P.; Guiet, P.; Nasser, V.; et al. *Brucella suis* biovar 2 infection in humans in France: Emerging infection or better recognition? *Epidemiol. Infect.* **2017**, *145*, 2711–2716. [CrossRef] [PubMed]

23. Munckhof, W.J.; Jennison, A.V.; Bates, J.R.; Gassiep, I. First report of probable neurobrucellosis in Australia. *Med. J. Aust.* **2013**, *199*, 423–425. [CrossRef] [PubMed]

24. Mor, S.M.; Wiethoelter, A.K.; Massey, P.D.; Robson, J.; Wilks, K.; Hutchinson, P. Pigs, pooches and pasteurisation: The changing face of brucellosis in Australia. *Aust. J. Gen. Pract.* **2018**, *47*, 99–103. [CrossRef] [PubMed]

25. Orr, B.; Malik, R.; Norris, J.; Westman, M. The Welfare of Pig-Hunting Dogs in Australia. *Animals* **2019**, *9*, 853. [CrossRef]

26. Orr, B.; Westman, M.E.; Norris, J.M.; Repousis, S.; Ma, G.; Malik, R. Detection of *Brucella* spp. during a serosurvey of pig-hunting and regional pet dogs in eastern Australia. *Aust. Vet. J.* **2022**, *100*, 360–366. [CrossRef]

27. Kneipp, C.C.; Sawford, K.; Wingett, K.; Malik, R.; Stevenson, M.A.; Mor, S.M.; Wiethoelter, A.K. *Brucella suis* Seroprevalence and Associated Risk Factors in Dogs in Eastern Australia, 2016 to 2019. *Front. Vet. Sci.* **2021**, *8*, 727641. [CrossRef]

28. Fabbri, M.C.; Crovetti, A.; Tinacci, L.; Bertelloni, F.; Armani, A.; Mazzei, M.; Fratini, F.; Bozzi, R.; Cecchi, F. Identification of candidate genes associated with bacterial and viral infections in wild boars hunted in Tuscany (Italy). *Sci. Rep.* **2022**, *12*, 8145. [CrossRef]

29. Cilia, G.; Fratini, F.; Turchi, B.; Ebani, V.V.; Turini, L.; Bilei, S.; Bossù, T.; De Marchis, M.L.; Cerri, D.; Bertelloni, F. Presence and Characterization of Zoonotic Bacterial Pathogens in Wild Boar Hunting Dogs (Canis Lupus Familiaris) in Tuscany (Italy). *Animals* **2021**, *11*, 1139. [CrossRef]

30. Montagnaro, S.; D'ambrosi, F.; Petruccelli, A.; Ferrara, G.; D'alessio, N.; Iovane, V.; Veneziano, V.; Fioretti, A.; Pagnini, U. A serological survey of brucellosis in Eurasian wild boar (*Sus scrofa*) in Campania region, Italy. *J. Wildl. Dis.* **2020**, *56*, 424–428. [CrossRef]

31. Bertelloni, F.; Mazzei, M.; Cilia, G.; Forzan, M.; Felicioli, A.; Sagona, S.; Bandecchi, P.; Turchi, B.; Cerri, D.; Fratini, F. Serological Survey on Bacterial and Viral Pathogens in Wild Boars Hunted in Tuscany. *EcoHealth* **2020**, *17*, 85–93. [CrossRef]

32. Pilo, C.; Addis, G.; Deidda, M.; Tedde, M.T.; Liciardi, M. A serosurvey for brucellosis in wild boar (*Sus scrofa*) in Sardinia, Italy. *J. Wildl. Dis.* **2015**, *51*, 885–888. [CrossRef]

33. Di Nicola, U.; Scacchia, M.; Marruchella, G. Pathological and serological findings in wild boars (*Sus scrofa*) from Gran Sasso and Monti della Laga National Park (Central Italy). *Large Anim. Rev.* **2015**, *21*, 167–171.

34. Risco, D.; García, A.; Serrano, E.; Fernandez-Llario, P.; Benítez, J.M.; Martínez, R.; García, W.L.; de Mendoza, J.H. High-Density dependence but low impact on selected reproduction parameters of *Brucella suis* biovar 2 in wild boar hunting estates from south-western Spain. *Transbound. Emerg. Dis.* **2014**, *61*, 555–562. [CrossRef] [PubMed]

35. Zurovac Sapundzic, Z.; Zutic, J.; Stevic, N.; Milicevic, V.; Radojicic, M.; Stanojevic, S.; Radojicic, S. First Report of *Brucella* Seroprevalence in Wild Boar Population in Serbia. *Vet. Sci.* **2022**, *9*, 575. [CrossRef]

36. Cvetnic, Ž.; Duvnjak, S.; Zdelar-Tuk, M.; Reil, I.; Mikulić, M.; Cvetnić, M.; Špičić, S. Swine brucellosis caused by *Brucella suis* biovar 2 in Croatia. *Slov. Vet. Res.* **2017**, *54*, 149–154. [CrossRef]

37. van Tulden, P.; Gonzales, J.L.; Kroese, M.; Engelsma, M.; de Zwart, F.; Szot, D.; Bisselink, Y.; van Setten, M.; Koene, M.; Willemsen, P.; et al. Monitoring results of wild boar (*Sus scrofa*) in The Netherlands: Analyses of serological results and the first identification of *Brucella suis* biovar 2. *Infect. Ecol. Epidemiol.* **2020**, *10*, 1794668. [CrossRef] [PubMed]

38. Malmsten, A.; Magnusson, U.; Ruiz-Fons, F.; González-Barrio, D.; Dalin, A.-M. A serologic survey of pathogens in wild boar (*Sus scrofa*) in Sweden. *J. Wildl. Dis.* **2018**, *54*, 229–237. [CrossRef]

39. Fredriksson-Ahomaa, M.; London, L.; Skrzypczak, T.; Kantala, T.; Laamanen, I.; Biström, M.; Maunula, L.; Gadd, T. Foodborne Zoonoses Common in Hunted Wild Boars. *EcoHealth* **2020**, *17*, 512–522. [CrossRef]

40. Neiffer, D.; Hewlett, J.; Buss, P.; Rossouw, L.; Hausler, G.; Deklerk-Lorist, L.-M.; Roos, E.; Olea-Popelka, F.; Lubisi, B.; Heath, L.; et al. Antibody Prevalence to African Swine Fever Virus, Mycobacterium bovis, Foot-and-Mouth Disease Virus, Rift Valley Fever Virus, Influenza A Virus, and *Brucella* and *Leptospira* spp. in Free-Ranging Warthog (*Phacochoerus africanus*) Populations in South Africa. *J. Wildl. Dis.* **2021**, *57*, 60–70. [CrossRef]

41. Gakuya, F.; Akoko, J.; Wambua, L.; Nyamota, R.; Ronoh, B.; Lekolool, I.; Mwatondo, A.; Muturi, M.; Ouma, C.; Nthiwa, D.; et al. Evidence of co-exposure with *Brucella* spp., *Coxiella burnetii*, and Rift Valley fever virus among various species of wildlife in Kenya. *PLoS Negl. Trop. Dis.* **2022**, *16*, e0010596. [CrossRef]

42. Kmetiuk, L.B.; Paulin, L.M.S.; Villalobos, E.M.C.; Lara, M.D.C.C.S.H.; Filho, I.R.B.; Pereira, M.S.; Bach, R.W.; Lipinski, L.C.; Fávero, G.M.; Dos Santos, A.P.; et al. Seroprevalence of anti-brucella spp. Antibodies in wild boars (*Sus scrofa*), hunting dogs, and hunters of Brazil. *J. Wildl. Dis.* **2021**, *57*, 974–976. [CrossRef] [PubMed]

43. Severo, D.R.T.; Werlang, R.A.; Mori, A.P.; Baldi, K.R.A.; Mendes, R.E.; Surian, S.R.S.; Coldebella, A.; Kramer, B.; Trevisol, I.M.; Gomes, T.M.A.; et al. Health profile of free-range wild boar (*Sus scrofa*) subpopulations hunted in Santa Catarina State, Brazil. *Transbound. Emerg. Dis.* **2021**, *68*, 857–869. [CrossRef] [PubMed]

44. Zimmermann, N.P.; Schabib Peres, I.A.H.F.; Braz, P.H.; Juliano, R.S.; Mathias, L.A.; Pellegrin, A.O. Serological prevalence of *Brucella* spp. in feral pigs and sympatric cattle in the pantanal of mato grosso do sul, Brazil [Prevalência sorológica de *Brucella* spp. em porcos ferais e bovinos em simpatria no pantanal do mato grosso do sul, brasil]. *Semin. Cienc. Agrar.* **2018**, *39*, 2437–2442. [CrossRef]

45. Montenegro, O.L.; Roncancio, N.; Soler-Tovar, D.; Cortés-Duque, J.; Contreras-Herrera, J.; Sabogal, S.; Acevedo, L.D.; Navas-Suárez, P.E. Serologic survey for selected viral and bacterial swine pathogens in colombian collared peccaries (*Pecari tajacu*) and feral pigs (*Sus scrofa*). *J. Wildl. Dis.* **2018**, *54*, 700–707. [CrossRef] [PubMed]

46. Cleveland, C.A.; DeNicola, A.; Dubey, J.P.; Hill, D.E.; Berghaus, R.D.; Yabsley, M.J. Survey for selected pathogens in wild pigs (*Sus scrofa*) from Guam, Marianna Islands, USA. *Vet. Microbiol.* **2017**, *205*, 22–25. [CrossRef]

47. Gaskamp, J.A.; Gee, K.L.; Campbell, T.A.; Silvy, N.J.; Webb, S.L. Pseudorabies virus and brucella abortus from an expanding wild pig (*Sus scrofa*) population in southern Oklahoma, USA. *J. Wildl. Dis.* **2016**, *52*, 383–386. [CrossRef]

48. Ridoutt, C.; Lee, A.; Moloney, B.; Massey, P.; Charman, N.; Jordan, D. Detection of brucellosis and leptospirosis in feral pigs in New South Wales. *Aust. Vet. J.* **2014**, *92*, 343–347. [CrossRef]

49. Pearson, H.E.; Toribio, J.-A.L.M.L.; Hernandez-Jover, M.; Marshall, D.; Lapidge, S.J. Pathogen presence in feral pigs and their movement around two commercial piggeries in Queensland, Australia. *Vet. Rec.* **2014**, *174*, 325. [CrossRef]

50. Hälli, O.; Ala-Kurikka, E.; Nokireki, T.; Skrzypczak, T.; Raunio-Saarnisto, M.; Peltoniemi, O.A.T.; Heinonen, M. Prevalence of and risk factors associated with viral and bacterial pathogens in farmed European wild boar. *Vet. J.* **2012**, *194*, 98–101. [CrossRef]

51. Grégoire, F.; Mousset, B.; Hanrez, D.; Michaux, C.; Walravens, K.; Linden, A. A serological and bacteriological survey of brucellosis in wild boar (*Sus scrofa*) in Belgium. *BMC Vet. Res.* **2012**, *8*, 80. [CrossRef]

52. Petersen, H.H.; Takeuchi-Storm, N.; Enemark, H.L.; Nielsen, S.T.; Larsen, G.; Chriél, M. Surveillance of Important Bacterial and Parasitic Infections in Danish Wild Boars (*Sus scrofa*). *Acta Vet. Scand.* **2020**, *62*, 41. [CrossRef]

53. Lama, J.K.; Bachoon, D.S. Detection of *Brucella Suis*, *Campylobacter Jejuni*, and *Escherichia coli* Strains in Feral Pig (*Sus scrofa*) Communities of Georgia. *Vector-Borne Zoonotic Dis.* **2018**, *18*, 350–355. [CrossRef] [PubMed]

54. Di Sabatino, D.; Garofolo, G.; Di Provvido, A.; Zilli, K.; Foschi, G.; Di Giannatale, E.; Ciuffetelli, M.; De Massis, F. *Brucella suis* biovar 2 multi locus sequence type ST16 in wild boars (*Sus scrofa*) from Abruzzi region, Italy. Introduction from Central-Eastern Europe? *Infection, Genet. Evol.* **2017**, *55*, 63–67. [CrossRef] [PubMed]

55. WHO. Available online: https://www.who.int/news-room/fact-sheets/detail/campylobacter (accessed on 1 March 2023).

56. Tomino, Y.; Andoh, M.; Horiuchi, Y.; Shin, J.; Ai, R.; Nakamura, T.; Toda, M.; Yonemitsu, K.; Takano, A.; Shimoda, H.; et al. Surveillance of Shiga Toxin-Producing *Escherichia coli* and *Campylobacter* spp. in Wild Japanese Deer (*Cervus nippon*) and Boar (*Sus scrofa*). *J. Vet. Med. Sci.* **2020**, *82*, 1287–1294. [CrossRef] [PubMed]

57. Díaz-Sánchez, S.; Sánchez, S.; Herrera-León, S.; Porrero, C.; Blanco, J.; Dahbi, G.; Blanco, J.E.; Mora, A.; Mateo, R.; Hanning, I.; et al. Prevalence of Shiga Toxin-Producing *Escherichia coli*, *Salmonella* spp. and *Campylobacter* spp. in Large Game Animals Intended for Consumption: Relationship with Management Practices and Livestock Influence. *Vet. Microbiol.* **2013**, *163*, 274–281. [CrossRef]

58. Carbonero, A.; Paniagua, J.; Torralbo, A.; Arenas-Montes, A.; Borge, C.; García-Bocanegra, I. Campylobacter Infection in Wild Artiodactyl Species from Southern Spain: Occurrence, Risk Factors and Antimicrobial Susceptibility. *Comp. Immunol. Microbiol. Infect. Dis.* **2014**, *37*, 115–121. [CrossRef]

59. Castillo-Contreras, R.; Marín, M.; López-Olvera, J.R.; Ayats, T.; Fernandez Aguilar, X.; Lavín, S.; Mentaberre, G.; Cerdà-Cuéllar, M. Zoonotic Campylobacter Spp. and Salmonella Spp. Carried by Wild Boars in a Metropolitan Area: Occurrence, Antimicrobial Susceptibility and Public Health Relevance. *Sci. Total Environ.* **2022**, *822*, 153444. [CrossRef]

60. Morita, S.; Sato, S.; Maruyama, S.; Miyagawa, A.; Nakamura, K.; Nakamura, M.; Asakura, H.; Sugiyama, H.; Takai, S.; Maeda, K.; et al. Prevalence and Whole-Genome Sequence Analysis of Campylobacter Spp. Strains Isolated from Wild Deer and Boar in Japan. *Comp. Immunol. Microbiol. Infect. Dis.* **2022**, *82*, 101766. [CrossRef]

61. Navarro-Gonzalez, N.; Casas-Díaz, E.; Porrero, C.M.; Mateos, A.; Domínguez, L.; Lavín, S.; Serrano, E. Food-Borne Zoonotic Pathogens and Antimicrobial Resistance of Indicator Bacteria in Urban Wild Boars in Barcelona, Spain. *Vet. Microbiol.* **2013**, *167*, 686–689. [CrossRef]

62. Peruzy, M.F.; Murru, N.; Smaldone, G.; Proroga, Y.T.R.; Cristiano, D.; Fioretti, A.; Anastasio, A. Hygiene Evaluation and Microbiological Hazards of Hunted Wild Boar Carcasses. *Food Control* **2022**, *135*, 108782. [CrossRef]

63. Stella, S.; Tirloni, E.; Castelli, E.; Colombo, F.; Bernardi, C. Microbiological Evaluation of Carcasses of Wild Boar Hunted in a Hill Area of Northern Italy. *J. Food Prot.* **2018**, *81*, 1519–1525. [CrossRef] [PubMed]

64. Peruzy, M.F.; Cristiano, D.; Delibato, E.; D'Alessio, N.; Proroga, Y.T.R.; Capozza, R.L.; Rippa, A.; Murru, N. Presence of Enteric Bacterial Pathogens in Meat Samples of Wild Boar Hunted in Campania Region, Southern Italy. *Ital. J. Food Saf.* **2022**, *11*, 9967. [CrossRef] [PubMed]

65. Commission Regulation (EC) No 2073/2005 of 15 November 2005 on Microbiological Criteria for Foodstuffs. *Off. J. Eur. Union* **2005**, *L338*, 1–26.

66. Kerkhof, P.J.; Peruzy, M.F.; Murru, N.; Houf, K. Wild boars as reservoir for Campylobacter and Arcobacter. *Vet. Microbiol.* **2022**, *270*, 109462. [CrossRef] [PubMed]

67. Sasaki, Y.; Goshima, T.; Mori, T.; Murakami, M.; Haruna, M.; Ito, K.; Yamada, Y. Prevalence and Antimicrobial Susceptibility of Foodborne Bacteria in Wild Boars (*Sus scrofa*) and Wild Deer (*Cervus Nippon*) in Japan. *Foodborne Pathog. Dis.* **2013**, *10*, 985–991. [CrossRef]

68. Cummings, K.J.; Rodriguez-Rivera, L.D.; McNeely, I.; Suchodolski, J.S.; Mesenbrink, B.T.; Leland, B.R.; Bodenchuk, M.J. Fecal Shedding of Campylobacter Jejuni and Campylobacter Coli among Feral Pigs in Texas. *Zoonoses Public Health* **2018**, *65*, 215–217. [CrossRef]

69. Zendoia, I.I.; Cevidanes, A.; Hurtado, A.; Vázquez, P.; Barral, M.; Barandika, J.F.; García-Pérez, A.L. Stable prevalence of *Coxiella burnetii* in wildlife after a decade of surveillance in northern Spain. *Vet. Microbiol.* **2022**, *268*, 109422. [CrossRef]

70. Espí, A.; Del Cerro, A.; Oleaga, Á.; Rodríguez-Pérez, M.; López, C.M.; Hurtado, A.; Rodríguez-Martínez, L.D.; Barandika, J.F.; García-Pérez, A.L. One health approach: An overview of q fever in livestock, wildlife and humans in Asturias (northwestern spain). *Animals* **2021**, *11*, 1395. [CrossRef]

71. Sgroi, G.; Iatta, R.; Lia, R.P.; Napoli, E.; Buono, F.; Bezerra-Santos, M.A.; Veneziano, V.; Otranto, D. Tick exposure and risk of tick-borne pathogens infection in hunters and hunting dogs: A citizen science approach. *Transbound. Emerg. Dis.* **2022**, *69*, e386–e393. [CrossRef]

72. Ebani, V.V.; Nardoni, S.; Fognani, G.; Mugnaini, L.; Bertelloni, F.; Rocchigiani, G.; Papini, R.A.; Stefani, F.; Mancianti, F. Molecular detection of vector-borne bacteria and protozoa in healthy hunting dogs from Central Italy. *Asian Pac. J. Trop. Biomed.* **2015**, *5*, 108–112. [CrossRef]

73. Ebani, V.; Bertelloni, F.; Cecconi, G.; Sgorbini, M.; Cerri, D. Zoonotic tick-borne bacteria among wild boars (Sus scrofa) in Central Italy. *Asian Pac. J. Trop. Dis.* **2017**, *7*, 141–143. [CrossRef]

74. Sgroi, G.; Iatta, R.; Lia, R.P.; D'Alessio, N.; Manoj, R.R.S.; Veneziano, V.; Otranto, D. Spotted fever group rickettsiae in *Dermacentor marginatus* from wild boars in Italy. *Transbound. Emerg. Dis.* **2021**, *68*, 2111–2120. [CrossRef] [PubMed]

75. Del Cerro, A.; Oleaga, A.; Somoano, A.; Barandika, J.F.; García-Pérez, A.L.; Espí, A. Molecular identification of tick-borne pathogens (*Rickettsia* spp., *Anaplasma phagocytophilum*, *Borrelia burgdorferi* sensu lato, *Coxiella burnetii* and piroplasms) in questing and feeding hard ticks from North-Western Spain. *Ticks Tick-Borne Dis.* **2022**, *13*, 101961. [CrossRef]

76. Castillo-Contreras, R.; Magen, L.; Birtles, R.; Varela-Castro, L.; Hall, J.L.; Conejero, C.; Aguilar, X.F.; Colom-Cadena, A.; Lavín, S.; Mentaberre, G.; et al. Ticks on wild boar in the metropolitan area of Barcelona (Spain) are infected with spotted fever group rickettsiae. *Transbound. Emerg. Dis.* **2022**, *69*, e82–e95. [CrossRef] [PubMed]

77. Santana, M.D.S.; Hoppe, E.G.L.; Carraro, P.E.; Calchi, A.C.; de Oliveira, L.B.; do Amaral, R.B.; Mongruel, A.C.B.; Machado, D.M.R.; Burger, K.P.; Barros-Batestti, D.M.; et al. Molecular detection of vector-borne agents in wild boars (Sus scrofa) and associated ticks from Brazil, with evidence of putative new genotypes of *Ehrlichia*, *Anaplasma*, and haemoplasmas. *Transbound. Emerg. Dis.* **2022**, *69*, e2808–e2831. [CrossRef] [PubMed]

78. Sumrandee, C.; Baimai, V.; Trinachartvanit, W.; Ahantarig, A. Molecular detection of *Rickettsia*, *Anaplasma*, *Coxiella* and *Francisella* bacteria in ticks collected from Artiodactyla in Thailand. *Ticks Tick-Borne Dis.* **2016**, *7*, 678–689. [CrossRef]

79. Orr, B.; Malik, R.; Westman, M.E.; Norris, J.M. Seroprevalence of *Coxiella burnetii* in pig-hunting dogs from north Queensland, Australia. *Aust. Vet. J.* **2022**, *100*, 230–235. [CrossRef]

80. WOAH. Available online: https://www.woah.org/app/uploads/2022/02/listeria-monocytogenes-infection-with.pdf (accessed on 1 March 2023).

81. Weindl, L.; Frank, E.; Ullrich, U.; Heurich, M.; Kleta, S.; Ellerbroek, L.; Gareis, M. *Listeria Monocytogenes* in Different Specimens from Healthy Red Deer and Wild Boars. *Foodborne Pathog. Dis.* **2016**, *13*, 391–397. [CrossRef]

82. Palacios-Gorba, C.; Moura, A.; Leclercq, A.; Gómez-Martín, Á.; Gomis, J.; Jiménez-Trigos, E.; Mocé, M.L.; Lecuit, M.; Quereda, J.J. *Listeria* spp. Isolated from Tonsils of Wild Deer and Boars: Genomic Characterization. *Appl. Environ. Microbiol.* **2021**, *87*, e02651-20. [CrossRef]

83. Roila, R.; Branciari, R.; Primavilla, S.; Miraglia, D.; Vercillo, F.; Ranucci, D. Microbial, Physicochemical and Sensory Characteristics of Salami Produced from Wild Boar (*Sus scrofa*). *Slovak J. Food Sci.* **2021**, *15*, 475–483. [CrossRef] [PubMed]

84. Lucchini, R.; Armani, M.; Novelli, E.; Rodas, S.; Masiero, A.; Minenna, J.; Bacchin, C.; Drigo, I.; Piovesana, A.; Favretti, M.; et al. Listeria monocytogenes in game meat cured sausages. In *Trends in Game Meat Hygiene: From Forest to Fork*; Paulsen, P., Bauer, A., Smulders, F.J.M., Eds.; Wageningen Academic Publishers: Wageningen, The Netherlands, 2014; pp. 167–174.

85. Cilia, G.; Turchi, B.; Fratini, F.; Bilei, S.; Bossù, T.; De Marchis, M.L.; Cerri, D.; Pacini, M.I.; Bertelloni, F. Prevalence, Virulence and Antimicrobial Susceptibility of Salmonella Spp., Yersinia Enterocolitica and Listeria Monocytogenes in European Wild Boar (*Sus scrofa*) Hunted in Tuscany (Central Italy). *Pathogens* **2021**, *10*, 93. [CrossRef] [PubMed]

86. Avagnina, A.; Nucera, D.; Grassi, M.A.; Ferroglio, E.; Dalmasso, A.; Civera, T. The Microbiological Conditions of Carcasses from Large Game Animals in Italy. *Meat Sci.* **2012**, *91*, 266–271. [CrossRef] [PubMed]

87. Kanabalan, R.D.; Lee, L.J.; Lee, T.Y.; Chong, P.P.; Hassan, L.; Ismail, R.; Chin, V.K. Human Tuberculosis and Mycobacterium Tuberculosis Complex: A Review on Genetic Diversity, Pathogenesis and Omics Approaches in Host Biomarkers Discovery. *Microbiol. Res.* **2021**, *246*, 126674. [CrossRef] [PubMed]

88. Prodinger, W.M.; Indra, A.; Koksalan, O.K.; Kilicaslan, Z.; Richter, E. Mycobacterium Caprae Infection in Humans. *Expert Rev. Anti-Infect. Ther.* **2014**, *12*, 1501–1513. [CrossRef]

89. Varela-Castro, L.; Gerrikagoitia, X.; Alvarez, V.; Geijo, M.V.; Barral, M.; Sevilla, I.A. A Long-Term Survey on Mycobacterium Tuberculosis Complex in Wild Animals from a Bovine Tuberculosis Low Prevalence Area. *Eur. J. Wildl. Res.* **2021**, *67*, 43. [CrossRef]

90. Reis, A.C.; Ramos, B.; Pereira, A.C.; Cunha, M.V. The Hard Numbers of Tuberculosis Epidemiology in Wildlife: A Meta-Regression and Systematic Review. *Transbound. Emerg. Dis.* **2021**, *68*, 3257–3276. [CrossRef]

91. Lopes, B.C.; Vidaletti, M.R.; Loiko, M.R.; da Andrade, J.S.; Maciel, A.L.G.; Doyle, R.L.; Bertagnolli, A.C.; Rodrigues, R.O.; Driemeier, D.; Mayer, F.Q. Investigation of Mycobacterium Bovis and Metastrongylus Sp. Co-Infection and Its Relationship to Tuberculosis Lesions' Occurrence in Wild Boars. *Comp. Immunol. Microbiol. Infect. Dis.* **2021**, *77*, 101674. [CrossRef]

92. Iovane, V.; Ferrara, G.; Petruccelli, A.; Veneziano, V.; D'Alessio, N.; Ciarcia, R.; Fioretti, A.; Pagnini, U.; Montagnaro, S. Prevalence of Serum Antibodies against the Mycobacterium Tuberculosis Complex in Wild Boar in Campania Region, Italy. *Eur. J. Wildl. Res.* **2020**, *66*, 20. [CrossRef]

93. Jamin, C.; Rivière, J. Assessment of Bovine Tuberculosis Surveillance Effectiveness in French Wildlife: An Operational Approach. *Prev. Vet. Med.* **2020**, *175*, 104881. [CrossRef]

94. Richomme, C.; Courcoul, A.; Moyen, J.-L.; Reveillaud, É.; Maestrini, O.; de Cruz, K.; Drapeau, A.; Boschiroli, M.L. Tuberculosis in the Wild Boar: Frequentist and Bayesian Estimations of Diagnostic Test Parameters When Mycobacterium Bovis Is Present in Wild Boars but at Low Prevalence. *PLoS ONE* **2019**, *14*, e0222661. [CrossRef]

95. Santos, N.; Nunes, T.; Fonseca, C.; Vieira-Pinto, M.; Almeida, V.; Gortázar, C.; Correia-Neves, M. Spatial Analysis of Wildlife Tuberculosis Based on a Serologic Survey Using Dried Blood Spots, Portugal. *Emerg. Infect. Dis.* **2018**, *24*, 2169–2175. [CrossRef]

96. Roos, E.O.; Olea-Popelka, F.; Buss, P.; Hausler, G.A.; Warren, R.; Van Helden, P.D.; Parsons, S.D.C.; De Klerk-Lorist, L.-M.; Miller, M.A. Measuring antigen-specific responses in Mycobacterium bovis-infected warthogs (Phacochoerus africanus) using the intradermal tuberculin test. *BMC Vet. Res.* **2018**, *14*, 360. [CrossRef] [PubMed]

97. Roos, E.O.; Olea-Popelka, F.; Buss, P.; de Klerk-Lorist, L.-M.; Cooper, D.; van Helden, P.D.; Parsons, S.D.C.; Miller, M.A. Seroprevalence of Mycobacterium bovis infection in warthogs (Phacochoerus africanus) in bovine tuberculosis-endemic regions of South Africa. *Transbound. Emerg. Dis.* **2018**, *65*, 1182–1189. [CrossRef] [PubMed]

98. Réveillaud, É.; Desvaux, S.; Boschiroli, M.-L.; Hars, J.; Faure, É.; Fediaevsky, A.; Cavalerie, L.; Chevalier, F.; Jabert, P.; Poliak, S.; et al. Infection of Wildlife by Mycobacterium Bovis in France Assessment Through a National Surveillance System, Sylvatub. *Front. Vet. Sci.* **2018**, *5*, 262. [CrossRef]

99. Maciel, A.L.G.; Loiko, M.R.; Bueno, T.S.; Moreira, J.G.; Coppola, M.; Dalla Costa, E.R.; Schmid, K.B.; Rodrigues, R.O.; Cibulski, S.P.; Bertagnolli, A.C.; et al. Tuberculosis in Southern Brazilian Wild Boars (*Sus scrofa*): First Epidemiological Findings. *Transbound. Emerg. Dis.* **2018**, *65*, 518–526. [CrossRef]

100. Amato, B.; Di Marco Lo Presti, V.; Gerace, E.; Capucchio, M.T.; Vitale, M.; Zanghì, P.; Pacciarini, M.L.; Marianelli, C.; Boniotti, M.B. Molecular Epidemiology of Mycobacterium Tuberculosis Complex Strains Isolated from Livestock and Wild Animals in Italy Suggests the Need for a Different Eradication Strategy for Bovine Tuberculosis. *Transbound. Emerg. Dis.* **2018**, *65*, e416–e424. [CrossRef]

101. Madeira, S.; Manteigas, A.; Ribeiro, R.; Otte, J.; Fonseca, A.P.; Caetano, P.; Abernethy, D.; Boinas, F. Factors That Influence Mycobacterium Bovis Infection in Red Deer and Wild Boar in an Epidemiological Risk Area for Tuberculosis of Game Species in Portugal. *Transbound. Emerg. Dis.* **2017**, *64*, 793–804. [CrossRef]

102. Jang, Y.; Ryoo, S.; Lee, H.; Kim, N.; Lee, H.; Park, S.; Song, W.; Kim, J.-T.; Lee, H.S.; Myung Kim, J. Isolation of Mycobacterium Bovis from Free-Ranging Wildlife in South Korea. *J. Wildl. Dis.* **2017**, *53*, 181–185. [CrossRef]

103. Che'Amat, A.; Armenteros, J.A.; González-Barrio, D.; Lima, J.F.; Díez-Delgado, I.; Barasona, J.A.; Romero, B.; Lyashchenko, K.P.; Ortiz, J.A.; Gortázar, C. Is Targeted Removal a Suitable Means for Tuberculosis Control in Wild Boar? *Prev. Vet. Med.* **2016**, *135*, 132–135. [CrossRef]

104. García-Jiménez, W.L.; Cortés, M.; Benítez-Medina, J.M.; Hurtado, I.; Martínez, R.; García-Sánchez, A.; Risco, D.; Cerrato, R.; Sanz, C.; Hermoso-de-Mendoza, M.; et al. Spoligotype Diversity and 5-Year Trends of Bovine Tuberculosis in Extremadura, Southern Spain. *Trop. Anim. Health Prod.* **2016**, *48*, 1533–1540. [CrossRef] [PubMed]

105. Matos, A.C.; Figueira, L.; Martins, M.H.; Pinto, M.L.; Matos, M.; Coelho, A.C. New Insights into Mycobacterium Bovis Prevalence in Wild Mammals in Portugal. *Transbound. Emerg. Dis.* **2016**, *63*, e313–e322. [CrossRef]

106. El Mrini, M.; Kichou, F.; Kadiri, A.; Berrada, J.; Bouslikhane, M.; Cordonnier, N.; Romero, B.; Gortázar, C. Animal Tuberculosis Due to Mycobacterium Bovis in Eurasian Wild Boar from Morocco. *Eur. J. Wildl. Res.* **2016**, *62*, 479–482. [CrossRef]

107. Miller, M.; Buss, P.; de Klerk-Lorist, L.-M.; Hofmeyr, J.; Hausler, G.; Lyashchenko, K.; Lane, E.P.; Botha, L.; Parsons, S.; van Helden, P. Application of Rapid Serologic Tests for Detection of Mycobacterium Bovis Infection in Free-Ranging Warthogs (Phacochoerus Africanus)—Implications for Antemortem Disease Screening. *J. Wildl. Dis.* **2015**, *52*, 180–182. [CrossRef] [PubMed]

108. Che' Amat, A.; González-Barrio, D.; Ortiz, J.A.; Díez-Delgado, I.; Boadella, M.; Barasona, J.A.; Bezos, J.; Romero, B.; Armenteros, J.A.; Lyashchenko, K.P.; et al. Testing Eurasian Wild Boar Piglets for Serum Antibodies against Mycobacterium Bovis. *Prev. Vet. Med.* **2015**, *121*, 93–98. [CrossRef]

109. Beerli, O.; Blatter, S.; Boadella, M.; Schöning, J.; Schmitt, S.; Ryser-Degiorgis, M.-P. Towards Harmonised Procedures in Wildlife Epidemiological Investigations: A Serosurvey of Infection with Mycobacterium Bovis and Closely Related Agents in Wild Boar (*Sus scrofa*) in Switzerland. *Vet. J.* **2015**, *203*, 131–133. [CrossRef] [PubMed]

110. Broughan, J.M.; Downs, S.H.; Crawshaw, T.R.; Upton, P.A.; Brewer, J.; Clifton-Hadley, R.S. Mycobacterium Bovis Infections in Domesticated Non-Bovine Mammalian Species. Part 1: Review of Epidemiology and Laboratory Submissions in Great Britain 2004–2010. *Vet. J.* **2013**, *198*, 339–345. [CrossRef] [PubMed]

111. Muñoz-Mendoza, M.; Marreros, N.; Boadella, M.; Gortázar, C.; Menéndez, S.; de Juan, L.; Bezos, J.; Romero, B.; Copano, M.F.; Amado, J.; et al. Wild boar tuberculosis in Iberian Atlantic Spain: A different picture from Mediterranean habitats. *BMC Vet. Res.* **2013**, *9*, 176. [CrossRef] [PubMed]

112. Boadella, M.; Vicente, J.; Ruiz-Fons, F.; de la Fuente, J.; Gortázar, C. Effects of Culling Eurasian Wild Boar on the Prevalence of Mycobacterium Bovis and Aujeszky's Disease Virus. *Prev. Vet. Med.* **2012**, *107*, 214–221. [CrossRef]

113. Zanella, G.; Bar-Hen, A.; Boschiroli, M.-L.; Hars, J.; Moutou, F.; Garin-Bastuji, B.; Durand, B. Modelling Transmission of Bovine Tuberculosis in Red Deer and Wild Boar in Normandy, France. *Zoonoses Public Health* **2012**, *59*, 170–178. [CrossRef]

114. Nugent, G.; Whitford, J.; Yockney, I.J.; Cross, M.L. Reduced Spillover Transmission of Mycobacterium Bovis to Feral Pigs (*Sus scofa*) Following Population Control of Brushtail Possums (Trichosurus Vulpecula). *Epidemiol. Infect.* **2012**, *140*, 1036–1047. [CrossRef]

115. García-Bocanegra, I.; de Val, B.P.; Arenas-Montes, A.; Paniagua, J.; Boadella, M.; Gortázar, C.; Arenas, A. Seroprevalence and Risk Factors Associated to Mycobacterium Bovis in Wild Artiodactyl Species from Southern Spain, 2006–2010. *PLoS ONE* **2012**, *7*, e34908. [CrossRef] [PubMed]

116. Orłowska, B.; Krajewska-Wędzina, M.; Augustynowicz-Kopeć, E.; Kozińska, M.; Brzezińska, S.; Zabost, A.; Didkowska, A.; Welz, M.; Kaczor, S.; Żmuda, P.; et al. Epidemiological Characterization of Mycobacterium Caprae Strains Isolated from Wildlife in the Bieszczady Mountains, on the Border of Southeast Poland. *BMC Vet. Res.* **2020**, *16*, 362. [CrossRef] [PubMed]

117. Csivincsik, Á.; Rónai, Z.; Nagy, G.; Svéda, G.; Halász, T. Surveillance of Mycobacterium Caprae Infection in a Wild Boar (*Sus scrofa*) Population in Southwestern Hungary. *Vet. Arh.* **2016**, *86*, 767–775.

118. Tagliapietra, V.; Boniotti, M.B.; Mangeli, A.; Karaman, I.; Alborali, G.; Chiari, M.; D'Incau, M.; Zanoni, M.; Rizzoli, A.; Pacciarini, M.L. Mycobacterium Microti at the Environment and Wildlife Interface. *Microorganisms* **2021**, *9*, 2084. [CrossRef] [PubMed]

119. Ghielmetti, G.; Hilbe, M.; Friedel, U.; Menegatti, C.; Bacciarini, L.; Stephan, R.; Bloemberg, G. Mycobacterial Infections in Wild Boars (*Sus scrofa*) from Southern Switzerland: Diagnostic Improvements, Epidemiological Situation and Zoonotic Potential. *Transbound. Emerg. Dis.* **2021**, *68*, 573–586. [CrossRef] [PubMed]

120. De Val, B.; Sanz, A.; Soler, M.; Allepuz, A.; Michelet, L.; Boschiroli, M.L.; Vidal, E. Mycobacterium Microti Infection in Free-Ranging Wild Boar, Spain, 2017–2019. *Emerg. Infect. Dis.* **2019**, *25*, 2152–2154. [CrossRef]

121. Bona, M.C.; Mignone, W.; Ballardini, M.; Dondo, A.; Mignone, G.; Ru, G. Tuberculosis in wild boar (Sus scrofa) in the Western Liguria Region. *Épidémiologie Et St. Anim.* **2018**, *74*, 151–158.

122. Chiari, M.; Ferrari, N.; Giardiello, D.; Avisani, D.; Pacciarini, M.L.; Alborali, L.; Zanoni, M.; Boniotti, M.B. Spatiotemporal and Ecological Patterns of Mycobacterium Microti Infection in Wild Boar (*Sus scrofa*). *Transbound. Emerg. Dis.* **2016**, *63*, e381–e388. [CrossRef]

123. Boniotti, M.B.; Gaffuri, A.; Gelmetti, D.; Tagliabue, S.; Chiari, M.; Mangeli, A.; Spisani, M.; Nassuato, C.; Gibelli, L.; Sacchi, C.; et al. Detection and molecular characterization of Mycobacterium microti isolates in wild boar from northern Italy. *J. Clin. Microbiol.* **2014**, *52*, 2834–2843. [CrossRef]

124. Schöning, J.M.; Cerny, N.; Prohaska, S.; Wittenbrink, M.M.; Smith, N.H.; Bloemberg, G.; Pewsner, M.; Schiller, I.; Origgi, F.C.; Ryser-Degiorgis, M.-P. Surveillance of Bovine Tuberculosis and Risk Estimation of a Future Reservoir Formation in Wildlife in Switzerland and Liechtenstein. *PLoS ONE* **2013**, *8*, e54253. [CrossRef]

125. Natarajan, A.; Beena, P.M.; Devnikar, A.V.; Mali, S. A Systemic Review on Tuberculosis. *Indian J. Tuberc.* **2020**, *67*, 295–311. [CrossRef] [PubMed]

126. Torres-Gonzalez, P.; Cervera-Hernandez, M.E.; Martinez-Gamboa, A.; Garcia-Garcia, L.; Cruz-Hervert, L.P.; Bobadilla-del Valle, M.; Ponce-de Leon, A.; Sifuentes-Osornio, J. Human Tuberculosis Caused by Mycobacterium Bovis: A Retrospective Comparison

with Mycobacterium Tuberculosis in a Mexican Tertiary Care Centre, 2000–2015. *BMC Infect. Dis.* **2016**, *16*, 657. [CrossRef] [PubMed]

127. Lekko, Y.M.; Che-Amat, A.; Ooi, P.T.; Omar, S.; Ramanoon, S.Z.; Mazlan, M.; Jesse, F.F.A.; Jasni, S.; Ariff Abdul-Razak, M.F. Mycobacterium Tuberculosis and Avium Complex Investigation among Malaysian Free-Ranging Wild Boar and Wild Macaques at Wildlife-Livestock-Human Interface. *Animals* **2021**, *11*, 3252. [CrossRef] [PubMed]

128. Mentaberre, G.; Romero, B.; de Juan, L.; Navarro-González, N.; Velarde, R.; Mateos, A.; Marco, I.; Olivé-Boix, X.; Domínguez, L.; Lavín, S.; et al. Long-Term Assessment of Wild Boar Harvesting and Cattle Removal for Bovine Tuberculosis Control in Free Ranging Populations. *PLoS ONE* **2014**, *9*, e88824. [CrossRef]

129. Aranha, J.; Abrantes, A.C.; Gonçalves, R.; Miranda, R.; Serejo, J.; Vieira-Pinto, M. GIS as an Epidemiological Tool to Monitor the Spatial–Temporal Distribution of Tuberculosis in Large Game in a High-Risk Area in Portugal. *Animals* **2021**, *11*, 2374. [CrossRef]

130. Pate, M.; Zajc, U.; Kušar, D.; Žele, D.; Vengušt, G.; Pirš, T.; Ocepek, M. *Mycobacterium* spp. in Wild Game in Slovenia. *Vet. J.* **2016**, *208*, 93–95. [CrossRef]

131. García-Jiménez, W.L.; Benítez-Medina, J.M.; Martínez, R.; Carranza, J.; Cerrato, R.; García-Sánchez, A.; Risco, D.; Moreno, J.C.; Sequeda, M.; Gómez, L.; et al. Non-Tuberculous Mycobacteria in Wild Boar (*Sus scrofa*) from Southern Spain: Epidemiological, Clinical and Diagnostic Concerns. *Transbound. Emerg. Dis.* **2015**, *62*, 72–80. [CrossRef]

132. Barasona, J.A.; Torres, M.J.; Aznar, J.; Gortázar, C.; Vicente, J. DNA Detection Reveals Mycobacterium Tuberculosis Complex Shedding Routes in Its Wildlife Reservoir the Eurasian Wild Boar. *Transbound. Emerg. Dis.* **2017**, *64*, 906–915. [CrossRef]

133. Barroso, P.; Barasona, J.A.; Acevedo, P.; Palencia, P.; Carro, F.; Negro, J.J.; Torres, M.J.; Gortázar, C.; Soriguer, R.C.; Vicente, J. Long-Term Determinants of Tuberculosis in the Ungulate Host Community of Doñana National Park. *Pathogens* **2020**, *9*, 445. [CrossRef]

134. Gortázar, C.; Fernández-Calle, L.M.; Collazos-Martínez, J.A.; Mínguez-González, O.; Acevedo, P. Animal Tuberculosis Maintenance at Low Abundance of Suitable Wildlife Reservoir Hosts: A Case Study in Northern Spain. *Prev. Vet. Med.* **2017**, *146*, 150–157. [CrossRef]

135. Ramos, B.; Pereira, A.C.; Reis, A.C.; Cunha, M.V. Estimates of the global and continental burden of animal tuberculosis in key livestock species worldwide: A meta-analysis study. *One Health* **2020**, *10*, 100169. [CrossRef] [PubMed]

136. Costa, P.; Ferreira, A.S.; Amaro, A.; Albuquerque, T.; Botelho, A.; Couto, I.; Cunha, M.V.; Viveiros, M.; Inácio, J. Enhanced detection of tuberculous mycobacteria in animal tissues using a semi-nested probe-based real-time PCR. *PLoS ONE* **2013**, *8*, e81337. [CrossRef] [PubMed]

137. Pedersen, K.; Miller, R.S.; Anderson, T.D.; Pabilonia, K.L.; Lewis, J.R.; Mihalco, R.L.; Gortazar, C.; Gidlewski, T. Limited antibody evidence of exposure to Mycobacterium bovis in feral swine (*Sus scrofa*) in the USA. *J. Wildl. Dis.* **2017**, *53*, 30–36. [CrossRef] [PubMed]

138. Santos, N.; Colino, E.F.; Arnal, M.C.; de Luco, D.F.; Sevilla, I.; Garrido, J.M.; Fonseca, E.; Valente, A.M.; Balseiro, A.; Queirós, J.; et al. Complementary Roles of Wild Boar and Red Deer to Animal Tuberculosis Maintenance in Multi-Host Communities. *Epidemics* **2022**, *41*, 100633. [CrossRef]

139. Palmer, M.V. Mycobacterium bovis: Characteristics of wildlife reservoir hosts. *Transb. Emerg. Dis.* **2013**, *60* (Suppl. 1), 1–13. [CrossRef]

140. Regulation (EC) No 853/2004 of the European Parliament and of the Council of 29 April 2004 Laying Down Specific Hygiene Rules for Food of Animal Origin. *Off. J. Eur. Union* **2003**, *L139*, 55.

141. Reis, A.C.; Tenreiro, R.; Albuquerque, T.; Botelho, A.; Cunha, M.V. Long-term molecular surveillance provides clues on a cattle origin for Mycobacterium bovis in Portugal. *Sci. Rep.* **2020**, *10*, 20856. [CrossRef] [PubMed]

142. Clausi, M.T.; Ciambrone, L.; Zanoni, M.; Costanzo, N.; Pacciarini, M.; Casalinuovo, F. Evaluation of the Presence and Viability of Mycobacterium bovis in Wild Boar Meat and Meat-Based Preparations. *Foods* **2021**, *10*, 2410. [CrossRef]

143. Varela-Castro, L.; Alvarez, V.; Sevilla, I.A.; Barral, M. Risk factors associated to a high Mycobacterium tuberculosis complex seroprevalence in wild boar (*Sus scrofa*) from a low bovine tuberculosis prevalence area. *PLoS ONE* **2020**, *15*, e0231559. [CrossRef]

144. Richomme, C.; Boadella, M.; Courcoul, A.; Durand, B.; Drapeau, A.; Corde, Y.; Hars, J.; Payne, A.; Fediaevsky, A.; Boschiroli, M.L. Exposure of Wild Boar to Mycobacterium Tuberculosis Complex in France since 2000 Is Consistent with the Distribution of Bovine Tuberculosis Outbreaks in Cattle. *PLoS ONE* **2013**, *8*, e77842. [CrossRef]

145. Albertti, L.A.G.; Souza-Filho, A.F.; Fonseca-Júnior, A.A.; Freitas, M.E.; de Oliveira-Pellegrin, A.; Zimmermann, N.P.; Tomás, W.M.; Péres, I.A.H.F.S.; Fontana, I.; Osório, A.L.A.R. Mycobacteria Species in Wild Mammals of the Pantanal of Central South America. *Eur. J. Wildl. Res.* **2015**, *61*, 163–166. [CrossRef]

146. Vicente, J.; Barasona, J.A.; Acevedo, P.; Ruiz-Fons, J.F.; Boadella, M.; Diez-Delgado, I.; Beltran-Beck, B.; González-Barrio, D.; Queirós, J.; Montoro, V.; et al. Temporal trend of tuberculosis in wild ungulates from Mediterranean Spain. *Transb. Emerg. Dis.* **2013**, *60*, 92–103. [CrossRef]

147. WOAH. Available online: https://www.woah.org/fileadmin/Home/fr/Health_standards/tahm/3.09.08_SALMONELLOSIS. pdf (accessed on 1 March 2023).

148. WOAH. Available online: https://www.woah.org/app/uploads/2021/05/salmonella-enterica-all-serovarsinfection-with.pdf (accessed on 1 March 2023).

149. Grimont, P.A.; Weill, F.X. Antigenic formulae of the Salmonella serovars. *WHO Collab. Cent. Ref. Res. Salmonella* **2007**, *9*, 1–166.

150. European Food Safety Authority (EFSA) and European Centre for Disease Prevention and Control (ECDC). The European Union One Health 2020 Zoonoses Report. *EFSA J.* **2021**, *19*, 6971. [CrossRef]

151. Torres, R.T.; Fernandes, J.; Carvalho, J.; Cunha, M.V.; Caetano, T.; Mendo, S.; Serrano, E.; Fonseca, C. Wild Boar as a Reservoir of Antimicrobial Resistance. *Sci. Total Environ.* **2020**, *717*, 135001. [CrossRef]

152. Shinoda, S.; Mizuno, T.; Miyoshi, S. General Review on Hog Cholera (Classical Swine Fever), African Swine Fever, and *Salmonella Enterica* Serovar Choleraesuis Infection. *J. Disaster Res.* **2019**, *14*, 1105–1114. [CrossRef]

153. Greig, J.; Rajić, A.; Young, I.; Mascarenhas, M.; Waddell, L.; LeJeune, J. A Scoping Review of the Role of Wildlife in the Transmission of Bacterial Pathogens and Antimicrobial Resistance to the Food Chain. *Zoonoses Public Health* **2015**, *62*, 269–284. [CrossRef]

154. Billinis, C. Wildlife Diseases That Pose a Risk to Small Ruminants and Their Farmers. *Small Rum. Res.* **2013**, *110*, 67–70. [CrossRef]

155. Paulsen, P.; Smulders, F.J.M.; Hilbert, F. Salmonella in Meat from Hunted Game: A Central European Perspective. *Food Res. Int.* **2012**, *45*, 609–616. [CrossRef]

156. Cano-Manuel, F.J.; López-Olvera, J.; Fandos, P.; Soriguer, R.C.; Pérez, J.M.; Granados, J.E. Long-Term Monitoring of 10 Selected Pathogens in Wild Boar (*Sus scrofa*) in Sierra Nevada National Park, Southern Spain. *Vet. Microbiol.* **2014**, *174*, 148–154. [CrossRef]

157. Flores Rodas, E.M.; Bogdanova, T.; Bossù, T.; Pecchi, S.; Tomassetti, F.; De Santis, P.; Tolli, R.; Condoleo, R.; Greco, S.; Brozzi, A.; et al. Microbiological Assessment of Freshly-Shot Wild Boars Meat in Lazio Region, Viterbo Territory: A Preliminary Study. *Ital. J. Food Saf.* **2014**, *3*, 1711. [CrossRef]

158. Touloudi, A.; Valiakos, G.; Athanasiou, L.V.; Birtsas, P.; Giannakopoulos, A.; Papaspyropoulos, K.; Kalaitzis, C.; Sokos, C.; Tsokana, C.N.; Spyrou, V.; et al. A Serosurvey for Selected Pathogens in Greek European Wild Boar. *Vet. Rec. Open* **2015**, *2*, e000077. [CrossRef]

159. McGregor, G.F.; Gottschalk, M.; Godson, D.L.; Wilkins, W.; Bollinger, T.K. Disease Risks Associated with Free-Ranging Wild Boar in Saskatchewan. *Can. Vet. J.* **2015**, *56*, 839–844. [PubMed]

160. Bassi, A.M.G.; Steiner, J.C.; Stephan, R.; Nüesch-Inderbinen, M. Seroprevalence of Toxoplasma Gondii and Salmonella in Hunted Wild Boars from Two Different Regions in Switzerland. *Animals* **2021**, *11*, 2227. [CrossRef] [PubMed]

161. Ortega, N.; Fanelli, A.; Serrano, A.; Martínez-Carrasco, C.; Escribano, F.; Tizzani, P.; Candela, M.G. Salmonella Seroprevalence in Wild Boar from Southeast Spain Depends on Host Population Density. *Res. Vet. Sci.* **2020**, *132*, 400–403. [CrossRef] [PubMed]

162. Zottola, T.; Montagnaro, S.; Magnapera, C.; Sasso, S.; De Martino, L.; Bragagnolo, A.; D'Amici, L.; Condoleo, R.; Pisanelli, G.; Iovane, G.; et al. Prevalence and Antimicrobial Susceptibility of Salmonella in European Wild Boar (*Sus scrofa*); Latium Region—Italy. *Comp. Immunol. Microbiol. Infect. Dis.* **2013**, *36*, 161–168. [CrossRef]

163. Petrović, J.; Mirčeta, J.; Babić, J.; Malešević, M.; Blagojević, B.; Radulović, J.P.; Antić, D. *Salmonella* in Wild Boars (*Sus scrofa*): Characterization and Epidemiology. *Acta Vet.* **2022**, *72*, 184–194. [CrossRef]

164. Piras, F.; Spanu, V.; Siddi, G.; Gymoese, P.; Spanu, C.; Cibin, V.; Schjørring, S.; De Santis, E.P.L.; Scarano, C. Whole-Genome Sequencing Analysis of Highly Prevalent Salmonella Serovars in Wild Boars from a National Park in Sardinia. *Food Control* **2021**, *130*, 108247. [CrossRef]

165. Bonardi, S.; Tansini, C.; Cacchioli, A.; Soliani, L.; Poli, L.; Lamperti, L.; Corradi, M.; Gilioli, S. Enterobacteriaceae and Salmonella Contamination of Wild Boar (*Sus scrofa*) Carcasses: Comparison between Different Sampling Strategies. *Eur. J. Wildl. Res.* **2021**, *67*, 88. [CrossRef]

166. Razzuoli, E.; Listorti, V.; Martini, I.; Migone, L.; Decastelli, L.; Mignone, W.; Berio, E.; Battistini, R.; Ercolini, C.; Serracca, L.; et al. Prevalence and Antimicrobial Resistances of Salmonella Spp. Isolated from Wild Boars in Liguria Region, Italy. *Pathogens* **2021**, *10*, 568. [CrossRef]

167. Plaza-Rodríguez, C.; Alt, K.; Grobbel, M.; Hammerl, J.A.; Irrgang, A.; Szabo, I.; Stingl, K.; Schuh, E.; Wiehle, L.; Pfefferkorn, B.; et al. Wildlife as Sentinels of Antimicrobial Resistance in Germany? *Front. Vet. Sci.* **2021**, *7*, 627821. [CrossRef] [PubMed]

168. Bonardi, S.; Bolzoni, L.; Zanoni, R.G.; Morganti, M.; Corradi, M.; Gilioli, S.; Pongolini, S. Limited Exchange of Salmonella Among Domestic Pigs and Wild Boars in Italy. *EcoHealth* **2019**, *16*, 420–428. [CrossRef] [PubMed]

169. Gil Molino, M.; García Sánchez, A.; Risco Pérez, D.; Gonçalves Blanco, P.; Quesada Molina, A.; Rey Pérez, J.; Martín Cano, F.E.; Cerrato Horrillo, R.; Hermoso-de-Mendoza Salcedo, J.; Fernández Llario, P. Prevalence of Salmonella Spp. in Tonsils, Mandibular Lymph Nodes and Faeces of Wild Boar from Spain and Genetic Relationship between Isolates. *Transbound. Emerg. Dis.* **2019**, *66*, 1218–1226. [CrossRef] [PubMed]

170. Cummings, K.J.; Rodriguez-Rivera, L.D.; Grigar, M.K.; Rankin, S.C.; Mesenbrink, B.T.; Leland, B.R.; Bodenchuk, M.J. Prevalence and Characterization of *Salmonella* Isolated from Feral Pigs Throughout Texas. *Zoonoses Public Health* **2016**, *63*, 436–441. [CrossRef]

171. Dias, D.; Torres, R.T.; Kronvall, G.; Fonseca, C.; Mendo, S.; Caetano, T. Assessment of Antibiotic Resistance of *Escherichia coli* Isolates and Screening of Salmonella Spp. in Wild Ungulates from Portugal. *Res. Microbiol.* **2015**, *166*, 584–593. [CrossRef]

172. Sannö, A.; Aspán, A.; Hestvik, G.; Jacobson, M. Presence of *Salmonella* Spp., *Yersinia Enterocolitica*, *Yersinia Pseudotuberculosis* and *Escherichia coli* O157:H7 in Wild Boars. *Epidemiol. Infect.* **2014**, *142*, 2542–2547. [CrossRef]

173. Favier, G.I.; Lucero Estrada, C.; Cortiñas, T.I.; Escudero, M.E. Detection and Characterization of Shiga Toxin Producing *Escherichia coli, Salmonella* spp., and *Yersinia* Strains from Human, Animal, and Food Samples in San Luis, Argentina. *Int. J. Microbiol.* **2014**, *2014*, 284649. [CrossRef]

174. Chiari, M.; Zanoni, M.; Tagliabue, S.; Lavazza, A.; Alborali, L.G. Salmonella Serotypes in Wild Boars (*Sus scrofa*) Hunted in Northern Italy. *Acta Vet. Scand.* **2013**, *55*, 42. [CrossRef]

175. Navarro-Gonzalez, N.; Mentaberre, G.; Porrero, C.M.; Serrano, E.; Mateos, A.; López-Martín, J.M.; Lavín, S.; Domínguez, L. Effect of Cattle on Salmonella Carriage, Diversity and Antimicrobial Resistance in Free-Ranging Wild Boar (*Sus scrofa*) in Northeastern Spain. *PLoS ONE* **2012**, *7*, e51614. [CrossRef]

176. Ranucci, D.; Roila, R.; Onofri, A.; Cambiotti, F.; Primavilla, S.; Miraglia, D.; Andoni, E.; Di Cerbo, A.; Branciari, R. Improving Hunted Wild Boar Carcass Hygiene: Roles of Different Factors Involved in the Harvest Phase. *Foods* **2021**, *10*, 1548. [CrossRef]

177. Orsoni, F.; Romeo, C.; Ferrari, N.; Bardasi, L.; Merialdi, G.; Barbani, R. Factors Affecting the Microbiological Load of Italian Hunted Wild Boar Meat (*Sus scrofa*). *Meat Sci.* **2020**, *160*, 107967. [CrossRef] [PubMed]

178. Peruzy, M.F.; Murru, N.; Yu, Z.; Kerkhof, P.-J.; Neola, B.; Joossens, M.; Proroga, Y.T.R.; Houf, K. Assessment of Microbial Communities on Freshly Killed Wild Boar Meat by MALDI-TOF MS and 16S RRNA Amplicon Sequencing. *Int. J. Food Microbiol.* **2019**, *301*, 51–60. [CrossRef] [PubMed]

179. Asakura, H.; Kawase, J.; Ikeda, T.; Honda, M.; Sasaki, Y.; Uema, M.; Kabeya, H.; Sugiyama, H.; Igimi, S.; Takai, S. Microbiological Quality Assessment of Game Meats at Retail in Japan. *J. Food Prot.* **2017**, *80*, 2119–2126. [CrossRef] [PubMed]

180. Mirceta, J.; Petrovic, J.; Malesevic, M.; Blagojevic, B.; Antic, D. Assessment of Microbial Carcass Contamination of Hunted Wild Boars. *Eur. J. Wildl. Res.* **2017**, *63*, 37. [CrossRef]

181. Russo, C.; Balloni, S.; Altomonte, I.; Martini, M.; Nuvoloni, R.; Cecchi, F.; Pedonese, F.; Salari, F.; Sant'ana Da Silva, A.M.; Torracca, B.; et al. Fatty Acid and Microbiological Profile of the Meat (Longissimus Dorsi Muscle) of Wild Boar (*Sus Scropha Scropha*) Hunted in Tuscany. *Ital. J. Anim. Sci.* **2017**, *16*, 1–8. [CrossRef]

182. Holmes, M.A.; Zadoks, R.N. Methicillin resistant S. aureus in human and bovine mastitis. *J. Mammary Gland. Biol. Neoplasia* **2011**, *16*, 373–382. [CrossRef] [PubMed]

183. Tong, S.Y.C.; Davis, J.S.; Eichenberger, E.; Holland, T.L.; Fowler, V.G. Staphylococcus Aureus Infections: Epidemiology, Pathophysiology, Clinical Manifestations, and Management. *Clin. Microbiol. Rev.* **2015**, *28*, 603–661. [CrossRef]

184. European Food Safety Authority (EFSA). Joint scientific report of ECDC, EFSA and EMEA on meticillin resistant Staphylococcus aureus (MRSA) in livestock, companion animals and food. *EFSA Sci. Rep.* **2009**, *301*, 1–10.

185. Sousa, M.; Silva, N.; Manageiro, V.; Ramos, S.; Coelho, A.; Gonçalves, D.; Caniça, M.; Torres, C.; Igrejas, G.; Poeta, P. First Report on MRSA CC398 Recovered from Wild Boars in the North of Portugal. Are We Facing a Problem? *Sci. Total Environ.* **2017**, *596–597*, 26–31. [CrossRef]

186. Porrero, M.C.; Mentaberre, G.; Sánchez, S.; Fernández-Llario, P.; Gómez-Barrero, S.; Navarro-Gonzalez, N.; Serrano, E.; Casas-Díaz, E.; Marco, I.; Fernández-Garayzabal, J.-F.; et al. Methicillin Resistant Staphylococcus Aureus (MRSA) Carriage in Different Free-Living Wild Animal Species in Spain. *Vet. J.* **2013**, *198*, 127–130. [CrossRef]

187. Porrero, M.C.; Mentaberre, G.; Sánchez, S.; Fernández-Llario, P.; Casas-Díaz, E.; Mateos, A.; Vidal, D.; Lavín, S.; Fernández-Garayzábal, J.-F.; Domínguez, L. Carriage of Staphylococcus Aureus by Free-Living Wild Animals in Spain. *Appl. Environ. Microbiol.* **2014**, *80*, 4865–4870. [CrossRef]

188. Traversa, A.; Gariano, G.R.; Gallina, S.; Bianchi, D.M.; Orusa, R.; Domenis, L.; Cavallerio, P.; Fossati, L.; Serra, R.; Decastelli, L. Methicillin Resistance in Staphylococcus Aureus Strains Isolated from Food and Wild Animal Carcasses in Italy. *Food Microbiol.* **2015**, *52*, 154–158. [CrossRef] [PubMed]

189. Seinige, D.; Von Altrock, A.; Kehrenberg, C. Genetic Diversity and Antibiotic Susceptibility of Staphylococcus Aureus Isolates from Wild Boars. *Comp. Immunol. Microbiol. Infect. Dis.* **2017**, *54*, 7–12. [CrossRef] [PubMed]

190. Ramos, B.; Rosalino, L.M.; Palmeira, J.D.; Torres, R.T.; Cunha, M.V. Antimicrobial Resistance in Commensal Staphylococcus Aureus from Wild Ungulates Is Driven by Agricultural Land Cover and Livestock Farming. *Environ. Pollut.* **2022**, *303*, 119116. [CrossRef] [PubMed]

191. Mama, O.M.; Ruiz-Ripa, L.; Fernández-Fernández, R.; González-Barrio, D.; Ruiz-Fons, J.F.; Torres, C. High Frequency of Coagulase-Positive Staphylococci Carriage in Healthy Wild Boar with Detection of MRSA of Lineage ST398-T011. *FEMS Microbiol. Lett.* **2019**, *366*, fny292. [CrossRef] [PubMed]

192. Meemken, D.; Blaha, T.; Hotzel, H.; Strommenger, B.; Klein, G.; Ehricht, R.; Monecke, S.; Kehrenberg, C. Genotypic and Phenotypic Characterization of Staphylococcus Aureus Isolates from Wild Boars. *Appl. Environ. Microbiol.* **2013**, *79*, 1739–1742. [CrossRef]

193. Rey Pérez, J.; Zálama Rosa, L.; García Sánchez, A.; Hermoso de Mendoza Salcedo, J.; Alonso Rodríguez, J.M.; Cerrato Horrillo, R.; Zurita, S.G.; Gil Molino, M. Multiple Antimicrobial Resistance in Methicillin-Resistant Staphylococcus Sciuri Group Isolates from Wild Ungulates in Spain. *Antibiotics* **2021**, *10*, 920. [CrossRef]

194. Holtmann, A.R.; Meemken, D.; Müller, A.; Seinige, D.; Büttner, K.; Failing, K.; Kehrenberg, C. Wild Boars Carry Extended-Spectrum β-Lactamase- and AmpC-Producing *Escherichia coli*. *Microorganisms* **2021**, *9*, 367. [CrossRef]

195. EFSA BIOHAZ Panel (EFSA Panel on Biological Hazards); Koutsoumanis, K.; Allende, A.; Alvarez-Ordonez, A.; Bover-Cid, S.; Chemaly, M.; Davies, R.; De Cesare, A.; Herman, L.; Hilbert, F.; et al. Pathogenicity assessment of Shiga toxin-producing *Escherichia coli* (STEC) and the public health risk posed by contamination of food with STEC. *EFSA J.* **2020**, *18*, e05967. [CrossRef]

196. Perrat, A.; Branchu, P.; Decors, A.; Turci, S.; Bayon-Auboyer, M.-H.; Petit, G.; Grosbois, V.; Brugère, H.; Auvray, F.; Oswald, E. Wild Boars as Reservoir of Highly Virulent Clone of Hybrid Shiga Toxigenic and Enterotoxigenic Escherichia coli Responsible for Edema Disease, France. *Emerg. Infect. Dis.* **2022**, *28*, 382–393. [CrossRef]

197. Topalcengiz, Z.; Danyluk, M.D. Assessment of Contamination Risk from Fecal Matter Presence on Fruit and Mulch in the tomato fields based on generic *Escherichia coli* population. *Food Microbiol.* **2022**, *103*, 103956. [CrossRef]

198. Topalcengiz, Z.; Jeamsripong, S.; Spanninger, P.M.; Persad, A.K.; Wang, F.; Buchanan, R.L.; Lejeune, J.; Kniel, K.E.; Jay-Russell, M.T.; Danyluk, M.D. Survival of shiga toxin–producing *Escherichia coli* in various wild animal feces that may contaminate produce. *J. Food Prot.* **2020**, *83*, 1420–1429. [CrossRef] [PubMed]

199. McKee, A.M.; Bradley, P.M.; Shelley, D.; McCarthy, S.; Molina, M. Feral swine as sources of fecal contamination in recreational waters. *Sci. Rep.* **2021**, *11*, 4212. [CrossRef]

200. Dias, D.; Costa, S.; Fonseca, C.; Baraúna, R.; Caetano, T.; Mendo, S. Pathogenicity of Shiga toxin-producing *Escherichia coli* (STEC) from wildlife: Should we care? *Sci. Total Environ.* **2022**, *812*, 152324. [CrossRef] [PubMed]

201. Morita, S.; Sato, S.; Maruyama, S.; Nagasaka, M.; Murakami, K.; Inada, K.; Uchiumi, M.; Yokoyama, E.; Asakura, H.; Sugiyama, H.; et al. Whole-genome sequence analysis of shiga toxin-producing escherichia coli O157 strains isolated from wild deer and boar in Japan. *J. Vet. Med. Sci.* **2021**, *83*, 1860–1868. [CrossRef] [PubMed]

202. Bertelloni, F.; Cilia, G.; Bogi, S.; Ebani, V.V.; Turini, L.; Nuvoloni, R.; Cerri, D.; Fratini, F.; Turchi, B. Pathotypes and antimicrobial susceptibility of *Escherichia coli* isolated from wild boar (*Sus scrofa*) in Tuscany. *Animals* **2020**, *10*, 744. [CrossRef]

203. Szczerba-Turek, A.; Socha, P.; Bancerz-Kisiel, A.; Platt-Samoraj, A.; Lipczynska-Ilczuk, K.; Siemionek, J.; Kończyk, K.; Terech-Majewska, E.; Szweda, W. Pathogenic potential to humans of Shiga toxin-producing *Escherichia coli* isolated from wild boars in Poland. *Int. J. Food Microbiol.* **2019**, *300*, 8–13. [CrossRef] [PubMed]

204. Dias, D.; Caetano, T.; Torres, R.T.; Fonseca, C.; Mendo, S. Shiga toxin-producing *Escherichia coli* in wild ungulates. *Sci. Total Environ.* **2019**, *651*, 203–209. [CrossRef]

205. Alonso, C.A.; Mora, A.; Díaz, D.; Blanco, M.; González-Barrio, D.; Ruiz-Fons, F.; Simón, C.; Blanco, J.; Torres, C. Occurrence and characterization of stx and/or eae-positive *Escherichia coli* isolated from wildlife, including a typical EPEC strain from a wild boar. *Vet. Microbiol.* **2017**, *207*, 69–73. [CrossRef]

206. Navarro-Gonzalez, N.; Porrero, M.C.; Mentaberre, G.; Serrano, E.; Mateos, A.; Cabal, A.; Domínguez, L.; Lavín, S. *Escherichia coli* O157:H7 in Wild Boars (*Sus scrofa*) and Iberian Ibex (*Capra pyrenaica*) Sharing Pastures with Free-Ranging Livestock in a Natural Environment in Spain. *Vet. Q.* **2015**, *35*, 102–106. [CrossRef]

207. Díaz-Sánchez, S.; Sánchez, S.; Sánchez, M.; Herrera-León, S.; Hanning, I.; Vidal, D. Detection and characterization of Shiga toxin-producing *Escherichia coli* in game meat and ready-to-eat meat products. *Int. J. Food Microbiol.* **2012**, *160*, 179–182. [CrossRef]

208. Rosner, B.M.; Werber, D.; Höhle, M.; Stark, K. Clinical Aspects and Self-Reported Symptoms of Sequelae of Yersinia Enterocolitica Infections in a Population-Based Study, Germany 2009–2010. *BMC Infect. Dis.* **2013**, *13*, 236. [CrossRef]

209. Fonnes, S.; Rasmussen, T.; Brunchmann, A.; Holzknecht, B.J.; Rosenberg, J. Mesenteric Lymphadenitis and Terminal Ileitis Is Associated With *Yersinia* Infection: A Meta-Analysis. *J. Surg. Res.* **2022**, *270*, 12–21. [CrossRef]

210. CDC. Yersinia Enterocolitica (Yersiniosis). Available online: www.cdc.gov/yersinia/index.html (accessed on 24 October 2016).

211. Platt-Samoraj, A. Toxigenic Properties of Yersinia Enterocolitica Biotype 1A. *Toxins* **2022**, *14*, 118. [CrossRef] [PubMed]

212. Fois, F.; Piras, F.; Torpdahl, M.; Mazza, R.; Ladu, D.; Consolati, S.G.; Spanu, C.; Scarano, C.; De Santis, E.P.L. Prevalence, Bioserotyping and Antibiotic Resistance of Pathogenic *Yersinia Enterocolitica* Detected in Pigs at Slaughter in Sardinia. *Int. J. Food Microbiol.* **2018**, *283*, 1–6. [CrossRef] [PubMed]

213. Argüello, H.; Estellé, J.; Leonard, F.C.; Crispie, F.; Cotter, P.D.; O'Sullivan, O.; Lynch, H.; Walia, K.; Duffy, G.; Lawlor, P.G.; et al. Influence of the Intestinal Microbiota on Colonization Resistance to *Salmonella* and the Shedding Pattern of Naturally Exposed Pigs. *mSystems* **2019**, *4*, e00021-19. [CrossRef] [PubMed]

214. Carella, E.; Romano, A.; Domenis, L.; Robetto, S.; Spedicato, R.; Guidetti, C.; Pitti, M.; Orusa, R. Characterisation of *Yersinia Enterocolitica* Strains Isolated from Wildlife in the Northwestern Italian Alps. *J. Vet. Res.* **2022**, *66*, 141–149. [CrossRef]

215. Selmi, R.; Tayh, G.; Srairi, S.; Mamlouk, A.; Ben Chehida, F.; Lahmar, S.; Bouslama, M.; Daaloul-Jedidi, M.; Messadi, L. Prevalence, Risk Factors and Emergence of Extended-Spectrum β-Lactamase Producing-, Carbapenem- and Colistin-Resistant Enterobacterales Isolated from Wild Boar (*Sus scrofa*) in Tunisia. *Microb. Pathog.* **2022**, *163*, 105385. [CrossRef]

216. Modesto, P.; De Ciucis, C.G.; Vencia, W.; Pugliano, M.C.; Mignone, W.; Berio, E.; Masotti, C.; Ercolini, C.; Serracca, L.; Andreoli, T.; et al. Evidence of Antimicrobial Resistance and Presence of Pathogenicity Genes in Yersinia Enterocolitica Isolate from Wild Boars. *Pathogens* **2021**, *10*, 398. [CrossRef]

217. Bonardi, S.; Brémont, S.; Vismarra, A.; Poli, I.; Diegoli, G.; Bolzoni, L.; Corradi, M.; Gilioli, S.; Le Guern, A.S. Is Yersinia Bercovieri Surpassing Yersinia Enterocolitica in Wild Boars (*Sus scrofa*)? *EcoHealth* **2020**, *17*, 388–392. [CrossRef]

218. Takahashi, T.; Kabeya, H.; Sato, S.; Yamazaki, A.; Kamata, Y.; Taira, K.; Asakura, H.; Sugiyama, H.; Takai, S.; Maruyama, S. Prevalence of *Yersinia* among Wild Sika Deer (*Cervus nippon*) and Boars (*Sus scrofa*) in Japan. *J. Wildl. Dis.* **2020**, *56*, 270. [CrossRef] [PubMed]

219. Sannö, A.; Rosendal, T.; Aspán, A.; Backhans, A.; Jacobson, M. Distribution of Enteropathogenic *Yersinia* spp. and *Salmonella* spp. in the Swedish Wild Boar Population, and Assessment of Risk Factors That May Affect Their Prevalence. *Acta Vet. Scand.* **2018**, *60*, 40. [CrossRef] [PubMed]

220. Syczyło, K.; Platt-Samoraj, A.; Bancerz-Kisiel, A.; Szczerba-Turek, A.; Pajdak-Czaus, J.; Łabuć, S.; Procajło, Z.; Socha, P.; Chuzhebayeva, G.; Szweda, W. The Prevalence of Yersinia Enterocolitica in Game Animals in Poland. *PLoS ONE* **2018**, *13*, e0195136. [CrossRef] [PubMed]

221. Lorencova, A.; Babak, V.; Lamka, J. Serological Prevalence of Enteropathogenic *Yersinia* Spp. in Pigs and Wild Boars from Different Production Systems in the Moravian Region, Czech Republic. *Foodborne Pathog. Dis.* **2016**, *13*, 275–279. [CrossRef] [PubMed]

222. Bancerz-Kisiel, A.; Socha, P.; Szweda, W. Detection and Characterisation of Yersinia Enterocolitica Strains in Cold-Stored Carcasses of Large Game Animals in Poland. *Vet. J.* **2016**, *208*, 102–103. [CrossRef]

223. Arrausi-Subiza, M.; Gerrikagoitia, X.; Alvarez, V.; Ibabe, J.C.; Barral, M. Prevalence of *Yersinia Enterocolitica* and *Yersinia Pseudotuberculosis* in Wild Boars in the Basque Country, Northern Spain. *Acta Vet. Scand.* **2016**, *58*, 4. [CrossRef]

224. Von Altrock, A.; Seinige, D.; Kehrenberg, C. *Yersinia enterocolitica* isolates from wild boars hunted in Lower Saxony, Germany. *Appl. Environ. Microbiol.* **2015**, *81*, 4835–4840. [CrossRef]

225. Arrausi-Subiza, M.; Ibabe, J.C.; Atxaerandio, R.; Juste, R.A.; Barral, M. Evaluation of different enrichment methods for pathogenic *Yersinia* species detection by real time PCR. *BMC Vet. Res.* **2014**, *10*, 192. [CrossRef]

226. Commission Implementing Regulation (EU) 2019/627 of 15 March 2019 laying Down Uniform Practical Arrangements for the Performance of Official Controls on Products of Animal Origin Intended for Human Consumption in Accordance with Regulation (EU) 2017/625 of the European Parliament and of the Council and amending Commission Regulation (EC) No 2074/2005 as Regards Official Controls. *Off. J. Eur. Union* **2004**, *L131*, 51.

227. Commission Notice on the Implementation of Food Safety Management Systems Covering Good Hygiene Practices and Procedures Based on the HACCP Principles, Including the Facilitation/Flexibility of the Implementation in Certain Food Businesses 2022/C 355/01—C/2022/5307. *Off. J. Eur. Union* **2002**, *C355*, 1–58.

228. Regulation (EC) No 852/2004 of the European Parliament and of the Council of 29 April 2004 on the Hygiene of Foodstuffs. *Off. J. Eur. Union* **2004**, *L139*, 1.

229. Regulation (EC) No 178/2002 of the European Parliament and of the Council of 28 January 2002 Laying Down the General Principles and Requirements of Food Law, Establishing the European Food Safety Authority and Laying Down Procedures in Matters of Food Safety. *Off. J. Eur. Union* **2002**, *L131*, 1–24.

230. Bandick, N.; Hensel, A. Zoonotic diseases and direct marketing of game meat: Aspects of consumer safety in Germany. In *Game Meat Hygiene in Focus: Microbiology, Epidemiology, Risk Analysis and Quality Assurance*; Paulsen, P., Bauer, A., Vodnansky, M., Winkelmayer, R., Smulders, F.J.M., Eds.; Wageningen Academic Publishers: Wageningen, The Netherlands, 2011; pp. 93–100.

231. Winkelmayer, R.; Stangl, P.-V.; Paulsen, P. Assurance of food safety along the game meat production chain: Inspection of meat from wild game and education of official veterinarians and "trained persons" in Austria. In *Game Meat Hygiene in Focus: Microbiology, Epidemiology, Risk Analysis and Quality Assurance*; Paulsen, P., Bauer, A., Vodnansky, M., Winkelmayer, R., Smulders, F.J.M., Eds.; Wageningen Academic Publishers: Wageningen, The Netherlands, 2011; pp. 245–258.

232. Citterio, C.V.; Bragagna, P.; Novelli, E.; Giaccone, V. Approaches to game hygine in the province Belluno (Italy): From training to meat microbiology. In *Game Meat Hygiene in focus: Microbiology, Epidemiology, Risk Analysis and Quality Assurance*; Paulsen, P., Bauer, A., Vodnansky, M., Winkelmayer, R., Smulders, F.J.M., Eds.; Wageningen Academic Publishers: Wageningen, The Netherlands, 2011; pp. 267–270.

 *foods*

Review

Foodborne Diseases in the Edible Insect Industry in Europe—New Challenges and Old Problems

Remigiusz Gałęcki [1],*, Tadeusz Bakuła [1] and Janusz Gołaszewski [2]

[1] Department of Veterinary Prevention and Feed Hygiene, Faculty of Veterinary Medicine, University of Warmia and Mazury in Olsztyn, 10-719 Olsztyn, Poland
[2] Center for Bioeconomy and Renewable Energies, Department of Genetics, Plant Breeding and Bioresource Engineering, Faculty of Agriculture and Forestry, University of Warmia and Mazury in Olsztyn, 10-719 Olsztyn, Poland
* Correspondence: remigiusz.galecki@uwm.edu.pl; Tel.: +48-89-523-49-08

Abstract: Insects play a key role in European agroecosystems. Insects provide important ecosystem services and make a significant contribution to the food chain, sustainable agriculture, the farm-to-fork (F2F) strategy, and the European Green Deal. Edible insects are regarded as a sustainable alternative to livestock, but their microbiological safety for consumers has not yet been fully clarified. The aim of this article is to describe the role of edible insects in the F2F approach, to discuss the latest veterinary guidelines concerning consumption of insect-based foods, and to analyze the biological, chemical, and physical hazards associated with edible insect farming and processing. Five groups of biological risk factors, ten groups of chemical risk factors, and thirteen groups of physical risks factors have been identified and divided into sub-groups. The presented risk maps can facilitate identification of potential threats, such as foodborne pathogens in various insect species and insect-based foods. Ensuring safety of insect-based foods, including effective control of foodborne diseases, will be a significant milestone on the path to maintaining a sustainable food chain in line with the F2F strategy and EU policies. Edible insects constitute a new category of farmed animals and a novel link in the food chain, but their production poses the same problems and challenges that are encountered in conventional livestock rearing and meat production.

Keywords: foodborne pathogens; entomophagy; biosecurity; microbiological safety; risk analysis; food chain

Citation: Gałęcki, R.; Bakuła, T.; Gołaszewski, J. Foodborne Diseases in the Edible Insect Industry in Europe—New Challenges and Old Problems. *Foods* **2023**, *12*, 770. https://doi.org/10.3390/foods12040770

Academic Editor: Frans J. M. Smulders

Received: 9 December 2022
Revised: 25 January 2023
Accepted: 7 February 2023
Published: 10 February 2023

1. Introduction

Insects (class Insecta) are ubiquitous in the world [1], and they come into direct contact with humans [2,3]. Social attitudes toward insects vary. In some countries, insects are regarded as ectoparasites and pests. However, in some cultures and ethnic groups, insects, as a source of protein and other nutrients, have been a part of the human and livestock diet for many centuries [4]. Many insect species are also used in traditional medicine around the world [5]. Insects are used in production of vaccines and protein preparations [6]. In 2004, extracts from *Lucilia sericata* larvae became the first insect-based treatment for chronic wounds that has been approved for use in the United States [7]. The venom of the Samsum ant (*Pseudomyrmex* sp.) has numerous medicinal properties. This powerful antioxidant has been shown to reduce inflammation, relieve pain, inhibit tumor growth, protect the liver, and aid hepatitis treatment [8,9]. Insects are also farmed animals [10]. Honey bees (*Apis mellifera*) have been exploited for honey for many millennia, whereas domestic silk moths (*Bombyx mori*) and Chinese oak silk moths (*Antheraea pernyi*) have long been reared for silk. Insects are also in human and animal diets.

Entomophagy, namely the practice of eating insects, continues to attract the interest of researchers, ecologists, and consumers as a potential solution to feeding the world's growing population in the coming decades [11,12]. In recent years, insects have emerged as one of the most innovative substrates in human and animal nutrition [13,14]. According to many scientists, edible insects are a major milestone in efforts aiming to diversify protein sources and guarantee global food security [15]. Edible insects are most widely consumed in subtropical and tropical regions, but entomophagy is not highly popular in Western culture [11]. Global insect consumption is difficult to estimate, but, according to the literature, around 2000 insect species are consumed in more than 80 countries [16,17]. The most widely consumed insects belong to the orders Coleoptera (31% of global consumption), Diptera (2%), Hemiptera (10%), Hymenoptera (14%), Isoptera (3%), Lepidoptera (18%), Odonata (3%), and Orthoptera (13%) [18]. Around 1500 species of wild and farmed edible insects are eaten in Africa [19]. Nearly 96 tons of edible insects are consumed in the Democratic Republic of Congo each year, and, in Kinshasa alone, an average family consumes around 300 caterpillars per week [20]. Latin America is the second largest market of edible insects, and entomophagy is most popular in Brazil, Ecuador, Colombia, Mexico, Peru, and Venezuela [21]. The Asian insect market is highly innovative [22]. In Asia, insects are not only popular substrates in food and feed production but are also used in the pharmaceutical industry [22]. Until recently, edible insects had not been regarded as a major food source in Europe. A breakthrough came on 20 December 2017, when a list of novel foods, including insects, was introduced by Commission Implementing Regulation (EU) 2017/2470 [23].

All arguments in favor and against entomophagy should be considered to promote introduction of long-term sustainable solutions on the European food market. Safety of edible insects should also be thoroughly analyzed before these products are authorized for human, companion animal, and livestock consumption. Numerous guidelines have been developed to ensure that edible insects are reared under safe conditions and can be safely used in food and feed production [24–26]. Despite the fact that most species of edible insects are harvested without proper biosecurity from the natural environment [27], farmed insects have to meet additional food safety standards and guidelines, including control of foodborne pathogens [28–30]. For this reason, microbiological safety of edible insects has to be thoroughly researched before they are approved for mass production. The optimal parameters for insect rearing, processing, and storage have already been described in the relevant regulations, but many edible insect species have not been tested for microbiological safety. Edible insects can be a source of biological hazards, including bacteria that cause foodborne diseases, and insect-based foods can become contaminated in all stages of production, delivery, and consumption. Other biological risks associated with insect farming, such as use of organic side-streams and food wastes in insect nutrition, are often disregarded.

The aim of this article was to: (i) discuss the role of edible insects in the farm-to-fork (F2F) strategy, (ii) present current veterinary guidelines relating to safe use of edible insects in food and feed production, and (iii) analyze biological, chemical, and physical risk factors in edible insect farming.

2. Edible Insects in the Farm-to-Fork Strategy

According to the Food and Agriculture Organization (FAO) of the United Nations, daily protein consumption per capita will reach around 54 g in 2030 and 57 g in 2050 [31]. Daily protein consumption per capita increased from 39 g in 1961 to 52 g in 2011. The global protein supply will have to increase by 76% to cater to the growing demand [32]. The rapid increase in protein demand can be attributed not only to global population growth but also to higher daily protein intake. It is estimated that around 30% of the world's land surface is used for cultivation of crops, whereas 7% of land is used for livestock production [11,33]. According to many researchers, livestock production has a significant impact on the environment by contributing to soil degradation, global warming, loss of biodiversity, greenhouse gas emissions, and air and water pollution [34]. These problems accentuate the need for a more sustainable approach to agricultural production. Many countries have committed to become carbon-neutral by 2050 as part of the European Green

Deal and the F2F strategy. According to the United Nations (UN), the global population will reach 8.5 billion in 2030 and 9.7 billion in 2050, which implies that the transition to carbon neutrality will be a highly challenging process [35]. Rapid population growth will increase demand for food, but it will also decrease availability of land for agricultural production [33,36]. Livestock production is one of the most rapidly growing agricultural sectors, and increased demand for animal-based products will also drive demand for feed. However, availability of feedstuffs on the global market could be compromised in the current geopolitical climate. Complete and balanced diets are essential for maintaining animal health and performance. Livestock diets should be characterized by high protein content (*Hermetia illucens* and *Tenebrio molitor* meal contains 40–60% protein), high palatability and digestibility (*H. illucens* and *T. molitor* meal digestibility has been estimated at 91–95%), an optimal amino acid profile (*H. illucens* and *T. molitor* meal contains more threonine, valine, isoleucine, leucine, and lysine than fish meal), and fatty acid profile [37–39]. Feeds should be free of antinutritional factors and pathogens, and they should be thoroughly tested to eliminate health risks for animals and ensure food chain safety [40]. High-quality ingredients should be used in feed production to maximize livestock performance. At present, fish meal and soybean meal are the main sources of protein in animal diets [41,42]. Fish meal is produced mainly from fish species that have high bone and fat content and are not suitable for direct human consumption. Fish meal is an excellent source of protein, minerals, and vitamins; it has a favorable composition of amino acids and fatty acids and is highly digestible [40,42]. However, overfishing, the environmental impact of fisheries, and legal regulations have reduced profits in the fish meal industry and have decreased the supply of fish meal for feed production [43–45]. Genetically modified (GM) soybeans dominate on the global market, and they are one of the leading sources of protein in food and feed production [41,46]. At present, soybean production meets the current demand for protein. In 2014, GM soybeans were cultivated on 82% of land under soybeans and on 50% of land under genetically engineered crops worldwide. According to estimates, 93–95% of soybean meal on the global market comes from GM plants [47–51]. As a result, industrial livestock production, particularly in Europe and North America, is highly dependent on feeds containing GM soybeans [3]. In Europe, soybean yields are low due to the harsh climate, and the EU is the world's second largest importer of feed protein. The EU imported 26 million tons of soybean meal and 15.9 million tons of soybeans in 2019 [47]. Innovative feed ingredients of comparable quality and profitability are being sought as part of the European Green Deal to minimize the EU's dependence on soybean imports. Various alternative protein sources have been considered, including distiller-dried grains with solubles, rapeseed meal, and legume seeds (lupin seeds and fava beans). However, these ingredients must be tested for protein content, nutritional value, and presence of antinutritional factors to ensure high productivity and profitability. To maintain continuity of feed production, feed ingredients characterized by uniform quality and composition should be available on the market.

In recent years, insects have emerged as a viable alternative in food and feed production. According to research, edible insects can replace or supplement other high-protein feed components. The experiences of cultures that practice entomophagy suggest that insect farming has considerable potential for improving food security and that edible insects can be farmed on an industrial scale. Research shows that *H. illucens*, *Musca domestica*, *T. molitor*, and fish proteins have similar amino acid compositions [48,49]. According to the UN, entomophagy could help to reduce world hunger. Insects are a sustainable and environmentally friendly source of protein for animals and humans [35].

Despite the fact that entomophagy is a controversial or even shocking practice for many Western consumers [50,51], insects could substantially contribute to global food security in the future [52]. In the EU, several insect species' protein has been approved for use in fish, poultry and pig feed, and pet food [10]. Several edible insect species, including *Acheta domesticus*, *Locusta migratoria*, and *T. molitor*, have also been approved for human consumption [10]. Insect farming is one of the most rapidly growing agricultural sectors [53]. In terms of volume, the edible insect market is projected to increase from 2000 tons in 2018 to around 200,000 tons in 2020 and 1.2 million tons in 2025 [54]. Insects make up a large part of diets consumed by wild animals [55], and insect protein is an important link in the food chain of many fish and poultry species under organic and natural conditions [56]. Insect-based feeds deliver health benefits and improve livestock welfare [57]. Edible insects are abundant in high-quality protein, and some insect species contain bioactive compounds with proven health benefits, including a beneficial amount of chitin (aids digestion), lauric acid (immunomodulatory properties), and antimicrobial peptides (bactericidal properties) [58]. In Europe, some insect farms cater specifically to the needs of the pet food industry. Numerous scientific and commercial initiatives suggest that popularity of edible insects will continue to rise. Insect larvae can be fed various organic and agricultural by-products, which suggests that insect farming is consistent with the F2F strategy [59]. Use of upcycled organic waste as a substrate for insect farming is a concept of strategic importance because it would help to alleviate the protein shortage in Europe and reduce the volume of agricultural wastes and by-products. Agricultural and food processing wastes and by-products can be effectively upcycled to recover valuable nutrients, and organic waste substrates can be converted into nutritious food products. However, edible insects, as well as other farm animals, are subject to the Feed Ban regulations, which means that use of some by-products in Europe is currently impossible. Insect protein from vertical farms can supplement vegetable protein sources in animal diets and increase availability of farmland for crop production. As a result, edible insect farms can substantially contribute to global food security.

In the EU, processed animal proteins (PAPs) have been approved for use in production of feeds for aquaculture, poultry, pigs, and companion animals [60]. The results of studies and analyses indicate that insect protein is safe for human consumption [61–63]. Insect farming is a new agricultural sector, and it can offer new livelihood opportunities for farmers whose livestock has been affected by avian influenza or African swine fever [58]. Insect farming will also contribute to emergence of a new food processing sector, and innovative marketing and production strategies will be required to eliminate negative attitudes to entomophagy and increase popularity of insect-based foods among consumers. As a result, insect farming will create new jobs, promote innovation and enterprise development, and increase food and feed production. Entomophagy is a relatively new concept for European consumers, which is why effective marketing campaigns are needed to increase awareness that insects can be a promising solution to overcoming global food insecurity [61–66].

The European Green Deal and the F2F strategy deliver synergistic effects by creating a legislative framework that supports waste recycling and reuse and minimizes the environmental impact of generated waste. In line with these guidelines, PAPs from slaughterhouse wastes should be used in animal nutrition to replace imported soybean meal. As a result, the EU has lifted the 2001 ban on use of PAPs in animal nutrition, excluding PAPs derived from the same species [60]. Use of PAPs of porcine origin is authorized in poultry feed and PAPs of poultry origin in pig feed. These changes should not increase risk of transmission of foodborne pathogens in the food chain. Lifting the species-to-species ban and use of insect protein in livestock nutrition can significantly contribute to development of protein sources alternative to soybean meal, thus improving animal performance and minimizing the environmental impact of livestock production [67]. Therefore, edible insects can be introduced to the human diet both indirectly (through livestock feed) and directly (through consumption) [67,68]. In light of the EU's agricultural policy, legislative solutions, and future investments in agriculture, edible insects are regarded as a new link in the food chain. However, insect farms should strictly adhere to biosecurity standards, and insect-based foods should be rigorously tested to ensure that foodborne pathogens are not transmitted to consumers.

3. The Role of Insects in Spread of Pathogenic Microorganisms and Foodborne Pathogens

Edible insects are regarded as a safe dietary alternative in livestock production [67–69]. However, microbiological safety of insect-based foods intended for human consumption is still under debate [29,70]. EFSA outputs on safety evaluation of such products have confirmed safety of edible insect consumption under certain conditions of use [71]. Insect farming can contribute to decreasing prevalence and spread of selected contagious diseases, including foodborne diseases, by eliminating pathogen carriers/reservoirs from the food chain. Due to species specificity and the specific physiology of insects, most entomopathogens do not play a role in epidemiology of zoonoses and do not pose a threat to humans [72]. Arthropods' ability to transmit foodborne pathogens and vector-borne diseases has been widely researched in the context of food production and the One Health approach [73–75]. Edible insects are highly unlikely to act as disease vectors [72,76]. Industrially farmed insects are fed agri-food by-products and plant-based products; therefore, the risk of transmission of zoonotic pathogens is low. Entomopathogens cannot cross the species barrier and cause disease in mammals, which is why edible insects are safe to use in food and feed [72]. It is worth noting that, in some cultures, insects infected with pathogens are regarded as a culinary delicacy or as medicinal products [77,78].

There is no evidence to suggest that edible insects harboring bacterial and viral entomopathogens pose a threat to vertebrates [61,79,80]. However, similar to other foods of animal origin, insect-based foods can raise safety concerns because problems can arise after death of insects and during their processing [81]. Companies that rear and process insects must implement strict sanitary rules to ensure microbiological safety of the end product [10,82,83]. Dedicated processing operations are put into place to eliminate any foodborne pathogens. However, the substrate and end product can become infected during processing. To minimize risk, insect farms should abide by the same biosecurity standards that are applied in the conventional food sector [10,24]. Work surfaces should be disinfected, farm workers should maintain good personal hygiene, farm premises should be regularly cleaned, and safe food preparation and delivery practices should be observed [84]. In farms that have not implemented biosecurity measures, insects and insect-based foods can become contaminated with pathogenic microorganisms transmitted by personnel and pests [53,85]. Therefore, legal regulations, in particular veterinary supervision procedures, should be introduced to guarantee safety of insects as a novel food [86]. Similar to other food products, edible insects are sensitive to deviations from approved production or distribution standards [87,88]. The end product can become contaminated when the required parameters are not observed during acquisition of raw materials, processing (such as drying), transport, storage, and distribution. The associated risks are presented in Table 1. Edible insects as final products should be regularly monitored for presence foodborne pathogens to ensure their safe implementation in the F2F strategy and the European food chain.

Table 1. Possible routes of contamination of edible insects and insect-based foods.

Stages of Contamination	Risks		Treatment	Reference
Substrate	1.	(Crickets) Minimal impact of external microbiota.		[89–91]
	2.	(Crickets) Bacterial endospore counts in crickets fed a standard + farm weed (S + W) diet were significantly lower and thus promising and could reduce risks associated with ready-to-eat insects.		
	3.	Risk of contamination with *Salmonella* spp. and *Campylobacter* spp. increases if materials such as used paper egg cartons are utilized in insect rearing. This risk is higher if cartons had been in contact with poultry feces.		

Table 1. *Cont.*

Stages of Contamination	Risks	Treatment	Reference
Rearing	1. (Crickets) *Aspergillus flavus* strains with low mycotoxigenic potential were identified in reared crickets, which could point to presence of mycotoxins in edible crickets.		[89]
Harvest	1. (Crickets) Starvation is not an effective method for reducing microbial loads in edible crickets.	Gut emptying by starvation prior to killing could reduce the microbial load in the insect gut, but it could also decrease fat and energy content and profitability of production.	[92]
Processing	1. (Crickets) High microbial loads of TAC and Enterobacteriaceae were detected in edible crickets, indicating a high risk of rapid spoilage. 2. (Crickets) Sporulating bacteria are a part of the cricket microbiome 3. Food safety risks associated with viruses are very low. 4. *Vibrio* spp., *Streptococcus* spp., *Staphylococcus* spp., *Clostridium* spp., and *Bacillus* spp. were identified in several studies on the microbiota of processed edible insects sold online.	Thermal treatments, novel processing methods (i.e., high-pressure processing), and additional post-processing treatments (acidification, addition of food preservatives, modified atmosphere packaging, etc.) should be applied to extend crickets' shelf-life.	[89,93,94]
Transport		https://ipiff.org/wp-content/uploads/2019/12/IPIFF-Guide-on-Good-Hygiene-Practices.pdf (accessed on: 13.November.2022)	
Preparation	1. Dried mopane worms, termites, and stink bugs sold at the Thohoyandou market were characterized by low contamination with coliforms, *Escherichia coli*, *Staphylococcus aureus*, *Salmonella* spp., TPC, yeasts, and molds.		[95]
Storage	1. (*T. molitor, Alphitobius diaperinus, Gryllus assimilis, Lo. Migratoria*) microbiological characteristics in different storage periods—safe for human consumption.	Insects intended for long-term storage should be killed in boiling water, dried at 103 °C for 12 h, and hermetically packed.	[96]
Consumption	1. The nutritional value and the microbiological and toxicological profiles of insects are influenced by composition of organic side streams. 2. The microbial risks associated with edible insects can be substantially reduced by observing good hygienic practices in rearing, handling, harvesting, processing, storage, and transport of insects and insect-based products. 3. Several spoilage-causing microbes that can alter food quality, including *Lysinibacillus* sp. and *Bacillus subtilis*, have been detected in edible insects. 4. Yeasts, including *Tetrapisispora* spp., *Candida* spp., *Pichia* spp., and *Debaryomyces* spp., and molds, including *Aspergillus* spp., *Alternaria* spp., *Cladosporium* spp., *Fusarium* spp., *Penicillium* spp., *Phycomycetes* spp., and *Wallemia* spp., are associated with the microbiota found on the body surface or in the gut of edible insects and may be harmful. 5. 38 samples of deep-fried and spiced *Ach. Domesticus, Lo. Migratoria*, and *Omphisa fuscidentalis* tested negative for *Salmonella* spp., *Listeria monocytogenes*, *E. coli*, and *S aureus*, but dried and powdered insects, as well as pollen, contained *Bacillus cereus*, coliforms, *Serratia liquefaciens*, *Listeria ivanovii*, *Mucor* spp., *Aspergillus* spp., *Penicillium* spp., and *Cryptococcus neoformans*.		[18,28,93,97]
R&D	1. (Crickets) Further efforts are needed to identify food-borne pathogens in edible crickets and define possible bacterial quality reference values.		[89]

Consumption of unprocessed insects may represent a significant risk factor. Insects can act as mechanical or biological vectors of pathogens [73], particularly critical priority pathogens in the food processing industry, including *Bacillus* spp., *Clostridium* spp., *E. coli*, *L. monocytogenes*, *Salmonella* spp., and *Staphylococcus* spp. [98–101]. Bacteriological hazards have been most widely investigated, but insects can also act as intermediate hosts or mechanical vectors for parasites in the natural environment [74]. Therefore, effective processing operations should be implemented and sanitary guidelines should be observed to minimize risk of contamination with foodborne pathogens [85].

Allergenicity of edible insects is yet another important safety concern. Similar to other food products, edible insects could pose certain risks to consumers with allergies. To date, 239 arthropod allergens have been identified by the Allergen Nomenclature Sub-committee of the World Health Organization (WHO) [102]. Edible insects may also cause cross-reactivity in people allergic to seafood. The following allergens are most frequently identified in edible insects: fructose-bisphosphate aldolase, phospholipase A, hyaluronidase, arginine kinase, myosin light chain, tropomyosin, α-tubulin, and β-tubulin [103]. A total of 116 allergic reactions to edible insects, mostly grasshoppers, locusts, and lentil weevils, have been identified in 2018 [102]. Insect allergens induce non-specific symptoms, such as anaphylaxis, allergic asthma, hypotension, gastrointestinal symptoms, loss of consciousness, urticaria, erythema, pruritus, and tachycardia. Employees of insect farms and insect processing plants can also develop allergic reactions [104,105]. Allergies also pose a threat to companion animals. Insects can also harbor foreign allergens [103,106], including mites and their metabolites. Direct contact with new proteins or symbiotic organisms can trigger heightened immune response. Presence of gluten in digestive tracts of insects fed grain [107] can pose a threat to people who suffer from celiac disease. Allergizing potential of edible insects should be monitored to eliminate these risks. Potential allergens in insect-based foods should be clearly listed on the product label.

Prions pose a significant biological hazard. Prions are one of the key hazards that have been identified by the European Food Safety Authority (EFSA) in the risk profile of edible insects [61,70]. Insect-specific prion diseases have not been identified because insects lack the gene encoding the prion protein PrP [70,108]. However, insects may act as vectors for prions from contaminated substrates derived from ruminants, which could pose a risk for humans, companion animals, and livestock [61,70].

At present, there is no scientific evidence to suggest that insects pose a viral risk to consumers [61,79,109,110]. Entomoviruses are not pathogenic to humans. Insects are commonly infected with viruses of the family Baculoviridae, which are not dangerous for humans or animals [72,111]. Humans do not harbor insect-specific viruses, and there is negligible risk that new mammalian-specific virus strains will evolve through recombination and reassortment and lead to host switching, as was the case with Swine flu [72]. Edible insects are unlikely to transmit foodborne viruses, such as Hepadnaviridae (hepatitis A and E), Reoviridae (reoviruses), and Caliciviridae (noroviruses) [53]. However, viruses could be transmitted to insects through feed or through contact with farm personnel. Viruses of the family Rhabdoviridae, which cause vesicular stomatitis, have been reported in edible insects [88]. Risk of SARS-CoV-2 transmission by edible insects is very low [72,110,112]. According to Doi et al. [72], risk of infection with SARS-CoV-2 as a foodborne pathogen is negligent in people who consume edible insects [72]. It should be noted that viruses causing foodborne diseases do not replicate in arthropods [79,109], but edible insects could become contaminated during processing and distribution.

Bacteria are presently regarded as the greatest safety hazard in production of edible insects [82]. Due to physiological, environmental, and behavioral differences, every species of edible insects intended for food and feed production harbors different bacteria [100,113]. According to the literature, the microbiome of edible insects poses a negligent risk to consumer safety [114,115]. Several bacteria that can act as opportunist pathogens in humans have been identified in edible insects, but these pathogens are specific to mammals [100]. The risks associated with bacterial symbionts in insects or their potential effects on vertebrates have not been evaluated to date. Insects can act as vectors and carriers of microorganisms that are harmful to humans, particularly when biosecurity and hygiene standards are not observed in insect farms. Insects can carry bacteria that are dangerous to humans, companion animals, and livestock and can act as vectors of foodborne pathogens [116]. Insect microbiota typically include the following bacterial families and genera: Enterobacteriaceae (*Proteus* spp., *Escherichia* spp.), *Pseudomonas* spp., *Staphylococcus* spp., *Streptococcus* spp.,

Bacillus spp., *Micrococcus* spp., *Lactobacillus* spp., and *Acinetobacter* spp. [117]. Some species of the above families and genera are potentially pathogenic to humans, whereas others are commonly encountered in healthy subjects. Unprocessed insects and insect-based foods can harbor *Campylobacter* spp., verotoxic *E. coli*, *Salmonella* spp., and *L. monocytogenes* if microbial inactivation techniques are not applied in production plants. Therefore, insects and insect-based foods should always be screened for these pathogens. Prevalence of some of these pathogens is lower in insects than in other animal protein sources. For example, *Campylobacter* spp. is not replicated in the digestive tract of insects [118–120]. Similar risks can be encountered during insect processing. Several bacterial species identified in edible insects can shorten the shelf-life of the final product. Presence of spore-forming bacteria in the end product poses one of the greatest bacteriological hazards [121]. Common sanitation practices, such as drying, boiling, or deep frying, may not be sufficient to eliminate these pathogens.

Entomopathogenic fungi are yet another group of potentially hazardous organisms. There is no scientific evidence to suggest that entomopathogenic fungi pose a risk to vertebrates. In some cultures, these fungi (such as *Ophiocordyceps sinensis*) have long been used in traditional medicine [77]. Mycosporidia could also pose a health threat to consumers [122], but their toxicity has not been analyzed to date. According to the literature, microsporidia *Trachipleistophora* spp. that probably originated from insects can infect vertebrates [123,124]. Due to specific insect rearing conditions and administered feeds, the end product can become contaminated with mycotoxins [125,126]. High concentrations of mycotoxins, such as deoxynivalenol, can lead to gastrointestinal dysfunction in mammals. Molds can also develop in insect-based products that have been stored and distributed in sub-optimal conditions. However, presence of molds in insect-based products has not been reported in the literature. Risks associated with fungi and mycotoxins in insect-derived foods are often disregarded, and further research is needed to guarantee safety of the end product.

Edible insects can potentially transmit parasitic diseases [74,127]. It appears that entomopathogenic parasites are unable to complete their full life cycle in humans or livestock due to biological specificity of the host. Entomopathogenic parasites cannot be transmitted between vertebrates either. However, there is evidence to suggest that some insect-specific parasites can cause digestive problems (such as horsehair worms, *Gordius* spp.) [128] or allergies (*Lophomonas blattarum*) [129]. Insects can also act as intermediate hosts for food-borne pathogens, including tapeworms (*Hymenolepis* spp.), lancet liver flukes (*Dicrocoelium dendriticum*), and nematodes (*Spirocerca lupi*) [127,130–132]. Insects can also act as mechanical vectors for different developmental stages of vertebrate parasites in different stages of their life cycle [74,133]. Insects can transmit parasites that colonize body surfaces (hairs, chitin exoskeletons) and digestive tracts. Mechanical transmission of parasites is a serious concern during insect farming. Research has demonstrated that insects can transmit protozoa [127,134,135]. It should also be noted that insects themselves can act as etiological factors of disease. Beetles of the family Tenebrionidae, such as yellow mealworms (*T. molitor*) and lesser mealworms (*A. diaperinus*), can cause canthariasis [136–138]. Insect farms can also be colonized by mites [139]. Table 2 provides a summary of potential biological hazards.

Table 2. Biological hazards associated with different species of edible insects.

Type of Hazard	Infectious Agent	Sensitive Species	Predisposing Factors	References
Prion vectors	Proteinaceous infectious particles	All species fed contaminated substrates of animal origin	• inadequate rearing practices • failure to observe legal regulations • contaminated feed and litter handling operations • absence of biosecurity measures • sanitation requirements are not observed by farm personnel	[70,108]

Table 2. *Cont.*

Type of Hazard	Infectious Agent	Sensitive Species	Predisposing Factors	References
Viruses	Caliciviridae Hepadnaviridae Vesicular stomatitis virus (VSV)	Migratory locust (*Lo. migratoria*), black soldier fly (*H. illucens*) Insects harvested from the natural environment	<ul><li>insects are reared with other animals</li><li>absence of biosecurity measures</li><li>sanitation requirements are not observed</li></ul>	[53,88]
Bacteria	*Aeromonas hydrophila, B. cereus, Clostridium difficile, Clostridium perfringens, Clostridium septicum, Clostridium sporogenes, E. coli, Enterococcus faecium, Enterococcus faecalis, Listeria* spp., *Salmonella* spp., *S. aureus.*	Migratory locust (*Lo. migratoria*) Yellow mealworm (*T. molitor*) Lesser mealworm (*A. diaperinus*) House cricket (*Ach. domesticus*) Domestic silk moth (*B. mori*) Insects harvested from the natural environment	<ul><li>handling operations</li><li>deviations from production standards</li><li>rearing conditions</li><li>inadequate rearing practices</li><li>contamination of feed and litter</li></ul>	[98–101,117]
Fungi and mycotoxins	*Aspergillus fumigatus, Aspergillus sclerotiorum, Cladosporium* spp. *Penicillium* spp., *Fusarium* spp., *Phycomycetes* spp. Microsporidia	Migratory locust (*Lo. migratoria*) Black soldier fly (*H. illucens*) Yellow mealworm (*T. molitor*)	<ul><li>high humidity</li><li>contamination of feed and litter</li><li>high water activity in the end product</li><li>inadequate storage conditions</li></ul>	[28,83,125,140]
Parasites	Protozoa (*Balantidium* spp., *Cryptosporidium* spp., *Entamoeba* spp.) Trematoda (*Dicrocoelium* spp., Lecithodendriidae) Cestoda (*Hymenolepis* spp., *Raillietina* spp.) Nematoda (*Gordius* spp., *Spirocerca* spp.)	Yellow mealworm (*T. molitor*) Lesser mealworm (*A. diaperinus*) House cricket (*Ach domesticus*) Insects harvested from the natural environment	<ul><li>insects as vectors of parasitic infections</li><li>insects as intermediate hosts</li><li>insects harvested in the natural environment</li><li>absence of biosecurity measures</li><li>dirty and contaminated feed (such as unwashed vegetables)</li><li>presence of pests</li><li>farm/processing personnel do not observe sanitation requirements</li><li>insects are reared with other animals</li></ul>	[4,127–135]
Mites	*Acarus* spp., *Dermatophagoides* spp., *Goheria* spp. *Tyrophagus* spp.	Mealworm (*T molitor*) Lesser mealworm (*A. diaperinus*) Black soldier fly (*H. illucens*) House cricket (*Ach. domesticus*)	<ul><li>feed substrates are contaminated with mites in different stages of the life cycle</li><li>biosecurity measures are not observed</li><li>sanitation requirements are not observed</li><li>high humidity</li><li>residual feed is not removed from farm premises</li></ul>	[139]

4. Risk Map

Microbiological safety of edible insects and insect-based foods is currently being extensively researched. The risk that insect-specific pathogens will adapt to new hosts cannot be predicted or ruled out [72]. Foodborne pathogens carried by insects can also pose a threat to immunocompromised and hyper-immunosensitive hosts [141,142]. Therefore, insect-specific microorganisms may turn out to be opportunistic pathogens. The gut microbiome of edible insects is species-specific [100,113], and its impact on mammals suffering from health problems cannot be reliably predicted. New pathogens could also be identified after insect-derived foods have been introduced to the food chain. Employees of livestock farms and food processing farms can be a potential source of infection [84,85,88]. In turn, insect farms require less personnel and can be automated in the future, which will significantly limit risk of pathogen transmission. This threat can be substantially minimized by implementing and rigorously observing biosecurity measures, ISO standards, and hazard analysis and critical control points (HACCP). Sanitary and veterinary supervision measures should also be developed and implemented [84,85,88] in insect farms to reduce risk of pathogen transmission to the level observed in conventional livestock farms. Insects and insect-based foods do not present greater risks than conventional foods because pathogenic microorganisms in both groups of products have low epizootic potential. Only local risks can be anticipated, for example in specific batches of contaminated products [72]. Unlike COVID-19, African swine fever, or avian influenza infections, which are associated with foodborne pathogens and livestock production, there is no evidence to suggest that edible insect farming could contribute to novel pandemic outbreaks [29,72,97,142].

To guarantee the safety of insect-derived foods, edible insects should be reared, processed, and stored according to the same sanitation requirements that are applied in conventional food and feed sectors [82,96]. In view of the biological composition of insect-based products, their microbiological safety, toxicity, palatability, and content of inorganic compounds should be analyzed. The overreaching goal of all processing operations should be to obtain end products that are safe for humans and animals, which can be achieved through implementation of HACCP systems [10,84,85,88]. Quality control measures in insect farms and the hazards associated with edible insects and insect-based foods should be addressed in the HACCP plan [10,84,85,88].

A hypothetical risk map listing the main threats for humans, animals, and insects associated with edible insect farming has been developed based on a review of the literature, the existing knowledge, veterinary regulations, and the authors' experience. The key risks were represented by groups of biological, chemical, and physical factors. Five groups of biological risk factors, ten groups of chemical risk factors, and thirteen groups of physical risk factors were identified and divided into sub-groups. The risks maps for each category of factors are presented in Supplementary Figures S1–S4. These maps can facilitate identification of the key risks in insect production and choice of the most effective methods for minimizing or eliminating these threats. It should be noted that risk maps present the widest possible range of threats associated with edible insect species classified as novel foods. Individual risk maps should also be developed for each species of edible insects. A combined map of the risks described is included in Supplementary File S1.

Viruses are the first group of biological factors in the risk map. Entomopathogenic viruses belonging to families Baculoviridae (*Granulovirus, Deltabaculovirus*), Iridoviridae (*Iridovirus*), and Reoviridae (*Cypovirus, Dinovernavirus*) pose a potential threat in insect farming. Edible insects can also play a certain role in transmission of pathogenic viruses. Therefore, the following viral families that play an important role in human and veterinary medicine were included in the risk map: Circoviridae, Coronaviridae, Flaviviridae, Herpesviridae, Orthomyxoviridae, Paramyxoviridae, Parvoviridae, and Picornaviridae.

Bacteria constitute the most important group of biological hazards. In the risk map, bacteria were divided into the following groups: symbionts, entomopathogens, and aerobic and anaerobic bacteria that are pathogenic to vertebrates. There is no evidence to suggest that bacterial symbionts in insects pose a health risk for mammals. Bacteria that are pathogenic to insects were divided into two groups: insect-specific (Morganellaceae—*Photorhabdus* spp. and *Xenorhabdus* spp.; Bacillaceae—*Paenibacillus* spp. and *Brevibacillus* spp.) and non-insect-specific pathogens (Pseudomonadaceae—*Pseudomonas* spp.; Streptomycetaceae—*Streptomyces* spp.; Enterobacteriaceae—*Yersinia* spp.). Bacteria that are specific to vertebrates were divided into two groups: anaerobic (Clostridiaceae—*Clostridium* spp.; Campylobacteraceae—*Campylobacter* spp.; Fusobacteriaceae—*Fusobacterium* spp.) and aerobic pathogens (Micrococcaceae—*Micrococcus* spp.; Listeriaceae—*L. monocytogenes*; Enterobacteriaceae—*Enterobacter* spp., *Yersinia* spp., and *Salmonella* spp.). Severity of these biological risks is determined mainly by the type of feed administered to insects.

Fungi are yet another biological risk factor. Inadequate rearing and feed storage conditions can contribute to fungal infections and contamination of the end product. This group of risks includes microsporidia that are pathogenic to mammals (*Encephalitozoon* spp.,

Trachipleistophora spp., and *Tubulinosema* spp.), as well as entomopathogenic microsporidia (*Nosema* spp. and *Paranosema* spp.). Entomopathogenic fungi are also present in the natural environment (Entomophthorales—*Conidiobolus* spp. and *Entomophthora* spp.) and in biological control agents (*Beauveria* spp. and *Metarhizium* spp.). Fungi and mycotoxins that occur commonly in the food chain could also pose biological risks.

Insects play an important role in the life cycle of many pathogens, which is why parasites could also pose a considerable risk in edible insect farming. The following groups of parasites were listed in the risk map: Protozoa, Trematoda, Cestoda, Nematoda, Acanthocephala, and mites. Insects, classified as farmed animals, can become infected with the following entomopathogenic parasites: Sporozoa (*Leidyana* spp., *Gregarine* spp., *Septatorina* spp.), Ciliates (*Tetrahymena* spp., *Nyctotherus* spp.), Nematoda (*Thelastoma* spp., *Steinernema* spp., *Heterorhabditis* spp.), as well as mites, including predatory mites (*Cheyletus eruditus*), opportunistic mites (*Dermatophagoides* spp., *Glicephagus* spp.), and storage mites (*Acarus* spp., *Rhizoglyphus* spp.). All parasites for which insects can act as intermediate or definitive hosts, carriers, mechanical vectors, and reservoirs are potentially harmful to vertebrates.

The last group of biological risk factors are pests that can pose a biosecurity threat in production of edible insects, such as wild animals and other insects. They can carry and transmit pathogens to the farm and lead to contamination of the end product. Attention should also be paid to parasitoids that can spread in the farm environment.

Severity and variation in biological threats are affected by numerous factors, including infectivity and virulence of pathogens, health status of hosts, presence of comorbidities/coinfections, immune status, physiological susceptibility, and history of previous infections. The map of biological risks is largely hypothetical because comprehensive epidemiological and epizootic data are required to fully characterize associated health risks. Edible insects are novel foods, and such detailed information is impossible to acquire at this point. The map of biological risks is presented in Supplementary Figure S1.

Chemical hazards can also be encountered in the edible insect industry [79]. Chemical contaminants can be introduced to insects and end products with the initial stock and feed, as well as by farm employees during biosecurity operations. Some substances can be introduced deliberately (for example, during veterinary treatment) or accidentally (with plant and animal substrates). Chemical substances can be accumulated by insects, which poses a significant threat to consumers. Insect metabolites, such as benzoquinones produced by beetles of the family Tenebrionidae, are also a potential risk factor, which is why stage of insect life cycle is an important consideration. Agri-food by-products can be effectively upcycled in insect rearing, and risk of chemical contamination is also influenced by observance of food safety regulations in crop production and conventional livestock farming. In automated insect rearing and processing systems, technical fluids, such as lubricants, are also a potential source of chemical contamination. Medicinal products used in both human and veterinary medicine, in particular antibiotics, hormones, antiparasitic agents, steroids, sedatives, and analgesics, also pose a risk of chemical contamination in insect farms. Disinfection, disinfestation, and deratting (DDD) measures involving disinfectants, rodenticides, and insecticides carry health risks for vertebrates and reared insects. Insect-based foods can be also contaminated with pesticides, such as acaricides, fungicides, and herbicides. Inadequate storage can lead to spoilage of final products and accumulation of toxic compounds, such as putrescine and indoles. Various chemical substances can be added to insect-derived foods to prevent spoilage, but high concentrations of food preservatives can have toxic effects. Some insects, such as *H. illucens*, can accumulate heavy metals, including arsenic, cadmium, and lead [143], which pose a health threat to consumers. The map of chemical risk factors also includes substances that exert adverse effects on consumers and insects, such as microplastics, bisphenol, and dioxins. Stimulant ingredients in foods, including theophylline, theobromine, nicotine, and caffeine, act as natural insecticides and constitute yet another group of hazardous chemical substances. Effective biosecurity measures should be introduced in insect rearing and processing to prevent or minimize accumulation of external toxins, drugs, and antinutritional factors. The map of chemical risks is presented in Supplementary Figure S2.

The identified physical risks in insect farming are presented in Supplementary Figure S3. Most of these factors do not pose a threat to consumers. Physical hazards, such as fluctuations in humidity and temperature, noise, suboptimal lighting, electromagnetic radiation, and vibration, compromise well-being of insects and influence productivity and profits. However, these factors are not dangerous for consumers. Particulate matter emissions, including PM10 and PM2.5, during insect rearing and processing can cause allergies in consumers and farm employees. Chitin can lead to gastrointestinal tract irritation in humans and animals. Moreover, dust containing chitin particles may pose a risk of airway irritation in farms where insects are fed agri-food by-products: there is a risk that the end product will contain hard particles (plant and animal residues, soil, or gravel). Therefore, insects and feed should be thoroughly cleaned before being converted into food products. Similar to conventional foods, insect-based products can also be contaminated with microplastics and micrometals that pose a threat to humans and animals.

5. Safety of Insects Reared for Food and Feed

Processed insects and insect-based foods have to adhere to food safety standards set forth by legal regulations [10,25,61,86,144]. Legal provisions play a key role in production and marketing of insect-based foods. Insect species that can be included in formulation of food and feed products have been listed in Commission Regulation (EU) 2017/893 of 24 May 2017 Amending Annexes I and IV to Regulation (EC) No. 999/2001 of the European Parliament and of the Council and Annexes X, XIV, and XV to Commission Regulation (EU) No. 142/2011 as Regards the Provisions on Processed Animal Protein [145]. In the EU, the risk profile and potential hazards associated with farmed insects used as food and feed were described in the EFSA opinion of 8 October 2015 [61–63]. The identified insect species do not transmit pathogens specific to plants and vertebrates. These insects are not invasive or pathogenic to mammals; they do not exert a negative impact on crops, and they are not protected [10]. It should also be noted that both whole edible insects and insect-based foods can be introduced to the EU market [23]. Similar to conventional livestock, edible insects have to be monitored to ensure the safety of the produced food and feed. Most legal regulations concern hygiene standards in food and feed production (Commission Regulation (EC) No. 1069/2009; Commission Regulation (EU) No. 142/2011; Commission Implementing Regulation (EU) 2017/2469) [23,146,147]. Observance of safety standards in the food processing sector is monitored by the respective authorities, including veterinary and sanitary inspectorates.

To eliminate foodborne pathogens, insect-based foods placed on the market must meet food and feed hygiene standards, good breeding practices, good hygiene practices, and good production practices [24,69,113,148,149]. Edible insects are classified as farmed animals; therefore, they can only be fed plant- and animal-based materials that have been approved for livestock nutrition [10]. Materials acquired outside the food chain may not be used as feed in insect farms. To minimize transmission of foodborne pathogens, commercial insect feeds must be purchased from certified manufacturers who adhere to HACCP requirements and European feed laws [150,151]. Insect producers must keep documents that confirm feed delivery dates and list feed manufacturers and initial feed parameters [150,151]. Products that do not meet safety standards (moldy feeds, feeds withdrawn from the market) cannot be fed to insects or processed into feed [150,152]. Each batch of insects placed on the food and feed market must conform to microbiological safety standards and maximum residue limits (MRL) stipulated in the relevant regulations [88]. Insects should be regularly monitored for presence of undesirable chemical substances, such as heavy metals, pesticides, and mycotoxins [85]. Each product batch should be clearly marked in every stage of the production process to ensure food traceability [10,152].

Applicability of animal-based substrates as insect feeds and the relevant processing requirements are set forth by Commission Notice (2018/C 133/02)—Guidelines for the Feed

Use of Food No Longer Intended for Human Consumption [153]. Most farmed insects will be processed into feed, which is why contamination with toxic compounds is a valid concern. Due to their specific physiology, edible insects can accumulate heavy metals [125,143]. Heavy metals have been identified in black soldier fly prepupae [143,154]. Microbiological contamination poses yet another threat. Insect farming conditions and food sources can promote development of specific pathogens. Some insects are capable of reducing microbial counts in digested food, but risk of microbiological contamination cannot be ruled out. For example, the black soldier fly can reduce microbial counts in alkaline poultry excreta but not in acidic pig manure [155–157]. In addition, not all pathogens (such as parasitic eggs) are effectively eliminated by the black soldier fly [158]. Effective treatments, such as high-temperature processing, are needed to minimize counts of pathogenic microorganisms in farmed insects [82,83,159]. Such treatments eliminate foodborne pathogens and microorganisms that cause food spoilage. Both insect feeds and end products must be free of pesticides, antibiotics, detergents, and other contaminants [150].

The described hazards and associated adverse health effects should be considered in qualitative and quantitative risk analyses [53]. Various microbiological hazards are associated with presence of pathogenic bacteria, such as *Campylobacter* spp., *S. aureus*, *B. cereus*, *E. coli*, *C. perfringens*, and *Enterococcus* spp. They should be monitored in production of insect-based foods, even if the relevant limits have not yet been introduced in the insect sector [53]. Samples of the end product have to meet guideline microbiological limits for *Salmonella* spp. (not detected in 25 g) and Enterobacteriaceae (up to 300 CFU in 1 g). Food products listed in Annex IV to Commission Regulation (EU) No. 142/2011 must be free of *C. perfringens* (1 g samples) [147]. According to Commission Regulation (EC) No. 2073/2005, *L. monocytogenes* counts in ready-to-eat foods may not exceed 100 CFU per 1 g of the product [160]. The above regulation also introduced microbiological limits for raw materials, minced meat and meat products (absence of *Salmonella* spp. in 10 g of minced meat and meat preparations that are made from species other than poultry and are intended to be eaten cooked; *E. coli*—up to 500 CFU/g in minced meat at the end of the manufacturing process), and cooked crustaceans and molluscan shellfish (absence of *Salmonella* spp. in 25 g of the product). If required, insects should also be periodically inspected for other pathogens and chemical substances, including pesticides, heavy metals, dioxins, and mycotoxins (Directive 2002/32/EC of the European Parliament and of the Council of 7 May 2002 on undesirable substances in animal feed). Insects should also be analyzed for presence of physical contaminants, such as plastic and metal components, and foreign particles [150], as well as physical parameters, such as water activity [61,94].

6. Conclusions

Insects have been long reared and consumed in many regions of the world (such as Southeast Asia, Mexico, and Africa), but little is known about their ability to transmit foodborne pathogens and their safety for consumers. In Europe, consumer attitudes toward entomophagy are gradually changing, and both whole insects and insect-based foods are gaining popularity. Therefore, food safety standards and veterinary inspection procedures targeting insect farms will have to be implemented to guarantee safety of European consumers. Despite the fact that the edible insect industry is a completely new sector of European agriculture, it will contribute to achievement of the main goals of the F2F strategy, which lies at the heart of the European Green Deal. The European Green Deal proposes a sustainable and inclusive growth strategy to improve consumers' health, care for the environment, and leave no one behind. The European Food Safety Authority has initiated a debate on strategic importance of edible insects for the European food and feed market (evaluation of insect-based foods, authorization of insect protein in poultry and pig feed). The F2F strategy and the resulting reforms in the EU's agricultural policy are major milestones on the path to a more sustainable food supply chain. Edible insects have been classified as farmed animals, but little remains known about their biology, physiology, biochemical pathways, specific pathogens, treatment options, and humanitarian rearing methods. Insect welfare and ethical criteria in insect farming are difficult to define. Veterinarians, in their daily practice, deal with insects, including pests or ectoparasites and bees, but edible insects are new and enigmatic because they have to compare them to conventional livestock. Action should be taken to educate veterinarians about farmed insects. Even though insects constitute a novel link in the food chain, scientists, veterinary practitioners, and breeders must face and solve the same old problems that are encountered in conventional livestock farming and food production.

Supplementary Materials: The following supporting information can be downloaded at: https://www.mdpi.com/article/10.3390/foods12040770/s1, Figure S1: Map of biological risk factors that pose a potential threat to vertebrates and edible insects. Figure S2: Map of chemical risk factors that pose a potential threat to vertebrates and edible insects. Figure S3: Map of physical risk factors that pose a potential threat to vertebrates and edible insects. Figure S4: Risk map for insect breeding, insect processing, and insect-based food and feed.

Author Contributions: Conceptualization, R.G.; methodology, R.G.; software, R.G.; validation, R.G.; formal analysis, R.G., T.B. and J.G.; investigation, R.G.; resources, R.G. and J.G.; data curation, R.G.; writing—original draft preparation, R.G.; writing—review and editing, R.G. and J.G.; visualization, R.G.; supervision, R.G. and J.G.; project administration, R.G. and J.G.; funding acquisition, R.G. and J.G. All authors have read and agreed to the published version of the manuscript.

Funding: This study was supported by a research project entitled "Sustainable up-cycling of agricultural residues: modular cascading waste conversion system", research grant agreement No. FACCE SURPLUS/III/UpWaste/02/2020. The research is supported by the National (Polish) Center for Research and Development (NCBiR) (Project FACCE SURPLUS/III/UpWaste/02/2020). This study was supported by a research project Lider XII entitled "Development of an insect protein food for companion animals with diet-dependent enteropathies", financed by the National Center for Research and Development (NCBiR) (LIDER/5/0029/L-12/20/NCBR/2021).

Institutional Review Board Statement: Not applicable.

Informed Consent Statement: Not applicable.

Data Availability Statement: Data are contained within the article.

Conflicts of Interest: The authors declare no conflict of interest. The funders had no role in the design of the study, in the collection, analyses, or interpretation of data, in the writing of the manuscript, or in the decision to publish the results.

References

1. Scudder, G.G. The importance of insects. *Insect Biodivers. Sci. Soc.* **2017**, *1*, 9–43.
2. Busvine, J.R. Insects and hygiene. In *Insects and Hygiene*; Springer: Berlin/Heidelberg, Germany, 1980; pp. 1–20.
3. Meyer-Rochow, V.B. Can Insects Help to Ease Problem of World Food Shortage. *Search* **1975**, *6*, 261–262.
4. Govorushko, S. *Human–Insect Interactions*; CRC Press: Boca Raton, FL, USA, 2018.
5. Dossey, A.T. Insects and their chemical weaponry: New potential for drug discovery. *Nat. Prod. Rep.* **2010**, *27*, 1737–1757. [CrossRef] [PubMed]
6. Qian, L.; Deng, P.; Chen, F.; Cao, Y.; Sun, H.; Liao, H. The exploration and utilization of functional substances in edible insects: A review. *Food Prod. Process. Nutr.* **2022**, *4*, 11. [CrossRef]
7. Heuer, H.; Heuer, L.; Saalfrank, V. Living Medication: Overview and Classification into Pharmaceutical Law. In *Nature Helps….*; Springer: Berlin/Heidelberg, Germany, 2011; pp. 349–367.
8. Altman, R.D.; Schultz, D.R.; Collins-Yudiskas, B.; Aldrich, J.; Arnold, P.I.; Arnold, P.I.; Brown, H.E. The effects of a partially purified fraction of an ant venom in rheumatoid arthritis. *Arthritis Rheum. Off. J. Am. Coll. Rheumatol.* **1984**, *27*, 277–284. [CrossRef]
9. Agarwal, S.; Sharma, G.; Verma, K.; Latha, N.; Mathur, V. Pharmacological potential of ants and their symbionts—A review. *Entomol. Exp. Et Appl.* **2022**, *170*, 1032–1048. [CrossRef]
10. Żuk-Gołaszewska, K.; Gałęcki, R.; Obremski, K.; Smetana, S.; Figiel, S.; Gołaszewski, J. Edible Insect Farming in the Context of the EU Regulations and Marketing—An Overview. *Insects* **2022**, *13*, 446. [CrossRef] [PubMed]
11. Van Huis, A. Potential of insects as food and feed in assuring food security. *Annu. Rev. Entomol.* **2013**, *58*, 563–583. [CrossRef]
12. Verbeke, W. Profiling consumers who are ready to adopt insects as a meat substitute in a Western society. *Food Qual. Prefer.* **2015**, *39*, 147–155. [CrossRef]

13. Rumpold, B.A.; Schlüter, O.K. Nutritional composition and safety aspects of edible insects. *Mol. Nutr. Food Res.* **2013**, *57*, 802–823. [CrossRef]

14. Patel, S.; Suleria, H.A.R.; Rauf, A. Edible insects as innovative foods: Nutritional and functional assessments. *Trends Food Sci. Technol.* **2019**, *86*, 352–359. [CrossRef]

15. Kinyuru, J.; Ndung'u, N. Promoting edible insects in Kenya: Historical, present and future perspectives towards establishment of a sustainable value chain. *J. Insects Food Feed.* **2020**, *6*, 51–58. [CrossRef]

16. Van Huis, A.; Halloran, A.; Van Itterbeeck, J.; Klunder, H.; Vantomme, P. How many people on our planet eat insects: 2 billion? *J. Insects Food Feed.* **2022**, *8*, 1–4. [CrossRef]

17. Jongema, Y. *List of Edible Insects of the World*; Laboratory of Entomology, Wageningen University: Wageningen, The Netherlands, 2017.

18. FAO. *Looking at Edible Insects from a Food Safety Perspective. Challenges and Opportunities for the Sector*; FAO: Rome, Italy, 2021.

19. Ebenebe, C.I.; Ibitoye, O.S.; Amobi, I.M.; Okpoko, V.O. African edible insect consumption market. In *African Edible Insects as Alternative Source of Food, Oil, Protein and Bioactive Components*; Springer: Berlin/Heidelberg, Germany, 2020; pp. 19–51.

20. Kitsa, K. Contribution des insectes comestibles à l'amélioration de la ration alimentaire au Kasaï-Occidental. *Zaïre-Afr. Économie Cult. Vie Soc.* **1989**, *29*, 511–519.

21. Costa-Neto, E.M. Anthropo-entomophagy in Latin America: An overview of the importance of edible insects to local communities. *J. Insects Food Feed.* **2015**, *1*, 17–23. [CrossRef]

22. Han, R.; Shin, J.T.; Kim, J.; Choi, Y.S.; Kim, Y.W. An overview of the South Korean edible insect food industry: Challenges and future pricing/promotion strategies. *Entomol. Res.* **2017**, *47*, 141–151. [CrossRef]

23. Commission Implementing Regulation (EU). *2017/2469 of 20 December 2017 Laying Down Administrative and Scientific Requirements for Applications Referred to in Article 10 of Regulation (EU) 2015/2283 of the European Parliament and of the Council on Novel Foods*; European Union: Luxemburg, 2017.

24. IPIFF Guide on Good Hygiene Practices for European Union Producers of Insects as Food and Feed. Available online: https://ipiff.org/wp-content/uploads/2019/12/IPIFF-Guide-on-Good-Hygiene-Practices.pdf (accessed on 6 January 2022).

25. Lähteenmäki-Uutela, A.; Marimuthu, S.; Meijer, N. Regulations on insects as food and feed: A global comparison. *J. Insects Food Feed.* **2021**, *7*, 849–856. [CrossRef]

26. Niassy, S.; Omuse, E.; Roos, N.; Halloran, A.; Eilenberg, J.; Egonyu, J.; Tanga, C.; Meutchieye, F.; Mwangi, R.; Subramanian, S. Safety, regulatory and environmental issues related to breeding and international trade of edible insects in Africa. *Rev. Sci. Tech.* **2022**, *41*, 117–131. [CrossRef]

27. Feng, Y.; Chen, X.M.; Zhao, M.; He, Z.; Sun, L.; Wang, C.Y.; Ding, W.F. Edible insects in China: Utilization and prospects. *Insect Sci.* **2018**, *25*, 184–198. [CrossRef]

28. Grabowski, N.T.; Klein, G. Microbiology of processed edible insect products–results of a preliminary survey. *Int. J. Food Microbiol.* **2017**, *243*, 103–107. [CrossRef]

29. Belluco, S.; Losasso, C.; Maggioletti, M.; Alonzi, C.; Ricci, A.; Paoletti, M.G. Edible insects: A food security solution or a food safety concern? *Anim. Front.* **2015**, *5*, 25–30.

30. Belluco, S.; Losasso, C.; Maggioletti, M.; Alonzi, C.C.; Paoletti, M.G.; Ricci, A. Edible insects in a food safety and nutritional perspective: A critical review. *Compr. Rev. Food Sci. Food Saf.* **2013**, *12*, 296–313. [CrossRef]

31. WHO Technical Report Series. *Protein and Amino Acid Requirements in Human Nutrition: Report of a Joint FAO/WHO/UNU Expert Consultation*; World Health Organization. 2007. Available online: http://whqlibdoc.who.int/trs/who_trs_935_eng.pdf (accessed on 3 May 2022).

32. WRAP. Food Futures. From Business as Usual to Business Unusual. In Proceedings of the World Economic Forum, Switzerland. 2021. Available online: http://www.wrap.org.uk/content/food-futures (accessed on 3 May 2022).

33. Premalatha, M.; Abbasi, T.; Abbasi, T.; Abbasi, S. Energy-efficient food production to reduce global warming and ecodegradation: The use of edible insects. *Renew. Sustain. Energy Rev.* **2011**, *15*, 4357–4360. [CrossRef]

34. Shafiullah, M.; Khalid, U.; Shahbaz, M. Does meat consumption exacerbate greenhouse gas emissions? Evidence from US data. *Environ. Sci. Pollut. Res.* **2021**, *28*, 11415–11429. [CrossRef]

35. United Nations. *World Population Prospects: The 2012 Revision*; Population division of the department of economic and social affairs of the United Nations Secretariat: New York, NY, USA, 2013; p. 18.

36. Mitsuhashi, J. The future use of insects as human food. *For. Insects Food: Hum. Bite Back* **2010**, *115*, 122.

37. Hong, J.; Han, T.; Kim, Y.Y. Mealworm (*Tenebrio molitor* Larvae) as an alternative protein source for monogastric animal: A review. *Animals* **2020**, *10*, 2068. [CrossRef]

38. Ojha, S.; Bekhit, A.E.-D.; Grune, T.; Schlüter, O.K. Bioavailability of nutrients from edible insects. *Curr. Opin. Food Sci.* **2021**, *41*, 240–248. [CrossRef]

39. Seyedalmoosavi, M.M.; Mielenz, M.; Veldkamp, T.; Daş, G.; Metges, C.C. Growth efficiency, intestinal biology, and nutrient utilization and requirements of black soldier fly (*Hermetia illucens*) larvae compared to monogastric livestock species: A review. *J. Anim. Sci. Biotechnol.* **2022**, *13*, 31. [CrossRef] [PubMed]

40. Barrows, F.T.; Bellis, D.; Krogdahl, Å.; Silverstein, J.T.; Herman, E.M.; Sealey, W.M.; Rust, M.B.; Gatlin III, D.M. Report of the plant products in aquafeed strategic planning workshop: An integrated, interdisciplinary research roadmap for increasing utilization of plant feedstuffs in diets for carnivorous fish. *Rev. Fish. Sci.* **2008**, *16*, 449–455. [CrossRef]

41. Banaszkiewicz, T. Nutritional value of soybean meal. *Soybean Nutr.* **2011**, *12*, 1–20.

42. Tacon, A.G.; Metian, M. Global overview on the use of fish meal and fish oil in industrially compounded aquafeeds: Trends and future prospects. *Aquaculture* **2008**, *285*, 146–158. [CrossRef]
43. Shepherd, C.; Jackson, A. Global fishmeal and fish-oil supply: Inputs, outputs and marketsa. *J. Fish Biol.* **2013**, *83*, 1046–1066. [CrossRef]
44. Jannathulla, R.; Rajaram, V.; Kalanjiam, R.; Ambasankar, K.; Muralidhar, M.; Dayal, J.S. Fishmeal availability in the scenarios of climate change: Inevitability of fishmeal replacement in aquafeeds and approaches for the utilization of plant protein sources. *Aquac. Res.* **2019**, *50*, 3493–3506. [CrossRef]
45. Olsen, R.L.; Hasan, M.R. A limited supply of fishmeal: Impact on future increases in global aquaculture production. *Trends Food Sci. Technol.* **2012**, *27*, 120–128. [CrossRef]
46. Dzwonkowski, W.; Rola, K.; Hanczakowska, E.; Niwińska, B.; Świątkiewicz, S. *Raport o Sytuacji na Światowym Rynku Roślin GMO i Możliwościach Substytucji Genetycznie Zmodyfikowanej Soi Krajowymi Roślinami Białkowymi w Aspekcie Bilansu Paszowego*; Instytut Ekonomiki Rolnictwa i Gospodarki Żywnościowej-Państwowy Instytut: Warszawa, Poland, 2015.
47. The Agricultural Market Information System. Supply and Demand. Available online: https://app.amis-outlook.org/#/market-database/supply-and-demand-overview (accessed on 17 November 2019).
48. Veldkamp, T.; Bosch, G. Insects: A protein-rich feed ingredient in pig and poultry diets. *Anim. Front.* **2015**, *5*, 45–50.
49. Katya, K.; Borsra, M.; Ganesan, D.; Kuppusamy, G.; Herriman, M.; Salter, A.; Ali, S.A. Efficacy of insect larval meal to replace fish meal in juvenile barramundi, Lates calcarifer reared in freshwater. *Int. Aquat. Res.* **2017**, *9*, 303–312. [CrossRef]
50. Clarkson, C.; Mirosa, M.; Birch, J. Consumer acceptance of insects and ideal product attributes. *Br. Food J.* **2018**, *120*, 2898–2911. [CrossRef]
51. Costa-Neto, E.M.; Dunkel, F. Insects as food: History, culture, and modern use around the world. In *Insects as Sustainable Food Ingredients*; Elsevier: Amsterdam, The Netherlands, 2016; pp. 29–60.
52. Van Huis, A. Edible insects contributing to food security? *Agric. Food Secur.* **2015**, *4*, 1–9. [CrossRef]
53. Bakuła, T.; Gałęcki, R. *Strategia Wykorzystania Alternatywnych Źródeł Białka w Żywieniu Zwierząt Oraz Możliwości Rozwoju Jego Produkcji na Terytorium Rzeczpospolitej Polski*; ERZET: Olsztyn, Poland, 2021; ISBN 978-83-961897-1-4.
54. International Platform of Insects for Food and Feed (IPIFF). *The European Insect Sector Today: Challenges, Opportunities and Regulatory Landscape. IPIFF Vision Paper on the Future of the Insect Sector towards 2030*; IPIFF: Brussels, Belgium, 2018.
55. Schoenly, K.; Beaver, R.; Heumier, T. On the trophic relations of insects: A food-web approach. *Am. Nat.* **1991**, *137*, 597–638. [CrossRef]
56. Józefiak, D.; Józefiak, A.; Kierończyk, B.; Rawski, M.; Świątkiewicz, S.; Długosz, J.; Engberg, R.M. Insects–a natural nutrient source for poultry–a review. *Ann. Anim. Sci.* **2016**, *16*, 297–313. [CrossRef]
57. Van Huis, A. New sources of animal proteins: Edible insects. In *New Aspects of Meat Quality*; Elsevier: Amsterdam, The Netherlands, 2017; pp. 443–461.
58. Gałęcki, R.; Zielonka, Ł.; Zasępa, M.; Gołębiowska, J.; Bakuła, T. Potential Utilization of Edible Insects as an Alternative Source of Protein in Animal Diets in Poland. *Front. Sustain. Food Syst.* **2021**, *5*, 675796. [CrossRef]
59. Oonincx, D.G.; Van Broekhoven, S.; Van Huis, A.; van Loon, J.J. Feed conversion, survival and development, and composition of four insect species on diets composed of food by-products. *PLoS ONE* **2015**, *10*, e0144601. [CrossRef] [PubMed]
60. Commission Regulation (EU). Commission Regulation (EU) 2021/1372 of 17 August 2021 amending Annex IV to Regulation (EC) No 999/2001 of the European Parliament and of the Council as Regards the Prohibition to Feed Non-Ruminant Farmed Animals, Other than Fur Animals, with Protein Derived from Animals. *Off. J. Eur. Union* **2021**, *295*, 1–17.
61. EFSA Scientific Committee. Scientific opinion on a risk profile related to production and consumption of insects as food and feed. *EFSA J.* **2015**, *13*, 4257. [CrossRef]
62. EFSA Panel on Nutrition, Novel Foods and Food Allergens (NDA). Safety of dried yellow mealworm (*Tenebrio molitor* larva) as a novel food pursuant to Regulation (EU) 2015/2283. *EFSA J.* **2021**, *19*, e06343.
63. EFSA. *Novel Food' Report: Opinion on the Risk Profile for House Cricket (Acheta domesticus) by the Swedish University of Agricultural Sciences (EFSA Funded Report, Adopted on 6 July 2018)*; EFSA: Parma, Italy, 2018.
64. Zielińska, E.; Zieliński, D.; Karaś, M.; Jakubczyk, A. Exploration of consumer acceptance of insects as food in Poland. *J. Insects Food Feed.* **2020**, *6*, 383–392. [CrossRef]
65. Kulma, M.; Tůmová, V.; Fialová, A.; Kouřimská, L. Insect consumption in the Czech Republic: What the eye does not see, the heart does not grieve over. *J. Insects Food Feed.* **2020**, *6*, 525–535. [CrossRef]
66. Mancini, S.; Sogari, G.; Espinosa Diaz, S.; Menozzi, D.; Paci, G.; Moruzzo, R. Exploring the Future of Edible Insects in Europe. *Foods* **2022**, *11*, 455. [CrossRef]
67. Gahukar, R. Edible insects farming: Efficiency and impact on family livelihood, food security, and environment compared with livestock and crops. In *Insects as Sustainable Food Ingredients*; Elsevier: Amsterdam, The Netherlands, 2016; pp. 85–111.
68. Dobermann, D.; Swift, J.; Field, L. Opportunities and hurdles of edible insects for food and feed. *Nutr. Bull.* **2017**, *42*, 293–308. [CrossRef]
69. Imathiu, S. Benefits and food safety concerns associated with consumption of edible insects. *NFS J.* **2020**, *18*, 1–11. [CrossRef]
70. Belluco, S.; Mantovani, A.; Ricci, A. Edible insects in a food safety perspective. In *Edible Insects in Sustainable Food Systems*; Springer: Berlin/Heidelberg, Germany, 2018; pp. 109–126.

71. Precup, G.; Ververis, E.; Azzollini, D.; Rivero-Pino, F.; Zakidou, P.; Germini, A. *The Safety Assessment of Insects and Products Thereof as Novel Foods in the European Union. Novel Foods and Edible Insects in the European Union*; EU: Brussels, Belgium, 2022; p. 123.

72. Doi, H.; Gałęcki, R.; Mulia, R.N. The merits of entomophagy in the post COVID-19 world. *Trends Food Sci. Technol.* **2021**, *110*, 849–854. [CrossRef] [PubMed]

73. Zurek, L.; Gorham, J.R. *Insects as Vectors of Foodborne Pathogens. Wiley Handbook of Science and Technology for Homeland Security*; Wiley: Hoboken, NJ, USA, 2008; p. 1.

74. Graczyk, T.K.; Knight, R.; Tamang, L. Mechanical transmission of human protozoan parasites by insects. *Clin. Microbiol. Rev.* **2005**, *18*, 128–132. [CrossRef]

75. Lounibos, L.P. Invasions by insect vectors of human disease. *Annu. Rev. Entomol.* **2002**, *47*, 233. [CrossRef] [PubMed]

76. Yates-Doerr, E. The world in a box? Food security, edible insects, and "One World, One Health" collaboration. *Soc. Sci. Med.* **2015**, *129*, 106–112. [CrossRef]

77. Cao, L.; Ye, Y.; Han, R. Fruiting body production of the medicinal Chinese caterpillar mushroom, Ophiocordyceps sinensis (Ascomycetes), in artificial medium. *Int. J. Med. Mushrooms* **2015**, *17*, 1107–1112. [CrossRef]

78. Evans, J.; Müller, A.; Jensen, A.; Dahle, B.; Flore, R.; Eilenberg, J.; Frøst, M. A descriptive sensory analysis of honeybee drone brood from Denmark and Norway. *J. Insects Food Feed.* **2016**, *2*, 277–283. [CrossRef]

79. van der Fels-Klerx, H.; Camenzuli, L.; Belluco, S.; Meijer, N.; Ricci, A. Food safety issues related to uses of insects for feeds and foods. *Compr. Rev. Food Sci. Food Saf.* **2018**, *17*, 1172–1183. [CrossRef]

80. Dall'Asta, C. *Why 'New' Foods Are Safe and How They Can Be Assessed. Novel Foods and Edible Insects in the European Union*; EU: Brussels, Belgium, 2022; p. 81.

81. Lacey, L.A.; Siegel, J.P. Safety and ecotoxicology of entomopathogenic bacteria. In *Entomopathogenic Bacteria: From Laboratory to Field Application*; Springer: Berlin/Heidelberg, Germany, 2000; pp. 253–273.

82. Klunder, H.; Wolkers-Rooijackers, J.; Korpela, J.M.; Nout, M.R. Microbiological aspects of processing and storage of edible insects. *Food Control* **2012**, *26*, 628–631. [CrossRef]

83. Mutungi, C.; Irungu, F.; Nduko, J.; Mutua, F.; Affognon, H.; Nakimbugwe, D.; Ekesi, S.; Fiaboe, K. Postharvest processes of edible insects in Africa: A review of processing methods, and the implications for nutrition, safety and new products development. *Crit. Rev. Food Sci. Nutr.* **2019**, *59*, 276–298. [CrossRef]

84. Rizou, M.; Galanakis, I.M.; Aldawoud, T.M.; Galanakis, C.M. Safety of foods, food supply chain and environment within the COVID-19 pandemic. *Trends Food Sci. Technol.* **2020**, *102*, 293–299. [CrossRef] [PubMed]

85. Kwiatek, K.; Bakuła, T.; Sieradzki, Z.; Osiński, Z.; Kowalczyk, E. *Wytyczne Dobrej Praktyki Higienicznej w Produkcji Owadów dla Celów Paszowych i Spożywczych*; Zakład Higieny Pasz Państwowy Instytut Weterynaryjny—Państwowy Instytut Badawczy: Puławy, Poland, 2021; p. 169. Available online: https://www.gov.pl/web/rolnictwo/wytyczne-dobrej-praktyki-higienicznej-w-produkcji-owadow-dla-celow-paszowych-i-spozywczych (accessed on 12 February 2022).

86. Belluco, S.; Halloran, A.; Ricci, A. New protein sources and food legislation: The case of edible insects and EU law. *Food Secur.* **2017**, *9*, 803–814. [CrossRef]

87. Vandeweyer, D.; Crauwels, S.; Lievens, B.; Van Campenhout, L. Metagenetic analysis of the bacterial communities of edible insects from diverse production cycles at industrial rearing companies. *Int. J. Food Microbiol.* **2017**, *261*, 11–18. [CrossRef]

88. Fraqueza, M.J.R.; Patarata, L. Constraints of HACCP application on edible insect for food and feed. *Future foods* **2017**, 89–113. Available online: https://books.google.co.jp/books?hl=pl&lr=&id=SW2PDwAAQBAJ&oi=fnd&pg=PA89&dq=Constraints+of+HACCP+application+on+edible+insect+for+food+and+feed.&ots=zxgVfrH4j8&sig=G6_KoH3yMYyO8Z62vfC2Jhws2nc&redir_esc=y#v=onepage&q=Constraints%20of%20HACCP%20application%20on%20edible%20insect%20for%20food%20and%20feed.&f=false (accessed on 16 February 2022).

89. Fernandez-Cassi, X.; Söderqvist, K.; Bakeeva, A.; Vaga, M.; Dicksved, J.; Vagsholm, I.; Jansson, A.; Boqvist, S. Microbial communities and food safety aspects of crickets (*Acheta domesticus*) reared under controlled conditions. *J. Insects Food Feed.* **2020**, *6*, 429–440. [CrossRef]

90. Ng'ang'a, J.; Imathiu, S.; Fombong, F.; Borremans, A.; Van Campenhout, L.; Broeck, J.V.; Kinyuru, J. Can farm weeds improve the growth and microbiological quality of crickets (*Gryllus bimaculatus*)? *J. Insects Food Feed.* **2020**, *6*, 199–209. [CrossRef]

91. Walia, K.; Kapoor, A.; Farber, J. Qualitative risk assessment of cricket powder to be used to treat undernutrition in infants and children in Cambodia. *Food Control* **2018**, *92*, 169–182. [CrossRef]

92. Inacio, A.C.; Vågsholm, I.; Jansson, A.; Vaga, M.; Boqvist, S.; Fraqueza, M. Impact of starvation on fat content and microbial load in edible crickets (*Acheta domesticus*). *J. Insects Food Feed.* **2021**, *7*, 1143–1147. [CrossRef]

93. Vandeweyer, D.; Lievens, B.; Van Campenhout, L. Identification of bacterial endospores and targeted detection of foodborne viruses in industrially reared insects for food. *Nat. Food* **2020**, *1*, 511–516. [CrossRef]

94. Fasolato, L.; Cardazzo, B.; Carraro, L.; Fontana, F.; Novelli, E.; Balzan, S. Edible processed insects from e-commerce: Food safety with a focus on the *Bacillus cereus* group. *Food Microbiol.* **2018**, *76*, 296–303. [CrossRef]

95. Ramashia, S.; Tangulani, T.; Mashau, M.; Nethathe, B. Microbiological quality of different dried insects sold at Thohoyandou open market, South Africa. *Food Res.* **2020**, *4*, 2247–2255. [CrossRef] [PubMed]

96. Adámek, M.; Mlček, J.; Adámková, A.; Suchánková, J.; Janalíková, M.; Borkovcová, M.; Bednářová, M. Effect of different storage conditions on the microbiological characteristics of insect. *Potravin. Slovak J. Food Sci.* **2018**, *12*, 248–253. [CrossRef] [PubMed]

97. Kooh, P.; Ververis, E.; Tesson, V.; Boué, G.; Federighi, M. Entomophagy and public health: A review of microbiological hazards. *Health* **2019**, *11*, 1272–1290. [CrossRef]

98. Borch, E.; Arinder, P. Bacteriological safety issues in red meat and ready-to-eat meat products, as well as control measures. *Meat Sci.* **2002**, *62*, 381–390. [CrossRef] [PubMed]

99. Balta, I.; Linton, M.; Pinkerton, L.; Kelly, C.; Stef, L.; Pet, I.; Stef, D.; Criste, A.; Gundogdu, O.; Corcionivoschi, N. The effect of natural antimicrobials against Campylobacter spp. and its similarities to *Salmonella* spp, *Listeria* spp., *Escherichia coli*, *Vibrio* spp., *Clostridium* spp. and *Staphylococcus* spp. *Food Control* **2021**, *121*, 107745. [CrossRef]

100. Garofalo, C.; Osimani, A.; Milanović, V.; Taccari, M.; Cardinali, F.; Aquilanti, L.; Riolo, P.; Ruschioni, S.; Isidoro, N.; Clementi, F. The microbiota of marketed processed edible insects as revealed by high-throughput sequencing. *Food Microbiol.* **2017**, *62*, 15–22. [CrossRef]

101. Chiang, Y.-C.; Tsen, H.-Y.; Chen, H.-Y.; Chang, Y.-H.; Lin, C.-K.; Chen, C.-Y.; Pai, W.-Y. Multiplex PCR and a chromogenic DNA macroarray for the detection of *Listeria monocytogens, Staphylococcus aureus, Streptococcus agalactiae, Enterobacter sakazakii, Escherichia coli* O157: H7, *Vibrio parahaemolyticus, Salmonella* spp. and *Pseudomonas fluorescens* in milk and meat samples. *J. Microbiol. Methods* **2012**, *88*, 110–116.

102. de Gier, S.; Verhoeckx, K. Insect (food) allergy and allergens. *Mol. Immunol.* **2018**, *100*, 82–106. [CrossRef]

103. Ribeiro, J.C.; Cunha, L.M.; Sousa-Pinto, B.; Fonseca, J. Allergic risks of consuming edible insects: A systematic review. *Mol. Nutr. Food Res.* **2018**, *62*, 1700030. [CrossRef]

104. Pener, M.P. Allergy to crickets: A review. *J. Orthoptera Res.* **2016**, *25*, 91–95. [CrossRef]

105. Pomés, A.; Mueller, G.A.; Randall, T.A.; Chapman, M.D.; Arruda, L.K. New insights into cockroach allergens. *Curr. Allergy Asthma Rep.* **2017**, *17*, 1–16. [CrossRef]

106. Pali-Schöll, I.; Meinlschmidt, P.; Larenas-Linnemann, D.; Purschke, B.; Hofstetter, G.; Rodríguez-Monroy, F.A.; Einhorn, L.; Mothes-Luksch, N.; Jensen-Jarolim, E.; Jäger, H. Edible insects: Cross-recognition of IgE from crustacean-and house dust mite allergic patients, and reduction of allergenicity by food processing. *World Allergy Organ. J.* **2019**, *12*, 100006. [CrossRef]

107. Mancini, S.; Fratini, F.; Tuccinardi, T.; Degl'Innocenti, C.; Paci, G. Tenebrio molitor reared on different substrates: Is it gluten free? *Food Control* **2020**, *110*, 107014. [CrossRef]

108. Van Raamsdonk, L.; Van der Fels-Klerx, H.; De Jong, J. New feed ingredients: The insect opportunity. *Food Addit. Contam. Part A* **2017**, *34*, 1384–1397. [CrossRef] [PubMed]

109. Eilenberg, J.; Vlak, J.; Nielsen-LeRoux, C.; Cappellozza, S.; Jensen, A. Diseases in insects produced for food and feed. *J. Insects Food Feed.* **2015**, *1*, 87–102. [CrossRef]

110. Dicke, M.; Eilenberg, J.; Salles, J.F.; Jensen, A.; Lecocq, A.; Pijlman, G.; Van Loon, J.; Van Oers, M. Edible insects unlikely to contribute to transmission of coronavirus SARS-CoV-2. *J. Insects Food Feed.* **2020**, *6*, 333–339. [CrossRef]

111. Paul, A.; Hasan, A.; Rodes, L.; Sangaralingam, M.; Prakash, S. Bioengineered baculoviruses as new class of therapeutics using micro and nanotechnologies: Principles, prospects and challenges. *Adv. Drug Deliv. Rev.* **2014**, *71*, 115–130. [CrossRef]

112. Roundy, C.M.; Hamer, S.A.; Zecca, I.B.; Davila, E.B.; Auckland, L.D.; Tang, W.; Gavranovic, H.; Swiger, S.L.; Tomberlin, J.K.; Fischer, R.S.B.; et al. No Evidence of SARS-CoV-2 Among Flies or Cockroaches in Households Where COVID-19 Positive Cases Resided. *J. Med. Entomol.* **2022**, *59*, 1479–1483. [CrossRef]

113. Osimani, A.; Milanović, V.; Garofalo, C.; Cardinali, F.; Roncolini, A.; Sabbatini, R.; De Filippis, F.; Ercolini, D.; Gabucci, C.; Petruzzelli, A. Revealing the microbiota of marketed edible insects through PCR-DGGE, metagenomic sequencing and real-time PCR. *Int. J. Food Microbiol.* **2018**, *276*, 54–62. [CrossRef]

114. Boemare, N.; Laumond, C.; Mauleon, H. The entomopathogenic nematode-bacterium complex: Biology, life cycle and vertebrate safety. *Biocontrol Sci. Technol.* **1996**, *6*, 333–346. [CrossRef]

115. Kikuchi, Y. Endosymbiotic bacteria in insects: Their diversity and culturability. *Microbes Environ.* **2009**, *24*, 195–204. [CrossRef]

116. Grabowski, N.T.; Klein, G. Bacteria encountered in raw insect, spider, scorpion, and centipede taxa including edible species, and their significance from the food hygiene point of view. *Trends Food Sci. Technol.* **2017**, *63*, 80–90. [CrossRef]

117. Amadi, E.; Ogbalu, O.; Barimalaa, I.; Pius, M. Microbiology and nutritional composition of an edible larva (*Bunaea alcinoe* Stoll) of the Niger Delta. *J. Food Saf.* **2005**, *25*, 193–197. [CrossRef]

118. Templeton, J.M.; De Jong, A.J.; Blackall, P.; Miflin, J.K. Survival of *Campylobacter* spp. in darkling beetles (*Alphitobius diaperinus*) and their larvae in Australia. *Appl. Environ. Microbiol.* **2006**, *72*, 7909–7911. [CrossRef] [PubMed]

119. Strother, K.O.; Steelman, C.D.; Gbur, E. Reservoir competence of lesser mealworm (Coleoptera: Tenebrionidae) for *Campylobacter jejuni* (Campylobacterales: Campylobacteraceae). *J. Med. Entomol.* **2005**, *42*, 42–47. [CrossRef] [PubMed]

120. Wales, A.; Carrique-Mas, J.; Rankin, M.; Bell, B.; Thind, B.; Davies, R. Review of the carriage of zoonotic bacteria by arthropods, with special reference to *Salmonella* in mites, flies and litter beetles. *Zoonoses Public Health* **2010**, *57*, 299–314.

121. Osimani, A.; Aquilanti, L. Spore-forming bacteria in insect-based foods. *Curr. Opin. Food Sci.* **2021**, *37*, 112–117. [CrossRef]

122. Stentiford, G.; Becnel, J.; Weiss, L.; Keeling, P.; Didier, E.; Bjornson, S.; Freeman, M.; Brown, M.; Roesel, K.; Sokolova, Y. Microsporidia–emergent pathogens in the global food chain. *Trends Parasitol.* **2016**, *32*, 336–348. [CrossRef]

123. Vávra, J.; Horák, A.; Modrý, D.; Lukeš, J.; Koudela, B. *Trachipleistophora extenrec n.* sp. a new microsporidian (Fungi: Microsporidia) infecting mammals. *J. Eukaryot. Microbiol.* **2006**, *53*, 464–476. [CrossRef]

124. Vávra, J.; Kamler, M.; Modrý, D.; Koudela, B. Opportunistic nature of the mammalian microsporidia: Experimental transmission of *Trachipleistophora extenrec* (Fungi: Microsporidia) between mammalian and insect hosts. *Parasitol. Res.* **2011**, *108*, 1565–1573. [CrossRef]

125. Schrögel, P.; Wätjen, W. Insects for food and feed-safety aspects related to mycotoxins and metals. *Foods* **2019**, *8*, 288. [CrossRef]

126. Evans, N.M.; Shao, S. Mycotoxin Metabolism by Edible Insects. *Toxins* **2022**, *14*, 217. [CrossRef] [PubMed]

127. Gałecki, R.; Sokół, R. A parasitological evaluation of edible insects and their role in the transmission of parasitic diseases to humans and animals. *PLoS ONE* **2019**, *14*, e0219303. [CrossRef] [PubMed]

128. Yamada, M.; Tegoshi, T.; Abe, N.; Urabe, M. Two Human Cases Infected by the Horsehair Worm, *Parachordodes* sp.(Nematomorpha: Chordodidae), in Japan. *Korean J. Parasitol.* **2012**, *50*, 263. [CrossRef] [PubMed]

129. Jorjani, O.; Bahlkeh, A.; Koohsar, F.; Talebi, B.; Bagheri, A. Chronic respiratory allergy caused by *Lophomonas blattarum*: A case report. *Med. Lab. J.* **2018**, *12*, 44–46. [CrossRef]

130. Pappas, P.W.; Barley, A.J. Beetle-to-beetle transmission and dispersal of *Hymenolepis diminuta* (Cestoda) eggs via the feces of *Tenebrio molitor*. *J. Parasitol.* **1999**, *85*, 384–385. [CrossRef] [PubMed]

131. Manga-González, M.Y.; González-Lanza, C.; Cabanas, E.; Campo, R. Contributions to and review of dicrocoeliosis, with special reference to the intermediate hosts of *Dicrocoelium dendriticum*. *Parasitology* **2001**, *123*, 91–114. [CrossRef]

132. Bailey, W.; Cabrera, D.; Diamond, D. Beetles of the family *Scarabaeidae* as intermediate hosts for *Spirocerca lupi*. *J. Parasitol.* **1963**, *49*, 485–488. [CrossRef] [PubMed]

133. Thyssen, P.J.; Moretti, T.d.C.; Ueta, M.T.; Ribeiro, O.B. The role of insects (Blattodea, Diptera, and Hymenoptera) as possible mechanical vectors of helminths in the domiciliary and peridomiciliary environment. *Cad. De Saúde Pública* **2004**, *20*, 1096–1102. [CrossRef]

134. Graczyk, T.K.; Cranfield, M.R.; Fayer, R.; Bixler, H. House flies (*Musca domestica*) as transport hosts of *Cryptosporidium parvum*. *Am. J. Trop. Med. Hyg.* **1999**, *61*, 500–504. [CrossRef]

135. Goodwin, M.A.; Waltman, W.D. Transmission of Eimeria, viruses, and bacteria to chicks: Darkling beetles (*Alphitobius diaperinus*) as vectors of pathogens. *J. Appl. Poult. Res.* **1996**, *5*, 51–55. [CrossRef]

136. Aelami, M.H.; Khoei, A.; Ghorbani, H.; Seilanian-Toosi, F.; Poustchi, E.; Hosseini-Farash, B.R.; Moghaddas, E. Urinary canthariasis due to *Tenebrio molitor* Larva in a ten-year-old boy. *J. Arthropod-Borne Dis.* **2019**, *13*, 416. [CrossRef]

137. Gałecki, R.; Michalski, M.M.; Wierzchosławski, K.; Bakuła, T. Gastric canthariasis caused by invasion of mealworm beetle larvae in weaned pigs in large-scale farming. *BMC Vet. Res.* **2020**, *16*, 1–7. [CrossRef] [PubMed]

138. Nguyen, N.; Yang, B.-K.; Lee, J.-S.; Yoon, J.U.; Hong, K.-J. Infestation status of the darkling beetle (*Alphitobius diaperinus*) in Broiler chicken houses of Korea. *Korean J. Appl. Entomol.* **2019**, *58*, 189–196.

139. Maciel-Vergara, G.; Jensen, A.; Lecocq, A.; Eilenberg, J. Diseases in edible insect rearing systems. *J. Insects Food Feed.* **2021**, *7*, 621–638. [CrossRef]

140. Gahukar, R.T. Edible insects collected from forests for family livelihood and wellness of rural communities: A review. *Glob. Food Secur.* **2020**, *25*, 100348. [CrossRef]

141. Henke, M.O.; De Hoog, G.S.; Gross, U.; Zimmermann, G.; Kraemer, D.; Weig, M. Human deep tissue infection with an entomopathogenic Beauveria species. *J. Clin. Microbiol.* **2002**, *40*, 2698–2702. [CrossRef]

142. Lange, K.W.; Nakamura, Y. Edible insects as future food: Chances and challenges. *J. Future Foods* **2021**, *1*, 38–46. [CrossRef]

143. Diener, S.; Zurbrügg, C.; Tockner, K. Bioaccumulation of heavy metals in the black soldier fly, *Hermetia illucens* and effects on its life cycle. *J. Insects Food Feed.* **2015**, *1*, 261–270. [CrossRef]

144. Skotnicka, M.; Karwowska, K.; Kłobukowski, F.; Borkowska, A.; Pieszko, M. Possibilities of the development of edible insect-based foods in Europe. *Foods* **2021**, *10*, 766. [CrossRef] [PubMed]

145. Commission Regulation (EU). *2017/893 of 24 May 2017 Amending Annexes I and IV to Regulation (EC) No. 999/2001 of the European Parliament and of the Council and Annexes X, XIV and XV to Commission Regulation (EU) No. 142/2011 as Regards the Provisions on Processed Animal Protein*; European Union: Luxemburg, 2017.

146. Commission Regulation (EC). *No. 1069/2009, Regulation (EC) No. 1069/2009 of the European Parliament and of the Council of 21 October 2009 Laying Down Health Rules as Regards Animal By-Products and Derived Products not Intended for Human Consumption and Repealing Regulation (EC) No. 1774/2002*; European Union: Luxemburg, 2002.

147. Commission Regulation (EU). *No. 142/2011, of 25 February 2011 Implementing Regulation (EC) No. 1069/2009 of the European Parliament and of the Council Laying Down Health Rules as Regards Animal By-Products and Derived Products Not Intended for Human Consumption and Implementing Council Directive 97/78/EC as Regards Certain Samples and Items Exempt from Veterinary Checks at the Border under that Directive*; European Union: Luxemburg, 2011.

148. Garofalo, C.; Milanović, V.; Cardinali, F.; Aquilanti, L.; Clementi, F.; Osimani, A. Current knowledge on the microbiota of edible insects intended for human consumption: Astate-of-the-art review. *Food Res. Int.* **2019**, *125*, 108527. [CrossRef] [PubMed]

149. Laurenza, E.C.; Carreño, I. Edible insects and insect-based products in the EU: Safety assessments, legal loopholes and business opportunities. *Eur. J. Risk Regul.* **2015**, *6*, 288–292. [CrossRef]

150. Commission Regulation (EC). *Directive 2002/32/EC of the European Parliament and of the Council of 7 May 2002 on Undesirable Substances in Animal Feed—Council statement; 2002 178/2002 of the European Parliament and of the Council of 28 January 2002*; European Union: Luxemburg, 2002.

151. Commission Regulation (EC). *Regulation (EC) No. 178/2002 of the European Parliament and of the Council of 28 January 2002 Laying Down the General Principles and Requirements of Food Law, Establishing the European Food Safety Authority and Laying Down Procedures in Matters of Food Safety*; European Union: Luxemburg, 2002.
152. Poma, G.; Cuykx, M.; Da Silva, K.M.; Iturrospe, E.; van Nuijs, A.L.; van Huis, A.; Covaci, A. Edible insects in the metabolomics era. *First steps towards the implementation of entometabolomics in food systems. Trends Food Sci. Technol.* **2021**, *119*, 371–377.
153. Commission Regulation (EC). *Commission Notice—Guidelines for the Feed Use of Food No Longer Intended for Human Consumption C/2018/2035*; European Union: Luxemburg, 2018.
154. Elechi, M.C.; Kemabonta, K.A.; Ogbogu, S.S.; Orabueze, I.C.; Adetoro, F.A.; Adebayo, H.A.; Obe, T.M. Heavy metal bioaccumulation in prepupae of black soldier fly *Hermetia Illucens* (Diptera: Stratiomyidae) cultured with organic wastes and chicken feed. *Int. J. Trop. Insect Sci.* **2021**, *41*, 2125–2131. [CrossRef]
155. Cai, M.; Ma, S.; Hu, R.; Tomberlin, J.K.; Thomashow, L.S.; Zheng, L.; Li, W.; Yu, Z.; Zhang, J. Rapidly mitigating antibiotic resistant risks in chicken manure by *Hermetia illucens* bioconversion with intestinal microflora. *Environ. Microbiol.* **2018**, *20*, 4051–4062. [CrossRef] [PubMed]
156. Awasthi, M.K.; Liu, T.; Awasthi, S.K.; Duan, Y.; Pandey, A.; Zhang, Z. Manure pretreatments with black soldier fly *Hermetia illucens* L. (*Diptera: Stratiomyidae*): A study to reduce pathogen content. *Sci. Total Environ.* **2020**, *737*, 139842. [CrossRef] [PubMed]
157. Zhang, X.; Zhang, J.; Jiang, L.; Yu, X.; Zhu, H.; Zhang, J.; Feng, Z.; Zhang, X.; Chen, G.; Zhang, Z. Black soldier fly (*Hermetia illucens*) larvae significantly change the microbial community in chicken manure. *Curr. Microbiol.* **2021**, *78*, 303–315. [CrossRef] [PubMed]
158. Muller, A.; Wiedmer, S.; Kurth, M. Risk evaluation of passive transmission of animal parasites by feeding of black soldier fly (*Hermetia illucens*) larvae and prepupae. *J. Food Prot.* **2019**, *82*, 948–954. [CrossRef]
159. Melgar-Lalanne, G.; Hernández-Álvarez, A.J.; Salinas-Castro, A. Edible insects processing: Traditional and innovative technologies. *Compr. Rev. Food Sci. Food Saf.* **2019**, *18*, 1166–1191. [CrossRef]
160. Commission Regulation (EC). *No. 2073/2005 of 15 November 2005 on Microbiological Criteria for Food-Stuffs*; European Union: Luxemburg, 2005.

Review

Foodborne Parasites and Their Complex Life Cycles Challenging Food Safety in Different Food Chains

Sarah Gabriël [1,*], Pierre Dorny [2], Ganna Saelens [1] and Veronique Dermauw [2]

[1]   Department of Translational Physiology, Infectiology and Public Health, Faculty of Veterinary Medicine, Ghent University, 9820 Merelbeke, Belgium
[2]   Department of Biomedical Sciences, Institute of Tropical Medicine, 2000 Antwerp, Belgium
*   Correspondence: sarah.gabriel@ugent.be

Abstract: Zoonotic foodborne parasites often represent complex, multi host life cycles with parasite stages in the hosts, but also in the environment. This manuscript aims to provide an overview of important zoonotic foodborne parasites, with a focus on the different food chains in which parasite stages may occur. We have chosen some examples of meat-borne parasites occurring in livestock (*Taenia* spp., *Trichinella* spp. and *Toxoplasma gondii*), as well as *Fasciola* spp., an example of a zoonotic parasite of livestock, but transmitted to humans via contaminated vegetables or water, covering the *'farm to fork'* food chain; and meat-borne parasites occurring in wildlife (*Trichinella* spp., *Toxoplasma gondii*), covering the *'forest to fork'* food chain. Moreover, fish-borne parasites (*Clonorchis* spp., *Opisthorchis* spp. and Anisakidae) covering the *'pond/ocean/freshwater to fork'* food chain are reviewed. The increased popularity of consumption of raw and ready-to-eat meat, fish and vegetables may pose a risk for consumers, since most post-harvest processing measures do not always guarantee the complete removal of parasite stages or their effective inactivation. We also highlight the impact of increasing contact between wildlife, livestock and humans on food safety. Risk based approaches, and diagnostics and control/prevention tackled from an integrated, multipathogen and multidisciplinary point of view should be considered as well.

Keywords: foodborne parasites; food chain; food safety; diagnostics; control; prevention; infection risk; meat-borne parasites; fish-borne parasites

Citation: Gabriël, S.; Dorny, P.; Saelens, G.; Dermauw, V. Foodborne Parasites and Their Complex Life Cycles Challenging Food Safety in Different Food Chains. *Foods* **2023**, *12*, 142. https://doi.org/10.3390/foods12010142

Academic Editor: Frans J.M. Smulders

Received: 23 November 2022
Revised: 20 December 2022
Accepted: 21 December 2022
Published: 27 December 2022

1. Introduction

Foodborne parasites (FBPs) have been long neglected, yet are slowly obtaining more attention, as diagnostic tools are improved and increasingly used, and burden data are becoming slowly available. Efforts made by the Food and Agriculture Organisation and the World Health Organisation (FAO/WHO) into the development of a multicriteria-based ranking for risk management of foodborne parasites have further placed the FBPs in the picture. This ranking was based on a number of criteria including amongst other number of global illnesses, morbidity, mortality; leading to a top four list related to the parasite's public health impact: *Taenia solium, Echinococcus granulosus, Echinococcus multilocularis, Toxoplasma gondii*; and a top four of *Trichinella spiralis, T. solium, Taenia saginata*, Anisakidae when assessing their trade impact [1]. In the WHO's burden assessment of foodborne pathogens initiative, conducted by the foodborne disease burden epidemiology reference group, the lack of knowledge of the burden of FBPs was acknowledged as well. The report lists the Disability Adjusted Life Years (DALYs) caused by 31 foodborne pathogens, including 14 parasites for which a total of 7,195,014 DALYs, 90,391,678 illnesses and 51,468 deaths were estimated for 2010. The estimates were judged conservative as often data were missing [2]. Foodborne parasites are notorious for their underreporting, most of them not having an obligatory notification.

On the level of the European Union (EU), several COST Actions including CYSTINET (TD1302) and EURO-FBP (FA1408) established or enforced Networks which have contributed greatly to the knowledge and management of FBPs in the EU.

While FBPs were previously majorly linked to endemic areas in the global south, this picture is rapidly changing due to globalisation, including increased international transports, distributions of food stuffs/products, increased trade, changing culinary behaviours moving towards less cooked, raw dishes and increased travel and migration of people. This, combined with an increase in susceptibility of a higher proportion of the global population due to, for example, a higher proportion of elderly people, leads to an increased number of people at higher risk for foodborne parasitic infections [3]. With improvements in diagnostic tools, increase in research efforts, knowledge and awareness on the impact of FBPs has increased, yet a lot of gaps are remaining.

A challenge typically related to FBPs is their long incubation times in humans, whereby clinical symptoms may take even years to appear (e.g., cyst stages of *Echinococcus* spp., *T. solium*) which complicates diagnosis, as well as the establishment of the original source of infection.

The FBPs often represent complex, multi host life cycles with parasite stages in the hosts, but also in the environment, where parasite stages may survive for many months or even years. Parasite stages may also contaminate other food stuffs such as fruits and vegetables. A human infection risk therefore may occur (even for a single parasite) at several points in different food chains.

Reduction of the risk can and should be envisaged at different points as well, following a One Health approach, accompanied by an efficient detection (and monitoring) of infection in different hosts. Yet, as will be exemplified below, currently, on a global level, many FBPs are not or insufficiently controlled, diagnostic tools not used or characterised by an insufficient performance. Furthermore, generally, levels of awareness by the different stakeholders are still too low.

In this Special Issue on safety of the food chain, this manuscript aims to provide an overview of a number of important zoonotic foodborne parasites, with a focus on the different food chains in which parasite stages may occur. This paper focusses on the human infection risk, on safety of the food chain, and as such does not cover diagnosis/treatment in human patients. We have chosen some examples of meat-borne parasites occurring in livestock (*Taenia* spp., *Trichinella* spp. and *Toxoplasma gondii*), as well as *Fasciola* spp., an example of a zoonotic parasite of livestock, but transmitted to humans via contaminated vegetables or water, covering the *'farm to fork'* food chain as well as meat-borne parasites occurring in wildlife (*Trichinella* spp., *Toxoplasma gondii*), covering the *'forest to fork'* food chain. Moreover, fish-borne parasites (*Clonorchis* spp., *Opisthorchis* spp. and Anisakidae) covering the *'pond/ocean/freshwater to fork'* food chain were added (Figure 1).

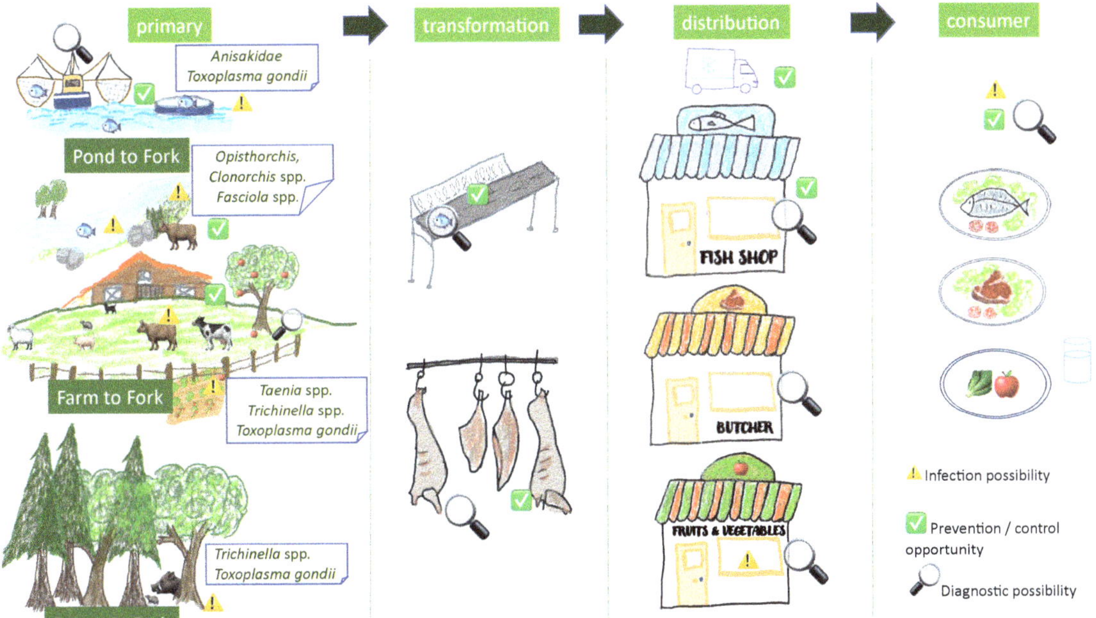

Figure 1. Simplified presentation of the foodborne zoonotic parasites in the different food chains, and identification of sites for diagnosis and control/prevention.

2. Farm to Fork and Forest to Fork

In this chapter, four important meat-borne zoonotic parasites from livestock will be discussed, including *T. solium*, *T. saginata*, *Trichinella* spp. and *Toxoplasma gondii*, as well as *Fasciola* spp. in the framework of the *Farm to Fork food chain*, while for *Trichinella* spp. and *Toxoplasma gondii* the *Forest to Fork food chain* will be highlighted as well (Figure 1).

2.1. Taenia solium, Taenia saginata

2.1.1. Introduction

Taenia solium and *T. saginata* are two human tapeworms. *Taenia saginata* has cattle as the intermediate hosts and primarily represents an economic burden to the meat sector. *Taenia solium* has pigs as intermediate hosts, but also humans may act as accidental intermediate host, developing neurocysticercosis, a main cause of epilepsy in endemic areas. *Taenia solium* therefore represents not only an economic problem in the human and veterinary sector, but also a serious public health problem. *Taenia saginata* has a global distribution, occurring in a high number of countries, with higher prevalences described in countries with raw meat consumption practices such as Belgium [4,5].

After consumption of undercooked infected pork/beef, a tapeworm develops in the human intestine (taeniosis). Eggs are shed from the gravid proglottids, which leave the host actively (only for *T. saginata*) or with the stool and subsequently contaminate the environment. Taeniosis generally leads to no/limited clinical signs and symptoms, though abdominal complaints, nausea and weight loss have been reported [4].

The metacestode larvae (cysticerci) develop in the (accidental) intermediate hosts after ingestion of the eggs (cysticercosis). In the natural intermediate host, the cysticerci lodge in the muscles, but also subcutaneously and in the brain (porcine cysticercosis, bovine cysticercosis). Generally speaking, the pig/cattle host shows no clinical signs or symptoms, though seizures and changes in behaviour have been described in heavily infected pigs [6,7]. For cattle, heavily infected carcasses detected at slaughter are condemned, while lightly infected carcasses require treatment, but can be sold afterwards. Nevertheless, the freezing of carcasses also entails a value loss. In Belgium, farmers pay an insurance to cover their losses due to bovine cysticercosis, the cattle owners bear an economic cost estimated at €3,408,455/year, while on the human side, an economic cost of €795,858/year was estimated for the taeniosis cases [8]. For porcine cysticercosis, infected carcasses should be condemned at slaughter, leading to economic losses for the farmer or trader. While meat inspection is often not implemented in rural areas in the Global South where most infections occur and therefore carcasses not condemned, infected carcasses do frequently represent an economic loss, as the value of the carcasses may be reduced by 50% [9]. For example, in Tanzania, a loss of nearly three million USD was estimated due to porcine cysticercosis in 2012 [10]. In the human accidental intermediate host, *T. solium* cysticerci may also lodge in the muscles, subcutaneously, and in the central nervous system. The latter is responsible for most clinical signs and symptoms including seizures, epilepsy, severe progressive chronic headache, vision problems etc. [4]. *Taenia solium* has been recognised as the FBP with the highest burden, estimated conservatively at 2,788,426 DALYs [2].

2.1.2. Localisation of the Infection Risks for the Consumer in the Food Chain

People develop taeniosis after consumption of undercooked infected pork/beef with viable cysticerci. When the infected pork/beef reaches the consumer in the food chain, the risk of exposure depends on the culinary practices of the consumer (see Section 2.1.4).

For *T. solium*, causing human cysticercosis, the routes of human infection are more complex, multi causal, and do not require the pig host as the transmission is human-to-human either direct or indirect. Eggs excreted by a human tapeworm carrier are immediately infective. Eggs may be ingested by tapeworm carriers or by close contacts of the carrier via the fecal oral route due to insufficient hand hygiene [4]. Eggs may also end up in the environment and contaminate fruits, vegetables, soil, water and surfaces [11], all potential sources of human infection. In areas with poor sanitation this happens in a more direct way via open defecation, equally so human stool may be used as a fertiliser for vegetable gardens. On the other hand, *Taenia* spp. eggs have been detected in sewage, in water purification plants, including effluent of these plants, indicating an insufficient clearing of the eggs from the dirty water entering the plants [11,12]. The role of insects in the dispersal of eggs in the environment needs clarification, and their importance for transmission of infective eggs to the human host is unknown [12]. The importance of the different routes of transmission to humans, either direct via contact with a carrier, or indirect via fruits, vegetables, water, etc., needs further research.

Also the survival time of the eggs in different environmental matrices (e.g., soil, water, . . .) under different climatic conditions, and as such their potential infectivity to humans/animals needs further investigation. Studies have been conducted, though primarily for *T. saginata*, indicating timespans of survival of up to one year and also the capability of the eggs to survive a European winter [13] (reviewed by [12]).

2.1.3. Diagnostic Options in the Food Chain

In the food chain, different points of diagnosis are possible. For taeniosis, infective cysticerci (viable) should not enter the food chain, therefore, infected carcasses should be picked up and removed from the food chain or should be treated (for *T. saginata*) [4]. There are two major hurdles in this detection, (1) the implementation of meat inspection and (2) the sensitivity of meat inspection. While meat inspection is implemented on most/all carcasses in industrialised countries, this is not the case in a lot of countries in the Global South, where backyard slaughter, or slaughter in illegal slaughterslabs is routine practice. Even though there seems to be an increasing level of knowledge and awareness regarding the risk of cysticercosis [14], the knowledge is still limited, and the risk perception insufficiently high. Farmers/butchers will still allow the infected meat to reach consumers, often for economic reasons [15]. As an alternative, farmers and traders may conduct a tongue inspection on the live pig, before purchasing. While this system works reasonably well in heavily infected pigs, the sensitivity is very low for other levels of infection, allowing a high number of infected carcasses to enter the food chain [16,17].

Meat inspection has a high specificity, but a notoriously low sensitivity, especially in lightly infected carcasses. For porcine cysticercosis, a sensitivity of 22% has been determined [16], for bovine cysticercosis, this has been estimated at 0.54% in Belgium, where primarily low levels of infection were observed [18].

Serological tools, detecting circulating antigens or specific antibodies have been developed, both for porcine and bovine cysticercosis. Yet, here also, though sensitivities are usually (somewhat) higher than for meat inspection, the tests have specificity problems. For circulating antigen detection, cross reactions with *Taenia hydatigena* are common, which is especially a problem for porcine cysticercosis, as in a number of endemic areas there is a co-occurrence of these parasites [16].

Most available molecular test are used for confirmation of detected lesions in the meat (reviewed by [19] for porcine cysticercosis).

Diagnostic options in the food chain related to human cysticercosis (*T. solium*) are currently not routinely implemented. Tests to detect eggs on fruits/vegetables or in environmental matrices such as water, soil have been reviewed recently by [11], with as main conclusions a general lack of sensitivity largely due to low recovery rates of the eggs from the matrices; specificity issues when only microscopic detection is included; and mostly a complete lack of standardisation of tests.

2.1.4. Prevention and Control Options in the Food Chain

Detection and removal of infected carcasses from the food chain (or treatment of lightly infected carcasses) is one of the main preventive targets for taeniosis. As mentioned above, this target is complicated by the lack of implementation of meat inspection in certain highly endemic areas, and even if implemented, especially lightly infected carcasses may still enter the food chain.

For bovine cysticercosis, the impact of meat inspection has been estimated in Belgium, where a prevalence of 42.5% was determined [18]. Meat inspection would only pick up 408 viable cysticerci from an estimated total of 213,344 viable cysticerci, present in the infected carcasses. Important here is that the performance of meat inspection is particularly low in the Belgian setting as most carcasses have very light infections. Nevertheless, in the same study, implementing a serological tool, in this case the circulating antigen detecting ELISA, would in a ten-year period greatly reduce the occurrence of bovine cysticercosis from >40% to 0.6%. The European legislation does allow for serological testing (EC 2019/627 [20]), but in practice, this is not yet implemented, probably due to the costs related to this approach.

In the EU, a more risk-based meat inspection approach is currently implemented considering age of the animals and their production system allowing a selection of animals to undergo a more in-depth meat inspection [21].

As for other FBPs, production systems are of major importance in allowing parasite access the animal hosts. In highly confined and controlled systems (see Section 2.2.4) pigs/cattle would not have access to *Taenia* eggs due to controlled feed and water access, high hygiene levels avoiding direct contamination via a potential human tapeworm carrier, etc. Nevertheless, while for pigs this is the case in a lot of commercial farming systems, for cattle this is much less so, as outdoor grazing is common; and with the trend towards more biological/organic farming, (renewed) access to potentially contaminated environment/feed/water is created [3].

Vaccination of cattle and pigs has been a subject of research for many years [22]. Especially for the pig host a number of vaccines have been developed, and have been evaluated in the field. The TSOL 18 vaccine is now commercially available, and has shown a very high efficacy, in combination with a single treatment with oxfendazole [23,24]. The latter, treatment with oxfendazole, has also been proposed as an effective option for the control of porcine cysticercosis [25], yet is not implemented outside scientific studies.

At the end of the food chain, the consumer may play a major role as well. Culinary habits will greatly influence the inactivation of viable cysticerci in pork/beef. As for other FBPs, good cooking practices avoiding raw/undercooked pork/beef consumption practices play an essential role in the level of risk of exposure of the consumer, leaving an important role to health education and awareness creation [25,26].

Of course, interventions directed towards the human host such as treatment of tapeworm carriers, stopping open defecation, or management of sludge on pastures are highly relevant [25], but not the focus of this food chain-oriented manuscript.

2.2. Trichinella spp.
2.2.1. Introduction

Larvae from the *Trichinella* genus present in different meat and meat products have been a source of infection for humans for centuries, with outbreaks still occurring regularly today, with a global distribution.

Transmission of *Trichinella* spp. among non-human animals occurs by predation or carrion consumption. Transmission to humans occurs via the consumption of raw or undercooked meat [27], whereby infective larvae situated within the muscle cells are released in the stomach to further mature in the small intestine into adult worms. Subsequently, after mating, adult females produce larvae that will migrate to the muscles. Myalgia, diarrhea, fever, facial edema and headaches were mostly reported as clinical signs and symptoms in infected people. Most of these disappeared within 2–8 weeks after treatment, nevertheless myalgia and fatigue may remain present for years. Early diagnosis and treatment in the last decades probably have contributed to a decrease in mortality due to trichinellosis [27]. *Trichinella* spp. were detected in domestic and wild animals in 66 countries and in humans in 55 countries [27,28]. Trichinellosis represents 550 DALYs, with an estimated 4470 illnesses globally in 2010 [2]. A review conducted to assess the global incidence and clinical impact of trichinellosis identified 65,818 cases from 41 countries between 1986 and 2009. Reporting of cases varies greatly though, and is mainly linked to hospitalised cases, representing a serious underestimated from the true number of cases, including mild or asymptomatic cases. The World health Organisation European Region accounted for the majority of the cases (87%) [27] occurring primarily in adults.

Two life cycles are described for *Trichinella* spp., a domestic and a sylvatic cycle. The domestic cycle, primarily involving pigs and rodents as hosts and the encapsulated species *Trichinella spiralis* (T1), relates in this manuscript to the *Farm to Fork food chain*.

The sylvatic cycle including a large range of wild mammals, birds and reptiles [29], relating to the *Forest to Fork food chain*, includes primarily the other *Trichinella* species and genotypes. The encapsulated clade with species infecting mammals only, e.g., *Trichinella nativa* (T2), *Trichinella britovi* (T3), *Trichinella murelli* (T5), *Trichinella nelsoni* (T7), *Trichinella patagoniensis* (T12), and *Trichinella chanchalensis* (T13), with three genotypes *Trichinella* T6; *Trichinella* T8 and *Trichinella* T9, and less common *T. spiralis* (T1). Species from the non-encapsulated clade infecting mammals and birds include *T. pseudospiralis* (T4) and infecting mammals and reptiles include *T. papuae* (T10) and *T. zimbabwensis* (T11) [28,29].

A recent paper by [30] reviews the presence of *Trichinella* in wildlife globally, identifying the polar bear (57.58%), martens (32.39%) as the species with the highest prevalence in the palearctic region. While most studies cover terrestrial mammals with as most studied the wild boar, red fox, raccoon dog, wolf, black and polar bears, *Trichinella* has been detected in marine mammals and birds as well. In addition, rodents and lagomorph species were found with *Trichinella*. These animal species are important for the maintenance of the transmission cycle, at the same time may be hunted for food consumption, yet are often not part of the food safety control systems. While lions and hyenas have been reported as hosts, and again play a role in the maintenance of the life cycle, infection detected in crocodiles presents a higher risk for unsafe food for humans. The zoonotic potential of *T. zimbabwensis* is still a matter of debate, an increased use of molecular tools to identify larvae either from a patient (after a muscle biopsy) or after tracing back the meat source will bring more clarity.

2.2.2. Localisation of the Infection Risks for the Consumer in the Food Chain

People are infected via the consumption of infected, undercooked meat. While most of the human infections/outbreaks are associated with pork and pork products from pigs from outdoor breeding farms, cases of infection via consumption of insufficiently cooked other meat and meat products including game meat, especially from wild boars, have been reported regularly [27]. Examples of outbreaks related to horse meat in France and Italy [31,32], dog meat in China [33] have been described. Two large trichinellosis outbreaks in France with direct parasitological evidence indicating horse meat as the source of infection included 128 and 407 cases (reviewed by [31]).

Pork and pork products originating from pigs kept under highly confined and controlled conditions is not a source of infection if the conditions are properly implemented and maintained (see below). Uncontrolled pork/meat/game meat often consumed at the household level, the latter linked to hunting activities, has been regularly reported as source of infection. As described above, a high number of wildlife species may be infected with *Trichinella* spp., and may therefore be potential sources of infection. Moreover, illegal import of uncontrolled meat from endemic areas has led to outbreaks. International travellers returning home from endemic areas may also develop disease upon their return [27].

Within the limitations of the diagnostic tests, controlled pork/meat/game meat (after slaughter) should prevent infected meat reaching the consumer, or at least at infection levels leading to clinical disease.

2.2.3. Diagnostic Options in the Food chain

Direct testing of muscle samples collected from pig carcasses at the slaughterhouse based on the artificial digestion method is now the most often implemented diagnostic technique, while previously, compression of a small amount of (porcine) muscle between glass slides followed by microscopic examination was routinely applied [34]. Pooled samples of pig muscles are analysed by the magnetic stirrer digestion method (in acidified pepsin). Allowing testing of multiple samples saves a substantial amount of time, especially in low prevalence situations, even though a positive batch result has to be followed by smaller batch testing. The method has a high specificity when conducted by well trained staff. As detected larvae can be recovered, subsequent molecular identification majorly helps epidemiological investigations. Lack of a high sensitivity is often mentioned as a major disadvantage, as infection levels of 3–5 larvae in one gram of muscle would be needed to obtain a sensitivity close to 100% [34]. Nevertheless, application of the test does allow detection of carcasses at this minimum infection level, which are exactly those carcasses that present the highest risk for causing clinical disease in humans [35]. As the food safety objective is to avoid clinical trichinellosis, the performance is satisfactory.

For game, hunters are advised (or obliged depending on the applying regulations) to collect muscle samples for testing.

Indirect detection of infection by specific antibody detection in serum via Enzyme Linked Immunosorbent Assays (ELISAs) is currently not advised for the determination of the infection status of individual carcasses in a food safety control system. Antibody detection suffers from insufficient sensitivity and specificity, though this has been a matter of debate [32,36]. Cross reactions may occur with other parasites infecting pigs, though the level is strongly dependant on the antigen used in the ELISA. Use of properly prepared excretory-secretory (ES) antigens, or ES products, recombinant and synthetic antigens with the same dominant epitopes have been claimed to provide a good specificity in ELISA [36]. Nevertheless, taking into consideration potential false positive results, positive serology would have to be followed by a direct test to confirm infection. The ELISA is usually characterised by a higher sensitivity than the artificial digestion, with a reported detection of one larva per gram of tissue [35,36]. However, this performance is related to the dose of infection, which will influence the level and timing of presence of detectable levels of specific antibodies in the serum, whereby low infection doses may be detectable only as early as six weeks after infection, allowing for a rather large diagnostic window of false negativity. This particularly renders the ELISA much less suitable for individual carcass assessment, as while the test result will be negative, *Trichinella* larvae may already be present in the muscles. Heavy infections (large infections dose) on the contrary, can be picked up as early as seven days post infection. However, as [35] debates, these low levels of infection may be missed by the direct detection methods as well, as these methods require a presence of 3–5 larvae per gram of tissue to be detectable. Nevertheless, given the challenges related to both the ELISA's diagnostic specificity and sensitivity, currently the artificial digestion method is still the method of choice recommended by the scientific community and described in regulatory documents.

On the contrary, ELISA can be useful for surveillance or monitoring in pigs and other animals [35] at primary level, in live animals.

Besides the diagnostic performance of the test, cost plays an important role as well. Barlow and colleagues [37] reviewed the available tests for assessment of safe pork and estimated the artificial digestion to be the most optimal choice. The possibility for analyses of pooled samples greatly contributed to a cost reduction, allowing for a system with a satisfactory diagnostic performance and an acceptable cost. ELISA was not selected for individual carcass assessment for reasons explained above. Molecular techniques, such as the conventional and real time PCR, while ideal to identify the *Trichinella* larva(e), are currently not cost efficient for large scale testing of carcasses. Nevertheless, with technological advances, molecular based methods may become more plausible replacements of the artificial digestion [37]. A promising example could be the lateral flow- recombinase polymerase amplification (LF-RPA) targeting the mitochondrial small subunit ribosomal RNA (rrnS) gene, which can be applied to fresh and frozen pork samples. Besides a very high sensitivity, it has a reported 100% specificity, quick time to result (less than 20 min) and relatively low equipment needs [38]. Nonetheless, a more large-scale evaluation on meat samples is needed [37].

2.2.4. Prevention and Control Options in the Food Chain

In the farm to fork food chain, direct systematic testing using the artificial digestion (see Section 2.2.3) of individual pig carcasses has been a cornerstone of clinical human trichinellosis prevention in many countries for many years. While testing removes a number of infected carcasses from the food chain, it does not detect the lightly infected carcasses and as such does not remove the parasite from the food chain, neither does it fully prevent human exposure to the parasite [35]. Still, the aim to avoid human clinical trichinellosis cases can be achieved implementing this system [39]. Moreover, detection of (heavily) infected carcasses and subsequent correct removal of these carcasses, avoids further potential transmission via animal feed for example. Of course, systematic testing comes with a high cost. In the European Union, this was estimated at an annual cost ranging from 25–400 million euro [40].

With the changes in farm management system, moving towards highly confined and highly controlled housing and farming systems (high levels of biosecurity), the risk of exposure of the pig host is removed and the occurrence of *Trichinella* spp. in the pig host subsequently dropped drastically. Indeed, under these conditions, the parasite may be removed from this particular farm to fork food chain. The development of international guidelines describing criteria for controlled housing and management, thereby prevented the risk of exposure of animals. Implementation of these systems actually reduces/removes the need for testing [35], as the parasite is absent from these particular commercial pork production systems. Indeed, summary data from the EU (European Centre for Disease Prevention and Control, [41]) for 2019 described that routine slaughter testing of 72.8 million pigs raised under controlled housing reported no *Trichinella* spp. infections. Considering the cost of testing, especially in low prevalence settings, with most animals testing negative, alternatives were searched to replace the systematic testing.

A set of criteria described by the World Organisation for Animal Health (OIE) to recognise free regions or countries was fairly rapidly replaced by a set of requirements to meet for a status of negligible risk as advised by the International Commission on trichinellosis [35,42]. The latter has as a purpose to ensure food safety (consumer health) and to obtain standardised requirements for international trade, removing the need for testing of animals originating from herds classified as negligible risk. The standards included amongst other good feed manufacturing and storage practices, rodent control, prevention of pigs accessing wildlife, removal of deceased pigs and controlled animal movement [35].

Of course, not all pigs are bred under these controlled management systems, and as such, are exposed to a higher risk of *Trichinella* spp. infection as exemplified by the EU summary data whereby 218 positive pigs were detected from 139.6 million pigs from non-controlled housing tested (European Centre for Disease Prevention and Control, [41]). Non-controlled housing may allow pigs access to potentially infected rodents and wildlife. The same counts for other animal species bred under non-controlled systems, such as horses. Carcasses from the latter also need to be tested systematically, like pigs. The trend towards biological/organic pig farming [3], moving away from these highly controlled/confined systems opens opportunities for the parasite to enter the farms. Whether there is a risk and how high this risk is, depends on the infection levels in the local wildlife, including rodents. In this interface area setting, both food chains, farm to fork and forest to fork merge. The expansion of wild boars into areas of free-range pig production in the Unites States represents a risk for pig infection, leading to the introduction of *Trichinella* spp. into the farm to fork food chain [43], at the same time increasing the availability of potentially infected wild boar meat via hunters.

For *Trichinella* spp., the forest to fork food chain is an important source of infection to humans, especially game meat consumers, hunters. Human behaviour is particularly important in this food chain, as the testing (or not) of game meat, and consumption practices are dependent on the hunter/consumer. Often home consumption of hunted game is not subject to regulations, therefore testing is frequently not performed. Risk of infection is subsequently dependant on the presence and level of infection as well as on the culinary practice of the household which may include cooking, freezing or curing, influencing the inactivation of the parasite. Health education on good preparation methods for meat that might contain *Trichinella* larvae, as described by the International Commission on Trichinellosis is encouraged, including cooking, freezing (for meat from domestic pigs), and irradiation [34].

Efforts have been ongoing in the development of vaccines targeting to reduce the larval or adult worm burdens, but to date, no sufficiently effective vaccine is available [44].

As highlighted in Section 2.2.3, ELISA detecting specific antibodies may be used for surveillance and monitoring in pigs and other animal species [36].

2.3. Toxoplasma gondii

2.3.1. Introduction

Toxoplasma gondii is an obligate intracellular protozoon that causes toxoplasmosis in man and animals. Toxoplasmosis is one of the most common parasitic zoonoses worldwide with an estimated one third of the global population being infected [45]. Toxoplasmosis is present in every country and seropositivity rates range from less than 10% to over 90% [45]. FAO and WHO have ranked *T. gondii* fourth out of 24 foodborne parasites of global importance [1]. The global annual incidence of congenital toxoplasmosis was estimated to be 190,100 cases. This was equivalent to a burden of 1.20 million DALYs (95% CI: 0.76–1.90). High burdens were seen in South America and in some Middle Eastern and low-income countries [46].

T. gondii uses felines as the definitive hosts, but is less specific when it comes to the intermediate hosts, which can be any warm-blooded animal, including mammals and birds. The lifecycle comprises an intestinal sexual phase that takes place only in the final hosts and results in the production of unsporulated oocysts that are faecally shed by infected felines and sporulate in the environment; and an extra-intestinal phase in intermediate hosts which comprises the formation of tachyzoites and tissue cysts that contain bradyzoites. In most hosts, *T. gondii* causes a lifelong latent infection in tissues such as skeletal and heart muscles, visceral organs and the central nervous system [47].

Most infections with *T. gondii* in humans and animals are subclinical or cause mild clinical signs and symptoms. Severe disease may occur in hosts with an immature or compromised immune system, or in case of higher pathogenicity of the parasite strain. In livestock, *T. gondii* is an important cause of abortion in sheep and goats. In humans, infection with *T. gondii* is particularly important in pregnant women and in immunocompromised people. When primary infection occurs during pregnancy *T. gondii* can cross the placenta and reach the foetus causing a mild to life threatening infection depending on the gestational stage. Disease- or medication-induced immunosuppression can lead to encephalitis or disseminated toxoplasmosis in adults. While toxoplasmosis in immunocompetent individuals is asymptomatic in around 80% of cases, in about 20% infection it may cause fever, mononucleosis-like symptoms, or ocular manifestations such as chorioretinitis [47].

2.3.2. Localisation of the Infection Risks for the Consumer in the Food Chain

Infection in humans may occur from the accidental ingestion of oocysts that can be present in food such as vegetables and fruits but also shellfish, or in water due to environmental contamination with cat faeces; or from the consumption of raw or undercooked meat containing tissue cysts. As mentioned above vertical transmission may occur when a parasite-naïve women gets infected during pregnancy. Less frequent ways of infection are caused by drinking goat milk containing tachyzoites during the acute phase of infection or by transplantation of tissues from a *T. gondii*-infected donor. Consumption of undercooked infected meat is considered a major risk factor for humans, especially in Europe, where it has been associated with 30–63% of infections [48,49]; Hill et al. [50] demonstrated the predominance of oocyst-caused infections in North America. Based on the relative proportion of the different animals in the overall meat consumption, the proportion of meat types consumed raw or undercooked, the exposure and susceptibility of different animal species to *T. gondii* and the subsequent establishment and survival of parasites in their tissues, meat from pigs and small ruminants are to be seen as the main sources of infection (farm to fork food chain). Meat from wildlife (forest to fork food chain), horses, poultry and cattle are less common sources [49].

Toxoplasma gondii is an example of a health issue that can be directly connected to outdoor animal husbandry [51]. Grazing ruminants mostly acquire infection by ingestion of oocysts from the environment. As a result, the seroprevalence in sheep tends to be high, e.g., between 27.8% and 87.4% in West-European countries [52]. Pigs may become infected both by ingesting oocysts and by the consumption of meat containing tissue cysts from infected rodents or kitchen leftovers. Consequently, a low prevalence (0–1%) is found in pig farms with well-managed controlled housing conditions that practise rodent control, keep cats away from the farm and the feed, and restrict the access to the farm. In contrast, a high prevalence of up to 60% is found in poorly managed or free-range farms [53–55]. Outdoor access of pigs considerably increases the risk for *T. gondii* infection such as in free-range organic farms [54,56]. Thomas et al. [57] studied the detailed anatomical distribution of *T. gondii* in naturally and experimentally infected lambs and found that parasite DNA could be detected in all the edible parts.

Environmental contamination with *T. gondii* oocysts is understudied and likely underestimated, which is partly due to the lack of suitable harmonized sampling approaches and detection methods. Oocysts can be spread in the soil by arthropods, earthworms, wind, and rain [58]. Sporulated oocysts are highly resistant and can remain infective in soil for up to two years [59]. Several waterborne infections associated with *T. gondii* oocysts have been described. Water in irrigation systems, rivers, lakes, beaches, and coasts, as well as wastewater and groundwater can be contaminated with the environmentally resistant oocysts. Moreover, oocysts can survive various inactivation procedures using chemical reagents, including sodium hypochlorite and chlorine [60]. Oocyst contamination of fresh vegetables may occur through cultivation in contaminated soil or using contaminated water for irrigation or washing. Consumption of raw vegetables and fruits are a risk factor. *T. gondii* oocysts can also enter the marine environment through disposal of sewage and water runoff, where they can cause infections in marine animals [61]. Oocysts have been detected in wild and commercial bivalve mollusks which are filter-feeders and can concentrate microorganisms. They can retain viable *T. gondii* oocysts for 85 days following uptake [62]. These shellfishes can pose another risk for consumers when consumed undercooked or raw [63].

2.3.3. Diagnostic Options in the Food Chain

Toxoplasma gondii infections can be detected by direct and indirect techniques. Coprological methods are used to detect oocysts in cat faeces. (Immuno-)histological and molecular methods and bioassay are used to detect the parasite in tissues, in most cases post-mortally. Recently, an improved molecular method using magnetic capture combined with RT-PCR has been developed that allows the detection of the parasite in larger portions of tissue [64]. Cat and mouse bioassays are the reference direct techniques to isolate *T. gondii*; however, these tests are not commonly used due to the long time it takes to obtain results, ethical issues, and high costs [47]. An alternative method is cell culture which is limited in use because of the variability of the results [65]. Many serological methods (indirect detection) have been developed and validated in humans and several animal species. Among these techniques are the immunofluorescent assay (IFAT), enzyme-linked immunosorbent assay (ELISA), latex agglutination tests (LAT) and the modified agglutination test (MAT). An important question is whether seropositivity can be linked to the infectivity of tissues. Opsteegh et al. [66] found that the correlation between antibody detection against *T. gondii* and direct parasite detection is high in pigs, small ruminants, and chickens. In these species, the use of serology can help determine the risk to the consumer, but it may not be as useful in other species, such as horses and cattle. In addition, a seronegative result does not necessarily mean that the meat is free of *T. gondii* [67].

2.3.4. Prevention and Control Options in the Food Chain

Although *T. gondii* is a high priority foodborne zoonotic parasite, it is not systematically controlled. At present, there are no specific regulations and no standardised methods for the detection of *T. gondii* in any food matrix. Because chronically infected animals are mostly asymptomatic and tissue cysts cannot be detected during routine meat inspection—the size of tissue cysts is less than 100 μm—most infected carcasses pass meat inspection and enter the food chain.

Currently, most of the control of *T. gondii* infection is carried out at home, especially in people that are most vulnerable to the parasite, such as pregnant women and immune-compromised persons. Primary prevention consisting of dietary recommendations, pet care measures, environmental measures, knowledge of risk factors and ways to control toxoplasmosis infection, has been found to be effective in reducing congenital toxoplasmosis [68].

At farm level preventive measures mostly apply to the pig industry where the parasite can be virtually eliminated by a set of hygienic measures, as discussed above. The serological prevalence of *T. gondii* in the pig population may be a useful indicator of the risk of human toxoplasmosis associated with the consumption of pork products. In the EU, the Commission Regulation No. 219/2014 modernised some specific requirements of the post-mortem inspection of pigs, favouring visual inspection instead of palpation and incisions [69]. However, it does not solve the *Toxoplasma* problem at slaughterhouse level. Therefore, the categorisation of risk for different types of farms (intensive systems and organic farms) would help the official veterinarian during ante-mortem and post-mortem visits [70].

Post-slaughter methods involve meat processing for human consumption and include mainly heat inactivation and freezing that are effective ways to kill the parasites if properly applied. Rani and Pradhan [71] studied the survival of *T. gondii* during cooking and low temperature storage and concluded that viable parasites were not found when the internal temperature of meat reached 64 °C and below −18 °C. Other meat curing procedures such as salting, smoking, or fermentation are less reliable for killing the parasites in meat because they are very much dependent of the concentration of the salt, the storage temperature, and the time of the processing [55]. The modern trend steered toward meat production from organic breeding, consumed raw or undercooked, with low concentrations of salt and additives (e.g., nitrites), may result in an increase of the zoonotic risk [54,56]. The current scientific knowledge is outdated and not sufficient for a full risk assessment. For this reason, innovative studies on *T. gondii* inactivation focusing on modern processing technologies may contribute to outline new preventive measures for consumers [55].

Currently, testing for parasite contamination in fresh produce is neither regulated nor mandatory. The increased popularity of consumption of raw and ready-to-eat vegetables may pose a new potential risk for consumers who could be accidentally exposed to oocysts, since most post-harvest processing measures do not guarantee the complete removal of oocysts or their effective inactivation [58].

Similar to other pathogens research is ongoing on the development of vaccines against toxoplasmosis. Currently, in animals, the only available vaccine is a live, attenuated, *T. gondii* S48 strain licensed for use in sheep in Europe and New Zealand for prevention of abortion [72]. However, using a live vaccine raises safety concerns for use in food-producing animals since the vaccine strain may revert to a wild type that might cause tissue cyst formation. DNA vaccination has been shown to determine long-lived humoral and cellular immune responses in vivo in animals [73]. Vaccination of cats has been proposed as the ultimate preventive measure because of the pivotal role of cats in the lifecycle of *T. gondii*. However, in case an effective vaccine for cats would be available, prospects on preventing oocyst-originated human toxoplasmosis by vaccination in large populations of cats are not favourable due to the large vaccination coverage needed [74].

2.4. Fasciola spp.

2.4.1. Introduction

The trematodes *Fasciola hepatica* and *Fasciola gigantica* are the causative agents of the disease fasciolosis. The life cycle of *Fasciola* spp. includes plant-eating mammals (mainly ruminants, but also pigs and others) as final hosts, aquatic, lymnaeid snails as intermediate hosts and aquatic plants as carriers. Humans can also act as final hosts, and may even contribute to the perpetuation of the life cycle in endemic areas, where poor sanitation occurs [75], although the extent to which this actually occurs has never been quantified.

For decades, fasciolosis was perceived to be a purely veterinary problem [76], however, it is now also seen as an important disease in humans. As a response, WHO has listed fasciolosis as one of the neglected tropical diseases to be prioritized for control [77]. Human fasciolosis is known to occur worldwide, although it mainly affects the poorest communities in rural areas across subtropical and tropical countries. No recent burden data are available for fasciolosis. In 2012, Fürst et al. [78] estimated the global burden of fasciolosis at 35,206 DALYs, whereas the 2015 Global Burden of Disease estimate amounted to 90,041 DALYs, however the latter estimate came with a wide uncertainty interval (58,050–209,097) [79]. Estimates for the number of people infected range between 2.4 and 17 million [78,80,81], whereas those for the population at risk range between 91 and 180 million people [82,83]. Based on reported infection intensities, it has been estimated that 14% of fasciolosis infections are symptomatic [78]. These numbers could underestimate the true occurrence and impact of the disease, as for many regions, such as for instance a number of countries on the African continent, few or no community-based epidemiological surveys have been performed [84], and considering that the current burden estimation does not yet account for account for the immunosuppression, neurological or ocular effects due to fasciolosis.

Overall, fasciolosis is considered an emerging disease [82,85,86]. Its emergence could be due to the increased attention it received since its designation as neglected tropical disease, combined with factors facilitating its expansion and importance. For instance, climate change is thought to cause an increased risk for its snail host survival and distribution, and thus spread of the disease [87,88]. Moreover, the increased consumption of raw vegetables and fruits as part of increasingly popular healthy life styles and the introduction of sylvatic reservoir animals can assist in the spread of the disease [89,90].

Infected individuals harbour the adult *Fasciola* spp. worms, and unlike in *Clonorchis/Opisthorchis* spp. infection, will shed unembryonated eggs with their faeces/stool [91]. Once in a favourable freshwater environment, the eggs will then embryonate after approximately two weeks. After hatching, the miracidia will actively search for a suitable freshwater snail to penetrate and infect. In the snail, the miracidia will undergo several developmental stages (i.e., sporocyst and redia stages) and various rounds of asexual multiplication [90]. Next, an exponential number of free-swimming cercariae will leave the snail and encyst into metacercariae (MC) on plants growing in the same aquatic environment. Susceptible final hosts can acquire the infection by ingesting contaminated raw water plants. Upon ingestion, the MC will excyst in the small intestine, and penetrate the intestinal wall. Next, the juvenile flukes will migrate through the liver where they will mature in the bile ducts and start producing eggs, completing the life cycle [92]. Ectopic infections can also occur, with the juvenile or immature flukes erroneously migrating to subcutaneous tissues, gastrointestinal tract, heart, lung, or rarely, brain and eye [93,94].

In humans, fasciolosis can cause fever, abdominal colic, digestive disorders, weight loss, anaemia and jaundice due to fluke migration and subsequent destruction of the liver tissue, inflammation and blockage of the bile ducts [94]. In ectopic infections, symptoms and signs are specific to the organ affected by the migration path and presence of the flukes. Recently, it has been shown that neurological, meningeal and ocular symptoms, such limb and facial paralysis, speech disorders, blindness can also occur in hepatic fasciolosis, due to leakages in blood-brain barrier [95]. Finally, for a long time, *Fasciola* spp. infection was thought to cause parasitic pharyngitis, called Halzoun. The proposed route for acquiring this condition was the ingestion of raw liver, contaminated with immature or young flukes, attaching to the pharyngeal mucosa and invoking pain and bleeding. However, it is now argued the condition is rather due to infection with other parasites, such as *Linguatula serrata*, and *Dicrocoelium dendriticum* [90].

2.4.2. Localisation of the Infection Risks for the Consumer in the Food Chain

Based on current knowledge, the most common pathway of infection for humans is the consumption of contaminated water plants. The best known infection source is watercress (*Nasturtium* spp.), an ubiquitous green leafy vegetable, with case reports around the globe mentioning the consumption of this plant in the anamnesis [96–98]. Furthermore, a wide range of other freshwater plants have been reported as infection sources, both wild and cultivated ones, such as dandelion (*Taraxacum* spp.) leaves, lamb's lettuce (*Valerianella locusta*), and water spinach (*Ipomoea aquatica*). Contamination of freshwater plants is mostly due to the direct shedding of eggs by an infected host, being domestic livestock, humans or even sylvatic hosts, such as the nutria (*Myocastor coypus*), into an aquatic environment where the suitable intermediate snail host is present as well as the water plant eventually serving as carrier for the infective stage, metacercariae [90].

Next to water plants, a variety of terrestrial plants can carry the infective stage of *Fasciola* spp. For instance, infections due the consumption of lettuce (*Lactuca sativa*), and parsley (*Petroselinum sativum*) have been reported [99,100], while in other cases, infection was presumed to be due to consumption of wild aromatic plants, such as mint

(*Mentha* spp.) [101]. Contamination of these plants can be due to them being submerged in water during a certain period of time, either in an environment where eggs had been shed, and thus snails were infected, or due to flooding of an adjacent field, with the runoff transporting infected snails or MC over a certain distance [102]. Other contamination pathways in terrestrial plants are washing in contaminated water bodies, and through irrigation with contaminated water.

Next to the ingestion of plants, the chewing and sucking contaminated plants are common infection routes for *Fasciola* spp. Well known examples are grass chewing, especially in children [103], as well as chewing on leaves of khat (*Catha edulis*), a popular tradition in the Horn of Africa and the Arabian Peninsula. Moreover, in certain regions, drinking beverages, such as herbal teas, and juices made from local plants are a known risk factor [104]. Drinking contaminated water is another suggested route of infection, although its importance is not well understood. Indeed, even though it is proven that unattached MC sink quite quickly to the bottom of water bodies, drinking contaminated water has been identified as the sole infection routes in a number of areas [105–107]. Naturally, ingestion of dishes and soups made with contaminated water, and washing of vegetables, fruits, tubercles, kitchen utensils or other objects with contaminated water can equally cause infection. Finally, in pig experiments, it was shown that the ingestion of raw liver with juvenile *Fasciola* spp. can lead to established liver infections [108], however to what extent this occurs in humans is not known.

2.4.3. Diagnostic Options in the Food Chain

Surprisingly little attention has been given to the development of diagnostic tools for MC detection on plants and in water. Moreover, the sensitivity and specificity of available techniques have never been assessed. Up to now, most techniques for plants entail the screening of plant surfaces by means of a stereomicroscope [109]. Not only is this is not a feasible option in a commercial context, due to the time consuming nature of the method, it also requires the necessary expertise, to recognize *Fasciola* spp. MC, especially on thicker leave and stem surfaces, and to differentiate them from other digenean MC (e.g., *Paramphistomum* spp.) [109]. In water, while originally developed to investigate the contamination of rice fields, floats or buoys could be used to catch MC present in important waterflows from irrigation channels and/or streams [110,111]. Overall, cost-effective techniques to detect metacercarial contamination of consumed plants and water are lacking.

2.4.4. Prevention and Control Options in the Food Chain

The most obvious method for fasciolosis prevention is avoiding oral contact with raw plants, whether it is the sucking/chewing or consumption of raw plants, especially those growing in an aquatic environment. However, culinary habits are often difficult to break, and therefore such recommendation might not be realistic for many communities and individuals. Another option would be the removal and/or destruction of MC attached on plant leaves. Unfortunately, washing vegetables using running water alone has proven to be only moderately successful in detaching MC [112]. On the contrary, briefly (5–10 min) soaking plants in vinegar (e.g., 120 mL/L) or liquid soap (e.g., 12 mL/L) solutions, seems effective in detaching and killing the MC [112,113], however these studies should be repeated to confirm the application of these methods in a commercial context. Some studies have investigated the effects of other chemical agents, of which potassium permanganate, and sodium hydroxide treatments seemed successful detachment/destruction methods [101,112,113], however, such agents have a considerable impact on plant palatability and appearance, and therefore their application seems unfeasible in a commercial context [113]. Cooking vegetables seems a more effective method to kill MC, however culinary traditions in raw plant consumption and in some regions, inadequate means to ensure sufficiently high temperatures while cooking, might hamper the application of this preventive measures.

Similarly, boiling potentially contaminated water might not be feasible due to lack of means, in such regions filtration with an appropriate mesh size might be more effective [90].

Currently there are few preventive measures for fasciolosis implemented by governments. In the European Union, all bovines and small ruminants processed through abattoirs will be inspected for *Fasciola* spp., by means of visual inspection and incisions of the liver, in line with the European meat inspection legislation (EC 2019/627, [20]), however this inspection will not stop infected livestock from contaminating fields and therefore indirectly causing human fasciolosis cases. In a wider preventive context, plants commonly consumed by humans, should be grown under controlled conditions, inaccessible to snails, ruminants and other animals. Appropriate legislation has been implemented by a number of countries, e.g., France and Australia [90,114]. Moreover, the risk of contamination of these fields, due to run-off or irrigation with water from adjacent areas where *Fasciola* spp. can be present, should be excluded. A One Health approach is needed to ensure that all stakeholders are involved and informed about the disease, the risks and responsibilities. In any given region, appropriate fasciolosis prevention and control may only be achieved once the local transmission dynamics of *Fasciola* spp. are fully understood.

3. Pond/Ocean/River to Fork: Fish-Borne Parasites

In this chapter, three important fish-borne zoonotic parasites will be discussed, including *Clonorchis* spp. and *Opisthorchis* spp., related to freshwater fish, as well as the group of Anisakidae, related to marine fish (Figure 1).

3.1. Fish Borne Trematodes: Clonorchis, Opisthorchis spp.

3.1.1. Introduction

Liver flukes of the Opisthorchiidae family, of which *Opisthorchis viverrini*, *O. felineus* and *Clonorchis sinensis* are the most important, cause opisthorchiasis and clonorchiasis in humans. The parasites have a complex lifecycle, requiring *Bithynia* spp. freshwater snails and cyprinid fish as primary and secondary intermediate hosts, respectively, and a fish-consuming mammal, such as humans, cats and dogs, as the final hosts [115]. Although cat and dogs are known to contribute to the life cycle of for instance *O. viverrini*, it has equally been shown for the same parasite, cats and dogs cannot sustain transmission in absence of humans as a final host [116]. It is not known, if the same is true for *O. felineus* and *C. sinensis*. The infected final host will discharge embryonated eggs in the biliary ducts and shed them via the stool/faeces. If shed in a suitable environment, the eggs will be ingested by the snail intermediate host. In the snail, the eggs will release miracidia that will subsequently develop into sporocysts, rediae and cercariae [117]. The latter developmental stage will be released from the snail host into the aquatic environment, where it will actively seek for a suitable secondary intermediate host, a freshwater cyprinid fish, penetrate its flesh or skin and encyst as metacercaria (MC) [118]. The final host acquires the infection by ingesting the raw or undercooked contaminated fish. Upon ingestion, the MC excysts in the small intestine and travels to the biliary tract via the ampulla of Vater. In the biliary ducts, the flukes will mature and start producing eggs, thereby restarting the life cycle [119].

Opistorchiosis and clonorchiosis can cause cholangitis, jaundice, cholecystitis, hepatomegaly and cholelithiasis [117]. Moreover, the International Agency for Research on Cancer has classified both *O. viverrini* and *C. sinensis* as Type I carcinogens. Indeed, in chronic infections, the (i) repeated mechanical damage due to the feeding and migrating flukes, combined with (ii) the secretion and excretion of metabolic products by the flukes as well as by *Helicobacter* spp. often co-infecting the final hosts, and (iii) the immunopathological response by the host, causing fibrosis and blockage of the bile ducts, may over time lead to oxidative damage to the epithelial cell DNA. Normal repair mechanisms and apoptosis are inhibited by certain fluke excretory/secretory products, and oncogenic mutations can occur as a consequence [117,120,121]. Eventually, the malignant transformations will lead to the development of cholangiocarcinoma (CCA), a highly lethal cancer of the bile duct, with an estimated median survival time of 4.3 months after diagnosis [121,122]. Up to now, while *O. felineus* has not been listed officially as a Type I carcinogen, there are indications both from animal experiments and reviews of the occurrence of cholangiocarcinoma in regions where the parasite is prevalent, that this fluke too has carcinogenic potential [123,124]. Opistorchiosis and clonorchiosis are two neglected yet emerging zoonotic diseases [82,125]. In the 1990s the total number of clonorchiosis and opistorchiosis cases was estimated at 7 and 10 million, respectively [83]. Nowadays, the total global number of infected people is estimated at about 20 million for *C. sinensis*, and at 10 million for *O. viverinni*, with most infections occurring in East Asia. Between 1.2 and 1.6 million people; mainly in Eastern Europe, are estimated to be infected with *O. felineus* [126]. Another 601 million people are thought to be at risk for *C. sinensis* infection, while 80 million for *Opisthorchis* spp. infection. The estimates for global burden due to clonorchiosis ranges between 275,370 and 522,863 Disability Adjusted Life Years (DALYS), whereas the estimates for opistorchiosis burden vary between 74,367 and 188,346 [78,79]. Based on reported infection intensities, it has been estimated that 8·2% of clonorchiosis, and 4·9% of opistorchiosis cases are symptomatic [78]. Increased importation of potentially contaminated fish, and newly acquired raw-fish eating habits might cause a future expansion in the distribution of the diseases [26].

3.1.2. Localisation of the Infection Risks for the Consumer in the Food Chain

Humans principally acquire opistorchiosis and clonorchiosis through the consumption of undercooked or raw infected freshwater cyprinid fish [91]. Moreover, other preparation styles, such as fermentation, pickling, inadequately freezing, or smoking of fish can pose a risk to the consumer [127]. In addition to cyprinid fish, some other fish types such eleotrids, cichlids, and osmerids are known to harbour MC [128]. Finally, some freshwater shrimp species have been reported to carry MC, and their consumption could thus pose the consumer at risk for infections [128].

While it is clear that the MC only encyst in fish/shrimp tissue, and thus the consumption of these tissues is the main infection source for acquiring the diseases, it is not well understood whether the preparation of contaminated fish can lead to contamination of the cooking environment, and thus subsequent human infection via contact with this environment. Indeed, *Opisthorchis* and *Clonorchis* spp. MC are distributed over different parts of the fish body, being the muscles, fins, heads and organs [129]. Therefore, it is not unimaginable that cutting boards might become contaminated with MC during the preparation of the fish dish. For instance, in their study performed in Laos, Araki et al. [130] reported on the observation that household cooks clean their chopping boards by scratching them using a knife and running water, and at times fish scales could still be found on the chopping boards after cleaning. Likewise, utensils and hands might become contaminated during food preparation or while eating [130]. Such contamination could explain infections in people reporting to never consume raw fish [131]. Finally, it has been hypothesized that people could become infected by drinking water contaminated by MC released from dead fish tissues, although this has never been proven to occur under field conditions [128].

3.1.3. Diagnostic Options in the Food Chain

All techniques to detect *Opisthorchis* spp. and *C. sinensis* MC in fish are based on postmortem investigations. Most consist of rather labour-intensive, traditional parasitological techniques. The most basic method is the direct compression method: fins, muscle, scales and subcutaneous tissues are collected, compressed between glass slides and examined for MC under a stereomicroscope [83]. In a second commonly used method, the artificial digestion method, the fish is divided in five parts (head, anterior and posterior trunk, tail and subcutaneous tissue). The tissues are subsequently ground, then digested using an artificial gastric juice (mostly a pepsin-HCl solution), and after sedimentation, MC are sought using a stereomicroscope [83]. A systematic review has shown that studies applying the compression method find higher prevalence estimates than those using the digestion method, however, their performance has not been compared directly [132]. Either way, both techniques require expert knowledge on the morphology of the different MC present in fish to allow for their correct differentiation (e.g., of Opisthorchiidae, Heterophyidae and Lecithodendriidae) [133].

Molecular techniques are however available to ensure correct differentiation of MC. For instance, a multiplex PCR was developed, targeting mitochondrial DNA, to allow differentiation of *Clonorchis* and *Opisthorchis* MC, particularly useful for regions where their distribution overlaps [134]. Moreover, a loop-mediated isothermal amplification (LAMP) was also developed for *C. sinensis* detection in fish, with a markedly higher sensitivity as compared to PCR [135]. Nevertheless, such molecular techniques remain expensive and not available in a commercial aquaculture context. Up to now, quick, easy, cheap diagnostic tools with adequate test performance are still lacking for *Clonorchis* and *Opisthorchis* detection in fish.

3.1.4. Prevention and Control Options in the Food Chain

Prevention of clonorchiosis and opisthorchis entails preventing the consumption of raw or undercooked contaminated fish/shrimps and avoiding contact with a contaminated environment. In many communities, the consumption of raw or undercooked fish is however deeply engrained in the culinary and social culture. For instance, the consumption of raw fish is often part of a social drinking events for males in many Asian countries [136–138]. Simply halting its consumption or changing the preparation style might therefore not be feasible. Moreover, MC are moderately tolerant to several preparation and preservation styles, which would usually be considered as effective in killing off germs. For instance, freezing at −12 °C had to be continued for 480 h to inactivate *Clonorchis* MC, whereas cooking at 50 °C required five hours to inactivate unspecified MC [127]. The only effective method is thoroughly heating the fish at a high temperature (e.g., 65 °C for 1 min) [127]. Another route for prevention in the intermediate host, namely a vaccine for freshwater fish, is currently being explored. Indeed, an oral vaccine based on *Bacillus subtilis* expressing enolase, was developed and is being tested [139,140]. In a wider preventive perspective, the life cycle can be broken by disconnecting human and animal faeces from the aquatic environment. At the moment, in some countries, the use of toilet types draining stool directly to ponds, still persist [141], and animal and human faeces continue being used as fish feed [139], therefore the transmission perpetuates. In Thailand, an opistorchiosis control program, called the "Lawa model" was developed. This EcoHealth/One Health-inspired approach combining human treatment with novel community-based and school health education, ecosystem monitoring and community participation, was implemented in the opistorchiosis endemic area at Lawa Lake, Khon Kaen province, Thailand [142]. The program successfully cut back infection rates in both humans, fish and snails, and will now be scaled up to other regions in Thailand and beyond [143].

3.2. Anisakidae

3.2.1. Introduction

Nematodes from the family Anisakidae are by far the most prevalent macroparasites in fish implicated in human disease. In their adult stage, anisakids are mostly found in the stomach of marine mammals as their definitive hosts, whereas the larval stages are found in smaller invertebrates, such as crustaceans as their first intermediate host, and in fish as their second intermediate host. Many commonly exploited marine fish species are infected, with the main zoonotic anisakid species being, although not limited to these, *Anisakis simplex* sensu stricto, *A. pegreffii*, *Pseudoterranova decipiens*, and *Contracaecum osculatum* [144]. Data on their prevalence in fish are known for many fish species with varying prevalences (e.g., up to 100% in herring) dependent on the geographical fishing grounds and seasons [145,146]. Yet ultimately, almost all species of teleost fish throughout the oceans may act as hosts where larvae can be found in the gastrointestinal tract or musculature of the fish [147,148].

Human infection, collectively named anisakidosis, takes place after consumption of undercooked fish containing a viable third-stage larva (L3), and may lead to several gastro-intestinal symptoms depending on the localisation of the larvae (acute, chronic, and ectopic) [149]. In addition to abdominal symptoms, a series of thermostable allergens present in *A. simplex* and *A. pegreffii* may also compromise human health with acute allergic manifestations ranging from urticaria, angioedema, asthma, conjunctivitis, to even a potential lethal anaphylactic shock [150]. The number of human health problems related to Anisakidae has not been fully quantified as global awareness of this foodborne parasite is still in its upsurge. Collection of more epidemiological data on the disease is therefore encouraged to provide better insights into the impact of Anisakidae on human disease. Nevertheless, specific country case studies from Spain and Norway have estimated 10,383–20,978 annual anisakidosis cases, as well as a 22% prevalence value of *Anisakis*-sensitization in certain regions [151,152]. Furthermore, it was stated that the most frequent cause of an anaphylactic episode due to a hidden allergen, is fish infected with *A. simplex* [153]. Finally, sufficient proof-of-principle has been provided in the past that demonstrates the transmissibility of anisakid allergenic peptides from fishmeal, as a feed component, to aquacultured fish and chicken meat [154–157]. These findings may significantly change the importance of these zoonotic nematodes from originally a purely fish borne food risk, to a much wider risk from several food sources (pork meat, chicken meat, aquacultured fish, etc.).

3.2.2. Localisation of the Infection Risks for the Consumer in the Food Chain

As abovementioned, consumers obtain an anisakid infection via consumption of infected fish that has not been sufficiently frozen or cooked [149]. Different preventive measurements can be taken to avoid this (see below), though these are primarily related to the infection risk with viable larvae. By adequate cooking or freezing, *Anisakis*-sensitized consumers remain at risk given the thermoresistant characteristics of the allergens [158]. Moreover, removal of the larva does not at all guarantee freedom of allergens since some of these anisakid allergens are excretory-secretory products excreted/secreted by the larval body during its migration through the fish flesh. As such, patients may be exposed to them when consuming fish, even though the larvae might have been removed during the quality control of the fish [159]. Finally, and different to other food allergies, occupational allergies in aquaculture and fishery workers after inhalation and/or skin contact with anisakid allergens have been reported [160].

3.2.3. Diagnostic Options in the Food Chain

Fish and fishery products can be examined for the presence of anisakid larvae and traces by a variety of detection methods. In the industrial setting (i.e., on the boat, in processing plants), candling is the most routinely used method for the detection of anisakids in commercial fish fillets. It entails a brief visual inspection of fish fillets on a light table to spot and manually remove parasites [161]. While this technique has the major advantage of not affecting the fish quality and thus allowing consumption afterwards, it is labour intensive and as such highly costly. Moreover, studies report a poor sensitivity with up to 76% of the larvae not being recovered, although this is dependent on the fish (colour, thickness, skin), the larvae (size, colour), and the skills of the inspector [161–164]. In laboratory settings on the other hand, a range of highly accurate alternatives such as UV press method, enzymatic digestion, and immunoassays are available and implemented, though the complete destruction of the fish tissue, renders these methods unfit for the application in the industry [165–167]. As a result, candling is still the standard method for the detection and removal of anisakids from fish fillets on an industrial scale. Future research should thus look into the development of more accurate, fast, non-destructive scanning methods to replace candling.

As dead larvae may still be responsible for allergic reactions in sensitized consumers, other tools directly targeting anisakid proteins have been developed [168]. However, the presence of anisakid proteins does not necessarily correlate with an allergic reaction since not all proteins are allergens. To deal with this, liquid chromatography tandem mass spectrometry methods and allergen specific enzyme-linked immunosorbent assays are available [169,170], but high equipment costs and their destructive entity still hamper their use in an industrial setting.

3.2.4. Prevention and Control Options in the Food Chain

Prevention of gastro-intestinal anisakidosis is simply based on avoiding the ingestion of a live L3 in raw/undercooked fish. In the food chain, this can be accomplished by maintaining a cold chain from boat to plate and by immediate gutting of the fish on the boat in order to avoid *post-mortem* migration from the fish gut to musculature. Care must be taken to destroy these viscera rather than disposing them in the sea water as this practice results in, once again, dissemination of the parasites [171]. Freezing of the fish has also been recommended by public health agencies and included in the current legislation of the European Union and Japan [172,173]. Specifically, food industry business that sell fish intended for raw, marinated, or salted consumption, must first freeze the fish in its entirety at −20 °C for >24 h, or −35 °C for >15 h to ensure killing of the larvae [174]. It is to be expected though, that in a big container of fish, not all parts of the fish will reach a temperature to kill all larvae. Ideally, freezing should therefore be followed by a period of storage in the frozen state to ensure complete elimination of the parasite. In addition to freezing/cooking, a visual check for larvae by the fish industry can be conducted by ways of candling (see above). The abovementioned limitations of this tool, however, make this tool insufficient for full clearance of the parasite and emphasizes the importance of adequate deep-freezing. Finally, at the stage of not only the consumer, but also the relevant governmental institutes and medical/veterinary staff, raising awareness regarding the presence of these parasites and possible preventive measures (e.g., cooking, consumption behaviour) is a principle preventive measure that should be taken.

While a variety of measures can be taken to reduce the incidence of human gastro-intestinal anisakidosis, it is important to consider that an *Anisakis*-sensitized individual may still develop an allergic reaction if the larva is dead/removed or if its traces are present [158]. So far, no allergen destruction process has been discovered, and standard testing for anisakid allergens is up to date not conducted on any food type. A suggestion, however, could be to label high-risk products for the possible presence of *Anisakis* allergens to warn sensitized patients who do not tolerate even properly cooked or canned fish.

4. Discussion and Conclusions

In this paper we have attempted to highlight the presence of a number of FBPs in different food chains, the risk of infection of humans by parasites, and their diagnostic and control challenges. We have also highlighted the complexity of their life cycles whereby different parasite stages, infective to the human host, may occur at different places within a food chain, for example, in the case of *T. solium* whereby the meat-borne component will place humans at risk when consuming pig carcasses infected with viable cysticerci, but humans may also be at risk when consuming vegetables or fruits contaminated with the parasite's eggs from the farm, or via contaminated water. *Toxoplasma gondii* is also a good example of multiple complex ways of transmission to humans.

As mentioned, the increased popularity of consumption of raw and ready-to-eat vegetables may pose a new potential risk for consumers, since most post-harvest processing measures do not guarantee the complete removal of oocysts/eggs or their effective inactivation [58]. The same can be said for raw fish/meat consumption, considering the generally low sensitivities or simply non implementation of detection tools.

We have also highlighted how interactions in interface areas may lead to the introduction of parasites from the *forest to fork* food chain into the *farm to fork* food chain. The impact of increasing contact between wildlife, livestock and humans on food safety needs to be considered carefully, especially bearing in mind the trend towards outdoor farming [3] and the increase in relevant wildlife species such as the wild boar [27].

The multidisciplinary One Health approach will be needed to deal with the impact of globalisation and climate change on the transmission of foodborne parasites to humans [26]. The One Health approach is increasingly applied by governments, yet mostly considering human and animal components. The environment is the most understudied component of parasite transmission, yet is very important. There is an urgent need for researchers to take up this work, e.g., to develop highly performing diagnostic tools to detect environmental stages, to assess their viability, etc.

The availability of cheap, easy to use, highly performing diagnostic tools fit for purpose is another challenge. While an increased amount of effort has been put into developments of new tests in different formats, and target product profiles are being developed (e.g., for *T. solium* [175]), their large-scale evaluation in relevant matrices is often lacking. Integrated systems looking into more integrated sample collections where relevant (e.g., environmental samples, meat samples from pig farms with outdoor access), and test systems allowing multipathogen detection are other aspects that need to be considered.

Monitoring and surveillance, and improved reporting based on proper diagnostics would be useful to implement for a number of FBPs. As resources are always limited, risk-based surveillance might help in the prioritisation, and lead to an efficient and effective allocation of, resources [21]. For *Trichinella* spp. and *T. saginata* this system has been put in place in the EU (see Sections 2.1.4 and 2.2.4).

Integration may be considered not only on a detection level, but also on a prevention/control level, where chosen interventions may impact on several pathogens. The implementation of highly controlled and confined farming systems has led to commercial pork originating from these farms to be free of *Trichinella* spp. These high levels of biosecurity not only impact the presence of *Trichinella* spp., but also *Taenia* spp. and *T. gondii* are rarely detected. Moreover, a focus on herd health and management, biosecurity at the farm level would tackle not only a number of FBPs, but also other pathogens such as Salmonella and Campylobacter, and this at a primary, pre-harvest level [176]. Nevertheless, the trend towards farming systems with more outdoor access, encouraged by animal welfare expectations, increase infection risk [3]. The latter should be compensated with either pre-harvest (e.g., improved detection) or post-harvest (e.g., inactivation via cooking) measures.

To finalise, the consumer also plays an essential role in risk of exposure, via human behaviour in choice of food to consume but also in the culinary practices in how to process food for consumption. While cooking at a sufficient temperature would deal with all parasites described above, often, this is not done. Other practices such as marinating, salting, drying, smoking are more often than not, insufficient to inactivate the parasite. From a different perspective, human migration and import of infected animals from endemic to non-endemic areas may lead to a (re)introduction of pathogens [3]. Consumer education regarding the risks and awareness creation would be highly beneficial.

Author Contributions: Conceptualization, S.G., P.D., V.D. and G.S.; writing S.G., P.D., G.S. and V.D. Original Draft Preparation, Writing—Review & Editing, S.G.; Visualization, S.G. All authors have read and agreed to the published version of the manuscript.

Funding: This research received no external funding.

Conflicts of Interest: The authors declare no conflict of interest.

References

1. FAO/WHO (Food and Agriculture Organization of the United Nations/World Health Organization). *Multicriteria-Based Ranking for Risk Management of Food-Borne Parasites*; Microbiological Risk Assessment Series No. 23, Rome; World Health Organization: Geneva, Switzerland, 2014; p. 302.
2. WHO. *World Health Organisation, WHO Estimates of the Global Burden of Foodborne Diseases Foodborne Disease Burden Epidemiology Reference Group 2007–2015*; World Health Organization: Geneva, Switzerland, 2015.
3. Gabriël, S.; Johansen, M.V.; Pozio, E.; Smit, G.S.; Devleesschauwer, B.; Allepuz, A.; Papadopoulos, E.; van der Giessen, J.; Dorny, P. Human migration and pig/pork import in the European Union: What are the implications for *Taenia solium* infections? *Vet. Parasitol.* **2015**, *213*, 38–45. [CrossRef] [PubMed]

4. Murrell, K.D.; Dorny, P.; Flisser, A.; Geerts, S.; Kyvsgaard, N.C.; McManus, D.; Nash, T.E.; Pawlowski, Z.S. *WHO/FAO/OIE Guidelines for the Surveillance, Prevention & Control of Taeniosis/Cysticercosis*; Murrell, K.D., Ed.; OIE: Paris, France, 2005.

5. Jansen, F.; Dorny, P.; Berkvens, D.; Van Hul, A.; Van den Broeck, N.; Makay, C.; Praet, N.; Eichenberger, R.M.; Deplazes, P.; Gabriël, S. High prevalence of bovine cysticercosis found during evaluation of different post-mortem detection techniques in Belgian slaughterhouses. *Vet. Parasitol.* **2017**, *244*, 1–6. [CrossRef] [PubMed]

6. Trevisan, C.; Mkupasi, E.M.; Ngowi, H.A.; Forkman, B.; Johansen, M.V. Severe seizures in pigs naturally infected with *Taenia solium* in Tanzania. *Vet. Parasitol.* **2016**, *220*, 67–71. [CrossRef] [PubMed]

7. Trevisan, C.; Johansen, M.V.; Mkupasi, E.M.; Ngowi, H.A.; Forkman, B. Disease behaviours of sows naturally infected with *Taenia solium* in Tanzania. *Vet. Parasitol.* **2017**, *235*, 69–74. [CrossRef] [PubMed]

8. Jansen, F.; Dorny, P.; Trevisan, C.; Dermauw, V.; Laranjo-González, M.; Allepuz, A.; Dupuy, C.; Krit, M.; Gabriël, S.; Devleesschauwer, B. Economic impact of bovine cysticercosis and taeniosis caused by *Taenia saginata* in Belgium. *Parasit. Vectors* **2018**, *11*, 241. [CrossRef]

9. Hobbs, E.C.; Mwape, K.E.; Devleesschauwer, B.; Gabriël, S.; Chembensofu, M.; Mambwe, M.; Phiri, I.K.; Masuku, M.; Zulu, G.; Colston, A.; et al. *Taenia solium* from a community perspective: Preliminary costing data in the Katete and Sinda districts in Eastern Zambia. *Vet. Parasitol.* **2018**, *251*, 63–67. [CrossRef]

10. Trevisan, C.; Devleesschauwer, B.; Schmidt, V.; Winkler, A.S.; Harrison, W.; Johansen, M.V. The societal cost of *Taenia solium* cysticercosis in Tanzania. *Acta Trop.* **2017**, *165*, 141–154. [CrossRef]

11. Saelens, G.; Robertson, L.; Gabriël, S. Diagnostic tools for the detection of taeniid eggs in different environmental matrices: A systematic review. *Food Waterborne Parasitol.* **2022**, *26*, e00145. [CrossRef]

12. Jansen, F.; Dorny, P.; Gabriël, S.; Dermauw, V.; Johansen, M.V.; Trevisan, C. The survival and dispersal of *Taenia* eggs in the environment: What are the implications for transmission? A systematic review. *Parasit. Vectors* **2021**, *14*, 88. [CrossRef]

13. Bucur, I.; Gabriël, S.; Van Damme, I.; Dorny, P.; Vang Johansen, M. Survival of *Taenia saginata* eggs under different environmental conditions. *Vet. Parasitol.* **2019**, *266*, 88–95. [CrossRef]

14. Gabriël, S.; Mwape, K.E.; Phiri, I.K.; Devleesschauwer, B.; Dorny, P. *Taenia solium* control in Zambia: The potholed road to success. *Parasite Epidemiol. Control* **2018**, *4*, e00082. [CrossRef]

15. Thys, S.; Mwape, K.E.; Lefèvre, P.; Dorny, P.; Phiri, A.M.; Marcotty, T.; Phiri, I.K.; Gabriël, S. Why pigs are free-roaming: Communities' perceptions, knowledge and practices regarding pig management and taeniosis/cysticercosis in a *Taenia solium* endemic rural area in Eastern Zambia. *Vet. Parasitol.* **2016**, *225*, 33–42. [CrossRef]

16. Dorny, P.; Phiri, I.K.; Vercruysse, J.; Gabriel, S.; Willingham, A.L., 3rd; Brandt, J.; Victor, B.; Speybroeck, N.; Berkvens, D. A Bayesian approach for estimating values for prevalence and diagnostic test characteristics of porcine cysticercosis. *Int. J. Parasitol.* **2004**, *34*, 569–576. [CrossRef] [PubMed]

17. Chembensofu, M.; Mwape, K.E.; Van Damme, I.; Hobbs, E.; Phiri, I.K.; Masuku, M.; Zulu, G.; Colston, A.; Willingham, A.L.; Devleesschauwer, B.; et al. Re-visiting the detection of porcine cysticercosis based on full carcass dissections of naturally *Taenia solium* infected pigs. *Parasit. Vectors* **2017**, *10*, 572. [CrossRef]

18. Jansen, F.; Dorny, P.; Berkvens, D.; Gabriël, S. Bovine cysticercosis and taeniosis: The effect of an alternative post-mortem detection method on prevalence and economic impact. *Prev. Vet. Med.* **2018**, *161*, 1–8. [CrossRef]

19. Waema, M.W.; Misinzo, G.; Kagira, J.M.; Agola, E.L.; Ngowi, H.A. DNA-Detection Based Diagnostics for *Taenia solium* Cysticercosis in Porcine. *J. Parasitol. Res.* **2020**, *2020*, 5706981. [CrossRef]

20. EU. Commission implementing regulation (EU) 2019/627 of 15 March 2019 laying down uniform practical arrangements for the performance of official controls on products of animal origin intended for human consumption in accordance with Regulation (EU) 2017/625 of the European Parliament and of the Council and amending Commission Regulation (EC) No 2074/2005 as regards official controls. *Off. J. Eur. Union* **2019**, *L 131*, 72.

21. Alban, L.; Häsler, B.; van Schaik, G.; Ruegg, S. Risk-based surveillance for meat-borne parasites. *Exp. Parasitol.* **2020**, *208*, 107808. [CrossRef]

22. Lightowlers, M.W. Vaccines for prevention of cysticercosis. *Acta Trop.* **2003**, *87*, 129–135. [CrossRef]

23. Lightowlers, M.W. Eradication of *Taenia solium* cysticercosis: A role for vaccination of pigs. *Int. J. Parasitol.* **2010**, *40*, 1183–1192. [CrossRef]

24. Gabriël, S.; Mwape, K.E.; Hobbs, E.C.; Devleesschauwer, B.; Van Damme, I.; Zulu, G.; Mwelwa, C.; Mubanga, C.; Masuku, M.; Mambwe, M.; et al. Evidence for potential elimination of active *Taenia solium* transmission in Africa? *N. Engl. J. Med.* **2020**, *383*, 396–397. [CrossRef] [PubMed]

25. De Coster, T.; Van Damme, I.; Baauw, J.; Gabriël, S. Recent advancements in the control of *Taenia solium*: A systematic review. *Food Waterborne Parasitol.* **2018**, *13*, e00030. [CrossRef] [PubMed]

26. Pozio, E. How globalization and climate change could affect foodborne parasites. *Exp. Parasitol.* **2020**, *208*, 107807. [CrossRef] [PubMed]

27. Murrell, K.D.; Pozio, E. Worldwide Occurrence and Impact of Human Trichinellosis, 1986–2009. *Emerg. Infect. Dis.* **2011**, *17*, 2194–2202. [CrossRef] [PubMed]

28. Pozio, E. World distribution of *Trichinella* spp. infections in animals and humans. *Vet. Parasitol.* **2007**, *149*, 3–21. [CrossRef]

29. Pozio, E.; Hoberg, E.; La Rosa, G.; Zarlenga, D.S. Molecular taxonomy, phylogeny and biogeography of nematodes belonging to the *Trichinella* genus. *Infect Genet. Evol.* **2009**, *9*, 606–616. [CrossRef]

30. Crisóstomo-Jorquera, V.; Landaeta-Aqueveque, C. The genus *Trichinella* and its presence in wildlife worldwide: A review. *Transbound. Emerg. Dis.* **2022**, *69*, e1269–e1279. [CrossRef]

31. Boireau, P.; Vallée, I.; Roman, T.; Perret, C.; Mingyuan, L.; Gamble, H.R.; Gajadhar, A. *Trichinella* in horses: A low frequency infection with high human risk. *Vet. Parasitol.* **2000**, *93*, 309–320. [CrossRef]

32. Gottstein, B.; Pozio, E.; Nöckler, K. Epidemiology, diagnosis, treatment, and control of trichinellosis. *Clin. Microbiol. Rev.* **2009**, *22*, 127–145. [CrossRef]

33. Liu, M.; Boireau, P. Trichinellosis in China: Epidemiology and control. *Trends Parasitol.* **2002**, *18*, 553–556. [CrossRef]

34. Noeckler, K.; Pozio, E.; van der Giessen, J.; Hill, D.E.; Gamble, H.R. International Commission on Trichinellosis: Recommendations on post-harvest control of *Trichinella* in food animals. *Food Waterborne Parasitol.* **2019**, *21*, e00041. [CrossRef]

35. Gamble, H.R. *Trichinella* spp. control in modern pork production systems. *Food Waterborne Parasitol.* **2022**, *28*, e00172. [CrossRef]

36. Bruschi, F.; Gómez-Morales, M.A.; Hill, D.E. International Commission on Trichinellosis: Recommendations on the use of serological tests for the detection of *Trichinella* infection in animals and humans. *Food Waterborne Parasitol.* **2019**, *5*, e00032. [CrossRef]

37. Barlow, A.; Roy, K.; Hawkins, K.; Ankarah, A.A.; Rosenthal, B. A review of testing and assurance methods for *Trichinella* surveillance programs. *Food Waterborne Parasitol.* **2021**, *9*, e00129. [CrossRef]

38. Li, X.; Liu, W.; Wang, J.; Zou, D.; Wang, X.; Yang, Z.; Yin, Z.; Cui, Q.; Shang, W.; Li, H.; et al. Rapid detection of *Trichinella spiralis* larvae in muscles by loop-mediated isothermal amplification. *Int. J. Parasitol.* **2012**, *42*, 1119–1126. [CrossRef]

39. Van Knapen, F. Control of trichinellosis by inspection and farm management practices. *Vet. Parasitol.* **2000**, *93*, 385–392. [CrossRef]

40. Kapel, C.M. Changes in the EU legislation on *Trichinella* inspection—New challenges in the epidemiology. *Vet. Parasitol.* **2005**, *132*, 189–194. [CrossRef]

41. European Centre for Disease Prevention and Control. Trichinellosis. In *ECDC. Annual Epidemiological Report for 2019*; ECDC: Stockholm, Sweden, 2021. Available online: https://www.ecdc.europa.eu/en/publications-data/trichinellosis-annual-epidemiological-report-2019 (accessed on 23 November 2022).

42. Dupouy-Camet, J.; Murrell, K.D. *FAO/WHO/OIE Guidelines for the Surveillance, Management. Prevention and Control of Trichinellosis*; OIE: Paris, France, 2007; pp. 101–110.

43. Burke, R.; Masuoka, P.; Murrell, K.D. Swine *Trichinella* infection and geographic information system tools. *Emerg. Infect Dis.* **2008**, *14*, 1109–1111. [CrossRef]

44. Tang, B.; Li, J.; Li, T.; Xie, Y.; Guan, W.; Zhao, Y.; Yang, S.; Liu, M.; Xu, D. Vaccines as a Strategy to Control Trichinellosis. *Front. Microbiol.* **2022**, *23*, 857786. [CrossRef]

45. Tenter, A.M.; Heckeroth, A.; Weiss, L. *Toxoplasma gondii*: From animals to humans. *Int. J. Parasitol.* **2000**, *30*, 1217–1258. [CrossRef]

46. Torgerson, P.R.; Mastroiacovo, P. The global burden of congenital toxoplasmosis: A systematic review. *Bull. WHO* **2013**, *91*, 501–508. [CrossRef] [PubMed]

47. Dubey, J.P. *Toxoplasmosis of Animals and Humans*; CRC Press: Boca Raton, FL, USA, 2010.

48. Cook, A.J.; Gilbert, R.E.; Buffolano, W.; Zufferey, J.; Petersen, E.; Jenum, P.A.; Foulon, W.; Semprini, A.E.; Dunn, D.T. Sources of toxoplasma infection in pregnant women: European multi- centre case-control study. European Research Network on Congenital Toxoplasmosis. *British Med. J.* **2000**, *321*, 142–147. [CrossRef] [PubMed]

49. Tenter, A.M. *Toxoplasma gondii* in animals used for human consumption. *Mem. Inst. Oswaldo Cruz.* **2009**, *104*, 364–369. [CrossRef] [PubMed]

50. Hill, D.; Coss, C.; Dubey, J.P.; Wroblewski, K.; Sautter, M.; Hosten, T.; Munoz-Zanzi, C.; Mui, E.; Withers, S.; Boyer, K.; et al. Identification of a sporozoite-specific antigen from *Toxoplasma gondii*. *J. Parasitol.* **2011**, *97*, 328–337. [CrossRef] [PubMed]

51. Kijlstra, A.; Meerburg, B.G.; Bos, A.P. Food safety in free-range and organic livestock systems: Risk management and responsibility. *J. Food Prot.* **2009**, *72*, 2629–2637. [CrossRef]

52. Verhelst, D.; De Craeye, S.; Vanrobaeys, M.; Czaplicki, G.; Dorny, P.; Cox, E. Seroprevalence of *Toxoplasma gondii* in domestic sheep in Belgium. *Vet. Parasitol.* **2014**, *205*, 57–61. [CrossRef]

53. Dubey, J.P.; Jones, J.L. *Toxoplasma gondii* infection in humans and animals in the United States. *Int. J. Parasitol.* **2008**, *38*, 1257–1278. [CrossRef]

54. Bacci, C.; Vismarra, A.; Mangia, C.; Bonardi, S.; Bruini, I.; Genchi, M.; Kramer, L.; Brindani, F. Detection of *Toxoplasma gondii* in free-range, organic pigs in Italy using serological and molecular methods. *Int. J. Food Microbiol.* **2015**, *202*, 54–56. [CrossRef]

55. De Berardinis, A.; Paludi, D.; Pennisi, L.; Vergara, A. *Toxoplasma gondii*, a foodborne pathogen in the swine production chain from a European perspective. *Foodborne Pathogens Dis.* **2017**, *14*, 637–648. [CrossRef]

56. Kijlstra, A.; Eissen, O.A.; Cornelissen, J.; Munniksma, K.; Eijck, I.; Kortbeek, T. *Toxoplasma gondii* infection in animal-friendly pig production systems. *Investig. Ophthalmol. Vis. Sci.* **2004**, *45*, 3165–3169. [CrossRef]

57. Thomas, M.; Aubert, D.; Escotte-Binet, S.; Durand, B.; Robert, C.; Geers, R.; Alliot, A.; Belbis, G.; Villena, I.; Blaga, R. Anatomical distribution of *Toxoplasma gondii* in naturally and experimentally infected lambs. *Parasite* **2022**, *29*, 3. [CrossRef]

58. Shapiro, K.; Kim, M.; Rajal, V.B.; Arrowood, M.J.; Packham, A.; Aguilar, B.; Wuertz, S. Simultaneous detection of four protozoan parasites on leafy greens using a novel multiplex PCR assay. *Food Microbiol.* **2019**, *84*, 103252. [CrossRef]

59. Dumètre, A.; Dardé, M.L. How to detect *Toxoplasma gondii* oocysts in environmental samples? *FEMS Microbiol. Rev.* **2003**, *27*, 651–661. [CrossRef]

60. Mirza Alizadeh, A.; Jazaeri, S.; Shemshadi, B.; Hashempour-Baltork, F.; Sarlak, Z.; Pilevar, Z.; Hosseini, H.A. Review on inactivation methods of *Toxoplasma gondii* in foods. *Pathog. Glob. Health* **2018**, *112*, 306–319. [CrossRef]

61. Shapiro, K.; Vanwormer, E.; Aguilar, B.; Conrad, P.A. Surveillance for *Toxoplasma gondii* in California mussels (*Mytilus californianus*) reveals transmission of atypical genotypes from land to sea. *Environ. Microbiol.* **2015**, *17*, 4177–4188. [CrossRef]

62. Lindsay, D.S.; Collins, M.V.; Mitchell, S.M.; Wetch, C.N.; Rosypal, A.C.; Flick, G.J.; Zajac, A.M.; Lindquist, A.; Dubey, J.P. Survival of *Toxoplasma gondii* oocysts in eastern oysters (*Crassostrea virginica*). *J. Parasitol.* **2004**, *90*, 1054–1057. [CrossRef]

63. Hohweyer, J.; Dumètre, A.; Aubert, D.; Azas, N.; Villena, I. Tools and methods for detecting and characterizing *Giardia*, *Cryptosporidium*, and *Toxoplasma* parasites in marine mollusks. *J. Food Prot.* **2013**, *76*, 1649–1657. [CrossRef]

64. Gisbert Algaba, I.; Geerts, M.; Jennes, M.; Coucke, W.; Opsteegh, M.; Cox, E.; Dorny, P.; Dierick, K.; De Craeye, S. A more sensitive, efficient and ISO 17025 validated Magnetic Capture real time PCR method for the detection of archetypal *Toxoplasma gondii* strains in meat. *Int. J. Parasitol.* **2017**, *47*, 875–884. [CrossRef]

65. Opsteegh, M.; Dam-Deisz, C.; de Boer, P.; De Craeye, S.; Faré, A.; Hengeveld, P.; Luiten, R.; Schares, G.; van Solt-Smits, C.; Verhaegen, B.; et al. Methods to assess the effect of meat processing on viability of *Toxoplasma gondii*: Towards replacement of mouse bioassay by in vitro testing. *Int. J. Parasitol.* **2020**, *50*, 357–369. [CrossRef]

66. Opsteegh, M.; Maas, M.; Schares, G.; van der Giessen, J. Relationship between seroprevalence in the main livestock species and presence of *Toxoplasma gondii* in meat (GP/EFSA/BIOHAZ/2013/01) An extensive literature review, Final report. *EFSA Support Publ.* **2016**, *13*, 1–294. [CrossRef]

67. Zdolec, N.; Kiš, M. Meat safety from farm to slaughter—Risk-based control of *Yersinia enterocolitica* and *Toxoplasma gondii*. *Processes* **2021**, *9*, 815. [CrossRef]

68. Wehbe, K.; Pencole, L.; Lhuaire, M.; Sibiude, J.; Mandelbrot, L.; Villena, I.; Picone, O. Hygiene measures as primary prevention of toxoplasmosis during pregnancy: A systematic review. *J. Gynecol. Obstet. Hum. Reprod.* **2022**, *51*, 102300. [CrossRef] [PubMed]

69. EU. Commission Regulation No. 219/2014 Commission Regulation (EU) No 219/2014 of 7 March 2014 amending Annex I to Regulation (EC) No 854/2004 of the European Parliament and of the Council as regards the specific requirements for post-mortem inspection of domestic swine. *Off. J. Eur. Union* **2014**, *L 69*, 99–100.

70. EFSA (European Food Safety Agency). Scientific Opinion on the public health hazard to be covered by inspection of meat (swine). *EFSA J.* **2011**, *9*, 2351. [CrossRef]

71. Rani, S.; Pradhan, A.K. Evaluating uncertainty and variability associated with *Toxoplasma gondii* survival during cooking and low temperature storage of fresh cut meats. *Int. J. Food Microbiol.* **2021**, *341*, 109031. [CrossRef]

72. Buxton, D.; Thomson, K.; Maley, S.; Wright, S.; Bos, H.J. Vaccination of sheep with a live incomplete strain (S48) of *Toxoplasma gondii* and their immunity to challenge when pregnant. *Vet. Rec.* **1991**, *129*, 89–93. [CrossRef]

73. Yuan, Z.G.; Zhang, X.X.; Lin, R.Q.; Petersen, E.; He, S.; Yu, M.; He, X.H.; Zhou, D.H.; He, Y.; Li, H.X.; et al. Protective effect against toxoplasmosis in mice induced by DNA immunization with gene encoding *Toxoplasma gondii* ROP18. *Vaccine* **2011**, *29*, 6614–6619. [CrossRef]

74. Bonačić Marinović, A.A.; Opsteegh, M.; Deng, H.; Suijkerbuijk, A.W.M.; van Gils, P.F.; van der Giessen, J. Prospects of toxoplasmosis control by cat vaccination. *Epidemics* **2020**, *30*, 100380. [CrossRef]

75. Mas-Coma, S.; Esteban, J.; Bargues, M. Epidemiology of human fascioliasis: A review and proposed new classification. *Bull. WHO* **1999**, *77*, 340–346.

76. Chen, M.G.; Mott, K.E. Progress in assessment of morbidity due to *Fasciola hepatica* infection: A review of recent literature. *Trop Dis. Bull* **1990**, *87*, 1–38.

77. WHO. *Working to Overcome the Global Impact of Neglected Tropical Diseases. First WHO Report on Neglected Tropical Diseases*; World Health Organization: Geneva, Switzerland, 2010. [CrossRef]

78. Fürst, T.; Keiser, J.; Utzinger, J. Global burden of human food-borne trematodiasis: A systematic review and meta-analysis. *Lancet Inf. Dis.* **2012**, *12*, 210–221. [CrossRef]

79. Havelaar, A.; Kirk, M.; Torgerson, P.; Gibb, H.; Hald, T.; Lake, R.; Praet, N.; Bellinger, D.; de Silva, N.; Gargouri, N.; et al. Disease Burden Epidemiology Reference Group. World Health Organization global estimates and regional comparisons of the burden of foodborne disease in 2010. *PLoS Med.* **2015**, *12*, e1001923. [CrossRef]

80. Hopkins, D. Homing in on helminths. *Am. J. Trop.* **1992**, *46*, 626–634. [CrossRef]

81. Rim, H.; Madsen, H.; Cross, J.H.; Rim, H.; Farag, H.F.; Sornmani, S. Food-borne trematodes: Ignored or emerging? Food-borne Ignored or Emerging? *Parasitol. Today* **1994**, *10*, 207–209. [CrossRef]

82. Keiser, J.; Utzinger, J. Emerging foodborne trematodiasis. *Emerg. Inf. Dis.* **2005**, *11*, 1507–1514. [CrossRef]

83. WHO. *Control of Foodborne Trematode infections: Report of a WHO Study Group*; 1995 WHO Technical Report Series 894; WHO: Geneva, Switzerland, 1995.

84. Dermauw, V.; Muchai, J.; Kappany, Y.; Al Fajardo, A.L.; Dorny, P. Human fascioliasis in Africa: A systematic review. *PLoS ONE* **2021**, *16*, e0261166. [CrossRef] [PubMed]

85. Dorny, P.; Praet, N.; Deckers, N.; Gabriel, S. Emerging food-borne parasites. *Vet. Parasitol.* **2009**, *163*, 196–206. [CrossRef] [PubMed]

86. Mas-Coma, S.; Valero, M.; Bargues, M. Chapter 2—Fasciola, lymnaeids and human fascioliasis, with a global overview on disease transmission, epidemiology, evolutionary genetics, molecular epidemiology and control. In *Advances in Parasitology*, 1st ed.; Elsevier: Amsterdam, The Netherlands, 2009. [CrossRef]

87. Fox, N.J.; White, P.C.L.; McClean, C.J.; Marion, G.; Evans, A.; Hutchings, M.R. Predicting impacts of climate change on *fasciola hepatica* risk. *PLoS ONE* **2011**, *6*, 19–21. [CrossRef] [PubMed]
88. Mas-Coma, S.; Valero, M.A.; Bargues, M.D. Climate change effects on trematodiases, with emphasis on zoonotic fascioliasis and schistosomiasis. *Vet. Parasitol.* **2009**, *163*, 264–280. [CrossRef]
89. Mas-Coma, S. Human fascioliasis emergence risks in developed countries: From individual patients and small epidemics to climate and global change impacts. *Enferm. Infec. y Microbiol. Clinic.* **2020**, *38*, 253–256. [CrossRef]
90. Mas-Coma, S.; Bargues, M.D.; Valero, M.A. Human fascioliasis infection sources, their diversity, incidence factors, analytical methods and prevention measures. *Parasitology* **2018**, *145*, 1665–1699. [CrossRef]
91. Keiser, J.; Utzinger, J. Food-borne trematodiases. *Clin. Microbiol. Rev.* **2009**, *22*, 466–483. [CrossRef]
92. Mas-Coma, S. Human Fascoliasis: Epidemiological patterns in human endemic areas of South America, Africa and Asia. *Southeast Asian J. Trop. Med. Public Health* **2004**, *35*, 1–11.
93. Taghipour, A.; Zaki, L.; Rostami, A.; Foroutan, M.; Ghaffarifar, F.; Fathi, A.; Abdoli, A. Highlights of human ectopic fascioliasis: A systematic review. *Inf. Dis.* **2019**, *51*, 785–792. [CrossRef]
94. Mas-Coma, S.; Agramunt, V.; Valero, M. Chapter 2. Neurological and Ocular Fascioliasis in Humans. In *Advances in Parasitology*; Academic Press: Cambridge, MA, USA, 2014; Volume 84. [CrossRef]
95. González-Miguel, J.; Valero, M.A.; Reguera-Gomez, M.; Mas-Bargues, C.; Bargues, M.D.; Simón, F.; Mas-Coma, S. Numerous *Fasciola* plasminogen-binding proteins may underlie blood-brain barrier leakage and explain neurological disorder complexity and heterogeneity in the acute and chronic phases of human fascioliasis. *Parasitol* **2019**, *146*, 284–298. [CrossRef]
96. Mailles, A.; Capek, A.; Ajana, F.; Schepens, C.; Ilef, D.; Vaillant, V. Commercial watercress as an emerging source of fascioliasis in Northern France in 2002: Results from an outbreak investigation. *Epidemiol. Inf.* **2006**, *134*, 942–945. [CrossRef]
97. Micic, D.; Oto, A.; Charlton, M.R.; Benoit, J.-L.; Siegler, M. Hiding in the Water. *N. Engl. J. Med.* **2020**, *382*, 1844–1849. [CrossRef]
98. Sah, R.; Khadka, S.; Khadka, M.; Gurubacharya, D.; Sherchand, J.B.; Parajuli, K.; Shah, N.P.; Kattel, H.P.; Pokharel, B.M.; Rijal, B. Human fascioliasis by *Fasciola hepatica*: The first case report in Nepal. *BMC Res. Notes* **2017**, *10*, 10–13. [CrossRef]
99. Goral, V.; Senturk, S.; Mete, O.; Cicek, M.; Ebik, B.; Kaya, B. A Case of Biliary Fascioliasis by *Fasciola gigantica* in Turkey. *Korean J. Parasitol.* **2011**, *49*, 65–68. [CrossRef]
100. Nino Incani, R.; Vieira, J.; Pacheco, M.; Planchart, S.; Amarista, M.; Lazdins, J. Human infection by *Fasciola hepatica* in Venezuela: Report of a geriatric case. *Investig. Clin.* **2003**, *44*, 255–260.
101. Ashrafi, K.; Valero, M.A.; Massoud, J.; Sobhani, A.; Solaymani-Mohammadi, S.; Conde, P.; Khoubbane, M.; Bargues, M.D.; Mas-Coma, S. Plant-borne human contamination by fascioliasis. *Am. J. Trop. Med. Hyg.* **2006**, *75*, 295–302. [CrossRef] [PubMed]
102. Milas, S.; Rossi, C.; Philippart, I.; Dorny, P.; Bottieau, E. Autochthonous human Fascioliasis, Belgium. *Em. Infect. Dis.* **2020**, *26*, 155–157. [CrossRef] [PubMed]
103. Rodríguez-Ulloa, C.; Rivera-jacinto, M.; del Valle-Mendoza, J.; Cerna, C.; Hoban, C.; Ortiz, P. Risk factors for human fascioliasis in schoolchildren in Baños del Inca. *Trans. Royal Soc. Trop. Med. Hyg.* **2018**, *112*, 1–7. [CrossRef] [PubMed]
104. Choi, S.; Park, S.; Hong, S.; Shin, H.; Jung, B.K.; Kim, M.J. Green vegetable juice as a potential source of human fascioliasis in Korea. *One Health* **2022**, *15*, 100441. [CrossRef] [PubMed]
105. Curtale, F.; Mas-Coma, S.; Hassanein, Y.A.W.; Barduagni, P.; Pezzotti, P.; Savioli, L. Clinical signs and household characteristics associated with human fascioliasis among rural population in Egypt: A case-control study. *Parassitologia* **2003**, *45*, 5–11.
106. Esteban, J.G.; Gonzalez, C.; Curtale, F.; Muñoz-Antoli, C.; Valero, M.; Bargues, M.; El Sayed, M.; El Wakeel, A.; Abdel-Wahab, Y.; Montresor, A.; et al. Hyperendemic fascioliasis associated with schistosomiasis in villages in the Nile Delta of Egypt. *Am J Trop. Med. Hyg.* **2003**, *69*, 429–437. [CrossRef]
107. Mas-Coma, S. Human fascioliasis. In *Waterborne Zoonoses: Identification, Causes and Control*; IWA Publish: London, UK, 2004.
108. Taira, N.; Yoshifuji, H.; Boray, J.C. Zoonotic potential of infection with *Fasciola* spp. by consumption of freshly prepared raw liver containing immature flukes. *Int. J. Parasitol.* **1997**, *27*, 775–779. [CrossRef]
109. Dreyfuss, G.; Vignoles, P.; Rondelaud, D. *Fasciola hepatica*: Epidemiological surveillance of natural watercress beds in central France. *Parasitol. Res.* **2005**, *95*, 278–282. [CrossRef]
110. Gaasenbeek, C.P.H.; Over, H.J.; Noorman, N.; Leeuw, W.A.; An, W.L. An epidemiological study of *Fasciola hepatica* in the Netherlands. *Vet. Q.* **2011**, *2176*, 140–144. [CrossRef]
111. Ueno, H. Metacercaria detecting buoy (MDB) method for estimating contamination of rice fields with *Fasciola* metacercariae. *Jarq* **1976**, *10*, 149–152.
112. El-Sayad, M.; Allam, A.; Osman, M. Prevention of human fascioliasis: A study on the role of acids detergents and potassium permenganate in clearing salads from metacercariae. *J. Egyptical. Soc. Parasitol.* **1997**, *27*, 163–169.
113. Hassan, A.; Shoukary, N.; El-Motayam, M.; Morsy, A. Efficacy of five chemicals on *Fasciola gigantica* encysted metacercariae infectivity. *J. Egyptical. Soc. Parasitol.* **2008**, *38*, 919–928.
114. Fennell, J. *Potential for Watercress Production in AUSTRALIA. Rural Industries Research and Development Corporation, Australian Government, Report*; RIRDC Publication No. 06/105, Project No. DAS-49A (Issue 06); RIRDC Publication: Canberra, Australia.
115. King, S.; Scholz, T. Trematodes of the family Opisthorchiidae: A minireview. *Korean J. Parasitol.* **2001**, *39*, 209–221. [CrossRef]
116. Bürli, C.; Harbrecht, H.; Odermatt, P.; Sayasone, S.; Chitnis, N. Mathematical analysis of the transmission dynamics of the liver fluke, *Opisthorchis viverrini. J. Theoretical. Biol.* **2018**, *439*, 181–194. [CrossRef]

117. Sripa, B.; Kaewkes, S.; Sithithaworn, P.; Mairiang, E.; Laha, T.; Smout, M.; Pairojkul, C.; Bhudhisawasdi, V.; Tesana, S.; Thinkamrop, B.; et al. Liver fluke induces cholangiocarcinoma. *PLoS Med.* **2007**, *4*, 1148–1155. [CrossRef]

118. Sithithaworn, P.; Andrews, R.H.; Van De, N.; Wongsaroj, T.; Sinuon, M.; Odermatt, P.; Nawa, Y.; Liang, S.; Brindley, P.J.; Sripa, B. The current status of opisthorchiasis and clonorchiasis in the Mekong Basin. *Parasitol. Int.* **2012**, *61*, 10–16. [CrossRef]

119. Kaewkes, S. Taxonomy and biology of liver flukes. *Acta Trop.* **2003**, *88*, 177–186. [CrossRef]

120. Na, B.K.; Pak, J.H.; Hong, S.J. *Clonorchis sinensis* and clonorchiasis. *Acta Trop.* **2020**, *203*, 105309. [CrossRef]

121. Sripa, B.; Tangkawattana, S.; Brindley, P.J. Update on Pathogenesis of Opisthorchiasis and Cholangiocarcinoma. In *Advances in Parasitology*; Academic Press: Cambridge, MA, USA, 2018; Volume 12. [CrossRef]

122. Woradet, S.; Promthet, S.; Songserm, N.; Parkin, D.M. Factors affecting survival time of cholangiocarcinoma patients: A prospective study in Northeast Thailand. *Asian Pacific. J. Cancer Prev.* **2013**, *14*, 1623–1627. [CrossRef]

123. Gouveia, M.J.; Pakharukova, M.Y.; Laha, T.; Sripa, B.; Maksimova, G.A.; Rinaldi, G.; Brindley, P.J.; Mordvinov, V.A.; Amaro, T.; Santos, L.L.; et al. Infection with *Opisthorchis felineus* induces intraepithelial neoplasia of the biliary tract in a rodent model. *Carcinogenesis* **2017**, *38*, 929–937. [CrossRef]

124. Pakharukova, M.Y.; Mordvinov, V.A. The liver fluke *Opisthorchis felineus*: Biology, epidemiology and carcinogenic potential. *Trans. Royal Soc. Trop. Med. Hyg.* **2015**, *110*, 28–36. [CrossRef] [PubMed]

125. WHO. *Accelerating Work to Overcome the Global Impact of Neglected Tropical Diseases—A Roadmap for Implementation*; WHO: Geneva, Switzerland, 2012.

126. Chai, J.; Jung, B.K. General overview of the current status of human foodborne trematodiasis. *Parasitol* **2022**, *149*, 1262–1285. [CrossRef] [PubMed]

127. Murrell, K.; Pozio, E. The liver flukes: *Clonorchis sinensis*, *Opisthorchis* spp., and *Metorchis* spp. In *Global Water Pathogen Project. Part Three. Specific Excreted Pathogens: Environmental and Epidemiology Aspects*; Michigan State University: East Lansing, MI, USA, 2017; pp. 3–17.

128. Rim, H. The current pathobiology and chemotherapy of clonorchiasis. *Korean J. Parasitol.* **1986**, *24*, 1–141. [CrossRef] [PubMed]

129. Manivong, K.; Komalamisra, C.; Waikagul, J.; Radomyos, P. *Opisthorchis viverrini* Metacercariae in Cyprinoid fish from three rivers in Khammouane Province, Lao PDR. *J. Trop. Med. Parasitol.* **2009**, *32*, 23–29. [CrossRef]

130. Araki, H.; Ong, K.I.C.; Lorphachan, L.; Soundala, P.; Iwagami, M.; Shibanuma, A.; Hongvanthong, B.; Brey, P.T.; Kano, S.; Jimba, M. Mothers' *Opisthorchis viverrini* infection status and raw fish dish consumption in Lao People's Democratic Republic: Determinants of child infection status. *Trop. Med. Health* **2018**, *46*, 1–7. [CrossRef]

131. Nguyen, T.T.B.; Dermauw, V.; Dahma, H.; Bui, D.T.; Le, T.T.H.; Phi, N.T.T.; Lempereur, L.; Losson, B.; Vandenberg, O.; Do, D.T.; et al. Prevalence and risk factors associated with *Clonorchis sinensis* infections in rural communities in northern Vietnam. *PLoS Negl. Trop. Dis.* **2020**, *14*, e0008483. [CrossRef]

132. Zhang, Y.; Gong, Q.L.; Lv, Q.B.; Qiu, Y.Y.; Wang, Y.C.; Qiu, H.Y.; Guo, X.R.; Gao, J.F.; Chang, Q.C.; Wang, C.R. Prevalence of *Clonorchis sinensis* infection in fish in South-East Asia: A systematic review and meta-analysis. *J. Fish Dis.* **2020**, *43*, 1409–1418. [CrossRef]

133. Sun, J.; Xin, H.; Jiang, Z.; Qian, M.; Duan, K.; Li, W.; Huang, S.; Gan, X.; Yang, Y.; Li, Z. High endemicity of *Clonorchis sinensis* infection in Binyang County, Southern China. *PLoS Negl. Trop. Dis.* **2020**, *14*, e0008540. [CrossRef]

134. Le, T.H.; Van De, N.; Blair, D.; Sithithaworn, P.; McManus, D.P. *Clonorchis sinensis* and *Opisthorchis viverrini*: Development of a mitochondrial-based multiplex PCR for their identification and discrimination. *Exp. Parasitol.* **2006**, *112*, 109–114. [CrossRef]

135. Cai, X.Q.; Xu, M.; Wang, W.H.; Qiu, D.Y.; Liu, G.X.; Lin, A.; Tang, J.D.; Zhang, R.L.; Zhu, X.Q. Sensitive and rapid detection of *Clonorchis sinensis* infection in fish by loop-mediated isothermal amplification (LAMP). *Parasitol. Res.* **2010**, *106*, 1379–1383. [CrossRef]

136. Lee, S.E.; Shin, H.E.; Lee, M.R.; Kim, Y.H.; Cho, S.H.; Ju, J.W. Risk factors of *clonorchis sinensis* human infections in endemic areas, Haman-gun, Republic of Korea: A case-control study. *Korean J. Parasitol.* **2020**, *58*, 647–652. [CrossRef]

137. Qian, M.B.; Li, H.M.; Jiang, Z.H.; Yang, Y.C.; Lu, M.F.; Wei, K.; Wei, S.L.; Chen, Y.; Zhou, C.H.; Chen, Y.D.; et al. Severe hepatobiliary morbidity is associated with *clonorchis sinensis* infection: The evidence from a cross-sectional community study. *PLoS Negl. Trop. Dis. 1* **2021**, *5*, e0009116. [CrossRef]

138. Steele, J.A.; Richter, C.H.; Echaubard, P.; Saenna, P.; Stout, V.; Sithithaworn, P.; Wilcox, B.A. Thinking beyond *Opisthorchis viverrini* for risk of cholangiocarcinoma in the lower Mekong region: A systematic review and meta-analysis. *Inf. Dis. Poverty* **2018**, *7*, 44. [CrossRef]

139. Qian, M.B.; Utzinger, J.; Keiser, J.; Zhou, X.N. Clonorchiasis. *Lancet* **2016**, *387*, 800–810. [CrossRef]

140. Wang, X.; Chen, W.; Tian, Y.; Mao, Q.; Lv, X.; Shang, M.; Li, X.; Yu, X.; Huang, Y. Surface display of *Clonorchis sinensis* enolase on *Bacillus subtilis* spores potentializes an oral vaccine candidate. *Vaccine* **2014**, *32*, 1338–1345. [CrossRef]

141. Huyen, T.; Thi, D.B.; Trung, D.D.; Phi, N.; Thuy, T. Knowledge, attitude and practices related to small liver fluke infection in rural communities of Ha Trung District, Thanh Hoa Province, Vietnam. *J. Kasetsart Vet.* **2021**, *31*, 15–32.

142. Sripa, B.; Tangkawattana, S.; Laha, T.; Kaewkes, S.; Mallory, F.F.; Smith, J.F.; Wilcox, B.A. Toward integrated opistorchiasis control in Northeast Thailand: The Lawa Project. *Acta Trop.* **2015**, *141*, 361–367. [CrossRef]

143. Sripa, B.; Tangkawattana, S.; Sangnikul, T. The Lawa model: A sustainable, integrated opisthorchiasis control program using the EcoHealth approach in the Lawa Lake region of Thailand. *Parasitol. Int.* **2017**, *66*, 346–354. [CrossRef]

144. Aibinu, I.E.; Smooker, P.M.; Lopata, A.L. *Anisakis* Nematodes in Fish and Shellfish- from infection to allergies. *Int. J. Parasitol. Parasites Wildl.* **2019**, *6*, 384–393. [CrossRef]

145. Baird, F.J.; Gasser, R.B.; Jabbar, A.; Lopata, A.L. Foodborne anisakiasis and allergy. *Mol. Cell Probes.* **2014**, *28*, 167–174. [CrossRef]

146. Mercken, E.; Van Damme, I.; Vangeenberghe, S.; Serradell, A.; De Sterck, T.; Lumain, J.P.L.; Gabriël, S. Ascaridoids in commercial fish: Occurrence, intensity and localization in whole fish and fillets destined for the Belgian market. *Int. J. Food Microbiol.* **2020**, *16*, 108657. [CrossRef] [PubMed]

147. Adroher-Auroux, F.J.; Benítez-Rodríguez, R. Anisakiasis and *Anisakis*: An underdiagnosed emerging disease and its main etiological agents. *Res. Vet. Sci.* **2020**, *132*, 535–545. [CrossRef] [PubMed]

148. Cipriani, P.; Acerra, V.; Bellisario, B.; Sbaraglia, G.L.; Cheleschi, R.; Nascetti, G.; Mattiucci, S. Larval migration of the zoonotic parasite *Anisakis pegreffii* (Nematoda: Anisakidae) in European anchovy, *Engraulis encrasicolus*: Implications to seafood safety. *Food Control* **2016**, *519*, 148–157. [CrossRef]

149. Bucci, C.; Gallotta, S.; Morra, I.; Fortunato, A.; Ciacci, C.; Iovino, P. *Anisakis*, just think about it in an emergency! *Int J Infect Dis* **2013**, *17*, e1071–e1072. [CrossRef] [PubMed]

150. Nieuwenhuizen, N.E. *Anisakis*-immunology of a foodborne parasitosis. *Parasite Immunol.* **2016**, *38*, 548–557. [CrossRef] [PubMed]

151. Mazzucco, W.; Raia, D.D.; Marotta, C.; Costa, A.; Ferrantelli, V.; Vitale, F.; Casuccio, A. *Anisakis* sensitization in different population groups and public health impact: A systematic review. *PLoS ONE* **2018**, *20*, e0203671. [CrossRef] [PubMed]

152. Rahmati, A.R.; Kiani, B.; Afshari, A.; Moghaddas, E.; Williams, M.; Shamsi, S. World-wide prevalence of *Anisakis* larvae in fish and its relationship to human allergic anisakiasis: A systematic review. *Parasitol. Res.* **2020**, *119*, 3585–3594. [CrossRef]

153. Añíbarro, B.; Seoane, F.J.; Múgica, M.V. Involvement of hidden allergens in food allergic reactions. *J. Investig. Allergol. Clin. Immunol.* **2007**, *17*, 168–172.

154. Armentia, A.; Martin-Gil, F.J.; Pascual, C.; Martín-Esteban, M.; Callejo, A.; Martínez, C. *Anisakis simplex* allergy after eating chicken meat. *J. Investig. Allergol. Clin. Immunol.* **2006**, *16*, 258–263.

155. Fæste, C.; Levsen, A.; Lin, A.H.; Larsen, N.; Plassen, C.; Moen, A.; Van Do, T.; Egaas, E. Fish feed as source of potentially allergenic peptides from the fish parasite *Anisakis simplex* (s.l.). *Anim. Feed Sci. Technol.* **2015**, *202*, 52–61. [CrossRef]

156. Polimeno, L.; Lisanti, M.T.; Rossini, M.; Giacovazzo, E.; Polimeno, L.; Debellis, L.; Ballini, A.; Topi, S.; Santacroce, L. *Anisakis* Allergy: Is aquacultured fish a safe and alternative food to wild-capture fisheries for *Anisakis simplex*-sensitized patients? *Biology* **2021**, *10*, 106. [CrossRef]

157. Saelens, G.; Planckaert, S.; Devreese, B.; Gabriël, S. Transmissibility of anisakid allergenic peptides from animal feed to chicken meat: Proof of concept. *J. Food Comp. Anal.* **2023**, *115*, 104939. [CrossRef]

158. Nieuwenhuizen, N.E.; Lopata, A.L. *Anisakis*-A food-borne parasite that triggers allergic host defences. *Int. J. Parasitol.* **2013**, *43*, 1047–1057. [CrossRef]

159. Ivanović, J.; Baltić, M.Ž.; Bošković, M.; Kilibarda, N.; Dokmanović, M.; Marković, R.; Janjić, J.; Baltić, B. Anisakis allergy in human. *Trends Food Sci. Technol.* **2017**, *59*, 25–29. [CrossRef]

160. Nieuwenhuizen, N.E.; Lopata, A. Allergic reactions to *Anisakis* found in fish. *Curr. Allergy Asthma Rep.* **2014**, *14*, 455. [CrossRef]

161. Levsen, A.; Lunestad, B.T.; Berland, B. Low detection dfficiency of candling as a commonly recommended inspection method for nematode larvae in the flesh of Pelagic fish. *J. Food Protect* **2005**, *68*, 828–832. [CrossRef]

162. Karl, H.; Leinemann, M. A fast and quantitative detection method for nematodes in fish fillets and fishery products. *Arch. Lebensmttelhyg.* **1993**, *44*, 124–125.

163. Petrie, A.; Wootten, R.; Bruno, D.; MacKenzie, K.; Bron, J. *A Survey of Anisakis and Pseudoterranova in Scottish Fisheries and the Efficacy of Current Detection Methods FSAS Project S14008*; Food Standards Scotland: Aberdeen, UK, 2007.

164. Mercken, E.; Van Damme, I.; Šoba Šparl, B.; Vangeenberghe, S.; Serradell, A.; De Sterck, T. Sensitivity of candling as routine method for the detection and recovery of ascaridoids in commercial fish fillets. *Sci. Rep.* **2022**, *12*, 1358. [CrossRef]

165. Pippy, J. Use of ultraviolet light to find parasitic nematodes in situ. *J. Fish Res. Board. Can.* **1970**, *27*, 963–965. [CrossRef]

166. Llarena-Reino, M.; Piñeiro, C.; Antonio, J.; Outeriño, L.; Vello, C.; González, Á.F.; Pascual, S. Optimization of the pepsin digestion method for anisakids inspection in the fishing industry. *Vet. Parasitol.* **2013**, *191*, 276–283. [CrossRef]

167. Kochanowski, M.; Różycki, M.; Dąbrowska, J.; Karamon, J.; Sroka, J.; Antolak, E.; Bełcik, A.; Cencek, T. Development and application of novel chemiluminescence immunoassays for highly sensitive detection of *Anisakis simplex* proteins in thermally processed seafood. *Pathogens* **2020**, *23*, 777. [CrossRef] [PubMed]

168. Werner, M.T.; Fæste, C.K.; Levsen, A.; Egaas, E. A quantitative sandwich ELISA for the detection of *Anisakis simplex* protein in seafood. *Eur. Food Res. Technol.* **2011**, *232*, 967–973. [CrossRef]

169. Fæste, C.K.; Moen, A.; Schniedewind, B.; Haug Anonsen, J.; Klawitter, J.; Christians, U. Development of liquid chromatography-tandem mass spectrometry methods for the quantitation of *Anisakis simplex* proteins in fish. *J. Chromatogr. A* **2016**, *5*, 58–72. [CrossRef] [PubMed]

170. Rodríguez-Mahillo, A.I.; González-Muñoz, M.; de las Heras, C.; Tejada, M.; Moneo, I. Quantification of *Anisakis simplex* allergens in fresh, long-term frozen, and cooked fish muscle. *Foodborne Pathog. Dis.* **2010**, *7*, 967–973. [CrossRef]

171. McClelland, G. Larval anisakine nematodes in various fish species from Sable Island Bank and vicinity. *Can. Bull Fish Aquat. Sci.* **1990**, *222*, 83–118.

172. EU. Commission regulation no 1276/2011 of 8 December 2011 amending annex III to regulation (EC) no 853/2004 of the European Parliament and of the council as regards the treatment to kill viable parasites in fishery products for human consumption. *Off. J. Eur. Union* **2011**, *L 327*, 39–41.
173. FDA. Parasites. In *Fish and Fishery Products Hazards and Controls Guidance*, 4th ed.; Food and Drug Administration, U.S. Department of Health and Human Services: Silver Spring, MD, USA, 2019; pp. 91–98.
174. Chen, H.Y.; Cheng, Y.S.; Grabner, D.S.; Chang, S.H.; Shih, H.H. Effect of different temperatures on the expression of the newly characterized heat shock protein 90 (*Hsp90*) in L3 of *Anisakis* spp. isolated from *Scomber australasicus*. *Vet. Parasitol.* **2014**, *205*, 540–550. [CrossRef]
175. Donadeu, M.; Fahrion, A.S.; Olliaro, P.L.; Abela-Ridder, B. Target product profiles for the diagnosis of *Taenia solium* taeniasis, neurocysticercosis and porcine cysticercosis. *PLoS Negl. Trop. Dis.* **2017**, *11*, e0005875. [CrossRef]
176. Rodrigues da Costa, M.; Pessoa, J.; Meemken, D.; Nesbakken, T. A systematic review on the effectiveness of pre-harvest meat safety interventions in pig herds to control *Salmonella* and other foodborne pathogens. *Microorganisms* **2021**, *9*, 1825. [CrossRef]

Perspective

The Environment, Farm Animals and Foods as Sources of *Clostridioides difficile* Infection in Humans

Declan Bolton * and Pilar Marcos

Teagasc Food Research Centre, Ashtown, D15 DY05 Dublin, Ireland
* Correspondence: declan.bolton@teagasc.ie; Tel.: +353-1-8059539

Abstract: The recent discovery of the same *Clostridioides difficile* ribotypes associated with human infection in a broad range of environments, animals and foods, coupled with an ever-increasing rate of community-acquired infections, suggests this pathogen may be foodborne. The objective of this review was to examine the evidence supporting this hypothesis. A review of the literature found that forty-three different ribotypes, including six hypervirulent strains, have been detected in meat and vegetable food products, all of which carry the genes encoding pathogenesis. Of these, nine ribotypes (002, 003, 012, 014, 027, 029, 070, 078 and 126) have been isolated from patients with confirmed community-associated *C. difficile* infection (CDI). A meta-analysis of this data suggested there is a higher risk of exposure to all ribotypes when consuming shellfish or pork, with the latter being the main foodborne route for ribotypes 027 and 078, the hypervirulent strains that cause most human illnesses. Managing the risk of foodborne CDI is difficult as there are multiple routes of transmission from the farming and processing environment to humans. Moreover, the endospores are resistant to most physical and chemical treatments. The most effective current strategy is, therefore, to limit the use of broad-spectrum antibiotics while advising potentially vulnerable patients to avoid high-risk foods such as shellfish and pork.

Keywords: *Clostridioides difficile*; ribotypes; environment; food; epidemiology

Citation: Bolton, D.; Marcos, P. The Environment, Farm Animals and Foods as Sources of *Clostridioides difficile* Infection in Humans. *Foods* **2023**, *12*, 1094. https://doi.org/10.3390/foods12051094

Academic Editor: Frans J.M. Smulders

Received: 14 November 2022
Revised: 21 February 2023
Accepted: 28 February 2023
Published: 4 March 2023

1. Introduction

Clostridioides difficile is a Gram-positive, endospore-forming anaerobic bacterium often carried asymptomatically in the human gastrointestinal tract [1–5]. However, when conditions are favourable, the endospores germinate in the colon, vegetative cells multiply, and toxins are produced [6], resulting in watery, non-bloody diarrhoea with abdominal pain, toxic megacolon and/or pseudomembranous colitis, which may be fatal [7–9].

The most common risk factor associated with CDI is the use/misuse of broad-spectrum antibiotics. *C. difficile* is often resistant to a wide range of antibiotics [10], and the administration of antibiotics like clindamycin, cephalosporins, penicillins and fluoroquinolones eliminate competitive bacteria in the colon and promote *C. difficile* outgrowth [11]. The elderly, infants, other immune compromised, and patients on antibiotic therapies are therefore most at risk [1,2,4], although the incidence of CDI in pregnant women, children and patients with inflammatory bowel disease (IBD) has also increased [12].

The generally accepted route for human CDI is transmission from the healthcare environment [13]. However, in recent years the proportion of community-acquired CDI, where the patient has no association with a healthcare facility, has increased [14]. At the same time, non-human reservoirs, including the natural environment (soil, rivers and lakes) [15] and animals, including domestic pets [16,17], food animals [18–20] and wild fauna [21] have been reported. Moreover, food may be contaminated [22,23].

The link between *C. difficile* and animals has been known for at least 60 years. In 1960, McBee [24] isolated this bacterium from the large intestine of a seal in Antarctica. By 1974 *C. difficile* had also been detected in animal faeces (donkeys, horses, cows and camels) and in the environment (hay, soil, sand and mud) [25]. In the early 1980s, *C. difficile* reservoirs were reported in healthy pigs and cattle [26,27] and in asymptomatic domestic pets, such as dogs, cats and birds, which had a prevalence of 21%, 30% and 33%, respectively [28]. Thus it was suggested that animals could be a vehicle of transmission to humans [29]. Interestingly, a common human pathogenic *C. difficile* ribotype (ribotype 078) was also isolated from pigs, cattle, and horses later, providing additional evidence of zoonotic transmission of *C. difficile* between animals and humans [30–33]. In more recent years, several studies have reported *C. difficile* in animals, on carcasses [21,34], in food processing facilities and in both raw and cooked foods [35–41].

Despite the increase in community-acquired CDI and data on *C. difficile* in the food chain, it is difficult to prove the source of infection in a given patient or outbreak as the same ribotypes and strains are common to both healthcare and food chain sources. Moreover, the patient may have acquired *C. difficile* sometime before the conditions in the colon changed to promote outgrowth. The objective of this review was to examine the evidence (CDI, virulence, ribotypes, environment, food animal and food sources and the current epidemiology of CDI in humans) supporting the hypothesis that *C. difficile* may be foodborne.

2. *C. difficile* Infection (CDI) in Humans

Elderly people are especially vulnerable to CDI, and cases are more likely to result in severe outcomes [42], possibly due to a decreased immune response or changes in the intestinal microbiota with age [43,44]. An underlying condition, chemotherapy or gastrointestinal surgery can increase susceptibility to CDI [45], which may become recurrent, leading to increased morbidity and mortality [46,47]. Broad-spectrum antibiotics significantly reduce the gut microflora diversity and alter the bile composition in the colon, facilitating CDI and recurrent infection in humans [48]. Treatment with acid suppression medication to prevent ulcers or treat acid-related diseases is also a risk factor for recurrence [49–51].

Metronidazole is used to treat mild to moderate CDI, while vancomycin is used in more severe cases, although the combination of both may be used when there are complications [52]. When these are ineffective, fidaxomicin has been proposed as an alternative to vancomycin [53,54] and has proven effective in preventing recurrent infection [55].

3. Virulence

Within the host, *C. difficile* endospores germinate into vegetative cells, colonise the intestinal tract and produce toxins resulting in disease [56,57], which causes intestinal inflammation, perforation, toxic megacolon and pseudomembranous colitis [58,59]. Mortality rates range from less than 2% to 17% [60,61]. The main virulence factors in *C. difficile* are toxin A and toxin B, encoded by the *tcdA* (308 kDa) and *tcdB* (270 kDa) genes located on a pathogenicity locus (PaLoc) (Figure 1 and Table 1). Both are large clostridial glycosylation toxins and are activated in response to environmental signals during the late log and stationary phases. In addition to the toxins, two regulatory proteins (TcdR and TcdC) and a protein whose function remains unclear (TcdE) complete the PaLoc [62,63]. TcdR (also referred to as TcdD) is a positive regulator activated in stationary phase growth, while TcdC is a negative regulator produced during the exponential phase. Mutations, such as deletions in the *tcdC* gene, may cause increased production of toxins A and B [62,64].

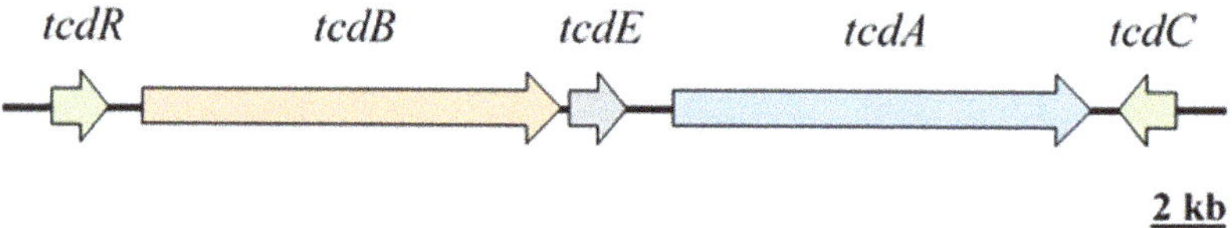

Figure 1. Illustration of the *C. difficile* Pathogenicity locus (PaLoc). Adapted from [65].

TcdA and TcdB possess the same biological activities, among which is the disruption of the cytoskeleton that leads to cytopathic effects in cultured cells. They also possess proinflammatory activity and can stimulate intestinal epithelial cells and immune cells to produce cytokines and chemokines [66,67]. Even low doses of toxins A and B damage the tight junctions of the gut epithelial barrier, facilitating the translocation of commensal bacteria, inflammation and cell apotheosis [66–68]. Sequence variations, deletions, and duplications within the pathogenicity locus account for different toxinotypes of *C. difficile*, with 27 currently identified. Certain strains can present only one of the toxins genes (A⁻B⁺ or A⁺B⁻), however, they reportedly still cause severe disease in humans [62]. In addition, the cytotoxicitybetween toxins that belong to different toxinotypes may vary, making the relation between strain type and CDI severity even more complex [59]. Strains lacking toxin A are more frequently reported due to deletions in the receptor-binding repetitive regions of TcdA caused by the recombination between short repetitive sequences highly conserved in this toxin gene [63]. Donta et al. [66] reported TcdB to be 4 to 200-fold more cytotoxic than TcdA in a mouse model. Therefore, strains producing toxin B have a higher severity in humans.

Up to a third of *C. difficile* isolates also produce the transferase *C. difficile* binary toxin (CDT) [69,70]. CDT, composed of CDTa (biological activity) and CDTb (binding), inhibits the protein actin, damaging the cytoskeleton of the gastrointestinal tract (GIT) cells [71]. The presence of the full-length CDT locus implies the potential expression of the binary toxin, and although some strains contain portions of the CDT locus, these are predicted as non-binary toxin-producing strains [68,70]. CDT-producing strains have been previously associated with a higher production of toxins A and B, leading to an increased disease severity [71,72]. However, CDT is not always present in severe cases [73,74]. In addition, CDT can also be produced by only B⁺ and non-toxigenic strains (A⁻B⁻) [72]. Although CDT production is commonly associated with higher severity of *C. difficile* infection, the role of this toxin during infection and its mechanism of secretion is still not well understood.

Table 1. The virulence factors in *C. difficile* and their function.

Virulence Factor	Encoding Genes	Role in CDI	References
Toxin A	*tcdA*	Multiple cytopathic and cytotoxic effects on the targeted cells include disruption of Rho, Rac and Cdc42-dependent signalling, the actin cytoskeleton and the tight adherence junctions, increasing epithelial permeability, allowing commensal bacterial translocation, inflammation, diarrhoea and sometimes death.	
Toxin B	*tcdB*		[66–68,75]
TcdR	*tcdR*	TcdR is a positive regulator (produced in response environmental conditions) that triggers the induction of transcription of the toxin genes (*tcdA* and *tcdB*).	[76,77]
TcdC	*tcdC*	TcdC is a negative regulator that inhibits the expression of *tcdA* and *tcdB*. Mutations may cause increased production of toxins A and B.	[62,64]
TcdE	*tcdE*	TcdE may function as a lytic protein to facilitate the release of toxins A and B to the extracellular environment by a phage-like system, as these toxins lack signal peptides.	[78,79]
CDT	*cdtA & cdtB*	*C. difficile* binary toxin (CDT) is a transferase that disrupts the normal cytoskeletal function of cells by inhibiting the protein actin. The altered actin cytoskeleton causes an imbalance between actin and microtubules.	[69–71]

4. Ribotypes

There are in excess of 800 *C. difficile* ribotypes (RT), some of which are associated with increased virulence [6,80,81], including RT027 and RT078 [82,83]. These ribotypes are also more prevalent in human cases. RT027 (toxinotype III) has a mutation in *tcdC*, resulting in significantly increased production of toxins A and B while also carrying the genes encoding CDT production and fluoroquinolone resistance [84,85]. Although prevalence has decreased in Europe in recent years, RT027 is associated with a higher mortality and morbidity rate than other ribotypes [86]. The fluoroquinolone resistance, which emerged in two genetically distinct epidemiological lineages (FQR1 and FQR2), was a key driver in the rapid emergence of RT027 [57]. Moreover, this is essential to the increased severity of this ribotype, as this strain typically infects elderly hospital patients on fluoroquinolone treatment [5].

RT078 carries a 39 bp deletion in the *tcdC* gene and therefore overproduces toxins A and B in addition to the binary toxin CDT. In contrast to RT027, which is mostly hospital-acquired, RT078 is more prevalent in younger people and is generally associated with the community [87]. RT078 strains are resistant to fluoroquinolones and erythromycin, which has contributed to their higher prevalence in CDI [88]. Ribotype 126 has the same mutation in its *tcdC* gene found in RT078, is resistant to moxifloxacin and tetracycline and is also considered hypervirulent [89–91]. Other significant ribotypes from a public health perspective include RT017 and RT018. Although the former only produces toxin B, it is resistant to fluoroquinolones and rifampicin and has been associated with numerous outbreaks [92–94]. RT018 has high toxin production capacity, increased cell adhesion, is multidrug-resistant (erythromycin, clindamycin and moxifloxacin) and has become endemic in several countries, including Italy, Spain, Austria and Slovenia [95–97].

5. *C. difficile* in the Environment, Farm Animals and Food

5.1. Water

Toxigenic *C. difficile* has been isolated from a variety of aquatic environments, including drinking water, rivers, sewage effluent and swimming pools [98,99]. Coastal beaches and river sediments are also contaminated [98,99], in some cases by runoff from fields or effluents from wastewater treatment plants [100]. Indeed, *C. difficile* is often detected in water from treatment plants [101], and contamination of drinking water was the source of at least one *C. difficile* outbreak in Finland [102]. Thus, *C. difficile* survives in water and through the effluent treatment process [100].

5.2. Soil, Manure and Silage

C. difficile is commonly found in soil on farms as well as in forests, recreational parks, residential gardens, etc. [103–107]. These authors reported the highest prevalence in urban settings (57%), followed by farms (31%) and forests (28%). Shivaperumal et al. [108] found prevalence rates of 62%, 13% and 15% in garden soil, manure and compost, respectively, while Fröschle et al. [109] reported *C. difficile* to be the most prevalent *Clostridium* spp. in grass silage and cattle manure.

5.3. Farm Environment and Animals

Marcos et al. [110] reported that *C. difficile* were widespread in soil, water and faeces on beef, sheep and broiler farms, with the prevalence ranging from 7% to 83% and counts from 2.9 to 8.4 $\log_{10}$ cfu/g or /mL, depending on the animal species and sample type being tested. Other studies also found *C. difficile* in the faeces of a range of farm animals, including cattle, sheep, poultry and pigs [111–116]. Of these, pigs are the most important source of *C. difficile* [113,116], with the relative prevalence by age being 45%, 3% and 1% in suckling piglets, post-weaning piglets and finishing pigs, respectively [114]. Although these animals may show symptoms (diarrhoea), most are asymptomatic [114]. Other similar studies have reported a prevalence of 37% [115] and 78% [111] in piglets and 4% [115], 62% [117] and 9% [16] in mature pigs.

C. difficile are also found in cattle, especially younger animals. Rodriguez et al. [113] reported a prevalence of 11% in calves and 6% in adult cattle. Other studies have found these bacteria in 11% [118], 14% [117] and 22% of calves [111] and 7% of mature animals [16]. Sheep, including lambs, are also potential carriers, with 0.6 to 2% in the former and 7% reported in the latter [16,119].

Toxigenic *C. difficile* strains have also been reported in poultry faeces in several countries, including the USA (2.3%) [120], the Netherlands (5.8%) [107], Egypt (11.5%) [121], India (14%) [122], Zimbabwe (29%) [123] and Slovenia (62.3%) [124].

5.4. C. difficile at the Animal Slaughter Stage

Pathogenic bacteria in faeces on the hide/fleece or in the gastrointestinal tract are readily transferred to the carcass during slaughter and dressing [125]. *C. difficile* was found in 1%, 3% and 28% of porcine gut contents at slaughter in Belgium [126], Austria [18] and the Netherlands [127], respectively. Reported carcass contamination rates include 7% in Belgium [126], 15% in Canada [128] and 23% in Taiwan [129]. The prevalence of bovine carcass contamination ranges from 7–8% [111,126] but may be as high as 34% [130]. Ovine carcass contamination rates of 15% and 25% have been reported in Iran and Turkey, respectively [130,131]. While poultry carcass data is lacking, Candel-Pérez et al. [132] found *C. difficile* in 28% of gizzard and 6% of liver samples collected in a poultry processing plant in Spain. In Ireland, beef, sheep and broiler carcass contamination rates ranged from 40% to 100%, 40% to 60% and 10% to 40%, respectively, depending on the sampling stage during carcass processing [16].

Ribotypes 002, 005, 013, 014, 015, 019, 035, 062, 081, 087 and 126 have been identified in porcine faeces and rectal swabs at slaughter plants in Europe [18,111,126,127,133]. The *C. difficile* ribotypes isolated from other animal carcasses include 027 from cattle and IR46 from ovine carcasses [131]. Poultry slaughter data is lacking, although Koene et al. [16] found toxigenic ribotypes 056, 014 and 003 in faecal samples from poultry in Dutch slaughter plants.

5.5. C. difficile in Retail Foods

C. difficile has been reported in a range of foods at the retail stage. Thus, the consumption of contaminated retail foods, especially ready-to-eat (RTE) foods, is a risk factor for human infection [134]. Marcos et al. tested meat, dairy and vegetable retail foods and detected *C. difficile* in 9 out of the 240 samples tested [110]. These include corned beef (1), spinach leaves (2), iceberg and little gem lettuce (1 sample each), wild rocket, coleslaw, whole milk yoghurt and cottage cheese (also 1 sample each). Of these samples, direct counts were obtained for the spinach leaves (5.8 $\log_{10}$ cfu/g), coleslaw (4.3 $\log_{10}$ cfu/g) and cottage cheese (6.8 $\log_{10}$ cfu/g).

5.6. C. difficile in Meat and Seafood

Both raw and RTE meat and seafood are frequently contaminated with *C. difficile* [35,118], and the prevalence, including toxin gene profiles and ribotypes, is summarised in Table 2. The reported contamination rates include 41% [35] and 20% [135] for raw pork meat, 12% for ground pork meat [36] and up to 29% for pork sausages and RTE pork products [135]. A beef contamination rate of 42% was reported by Rodriguez-Palacios et al. [118], while ground beef rates include 2% [37], 12% [36], 20% [118] and 50% [35]. In one study, de Boer et al. [38] detected *C. difficile* in 6% of raw lamb samples. Reported poultry contamination rates include 1% [38], 3% [39], 8% [136,137], 13% [36,120] and 44% [35]. *C. difficile* has also been detected in shellfish and fish in several countries, with prevalence ranging from 4% to 49% [138–141].

Table 2. Meat and seafood retail foods contaminated with *C. difficile*, including toxin gene profiles (toxins A, B and CDT) and ribotypes.

Product	Raw or RTE	Total No. (%) Positive	Toxin Gene Profile	Ribotype(s)	Reference
Ground pork	Raw	3/7 (41.3%)	$A^+ B^+ CDT^+$	027 078	[35]
Ground pork	Raw	14/115 (12%)	$A^+ B^+ CDT^+$	027 078	[36]
Ground pork	Raw	2/66 (3.0%)	$A^+ B^+ CDT^-$	029	[39]
Pork meat	Raw	35/303 (11.5%)	$A^+ B^+ CDT^+$	078	[136]
Pork sausages	RTE	10/16 (62.5%)	$A^+ B^+ CDT^+$	027 078	[35]
Ground beef	Raw	13/26 (42.4%)	$A^+ B^+ CDT^+$	027 078	[35]
Ground beef	Raw	11/53 (20.8%)	$A^+ B^+ CDT^+$ / $A^+ B^+ CDT^-$	M31 / 014 077	[118]
Ground beef	Raw	14/115 (12%)	$A^+ B^+ CDT^+$	027 078	[36]
Ground beef	Raw	2/105 (1.9%)	$A^+ B^+ CDT^{ND}$	012	[37]
Ground beef	Raw	21/303 (6.9%)	$A^+ B^+ CDT^+$	PA22	[136]
Beef	Raw	1/67 (1.5%)	$A^+ B^+ CDT^-$	029	[39]
Beef sausages	RTE	1/7 (14.3%)	$A^+ B^+ CDT^+$	027	[35]
Corned beef	RTE	1/10 (10%)	$A^{ND} B^+ CDT^{ND}$	ND [1]	[110]
Ground veal	Raw	1/7 (14.3%)	$A^+ B^+ CDT^+$	M31	[118]
Turkey	Raw	44/303 (14.5%)	$A^+ B^+ CDT^+$	PA01 PA05 PA16	[136]
Ground turkey	Raw	4/9 (44.4%)	$A^+ B^+ CDT^+$	078	[35]
Lamb	Raw	1/16 (6.3%)	$A^+ B^+ CDT^+$	045	[38]
Chicken	Raw	7/257 (2.7%)	$A^+ B^+ CDT^-$	001 003 071 087	[38]
Chicken	Raw	1/67 (1.5%)	$A^+ B^+ CDT^-$	029	[39]
Chicken	Raw	25/310 (8.0%)	$A^+ B^+ CDT^-$	ND [1]	[137]
Chicken	Raw	26/203 (12.8%)	$A^+ B^+ CDT^+$	078	[23]
Chicken	Raw	24/303 (7.8%)	$A^+ B^+ CDT^+$	PA05 PA16	[136]
Chicken	Raw	4/32 (12.5%)	$A^+ B^+ CDT^+$	078	[110]
Chicken	RTE	1/130 (0.8%)	$A^+ B^+ CDT^-$	014 020	[41]
Shellfish	Raw	118/702 (16.8%)	$A^+ B^+ CDT^+$	126 475	[141]

Table 2. *Cont.*

Product	Raw or RTE	Total No. (%) Positive	Toxin Gene Profile	Ribotype(s)	Reference
Bivalve molluscs	Raw	26/53 (49%)	A$^+$ B$^+$ CDT$^+$	078	[139]
			A$^+$ B$^+$ CDT$^-$	002 012 014/020 018 001	
Bivalve molluscs	Raw	36/925 (3.9%)	A$^+$ B$^+$ CDT$^+$	078 126 010	[140]
			A$^-$ B$^+$ CDT$^-$	017 001	

+: Positive; −: Negative; [1] ND: Not determined.

5.7. C. difficile in Vegetables

The information on *C. difficile* isolated from vegetables is summarised in Table 3, with overall prevalence rates of 2% to 5% [22,103,142]. Lim et al. detected *C. difficile* in 56% of organic and 50% of non-organic potatoes, 22% of organic beetroots, 56% of organic onions and 53% of organic carrots [143]. Tkalec et al. found this pathogen in 9% of leaf vegetables, 7% of ginger, 26% and 60% of potatoes, and 14.3% of homegrown leaf vegetables [144]. RTE salads contamination rates included 2% [41], 3% [142], 3.3% [145] and 8% (153).

Table 3. Vegetable retail foods contaminated with *C. difficile*, including toxin gene profiles (toxins A, B and CDT) and ribotypes.

Product	Raw or RTE	Total No. (%) Positive	Toxin Gene Profile	Ribotype(s)	Reference
Root vegetables (potatoes, beetroots, onions and carrots)	Raw	30/100 (30%)	A$^+$ B$^+$ CDT$^+$	QX 274	[143]
			A$^+$ B$^+$ CDT$^-$	002 137 QX519 QX049 101	
			A$^-$ B$^+$ CDT$^+$	070 237 584	
			A$^-$ B$^-$ CDT$^+$	033	
Root vegetables (potatoes, ginger) and leaf vegetables	Raw and RTE	28/154 (18.2%)	A$^+$ B$^+$ CDT$^-$	001/072 011/049 014/020 012 070 150 394 SLO129 SLO187 SLO279	[144]
			A$^+$B$^+$CDT$^+$	027 244 126 023	
Lettuce	RTE	1/54 (1.9%)	A$^+$ B$^+$ CDT$^+$	126	[41]

Table 3. *Cont.*

Product	Raw or RTE	Total No. (%) Positive	Toxin Gene Profile	Ribotype(s)	Reference
Vegetables (potato, onion, mushroom, carrot, radish and cucumber)	Raw	7/300 (2.4%)	A^+ B^{ND} CDT ND	ND [1]	[103]
Salad (lettuce, lamb's lettuce) and vegetable (pea sprouts)	RTE	3/104 (2.8%)	A^+ B^+ CDT$^-$	014/020 001 015	[142]
Vegetables (carrots, potatoes, garlic, ginger, beets, mushrooms, lettuce, green onions, radishes, etc.)	Raw and RTE	5/111 (4.5%)	A^+ B^+ CDT$^+$	078	[22]
			A^+ B^+ CDT$^-$	ND [1]	
Salad (baby leaf spinach)	RTE	2/60 (3.3%)	A^+ B^+ CDT$^+$	078 126	[145]
Salad (baby leaf spinach, organic mixed leaf salad, organic lettuce)	RTE	3/40 (7.5%)	A^+ B^+ CDT ND	001	[146]
			A^- B^+ CDT ND	017	
Spinach leaves	RTE	2/10 (20%)	$A^-$$B^+$ CDT$^-$	ND [1]	[110]
Iceberg lettuce leaves	RTE	1/10 (10%)	$A^-$$B^+$ CDT$^-$	ND [1]	[110]
Little Gem lettuce leaves	RTE	1/10 (10%)	$A^-$$B^+$ CDT$^-$	ND [1]	[110]
Wild rocket leaves	RTE	1/10 (10%)	$A^-$$B^+$ CDT$^+$	ND [1]	[110]
Coleslaw	RTE	1/10 (10%)	$A^-$$B^+$ CDT$^-$	ND [1]	[110]

+: Positive; −: Negative; [1] ND: Not determined.

All of these ribotypes have toxin genes associated with illness in humans. Many have been isolated directly from patients with CDI (Table 4), including 001, 002, 003, 010, 011, 012, 014, 015, 017, 018, 020, 023, 027, 029, 070, 071, 072, 077, 078, 087, 101, 126, 137 and 150. Of these, 002, 003, 012, 014, 027, 029, 070, 078 and 126 have been reported in confirmed community-acquired CDI, while 001, 017, 027, 072, 078 and 126 are hypervirulent.

Table 4. Further characterisation (pathogenicity, hypervirulence and association with community-acquired CDI) of the ribotypes isolated from foods (Tables 2 and 3).

Ribotype	Pathogenic			Hypervirulent			CA CDI [1]	Reference(s)
	yes	no	unk [2]	yes	no	unk		
001	✓			✓				[81,99,147–150]
002	✓				✓		✓	[81,99,148,149,151,152]
003	✓				✓		✓	[81,99]
010	✓				✓			[150]
011	✓				✓			[148]
012	✓				✓		✓	[81,148–150,153]
014	✓				✓		✓	[81,99,148,151,153]
015	✓				✓			[148,149,151]
017	✓			✓				[145,148,149,154]
018	✓				✓			[148,149]
020	✓				✓			[148,149,151]

Table 4. *Cont.*

Ribotype	Pathogenic	Hypervirulent	CA CDI [1]	Reference(s)
023	✓	✓		[148,149,151]
027	✓	✓	✓	[72,73,147–149,153,155]
029	✓	✓	✓	[99,153]
070	✓	✓	✓	[81,149]
071	✓	✓		[149]
072	✓	✓		[99,148,149,156]
077	✓	✓		[149]
078	✓	✓	✓	[72,73,148,149,153–155]
087	✓	✓		[148,149]
101	✓	✓		[149]
126	✓	✓	✓	[6,80,81,99,149,153]
137	✓	✓		[149]
150	✓	✓		[149]
033, 045, 049, 237, 244, 394, 475, 584, M31, PA01, PA05, PA16, PA22, QX049, QX274, QX519, SLO129, SLO187, SLO279		No information		

[1] CA CDI = community acquired *C. difficile* infection; [2] unk = unknown.

5.8. Meta-Analysis

The data presented in Tables 2 and 3 were analysed using Graphpad Prism version 9.3.1. The odds ratios (OR) (the odds of consuming a contaminated product) were calculated for each food type. Briefly, the OR was calculated as the number of positive over negative samples reported for each study. Turkey (with only two studies) was combined with the chicken data (poultry category), while the single lamb study was omitted. The medians and 95% confidence intervals were obtained and were then used to prepare the forest plots. In these Figures, the vertical line is set at an OR = 1 (50:50 chance of the food being contaminated). When all ribotypes are considered, shellfish and pork present a higher risk to the consumer (Figure 2). However, when the analysis is repeated, focusing exclusively on ribotypes 027 and 078 (the 2 hypervirulent strains most commonly associated with human infection), the increased risk is only associated with the consumption of pork (Figure 3).

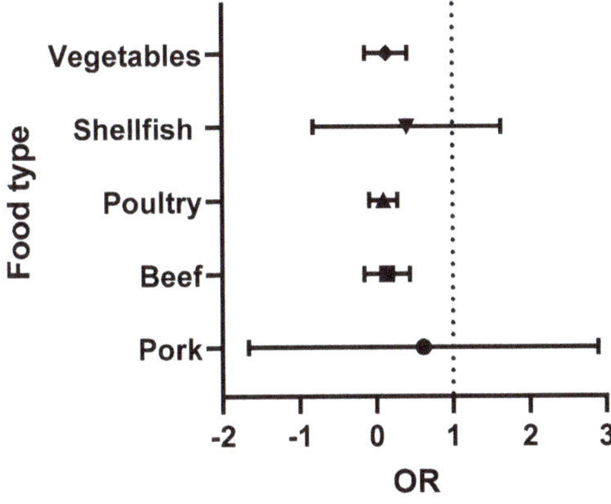

Figure 2. Forest plot of the OR of *C. difficile* (all ribotypes) in each food type.

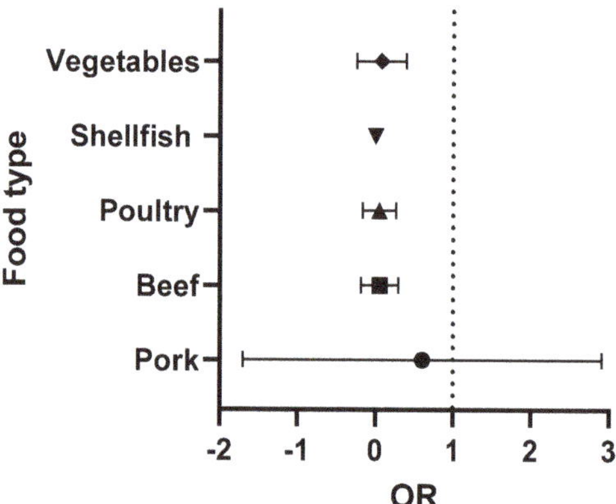

Figure 3. Forest plot of the OR of *C. difficile* 027 and 078 in each food type.

6. The Epidemiology of Foodborne Infection

In 1978, *C. difficile* was recognised as the causative agent of pseudomembranous colitis and diarrhoea in patients on antimicrobial therapy and it was a hospital-associated disease [157]. In the 1980s and 1990s, the incidence of CDI increased significantly, driven by the use of broad-spectrum third-generation cephalosporins (to which *C. difficile* is intrinsically resistant), but the disease was rarely fatal [158,159]. There was a further increase in CDI in the first 10 years of this century driven by the emergence and epidemic spread of the hypervirulent strain, ribotype 027 [160]. The epidemiology of CDI also changed in terms of clinical presentation, response to treatment, and disease outcome. Community-acquired CDI, defined as cases with symptom onset in the community with no history of hospitalisation in the previous 12 weeks or symptom onset within 48 h of hospital admission [161], also emerged. Since then, the incidence of CDI has remained high in developed countries [159,162], and rates of community acquired CDI have increased, accounting for 41%, 30% and 14% of total CDI in the USA, Australia and the EU, respectively [96,159,163]. Furthermore, community acquired CDI patients are generally younger, healthy, often female and lack the traditional risk factors of CDI, including a history of antimicrobial usage [164].

The natural habitat of *C. difficile* is the mammalian gastrointestinal tract (GIT). These bacteria colonise the neonatal GIT, proliferate and are excreted in the faeces to which other newborn animals are exposed, and the cycle recommences. As mammals develop, other bacterial species colonise the GIT, and the prevalence of *C. difficile* decreases [165]. The GIT microbiota inhibit germination, vegetative growth and toxin production, thus protecting against *C. difficile* [48]. However, in the 1990s, this protection was removed when cephalosporins were used in animal husbandry, and food animals became a major reservoir and amplification host for *C. difficile* [119,166], resulting in the contamination of the environment and a range of foods [100,119,166].

Once the environment is contaminated, there are multiple direct and indirect routes to humans, including via food (as illustrated in Figure 4). It is all but impossible to provide incontrovertible proof of foodborne transmission because of the ubiquitous nature of *C. difficile*, delayed onset of symptoms, ability to persist for extended periods as an endospore, etc. However, it has been shown that *C. difficile* endospores in animal waste, wastewater treatment sludge, soil, manure and compost may survive for extended periods of time, facilitating direct contamination of vegetables and fruit or meat via cross-contamination of carcasses during slaughter and processing [108,147]. Water also frequently contains *C. difficile* endospores [99,100,148], and food production may also be contaminated via water used for irrigation or food processing [100,144]. Moreover, the presence of endospores in rivers may contaminate fish and seafood [100,138,139,141]. Transfer from food and wild animals and from domestic pets has also been described [116,149].

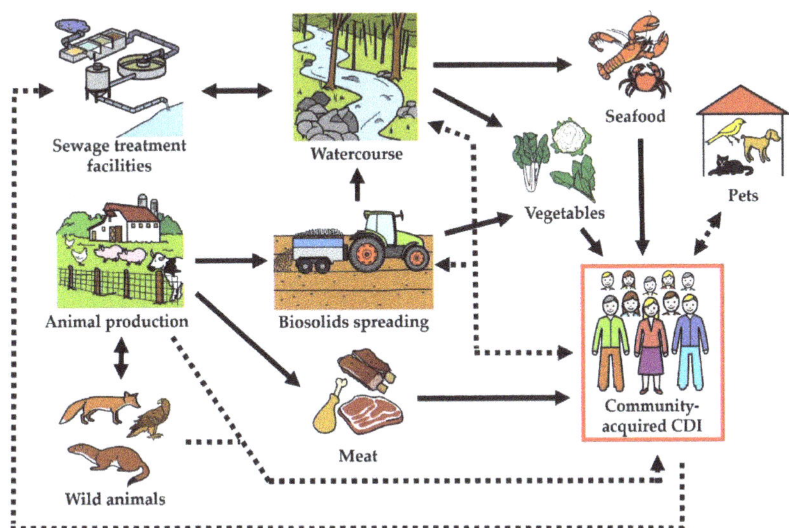

Figure 4. The cycle of community-associated CDI infections from zoonotic, environmental or food-borne sources. Adapted from [100] using ARASAAC pictograms.

Of particular interest, from the public health perspective, is the detection of similar *C. difficile* isolates in farm animals and in humans suffering from CDI, suggesting this pathogen may be zoonotic [150]. Whole genome sequencing (WGS) analysis has shown that ribotypes 078, 126 and 066, commonly found in pigs and/or cattle, are genetically identical to those in humans [151–155]. Although ribotype data for sheep is limited, ribotypes 014, 010 and 045 are common to both humans and ovine sources [119,156,167], while human-related ribotypes 001, 014 and 039 are also found in broilers [120–122,168].

7. Control Strategies

CDI can be controlled in hospitals using deep environmental cleaning, appropriate hand hygiene, stringent infection control and antimicrobial stewardship [169]. However, the same strategies cannot be used in agriculture and food processing [15]. Reduced usage of antibiotics in food animal production would reduce *C. difficile* amplification but is unlikely as increasing global food demand is driving increased antimicrobial usage in animal husbandry, which is projected to rise by 67% by 2030 [15,140]. In 2006 the EU banned the use of antibiotics as growth promoters, followed by the USA in 2017, but other major food-producing countries still allow this practice [170].

Preventing the recycling and dissemination of *C. difficile* endospores in animal slurries applied to land as organic fertilisers would also facilitate reduced environmental contamination and animal carriage. However, research is required to develop effective treatments [171]. Vaccination of food-producing animals is another possible control strategy, but an effective vaccine has not been developed yet [15]. Controlling *C. difficile* in food is dependent on reducing or eliminating the endospores, which are resistant to chilled (4 °C) and freezing (−18 °C and −80 °C) temperatures [172,173]. Although the endospores are resistant at 80 °C [172–174] and will survive the recommended cooking time temperature combinations recommended for meat [174], they are eliminated at 98 °C for 2 min [175]. The same authors suggested microwave irradiation (800 W/60 s) also achieved complete inactivation by denaturing the outer coat.

C. difficile endospores are also resistant to desiccation, hydrostatic pressure [37,176–179] and a range of food preservatives, including sodium nitrite, sodium nitrate and sodium metabisulfite, at permitted concentrations [180]. In contrast, nisin [181], black seed oil, myrrh water [182], garlic juice, peppermint oil, trans-cinnamaldehyde, allicin, menthol and zingerone [183] have a potential application, but validation studies are required before they can be used in controlling *C. difficile* in food.

8. Conclusions

Based on the information provided, it was concluded that *C. difficile* is widespread in the environment and along the food chain. Many food isolates carry the virulence factors required for human infection, and there is no conceivable reason why food is not a source of these pathogens. This conclusion is further supported by the presence of the same ribotypes in food and humans suffering from community-acquired CDI. Based on our analysis, potentially vulnerable consumers should be advised not to handle or consume shellfish or pork.

Author Contributions: Conceptualization, D.B.; investigation, D.B. and P.M.; data curation, P.M.; writing—original draft preparation, D.B.; writing—review and editing, D.B. and P.M.; supervision, D.B.; project administration, D.B.; funding acquisition, D.B. All authors have read and agreed to the published version of the manuscript.

Funding: This research was funded by the Food Institutional Research Measure (FIRM) administered by the Department for Agriculture, Food and the Marine (DAFM) (Grant number 17F206). Pilar Marcos was supported by the Teagasc Walsh Scholarship Scheme (number 2018210).

Data Availability Statement: Not applicable.

Acknowledgments: The pictographic symbols used are the property of the Government of Aragón and have been created by Sergio Palao for the Aragonese Portal of Augmentative and Alternative Communication (ARASAAC) (http://www.arasaac.org; accessed on 27 September 2022) that distributes them under Creative Commons License BY-NC-SA.

Conflicts of Interest: The authors declare no conflict of interest.

References

1. Heinlen, L.; Ballard, J.D. Clostridium difficile infection. *Am. J. Med. Sci.* **2010**, *340*, 247–252. [CrossRef] [PubMed]
2. Rousseau, C.; Poilane, I.; De Pontual, L.; Maherault, A.C.; Le Monnier, A.; Collignon, A. *Clostridium difficile* carriage in healthy infants in the community: A potential reservoir for pathogenic strains. *Clin. Inf. Dis.* **2012**, *55*, 1209–1215. [CrossRef] [PubMed]
3. Galdys, A.L.; Curry, S.R.; Harrison, L.H. Asymptomatic *Clostridium difficile* colonization as a reservoir for *Clostridium difficile* infection. *Expert Rev. Anti-Infect. Ther.* **2014**, *12*, 967–980. [CrossRef]
4. Martin, J.S.; Monaghan, T.M.; Wilcox, M.H. *Clostridium difficile* infection: Epidemiology, diagnosis and understanding transmission. *Nat. Rev. Gastroenterol. Hepatol.* **2016**, *13*, 206–216. [CrossRef]
5. Czepiel, J.; Dróżdż, M.; Pituch, H.; Kuijper, E.J.; Perucki, W.; Mielimonka, A.; Goldman, S.; Wultańska, D.; Garlicki, A.; Biesiada, G. *Clostridium difficile* infection: Review. *Eur. J. Clin. Microbiol. Infect. Dis.* **2019**, *38*, 1211–1221. [CrossRef]
6. Vedantam, G.; Clark, A.; Chu, M.; McQuade, R.; Mallozzi, M.; Viswanathan, V.K. *Clostridium difficile* infection: Toxins and non-toxin virulence factors, and their contributions to disease establishment and host response. *Gut Microbes* **2012**, *3*, 121–134. [CrossRef]
7. Manabe, Y.C.; Vinetz, J.M.; Moore, R.D.; Merz, C.; Charache, P.; Bartlett, J.G. Clostridium difficile colitis: An efficient clinical approach to diagnosis. *Ann. Intern. Med.* **1995**, *123*, 835–840. [CrossRef]
8. Borriello, S.P. Pathogenesis of Clostridium difficile infection. *J. Antimicrobial. Chemo.* **1998**, *41*, 13–19. [CrossRef]
9. Brazier, J.S. Clostridium difficile: From obscurity to superbug. *Brit. J. Biomed. Sci.* **2008**, *65*, 39–44. [CrossRef]
10. Peng, Z.; Jin, D.; Kim, H.B.; Stratton, C.W.; Wu, B.; Tang, Y.-W.; Sun, X. Update on Antimicrobial Resistance in *Clostridium difficile*. *J. Clin. Microbiol.* **2017**, *55*, 1998–2008. [CrossRef]
11. Owen, R.C.; Donskey, C.J.; Gaynes, R.P.; Loo, V.G.; Muto, C.A. Antimicrobial-associated risk factors for *Clostridium difficile* infection. *Clin. Infect. Dis.* **2008**, *46*, 19–31.
12. Ananthakrishnan, A. *Clostridium difficile* infection: Epidemiology, risk factors and management. *Nat. Rev. Gastroenterol. Hepatol.* **2011**, *8*, 17–26. [CrossRef] [PubMed]
13. Enoch, D.A.; Aliyu, S.H. Is *Clostridium difficile* infection still a problem for hospitals? *Can. Med. Assoc. J.* **2012**, *184*, 17–18. [CrossRef] [PubMed]
14. Kim, G.; Zhu, N.A. Community-acquired *Clostridium difficile* infection. *Can Fam Physician.* **2017**, *63*, 131–132.
15. Lim, S.C.; Knight, D.R.; Riley, T.V. *Clostridium difficile* and One Health. *Clin. Microbiol. Infect.* **2020**, *26*, 857–863. [CrossRef]

16. Koene, M.; Mevius, D.; Wagenaar, J.; Harmanus, C.; Hensgens, M.; Meetsma, A.; Putirulan, F.; van Bergen, M.; Kuijper, E. *Clostridium difficile* in Dutch animals: Their presence, characteristics and similarities with human isolates. *Clin. Microbiol. Infect.* **2012**, *18*, 778–784. [CrossRef]

17. Schneeberg, A.; Rupnik, M.; Neubauer, H.; Seyboldt, C. Prevalence and distribution of *Clostridium difficile* PCR ribotypes in cats and dogs from animal shelters in Thuringia, Germany. *Anaerobe* **2012**, *18*, 484–488. [CrossRef]

18. Indra, A.; Lassnig, H.; Baliko, N.; Much, P.; Fiedler, A.; Huhulescu, S. *Clostridium difficile*: A new zoonotic agent? *Wien. Klin. Wochenschr.* **2009**, *121*, 91–95. [CrossRef]

19. Schmid, A.; Messelhäusser, U.; Hörmansdorfer, S.; Sauter-Louis, C.; Mansfeld, R. Occurrence of zoonotic Clostridia and Yersinia in healthy cattle. *J. Food Protect.* **2013**, *76*, 1697–1703. [CrossRef]

20. Noren, T.; Johansson, K.; Unemo, M. *Clostridium difficile* PCR-ribotype 046 is common among neonatal pigs and humans in Sweden. *Clin. Microbiol. Infect.* **2014**, *20*, O2–O6. [CrossRef]

21. Rodriguez-Palacios, A.; Borgmann, S.; Kline, T.R.; LeJeune, J.T. *Clostridium difficile* in foods and animals: History and measures to reduce exposure. *Animal Health Res. Rev.* **2013**, *14*, 11–29. [CrossRef]

22. Metcalf, D.; Costa, M.C.; Dew, W.M.; Weese, J.S. *Clostridium difficile* in vegetables, Canada. *Lett. Appl. Microbiol.* **2010**, *51*, 600–602. [CrossRef] [PubMed]

23. Weese, J. *Clostridium difficile* in food—Innocent bystander or serious threat? *Clin. Microbiol. Infect.* **2010**, *16*, 3–10. [CrossRef] [PubMed]

24. McBee, R.H. Intestinal flora of some antarctic birds and mammals. *J. Bacteriol.* **1960**, *79*, 311–312. [CrossRef] [PubMed]

25. Hafiz, S. *Clostridium difficile* and Its Toxins. Ph.D. Thesis, Department of Microbiology, University of Leeds, Leeds, UK, 1974.

26. Princewell, T.J.T.; Agba, M.I. Examination of bovine faeces for the isolation and identification of *Clostridium* species. *J. Appl. Bacteriol.* **1982**, *52*, 97–102. [CrossRef]

27. Nagy, J.; Bilkei, G. Neonatal piglet losses associated with *Escherichia coli* and *Clostridium difficile* infection in a Slovakian outdoor production unit. *Vet. J.* **2003**, *166*, 98–100. [CrossRef]

28. Borriello, S.P.; Honour, P.; Turner, T.; Barclay, F. Household pets as a potential reservoir for *Clostridium difficile* infection. *J. Clin. Pathol.* **1983**, *36*, 84–87. [CrossRef]

29. Weber, A.; Kroth, P.; Heil, G. Domestic animals as excreters of *Clostridium difficile*. *Dtsch. Med. Wochenschr.* **1988**, *113*, 1617–1618.

30. Goorhuis, A.; Debast, S.B.; van Leengoed, L.A.; Harmanus, C.; Notermans, D.W.; Bergwerff, A.A.; Kuijper, E.J. *Clostridium difficile* PCR ribotype 078: An emerging strain in humans and in pigs? *J. Clin. Microbiol.* **2008**, *46*, 1157. [CrossRef]

31. Jhung, M.A.; Thompson, A.D.; Killgore, G.E.; Zukowski, W.E.; Songer, G.; Warny, M.; Johnson, S.; Gerding, D.N.; McDonald, L.C.; Limbago, B.M. Toxinotype V *Clostridium difficile* in humans and food animals. *Emer. Infect. Dis.* **2008**, *14*, 1039–1045. [CrossRef]

32. Rupnik, M.; Widmer, A.; Zimmermann, O.; Eckert, C.; Barbut, F. *Clostridium difficile* toxinotype V, ribotype 078, in animals and humans. *J. Clin. Microbiol.* **2008**, *46*, 2146. [CrossRef] [PubMed]

33. Debast, S.B.; van Leengoed, L.A.; Goorhuis, A.; Harmanus, C.; Kuijper, E.J.; Bergwerff, A.A. *Clostridium difficile* PCR ribotype 078 toxinotype V found in diarrhoeal pigs identical to isolates from affected humans. *Environ. Microbiol.* **2009**, *11*, 505–511. [CrossRef] [PubMed]

34. Susick, E.K.; Putnam, M.; Bermudez, D.M.; Thakur, S. Longitudinal study comparing the dynamics of *Clostridium difficile* in conventional and antimicrobial free pigs at farm and slaughter. *Vet. Microbiol.* **2012**, *157*, 172–178. [CrossRef]

35. Songer, J.G.; Trinh, H.T.; Killgore, G.E.; Thompson, A.D.; McDonald, L.C.; Limbago, B.M. *Clostridium difficile* in Retail Meat Products, USA, 2007. *Emer. Infect. Dis.* **2009**, *15*, 819–821. [CrossRef]

36. Weese, J.; Avery, B.P.; Rousseau, J.; Reid-Smith, J. Detection and Enumeration of *Clostridium difficile* Spores in Retail Beef and Pork. *Appl. Environ. Microbiol.* **2009**, *15*, 5009–5011. [CrossRef]

37. Bouttier, S.; Barc, M.C.; Felix, B.; Lambert, S.; Collignon, A.; Barbut, F. *Clostridium difficile* in ground meat, France. *Emer. Infect. Dis.* **2010**, *16*, 733–735. [CrossRef]

38. de Boer, E.; Zwartkruis-Nahuis, A.; Heuvelink, A.E.; Harmanus, C.; Kuijper, E.J. Prevalence of *Clostridium difficile* in retailed meat in The Netherlands. *Int. J. Food Microbiol.* **2011**, *144*, 561–564. [CrossRef] [PubMed]

39. Quesada-Gomez, C.; Mulvey, M.R.; Vargas, P.; Gamboa-Coronado Mde, l.M.; Rodriguez, C.; Rodriguez-Cavillini, E. Isolation of a toxigenic and clinical 25 genotype of *Clostridium difficile* in retail meats in Costa Rica. *J. Food Pro.* **2013**, *76*, 348–351. [CrossRef]

40. Kouassi, K.A.; Dadie, A.T.; N'Guessan, K.F.; Dje, K.M.; Loukou, Y.G. *Clostridium perfringens* and *Clostridium difficile* in cooked beef sold in Côte d'Ivoire and their antimicrobial susceptibility. *Anaerobe* **2014**, *28*, 90–94. [CrossRef]

41. Primavilla, S.; Farneti, S.; Petruzzelli, A.; Drigo, I.; Scuota, S. Contamination of hospital food with *Clostridium difficile* in Central Italy. *Anaerobe* **2019**, *55*, 8–10. [CrossRef]

42. Miller, M.; Gravel, D.; Mulvey, M.; Taylor, G.; Boyd, D.; Simor, A.; Gardam, M.; McGeer, A.; Hutchinson, J.; Moore, D.; et al. Health Care-Associated *Clostridium difficile* Infection in Canada: Patient Age and Infecting Strain Type Are Highly Predictive of Severe Outcome and Mortality. *Clin. Infect. Dis* **2010**, *50*, 194–201. [CrossRef] [PubMed]

43. Kelly, C.P. Immune response to Clostridium difficile infection. *Eur. J. Gastroenterol. Hepatol* **1996**, *8*, 1048–1053. [CrossRef] [PubMed]

44. Ogra, P.L. Ageing and its possible impact on mucosal immune responses. *Ageing Res. Rev.* **2010**, *9*, 101–106. [CrossRef] [PubMed]

45. Cole, S.A.; Stahl, T.J. Persistent and Recurrent *Clostridium difficile* Colitis. *Clin. Colon Rectal Sur.* **2015**, *28*, 65–69. [CrossRef] [PubMed]

46. Dharbhamulla, N.; Abdelhady, A.; Domadia, M.; Patel, S.; Gaughan, J.; Roy, S. Risk Factors Associated with Recurrent *Clostridium difficile* Infection. *J. Clin. Med. Res.* **2019**, *11*, 1–6. [CrossRef]

47. Song, J.H.; Kim, Y.S. Recurrent *Clostridium difficile* Infection: Risk Factors, Treatment, and Prevention. *Gut Liver* **2019**, *13*, 16–24. [CrossRef]

48. Britton, R.A.; Young, V.B. Role of the intestinal microbiota in resistance to colonization by *Clostridium difficile*. *Gastroenterology* **2014**, *146*, 1547–1553. [CrossRef]

49. Min, J.H.; Kim, Y.S. Proton pump inhibitors should be used with caution in critically Ill patients to prevent the risk of *Clostridium difficile* infection. *Gut Liver* **2016**, *10*, 493–494. [CrossRef]

50. Tariq, R.; Singh, S.; Gupta, A.; Pardi, D.S.; Khanna, S. Association of gastric acid suppression with recurrent *Clostridium difficile* infection: A systematic review and meta-analysis. *JAMA Internal Med.* **2017**, *177*, 784–791. [CrossRef]

51. Mehta, P.; Nahass, R.G.; Brunetti, L. Acid suppression medications during hospitalization as a risk factor for recurrence of *Clostridioides difficile* infection: Systematic review and meta-analysis. *Clin. Infect. Dis.* **2021**, *545*, 1–7. [CrossRef]

52. Surawicz, C.M.; Brandt, L.J.; Binion, D.G.; Ananthakrishnan, A.N.; Curry, S.R.; Gilligan, P.H.; McFarland, L.V.; Mellow, M.; Zuckerbraun, B.S. Guidelines for diagnosis, treatment, and prevention of *Clostridium difficile* infections. *Am. J. Gastroenterol.* **2013**, *108*, 478–498. [CrossRef] [PubMed]

53. Louie, T.J.; Miller, M.A.; Mullane, K.M.; Weiss, K.; Lentnek, A.; Golan, Y.; Gorbach, S.; Sears, P.; Shue, Y.-K. Fidaxomicin versus vancomycin for *Clostridium difficile* infection. *N. Eng. J. Med.* **2011**, *364*, 422–431. [CrossRef] [PubMed]

54. Chaparro-Rojas, F.; Mullane, K.M. Emerging therapies for *Clostridium difficile* infection—Focus on fidaxomicin. Infect. *Drug Res.* **2013**, *6*, 41–53.

55. Iv, E.C.O.; Iii, E.C.O.; Johnson, D.A. Clinical update for the diagnosis and treatment of *Clostridium difficile* infection. *World J. Gastro. Pharmacol. Ther.* **2014**, *5*, 1–26. [CrossRef] [PubMed]

56. Dyer, C.; Hutt, L.P.; Burky, R.; Joshi, L.T. Biocide Resistance and Transmission of *Clostridium difficile* Spores Spiked onto Clinical Surfaces from an American Health Care Facility. *Appl. Environ. Microbiol.* **2019**, *85*, e01090-19. [CrossRef] [PubMed]

57. Depestel, D.D.; Aronoff, D.M. Epidemiology of *Clostridium difficile* infection. *J. Pharm. Pract.* **2013**, *26*, 464–475. [CrossRef] [PubMed]

58. Barbut, F.; Petit, J.C. Epidemiology of *Clostridium difficile*-associated infections. *Clin. Microbiol. Infect.* **2001**, *7*, 405–410. [CrossRef]

59. Akerlund, T.; Svenungsson, B.; Lagergren, A.; Burman, L.G. Correlation of disease severity with fecal toxin levels in patients with *Clostridium difficile*-associated diarrhea and distribution of PCR ribotypes and toxin yields in vitro of corresponding isolates. *J. Clin. Microbiol.* **2006**, *44*, 353–358. [CrossRef]

60. Evans, C.T.; Safdar, N. Current trends in the epidemiology and outcomes of *Clostridium difficile* infection. *Clin Infect Dis.* **2015**, *60* (Suppl. S2), S66–S71. [CrossRef]

61. Czepiel, J.; Krutova, M.; Mizrahi, A.; Khanafer, N.; Enoch, D.A.; Patyi, M.; Deptuła, A.; Agodi, A.; Nuvials, X.; Pituch, H.; et al. Mortality Following *Clostridioides difficile* Infection in Europe: A Retrospective Multicenter Case-Control Study. *Antibiotic* **2021**, *10*, 299. [CrossRef]

62. Voth, D.; Ballard, J.D. *Clostridium difficile* Toxins: Mechanism of Action and Role in Disease. *Clin. Microbiol. Rev.* **2005**, *2*, 247–263. [CrossRef] [PubMed]

63. Rupnik, M.; Janezic, S. An update on *Clostridium difficile* toxinotyping. *J. Clin. Microbiol.* **2016**, *54*, 13–18. [CrossRef] [PubMed]

64. Hundsberger, T.; Braun, V.; Weidmann, M.; Leuke, P.; Sauerborn, M.; von Eichel-Streiber, C. Transcription analysis of the genes *tcdA-E* of the pathogenicity locus of *Clostridium difficile*. *Eur. J. Biochem.* **1997**, *244*, 735–742. [CrossRef] [PubMed]

65. Matamouros, S.; England, P.; Dupuy, B. *Clostridium difficile* toxin expression is inhibited by the novel regulator TcdC. *Mol. Microbiol.* **2007**, *64*, 1274–1288. [CrossRef] [PubMed]

66. Donta, S.T.; Sullivan, N.; Wilkins, T.D. Differential effects of *Clostridium difficile* toxins on tissue-cultured cells. *J. Clin. Microbiol.* **1982**, *15*, 1157–1158. [CrossRef]

67. Feltis, B.A.; Wiesner, S.M.; Kim, A.S.; Erlandsen, S.L.; Lyerly, D.L.; Wilkins, T.D.; Wells, C.L. *Clostridium difficile* toxins A and B can alter epithelial permeability and promote bacterial paracellular migration through HT-29 enterocytes. *Shock* **2000**, *14*, 629–634. [CrossRef]

68. Riggs, M.M.; Sethi, A.K.; Zabarsky, T.F.; Eckstein, E.C.; Jump, R.L.P.; Donskey, C.J. Asymptomatic carriers are a potential source for transmission of epidemic and nonepidemic *Clostridium difficile* strains among long-term care facility residents. *Clin. Infect. Dis.* **2007**, *45*, 992–998. [CrossRef]

69. Stiles, B.G.; Wigelsworth, D.J.; Popoff, M.R.; Barth, H. Clostridial binary toxins: Iota and c2 family portraits. *Front. Cell. Infect. Microbiol.* **2011**, *1*, 11. [CrossRef]

70. Gerding, D.; Johnson, S.; Rupnik, M.; Aktories, K. *Clostridium difficile* binary toxin CDT: Mechanism, epidemiology, and potential clinical importance. *Gut Microbes* **2014**, *5*, 15–27. [CrossRef]

71. Aktories, K.; Papatheodorou, P.; Schwan, C. Binary *Clostridium difficile* toxin (CDT)—A virulence factor disturbing the cytoskeleton. *Anaerobe* **2018**, *53*, 21–29. [CrossRef]

72. Kuehne, S.A.; Collery, M.M.; Kelly, M.L.; Cartman, S.T.; Cockayne, A.; Minton, N.P. Importance of toxin A, toxin B, and CDT in virulence of an epidemic *Clostridium difficile* strain. *J. Infect. Dis.* **2014**, *209*, 83–86. [CrossRef] [PubMed]

73. Wilson, V.; Cheek, L.; Satta, G.; Walker-Bone, K.; Cubbon, M.; Citron, D. Predictors of death after *Clostridium difficile* infection: A report on 128 strain-typed cases from a teaching hospital in the United Kingdom. *Clin. Infect. Dis.* **2010**, *50*, 77–81. [CrossRef] [PubMed]

74. Goldenberg, S.D.; French, G.L. Lack of association of *tcdC* type and binary toxin status with disease severity and outcome in toxigenic *Clostridium Difficile*. *J. Infect.* **2011**, *62*, 355–362. [CrossRef] [PubMed]

75. Di Bella, S.; Ascenzi, P.; Siarakas, S.; Petrosillo, N.; di Masi, A. *Clostridium difficile* Toxins A and B: Insights into Pathogenic Properties and Extraintestinal Effects. *Toxins* **2016**, *8*, 134. [CrossRef] [PubMed]

76. Mizuno, T.; Kaibuchi, K.; Yamamoto, T.; Kawamura, M.; Sakoda, T.; Fujioka, H.; Matsuura, Y.; Takai, Y. A stimulatory GDP/GTP exchange protein for smg p21 is active on the post-translationally processed form of c-Ki-ras p21 and rhoA p21. *Proc. Natl. Acad. Sci. USA* **1991**, *88*, 6442–6446. [CrossRef] [PubMed]

77. Mani, N.; Dupuy, B. Regulation of toxin synthesis in *Clostridium difficile* by an alternative RNA polymerase sigma factor. *Proc. Natl. Acad. Sci. USA* **2001**, *98*, 5844–5849. [CrossRef]

78. Tan, K.S.; Wee, B.Y.; Song, K.P. Evidence for holin function of tcdE gene in the pathogenicity of *Clostridium difficile*. *J. Med. Microbiol.* **2001**, *50*, 613–619. [CrossRef]

79. Govind, R.; Dupuy, B. Secretion of *Clostridium difficile* toxins A and B requires the holin-like protein TcdE. *PLoS Pathog.* **2012**, *8*, e1002727. [CrossRef]

80. Dayananda, P.; Wilcox, M.H. A Review of Mixed Strain *Clostridium difficile* colonization and infection. *Front. Microbiol.* **2019**, *10*, 692. [CrossRef]

81. Azimirad, M.; Krutova, M.; Yadegar, A.; Shahrokh, S.; Olfatifar, M.; Aghdaei, H.A.; Fawley, W.N.; Wilcox, M.H.; Zali, M.R. *Clostridioides difficile* ribotypes 001 and 126 were predominant in Tehran healthcare settings from 2004 to 2018: A 14-year-long cross-sectional study. *Emerg. Microbes Infect.* **2020**, *1*, 1432–1443. [CrossRef]

82. Neely, F.; Lambert, M.L.; Van Broeck, J.; Delmée, M. Clinical and laboratory features of the most common *Clostridium difficile* ribotypes isolated in Belgium. *J. Hosp. Infect.* **2017**, *4*, 394–399. [CrossRef] [PubMed]

83. Stein, K.; Egan, S.; Lynch, H.; Harmanus, C.; Kyne, L.; Herra, C.; McDermott, S.; Kuijper, E.; Fitzpatrick, F.; FitzGerald, S.; et al. PCR-ribotype distribution of *Clostridium difficile* in Irish pigs. *Anaerobe* **2017**, *48*, 237–241. [CrossRef]

84. Kuijper, E.J.; Coignard, B.; Tüll, P. Emergence of *Clostridium difficile*-associated disease in North America and Europe. Clinical microbiology and infection: The official publication of the European Society of Clinical Microbiology and Infectious Diseases. *Clin. Microbiol. Infect.* **2006**, *6*, 2–18. [CrossRef] [PubMed]

85. Hookman, P.; Barkin, J.S. *Clostridium difficile* associated infection, diarrhea and colitis. *World J. Gastroenterol.* **2009**, *15*, 1554–1580. [CrossRef] [PubMed]

86. Janezic, S.; Garneau, J.R.; Monot, M. Comparative Genomics of *Clostridium difficile*. *Adv. Exp. Med. Biol.* **2018**, *1050*, 59–75. [PubMed]

87. Jones, A.M.; Kuijper, E.J.; Wilcox, M.H. *Clostridium difficile*: A European perspective. *J. Infect.* **2013**, *66*, 115–128. [CrossRef]

88. Baldan, R.; Trovato, A.; Bianchini, V.; Biancardi, A.; Cichero, P.; Mazzotti, M.; Nizzero, P.; Moro, M.; Ossi, C.; Scarpellini, P.; et al. *Clostridium difficile* PCR ribotype 018, a successful epidemic genotype. *J. Clin. Microbiol.* **2015**, *53*, 2575–2580. [CrossRef]

89. Knetsch, C.W.; Hensgens, M.P.M.; Harmanus, C.; van der Bijl, M.W.; Savelkoul, P.H.M.; Kuijper, E.J.; Corver, J.; van Leeuwen, H.C. Genetic markers for *Clostridium difficile* lineages linked to hypervirulence. *Microbiology* **2011**, *157*, 3113–3123. [CrossRef]

90. Schneeberg, A.; Neubauer, H.; Schmoock, G.; Baier, S.; Harlizius, J.; Nienhoff, H.; Brase, K.; Zimmermann, S.; Seyboldt, C. *Clostridium difficile* genotypes in piglet populations in Germany. *J. Clin. Microbiol.* **2013**, *51*, 3796–3803. [CrossRef]

91. Álvarez-Pérez, S.; Blanco, J.L.; Harmanus, C.; Kuijper, E.; Garcia, M.E. Subtyping and antimicrobial susceptibility of *Clostridium difficile* PCR ribotype 078/126 isolates of human and animal origin. *Vet. Microbiol.* **2017**, *199*, 15–22. [CrossRef]

92. Cairns, M.D.; Stabler, R.A.; Shetty, N.; Wren, B.W. The continually evolving *Clostridium difficile* species. *Future Microbiol.* **2012**, *7*, 945–957. [CrossRef] [PubMed]

93. Cairns, M.D.; Preston, M.D.; Lawley, T.D.; Clark, T.G.; Stabler, R.A.; Wren, B.W. Genomic epidemiology of a protracted hospital outbreak caused by a toxin A-negative *Clostridium difficile* sublineage PCR ribotype 017 strain in London, England. *J. Clin. Microbiol.* **2015**, *53*, 3141–3147. [CrossRef] [PubMed]

94. Collins, D.A.; Hawkey, P.M.; Riley, T.V. Epidemiology of *Clostridium difficile* infection in Asia. *Antimicrob. Resist. Infect. Control.* **2013**, *2*, 21. [CrossRef] [PubMed]

95. Spigaglia, P.; Barbanti, F.; Dionisi, A.M.; Mastrantonio, P. *Clostridium difficile* isolates resistant to fluoroquinolones in Italy: Emergence of PCR ribotype 018. *J. Clin. Microbiol.* **2010**, *48*, 2892–2896. [CrossRef]

96. Bauer, M.P.; Notermans, D.W.; van Benthem, B.H.; Brazier, J.S.; Wilcox, M.H.; Rupnik, M.; Monnet, D.L.; van Dissel, J.T.; Kuijper, E.J. *Clostridium difficile* infection in Europe: A hospital-based survey. *Lancet* **2011**, *377*, 63–73. [CrossRef]

97. Barbanti, F.; Spigaglia, P. Characterization of *Clostridium difficile* PCR-ribotype 018: A problematic emerging type. *Anaerobe* **2016**, *42*, 123–129. [CrossRef]

98. Zidaric, V.; Beigot, S.; Lapajne, S.; Rupnik, M. The occurrence and high diversity of *Clostridium difficile* genotypes in rivers. *Anaerobe* **2010**, *16*, 371–375. [CrossRef]

99. Bazaid, F. *Distribution and Sources of Clostridium difficile Present in Water Sources*; MSc University of Guelph: Guelph, Canada, 2012.

100. Warriner, K.; Xu, C.; Habash, M.; Sultan, S.; Weese, S.J. Dissemination of *Clostridium difficile* in food and the environment: Significant sources of *C. difficile* community acquired infection? *J. Appl. Microbiol.* **2016**, *122*, 542–555. [CrossRef]

101. Moradigaravand, D.; Gouliouris, T.; Ludden, C.; Reuter, S.; Jamrozy, D.; Blane, B.; Naydenova, P.; Judge, K.H.; Aliyu, S.F.; Hadjirin, N.A.; et al. Genomic survey of *Clostridium difficile* reservoirs in the East of England implicates environmental contamination of wastewater treatment plants by clinical lineages. *Microb. Genom.* **2018**, *4*, e000162. [CrossRef]
102. Kotila, S.M.; Pitkänen, T.; Brazier, J.; Eerola, E.; Jalava, J.; Kuusi, M.; Könönen, E.; Laine, J.; Miettinen, I.T.; Vuento, R.; et al. *Clostridium difficile* contamination of public tap water distribution system during a waterborne outbreak in Finland. *Scand. J. Public Health* **2013**, *41*, 541–545. [CrossRef]
103. Al Saif, N.; Brazier, J.S. The distribution of *Clostridium difficile* in the environment of South Wales. *J. Med. Microbiol.* **1996**, *45*, 133–137. [CrossRef] [PubMed]
104. Baverud, V.; Gustafsson, A.; Franklin, A.; Aspan, A.; Gunnarsson, A. *Clostridium difficile*: Prevalence in horses and environment, and antimicrobial susceptibility. *Equine Vet. J.* **2003**, *35*, 465–471. [CrossRef] [PubMed]
105. Higazi, T.B.; AL-Saghir, M.; Burkett, M.; Pusok, R. PCR detection of *Clostridium difficile* and its toxigenic strains in public places in Southeast Ohio. *Int. J. Microbiol. Res.* **2011**, *2*, 105–111.
106. Rodriguez, C.; Bouchafa, L.; Soumillion, K.; Ngyuvula, E.; Taminiau, B.; VanBroeck, J.; Delmée, M.; Daube, G. Seasonality of *Clostridium difficile* in the natural environment. *Transbound. Emerg. Dis.* **2019**, *66*, 2440–2449. [CrossRef] [PubMed]
107. Dharmasena, M.; Jiang, X. Isolation of toxigenic *Clostridium difficile* from animal manure and composts being used as biological soil amendments. *Appl. Environ. Microbiol.* **2018**, *84*, e00738-18. [CrossRef]
108. Shivaperumal, N.; Chang, B.J.; Riley, T.V. High Prevalence of *Clostridium difficile* in Home Gardens in Western Australia. *Appl. Environ. Microbiol. J.* **2020**, *87*, e01572-20. [CrossRef]
109. Fröschle, B.; Messelhäusser, U.; Höller, C.; Lebuhn, M. Fate of *Clostridium botulinum* and incidence of pathogenic clostridia in biogas processes. *J. Appl. Microbiol.* **2015**, *119*, 936–947. [CrossRef]
110. Marcos, P.; Whyte, P.; Rogers, T.; McElroy, M.; Fanning, S.; Frias, J.; Bolton, D. The prevalence of *Clostridioides difficile* on farms, in abattoirs and in retail foods in Ireland. *Food Microbiol.* **2021**, *98*, 103781. [CrossRef]
111. Rodriguez, C.; Taminiau, B.; Van Broeck, J.; Avesani, V.; Delmee, M.; Daube, G. *Clostridium difficile* in young farm animals and slaughter animals in Belgium. *Anaerobe* **2012**, *18*, 621–625. [CrossRef]
112. Wu, Y.C.; Lee, J.J.; Tsai, B.Y.; Liu, Y.F.; Chen, C.M.; Tien, N.; Tsai, P.J.; Chen, T.H. Potentially hypervirulent *Clostridium difficile* PCR ribotype 078 lineage isolates in pigs and possible implications for humans in Taiwan. *Int. J. Med. Microbiol.* **2016**, *306*, 115–122. [CrossRef]
113. Rodriguez, C.; Hakimi, D.E.; Vanleyssem, R.; Taminiau, B.; Van Broeck, J.; Delmee, M.; Korsak, N.; Daube, G. *Clostridium difficile* in beef cattle farms, farmers and their environment: Assessing the spread of the bacterium. *Vet. Microbiol.* **2017**, *210*, 183–187. [CrossRef] [PubMed]
114. Kim, H.Y.; Cho, A.; Kim, J.W.; Kim, H.; Kim, B. High prevalence of *Clostridium difficile* PCR ribotype 078 in pigs in Korea. *Anaerobe* **2018**, *51*, 42–46. [CrossRef]
115. Krutova, M.; Zouharova, M.; Matejkova, J.; Tkadlec, J.; Krejci, J.; Faldyna, M.; Nyc, O.; Bernardy, J. The emergence of *Clostridium difficile* PCR ribotype 078 in piglets in the Czech Republic clusters with *Clostridium difficile* PCR ribotype 078 isolates from Germany, Japan and Taiwan. *Int. J. Med. Microbiol.* **2018**, *308*, 770–775. [CrossRef] [PubMed]
116. Rodriguez, C.; Taminiau, B.; van Broeck, J.; Daube, G. *Clostridium difficile* in Food and Animals: A Comprehensive Review. *Adv. Exp. Med. Biol.* **2016**, *932*, 65–92. [PubMed]
117. Keel, K.; Brazier, J.S.; Post, K.W.; Weese, S.; Songer, J.G. Prevalence of PCR Ribotypes among *Clostridium difficile* Isolates from Pigs, Calves, and Other Species. *J. Clin. Microbiol.* **2007**, *6*, 1963–1964. [CrossRef]
118. Rodriguez-Palacios, A.; Staempfli, H.R.; Duffield, T.; Weese, J.S. *Clostridium difficile* in Retail Ground Meat, Canada. *Emerg. Infect. Dis.* **2007**, *3*, 485–487. [CrossRef]
119. Knight, D.R.; Riley, T.V. Prevalence of Gastrointestinal *Clostridium difficile* Carriage in Australian Sheep and Lambs. *Appl. Environ. Microbiol.* **2013**, *79*, 5689–5692. [CrossRef]
120. Harvey, R.; Norman, K.; Andrews, K.; Hume, M.; Scanlan, C.; Callaway, T. *Clostridium difficile* in poultry and poultry meat. *Foodborne Path. Dis.* **2011**, *8*, 1321–1323. [CrossRef]
121. Abdel-Glil, M.Y.; Thomas, P.; Schmoock, G.; Abou-El-Azm, K.; Wieler, L.H.; Neubauer, H.; Seyboldt, C. Presence of *Clostridium difficile* in poultry and poultry meat in Egypt. *Anaerobe* **2018**, *51*, 21–25. [CrossRef]
122. Hussain, I.; Borah, P.; Sharma, R.K.; Rajkhowa, S.; Rupnik, M.; Saikia, D.P. Molecular characteristics of *Clostridium difficile* isolates from human and animals in the North Eastern region of India. *Mol. Cell Probes* **2016**, *30*, 306–311. [CrossRef]
123. Simango, C.; Mwakurudza, S. *Clostridium difficile* in broiler chickens sold at market places in Zimbabwe and their antimicrobial susceptibility. *Int. J. Food Microbiol.* **2008**, *124*, 268–270. [CrossRef] [PubMed]
124. Zidaric, V.; Zemljic, M.; Janezic, S.; Kocuvan, A.; Rupnik, M. High diversity of *Clostridium difficile* genotypes isolated from a single poultry farm producing replacement laying hens. *Anaerobe* **2008**, *14*, 325–327. [CrossRef] [PubMed]
125. European Food Safety Authority (EFSA). Scientific Opinion on the public health hazards to be covered by inspection of meat (bovine animals). *Eur. Food Saf. Aut. J.* **2013**, *11*, 1–261.
126. Rodriguez, C.; Avesani, V.; Van Broeck, J.; Taminiau, B.; Delmée, M.; Daube, G. Presence of *Clostridium difficile* in pigs and cattle intestinal contents and carcass contamination at the slaughterhouse in Belgium. *Int. J. Food Microbiol.* **2013**, *166*, 256–262. [CrossRef]

127. Hopman, N.E.; Oorburg, D.; Sanders, I.; Kuijper, E.J.; Lipman, L.J. High occurrence of various *Clostridium difficile* PCR ribotypes in pigs arriving at the slaughterhouse. *Vet. Quart.* **2011**, *31*, 179–181. [CrossRef]

128. Hawken, P.; Weese, J.S.; Friendship, R.; Warriner, K. Carriage and dissemination of *Clostridium difficile* and methicillin resistant *Staphylococcus aureus* in pork processing. *Food Control* **2013**, *31*, 433–437. [CrossRef]

129. Wu, Y.C.; Chen, C.M.; Kuo, C.J.; Lee, J.J.; Chen, P.C.; Chang, Y.C. Prevalence and molecular characterization of *Clostridium difficile* isolates from a pig slaughterhouse, pork, and humans in Taiwan. *Int. J. Food Microbiol.* **2013**, *242*, 37–44. [CrossRef]

130. Hampikyan, H.; Bingol, E.B.; Muratoglu, K.; Akkaya, E.; Cetin, O.; Colak, H. The prevalence of *Clostridium difficile* in cattle and sheep carcasses and the antibiotic susceptibility of isolates. *Meat Sci.* **2018**, *139*, 120–124. [CrossRef]

131. Esfandiari, Z.; Weese, J.S.; Ezzatpanah, H.; Chamani, M.; Shoaei, P.; Yaran, M.; Ataei, B.; Maracy, M.R.; Ansariyan, A.; Ebrahimi, F.; et al. Isolation and characterization of *Clostridium difficile* in farm animals from slaughterhouse to retail stage in Isfahan, Iran. *Food. Path. Dis.* **2015**, *10*, 864–866. [CrossRef]

132. Candel-Pérez, C.; Santaella-Pascual, J.; Ros-Berruezo, G.; Martínez-Gracia, C. Occurrence of *Clostridioides* [*Clostridium*] *difficile* in poultry giblets at slaughter and retail pork and poultry meat in southeastern Spain. *J. Food Prot.* **2021**, *84*, 310–314. [CrossRef]

133. Keessen, E.C.; van den Berkt, A.J.; Haasjes, N.H.; Hermanus, C.; Kuijper, E.J.; Lipman, L.J. The relation between farm specific factors and prevalence of *Clostridium difficile* in slaughter pigs. *Vet. Microbiol.* **2011**, *54*, 130–134. [CrossRef] [PubMed]

134. Lund, B.M.; Peck, M.W. A possible route for foodborne transmission of *Clostridium difficile*? *Food. Path. Dis.* **2015**, *12*, 177–182. [CrossRef] [PubMed]

135. Harvey, R.; Norman, K.; Andrews, K.; Norby, B.; Hume, M.; Scanlan, C.; Hardin, M.; Scott, H. *Clostridium difficile* in retail meat and processing plants in Texas. *J. Vet. Diag. Investig.* **2011**, *23*, 807–811. [CrossRef]

136. Varshney, J.B.; Very, K.J.; Williams, J.L.; Hegarty, J.P.; Stewart, D.B.; Lumadue, J. Characterization of *Clostridium difficile* isolates from human fecal samples and retail meat from Pennsylvania. *Food. Path. Dis.* **2014**, *11*, 822–829. [CrossRef]

137. Guran, H.S.; Ilhak, O.I. *Clostridium difficile* in retail chicken meat parts and liver in the Eastern Region of Turkey. *J. Con. Prot. Food Saf.* **2015**, *10*, 359–364. [CrossRef]

138. Metcalf, D.; Avery, B.P.; Janecko, N.; Matic, N.; Reid-Smith, R.; Weese, J.S. *Clostridium difficile* in seafood and fish. *Anaerobe* **2011**, *17*, 85–86. [CrossRef] [PubMed]

139. Pasquale, V.; Romano, V.; Rupnik, M.; Capuano, F.; Bove, D.; Aliberti, F.; Krovacek, K.; Dumontet, S. Occurrence of toxigenic *Clostridium difficile* in edible bivalve molluscs. *Food Microbiol.* **2012**, *31*, 309–312. [CrossRef]

140. Troiano, T.; Harmanus, C.; Sanders, I.; Pasquale, V.; Dumontet, S.; Capuano, F.; Romano, V.; Kuijper, E.J. Toxigenic *Clostridium difficile* PCR ribotypes in ediblemarine bivalve molluscs in Italy. *Int. J. Food Microbiol.* **2015**, *208*, 30–34. [CrossRef]

141. Agnoletti, F.; Arcangelia, G.; Barbantib, F.; Barcoa, L.; Brunettaa, R.; Cocchia, M.; Conederaa, G.; D'Estea, L.; Drigoa, I.; Spigagliab, P.; et al. Survey, characterization and antimicrobial susceptibility of *Clostridium difficile* from marine bivalve shellfish of North Adriatic Sea. *Int. J. Food Microbiol.* **2019**, *298*, 74–80. [CrossRef]

142. Eckert, C.; Burghoffer, B.; Barbut, F. Contamination of ready-to-eat raw vegetables with *Clostridium difficile*, France. *J. Med. Microbiol.* **2013**, *62*, 1435–1438. [CrossRef]

143. Lim, S.C.; Foster, N.F.; Elliott, B.; Riley, T.V. High prevalence of *Clostridium difficile* on retail root vegetables, Western Australia. *J. Appl. Microbiol.* **2018**, *124*, 585–590. [CrossRef]

144. Tkalec, V.; Janezica, S.; Skoka, B.; Simonica, T.; Mesarica, S.; Vrabica, T.; Rupnik, M. High *Clostridium difficile* contamination rates of domestic and imported potatoes compared to some other vegetables in Slovenia. *Food Microbiol.* **2019**, *78*, 194–200. [CrossRef]

145. Marcos, P.; Whyte, P.; Burgess, C.; Ekhlas, D.; Bolton, D. Detection and Genomic Characterisation of *Clostridioides difficile* from Spinach Fields. *Pathogens* **2022**, *11*, 1310. [CrossRef]

146. Bakri, M.M.; Brown, D.J.; Butcher, J.P.; Sutherland, A.D. *Clostridium difficile* in ready-to-eat salads, Scotland. *Emerg. Infect. Dis.* **2009**, *15*, 817–818. [CrossRef] [PubMed]

147. Dharmasena, M.; Wang, H.; Wei, T.; Bridges, W.C., Jr.; Jiang, X. Survival of *Clostridioides difficile* in finished dairy compost under controlled conditions. *J. Appl. Microbiol.* **2021**, *131*, 996–1006. [CrossRef] [PubMed]

148. Xu, C.; Weese, J.S.; Flemming, C.; Odumeru, J.; Warriner, K. Fate of *Clostridium difficile* during wastewater treatment and incidence in Southern Ontario watersheds. *J. Appl. Microbiol.* **2014**, *117*, 891–904. [CrossRef] [PubMed]

149. Wei, Y.; Sun, M.; Zhang, Y.; Gao, J.; Kong, F.; Liu, D.; Yu, H.; Du, J.; Tang, R. Prevalence, genotype and antimicrobial resistance of *Clostridium difficile* isolates from healthy pets in Eastern China. *BMC Infect. Dis.* **2019**, *19*, 46. [CrossRef]

150. Weese, J.S. Clostridium (Clostridioides) difficile in animals. *J. Vet. Diagn.Investig. Off. Publ. Am. Assoc. Vet. Lab. Diag.* **2020**, *32*, 213–221. [CrossRef]

151. Hammitt, M.C.; Bueschel, D.M.; Keel, M.K.; Glock, R.D.; Cuneo, P.; DeYoung, D.W.; Reggiardo, C.; Trinh, H.T.; Songer, J.G. A possible role for *Clostridium difficile* in the etiology of calf enteritis. *Vet. Microbiol.* **2008**, *127*, 343–352. [CrossRef]

152. Costa, M.C.; Stämpfli, H.R.; Arroyo, L.G.; Pearl, D.L.; Weese, J.S. Epidemiology of *Clostridium difficile* on a veal farm: Prevalence, molecular characterization and tetracycline resistance. *Vet. Microbiol.* **2011**, *152*, 379–384. [CrossRef]

153. Costa, M.C.; Reid-Smith, R.; Gow, S.; Hannon, S.J.; Booker, C.; Rousseau, J.; Benedict, K.M.; Morley, P.S.; Weese, J.S. Prevalence and molecular characterization of *Clostridium difficile* isolated from feedlot beef cattle upon arrival and mid-feeding period. *BMC Vet. Res.* **2012**, *8*, 38. [CrossRef] [PubMed]

154. Keessen, E.C.; Harmanus, C.; Dohmen, W.; Kuijper, E.J.; Lipman, L.J. *Clostridium difficile* infection associated with pig farms. *Emerg. Infect. Dis.* **2013**, *19*, 1032–1034. [CrossRef] [PubMed]

155. Magistrali, C.F.; Maresca, C.; Cucco, L.; Bano, L.; Drigo, I.; Filippini, G.; Dettori, A.; Broccatelli, S.; Pezzotti, G. Prevalence and risk factors associated with *Clostridium difficile* shedding in veal calves in Italy. *Anaerobe* **2015**, *33*, 42–47. [CrossRef] [PubMed]

156. Romano, V.; Albanese, F.; Dumontet, S.; Krovacek, K.; Petrini, O.; Pasquale, V. Prevalence and genotypic characterization of *Clostridium difficile* from ruminants in Switzerland. *Zoo. Pub. Health* **2012**, *59*, 545–548. [CrossRef] [PubMed]

157. Larson, H.E.; Price, A.B.; Honour, P.; Borriello, S.P. *Clostridium difficile* and the aetiology of pseudomembranous colitis. *Lancet* **1978**, *1*, 1063–1066. [CrossRef] [PubMed]

158. Olson, M.M.; Shanholtzer, C.J.; Lee, J.T., Jr.; Gerding, D.N. Ten years of prospective *Clostridium difficile*-associated disease surveillance and treatment at the Minneapolis VA Medical Center, 1982–1991. *Infect. Control Hosp. Epidemiol.* **1994**, *15*, 371–381. [CrossRef]

159. Slimings, C.; Riley, T. Antibiotics and hospital-acquired *Clostridium difficile* infection: Update of systematic review and meta-analysis. *J. Antimicrob. Chemother.* **2014**, *69*, 881–891. [CrossRef]

160. Freeman, J.; Vernon, J.; Morris, K.; Nicholson, S.; Todhunter, S.; Longshaw, C.; Wilcox, M.H. Pan-European Longitudinal Surveillance of Antibiotic Resistance among Prevalent *Clostridium difficile* Ribotypes' Study Group. *Clin. Microbiol. Infect.* **2015**, *21*, 248.e9–248.e16. [CrossRef]

161. McDonald, L.C.; Coignard, B.; Dubberke, E.; Song, X.; Horan, T.; Kutty, P.K.; Ad Hoc *Clostridium difficile* Surveillance Working Group. Recommendations for surveillance of *Clostridium difficile*-associated disease. *Infect. Control. Hosp. Epidemiol.* **2007**, *28*, 140–145. [CrossRef]

162. Lessa, F.C.; Mu, Y.; Bamberg, W.M.; Beldavs, Z.G.; Dumyati, G.K.; Dunn, J.R.; Farley, M.M.; Holzbauer, S.M.; Meek, J.I.; Phipps, E.C.; et al. Burden of *Clostridium difficile* infection in the United States. *N. Engl. J. Med.* **2015**, *372*, 825–834. [CrossRef]

163. Khanna, S. Microbiota restoration for recurrent *Clostridioides difficile*: Getting one step closer every day! *J. Intern. Med.* **2021**, *290*, 294–309. [CrossRef] [PubMed]

164. Kuntz, J.L.; Chrischilles, E.A.; Pendergast, J.F.; Herwaldt, L.A.; Polgreen, P.M. Incidence of and risk factors for community-associated Clostridium difficile infection: A nested case-control study. *BMC Infect. Dis.* **2011**, *11*, 194. [CrossRef] [PubMed]

165. Moono, P.; Lim, S.C.; Riley, T.V. High prevalence of toxigenic *Clostridium difficile* in public space lawns in Western Australia. *Sci. Rep.* **2017**, *7*, 41196. [CrossRef]

166. Rodriguez Diaz, C.; Seyboldt, C.; Rupnik, M. Non-human *C. difficile* reservoirs and sources: Animals, food, environment. *Adv. Exp. Med. Biol.* **2018**, *1050*, 227–243.

167. Avberšek, J.; Pirš, T.; Pate, M.; Rupnik, M.; Ocepek, M. *Clostridium difficile* in goats and sheep in Slovenia: Characterisation of strains and evidence of age-related shedding. *Anaerobe* **2014**, *28*, 163–167. [CrossRef] [PubMed]

168. Rodriguez-Palacios, A.; Barman, T.; LeJeune, J.T. Three-week summer period prevalence of *Clostridium difficile* in farm animals in a temperate region of the United States (Ohio). *Can. Vet. J.* **2014**, *55*, 786–789. [PubMed]

169. Cataldo, M.A.; Granata, G.; Petrosillo, N. *Clostridium difficile* infection: New approaches to prevention, non-antimicrobial treatment, and stewardship. *Expert Rev. Anti-Infect. Ther.* **2017**, *15*, 1027–1040. [CrossRef]

170. Maron, D.F.; Smith, T.J.; Nachman, K.E. Restrictions on antimicrobial use in food animal production: An international regulatory and economic survey. *Glob. Health* **2013**, *9*, 48. [CrossRef]

171. Squire, M.M.; Riley, T.V. *Clostridium difficile* infection in humans and piglets: A 'One Health' opportunity. *Curr. Top. Microbiol. Immunol.* **2013**, *365*, 299–314.

172. Deng, K.; Plaza-Garrido, A.; Torres, J.A.; Paredes-Sabja, D. Survival of *Clostridium difficile* spores at low temperatures. *Food Microbiol.* **2015**, *46*, 218–221. [CrossRef]

173. Flock, G.; Chen, C.-H.; Yin, H.-B.; Fancher, S.; Mooyottu, S.; Venkitanarayanan, K. Effect of chilling, freezing and cooking on survivability of *Clostridium difficile* spores in ground beef. *Meat Sci.* **2016**, *112*, 161. [CrossRef]

174. Rodriguez-Palacios, A.; Reid-Smith, R.J.; Staempfli, H.R.; Weese, J.S. *Clostridium difficile* survives minimal temperature recommended for cooking ground meats. *Anaerobe* **2010**, *16*, 540–542. [CrossRef] [PubMed]

175. Ojha, S.C.; Chankhamhaengdecha, S.; Singhakaew, S.; Ounjai, P.; Janvilisri, T. Inactivation of *Clostridium difficile* spores by microwave irradiation. *Anaerobe* **2016**, *38*, 14–20. [CrossRef] [PubMed]

176. Doona, C.J.; Feeherry, F.E.; Setlow, B.; Wang, S.; Li, W.; Nichols, F.C. Effects of high-pressure treatment on spores of *Clostridium* species. *Appl. Environ. Microbiol.* **2016**, *82*, 5287–5297. [CrossRef]

177. Deng, K.; Talukdar, P.K.; Sarker, M.R.; Paredes-Sabja, D.; Torres, J.A. Survival of *Clostridium difficile* spores at low water activity. *Food Microbiol.* **2017**, *65*, 274–278. [CrossRef]

178. Broda, D.M.; Delacy, K.M.; Bell, R.G.; Terry, J.; Cook, R.L. Psychrotrophic *Clostridium* spp. associated with 'blown pack' spoilage of chilled vacuum-packed red meats and dog rolls in gas-impermeable plastic casings. *Int. J. Food Microbiol.* **1996**, *29*, 335–352. [CrossRef]

179. Atasoy, F.; Gücükoğlu, A. Detection of *Clostridium difficile* and toxin genes in samples of modified atmosphere packaged (MAP) minced and cubed beef meat. *Ank. ÜNiversitesi Vet. Fakültesi Derg.* **2017**, *64*, 165–170.

180. Lim, S.C.; Foster, N.F.; Riley, T.V. Susceptibility of *Clostridium difficile* to the food preservatives sodium nitrite, sodium nitrate and sodium metabisulphite. *Anaerobe* **2016**, *37*, 67–71. [CrossRef]

181. Le Lay, C.; Dridi, L.; Bergeron, M.G.; Ouellette, M.; Fliss, I. Nisin is an effective inhibitor of *Clostridium difficile* vegetative cells and spore germination. *J. Med. Microbiol.* **2016**, *65*, 169–175. [CrossRef]

182. Aljarallah, K.M. Inhibition of *Clostridium difficile* by natural herbal extracts. *J. Taibah Univ. Med. Sci.* **2016**, *11*, 427–431. [CrossRef]
183. Roshan, N.; Riley, T.V.; Hammer, K.A. Antimicrobial activity of natural products against *Clostridium difficile* in vitro. *J. Appl. Microbiol.* **2017**, *123*, 92–103. [CrossRef] [PubMed]

Review

Asymptomatic Carriage of *Listeria monocytogenes* by Animals and Humans and Its Impact on the Food Chain

Dagmar Schoder [1,2,*], **Claudia Guldimann** [3] **and Erwin Märtlbauer** [4]

[1] Department of Veterinary Public Health and Food Science, Institute of Food Safety,
University of Veterinary Medicine, 1210 Vienna, Austria
[2] Veterinarians without Borders Austria, 1210 Vienna, Austria
[3] Department of Veterinary Sciences, Faculty of Veterinary Medicine, Institute of Food Safety and Analytics,
Ludwig-Maximilians-University Munich, 85764 Oberschleißheim, Germany
[4] Department of Veterinary Sciences, Faculty of Veterinary Medicine, Institute of Milk Hygiene,
Ludwig-Maximilians-University Munich, 85764 Oberschleißheim, Germany
[*] Correspondence: dagmar.schoder@vetmeduni.ac.at

Abstract: Humans and animals can become asymptomatic carriers of *Listeria monocytogenes* and introduce the pathogen into their environment with their feces. In turn, this environmental contamination can become the source of food- and feed-borne illnesses in humans and animals, with the food production chain representing a continuum between the farm environment and human populations that are susceptible to listeriosis. Here, we update a review from 2012 and summarize the current knowledge on the asymptomatic carrier statuses in humans and animals. The data on fecal shedding by species with an impact on the food chain are summarized, and the ways by which asymptomatic carriers contribute to the risk of listeriosis in humans and animals are reviewed.

Keywords: domestic animals; ruminants; wildlife; human; crop; vegetable; environmental contamination; asymptomatic carriers; *Listeria monocytogenes*

Citation: Schoder, D.; Guldimann, C.; Märtlbauer, E. Asymptomatic Carriage of *Listeria monocytogenes* by Animals and Humans and Its Impact on the Food Chain. *Foods* **2022**, *11*, 3472. https://doi.org/10.3390/foods11213472

Academic Editors: Evandro Leite de Souza and Frans J.M. Smulders

Received: 13 September 2022
Accepted: 26 October 2022
Published: 1 November 2022

Publisher's Note: MDPI stays neutral with regard to jurisdictional claims in published maps and institutional affiliations.

1. Introduction

Food-borne listeriosis caused by *Listeria monocytogenes* accounted for 1876 human cases in the EU in 2020. It is also the zoonosis with the highest case fatality rate of 10% in the EU [1]. Combined with the often severe neurological symptoms, this makes listeriosis a high priority for food safety efforts worldwide. *L. monocytogenes* has a broad host range in humans, as well as wild and domestic animals that typically become infected by the ingestion of food or feed that has been contaminated with *L. monocytogenes*.

Potential sources for *L. monocytogenes* in feed and food result from the ubiquitous presence of *L. monocytogenes* in the environment [2], fecal shedding by hosts and the ability of *L. monocytogenes* to establish itself in suitable niches in the farm or food-processing environment due to its capacity to adapt to a broad range of environmental stresses [3].

The food chain provides a direct link between the farm environment and human hosts. *L. monocytogenes* gains access to food production facilities through either raw materials of animal origin (meat and milk) via produce that are contaminated with *L. monocytogenes* from soil or feces or from other sources through a lack in hygiene management. A subset of strains of *L. monocytogenes* (e.g., clonal complex (CC) 9 or 121) have shown a higher propensity to persist in the food production environment, mainly through increased resistance against disinfectants such as quaternary ammonium compounds [4]. It is not uncommon for *L. monocytogenes* to persist in niches in food processing facilities for years or even decades [5].

Taken together, food intended for human consumption may become contaminated with *L. monocytogenes* at any level: (i) during primary production at the farm level, (ii) during processing or (iii) at the retail or (iv) consumer level due to insufficient hygiene measures during food handling. If *L. monocytogenes* is able to grow in food or feed matrices that are consumed without an inactivation step (e.g., heating), the basic conditions for an outbreak of listeriosis are met.

Infections of human or animal hosts result in clinical presentations that range from asymptomatic carriers to septicemia, encephalitis or abortions [6]. While the pathomechanisms in the host and the bacterial virulence factors in *L. monocytogenes* are well-understood, it remains largely unclear why some individuals become asymptomatic carriers. In some cases, truncated alleles of the gene *inlA* encoding surface protein internalin A were found in strains isolated from asymptomatic human carriers [7]. Truncated forms of the *inlA* gene were associated with a loss of virulence [8], which may lead to asymptomatic carriage, and were also overrepresented in isolates from food compared to clinical isolates [9], suggesting the potential exposure of consumers to these isolates. Additionally, a contribution of viable but nonculturable (VBNC) forms of *L. monocytogenes* [10] to asymptomatic carriage has been hypothesized [11,12].

Among other bacteria, *L. monocytogenes* has developed various mechanisms for switching from a vegetative to a metabolically inactive state. However, *L. monocytogenes* in the VBNC state represent a diagnostic challenge, because the majority of the current tests need at least one cultivation step, thus failing to detect nongrowing VBNC cells. Fortunately, in recent years, a variety of PCR and qPCR applications combined with DNA intercalating dyes have been established for detecting viable and VBNC cells [12].

Asymptomatic carriers present a major challenge to food safety (Figure 1). On the other hand, according to a recent study, the gut microbiota itself is a major line of defense against foodborne pathogens [13]. The authors point out that a specific microbiota signature is associated with the asymptomatic shedding of *L. monocytogenes*. They conclude that fecal carriage of this pathogen is a common phenomenon in healthy individuals and is very much influenced by the gut microbiota [13].

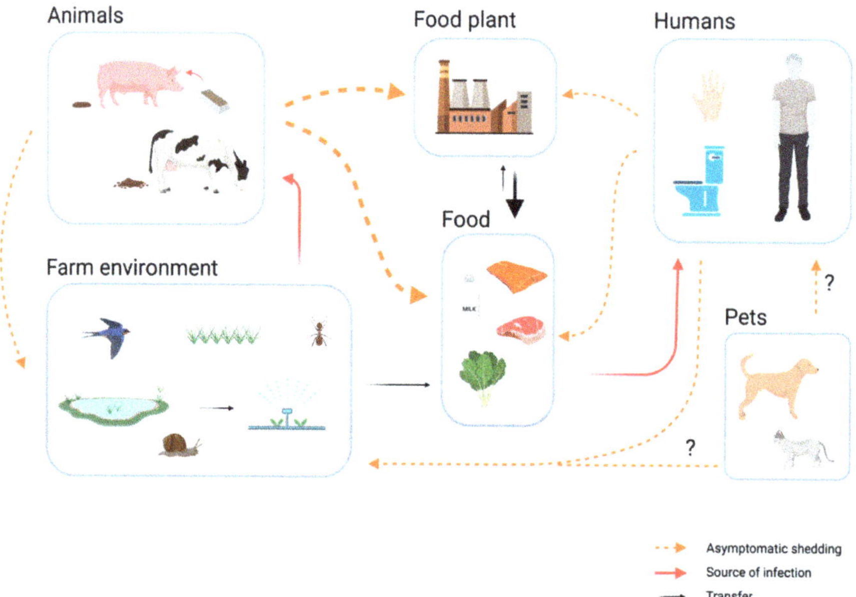

Figure 1. Role of asymptomatic fecal shedding of *L. monocytogenes* by humans and animals in the spread between habitats with a focus on food safety. Bold arrows indicate the most relevant contamination pathways in food production chains in Europe today—in particular, along the farm–food plant–food continuum. Figure was created with BioRender.com.

Efficient controls are in place to exclude animals that are obviously ill from milk or meat production. In contrast, the intermittent fecal shedding of asymptomatic carriers often remains invisible in humans and animals. [14]. With regard to farm animals, the fecal shedding of *L. monocytogenes* can lead to three contamination scenarios with potential implications for food safety: (i) It contributes to a higher load of *L. monocytogenes* in the immediate barn environment, increasing the risk of additional animals becoming infected. (ii) Manure from these animals may be used to fertilize fields, and runoff from farms may contaminate water sources, both risk factors for the contamination of feed and crops with *L. monocytogenes* [15,16]. (iii) Unrecognized carrier animals may lead to raw milk and meat contamination due to insufficient hygiene practices during milking or slaughtering. Finally, asymptomatic human colonization with *L. monocytogenes* may result in the direct contamination of food or the food processing environment due to insufficient hand hygiene.

The aim of this review is to update a 2012 review on the asymptomatic carrier statuses in different species [17] and to summarize the current knowledge on risk factors associated with fecal shedding in different species.

2. Domestic Animals as Asymptomatic Carriers

From studies on Listeria ecology, the asymptomatic carriage of *L. monocytogenes* seems to be evident worldwide, and most domestic animal species, including dogs and cats, can shed *L. monocytogenes* intermittently via feces (Table 1). In addition, the prevalence data showed significant counts of *L. monocytogenes* in tonsil samples from healthy domestic animals (Table 1). The role of household pets in spreading *L. monocytogenes* is not well-studied. To our knowledge, there is no documented clinical case of the transmission of *L. monocytogenes* from pets to humans. However, in recent years, it has become increasingly popular for dog and cat owners to feed their pets raw meat-based diets (RMBDs) instead of the more conventional dry or canned pet foods. A Dutch research team demonstrated that RMBDs may be a possible source of *L. monocytogenes* infection in pet animals and, if transmitted, pose a risk for human beings. They analyzed 35 commercial RMBDs from eight different brands. Alarmingly, *L. monocytogenes* was present in 54% of all tested samples [18].

Pigs can be important reservoirs for *L. monocytogenes* (Table 1), and in particular, younger animals are at risk for asymptomatic carriage. For example, the prevalence of *L. monocytogenes* in the tonsils of fattening pigs (22%) was significantly higher than in sows (6%) [19]. Hypervirulent clones of *L. monocytogenes* such as CC6 [9] were found in pig tonsils, and due to the presence of closely related isolates along the production chain, the cross-contamination or recontamination of meat from a specific source in the slaughterhouse seems to play an important role [20].

Housing conditions significantly influence the risk of *L. monocytogenes* detection in healthy pigs: According to Hellstrom et al. [21], there is a higher prevalence of *L. monocytogenes* in animals from organic production compared to conventional farms. The EU regulation on organic production (EU 2018/848) stipulates that pigs in organic production systems must have straw bedding and must have outdoor access. Additionally, pigs in organic production systems are typically housed in larger groups [21], all of which may contribute to a higher exposure of the animals to *L. monocytogenes* from the environment or from other animals within the same group. On the other hand, it cannot be denied that pigs of intensive indoor farming are often exposed to prolonged social, environmental and metabolic stress [22], which may also enhance the shedding of *L. monocytogenes*.

A highly variable prevalence has been found in fecal samples of healthy dairy cattle ranging from ±1.9% in individual animals to ≥46% of beef herds [23]. Antibody titers to specific *L. monocytogenes* virulence proteins, such as listeriolysin O and internalin A, were demonstrated in 11% of 1652 healthy dairy cows in Switzerland, suggesting that contact with *L. monocytogenes* is relatively frequent in this animal species [24]. A large-scale longitudinal study conducted to monitor *Listeria* spp. in dairy farms during three consecutive seasons in Spain showed that the prevalence of *L. monocytogenes* was affected by season and age: a higher prevalence was observed during the winter in cattle, and cows in their second lactation had the highest probability of *L. monocytogenes* fecal shedding [25].

In all likelihood, the fecal shedding of *L. monocytogenes* by cattle depends on extraneous factors, including feedstuff contamination and season. Shedding appears to be directly associated with feeding practices. A higher prevalence of *L. monocytogenes* in feces has occurred in farms with contaminated feed. Generally, *Listeria* spp. and *L. monocytogenes* prevalence were higher during the indoor season compared to the pasture season [26,27]. *L. monocytogenes* shedding by cows on a study farm was (i) dependent on the subtype of *L. monocytogenes*, (ii) highly associated with silage contamination and (iii) related to animal stress [28]. Poor-quality silage with fermentation defects can have pH values that are permissive to the growth of *L. monocytogenes* and act as a major risk factor for listeriosis in ruminants [29]. This may also explain the seasonal shedding patterns of *L. monocytogenes* by ruminants [14,30,31]. Additionally, for sheep and goats, the likelihood of the isolation of *L. monocytogenes* was three to seven times higher in farms that relied on silage feeding compared to those without [32]. Finally, it is interesting to note that *L. monocytogenes* clonal complex 1 is the most prevalent clonal group associated with human listeriosis and is strongly associated with cattle and dairy products [33].

Given the general proclivity of *L. monocytogenes* for most vertebrates, the special association of *L. monocytogenes* with ruminants may be a specific host adaptation that reflects the unique conditions in the pre-fermentative ruminant fore-stomach. It is the voluminous rumen that may favor the rapid multiplication of *L. monocytogenes* at a pH between 6.5 and 7.2 and at body temperatures between 38.0 and 40.5 °C before confrontation with the acidic environment of the abomasum. This hypothesis is supported by findings that brief, and the low-level fecal excretion of *L. monocytogenes* in sheep is concomitant with a transitory asymptomatic infection after translocation from the gastrointestinal tract (GIT), with the rumen digesta serving as a reservoir. In this study, the asymptomatic carriage of *L. monocytogenes* in sheep was not simply a case of passive passage of the bacteria but was associated with transitory multiplication in the rumen, depending on the dose of *L. monocytogenes* ingested and the age of the animal [34].

Table 1. Isolation of *Listeria monocytogenes* from healthy domestic animals.

Country	Animal	Target	Sample (*n*)	Positive (%)	Ref.
Austria	Sheep/Goat	*L.* spp. *L. monocytogenes*	Feces (53)	42.6 13	[32]
Egypt	Cattle	*L. monocytogenes*	Feces (660) Milk (660)	6.8 5.9	[37]
Finland	Chicken	*L monocytogenes*	Cloacal swabs (457)	1.3	[36]
Germany	Cattle Sheep Hens Pigs Horses Dogs Cats	*L. monocytogenes*	Feces (138) (100) (100) (34) (400) (300) (275)	33 8 8 5.9 4.8 1.3 0.4	[38]
Germany	Pigs	*L. monocytogenes* *L. innocua*	Tonsils (430)	1.6 1.2	[39]

Table 1. *Cont.*

Country	Animal	Target	Sample (*n*)	Positive (%)	Ref.
Italy	Pigs	*L. monocytogenes*	Salivary gland, lymph nodes, tonsils (189)	13.2	[40]
Japan	Cattle Pigs Dogs Rats	*L. monocytogenes*	Feces (1705)	1.9 0.6 0.9 6.5	[41]
Japan	Cattle	*L. monocytogenes*	Feces (1738)	6	[42]
Jordan	Cattle	*L. monocytogenes*	Feces (610)	1.5	[43]
Qatar	Camel	*L. monocytogenes*	Feces (50)	4	[44]
Slovenia	Cows Calves	*L. monocytogenes*	Feces (540) (511)	18.2 8.4	[45] *
Spain	Cattle (beef) Cattle (dairy) Sheep	*L. monocytogenes*	Feces (301)	42.3 46.3 23.5	[23]
Spain	Cattle Sheep Goat	*L. monocytogenes*	Feces (953) Feces (483) Feces (333)	4.3 5.8 0.3	[25]
Switzerland	Cattle	Ab to LO and IA **	Serum (1652)	11	[24]
Taiwan	Chicken	*L. monocytogenes*	Carcass rinse (246)	11.4	[35]
USA	Cattle	*L. monocytogenes*	Feces (825)	31	[14] ***
USA	Cattle	*L. monocytogenes*	Feces (528)	20.2	[31] ****
USA	Broiler	*L. monocytogenes*	Feces (555)	14.9	[46]
USA (Central NY State)	Cattle	*L. monocytogenes*	Milk (1412) Udder swab (1408) Feces (1414)	13 19 43	[47]

* Fecal samples were collected from cows and calves on 20 family dairy farms in 2-week intervals for a period of 1 year. ** Antibodies to listeriolysin O and internalin A. *** Twenty-five fecal samples were collected daily for two 2-week periods and one 5-day period. **** A case–control study involving 24 case farms with at least one recent case of listeriosis and 28 matched control farms with no listeriosis cases was conducted to study the transmission and ecology of *Listeria monocytogenes* on farms.

Poultry, turkeys, ducks and geese can asymptomatically carry *L. monocytogenes* [17]. Recently, carcass rinses and cloacal swabs were reported to be positive at a level of 11 and 1.3%, respectively [35,36]. In addition, there are numerous reports about contamination rates in poultry production establishments and poultry meat and meat products. Stress, such as transport, is plausibly one important factor that exacerbates shedding and thus contributes to the contamination of production lines.

In summary, healthy domestic animals can be asymptomatic carriers of *L. monocytogenes*. While the prevalence of fecal shedding tends to be low, husbandry practices involving silage feeding, as well as stressors associated with housing conditions, group sizes and transport are risk factors that can increase fecal shedding.

3. Carriage of *L. monocytogenes* by Wild Animals

Typically, studies on wildlife shedding *L. monocytogenes* provide no metadata on the health status of the animals, either because it was not evaluated, because animal droppings were sampled in the absence of the animal or because the symptoms of listeriosis were difficult to spot or unknown in a species. Additionally, catch and release studies may be biased towards animals with an impaired health status that may render them more likely to be caught. This makes a classification as "asymptomatic" carriers in wild animals problematic. However, for the purpose of this review, we assume that wild animals that are fecal shedders, symptomatic or not, contribute to the distribution of *L. monocytogenes* in and between environments and should therefore be considered in food safety risk assessments.

A variety of birds, including pheasants, pigeons, gulls, crows, rooks and sparrows, have been shown to be asymptomatic carriers of *L. monocytogenes* (Table 2). A comprehensive prevalence study in Japan looked at fecal or intestinal samples from 996 birds across 18 species and found *Listeria* spp. in 13.4% of all samples, most commonly in samples from crows [48]. Additionally, the fecal presence of *L. monocytogenes* was shown in 33% of urban rooks [49]. According to Hellstrom et al. [50], feces from wild birds (mostly from gulls, pigeons and sparrows) collected in Finland exhibited an overall *L. monocytogenes* prevalence of 36%. Pulsotypes obtained from the birds were often similar to those collected from the food chain, suggesting a possible role of birds in the spread of *L. monocytogenes* strains that are relevant in the context of human infections.

The carriage of *L. monocytogenes* in wildlife is not confined to wild birds. Table 2 shows that *Listeria* spp., including *L. monocytogenes*, have been isolated from a broad variety of mammals (e.g., deer, rodents and wild boars) and also other vertebrates such as reptiles. A Japanese study [48] that included fecal or intestinal samples from 623 wild mammals from eleven species identified *Listeria* spp. in 38 (6.1%) of the tested animals. The highest number of *Listeria* spp. isolates (16/38) were from monkeys, which resulted in a prevalence of 20.0% (16/80) in the monkey samples. A similar study conducted in Canada analyzed 268 fecal samples from a variety of animals, 112 of which were from wildlife, including deer, moose, otters and raccoons. Among these, 35 samples were positive for *L. monocytogenes* (29%) [51]. In samples of 45 red deer and 49 wild boars hunted in Austria and Germany during 2011/12, a total of 19 (42.2%) red deer were positive for *L. monocytogenes*, as were 4 (18.2%) out of 22 pooled feed samples and 12 (24.5%) boars [52]. In several samples, *L. monocytogenes* was isolated from the tonsils and ruminal or cecal contents without its presence in feces, implying that game can carry *L. monocytogenes* even if it is not detectable in their feces. The highest counts for *L. monocytogenes* were found in the rumen of deer and in the tonsils of boars. Pulsed-field gel electrophoresis showed a wide variety of strains, but the serotypes were predominantly 1/2a and 4b. A Polish study examining free-living carnivores as potential sources of infection [53] isolated *L. monocytogenes* from approximately 5% of animals, which included red foxes, beech martens and raccoons. A full set of intact virulence genes was present in 35% of the isolates; the remainder contained varying numbers and configurations of the genes.

Table 2. Isolation of *Listeria monocytogenes* from healthy wild animals.

Country	Animal	Target	Sample (n)	Positive (%)	Ref.
Austria/ Germany	Red deer Wild boar	*L. monocytogenes*	samples* (45) (49)	42 25	[52]
Bulgaria	Birds (*Riparia riparia, Motacilla flava*)	*L. monocytogenes*	Feces (673)	0.6	[54]
Canada	Geese	*L.* spp. *L. monocytogenes*	Feces (495)	9.5 4.0	[55]
Canada	Gulls (*Laurus delawarensis*)	*L. monocytogenes*	Cloacal swabs (264)	9.5	[56]
China	Rodents	*L. m L. ivanovii L. innocua*	Feces (702)	0.3 3.7 6.7	[57]

Table 2. *Cont.*

Country	Animal	Target	Sample (n)	Positive (%)	Ref.
China	Rodents	*L.* spp. *L. monocytogenes* *L. innocua*	Feces (341)	9 3.2 2.9	[58]
Finland	Birds	*L. monocytogenes*	Feces (212)	36	[50]
Finland/Norway	Reindeer	*L. monocytogenes*	Feces (470)	3.2	[59]
France	Rooks	*L. monocytogenes* *L. innocua* *L. seeligeri*	Feces (112)	33 24 8	[49]
Germany	Pigeons	*L. monocytogenes* *L. innocua* *L. seeligeri*	Feces (350)	0.8 2.3 0.6	[60]
Japan	Crows	*L. monocytogenes* *L. innocua*	Feces (301)	1.7 43	[48]
Kenya	Nile tilapia	*L. monocytogenes*	Muscle (167)	1.2	[61]
Poland	Red deer	*L. monocytogenes*	Feces (120)	1.75	[62]
Poland	Red fox, beech marten, racoon	*L. monocytogenes*	Rectal swab (286) (65) (70)	3.5 6.1 4.3	[53]
Switzerland	Wild boars	*L. monocytogenes*	Tonsils (153) Feces (153)	17 1	[63]
USA (central New York)	Reptiles Mammals Birds	*L. monocytogenes*	Feces (17) (64) (242)	12 8 4.5	[64]

* Tonsils and content of the rumen or the stomach, liver, intestinal lymph nodes, cecum content and feces.

Ready-to-eat fish and seafood products—in particular, cold smoked salmon—are frequent sources of human listeriosis [1]. The majority of cases are likely a consequence of post-harvest contamination by *L. monocytogenes* strains that persist in food processing facilities [65]. For farmed fish, factors such as water pollution, agricultural runoff and seagull feces are important contributing factors to the presence of *L. monocytogenes* in the fish and the farm environment [66]. Another alternative is the fish, such as salmon from wild catch. According to a recent Norwegian study, freshly slaughtered salmon contaminated with *L. monocytogenes* was a likely source for the introduction and subsequent persistence in a salmon processing plant [67]. However, fish themselves do not seem to be very susceptible to *L. monocytogenes*. After a gavage of *L. monocytogenes* into the stomachs of live salmon, they were readily cleared without pathologic changes to the animals within three days [66], and rainbow trout held in fish farms where *L. monocytogenes* was detected in the water were rarely positive for *L. monocytogenes* [68]. A recent study [61] demonstrated an increase in the *L. monocytogenes* contamination level in tilapia from capture (1.2%) to the domestic market (5.8%). Taken together, these data suggest that fish may become transient asymptomatic carriers of *L. monocytogenes* after exposure but are not likely to be long-term spreaders of the pathogen.

Reptiles [64], insects [69] and even protozoa [70] may also harbor *L. monocytogenes*. The ongoing trend to keep reptiles, such as snakes, turtles and geckos, as exotic pets should not go unmentioned. In Europe alone, it is estimated that more than 11 million reptiles were kept as pets in 2021 [71]. Further studies are required to assess the possible risk of infection for reptile keepers.

Lately, invading Spanish slugs (*Arion vulgaris*) have been implicated as vectors for *L. monocytogenes* [72]. Of the pooled samples of 710 slugs, 43% were positive, and 16% of them had mean counts of 405 CFU/g of slug tissue. Of 62 slugs cultured, 11% had a positive surface or mucus. Additionally, when the slugs were fed with *L. monocytogenes*, they shed viable bacteria in their feces for up to 22 days. Recently, ants were found to harbor *L. monocytogenes* sporadically, and their potential to transmit pathogenic microorganisms from contaminated environments to food has been demonstrated [69].

Overall, these data show that a wide range of vertebrates, including reptiles, birds and mammals, as well as some invertebrates, can act as carriers of *L. monocytogenes* and contribute to its spread between habitats through asymptomatic carriage.

4. Asymptomatic Carriage of *L. monocytogenes* in Humans

The fecal transmission of *L. monocytogenes* is not only confined to domestic and wild animals. Humans have been shown to shed *L. monocytogenes* intermittently, with the prevalence of fecal shedding in healthy individuals determined by cultures typically ranging below 5% (Table 3); for older studies, see [73].

Interestingly, although low levels of carriage were found in Austria [74] and the USA [75] for healthy people, a later study in Austria compared feces from three individuals sampled over a three-year period. They found that ten (1.2%) out of 868 samples proved positive for *L. monocytogenes*, all of which were serotypes 1/2a and 1/2b. A closer analysis revealed that there were five independent asymptomatic exposures to the bacterium, corresponding to an average of two exposures per person per year [76]. According to the scientific opinion of the European Food Safety Authority on *L. monocytogenes* contamination of RTE foods and the risk for human health in the EU, there is an increasing number of clinical cases for the over 75 years of age group and female age group between 25 and 44 years old. Quantitative modeling demonstrated that more than 90% of invasive listeriosis is caused by the ingestion of RTE food containing > 2000 CFU/g and that one-third of cases are due to growth of the organism in the consumer phase [77].

Underlying medical conditions may also be a predisposing factor for the asymptomatic carriage of *L. monocytogenes* in humans—for instance, in patients on renal dialysis who received the H_2 receptor antagonist antacid ranitidine [78]. On the other hand, the fecal prevalence of *Listeria* spp. or *L. monocytogenes* was the same between HIV-infected pregnant women receiving antiretrovirals and uninfected controls [79]. Pregnancy itself does not seem to affect the rate of human asymptomatic carriage. It was shown that 51 women in their 10–16th weeks of pregnancy had a fecal carriage rate of only 2% [80]. In comparison, the same fecal carriage was confirmed for 3.4% out of 59 nonpregnant controls. Moreover, when the fecal carriage rates in pregnant women with listeriosis were compared with matched, nonpregnant controls following an outbreak in Los Angeles, similar carriage rates were found [81]. A recent study [82] indicated that dysbiosis of breast milk microbiota may result in an increased relative abundance of *L. monocytogenes* in the milk of the mothers of children showing severe acute malnutrition (SAM).

Table 3. Humans as carriers of *Listeria* spp.

Country	Target	Sample (n)	Positive (%)	Reference
Austria	*L. monocytogenes*	Feces, healthy people (505)	0.2	[74]
Brazil	*L. monocytogenes* *L.* spp.	Feces, pregnant women (213)	2.4 7.5	[79]
Egypt	*L. monocytogenes* *L. innocua*	Hand swabs, farm workers (100)	16 2	[83]
France	*hly* gene	Feces (900)	10	[13]
Germany	*L. monocytogenes* *L. innocua*	Feces, patients with diarrhea (1000)	0.6 1.7	[84]

Table 3. *Cont.*

Country	Target	Sample (n)	Positive (%)	Reference
Germany	*L. monocytogenes* *L. innocua*	Feces, healthy people (2000)	0.8 2	[84]
Iran	*L. monocytogenes*	Feces (80) Vaginal swabs (80 samples from women with at least two abortions)	7.5 11.3	[85]
Senegal	*L. monocytogenes*	Breast milk, mothers of SAM children (120) Breast milk, mothers of healthy children (32)	100 37.5	[82]
Turkey	*L. monocytogenes*	Hand and clothes swabs, abattoir workers (70)	5.7	[86]
Turkey	*L. monocytogenes*	Feces (1061)	0.9	[87]
UK	*L. monocytogenes*	Feces, patients with gastroenteritis (171)	1.8	[78]
USA	*L. monocytogenes*	Feces (827)	0.12	[75]

A culture-independent approach based on molecular methods detected *L. monocytogenes* in 173/3338 (5.2%) human microbiome datasets on MG-RAST (16S sequencing) and in 90/900 (10%) stool samples from healthy individuals using PCR [13]. The interpretation of these data should bear in mind that DNA-based detection methods do not differentiate between live and dead organisms. The same study also showed a correlation between specific gut microbiota and the presence of *L. monocytogenes*. Interestingly, a study in mice indicated that aging may cause significant dysbiosis of the commensal microbiota, resulting in increased *L. monocytogenes* colonization of the gut [88]. Additionally, occupational groups encountering animals, feces and meat and those who undergo work-related exposure to the bacterium are anticipated to be at an increased risk of asymptomatic infection. For example, the cumulative prevalence of *L. monocytogenes* in hand swabs from farm workers and hand and clothes swabs from abattoir workers was 16% and 6%, respectively [83,86], which is higher than the average prevalence typically found in fecal samples from healthy people (Table 3).

In order to grasp the extent of *L. monocytogenes* exposure in the wider human community, a European Union-wide baseline survey was carried out in 2010 and 2011. All in all, 13,088 food samples were examined for the presence of *L. monocytogenes*. The prevalence across the entire European Union in fish samples was 10.4%, while, for meat and cheese samples, the prevalence were 2.07% and 0.47%, respectively [89].

Wagner et al. [90] sampled ready-to-eat foods in Austria. Out of 946 food samples collected from food retailers in Vienna, 124 (13.1%) and 45 (4.8%) tested positive for *Listeria* spp. and *L. monocytogenes*, respectively. Products showing the highest contamination were fish and seafood (19.4%), followed by raw meat sausages (6.3%), soft cheese (5.5%) and cooked meats (4.5%). The samples were also collected from households in the same region, and 5.6% and 1.7% out of 640 foodstuffs tested positive for *Listeria* spp. and *L. monocytogenes*, respectively. Alarmingly, the same isolates from the latter products could be detected from pooled fecal samples of household members, suggesting that even low-level contaminated foods (<100 CFU/g) may result in fecal shedding.

5. The Impact of Asymptomatic Carriers on the Presence of *L. monocytogenes* on Farms and in the Food Processing Environment

As discussed above, asymptomatic fecal shedding of *L. monocytogenes* by farm animals contributes to an increased presence of the pathogen in the farm environment with an associated risk to food and feed safety. A systems approach to food safety therefore should include a thorough analysis of the ecology of *L. monocytogenes* in the agricultural environment, and the identification and elimination of farm reservoirs for *L. monocytogenes* is a prerequisite for the implementation of farm-specific pathogen reduction programs [91]. Table 4 summarizes the recent studies that were performed in this context.

We can conclude that asymptomatic fecal shedding by farm animals is linked to diet, particularly to silage [92]. The incidence of *L. monocytogenes* silages was reported to range from 2.5% (clamp silage) to 22.2% (large bales) and to be even higher (44%) in moldy samples [93]. *L. monocytogenes* is thought to initially access silage from the contamination of raw grass via soil. Insufficient acidification of the silage caused by inadequate fermentation then allows the growth of *L. monocytogenes* to levels that can cause disease. Surprisingly, *L. monocytogenes* was rarely detected on grass and vegetables prior to processing [92], which may reflect the low numbers of bacteria needed as an initial contamination. However, once plant material is contaminated, the bacteria can survive for weeks, with implications for feed safety if the grass is contaminated and food safety when the crops are contaminated [94]. Importantly, recent studies in the USA demonstrated that the use of surface water for irrigation could be a major source of contamination [95–97]. *L. monocytogenes* was found in up to 27% of the samples of pond water [98] and in up to up to 99% in the waste water of stabilization ponds in the arctic region of Canada [99]. In addition, the ability of *L. monocytogenes* to enter the VBNC state may contribute to adaptation, persistence and transmission between different ecological niches [11].

Besides silage, *L. monocytogenes* was regularly isolated from samples obtained from feed bunks, water troughs and bedding [47,100], which is consistent with its ubiquitous presence in soil and subsequent spread through feed and animals. Most interestingly, recent studies indicated that dairy farms may favor the selection of hypervirulent *L. monocytogenes* clones, which can then enter the food chain [4,25].

Taken together, the persistence of *L. monocytogenes* in the ruminant farm environment may be supported by a cycle of ingestion of *L. monocytogenes* with contaminated feed, multiplication in animal hosts and subsequent fecal contamination of the environment [101].

Although pigs seldom develop clinical listeriosis, pork products have consistently been linked to human infection [1,102,103]. Slaughter and processing environment contaminations have been traced back to healthy carrier pigs [104]. As for in–out or empty and clean finishing pig facilities, when the duration of the empty period prior to the introduction of growing pigs was less than one day in the fattening section, the risk of *L. monocytogenes* contamination was significantly increased [105]. This same group also proposed that wet feeding is a risk factor for *L. monocytogenes* colonization of a finishing batch, likely because of feed residue layers and biofilm formation within the pipes and valves. The prevalence of fecal shedding of *L. monocytogenes* in healthy swine generally increases from the farm to food manufacturing plants [91]. However, the main source for *L. monocytogenes* contamination in food appears to be at the slaughter and processing steps, where the bacterium can survive for very long periods [104].

Table 4. *Listeria* spp. isolated from the farm and environment.

Country	Target	Sample (n)	Positive (%)	Ref.
Austria	*L.* spp.	Working Boots (53)	51	[32]
		Floor (53)	39.3	
	L. monocytogenes	Working Boots (53)	15.7	
		Floor (53)	7.9	
Canada	*L. monocytogenes*	Irrigation water (223)	10.3	[106]
Canada (Arctic region)	*L. monocytogenes*	Wastewater stabilization ponds (109)	99	[99]
Denmark	*L. monocytogenes*	Abattoir poultry (3080)	8.0	[107]

Table 4. *Cont.*

Country	Target	Sample (n)	Positive (%)	Ref.
Egypt	*L. monocytogenes*	Water (36) Silage (36) Manure (36) Soil (36) Milking equipment (432)	8.3 27.8 19.4 8.3 6.9	[37]
Germany	*L. monocytogenes*	Slaughterhouse (environment and equipment, 77)	0.9	[20]
Iran	*L. monocytogenes*	Water (180)	16.7	[85]
Iran	*L. monocytogenes*	Iranian currency (108)	0.93	[108]
Jordan	*L. monocytogenes*	Bulk tank milk (305)	7.5	[43]
New Zealand	*L. monocytogenes*	Bulk tank milk (400)	4.0	[109] *
South Africa	*L. monocytogenes*	Roof-harvested rainwater (264)	22	[110]
Taiwan	*L. monocytogenes*	Abattoir environment (246)	0	[35]
USA	*L. monocytogenes*	Soil (555)	15.3	[46]
USA	*L.* spp.** *L. monocytogenes*	Stream water (196)	28 10	[95]
USA (Colorado, wilderness area)	*L. monocytogenes* *L.* spp.	Soil, water, sediment, surface soil and wildlife fecal samples (572)	0.23 1.5	[111]
USA (Idaho)	*L. monocytogenes*	Dairy wastewater ponds (30)	6.7	[112]
USA (New York State)	*L. monocytogenes* *L. innocua*	Water (132)	48 10.6	[113]
	L. monocytogenes *L. innocua*	Feces (77)	29 8	
USA (NYS)	*L.* spp. *L. monocytogenes*	Spinach field soil (1092)	12 7.8	[96]
USA (NYS)	*L.* spp. *L. monocytogenes*	Pond/river water used for irrigation (9)	44 22	[97]
USA (Virginia)	*L. monocytogenes*	irrigation water pond (48) well (48)	27.1 4.2	[98]

* Survey from November 2011 to August 2012 during which 25-mL milk samples were collected five times from each of 80 randomly selected dairy farms and tested for the presence of *L. monocytogenes*. ** Excluding *L. monocytogenes*.

Taken together, the asymptomatic shedding of humans and animals, as well as *L. monocytogenes* persistence in the farm environment, present a risk to animal and human health. Since *L. monocytogenes* may access food production facilities from these primary sources, preventative strategies at this level of the food production chain should focus on a high standard of feed and animal hygiene, sanitary milk production and good farming practices. Poor hygiene on farms, such as inattention to boot cleaning, hand washing, failure to wear protective clothing and indifference to silage quality, increases the risk of animals becoming colonized with *L. monocytogenes*, including the downstream risk to the human consumer.

The colonization of food processing equipment and facilities can originate from raw food sources or introduction by poor hygiene practices or fomites [114]. Persistent strains of *L. monocytogenes* isolated from the food processing environment show enhanced adherence with short contact times, promoting survival and possibly initiating the establishment of a strain as a "house strain" in a food processing plant [115]. Sodium chloride, which is often used in food production, induces autoaggregation and increases *L. monocytogenes* adhesion to plastic [114]. The same authors found that persistent strains might have a lower virulence potential than clinical strains. Others have also observed that *L. monocytogenes* strains responsible for persistent contamination differ from sporadic strains, but there does not seem to be any specific evolutionary lineage of persistent strains [19]. Disturbingly, this lower virulence may change following exposure to disinfectants [116,117].

Asymptomatic fecal shedding of *L. monocytogenes* by cows can lead to the entry of *L. monocytogenes* into dairy processing plants via contaminated raw milk and result in persistence as disinfectant-tolerant biofilms on surfaces and the subsequent contamination of processed products [101]. While *L. monocytogenes* is killed by short-term, high-temperature pasteurization, it can survive and thrive in post-pasteurization processing environments and thereby recontaminate dairy products [118–120]. The unique growth and survival properties of *L. monocytogenes* and its ability to adhere to surfaces contribute to the difficulty of eliminating it entirely [121].

In a poultry processing facility in Northern Ireland, a particular genotype of *L. monocytogenes*, considered to originate from incoming birds and prevalent in the raw meat processing area, was found to be widespread on food contact surfaces, floors and drains [122]. One year later, the strains isolated from cooked poultry products and the cooked poultry-processing environment contained only that genotype, plus one other, common to both raw and cooked meat areas. This highlights the potential for persistent strains to cross-contaminate processed foods in the same facility.

6. Future Implications

Since it is evident that animals and humans can persistently carry *L. monocytogenes* and thereby act as a source for the contamination of processed food, control measures must begin on the farm and include humans, animals and the environment in a one-health approach. An awareness of asymptomatic carriage should inform hygiene regulations with respect to the animal and food handlers at all stages of food production. In the farm context, proactive farm hygiene practices to lower the bacterial burden on crops and in animal feed can reduce the root causes of *L. monocytogenes* access to animal and human hosts. In particular, sewage handling and irrigation techniques for crops should take into account the possibility of spreading *L. monocytogenes* to growing plants. Attention to feed hygiene and to correct fermentation during silage production helps interrupt the cycle between the shedding of *L. monocytogenes* from asymptomatic ruminant carriers to grass via manure and the subsequent colonization of more animals from contaminated feed.

Meaningful preventive measures include the adequate compartmentalizing of the raw food processing steps, the critical rethinking of the need for silage feeding, avoiding irrigation close to the harvest, the scrupulous cleanliness of food contact surfaces and equipment, the strict personal hygiene of food handlers and regular monitoring for the persistent colonization of the food processing environment with *L. monocytogenes*.

Author Contributions: Conceptualization, D.S., E.M. and C.G.; writing—original draft preparation, D.S., E.M. and C.G. and writing—review and editing, D.S., E.M. and C.G. All authors have read and agreed to the published version of the manuscript.

Funding: This research received no external funding.

Data Availability Statement: Not applicable.

Conflicts of Interest: The authors declare no conflict of interest.

References

1. Anonymous. The European Union One Health 2020 Zoonoses Report. *EFSA J.* **2021**, *19*, e06971. [CrossRef]
2. Haase, J.K.; Didelot, X.; Lecuit, M.; Korkeala, H.; Group, L.m.M.S.; Achtman, M. The ubiquitous nature of *Listeria monocytogenes* clones: A large-scale Multilocus Sequence Typing study. *Environ. Microbiol.* **2014**, *16*, 405–416. [CrossRef] [PubMed]
3. Belias, A.; Sullivan, G.; Wiedmann, M.; Ivanek, R. Factors that contribute to persistent Listeria in food processing facilities and relevant interventions: A rapid review. *Food Control* **2022**, *133*, 108579. [CrossRef]

4. Maury, M.M.; Bracq-Dieye, H.; Huang, L.; Vales, G.; Lavina, M.; Thouvenot, P.; Disson, O.; Leclercq, A.; Brisse, S.; Lecuit, M. Hypervirulent *Listeria monocytogenes* clones' adaption to mammalian gut accounts for their association with dairy products. *Nat. Commun.* **2019**, *10*, 2488. [CrossRef]

5. Harrand, A.S.; Jagadeesan, B.; Baert, L.; Wiedmann, M.; Orsi, R.H. Evolution of *Listeria monocytogenes* in a Food Processing Plant Involves Limited Single-Nucleotide Substitutions but Considerable Diversification by Gain and Loss of Prophages. *Appl. Environ. Microbiol.* **2020**, *86*, e02493-19. [CrossRef]

6. Schlech, W.F.; Fischetti, V.A.; Novick, R.P.; Ferretti, J.J.; Portnoy, D.A.; Miriam Braunstein, M.; Rood, J.I. Epidemiology and Clinical Manifestations of *Listeria monocytogenes* Infection. *Microbiol. Spectr.* **2019**, *7*, 3. [CrossRef]

7. Olier, M.; Garmyn, D.; Rousseaux, S.; Lemaître, J.P.; Piveteau, P.; Guzzo, J. Truncated Internalin A and Asymptomatic *Listeria monocytogenes* Carriage: In Vivo Investigation by Allelic Exchange. *Infect. Immun.* **2005**, *73*, 644–648. [CrossRef]

8. Nightingale, K.K.; Windham, K.; Martin, K.E.; Yeung, Y.; Wiedmann, M. Select *Listeria monocytogenes* Subtypes Commonly Found in Foods Carry Distinct Nonsense Mutations in *inlA*, Leading to Expression of Truncated and Secreted Internalin A, and Are Associated with a Reduced Invasion Phenotype for Human Intestinal Epithelial Cells. *Appl. Environ. Microbiol.* **2005**, *71*, 8764–8772. [CrossRef]

9. Maury, M.M.; Tsai, Y.H.; Charlier, C.; Touchon, M.; Chenal-Francisque, V.; Leclercq, A.; Criscuolo, A.; Gaultier, C.; Roussel, S.; Brisabois, A.; et al. Uncovering *Listeria monocytogenes* hypervirulence by harnessing its biodiversity. *Nat. Genet.* **2016**, *48*, 308–313. [CrossRef]

10. Besnard, V.; Federighi, M.; Declerq, E.; Jugiau, F.; Cappelier, J.M. Environmental and physico-chemical factors induce VBNC state in *Listeria monocytogenes*. *Vet. Res.* **2002**, *33*, 359–370. [CrossRef]

11. Lotoux, A.; Milohanic, E.; Bierne, H. The Viable But Non-Culturable State of *Listeria monocytogenes* in the One-Health Continuum. *Front. Cell. Infect. Microbiol.* **2022**, *12*, 849915. [CrossRef] [PubMed]

12. Fleischmann, S.; Robben, C.; Alter, T.; Rossmanith, P.; Mester, P. How to Evaluate Non-Growing Cells—Current Strategies for Determining Antimicrobial Resistance of VBNC Bacteria. *Antibiotics* **2021**, *10*, 115. [CrossRef] [PubMed]

13. Hafner, L.; Pichon, M.; Burucoa, C.; Nusser, S.H.A.; Moura, A.; Garcia-Garcera, M.; Lecuit, M. *Listeria monocytogenes* faecal carriage is common and depends on the gut microbiota. *Nat. Commun.* **2021**, *12*, 6826. [CrossRef]

14. Ho, A.J.; Ivanek, R.; Grohn, Y.T.; Nightingale, K.K.; Wiedmann, M. *Listeria monocytogenes* fecal shedding in dairy cattle shows high levels of day-to-day variation and includes outbreaks and sporadic cases of shedding of specific *L. monocytogenes* subtypes. *Prev. Vet. Med.* **2007**, *80*, 287–305. [CrossRef] [PubMed]

15. Wilkes, G.; Edge, T.A.; Gannon, V.P.J.; Jokinen, C.; Lyautey, E.; Neumann, N.F.; Ruecker, N.; Scott, A.; Sunohara, M.; Topp, E.; et al. Associations among pathogenic bacteria, parasites, and environmental and land use factors in multiple mixed-use watersheds. *Water Res.* **2011**, *45*, 5807–5825. [CrossRef] [PubMed]

16. Lyautey, E.; Lapen, D.R.; Wilkes, G.; McCleary, K.; Pagotto, F.; Tyler, K.; Hartmann, A.; Piveteau, P.; Rieu, A.; Robertson, W.J.; et al. Distribution and characteristics of *Listeria monocytogenes* isolates from surface waters of the South Nation River watershed, Ontario, Canada. *Appl. Environ. Microbiol.* **2007**, *73*, 5401–5410. [CrossRef] [PubMed]

17. Schoder, D.; Wagner, M. Growing awareness of asymptomatic carriage of *Listeria monocytogenes*. *Wien. Tierarztl. Mon.* **2012**, *99*, 322–329.

18. van Bree, F.P.J.; Bokken, G.C.A.M.; Mineur, R.; Franssen, F.; Opsteegh, M.; van der Giessen, J.W.B.; Lipman, L.J.A.; Overgaauw, P.A.M. Zoonotic bacteria and parasites found in raw meat-based diets for cats and dogs. *Vet. Rec.* **2018**, *182*, 50. [CrossRef]

19. Autio, T.; Markkula, A.; Hellstrom, S.; Niskanen, T.; Lunden, J.; Korkeala, H. Prevalence and genetic diversity of *Listeria monocytogenes* in the tonsils of pigs. *J. Food Prot.* **2004**, *67*, 805–808. [CrossRef]

20. Oswaldi, V.; Luth, S.; Dzierzon, J.; Meemken, D.; Schwarz, S.; Fessler, A.T.; Felix, B.; Langforth, S. Distribution and Characteristics of *Listeria* spp. in Pigs and Pork Production Chains in Germany. *Microorganisms* **2022**, *10*, 512. [CrossRef]

21. Hellstrom, S.; Laukkanen, R.; Siekkinen, K.M.; Ranta, J.; Maijala, R.; Korkeala, H. *Listeria monocytogenes* contamination in pork can originate from farms. *J. Food Prot.* **2010**, *73*, 641–648. [CrossRef] [PubMed]

22. Martínez-Miró, S.; Tecles, F.; Ramón, M.; Escribano, D.; Hernández, F.; Madrid, J.; Orengo, J.; Martínez-Subiela, S. Causes, consequences and biomarkers of stress in swine: An update. *BMC Vet. Res.* **2016**, *12*, 171. [CrossRef] [PubMed]

23. Hurtado, A.; Ocejo, M.; Oporto, B. *Salmonella* spp. and *Listeria monocytogenes* shedding in domestic ruminants and characterization of potentially pathogenic strains. *Vet. Microbiol.* **2017**, *210*, 71–76. [CrossRef] [PubMed]

24. Boerlin, P.; Boerlin-Petzold, F.; Jemmi, T. Use of listeriolysin O and internalin A in a seroepidemiological study of listeriosis in Swiss dairy cows. *J. Clin. Microbiol.* **2003**, *41*, 1055–1061. [CrossRef] [PubMed]

25. Palacios-Gorba, C.; Moura, A.; Gomis, J.; Leclercq, A.; Gomez-Martin, A.; Bracq-Dieye, H.; Moce, M.L.; Tessaud-Rita, N.; Jimenez-Trigos, E.; Vales, G.; et al. Ruminant-associated *Listeria monocytogenes* isolates belong preferentially to dairy-associated hypervirulent clones: A longitudinal study in 19 farms. *Environ. Microbiol.* **2021**, *23*, 7617–7631. [CrossRef]

26. Husu, J.R. Epidemiological studies on the occurrence of *Listeria monocytogenes* in the feces of dairy cattle. *Zent. Vet. B* **1990**, *37*, 276–282. [CrossRef]

27. Husu, J.R.; Seppanen, J.T.; Sivela, S.K.; Rauramaa, A.L. Contamination of raw milk by *Listeria monocytogenes* on dairy farms. *Zent. Vet. B* **1990**, *37*, 268–275. [CrossRef]

28. Ivaneka, R.; Grohn, Y.T.; Ho, A.J.J.; Wiedmann, M. Markov chain approach to analyze the dynamics of pathogen fecal shedding— Example of *Listeria monocytogenes* shedding in a herd of dairy cattle. *J. Theor. Biol.* **2007**, *245*, 44–58. [CrossRef]

29. Vilar, M.J.; Yus, E.; Sanjuán, M.L.; Diéguez, F.J.; Rodríguez-Otero, J.L. Prevalence of and Risk Factors for Listeria Species on Dairy Farms. *J. Dairy Sci.* **2007**, *90*, 5083–5088. [CrossRef]
30. Nightingale, K.K.; Fortes, E.D.; Ho, A.J.; Schukken, Y.H.; Grohn, Y.T.; Wiedmann, M. Evaluation of farm management practices as risk factors for clinical listeriosis and fecal shedding of *Listeria monocytogenes* in ruminants. *J. Am. Vet. Med. Assoc.* **2005**, *227*, 1808–1814. [CrossRef]
31. Nightingale, K.K.; Schukken, Y.H.; Nightingale, C.R.; Fortes, E.D.; Ho, A.J.; Her, Z.; Grohn, Y.T.; McDonough, P.L.; Wiedmann, M. Ecology and transmission of *Listeria monocytogenes* infecting ruminants and in the farm environment. *Appl. Environ. Microbiol.* **2004**, *70*, 4458–4467. [CrossRef] [PubMed]
32. Schoder, D.; Melzner, D.; Schmalwieser, A.; Zangana, A.; Winter, P.; Wagner, M. Important vectors for *Listeria monocytogenes* transmission at farm dairies manufacturing fresh sheep and goat cheese from raw milk. *J. Food Prot.* **2011**, *74*, 919–924. [CrossRef] [PubMed]
33. Moura, A.; Lefrancq, N.; Wirth, T.; Leclercq, A.; Borges, V.; Gilpin, B.; Dallman, T.J.; Frey, J.; Franz, E.; Nielsen, E.M.; et al. Emergence and global spread of *Listeria monocytogenes* main clinical clonal complex. *Sci. Adv.* **2021**, *7*, eabj9805. [CrossRef] [PubMed]
34. Zundel, E.; Bernard, S. *Listeria monocytogenes* translocates throughout the digestive tract in asymptomatic sheep. *J. Med. Microbiol.* **2006**, *55*, 1717–1723. [CrossRef]
35. Lin, C.H.; Adams, P.J.; Huang, J.F.; Sun, Y.F.; Lin, J.H.; Robertson, I.D. Contamination of chicken carcasses and the abattoir environment with *Listeria monocytogenes* in Taiwan. *Br. Poult. Sci.* **2021**, *62*, 701–709. [CrossRef]
36. Pohjola, L.; Nykäsenoja, S.; Kivistö, R.; Soveri, T.; Huovilainen, A.; Hänninen, M.L.; Fredriksson-Ahomaa, M. Zoonotic Public Health Hazards in Backyard Chickens. *Zoonoses Public Health* **2016**, *63*, 420–430. [CrossRef]
37. Elsayed, M.M.; Elkenany, R.M.; Zakaria, A.I.; Badawy, B.M. Epidemiological study on *Listeria monocytogenes* in Egyptian dairy cattle farms' insights into genetic diversity of multi-antibiotic-resistant strains by ERIC-PCR. *Environ. Sci. Pollut. Res.* **2022**, *29*, 54359–54377. [CrossRef]
38. Weber, A.; Potel, J.; SchaferSchmidt, R.; Prell, A.; Datzmann, C. Investigations on the occurrence of *Listeria monocytogenes* in fecal samples of domestic and companion animals. *Zent. Hyg. Umw.* **1995**, *198*, 117–123.
39. Oswaldi, V.; Dzierzon, J.; Thieme, S.; Merle, R.; Meemken, D. Slaughter pigs as carrier of *Listeria monocytogenes* in Germany. *J. Consum. Prot. Food Saf.* **2021**, *16*, 109–115. [CrossRef]
40. Adorisio, E.; Tassone, D.D.; Conte, A.; Annicchiarico, L.S. *Lysteria monocytogenes* in samples of swines sent to the human alimentary chain. *Ann. Ig. Med. Prev. Comunita* **2003**, *15*, 575–581.
41. Iida, T.; Kanzaki, M.; Maruyama, T.; Inoue, S.; Kaneuchi, C. Prevalence of *Listeria monocytogenes* in intestinal contents of healthy animals in japan. *J. Vet. Med. Sci.* **1991**, *53*, 873–875. [CrossRef] [PubMed]
42. Hasegawa, M.; Iwabuchi, E.; Yamamoto, S.; Muramatsu, M.; Takashima, I.; Hirai, K. Prevalence and characteristics of *Listeria monocytogenes* in feces of black beef cattle reared in three geographically distant areas in Japan. *Foodborne Pathog. Dis.* **2014**, *11*, 96–103. [CrossRef] [PubMed]
43. Obaidat, M.M.; Stringer, A.P. Prevalence, molecular characterization, and antimicrobial resistance profiles of *Listeria monocytogenes*, Salmonella enterica, and Escherichia coli O157:H7 on dairy cattle farms in Jordan. *J. Dairy Sci.* **2019**, *102*, 8710–8720. [CrossRef]
44. Stipetic, K.; Chang, Y.C.; Peters, K.; Salem, A.; Doiphode, S.H.; McDonough, P.L.; Chang, Y.F.; Sultan, A.; Mohammed, H.O. The rIsk of carriage of *Salmonella* spp. and *Listeria monocytogenes* in food animals in dynamic populations. *Vet. Med. Sci.* **2016**, *2*, 246–254. [CrossRef]
45. Bandelj, P.; Jamnikar-Ciglenecki, U.; Ocepek, M.; Blagus, R.; Vengust, M. Risk factors associated with fecal shedding of *Listeria monocytogenes* by dairy cows and calves. *J. Vet. Intern. Med.* **2018**, *32*, 1773–1779. [CrossRef]
46. Locatelli, A.; Lewis, M.A.; Rothrock, M.J., Jr. The Distribution of Listeria in Pasture-Raised Broiler Farm Soils Is Potentially Related to University of Vermont Medium Enrichment Bias toward Listeria innocua over *Listeria monocytogenes*. *Front. Vet. Sci.* **2017**, *4*, 227. [CrossRef]
47. Mohammed, H.O.; Stipetic, K.; McDonough, P.L.; Gonzalez, R.N.; Nydam, D.V.; Atwill, E.R. Identification of potential on-farm sources of *Listeria monocytogenes* in herds of dairy cattle. *Am. J. Vet. Res.* **2009**, *70*, 383–388. [CrossRef]
48. Yoshida, T.; Sugimoto, T.; Sato, M.; Hirai, K. Incidence of *Listeria monocytogenes* in wild animals in Japan. *J. Vet. Med. Sci.* **2000**, *62*, 673–675. [CrossRef]
49. Bouttefroy, A.; Lemaitre, J.P.; Rousset, A. Prevalence of Listeria sp. in droppings from urban rooks (*Corvus frugilegus*). *J. Appl. Microbiol.* **1997**, *82*, 641–647. [CrossRef]
50. Hellstrom, S.; Kiviniemi, K.; Autio, T.; Korkeala, H. *Listeria monocytogenes* is common in wild birds in Helsinki region and genotypes are frequently similar with those found along the food chain. *J. Appl. Microbiol.* **2008**, *104*, 883–888. [CrossRef]
51. Lyautey, E.; Hartmann, A.; Pagotto, F.; Tyler, K.; Lapen, D.R.; Wilkes, G.; Piveteau, P.; Rieu, A.; Robertson, W.J.; Medeiros, D.T.; et al. Characteristics and frequency of detection of fecal *Listeria monocytogenes* shed by livestock, wildlife, and humans. *Can. J. Microbiol.* **2007**, *53*, 1158–1167. [CrossRef] [PubMed]
52. Weindl, L.; Frank, E.; Ullrich, U.; Heurich, M.; Kleta, S.; Ellerbroek, L.; Gareis, M. *Listeria monocytogenes* in Different Specimens from Healthy Red Deer and Wild Boars. *Foodborne Pathog. Dis.* **2016**, *13*, 391–397. [CrossRef]

53. Nowakiewicz, A.; Zieba, P.; Ziolkowska, G.; Gnat, S.; Muszynska, M.; Tomczuk, K.; Dziedzic, B.M.; Ulbrych, L.; Troscianczyk, A. Free-Living Species of Carnivorous Mammals in Poland: Red Fox, Beech Marten, and Raccoon as a Potential Reservoir of Salmonella, Yersinia, *Listeria* spp. and Coagulase-Positive Staphylococcus. *PLoS ONE* **2016**, *11*, e0155533. [CrossRef] [PubMed]

54. Najdenski, H.; Dimova, T.; Zaharieva, M.M.; Nikolov, B.; Petrova-Dinkova, G.; Dalakchieva, S.; Popov, K.; Hristova-Nikolova, I.; Zehtindjiev, P.; Peev, S.; et al. Migratory birds along the Mediterranean–Black Sea Flyway as carriers of zoonotic pathogens. *Can. J. Microbiol.* **2018**, *64*, 915–924. [CrossRef]

55. Gorham, T.J.; Lee, J. Pathogen Loading From Canada Geese Faeces in Freshwater: Potential Risks to Human Health Through Recreational Water Exposure. *Zoonoses Public Health* **2016**, *63*, 177–190. [CrossRef]

56. Quessy, S.; Messier, S. Prevalence of *Salmonella* spp., *Campylobacter* spp. and *Listeria* spp. in ring-billed gulls (*Larus delawarensis*). *J. Wildl. Dis.* **1992**, *28*, 526–531. [CrossRef] [PubMed]

57. Cao, X.; Wang, Y.; Wang, Y.; Li, H.; Luo, L.; Wang, P.; Zhang, L.; Li, H.; Liu, J.; Lu, L.; et al. Prevalence and Characteristics of Listeria ivanovii Strains in Wild Rodents in China. *Vector-Borne Zoonotic Dis.* **2019**, *19*, 8–15. [CrossRef] [PubMed]

58. Wang, Y.; Lu, L.; Lan, R.; Salazar, J.K.; Liu, J.; Xu, J.; Ye, C. Isolation and characterization of Listeria species from rodents in natural environments in China. *Emerg. Microbes Infect.* **2017**, *6*, e44. [CrossRef]

59. Laaksonen, S.; Oksanen, A.; Julmi, J.; Zweifel, C.; Fredriksson-Ahomaa, M.; Stephan, R. Presence of foodborne pathogens, extended-spectrum beta-lactamase -producing Enterobacteriaceae, and methicillin-resistant Staphylococcus aureus in slaughtered reindeer in northern Finland and Norway. *Acta Vet. Scand.* **2017**, *59*, 2. [CrossRef]

60. Weber, A.; Potel, J.; Schafer-Schmidt, R. The occurrence of *Listeria monocytogenes* in fecal samples of pigeons. *Berl. Und Muenchener Tieraerztliche Wochenschr.* **1995**, *108*, 26–27.

61. Onjong, H.A.; Ngayo, M.O.; Mwaniki, M.; Wambui, J.; Njage, P.M.K. Microbiological Safety of Fresh Tilapia (Oreochromis niloticus) from Kenyan Fresh Water Fish Value Chains. *J. Food Prot.* **2018**, *81*, 1973–1981. [CrossRef] [PubMed]

62. Gnat, S.; Troscianczyk, A.; Nowakiewicz, A.; Majer-Dziedzic, B.; Ziolkowska, G.; Dziedzic, R.; Zieba, P.; Teodorowski, O. Experimental studies of microbial populations and incidence of zoonotic pathogens in the faeces of red deer (*Cervus elaphus*). *Lett. Appl. Microbiol.* **2015**, *61*, 446–452. [CrossRef] [PubMed]

63. Wacheck, S.; Fredriksson-Ahomaa, M.; Konig, M.; Stolle, A.; Stephan, R. Wild boars as an important reservoir for foodborne pathogens. *Foodborne Pathog. Dis.* **2010**, *7*, 307–312. [CrossRef] [PubMed]

64. Chen, T.; Orsi, R.H.; Chen, R.; Gunderson, M.; Roof, S.; Wiedmann, M.; Childs-Sanford, S.E.; Cummings, K.J. Characterization of *Listeria monocytogenes* isolated from wildlife in central New York. *Vet. Med. Sci.* **2022**, *8*, 1319–1329. [CrossRef]

65. Ferreira, V.; Wiedmann, M.; Teixeira, P.; Stasiewicz, M.J. *Listeria monocytogenes* persistence in food-associated environments: Epidemiology, strain characteristics, and implications for public health. *J. Food Prot.* **2014**, *77*, 150–170. [CrossRef] [PubMed]

66. Hsu, J.-L.; Opitz, H.M.; Bayer, R.C.; Kling, L.J.; Halteman, W.A.; Martin, R.E.; Slabyj, B.M. *Listeria monocytogenes* in an Atlantic Salmon (Salmo salar) Processing Environment. *J. Food Prot.* **2005**, *68*, 1635–1640. [CrossRef]

67. Fagerlund, A.; Wagner, E.; Møretrø, T.; Heir, E.; Moen, B.; Rychli, K.; Langsrud, S. Pervasive *Listeria monocytogenes* Is Common in the Norwegian Food System and Is Associated with Increased Prevalence of Stress Survival and Resistance Determinants. *Appl. Environ. Microbiol.* **2022**, *88*, e0086122. [CrossRef] [PubMed]

68. Miettinen, H.; Wirtanen, G. Ecology of *Listeria* spp. in a fish farm and molecular typing of *Listeria monocytogenes* from fish farming and processing companies. *Int. J. Food Microbiol.* **2006**, *112*, 138–146. [CrossRef]

69. Simothy, L.; Mahomoodally, F.; Neetoo, H. A study on the potential of ants to act as vectors of foodborne pathogens. *AIMS Microbiol.* **2018**, *4*, 319–333. [CrossRef]

70. Pushkareva, V.I.; Ermolaeva, S.A. *Listeria monocytogenes* virulence factor Listeriolysin O favors bacterial growth in co-culture with the ciliate Tetrahymena pyriformis, causes protozoan encystment and promotes bacterial survival inside cysts. *BMC Microbiol.* **2010**, *10*, 26. [CrossRef]

71. Statista GmbH. Anzahl der Reptilien in Europa Nach Ländern im Jahr 2021. Available online: https://de.statista.com/statistik/daten/studie/454104/umfrage/reptilien-in-europa-nach-laendern/ (accessed on 7 October 2022).

72. Gismervik, K.; Aspholm, M.; Rorvik, L.M.; Bruheim, T.; Andersen, A.; Skaar, I. Invading slugs (*Arion vulgaris*) can be vectors for *Listeria monocytogenes*. *J. Appl. Microbiol.* **2015**, *118*, 809–816. [CrossRef] [PubMed]

73. Schuchat, A.; Swaminathan, B.; Broome, C.V. Epidemiology of human listeriosis. *Clin. Microbiol. Rev.* **1991**, *4*, 169–183. [CrossRef]

74. Grif, K.; Hein, I.; Wagner, M.; Brandl, E.; Mpamugo, O.; McLauchlin, J.; Dierich, M.P.; Allerberger, F. Prevalence and characterization of *Listeria monocytogenes* in the feces of healthy Austrians. *Wien. Klin. Wochenschr.* **2001**, *113*, 737–742. [PubMed]

75. Sauders, B.D.; Pettit, D.; Currie, B.; Suits, P.; Evans, A.; Stellrecht, K.; Dryja, D.M.; Slate, D.; Wiedmann, M. Low prevalence of *Listeria monocytogenes* in human stool. *J. Food Prot.* **2005**, *68*, 178–181. [CrossRef] [PubMed]

76. Grif, K.; Patscheider, G.; Dierich, M.P.; Allerberger, F. Incidence of fecal carriage of *Listeria monocytogenes* in three healthy volunteers: A one-year prospective stool survey. *Eur. J. Clin. Microbiol. Infect. Dis.* **2003**, *22*, 16–20. [CrossRef]

77. EFSA Panelon Biological Hazards (BIOHAZ); Ricci, A.; Allende, A.; Bolton, D.; Chemaly, M.; Davies, R.; Fernández Escámez, P.S.; Girones, R.; Herman, L.; Koutsoumanis, K.; et al. Listeria monocytogenes contamination of ready-to-eat foods and the risk for human health in the EU. *EFSA J.* **2018**, *16*, e05134. [CrossRef]

78. MacGowan, A.P.; Marshall, R.J.; MacKay, I.M.; Reeves, D.S. Listeria faecal carriage by renal transplant recipients, haemodialysis patients and patients in general practice: Its relation to season, drug therapy, foreign travel, animal exposure and diet. *Epidemiol. Infect.* **1991**, *106*, 157–166. [CrossRef]

79. Freitag, I.G.R.; de Castro Lisboa Pereira, R.; Machado, E.S.; Hofer, E.; Vallim, D.C.; Hofer, C.B. Prevalence of *Listeria monocytogenes* fecal carriers in HIV-infected and -uninfected pregnant women from Brazil. *Braz. J. Microbiol.* **2021**, *52*, 2081–2084. [CrossRef]

80. Lamont, R.J.; Postlethwaite, R. Carriage of *Listeria monocytogenes* and related species in pregnant and non-pregnant women in Aberdeen, Scotland. *J. Infect.* **1986**, *13*, 187–193. [CrossRef]

81. Mascola, L.; Ewert, D.P.; Eller, A. Listeriosis: A previously unreported medical complication in women with multiple gestations. *Am. J. Obstet. Gynecol.* **1994**, *170*, 1328–1332. [CrossRef]

82. Sarr, M.; Alou, M.T.; Delerce, J.; Khelaifia, S.; Diagne, N.; Diallo, A.; Bassene, H.; Bréchard, L.; Bossi, V.; Mbaye, B.; et al. A *Listeria monocytogenes* clone in human breast milk associated with severe acute malnutrition in West Africa: A multicentric case-controlled study. *PLoS Negl. Trop. Dis.* **2021**, *15*, e0009555. [CrossRef] [PubMed]

83. Tahoun, A.; Abou Elez, R.M.M.; Abdelfatah, E.N.; Elsohaby, I.; El-Gedawy, A.A.; Elmoslemany, A.M. *Listeria monocytogenes* in raw milk, milking equipment and dairy workers: Molecular characterization and antimicrobial resistance patterns. *J. Glob. Antimicrob. Resist.* **2017**, *10*, 264–270. [CrossRef] [PubMed]

84. Muller, H.E. Listeria isolations from feces of patients with diarrhea and from healthy food handlers. *Infection* **1990**, *18*, 97–99. [CrossRef] [PubMed]

85. Meghdadi, H.; Khosravi, A.D.; Sheikh, A.F.; Alami, A.; Nassirabady, N. Isolation and characterization of *Listeria monocytogenes* from environmental and clinical sources by culture and PCR-RFLP methods. *Iran. J. Microbiol.* **2019**, *11*, 7–12. [CrossRef]

86. Akkaya, L.; Alisarli, M.; Cetinkaya, Z.; Kara, R.; Telli, R. Occurrence of Escherichia coli O157: H7/O157, *Listeria monocytogenes* and *Salmonella* spp. in beef slaughterhouse environments, equipment and workers. *J. Muscle. Foods* **2008**, *19*, 261–274. [CrossRef]

87. Pala, K.; Oezakin, C.; Akis, N.; Sinirtas, M.; Gedikoglu, S.; Aytekin, H. Asymptomatic carriage of bacteria in food workers in Nilufer district, Bursa, Turkey. *Turk. J. Med. Sci.* **2010**, *40*, 133–139. [CrossRef]

88. Alam, M.S.; Gangiredla, J.; Hasan, N.A.; Barnaba, T.; Tartera, C. Aging-Induced Dysbiosis of Gut Microbiota as a Risk Factor for Increased *Listeria monocytogenes* Infection. *Front. Immunol.* **2021**, *12*, 672353. [CrossRef]

89. European Food Safety Authority. Analysis of the baseline survey on the prevalence of *Listeria monocytogenes* in certain ready-to-eat (RTE) foods in the EU, 2010-2011 Part A: *Listeria monocytogenes* prevalence estimates. *EFSA J.* **2013**, *11*, 3241. [CrossRef]

90. Wagner, M.; Auer, B.; Trittremmel, C.; Hein, I.; Schoder, D. Survey on the Listeria contamination of ready-to-eat food products and household environments in Vienna, Austria. *Zoonoses Public Health* **2007**, *54*, 16–22. [CrossRef]

91. Esteban, J.I.; Oporto, B.; Aduriz, G.; Juste, R.A.; Hurtado, A. Faecal shedding and strain diversity of *Listeria monocytogenes* in healthy ruminants and swine in Northern Spain. *BMC Vet. Res.* **2009**, *5*, 2. [CrossRef]

92. Fenlon, D.R.; Wilson, J.; Donachie, W. The incidence and level of *Listeria monocytogenes* contamination of food sources at primary production and initial processing. *J. Appl. Bacteriol.* **1996**, *81*, 641–650. [CrossRef] [PubMed]

93. Fenlon, D.R. Wild birds and silage as reservoirs of Listeria in the agricultural environment. *J. Appl. Bacteriol.* **1985**, *59*, 537–543. [CrossRef] [PubMed]

94. Honjoh, K.; Iwazako, Y.; Lin, Y.; Kijima, N.; Miyamoto, T. Possibilities for Contamination of Tomato Fruit by *Listeria monocytogenes* during Cultivation. *Food Sci. Technol. Res.* **2016**, *22*, 349–357. [CrossRef]

95. Weller, D.; Belias, A.; Green, H.; Roof, S.; Wiedmann, M. Landscape, Water Quality, and Weather Factors Associated With an Increased Likelihood of Foodborne Pathogen Contamination of New York Streams Used to Source Water for Produce Production. *Front. Sustain. Food Syst.* **2020**, *3*, 124. [CrossRef]

96. Weller, D.; Wiedmann, M.; Strawn, L.K. Spatial and Temporal Factors Associated with an Increased Prevalence of *Listeria monocytogenes* in Spinach Fields in New York State. *Appl. Environ. Microbiol.* **2015**, *81*, 6059–6069. [CrossRef]

97. Weller, D.; Wiedmann, M.; Strawn, L.K. Irrigation Is Significantly Associated with an Increased Prevalence of *Listeria monocytogenes* in Produce Production Environments in New York State. *J. Food Prot.* **2015**, *78*, 1132–1141. [CrossRef]

98. Gu, G.; Strawn, L.K.; Ottesen, A.R.; Ramachandran, P.; Reed, E.A.; Zheng, J.; Boyer, R.R.; Rideout, S.L. Correlation of Salmonella enterica and *Listeria monocytogenes* in Irrigation Water to Environmental Factors, Fecal Indicators, and Bacterial Communities. *Front. Microbiol.* **2020**, *11*, 557289. [CrossRef]

99. Huang, Y.; Hansen, L.T.; Ragush, C.M.; Jamieson, R.C. Disinfection and removal of human pathogenic bacteria in arctic waste stabilization ponds. *Environ. Sci. Pollut. Res. Int.* **2018**, *25*, 32881–32893. [CrossRef]

100. Walland, J.; Lauper, J.; Frey, J.; Imhof, R.; Stephan, R.; Seuberlich, T.; Oevermann, A. *Listeria monocytogenes* infection in ruminants: Is there a link to the environment, food and human health? A review. *Schweiz Arch. Tierheilkd* **2015**, *157*, 319–328. [CrossRef]

101. Oliver, S.P.; Jayarao, B.M.; Almeida, R.A. Foodborne pathogens in milk and the dairy farm environment: Food safety and public health implications. *Foodborne Pathog. Dis.* **2005**, *2*, 115–129. [CrossRef]

102. de Valk, H.; Vaillant, V.; Jacquet, C.; Rocourt, J.; Le Querrec, F.; Stainer, F.; Quelquejeu, N.; Pierre, O.; Pierre, V.; Desenclos, J.C.; et al. Two consecutive nationwide outbreaks of Listeriosis in France, October 1999-February 2000. *Am. J. Epidemiol.* **2001**, *154*, 944–950. [CrossRef] [PubMed]

103. McLauchlin, J.; Hall, S.M.; Velani, S.K.; Gilbert, R.J. Human listeriosis and pate: A possible association. *BMJ* **1991**, *303*, 773–775. [CrossRef] [PubMed]

104. Giovannacci, I.; Ragimbeau, C.; Queguiner, S.; Salvat, G.; Vendeuvre, J.L.; Carlier, V.; Ermel, G. *Listeria monocytogenes* in pork slaughtering and cutting plants: Use of RAPD, PFGE and PCR-REA for tracing and molecular epidemiology. *Int. J. Food Microbiol.* **1999**, *53*, 127–140. [CrossRef]

105. Beloeil, P.A.; Chauvin, C.; Toquin, M.T.; Fablet, C.; Le Notre, Y.; Salvat, G.; Madec, F.; Fravalo, P. *Listeria monocytogenes* contamination of finishing pigs: An exploratory epidemiological survey in France. *Vet. Res.* **2003**, *34*, 737–748. [CrossRef]
106. Falardeau, J.; Johnson, R.P.; Pagotto, F.; Wang, S. Occurrence, characterization, and potential predictors of verotoxigenic Escherichia coli, *Listeria monocytogenes*, and Salmonella in surface water used for produce irrigation in the Lower Mainland of British Columbia, Canada. *PLoS ONE* **2017**, *12*, e0185437. [CrossRef]
107. Ojeniyi, B.; Wegener, H.C.; Jensen, N.E.; Bisgaard, M. *Listeria monocytogenes* in poultry and poultry products: Epidemiological investigations in seven Danish abattoirs. *J. Appl. Bacteriol.* **1996**, *80*, 395–401. [CrossRef]
108. Moosavy, M.H.; Shavisi, N.; Warriner, K.; Mosta-Favi, E. Bacterial Contamination of Iranian Paper Currency. *Iran. J. Public Health* **2013**, *42*, 1067–1070.
109. Marshall, J.C.; Soboleva, T.K.; Jamieson, P.; French, N.P. Estimating Bacterial Pathogen Levels in New Zealand Bulk Tank Milk. *J. Food Prot.* **2016**, *79*, 771–780. [CrossRef]
110. Jongman, M.; Korsten, L. Microbial quality and suitability of roof-harvested rainwater in rural villages for crop irrigation and domestic use. *J. Water Health* **2016**, *14*, 961–971. [CrossRef]
111. Ahlstrom, C.A.; Manuel, C.S.; Den Bakker, H.C.; Wiedmann, M.; Nightingale, K.K. Molecular ecology of *Listeria* spp., Salmonella, Escherichia coli O157:H7 and non-O157 Shiga toxin-producing E. coli in pristine natural environments in Northern Colorado. *J. Appl. Microbiol.* **2018**, *124*, 511–521. [CrossRef]
112. Dungan, R.S.; Klein, M.; Leytem, A.B. Quantification of bacterial indicators and zoonotic pathogens in dairy wastewater ponds. *Appl. Environ. Microbiol.* **2012**, *78*, 8089–8095. [CrossRef] [PubMed]
113. Belias, A.; Strawn, L.K.; Wiedmann, M.; Weller, D. Small Produce Farm Environments Can Harbor Diverse *Listeria monocytogenes* and *Listeria* spp. Populations. *J. Food Prot.* **2021**, *84*, 113–121. [CrossRef]
114. Jensen, A.; Thomsen, L.E.; Jorgensen, R.L.; Larsen, M.H.; Roldgaard, B.B.; Christensen, B.B.; Vogel, B.F.; Gram, L.; Ingmer, H. Processing plant persistent strains of *Listeria monocytogenes* appear to have a lower virulence potential than clinical strains in selected virulence models. *Int. J. Food Microbiol.* **2008**, *123*, 254–261. [CrossRef] [PubMed]
115. Lunden, J.M.; Autio, T.J.; Sjoberg, A.M.; Korkeala, H.J. Persistent and nonpersistent *Listeria monocytogenes* contamination in meat and poultry processing plants. *J. Food Prot.* **2003**, *66*, 2062–2069. [CrossRef] [PubMed]
116. Kastbjerg, V.G.; Larsen, M.H.; Gram, L.; Ingmer, H. Influence of sublethal concentrations of common disinfectants on expression of virulence genes in *Listeria monocytogenes*. *Appl. Environ. Microbiol.* **2010**, *76*, 303–309. [CrossRef]
117. Rodrigues, D.; Cerca, N.; Teixeira, P.; Oliveira, R.; Ceri, H.; Azeredo, J. *Listeria monocytogenes* and Salmonella enterica enteritidis biofilms susceptibility to different disinfectants and stress-response and virulence gene expression of surviving cells. *Microb. Drug Resist.* **2011**, *17*, 181–189. [CrossRef]
118. Prates, D.D.; Haubert, L.; Wurfel, S.D.; Cavicchioli, V.Q.; Nero, L.A.; da Silva, W.P. *Listeria monocytogenes* in dairy plants in Southern Brazil: Occurrence, virulence potential, and genetic diversity. *J. Food Saf.* **2019**, *39*, e12695. [CrossRef]
119. Alessandria, V.; Rantsiou, K.; Dolci, P.; Cocolin, L. Molecular methods to assess *Listeria monocytogenes* route of contamination in a dairy processing plant. *Int. J. Food Microbiol.* **2010**, *141*, S156–S162. [CrossRef]
120. Ibba, M.; Cossu, F.; Spanu, V.; Virdis, S.; Spanu, C.; Scarano, C.; De Santis, E.P.L. *Listeria monocytogenes* contamination in dairy plants: Evaluation of *Listeria monocytogenes* environmental contamination in two cheese-making plants using sheeps milk. *Ital. J. Food Saf.* **2013**, *2*, 109–112. [CrossRef]
121. Mafu, A.A.; Roy, D.; Goulet, J.; Magny, P. Attachment of *Listeria monocytogenes* to Stainless Steel, Glass, Polypropylene, and Rubber Surfaces After Short Contact Times. *J. Food Prot.* **1990**, *53*, 742–746. [CrossRef]
122. Lawrence, L.M.; Gilmour, A. Characterization of *Listeria monocytogenes* isolated from poultry products and from the poultry-processing environment by random amplification of polymorphic DNA and multilocus enzyme electrophoresis. *Appl. Environ. Microbiol.* **1995**, *61*, 2139–2144. [CrossRef] [PubMed]

Review

Microbiological Safety and Shelf-Life of Low-Salt Meat Products—A Review

Coral Barcenilla [1,*], Avelino Álvarez-Ordóñez [1,2], Mercedes López [1,2], Ole Alvseike [3] and Miguel Prieto [1,2]

[1] Department of Food Hygiene and Technology, Faculty of Veterinary Medicine, University of León, 24071 León, Spain
[2] Institute of Food Science and Technology, University of León, 24007 León, Spain
[3] Animalia—Norwegian Meat and Poultry Research Centre, NO-0513 Oslo, Norway
*   Correspondence: cbarc@unileon.es; Tel.: +34-987-291245

Abstract: Salt is widely employed in different foods, especially in meat products, due to its very diverse and extended functionality. However, the high intake of sodium chloride in human diet has been under consideration for the last years, because it is related to serious health problems. The meat-processing industry and research institutions are evaluating different strategies to overcome the elevated salt concentrations in products without a quality reduction. Several properties could be directly or indirectly affected by a sodium chloride decrease. Among them, microbial stability could be shifted towards pathogen growth, posing a serious public health threat. Nonetheless, the majority of the literature available focuses attention on the sensorial and technological challenges that salt reduction implies. Thereafter, the need to discuss the consequences for shelf-life and microbial safety should be considered. Hence, this review aims to merge all the available knowledge regarding salt reduction in meat products, providing an assessment on how to obtain low salt products that are sensorily accepted by the consumer, technologically feasible from the perspective of the industry, and, in particular, safe with respect to microbial stability.

Keywords: microbiological safety; low-salt meat products; shelf-life; water activity

Citation: Barcenilla, C.; Álvarez-Ordóñez, A.; López, M.; Alvseike, O.; Prieto, M. Microbiological Safety and Shelf-Life of Low-Salt Meat Products—A Review. *Foods* **2022**, *11*, 2331. https://doi.org/10.3390/foods11152331

Academic Editors: Frans J.M. Smulders and Arun K. Bhunia

Received: 29 June 2022
Accepted: 29 July 2022
Published: 4 August 2022

Publisher's Note: MDPI stays neutral with regard to jurisdictional claims in published maps and institutional affiliations.

1. Introduction

Consumers are increasingly demanding foods with a low salt content, which are perceived as healthier and fresher. Reducing salt intake may lower many commonly associated risks, including high blood pressure, cardiovascular disease, stroke, and coronary heart attack, and has been identified by the World Health Organization (WHO) as one of the most cost-effective measures that countries can take to improve health outcomes [1].

The use of salt in food manufacturing has been traditionally related to its ability to keep foods edible for longer periods of time, which would allow manufacturers to extend their consumption in periods of scarcity [2]. In ancient times, salt was used as a preservative in many foods such as fish, meat, and dairy products, due to its preservative quality and sensorial properties [3]. Indeed, as a highly estimated trading commodity, it is argued that the word salary derives from the Latin salarium, i.e., money given to Roman legionnaires to buy salt [4].

To ensure safe foods with an extended shelf-life, classical food preservation processes (thermal processing, drying, salting, freezing, etc.) usually impose extreme physical and/or chemical barriers to prevent the growth of, or to inactivate, spoilage and pathogenic bacteria [5]. Salting has been traditionally used in many instances as a simple and inexpensive method of preservation since it does not need sophisticated equipment and imparts suitable sensory properties to the product. It was not until the 1950s that cold storage was introduced in private households, reducing the need for salt as a basic additive. Still, this technological advance did not cause a reduction in the amount of salt in the diet. Indeed, the consumption of salt increased simultaneously with the expansion of industrially processed foods, illustrating consumers' strong appreciation and preference for a salty taste [6–8]. These dietary changes resulted in less individual-level control of sodium intake and, for many, a chronic excess of sodium intake from non-discretionary sources [4]. In addition, salt is very appreciated from a technological point of view due to its various properties. Considered as a multifunctional ingredient, salt is used to enhance flavor, mask off-flavors, improve food structure and texture, promote water holding capacity, and reduce water activity, which finally leads to a preservation action [9].

The issue of salt and how it relates to health and food safety has been frequently dealt with by regulatory agencies [10–15]. The food industry faces big challenges derived from salt reduction, and, although it has met the challenge, there is still a need to balance salt reduction in products with consumers' taste preferences, the microbiological safety of the products, and food technological constraints [16,17]. Some producers have already reduced the salt content successfully [18–20], but further reductions depend on salt replacers or additives that work as effective substitutes [16,21,22]. However, salt concentrations commonly used in fermented meats inhibit the growth of undesired microorganisms, and, at the same time, promote the growth of more salt-tolerant lactic acid bacteria [23–25], while high salt concentrations may also inhibit the growth of starter cultures in fermented meats, as happened in a sausage made with 5% instead of 2.5% NaCl [26] or with 2.4% instead of 1.01% [27]. Paradoxically, food scares regarding E-numbers and the "clean label" trends are significant obstacles for the development of healthier low-sodium products [14,27]. Thus, research is still on-going regarding low-sodium meat products, which indicates that the direct reduction of this component to reach authorities' recommendations in this type of products is not an easy target [28].

The reduction of salt in food products and the increasing use of replacers (e.g., KCl) as an alternative to NaCl may represent potential safety risks arising from such reformulations. While the consequences of sensory and technological properties have received considerably more attention [9,29–34], salt reduction and the use of replacers as an alternative to NaCl may represent potential safety risks [17]. Before reducing the salt content of meat products or replacing it with alternative ingredients, it is necessary to assess the microbiological stability of original and reformulated meat products by studying the consequences for safety and quality [8]. Therefore, the aim of this article is to review the mechanisms of microbial inhibition by salt, to evaluate the microbiological risks deriving from a low level of NaCl in meat products and to critically review the available management measures aimed at minimizing the risks associated with this type of reformulated product.

2. Health Risks Associated with Salt Consumption

The ingestion of elevated levels of sodium chloride is associated with increased blood pressure and hypertension, induced cardiovascular disease (CVD) and stroke [28,29]. The increased risk of cardiovascular events associated with a higher sodium intake (>5 g/d) is most prominent in those with hypertension [30]. However, increasing evidence has shown that a high intake of salt is also a risk factor for otherwise healthy people [31,32]. Current evidence from prospective cohort studies suggests that the lowest risk of cardiovascular events and death occurs in populations consuming an average sodium intake ranging from 3 to 5 g/d. Sodium reduction seems to increase heart rate independently of the reducing effect on the baseline blood pressure of sodium. Hence, lowering the sodium intake is best targeted at populations with hypertension who consume high-sodium diets [33–35]. The long-term effect of salt intake in doses higher than the physiological need is mainly increased blood pressure with age. There is also strong evidence that risk is reduced when salt intake is lowered, independent of age. Therefore, there is a health motivation for almost everybody to control and reduce their salt intake [29].

Physiologically, sodium levels are strictly controlled in the bodies of humans and animals. Sodium is the most important and prevalent metal ion in the body's tissues. Indeed, it is essential for homeostasis, blood pressure maintenance, water holding, and neural transmission, among others [36]. The salty taste reflects the Na^+ ion concentration in the mouth, which in turn causes a positive response in the brain, and hypertonic blood concentrations seem to be preferred [37]. The daily salt intake across the world has typically varied between 9 and 12 g NaCl [38]. However, the World Health Organization recommends a daily consumption of 5 g NaCl, equivalent to approx. 2 g/day of sodium [13].

Estimates of the global burden of disease from high systolic blood pressure are receiving increased attention [39]. In the USA, it was estimated that a 3 g/day salt reduction would save 194,000–392,000 quality-adjusted life-years and $10–24 billion in healthcare costs annually [40]. In Europe, the Framework for National Salt Initiatives was developed in 2008 with the overall goal of contributing towards a reduced salt intake at population level. Thus, the initiative identified five key elements to focus action on, which included (i) data availability, (ii) benchmarks and major food categories, (iii) raising public awareness, (iv) developing reformulation actions with industry/catering, and (v) monitoring and evaluation of actions. Overall, the EU framework set a realistic benchmark of a minimum 16% reduction over a 4-year period in all food categories [41]. Remarkably, countries with a higher sodium intake, i.e., the Czech Republic, Slovenia, and Hungary, exhibit a higher prevalence of hypertension [42]. A decrease of salt intake to 5 g per day is expected to substantially reduce the burden of cardiovascular disease and mortality. In Finland, the salt consumption has been considerably reduced, while the salt intake in other countries such as Poland has remained relatively high. Indeed, a reduction in salt intake to reach the WHO population nutrient goal would reduce the prevalence of stroke around 10.1% in Finland and 23.1% in Poland [43]. Regarding cardiovascular diseases, a 17% decrease is expected if the WHO target is reached, which will prevent an estimated 4 million deaths annually worldwide [42]. The UK and Ireland are other successful examples that have been following salt-reduction strategies among different food products [8]. Japan, which is a country well-known for its low mortality rates, reduced its salt intake from 14.5 g in 1973 to 9.5 g in 2017, which has led to the reduction in stomach cancer and other cerebrovascular diseases [44]. However, this reduction is still far from the 5 g NaCl recommendation [13]. Nevertheless, the socioeconomic status of individuals plays an important role in the type of diet followed and thereby the amount of sodium ingested [42].

3. Functions and Content of Salt in Meat Products

Salt (NaCl) has three main functions in meat products: a preservative effect, the contribution to organoleptic quality attributes (flavor and texture), and a technological function as to provide binding between meat and fat [8,38,45,46].

The preservation (extended shelf-life and microbiological stability) of meat products can be attained by lowering the water activity by the addition of a solute and through dehydration by removal of water by simple evaporation. Water molecules are retained among Na^+ and Cl^- ions and thus become unavailable for other functions, such as chemical or enzymatic reactions or to be used by microorganisms [47]. This provokes the inhibition of the spoilage and pathogenic microbiota, which in turn increases the shelf-life and safety of foods [8]. Stringer et al. (2005), evaluated this by modelling the growth capacity of *Aeromonas hydrophila*, *Clostridium botulinum*, *Listeria monocytogenes*, *Yersinia enterocolitica*, and *Bacillus cereus* in two NaCl concentrations in chicken rolls (3.04 and 1.61%), ham (5.57 and 2.81%), and bacon (5.5 and 2.85%). All bacteria presented a greater growth rate under a reduced salt content in all products in the model. Moreover, *C. botulinum* was able to grow in products with 2.85% salt, whereas it was not able to in 5.5% salt [48].

Salt is widely used as a **flavor and palatability enhancer**, since it improves the positive sensory attributes of most foods and helps attenuate bitterness and sweetness, thus being a major determinant for consumer acceptance. When different salt concentrations ranging from 0.8 to 2.2% w/w were added to pork breakfast sausages, consumers found the most acceptable samples had 1.4% salt [49]. Salting also helps the volatilization of certain molecules in foods, thus intensifying the aroma of the food [45]. Salted meat products such as dry-cured ham are quite popular, because they have unique sensory characteristics [47].

In meat products, salt operates as a **binding agent** between meat and fat. It increases the water-binding capacity, so that the final product shows improved yield, texture, tenderness, and palatability. The salt in meat products at 1.5–2.5% promotes the solubilization and extraction of myofibrillar proteins (actin and myosin) that are insoluble in water alone, therefore being fundamental in the gelation and binding of many restructured meat products [46]. Salting, in combination with processing steps such as blending and tumbling, helps extract these salt-soluble myofibrillar proteins to the meat surface, which is essential for holding pieces of meat together in batters and restructured meats and contributes to forming a gel between meat particles and between meat and fat particles [50]. The solubility degree of myofibrillar proteins is directly depending on the amount of NaCl in the meat product [51].

The range of the salt content in meat products is very large (Table 1), and even similar or equivalent meat products can be elaborated with different concentrations depending on the particular formulations. This suggests that it is feasible to generally reduce the salt content of foods. Meat products are considered the second biggest source, after bakery products, of salt intake in Europe [42]. This is quite outstanding, especially considering that the amount of salt naturally present in fresh meat is very low compared to meat products after processing [8,42]. In an analysis performed on a series of meat products, the results showed an average of 2.14 g NaCl/100 g of product, with the lowest value of 0.84 for turkey breast and the highest of 7.81 for ham [52]. In relation to the intake, in industrialized countries, about 75–80% of salt is ingested in processed foods, and especially meat products, which constitute one of the major sources of sodium in the form of salt or other additives [42,52]. Four food groups include almost 60% of the total ingested salt, i.e., cured meat products (26.2%), bread products (19.1%), cheese (6.7%), and processed ready-to-eat (RTE) foods (4.9%). It is estimated that, in countries such as Ireland or the United States, processed meat products contribute to more than 20% of the daily sodium intake [15,53]. It should be noted that, in cured meat products, sodium can stem not only from common salt, but also from sodium nitrite and the additives sodium ascorbate and sodium erythorbate, which are used as reducing agents. Other possible sources are sodium tripolyphosphate, monosodium glutamate, and hydrolyzed vegetable protein.

Table 1. Salt content in a selection of meat products from different countries.

Product	NaCl Content	Country	Reference
Beef, cured, dried beef	8.68	USA	
Pork, cured, bacon, cooked, broiled, pan-fried, or roasted	3.99	USA	[54]
Pork sausage	3.23	USA	
Canned meat chop	3.44	Serbia	
Cooked sausages	2.95	Serbia	[55]
Smoked products	3.44	Serbia	
Hard pork sausage	3.18	Spain	[56]
Cooked ham	2.45	Czech Republic	
Frankfurters	2.44	Czech Republic	
Knackauer	2.34	Germany	[57]
Schinkenwurst	2.03	Germany	
Bierschinken	2.2	Germany	
Pancetta	2.94	Serbia	
Kulen sausage	4.24	Serbia	[58]
Chorizo	3.58	Spain	
Fuet-type sausage	3.94	Spain	[52]
Mortadella sausage	1.97	Spain	

3.1. Salt as a Chemical Preservative

Meat products are made primarily of muscle meat added with salt and nitrites, which are responsible for the curing. The curing process changes the flavor and color of the meat and improves the shelf-life and safety of the product by inhibiting spoiling microbiota and pathogens, while certain microbial groups such as lactic acid bacteria (LAB) and *Micrococcaceae* are promoted [17,59,60]. The addition of salt and other solutes (salt replacers, sugars, humectants, proteins, etc.), together with the dehydration caused by the removal of water by simple evaporation, decreases water activity (a_w), reducing or inhibiting microbial and enzymatic activity and, therefore, accomplishing a significant preservation effect. Meat products manufactured with NaCl levels below those typically formulated have usually a shorter shelf-life [8]. For instance, the shelf-life of a reduced-salt bacon (2.3% *w/w* NaCl) was 28 days, whereas for the control bacon (3.5% *w/w* NaCl) it was 56 days [48]. Certainly, salt is not the only barrier to spoilage microbiota and pathogens present in meat products. The combined use of other preservation hurdles, such as low temperature, acidification [60], antimicrobial compounds [38,61–63], limited oxygen availability or HPP treatment [64], contributes to the inhibition of certain microbial groups and causes a shift in the prevailing microbial populations. For example, Michelakou et al. (2021) stated that abusive storage temperatures somehow limited the effect of salt, thus indicating that low temperatures help to hold certain microbial growth [65]. Moreover, potassium lactate allowed a 30% reduction from 4 to 2.8% salt in salami without repercussions in the antimicrobial capacity [62]. Several antimicrobial compounds are under research for future application as antimicrobial food cultures in different products, among them meat products, and quite promising results are being achieved using lactic acid bacteria and their bacteriocins [66–68].

3.2. Low Water Activity as an Environmental Stress Factor—Molecular Basis of NaCl Action as a Preservative and the Bacterial Adaptive Response against NaCl

Water activity is an indicator of the amount of water that is available for microbial growth and other chemical reactions in a particular food and can be defined as the ratio between the partial vapor pressure of a given food in relation to the vapor pressure of pure water at the same temperature. Considering Raoult's law, where the vapor pressure is related to the molar fraction of a solute in a solution, the higher the concentration of salt, the lower the a_w. Generally, an a_w of 0.85 is considered the lowest value for any human pathogen bacteria to grow, in accordance with the requirement for *Staphylococcus aureus* toxin production (Table 2 shows the a_w required by different bacterial pathogens). However, yeast and mold are able to grow at lower a_w levels, and even some rare xerophilous bacteria, which undergo a phenomenon known as anhydrobiosis, can persist in extreme dehydration. Nonetheless, for most foods, the a_w is between 0.95 and 0.99 [69]. The a_w of a food can be reduced by increasing the concentration of solutes in its aqueous phase through drying, water extraction, freeze drying, etc., or by adding new solutes. Salt is one of those solutes able to reduce a_w due to the association of sodium [Na^+] and chloride ions [Cl^-] ions with water molecules. Water dissolves salts due to its marked polarity and capacity to form weak hydrogen bonds; the electronegative pole of the H_2O molecule (oxygen) is attracted to the positively charged [Na^+], and the electropositive pole (hydrogen) is attracted to the negatively charged [Cl^-] [70].

Table 2. Limiting conditions regarding a_w for the growth of bacterial pathogens (adapted from Table A-1 in (FDA, 2011) [71]).

Bacterial Pathogens	Min. a_w	Max. % Water-Phase NaCl	Min pH
Bacillus cereus	0.92	10	4.3
Campylobacter jejuni	0.987	1.7	4.9
Clostridium botulinum, type A, and proteolytic types B and F	0.935	10	4.6
Clostridium botulinum, type E, and nonproteolytic types B and F	0.97	5	5.0
Clostridium perfringens	0.93	7	5
Escherichia coli	0.95	6.5	4
Listeria monocytogenes	0.92	10	4.4
Salmonella spp.	0.94	8	3.7
Shigella spp.	0.96	5.2	4.8
Staphylococcus aureus growth	0.83	20	4
Staphylococcus aureus toxin formation	0.85	10	4
Vibrio cholerae	0.97	6	5
Vibrio parahaemolyticus	0.94	10	4.8
Vibrio vulnificus	0.96	5	5
Yersinia enterocolitica	0.945	7	4.2

Salt produces an osmotic imbalance in microbial cells by extracting water from the cells through the membranes of both food tissues and microorganisms through the process of osmosis. The retraction and reduction of the cytoplasmic volume is a phenomenon known as plasmolysis. Osmotic stress due to a high concentration of solutes can occur in food during manufacturing, forcing microbial cells to cope internally with this adverse environment in order to survive. This event can occur suddenly, after rapid exposure to highly concentrated solutions (e.g., osmotic shock during salting), or progressively, as the product slowly dehydrates (e.g., during ripening). Osmotic shock provokes water efflux and dehydration in the cells, the release of low molecular-weight compounds and cell proteins, and the perturbation of many cellular physiological functions [72,73]. However, the progressive exposure to osmotic stress can allow cells to adapt gradually and maintain homeostasis by means of an adaptive response.

Cellular adaptive response. To protect the cell against the damage inflicted on functions and key molecules such as enzymes and macromolecular structures, bacteria use common adaptive stress pathways in response to a diverse range of adverse environmental conditions (such as dehydration) [74–76]. For this purpose, they have evolved adaptive networks, such as biofilm formation, shifts in metabolism, or changes in the cell membranes, that allow them to cope with the challenges of a changing environment [69,77]. There are three basic microbial strategies used to overcome exposure to a low a_w environment [78]:

- some cells counterbalance the levels of inorganic ions (usually KCl) to achieve osmotic stability;
- some are able to modify the membrane permeability, structure, and/or composition to protect the cells; and
- some produce or accumulate low-molecular-weight compounds that have the osmotic capacity to counteract the extreme external osmotic pressure. These osmolytes are defined as compatible solutes and are polar, normally uncharged, molecules. Compatible solutes act in the cytosol to counterbalance high external osmolarity, thus preventing water loss from the cell and plasmolysis, without adversely affecting the

macromolecular structure and enzymatic functions. These compatible solutes belong to several classes of compounds with some common structural motifs, some amino acid derivatives being particularly important [79].

Bacterial cells usually activate several synergistic response pathways to survive low water-activity environments or hyperosmolarity conditions [69]. It is generally acknowledged that bacteria react to environments of elevated osmolarity by means of a biphasic response, which involves the stimulation of potassium uptake (and its counter-ion glutamate) followed by a dramatic increase in the cytoplasmic concentration (by synthesis and/or uptake) of compatible solutes or osmoprotectants [80]. Nevertheless, it is important to highlight that bacteria do not always activate the same responses or genetic factors when facing osmotic stress produced by different solutes (NaCl, KCl, sucrose, etc.) [81]. For potassium uptake, microorganisms have an inducible high-affinity system (Kdp) and low-affinity systems (Trk, Kup) [82]. On the other hand, the main compatible solutes are glycine-betaine, carnitine, ectoine, proline, and trehalose, and several genetic loci are responsible for their synthesis or uptake [78,79,81]. For example, *L. monocytogenes* at high salt concentrations (10–20%) survives mainly due to the accumulation of glycine-betaine, carnitine, and proline taken up from the environment. The accumulation of glycine-betaine and carnitine occurs via two glycine-betaine transporters which are encoded by the *BetL* gene and the *Gbu* operon. On the other hand, carnitine is internalized via a carnitine transporter encoded by the *OpuC* operon [77]. Some master regulators of the bacterial stress response, such as σB, which are induced by a wide spectrum of stress conditions and which control the expression of numerous genes that mediate the adaptation to suboptimal environments, seem to be involved in the regulation of this complex response to hyperosmotic environments. In fact, for example, both *BetL* and *OpuC* have putative σB-dependent promoters [73].

Some bacteria can display additional survival strategies such as the over-expression of sodium efflux systems, the induction of modifications in cell morphology or membrane fatty acids composition, and the synthesis of specific stress proteins [69,77,79]. In fact, it has been described that cold-shock proteins (Csp), salt-shock proteins (Ssp), and stress-acclimatation proteins (Sap) can contribute to osmotic stress resistance [73].

<u>Additional effects of salt</u>. Sodium chloride exerts its preservative action primarily by making water unavailable for microorganisms and enzymatic reactions, but it also operates a direct antimicrobial effect [47]. High salt concentrations may interfere with the action of several cellular enzymes and force cells to expel Na^+ ions, which can be very effective in the inhibition of some microorganisms that do not have the necessary tools to counteract those effects. In addition, salt can favor lipid oxidation and thus can affect the quality of meat. Salt (0.5–2%) was found to be pro-oxidant on ground beef and in pork (1.5% salt). The pro-oxidant action of NaCl could be explained because NaCl can disrupt the cell-membrane-liberating ions that form the molecules and finally inhibit the enzymes that are in charge of antioxidant activity [83]. Enzymatic activity can also be affected by low a_w values caused by the addition of salt. It was observed that cathepsins, aminopeptidases, and neutral lipases were strongly affected as a_w values were lowered from 1 to 0.85 during ham dry curing. Nevertheless, other enzymes like calpain, acid lipase, or acid phospholipase were less or not affected at all [84]. It should also be considered that salt is usually formulated not as pure NaCl but as curing salt (sodium chloride containing around 12% of sodium nitrite) [9,57]. Sodium nitrite has a particular inhibitory effect on some pathogens, such as *C. botulinum*, and helps modulate the microbiota of meat products [85,86].

<u>Consequences for microbial behavior (growth, survival and inactivation)</u>. When exposed to high salt concentrations, microbial cells are forced to consume energy for homeostasis, maintenance, and reparation tasks, to the detriment of other energy-demanding cellular processes, such as growth and multiplication [87]. As a rule of thumb, the lag time increases and the growth rate progressively slows down as the conditions in the surrounding environment deviate from the optimal situation [88].

Another consequence is that the susceptibility of microbial cells to other stress factors can be modified [89]. The induction of cross-protection responses by exposure to low

a_w environments can modulate the fate of pathogenic microorganisms in food products, impacting shelf-life and food safety. Indeed, it has been reported that microbial cells are more resistant to different processing technologies, such as thermal treatments or high hydrostatic pressure (HHP) at low a_w conditions [90,91], and that previous exposure to salt or osmotic stress conditions can increase microbial tolerance to them [92–94]. In meat processing, the co-existence or succession of two or more stresses is common, and the sequence of events can be an important factor. For example, an acid stress placing a large energy demand on the cell could greatly sensitize the cell to successive treatments, such as a_w stress. Certain cellular components such as cold-shock proteins (Csps) have been shown to contribute to resistance to osmotic stress by means of an adaptive response [95]. Therefore, the combined or sequential exposure of cells to two or more stresses in food environments might induce cross-protection responses [96,97].

4. Changes in the Microbiota of Meat Products Due to Salt

Fresh meat is particularly prone to microbial spoilage due to the abundance of nutrients and favorable intrinsic properties that do not hamper the metabolism of bacteria. When microbial activity impairs organoleptic properties, such as odor, taste, texture, or appearance, the food is considered unfit for human consumption and rejected [98–100].

The manufacture of meat products can be seen as a successful preservation method based on the shifting of microbial populations from spoilage and (sometimes) pathogenic Gram-negative bacteria to desirable and beneficious Gram-positive ones (such as many species of *Lactobacillaceae* and *Micrococcaceae*), which confer attractive sensory properties and achieve long shelf-lives [101]. This shift is attained mainly by processes and factors such as salting/curing, temperature control, atmosphere modification, acidification, drying, or the use of antimicrobials and can be reinforced by the addition of selected starter cultures [60,102,103]. Thus, salt contributes to enlarging the shelf-life of meat products and improving safety because it fosters a shift in the dominant microbial populations originally occurring in the raw material [104].

Salt is able to inhibit the growth of many spoilage and pathogenic bacteria, yeasts, and molds, albeit to a different extent depending on the microbial group [77]. Halophiles have been shown to contain enzymes active in solutions of very high ionic strength [105], while the salt tolerance of non-halophiles is related to their ability to accumulate potassium and other compatible solutes within the cells [106]. In general, Gram-positive bacteria isolated from foods are more tolerant to salt than Gram-negatives.

The inhibitory effect on microorganisms takes place in the aqueous phase of the product. Therefore, data such as salt concentration in the aqueous phase and/or a_w are preferable for estimating the inhibitory effect on the microbiota. The majority of spoilage bacteria grow at a_w above 0.90, but some can grow at a_w as low as 0.85 and in extreme cases even lower [69,91]. Yeasts and molds in general tolerate a lower a_w and many can grow at an a_w down to about 0.7 (down to 0.6 for some xerophile species) [69]. Among them, some fungi can produce mycotoxins under low a_w conditions. Salt is a powerful inhibitor of Gram-negative bacteria, which commonly colonize the surface of aerobically-stored refrigerated fresh meat and degrade low-molecular-weight compounds with the ultimate production of substances that contribute to off flavors and tastes [107,108]. Non-motile aerobic rods and coccobacilli belonging to *Pseudomonas*, *Moraxella*, *Psychrobacter*, *Acinetobacter*, and psychrotrophic *Enterobacteriaceae* are major components of the spoilage microbiota of refrigerated raw meat stored aerobically [107,108]. Fresh meat spoilage is preceded by a time-variable phase in which bacteria use low-molecular-weight compounds such as glucose and glucose-6-P as a carbon and energy source. Later on, Gram-negative bacteria use amino acids at refrigeration temperatures and typical pH conditions in post-mortem meat to obtain energy [109,110], which occurs when high amounts of bacteria (more than 10^7 CFU/cm^2) are present.

If environmental conditions change, a selection pressure on the bacterial community is exerted, and certain groups become dominant. The prevailing conditions during ripening (a$_w$, pH, Eh, NaCl concentration, etc.) favor the growth of microbial groups such as lactic acid bacteria (LAB), *Micrococcaceae*, certain yeasts, and molds, which dominate the microbiota of meat products [102]. These microbial groups are able to decrease the pH and convert lipids and proteins into desirable substances during the curing process [111,112]. The spoilage capacity of these groups is usually limited, and the end-products of their metabolism are not overtly offensive, although some Gram-positive species can be very detrimental [110]. In the industrial manufacture of some fermented meat products, ad hoc starter cultures from the above-mentioned groups are used to improve the quality and safety of the product by accelerating the change in microbial populations, displacing the spoiling microbiota [112–115].

4.1. Microbiological Safety Assessment and Shelf-Life of Low Salt Meat Products

Reducing the salt content of a particular meat product can represent a challenging task for the industry. The contribution of salt to the technological and sensory quality of the final product can be replaced, but the potential safety risks linked to reformulations should be assessed more carefully [20,53,116]. When microbiological safety is the major issue considered, the straightforward approach is to determine the likelihood of the growth or survival of spoilage and pathogenic microorganisms in the specific low-salt meat product [16,48]. For this purpose, it is necessary to initially identify the hazards and other microbiota occurring in the product; secondly, to evaluate the inhibitory capacity of other intrinsic and extrinsic conditions (hurdles) in the product; and, finally, to consider the technological steps that are feasible and suitable for achieving microbial stability. Predictive models, challenge testing, and shelf-life tests are suitable tools to obtain accurate estimates of the likelihood of the growth, survival, or inactivation of microorganisms in the product [117–119]. The implementation of an efficient strategy to guarantee the microbiological safety of low-salt meat products is achievable once all data are gathered and the adequate tools are suitably used.

Most of the scientific literature published on low-salt meat products deals with the sensory and technological issues, although there are also some articles that investigate the microbiological aspects [17,65,120,121]. Numerous articles show that, when NaCl levels used are below those typically formulated in meat products, the shelf-life is significantly reduced [53,104]. For example, the shelf-life of a typical Greek pork meat product (4.85% *w/w* NaCl) was reduced by between 10 and 78 days when the salt content was 50% reduced and samples were stored between 15 and 0 °C, respectively. In this research article, Michelakou et al. (2021) evaluated the influence of 50% reduced salt and incubation temperatures (0, 5, 10, and 15 °C) on a pork product. It is important to consider that, at higher incubation temperatures, the incubation period was reduced, both for the control samples and the reduced-salt samples [65]. A salt reduction promoted faster spoilage of raw sausages by lowering the overall bacterial diversity (both richness and evenness) in the product, including Gram-negative as well as Gram-positive bacteria [104]. Although no apparent changes were noticed in shelf-life, both aerobic and LAB counts increased significantly after 60 days of storage in low-salt turkey sausage formulations [122]. The authors concluded that a salt replacer could effectively and completely substitute NaCl in smoked turkey sausages, although some sensorial optimization may be required. The results obtained by Charmpi et al. (2020) suggest that the salt level (between 1 and 4%) influenced the diversity of microbial communities during the fermentation of pork minced meat. The highest salt concentration lowered the bacterial diversity, as Enterobacterales were detrimentally affected. LAB and coagulase-negative staphylococci predominated during the fermentation process, as they are well adapted to higher-salt environments (6% NaCl in regular sausage products) [60,115].

4.2. Microbial Hazards Associated with Low-Salt Cured Meat Products (Hazard Identification)

Bacterial pathogens can occur in the meat product because the raw materials are contaminated (meat, offal, spices, and other ingredients) or as a consequence of nonhygienic manipulation, bad manufacturing practices, and cross-contamination from utensils and equipment at the processing facilities. Once they have reached the products, the manufacturing and storage conditions (including temperature-time combinations, a_w of the foodstuff, and preservative concentration) dictate whether bacteria can grow, survive, or are inactivated [99,100,123]. If a significant inhibitory barrier, such as the salt, is reduced, the ability of pathogens to survive and grow increases, but the risk is likely greater for those microorganisms more susceptible to salt inhibition. Table 2 shows the tolerance limits for a_w of selected bacteria.

The identification of microbiological hazards linked to a product as one of the steps in a HACCP plan is a necessary phase that the industry should complete. The assessment has to consider the prevalence and concentration of biological hazards potentially present in the raw materials or introduced during food handling and processing, the intrinsic and extrinsic conditions during processing, and the conditions of storage and distribution [124]. The safety assurance is improved when, in addition to a proper risk evaluation, procedures for assessing the lethal effect of the treatments are included, as well as mechanisms to monitor, evaluate, optimize, and validate the lethal burden of the process. Epidemiological data from outbreaks linked to meat products [125], expert elicitation, and data from source attribution are all useful in identifying and ranking the main pathogens associated with meat products [126]. The published scientific literature and risk assessments constitute valuable data to carry out a hazard identification, but the most important data are aspects related to the hygienic and manufacturing conditions in a given factory (e.g., hygienic quality of raw materials, intrinsic factors such as fermentation temperature, storage time, etc.) [26].

Salmonella has usually been reported as a causative agent in outbreaks and cases of infection linked to the consumption of meat products [127]. The contamination with *Salmonella* of raw products of animal origin (meat, fat, spices, etc.) can occur very frequently and has been reported in many research articles [127,128]. Warm-blooded animals are frequent *Salmonella* reservoirs, and, therefore, *Salmonella* can contaminate the meat during slaughtering and meat processing. Cross-contamination and recontamination events linked to *Salmonella* during meat processing and preparation are also recurrently described in the literature [129,130]. The low a_w values achieved during the curing of traditional dry-cured salami or loins have been linked with a reduction in *Salmonella* presence [131,132], although the NaCl content did not significantly affect the probability of finding *Salmonella* [131]. *Salmonella enterica* can survive hyperosmotic stress conditions due to high NaCl concentration (6%), and its survival ability is influenced, as in other Gram-negative foodborne pathogens, by the alternative sigma factor RpoS [133]. Raybaudi-Massilia et al. (2019) did not find any significant differences regarding the occurrence of *Salmonella enteritidis* when comparing control samples from three meat products (cooked ham, 1.14% g Na: turkey breast, 1% g Na: and Deli type sausage, 1.33% g Na) with samples with up to a 30% reduction in salt content [19].

Shiga-toxin-producing *Escherichia coli* (STEC) are zoonotic agents characterized by the production of Shiga-like toxins commonly associated with foodborne disease episodes that can lead to severe health complications and sometimes death. Although the majority of reported STEC cases have been linked to strains of serotype O157, other serotypes, such as O45, O26, O91, O103, O111, O121, and O145, are emerging as causative agents of foodborne disease [127]. In the European Union, 4446 confirmed cases of STEC infections were reported in 2020, with a notification rate of 1.49 cases per 100,000 population. STEC has usually been associated with meat from ruminants, and the main food vehicles implicated as the source of outbreaks are bovine meat and meat products, together with other types of food and water [127]. As with *Salmonella*, *E. coli* is susceptible to low a_w values, and the numbers decrease as the curing process advances [134]. A short curing period has been identified as one of the factors responsible for an outbreak attributed to fermented sausages due to STEC [135].

Listeria monocytogenes also represents a hazard in meat and meat products. In 2020, 1876 confirmed cases of listeriosis were reported in the EU. It was the zoonosis that had the highest case-fatality rate (13.0%). No statistically significant increasing trend was observed during the 2016–2020 period [127], although a significant increasing trend was observed in previous years (2008–2016) [136]. *L. monocytogenes* is considered salt tolerant [137]. Indeed, it has been reported that growth can occur at NaCl concentrations as high as 10% and even more in the case of adapted strains [73]. Thus, microbial reduction in response to a low a_w is less accentuated when compared to other pathogens [92,132]. In sliced chouriço, salt acted as an effective hurdle to control *L. monocytogenes* growth, and manufacturing meat products with lower salt content (1.5% as compared to 3%) allowed the growth of the pathogen [67].

In general, Gram-positive bacteria are able to grow at lower a_w conditions compared to Gram-negative bacteria, and *Staphylococcus aureus* is the pathogenic bacteria with the lowest minimum growth a_w. *S. aureus* can grow in conditions of high salt concentration (10–20%) (a_w = 0.83 to 0.86) and performs better than other competitive flora under low a_w due to its great adaptative response to osmotic stress [138], although it does not produce enterotoxin in such conditions (it only produces enterotoxin at $a_w > 0.90$). Stress conditions, such as NaCl stress (4.5%), were shown to decrease *seb* (staphylococcal enterotoxin B) promoter activity [139].

In meat products, botulism outbreaks have usually been associated with food processing failures (thermal treatment) and the irregular distribution of curing salts, which allows spore outgrowth and botulism toxin production [140,141]. The products that are often implicated are home-made canned meat products and cured hams with curing defects and anaerobic conditions in the inner parts of the product that allow the germination of *C. botulinum* spores. Curing salts (nitrate and/or nitrite), independently of the salt formulation, have been shown to be adequate preservatives for the control of *C. botulinum* in dry-cured hams salted with formulations including replacers such as KCl and/or $CaCl_2$ and $MgCl_2$ [85].

A lower level of biogenic amines (particularly cadaverine, histamine and tyramine) has been reported in blood dry-cured sausages and traditional Portuguese sausages manufactured with a level of 3% salt as compared to 6% [115,142].

4.3. Processing Intrinsic and Extrinsic Hurdles Affecting Microbial Hazards in Meat Products

Salting, together with other classical food preservation processes (drying, freezing, thermal treatment, etc.), imposes extreme physical and chemical barriers on microbial growth and has traditionally been used in many instances as common methods of preservation (Table 3). For all these preservation processes, microbial stability for long periods of time is achieved using stringent conditions that constitute robust obstacles or "hurdles" for bacteria, even though they dramatically change the organoleptic characteristics of fresh meat. Meat products manufactured in this way are very different from fresh meat in their organoleptic characteristics [143].

In contrast, modern strategies in food preservation seek to apply mild treatments to inactivate or permanently inhibit injured microorganisms by using multiple barriers, especially in the case of minimally processed foods [144]. This is the reaction of the food industry to the demands and preferences of modern consumers in relation to quality, healthiness, nutrition, convenience, and hedonic perception [144,145]. Low-salt meat products are a perfect example, in which sensory attributes and microbial stability should be achieved using a combination of methods or preservation technologies and favorable intrinsic and extrinsic factors. This approach has been visualized as a sequential or simultaneous group of hurdles acting synergistically to inhibit the growth of or inactivate microbes [20,144]. An effective and stable system capable of prolonging the shelf-life and assuring the safety of the end-product is accomplished when the combined hurdles inactivate most of the spoilage and pathogenic microorganisms while survivors are inhibited. Microbial growth and pH remained within the normal range in sausages when the sodium chloride was replaced with 20% potassium chloride and 38% calcium chloride in combination with an olive oil emulsified alginate [27]. Some hurdles can achieve a complete microbial inactivation and thus have a bactericidal effect, (e.g., thermal treatment, HHP, acidification), while others only slow down or arrest the microbial growth (bacteriostatic effect) depending on the intensity of the hurdle (salting, refrigeration, use of modified atmospheres) [146]. As part of the recent methods to be used in hurdle technology, the use of food cultures and/or their metabolites (i.e., bacteriocins) as natural preservatives in food should be highlighted [68,147–150].

From a safety viewpoint, when meat processing does not include a killing step, the use of cumulative and synergistic hurdles is strictly necessary to maintain the safety and stability of the meat product since the inactivation of pathogens or spoilage agents cannot be completely guaranteed. In general, a number of microorganisms would be able to grow at the NaCl concentrations (<5%) encountered in most meat products in conditions of optimal temperature, pH, and nutrient availability. However, the presence of these additional growth barriers (acidity, refrigeration storage, vacuum or modified atmosphere packaging, the presence of other antimicrobial compounds such as nitrites, preservatives, food cultures, etc.) can slow down or stop microbial metabolism when combined with the relatively mild salt concentrations prevailing in this type of meat product.

The restrictions in the application and intensity of those hurdles come from constraints such as the maintenance of the sensory quality of the product (flavor, texture …), its conformity with legislative requirements (additive maximum limits), and the ability to meet the economic industrial demands (reduced costs, water loss …) [53,146]. Some non-thermal preservation technologies (pulsed electric fields, irradiation, electromagnetic fields, etc.) are not appropriate for the manufacture of meat products for technological reasons or limited consumer acceptance. On the other hand, technologies such as HHP are very suitable for the manufacture of low-salt meat products due to their capacity to inactivate the microbiota while contributing to protein solubilization [20,151–156].

Table 3. Use of cumulative and synergistic hurdles with salt to achieve the safety and quality of meat products.

Product	Combined Hurdles	Results	Reference
Raw pork meat	350 MPa HPP + 1, 1.5 or 3% NaCl	Synergism between HPP and salt showed to control bacteria recovery (aerobic mesophiles, LAB and *Enterobacteriaceae*) more than each hurdle alone during storage.	[156]
Sliced dry cured ham	2.8% NaCl + 600 MPa	Combined hurdles achieved *Salmonella* and *L. monocytogenes* inactivation 14 and 42 days earlier than HPP alone.	[157]
Sausage (*chorizo*)	1.01% NaCl + 0.48% KCl + 0.91% CaCl$_2$ + olive oil emulsified alginate replacing pork fat	A reduction of 58% NaCl in sausages seems to be feasible since no pH and microbial counts remained in the normal values.	[27]
Pork sausage	600 Mpa HPP + carrot fibers or potato starch + 1.2% NaCl	Reducing salt content from 1.8% to 1.2% with the addition of HPP and hydrocolloids did not negatively influence the water binding capacity, color, or texture of sausages.	[154]
Sheep natural sausage casings	0, 4, 7 or 12% NaCl + 0, 100, 150, 200 ug/g nisin	Combined hurdles greatly controlled *L. monocytogenes* than salt and nisin alone.	[158]
Pork	Ultrasound (9 and 54.9 W/cm^2) + 5% NaCl or a commercial salt replacer	Ultrasound only enhanced NaCl diffusion into the meat but did not influence the replacer diffusion.	[159]

5. Strategies to Guarantee the Microbiological Safety of Low-Salt Meat Products

Several strategies have been devised to guarantee the microbiological safety of low-salt meat products (salt replacers, antimicrobial compounds, flavor enhancers, improved salt application techniques, processing technologies) [20]. The strategies can be classified using different approaches (Table 4).

Table 4. Approaches to guarantee the microbiological safety of low-salt meat products.

Approach	Main Mechanism	Advantages *	Disadvantages
Use of preservatives that supplement or replace inhibitory power of salt	Low a_w and inhibition power of preservatives (KCl, $MgCl_2$, $CaCl_2$, $MgSO_4$, food cultures, bacteriocins, etc.)	Characteristics of the product remain (almost) unchanged.	Need to evaluate inhibitory power. Synergy with other hurdles absent. Sensory and technological properties of replacer should be assessed. No green label.
Increase intensity of remaining hurdles	Stricter conditions that inhibit microbiota (acidification, drying, freezing)	Green label. No need to change formulation, processing equipment.	Products of quite different sensory quality. Economic constraints (e.g., water loss).
Processing technologies. Decontamination	Inactivation of microbiota (HHP)	Avoids recontamination (product packaged). Green label. Useful also for technological properties.	No application to raw materials. Efficacy depends on the characteristics of product. High initial investment.
High level of hygiene in production	Raw materials of good quality with low numbers of spoilage and absence of pathogenic microorganisms (Hygiene, HACCP, GMP)	Strategy that it is beneficial and needed in any event.	Insufficient on its own, needs supplementation with other strategies. Depends on the raw material supplier.

* In addition to those linked to health due to salt reduction.

First approach is to replace NaCl (totally or partially) using other additive(s) with similar properties [38,61,143,160]. A 20% sodium reduction was obtained in turkey breast by replacing NaCl with Na_2HPO_4 prepared in a 50:50 blend [161]. The replacement needs to be carefully adjusted since the inhibitory barrier of the substitute may be lower than that of NaCl [162–164]. In addition, there is a synergistic effect of NaCl with other hurdles, which may be absent with the replacer [165]. The reduction of the sodium content (by reducing both salt and sodium nitrite) allows a rapid growth of lactic acid bacteria and proteolytic microorganisms in cured meats, resulting in a product that spoils more rapidly [48]. Dry fermented sausages with a 58% NaCl substitution with KCl and $CaCl_2$ showed a more pronounced growth of *Lactobacillus* than the control sample (2.4% NaCl). Moreover, *Lactobacillus* counts in the control decreased (2–3 log cfu/g) during ripening, while they maintained more or less stable levels in reduced sausages [27]. NaCl is very effective in controlling pathogens and spoilage organisms, thus it can be necessary to substitute its inhibitory action by using some other preservatives in case of replacement [121]. A higher amount of yeast (4.7–5.4 log cfu/g) was found in a 10% salt content bacon, while lower counts (1.3–3.9 log cfu/g) were found in 1.4% salt content bacon, which might indicate that a higher salt content is expected to suppress the growth of bacteria, enabling the slower-growing yeast to better grow in the product [16]. Alternative salts (e.g., KCl, MgCl$_2$, CaCl$_2$, MgSO$_4$, etc.), sugars, proteins, and humectants decrease the a$_w$ in foods, but they usually have an inferior inhibitory action. Replacers of salt and other sodium-containing preservatives, such as KCl [20], mixtures of potassium lactate and sodium diacetate [166], or mixtures of KCl, MgCl$_2$, and CaCl$_2$ [167,168] are not as effective in the inhibition of undesirable microbiota. Other a$_w$ depressors have no inhibitory effect at all. The replacement should also consider the other functions of salt (sensory, technological role, flavoring agent) in the meat product [9,143,169–172]. In a mortadella product, lower sensory acceptability, especially regarding flavor, was found when blends containing 1% NaCl, 0.5% KCl, and 0.5% MgCl$_2$ and 0.5% NaCl, 1% KCl, and MgCl$_2$ 0.5% were used in the formulation compared to the 2% NaCl control [173].

Another strategy is to increase the intensity of the remaining (intrinsic and extrinsic) hurdles, so they compensate for the reduction of salt and its inhibitory potential. Examples of these meat products are commercially-sterile canned products, products with a pH under 3.8 (submitted to an intense acidification), frozen products intended for immediate consumption after thawing, products with a low a$_w$ achieved by other means (e.g., extensive drying that increases concentrations of solutes in the final product), and natural seasonings [174]. García-Lomillo et al. (2017) achieved a 1% salt reduction when using 2% red wine pomace seasoning. However, these types of products may present a strong or defective sensory profile, due to an extreme application of one single barrier.

One more procedure is to include a further processing step, i.e., a decontamination treatment, applied either to the raw materials before processing or to the final meat product once it has been manufactured and packaged [8]. The introduction of an inactivation step for raw materials (thermal treatment, HHP, light pulses, chemical decontamination, etc.) may be difficult to put into practice, due to unwanted modifications that occur in fresh meat and sensory changes in the final product [175]. The second option (treatment of the final product) is very effective since the process's safety assurance is enhanced, as recontamination is prevented [176–178]. In any case, the introduction of an inactivation step in the food-manufacturing process requires a careful assessment of microbiological risks, including an adequate calculation of the lethality effect. There is also a need to have tools and instruments to monitor, optimize, and validate the process on-line and procedures to model the lethal effect of the treatment [179–181]. A combination of this option and a replacer or other additional hurdles in the formulation have also been proposed, with HHP as the most favored choice [20,155,182].

A final option is to use raw materials of optimal microbiological quality in the manufacture of low-salt meat products by increasing the hygienic standards at the slaughterhouse and cutting plant and strictly adhering to HACCP and GMP (good manufacturing practices), including environmental monitoring and sanitation. On its own, this procedure is considered insufficient to produce stable low-salt meat products and should be complemented by other methods, such as those listed above. In any circumstance, it is always necessary that the processing of meat products should adhere in all circumstances to the strictest conditions of process hygiene [175].

Either way, the safety of the whole process (formulation, hurdle combination, and processing steps) should preferably be verified by challenge testing and aided by mathematical modelling.

6. Use of Challenge Testing and Shelf-Life Tests

Challenge testing and shelf-life tests are useful tools that help food processors determine the quality of foods and estimate the ability of foodborne pathogens to grow during the foreseeable conditions of distribution and storage. This is especially necessary when changes in the product formulation (e.g., lowering salt content) are introduced, as the possibility of reformulated low-salt meat products having shorter shelf-life or causing foodborne illness has already been emphasized. Salt replacement or reduction has an impact on the a$_w$ of the food and, therefore, will undoubtedly modify the growth behavior of pathogenic and spoilage microorganisms; therefore, there is a need for effective tools ensuring the manufacture of safe foods with changes in the shelf-life [62,183].

The microbial growth ability in food products can be estimated based on specifications of the physico–chemical characteristics of the product, consultation of the available scientific literature, or predictive mathematical modelling (see below). In many cases, a growth assessment will have to involve laboratory-based studies, so-called challenge tests, and shelf-life studies [75]. Challenge and shelf-life testing is normally performed on a case-by case basis, which means that it can be very expensive and time-consuming, particularly if a range of products, formulations, and different bacteria have to be tested. Results can take many days until they are available, since they are usually obtained by classical microbiological analysis. Nonetheless, both tests can provide valuable information on microbial stability to food processors.

Challenge testing. As a primary objective, challenge tests aim to determine whether a particular food product has the ability to support the growth of a particular microorganism. Simulation of conditions in an artificially contaminated product allows us to study the fate of pathogens or spoilage microbiota. In any case, results should be analyzed with care (including fail-safe approach), considering all the constraints and assumptions introduced in simulating the natural contamination present in foods and the accurate reproduction of conditions of foods during storage, distribution, sale, and preparation. Challenge testing is a technique commonly employed in research [62,184–188]. Up to a 40% NaCl reduction was achieved during a challenge test in a pre-packed cooked meat product when it was replaced with a commercial mixture of potassium lactate and sodium diacetate without statistically affecting the shelf-life [166]. In a challenge test carried out in salami with 4% NaCl and 2.8% NaCl plus 1.6% potassium lactate, the reduced and replaced sample showed to be effective with respect to microbial benefits without compromising the product quality [62]. According to the authors, a limitation of this challenge test could be the absence of exposure to abusive temperatures, which does not allow the interpretation to be further extended to other storage temperatures.

Shelf-life studies. In the European Union, Regulation (EC) No 2073/2005 allows food-business operators to carry out shelf-life studies, as necessary, to investigate compliance with the food-safety criteria throughout the product's lifespan. They are conducted to study whether particular microorganisms are able to survive and grow in naturally contaminated foods during storage and distribution beyond the limits imposed by the Regulation. The consultation of the available scientific literature and specifications of physico-chemical characteristics of the product is encouraged. For example, referring to *L. monocytogenes*, according to the EURL Lm technical guidance document for conducting shelf-life studies on *Listeria monocytogenes* in ready-to-eat foods [183], shelf-life tests for *L. monocytogenes* would not be needed for the following meat products [189]:

- foods produced for immediate consumption (with a shelf-life of less than five days);
- foods (meat products) which are intended to be cooked or subjected to any other bacterial inactivation step before human consumption;
- foods which have received a heat treatment or other processing effective to eliminate *L. monocytogenes*, when recontamination is not possible after this treatment (e.g., products treated in their final package);
- meat products with pH $\leq$ 4.4, or $a_w \leq 0.92$, or pH ≤ 5.0 and $a_w \leq 0.94$, conditions which are already known as unable to support the growth of *L. monocytogenes*; and
- other categories of product can also belong to this category, subject to scientific justification (e.g., frozen products).

Moreover, historical data on the prevalence of the particular microbial species in the specific food product at the end of its shelf-life and particularly on results of durability studies (the number of samples exceeding 100 CFU/g) and outputs of predictive microbiology modules may be useful in deciding whether a test is required or not for a particular foodstuff.

Meat products contain several hurdles that impose a series of restrictions affecting the microbial growth potential of a given pathogen (see above). This growth potential can serve to classify foods or to evaluate particular food products with regards to shelf-life and safety. When the growth potential is lower than 0.5, it is considered that the intrinsic and extrinsic properties of the product are able to restrict pathogen growth during shelf-life, in case an accidental contamination of the product has taken place. Nonetheless, this aspect does not eliminate the risk or probability of diseases associated with these products, as the sole presence of the pathogen in the product implies a certain degree of exposure.

7. Predictive Microbiology in the Safety Assessment of Low-Salt Meat Products

Predictive microbiology uses mathematical functions to describe the behavior of microorganisms subjected to intrinsic and extrinsic factors in foods. For this purpose, diverse software tools (ComBase, Monte Carlo simulation, Decision Tools @Risk, MicroHibro, etc.) are available that allow to users calculate the growth, survival, and inactivation of bacteria in foods. Models attempt to estimate the quantitative or qualitative evolution of microbial populations over time and, therefore, the food shelf-life and pathogen fate.

Models that describe the growth of a population of microorganisms are being increasingly used to adopt strategies to improve food safety. From such a point of view, models have to be able to calculate and describe the growth, survival, or inactivation of spoilage or pathogenic bacteria in the food matrix under a defined set of extrinsic and intrinsic conditions and, eventually, when microbial numbers might reach a level compromising human health [117]. A variety of deterministic models describing the bacterial growth, survival, and inactivation in meats in response to environmental factors (temperature, pH, water activity, etc.) have been proposed [92,190,191]. Models are based on variations of the Bigelow, Baranyi, Gompertz, Logistic, or Richards models, with the environmental effects being expressed through changes in the equation parameters (Lopez et al., 2004). In addition, some models have also been published describing the fate of pathogens (growth, inactivation, and survival) under (static or dynamic) conditions of processing, studying the impact of extrinsic and intrinsic factors on meat products [92,179–181,192–195]. A model to describe the combined effect of salt and heating temperature on the heat tolerance of *L. monocytogenes* was described for meat and seafood products to achieve $\geq 3 \log_{10}$ reductions. Only the products with salt influenced the model, thus being independent from strains, temperatures, and type of food [196]. Nevertheless, the authors are aware that other intrinsic factors might influence the model, and that it needs deeper research. In another modelling study, *L. monocytogenes* growth was stimulated at 0.92–0.94 a_w when 4% NaCl was applied [197].

New genetic, physiological, and molecular information is increasingly available, which will improve the prediction capacity of models. In any case, the use of this methodology requires a high level of expertise [198]. Assumptions and limitations should be taken into account, e.g., information available is often obtained from studies carried out under optimal conditions (37 °C, neutral pH, etc.) and in laboratory-based rich media.

Using software with growth/no growth boundary modules, it is possible to obtain information on the growth probability of pathogens according to pH, a_w, and temperature. The models that investigate the growth–no growth interface of target microorganisms are particularly useful for these purposes, since they can afford information on the impact of intrinsic and extrinsic factors to determine the behavior of pathogens or spoilage microorganisms in the final product. Similarly to other processes, the validation of mathematical models in foods is necessary, together with challenge tests and shelf-life tests, especially if the assessments are performed in model systems.

8. Conclusions

Most of the scientific literature published on low-salt meat products deals with sensory and technological issues, while the safety viewpoint has been somehow overlooked. There is a need to further investigate the microbiological implications of salt reduction in meat products, since the inhibitory barrier offered by salt may not be adequately replaced. The assessment of the safety risk associated with meat products with a low salt concentration should be unavoidably performed on a case-by-case basis. To achieve this aim, mathematical modelling, challenge tests, and shelf-life studies are very useful tools that should be used by experienced personnel. In the industry setting, all this information should be assessed and integrated into an HACCP plan that includes a comprehensive hazard-identification phase and adequate tools able to control and monitor the critical points. To guarantee the microbiological safety of low-salt meat products, approaches can be addressed towards finding suitable replacers (salts or other depressors of a_w), processing changes that increase the intensity of remaining hurdles (intrinsic and extrinsic factors), the use of processing technologies able to decontaminate the end product, the use of more than one strategy as a part of hurdle technology (HPP, use of food cultures and/or natural antimicrobial compounds, etc.), or the (always) necessary hygienic strategies able to obtain raw materials with a low amount of microbial contaminants. The reduction or elimination of salt associated with a product reformulation has to ensure the same safety level, must be economically and technologically viable, and must be accepted by the consumer from a sensory point of view.

Author Contributions: The authors C.B., A.Á.-O., M.L., O.A. and M.P. contributed equally to the article. All authors have read and agreed to the published version of the manuscript.

Funding: The authors acknowledge the financial support of the European Union's Horizon 2020 Research and Innovation Program under grant agreement No. 818368 (MASTER). Coral Barcenilla is also grateful to *Junta de Castilla y León* and the European Social Fund for awarding her a pre-doctoral grant (BOCYL-D-07072020-6).

Institutional Review Board Statement: Not applicable.

Informed Consent Statement: Not applicable.

Data Availability Statement: Not applicable.

Conflicts of Interest: The authors declare no conflict of interest.

References

1. Halagarda, M.; Wójciak, K.M. Health and safety aspects of traditional European meat products. A review. *Meat Sci.* **2022**, *184*, 108623. [CrossRef] [PubMed]
2. Vidal, V.A.S.; Lorenzo, J.M.; Munekata, P.E.S.; Pollonio, M.A.R. Challenges to reduce or replace NaCl by chloride salts in meat products made from whole pieces—A review. *Crit. Rev. Food Sci. Nutr.* **2021**, *61*, 2194–2206. [CrossRef] [PubMed]
3. Vidal, V.A.S.; Paglarini, C.S.; Lorenzo, J.M.; Munekata, P.E.S.; Pollonio, M.A.R. Salted Meat Products: Nutritional Characteristics, Processing and Strategies for Sodium Reduction. *Food Rev. Int.* **2021**, 1–20. [CrossRef]
4. O'Donnell, M.; Mente, A.; Yusuf, S. Sodium Intake and Cardiovascular Health. *Circ. Res.* **2015**, *116*, 1046–1057. [CrossRef]
5. Coombs, C.E.; Holman, B.W.; Friend, M.A.; Hopkins, D.L. Long-term red meat preservation using chilled and frozen storage combinations: A review. *Meat Sci.* **2017**, *125*, 84–94. [CrossRef]
6. Gaudette, N.J.; Pietrasik, Z.; Johnston, S.P. Application of taste contrast to enhance the saltiness of reduced sodium beef patties. *LWT* **2019**, *116*, 108585. [CrossRef]
7. Pietrasik, Z.; Gaudette, N. The impact of salt replacers and flavor enhancer on the processing characteristics and consumer acceptance of restructured cooked hams. *Meat Sci.* **2014**, *96*, 1165–1170. [CrossRef]
8. Inguglia, E.S.; Zhang, Z.; Tiwari, B.K.; Kerry, J.P.; Burgess, C. Salt reduction strategies in processed meat products—A review. *Trends Food Sci. Technol.* **2017**, *59*, 70–78. [CrossRef]
9. Cruz-Romero, M.C.; O'Flynn, C.C.; Troy, D.; Mullen, A.M.; Kerry, J.P. The Use of Potassium Chloride and Tapioca Starch to Enhance the Flavour and Texture of Phosphate- and Sodium-Reduced Low Fat Breakfast Sausages Manufactured Using High Pressure-Treated Meat. *Foods* **2021**, *11*, 17. [CrossRef]
10. Kromhout, D.; Spaaij, C.J.K.; de Goede, J.; Weggemans, R.M. The 2015 Dutch food-based dietary guidelines. *Eur. J. Clin. Nutr.* **2016**, *70*, 869–878. [CrossRef]
11. WHO. *Mapping Salt Reduction Initiatives in the WHO European Region*; WHO: Geneva, Switzerland, 2013.
12. Palou, A.; Badiola, J.J.; Anadón, A.; Bosch, A.; Cacho, J.F.; Cameán, A.M. Informe Del Comité Científico de La Agencia Española de Seguridad Alimentaria y Nutrición (AESAN) En Relación Al Efecto de La Reducción de La Sal En La Seguridad Microbiológica de Los Productos Cárnicos Curados. *Rev. Del Com. Científico La AESAN* **2010**, *13*, 59–87.

13. WHO. Sodium Intake for Adults and Children. In *Guideline: Sodium Intake for Adults and Children*; WHO: Geneva, Switzerland, 2012.

14. Van Gunst, A.; Roodenburg, A.J.C.; Steenhuis, I.H.M. Reformulation as an Integrated Approach of Four Disciplines: A Qualitative Study with Food Companies. *Foods* **2018**, *7*, 64. [CrossRef]

15. Food Safety Authority of Ireland. *Salt and Health: Review of the Scientific Evidence and Recommendations for Public Policy in Ireland (Revision 1)*; Food Safety Authority of Ireland: Dublin, Ireland, 2016.

16. Aaslyng, M.D.; Vestergaard, C.; Koch, A.G. The effect of salt reduction on sensory quality and microbial growth in hotdog sausages, bacon, ham and salami. *Meat Sci.* **2014**, *96*, 47–55. [CrossRef]

17. Fraqueza, M.J.; Laranjo, M.; Elias, M.; Patarata, L. Microbiological hazards associated with salt and nitrite reduction in cured meat products: Control strategies based on antimicrobial effect of natural ingredients and protective microbiota. *Curr. Opin. Food Sci.* **2021**, *38*, 32–39. [CrossRef]

18. Zhang, X.; Yang, J.; Gao, H.; Zhao, Y.; Wang, J.; Wang, S. Substituting sodium by various metal ions affects the cathepsins activity and proteolysis in dry-cured pork butts. *Meat Sci.* **2020**, *166*, 108132. [CrossRef]

19. Raybaudi-Massilia, R.; Mosqueda-Melgar, J.; Rosales-Oballos, Y.; de Petricone, R.C.; Frágenas, N.; Zambrano-Durán, A.; Sayago, K.; Lara, M.; Urbina, G. New alternative to reduce sodium chloride in meat products: Sensory and microbiological evaluation. *LWT* **2019**, *108*, 253–260. [CrossRef]

20. Tamm, A.; Bolumar, T.; Bajovic, B.; Toepfl, S. Salt (NaCl) reduction in cooked ham by a combined approach of high pressure treatment and the salt replacer KCl. *Innov. Food Sci. Emerg. Technol.* **2016**, *36*, 294–302. [CrossRef]

21. Stelmasiak, A.; Damaziak, K.; Wierzbicka, A. The Impact of Low Salt Level and Addition of Selected Herb Extracts on the Quality of Smoked Pork. *Anim. Sci. Pap. Rep.* **2021**, *39*, 47–60.

22. Kehlet, U.; Pagter, M.; Aaslyng, M.D.; Raben, A. Meatballs with 3% and 6% dietary fibre from rye bran or pea fibre—Effects on sensory quality and subjective appetite sensations. *Meat Sci.* **2017**, *125*, 66–75. [CrossRef]

23. Thao, T.T.P.; Thoa, L.T.K.; Ngoc, L.M.T.; Lan, T.T.P.; Phuong, T.V.; Truong, H.T.H.; Khoo, K.S.; Manickam, S.; Hoa, T.T.; Tram, N.D.Q.; et al. Characterization halotolerant lactic acid bacteria *Pediococcus pentosaceus* HN10 and in vivo evaluation for bacterial pathogens inhibition. *Chem. Eng. Process. Process Intensif.* **2021**, *168*, 108576. [CrossRef]

24. Lee, Y.; Cho, Y.; Kim, E.; Kim, H.-J.; Kim, H.-Y. Identification of lactic acid bacteria in galchi- and myeolchi-jeotgalby 16S rRNA sequencing, MALDI TOF mass spectrometry, and PCR-DGGE. *J. Microbiol. Biotechnol.* **2018**, *28*, 1112–1121. [CrossRef] [PubMed]

25. Mani, A. Food Preservation by Fermentation and Fermented Food Products. *Int. J. Acad. Res. Dev.* **2018**, *1*, 51–57.

26. Holck, A.L.; Axelsson, L.; Rode, T.M.; Høy, M.; Måge, I.; Alvseike, O.; L'Abée-Lund, T.M.; Omer, M.K.; Granum, P.E.; Heir, E. Reduction of verotoxigenic *Escherichia coli* in production of fermented sausages. *Meat Sci.* **2011**, *89*, 286–295. [CrossRef] [PubMed]

27. Beriain, M.; Gómez, I.; Petri, E.; Insausti, K.; Sarriés, M. The effects of olive oil emulsified alginate on the physico-chemical, sensory, microbial, and fatty acid profiles of low-salt, inulin-enriched sausages. *Meat Sci.* **2011**, *88*, 189–197. [CrossRef]

28. Wyness, L.A.; Butriss, J.L.; Stanner, S.A. Reducing the population's sodium intake: The UK Food Standards Agency's salt reduction programme. *Public Health Nutr.* **2012**, *15*, 254–261. [CrossRef]

29. Trieu, K.; Neal, B.; Hawkes, C.; Dunford, E.; Campbell, N.R.C.; Rodriguez-Fernandez, R.; Legetic, B.; McLaren, L.; Barberio, A.; Webster, J. Salt Reduction Initiatives around the World—A Systematic Review of Progress towards the Global Target. *PLoS ONE* **2015**, *10*, e0130247. [CrossRef]

30. Graudal, N.A.; Hubeck-Graudal, T.; Jürgens, G. Reduced Dietary Sodium Intake Increases Heart Rate. A Meta-Analysis of 63 Randomized Controlled Trials Including 72 Study Populations. *Front. Physiol.* **2016**, *7*, 111. [CrossRef]

31. Aburto, N.J.; Ziolkovska, A.; Hooper, L.; Elliott, P.; Cappuccio, F.P.; Meerpohl, J.J. Effect of lower sodium intake on health: Systematic review and meta-analyses. *BMJ* **2013**, *346*, f1326. [CrossRef]

32. Poggio, R.; Gutierrez, L.; Matta, M.G.; Elorriaga, N.; Irazola, V.; Rubinstein, A. Daily sodium consumption and CVD mortality in the general population: Systematic review and meta-analysis of prospective studies. *Public Health Nutr.* **2015**, *18*, 695–704. [CrossRef]

33. Mente, A.; O'Donnell, M.; Rangarajan, S.; Dagenais, G.; Lear, S.; McQueen, M.; Diaz, R.; Avezum, A.; Lopez-Jaramillo, P.; Lanas, F.; et al. Associations of urinary sodium excretion with cardiovascular events in individuals with and without hypertension: A pooled analysis of data from four studies. *Lancet* **2016**, *388*, 465–475. [CrossRef]

34. Guallar-Castillón, P.; Muñoz-Pareja, M.; Aguilera, M.T.; León-Muñoz, L.M.; Rodríguez-Artalejo, F. Food sources of sodium, saturated fat and added sugar in the Spanish hypertensive and diabetic population. *Atherosclerosis* **2013**, *229*, 198–205. [CrossRef]

35. Cook, N.R.; Appel, L.J.; Whelton, P.K. Sodium Intake and All-Cause Mortality Over 20 Years in the Trials of Hypertension Prevention. *J. Am. Coll. Cardiol.* **2016**, *68*, 1609–1617. [CrossRef]

36. Sterns, R.H. Disorders of Plasma Sodium—Causes, Consequences, and Correction. *N. Engl. J. Med.* **2015**, *372*, 55–65. [CrossRef]

37. Bigiani, A. Electrophysiology of Sodium Receptors in Taste Cells. *J. Biomed. Sci. Eng.* **2016**, *9*, 367–383. [CrossRef]

38. Pateiro, M.; Munekata, P.E.S.; Cittadini, A.; Domínguez, R.; Lorenzo, J.M. Metallic-based salt substitutes to reduce sodium content in meat products. *Curr. Opin. Food Sci.* **2021**, *38*, 21–31. [CrossRef]

39. Huffman, M.D.; Lloyd-Jones, D.M. Global Burden of Raised Blood Pressure: Coming Into Focus. *JAMA* **2017**, *317*, 142–143. [CrossRef]

40. Coxson, P.G.; Cook, N.R.; Joffres, M.; Hong, Y.; Orenstein, D.; Schmidt, S.M.; Bibbins-Domingo, K. Mortality Benefits from US Population-Wide Reduction in Sodium Consumption: Projections from 3 Modeling Approaches. *Hypertension* **2013**, *61*, 564–570. [CrossRef]

41. European Commision Salt Campaign. Available online: https://ec.europa.eu/health/other-pages/basic-page/salt-campaign_en (accessed on 20 February 2022).

42. Kloss, L.; Meyer, J.D.; Graeve, L.; Vetter, W. Sodium intake and its reduction by food reformulation in the European Union—A review. *NFS J.* **2015**, *1*, 9–19. [CrossRef]

43. Hendriksen, M.A.H.; Van Raaij, J.M.A.; Geleijnse, J.M.; Breda, J.; Boshuizen, H.C. Health Gain by Salt Reduction in Europe: A Modelling Study. *PLoS ONE* **2015**, *10*, e0118873. [CrossRef]

44. Tsugane, S. Why has Japan become the world's most long-lived country: Insights from a food and nutrition perspective. *Eur. J. Clin. Nutr.* **2021**, *75*, 921–928. [CrossRef]

45. Verma, A.K.; Banerjee, R. Low-Sodium Meat Products: Retaining Salty Taste for Sweet Health. *Crit. Rev. Food Sci. Nutr.* **2011**, *52*, 72–84. [CrossRef]

46. Balestra, F.; Petracci, M. Technofunctional Ingredients for Meat Products: Current Challenges. In *Sustainable Meat Production and Processing*, 1st ed.; Charis, M., Ed.; Academic Press: Cambridge, MA, USA, 2018; pp. 45–68. [CrossRef]

47. Zhang, Y.; Guo, X.; Peng, Z.; Jamali, M.A. A review of recent progress in reducing NaCl content in meat and fish products using basic amino acids. *Trends Food Sci. Technol.* **2022**, *119*, 215–226. [CrossRef]

48. Stringer, S.; Pin, C. *Microbial Risks Associated with Salt Reduction in Certain Foods and Alternative Options for Preservation*; IFR Enterprises: Norwich, UK, 2005.

49. Tobin, B.D.; O'Sullivan, M.G.; Hamill, R.M.; Kerry, J.P. The impact of salt and fat level variation on the physiochemical properties and sensory quality of pork breakfast sausages. *Meat Sci.* **2013**, *93*, 145–152. [CrossRef]

50. Kim, T.-K.; Yong, H.-I.; Jung, S.; Kim, H.-W.; Choi, Y.-S. Technologies for the Production of Meat Products with a Low Sodium Chloride Content and Improved Quality Characteristics—A Review. *Foods* **2021**, *10*, 957. [CrossRef]

51. Weiss, J.; Gibis, M.; Schuh, V.; Salminen, H. Advances in ingredient and processing systems for meat and meat products. *Meat Sci.* **2010**, *86*, 196–213. [CrossRef]

52. Spanish Agency for Consumer Affairs Food Safety and Nutrition. *Salt Content of Foods in Spain*; Spanish Agency for Consumer Affairs Food Safety and Nutrition: Madrid, Spain, 2012.

53. Desmond, E. Reducing salt: A challenge for the meat industry. *Meat Sci.* **2006**, *74*, 188–196. [CrossRef]

54. The University of Maine. Sodium Content of Your Food. Available online: https://extension.umaine.edu/publications/wp-content/uploads/sites/52/2021/01/4059_updated2021.pdf (accessed on 22 July 2022).

55. Nadežda, P.; Milica, Ž.-B.; Željko, M.; Sandra, J.; Stojanov, I. Sodium Chloride Content in Meat Products. *Arh. Vet. Med.* **2013**, *6*, 71–79.

56. Perez-Palacios, T.; Salas, A.; Muñoz, A.; Ocaña, E.-R.; Antequera, T. Sodium chloride determination in meat products: Comparison of the official titration-based method with atomic absorption spectrometry. *J. Food Compos. Anal.* **2022**, *108*, 104425. [CrossRef]

57. Kameník, J.; Saláková, A.; Vyskočilová, V.; Pechová, A.; Haruštiaková, D. Salt, sodium chloride or sodium? Content and relationship with chemical, instrumental and sensory attributes in cooked meat products. *Meat Sci.* **2017**, *131*, 196–202. [CrossRef]

58. Stamenic, T.; Petricevic, M.; Samolovac, L.; Sobajic, S.; Cekic, B.; Gogic, M.; Zivkovic, V. Contents of sodium-chloride in various groups of locally manufactured meat. *Biotechnol. Anim. Husb.* **2021**, *37*, 223–234. [CrossRef]

59. Rai, A.K.; Tamang, J.P. Prevalence of *Staphylococcus* and *Micrococcus* in Traditionally Prepared Meat Products. *J. Sci. Ind. Res.* **2017**, *1*, 351–354.

60. Charmpi, C.; Van der Veken, D.; Van Reckem, E.; De Vuyst, L.; Leroy, F. Raw meat quality and salt levels affect the bacterial species diversity and community dynamics during the fermentation of pork mince. *Food Microbiol.* **2020**, *89*, 103434. [CrossRef] [PubMed]

61. Kumar, S.; Yadav, S.; Pal, A. Studies on the Impact of Partial Replacement of Sodium Chloride with Potassium Lactate on Quality Attributes of Buffalo Calf Meat Rolls. *Asian J. Dairy Food Res.* **2021**, *40*, 206–212. [CrossRef]

62. Muchaamba, F.; Stoffers, H.; Blase, R.; Von Ah, U.; Tasara, T. Potassium Lactate as a Strategy for Sodium Content Reduction without Compromising Salt-Associated Antimicrobial Activity in Salami. *Foods* **2021**, *10*, 114. [CrossRef]

63. Yadav, B.; Spinelli, A.C.; Govindan, B.N.; Tsui, Y.Y.; McMullen, L.M.; Roopesh, M. Cold plasma treatment of ready-to-eat ham: Influence of process conditions and storage on inactivation of *Listeria innocua*. *Food Res. Int.* **2019**, *123*, 276–285. [CrossRef]

64. Ramaroson, M.; Guillou, S.; Rossero, A.; Rezé, S.; Anthoine, V.; Moriceau, N.; Martin, J.-L.; Duranton, F.; Zagorec, M. Selection procedure of bioprotective cultures for their combined use with High Pressure Processing to control spore-forming bacteria in cooked ham. *Int. J. Food Microbiol.* **2018**, *276*, 28–38. [CrossRef]

65. Michelakou, E.; Giaouris, E.; Doultsos, D.; Nasopoulou, C.; Skandamis, P. Evaluation of the microbial stability and shelf life of 50% NaCl-reduced traditional Greek pork meat product "Syglino of Monemvasia" stored under vacuum at different temperatures. *Heliyon* **2021**, *7*, e08296. [CrossRef]

66. Da Costa, R.J.; Voloski, F.L.S.; Mondadori, R.G.; Duval, E.H.; Fiorentini, M. Preservation of Meat Products with Bacteriocins Produced by Lactic Acid Bacteria Isolated from Meat. *J. Food Qual.* **2019**, *2019*, 1–12. [CrossRef]

67. García-Díez, J.; Patarata, L. Influence of salt level, starter culture, fermentable carbohydrates, and temperature on the behaviour of *L. monocytogenes* in sliced chouriço during storage. *Acta Aliment.* **2017**, *46*, 206–213. [CrossRef]

68. Barcenilla, C.; Ducic, M.; López, M.; Prieto, M.; Álvarez-Ordóñez, A. Application of lactic acid bacteria for the biopreservation of meat products: A systematic review. *Meat Sci.* **2022**, *183*, 108661. [CrossRef]

69. Esbelin, J.; Santos, T.; Hébraud, M. Desiccation: An environmental and food industry stress that bacteria commonly face. *Food Microbiol.* **2018**, *69*, 82–88. [CrossRef]

70. Cheung, P.C.K.; Mehta, B.M. *Handbook of Food Chemistry*; Springer: Hong Kong, China, 2015; ISBN 9783642366055.

71. FDA. *Fish and Fishery Products Hazards and Controls Guidance*; FDA: Silver Spring, MD, USA, 2011.

72. Wood, J.M. Bacterial responses to osmotic challenges. *J. Gen. Physiol.* **2015**, *145*, 381–388. [CrossRef]

73. Thakur, M.; Asrani, R.K.; Patial, V. Listeria monocytogenes: A Food-Borne Pathogen. *Foodborne Dis.* **2018**, *15*, 157–192. [CrossRef]

74. Li, F.; Xiong, X.-S.; Yang, Y.-Y.; Wang, J.-J.; Wang, M.-M.; Tang, J.-W.; Liu, Q.-H.; Wang, L.; Gu, B. Effects of NaCl Concentrations on Growth Patterns, Phenotypes Associated with Virulence, and Energy Metabolism in *Escherichia coli* BW25113. *Front. Microbiol.* **2021**, *12*, 705326. [CrossRef]

75. Alvarez-Ordóñez, A.; Broussolle, V.; Colin, P.; Nguyen-The, C.; Prieto, M. The adaptive response of bacterial food-borne pathogens in the environment, host and food: Implications for food safety. *Int. J. Food Microbiol.* **2015**, *213*, 99–109. [CrossRef]

76. Begley, M.; Hill, C. Stress Adaptation in Foodborne Pathogens. *Annu. Rev. Food Sci. Technol.* **2015**, *6*, 191–210. [CrossRef]

77. Burgess, C.M.; Gianotti, A.; Gruzdev, N.; Holah, J.; Knøchel, S.; Lehner, A.; Margas, E.; Esser, S.S.; Sela Saldinger, S.; Tresse, O. The response of foodborne pathogens to osmotic and desiccation stresses in the food chain. *Int. J. Food Microbiol.* **2016**, *221*, 37–53. [CrossRef]

78. Zeidler, S.; Müller, V. Coping with Low Water Activities and Osmotic Stress in *Acinetobacter baumannii*: Significance, Current Status and Perspectives. *Environ. Microbiol.* **2019**, *21*, 2212–2230. [CrossRef]

79. Bremer, E.; Krämer, R. Responses of Microorganisms to Osmotic Stress. *Annu. Rev. Microbiol.* **2019**, *73*, 313–334. [CrossRef]

80. Booth, I.R. Bacterial mechanosensitive channels: Progress towards an understanding of their roles in cell physiology. *Curr. Opin. Microbiol.* **2014**, *18*, 16–22. [CrossRef]

81. Schuster, C.F.; Wiedemann, D.M.; Kirsebom, F.C.M.; Santiago, M.; Walker, S.; Gründling, A. High-throughput Transposon Sequencing Highlights the Cell Wall as an Important Barrier for Osmotic Stress in Methicillin Resistant *Staphylococcus aureus* and Underlines a Tailored Response to Different Osmotic Stressors. *Mol. Microbiol.* **2020**, *113*, 699–717. [CrossRef]

82. Liu, Y.; Ho, K.K.; Su, J.; Gong, H.; Chang, A.C.; Lu, S. Potassium transport of *Salmonella* is important for type III secretion and pathogenesis. *Microbiology* **2013**, *159*, 1705–1719. [CrossRef]

83. Mariutti, L.R.; Bragagnolo, N. Influence of salt on lipid oxidation in meat and seafood products: A review. *Food Res. Int.* **2017**, *94*, 90–100. [CrossRef]

84. Toldrá, F. The role of muscle enzymes in dry-cured meat products with different drying conditions. *Trends Food Sci. Technol.* **2006**, *17*, 164–168. [CrossRef]

85. Armenteros, M.; Aristoy, M.-C.; Toldrá, F. Evolution of nitrate and nitrite during the processing of dry-cured ham with partial replacement of NaCl by other chloride salts. *Meat Sci.* **2012**, *91*, 378–381. [CrossRef]

86. Flores, M.; Toldrá, F. Chemistry, safety, and regulatory considerations in the use of nitrite and nitrate from natural origin in meat products—Invited review. *Meat Sci.* **2020**, *171*, 108272. [CrossRef]

87. Kempes, C.P.; van Bodegom, P.; Wolpert, D.; Libby, E.; Amend, J.; Hoehler, T. Drivers of Bacterial Maintenance and Minimal Energy Requirements. *Front. Microbiol.* **2017**, *8*, 31. [CrossRef]

88. Freitag, A.; Cluff, M.; Hitzeroth, A.C.; van Wyngaardt, L.; Hugo, A.; Hugo, C.J. Community level physiological profiling of reduced or replaced salt fresh sausage inoculated with *Escherichia coli* ATCC 25922. *LWT* **2021**, *148*, 111786. [CrossRef]

89. Shah, M.; Bergholz, T. Variation in growth and evaluation of cross-protection in *Listeria monocytogenes* under salt and bile stress. *J. Appl. Microbiol.* **2020**, *129*, 367–377. [CrossRef] [PubMed]

90. Finn, S.; Condell, O.; McClure, P.; Amézquita, A.; Fanning, S. Mechanisms of survival, responses and sources of *Salmonella* in low-moisture environments. *Front. Microbiol.* **2013**, *4*, 331. [CrossRef] [PubMed]

91. Hennekinne, J.A.; De Buyser, M.L.; Dragacci, S. *Staphylococcus aureus* and Its Food Poisoning Toxins: Characterization and Outbreak Investigation. *FEMS Microbiol. Rev.* **2012**, *36*, 815–836. [CrossRef]

92. Valdramidis, V.; Patterson, M.; Linton, M. Modelling the recovery of *Listeria monocytogenes* in high pressure processed simulated cured meat. *Food Control* **2015**, *47*, 353–358. [CrossRef]

93. Kalburge, S.S.; Whitaker, W.B.; Boyd, E.F. High-Salt Preadaptation of *Vibrio parahaemolyticus* Enhances Survival in Response to Lethal Environmental Stresses. *J. Food Prot.* **2014**, *77*, 246–253. [CrossRef]

94. Ma, J.; Xu, C.; Liu, F.; Hou, J.; Shao, H.; Yu, W. Stress adaptation and cross-protection of *Lactobacillus plantarum* KLDS 1.0628. *CyTA J. Food* **2021**, *19*, 72–80. [CrossRef]

95. Keto-Timonen, R.; Hietala, N.; Palonen, E.; Hakakorpi, A.; Lindström, M.; Korkeala, H. Cold Shock Proteins: A Minireview with Special Emphasis on Csp-Family of Enteropathogenic *Yersinia*. *Front. Microbiol.* **2016**, *7*, 1–7. [CrossRef]

96. Schmid, B.; Klumpp, J.; Raimann, E.; Loessner, M.J.; Stephan, R.; Tasara, T. Role of Cold Shock Proteins in Growth of *Listeria monocytogenes* under Cold and Osmotic Stress Conditions. *Appl. Environ. Microbiol.* **2009**, *75*, 1621–1627. [CrossRef]

97. Bergholz, T.M.; Bowen, B.; Wiedmann, M.; Boor, K.J. *Listeria monocytogenes* Shows Temperature-Dependent and -Independent Responses to Salt Stress, Including Responses That Induce Cross-Protection against Other Stresses. *Appl. Environ. Microbiol.* **2012**, *78*, 2602–2612. [CrossRef]

98. De Filippis, F.; La Storia, A.; Villani, F.; Ercolini, D. Exploring the Sources of Bacterial Spoilers in Beefsteaks by Culture-Independent High-Throughput Sequencing. *PLoS ONE* **2013**, *8*, e70222. [CrossRef]

99. Marmion, M.; Ferone, M.; Whyte, P.; Scannell, A. The changing microbiome of poultry meat; From farm to fridge. *Food Microbiol.* **2021**, *99*, 103823. [CrossRef]

100. Pellissery, A.J.; Vinayamohan, P.G.; Amalaradjou, M.A.R.; Venkitanarayanan, K. Spoilage Bacteria and Meat Quality. In *Meat Quality Analysis*; Elsevier Inc.: Amsterdam, The Netherlands, 2019; pp. 307–334; ISBN 9780128192337.

101. Kumar, P.; Chatli, M.K.; Verma, A.K.; Mehta, N.; Malav, O.P.; Kumar, D.; Sharma, N. Quality, functionality, and shelf life of fermented meat and meat products: A review. *Crit. Rev. Food Sci. Nutr.* **2017**, *57*, 2844–2856. [CrossRef]

102. Quijada, N.M.; De Filippis, F.; Sanz, J.J.; García-Fernández, M.D.C.; Rodríguez-Lázaro, D.; Ercolini, D.; Hernández, M. Different *Lactobacillus* populations dominate in "Chorizo de León" manufacturing performed in different production plants. *Food Microbiol.* **2018**, *70*, 94–102. [CrossRef]

103. Juárez-Castelán, C.; García-Cano, I.; Escobar-Zepeda, A.; Azaola-Espinosa, A.; Alvarez-Cisneros, Y.M.; Ponce-Alquicira, E. Evaluation of the bacterial diversity of Spanish-type chorizo during the ripening process using high-throughput sequencing and physicochemical characterization. *Meat Sci.* **2019**, *150*, 7–13. [CrossRef]

104. Fougy, L.; Desmonts, M.-H.; Coeuret, G.; Fassel, C.; Hamon, E.; Hézard, B.; Champomier-Vergès, M.-C.; Chaillou, S. Reducing Salt in Raw Pork Sausages Increases Spoilage and Correlates with Reduced Bacterial Diversity. *Appl. Environ. Microbiol.* **2016**, *82*, 3928–3939. [CrossRef]

105. Ortega, G.; Laín, A.; Tadeo, X.; López-Méndez, B.; Castaño, D.; Millet, O. Halophilic enzyme activation induced by salts. *Sci. Rep.* **2011**, *1*, 6. [CrossRef]

106. Sekar, A.; Kim, K. Halophilic Bacteria in the Food Industry. In *Encyclopedia of Marine Biotechnology*, 1st ed.; Kim, S., Ed.; John Wiley & Sons Ltd.: Hoboken, NJ, USA, 2020; pp. 2061–2070. [CrossRef]

107. Casaburi, A.; Piombino, P.; Nychas, G.-J.; Villani, F.; Ercolini, D. Bacterial populations and the volatilome associated to meat spoilage. *Food Microbiol.* **2015**, *45*, 83–102. [CrossRef]

108. Doulgeraki, A.I.; Ercolini, D.; Villani, F.; Nychas, G.-J.E. Spoilage microbiota associated to the storage of raw meat in different conditions. *Int. J. Food Microbiol.* **2012**, *157*, 130–141. [CrossRef]

109. Stellato, G.; La Storia, A.; De Filippis, F.; Borriello, G.; Villani, F.; Ercolini, D. Overlap of Spoilage-Associated Microbiota between Meat and the Meat Processing Environment in Small-Scale and Large-Scale Retail Distributions. *Appl. Environ. Microbiol.* **2016**, *82*, 4045–4054. [CrossRef]

110. Hwang, B.K.; Choi, H.; Choi, S.H.; Kim, B.-S. Analysis of Microbiota Structure and Potential Functions Influencing Spoilage of Fresh Beef Meat. *Front. Microbiol.* **2020**, *11*, 1657. [CrossRef]

111. Fadda, S.; López, C.; Vignolo, G. Role of lactic acid bacteria during meat conditioning and fermentation: Peptides generated as sensorial and hygienic biomarkers. *Meat Sci.* **2010**, *86*, 66–79. [CrossRef]

112. Cui, S.; Fan, Z. Lactic Acid Bacteria and Fermented Meat Products. In *Lactic Acid Bacteria*; Chen, W., Ed.; Springer: Berlin/Heidelberg, Germany, 2019; pp. 211–226; ISBN 9789811372827.

113. Castilho, N.P.A.; Colombo, M.; De Oliveira, L.L.; Todorov, S.D.; Nero, L.A. *Lactobacillus* curvatus UFV-NPAC1 and other lactic acid bacteria isolated from calabresa, a fermented meat product, present high bacteriocinogenic activity against *Listeria monocytogenes*. *BMC Microbiol.* **2019**, *19*, 63. [CrossRef]

114. Komora, N.; Maciel, C.; Amaral, R.A.; Fernandes, R.; Castro, S.M.; Saraiva, J.A.; Teixeira, P. Innovative hurdle system towards *Listeria monocytogenes* inactivation in a fermented meat sausage model—High pressure processing assisted by bacteriophage P100 and bacteriocinogenic Pediococcus acidilactici. *Food Res. Int.* **2021**, *148*, 110628. [CrossRef]

115. Laranjo, M.; Gomes, A.; Agulheiro-Santos, A.C.; Potes, M.E.; Cabrita, M.J.; Garcia, R.; Rocha, J.M.; Roseiro, L.C.; Fernandes, M.J.; Fraqueza, M.J.; et al. Impact of salt reduction on biogenic amines, fatty acids, microbiota, texture and sensory profile in traditional blood dry-cured sausages. *Food Chem.* **2017**, *218*, 129–136. [CrossRef] [PubMed]

116. Sleator, R.D.; Hill, C. Food reformulations for improved health: A potential risk for microbial food safety? *Med. Hypotheses* **2007**, *69*, 1323–1324. [CrossRef]

117. Carrasco, E.; Valero, A.; Pérez-Rodríguez, F.; García-Gimeno, R.; Zurera, G. Management of microbiological safety of ready-to-eat meat products by mathematical modelling: *Listeria monocytogenes* as an example. *Int. J. Food Microbiol.* **2007**, *114*, 221–226. [CrossRef] [PubMed]

118. Uyttendaele, M.; Busschaert, P.; Valero, A.; Geeraerd, A.; Vermeulen, A.; Jacxsens, L.; Goh, K.; De Loy, A.; Van Impe, J.; Devlieghere, F. Prevalence and challenge tests of *Listeria monocytogenes* in Belgian produced and retailed mayonnaise-based deli-salads, cooked meat products and smoked fish between 2005 and 2007. *Int. J. Food Microbiol.* **2009**, *133*, 94–104. [CrossRef] [PubMed]

119. Iannetti, L.; Salini, R.; Sperandii, A.F.; Santarelli, G.A.; Neri, D.; Di Marzio, V.; Romantini, R.; Migliorati, G.; Baranyi, J. Predicting the kinetics of *Listeria monocytogenes* and *Yersinia enterocolitica* under dynamic growth/death-inducing conditions, in Italian style fresh sausage. *Int. J. Food Microbiol.* **2017**, *240*, 108–114. [CrossRef] [PubMed]

120. Leães, Y.S.V.; Silva, J.S.; Robalo, S.S.; Pinton, M.B.; dos Santos, S.P.; Wagner, R.; Brasil, C.C.B.; de Menezes, C.R.; Barin, J.S.; Campagnol, P.C.B.; et al. Combined effect of ultrasound and basic electrolyzed water on the microbiological and oxidative profile of low-sodium mortadellas. *Int. J. Food Microbiol.* **2021**, *353*, 109310. [CrossRef]

121. Perea-Sanz, L.; Montero, R.; Belloch, C.; Flores, M. Nitrate reduction in the fermentation process of salt reduced dry sausages: Impact on microbial and physicochemical parameters and aroma profile. *Int. J. Food Microbiol.* **2018**, *282*, 84–91. [CrossRef]

122. Pietrasik, Z.; Gaudette, N.J. The effect of salt replacers and flavor enhancer on the processing characteristics and consumer acceptance of turkey sausages. *J. Sci. Food Agric.* **2015**, *95*, 1845–1851. [CrossRef]

123. Peruzy, M.F.; Houf, K.; Joossens, M.; Yu, Z.; Proroga, Y.T.R.; Murru, N. Evaluation of microbial contamination of different pork carcass areas through culture-dependent and independent methods in small-scale slaughterhouses. *Int. J. Food Microbiol.* **2021**, *336*, 108902. [CrossRef]

124. Fraqueza, M.J.; Barreto, A.S. HACCP: Hazard Analysis and Critical Control Points. In *Handbook of Fermented Meat and Poultry*; Wiley: Hoboken, NJ, USA, 2015; pp. 469–485.

125. Omer, M.K.; Alvarez-Ordóñez, A.; Prieto, M.; Skjerve, E.; Asehun, T.; Alvseike, O.A. A Systematic Review of Bacterial Foodborne Outbreaks Related to Red Meat and Meat Products. *Foodborne Pathog. Dis.* **2018**, *15*, 598–611. [CrossRef]

126. Hazards, E.P.O.B. Scientific Opinion on the development of a risk ranking toolbox for the EFSA BIOHAZ Panel. *EFSA J.* **2015**, *13*, 3939. [CrossRef]

127. EFSA. The European Union One Health 2020 Zoonoses Report. *EFSA J.* **2021**, *19*, 6971. [CrossRef]

128. Arguello, H.; Álvarez-Ordoñez, A.; Carvajal, A.; Rubio, P.; Prieto, M. Role of Slaughtering in Salmonella Spreading and Control in Pork Production. *J. Food Prot.* **2013**, *76*, 899–911. [CrossRef]

129. Akil, L.; Ahmad, H.A. Quantitative Risk Assessment Model of Human Salmonellosis Resulting from Consumption of Broiler Chicken. *Diseases* **2019**, *7*, 19. [CrossRef]

130. Thames, H.T.; Sukumaran, A.T. A Review of *Salmonella* and *Campylobacter* in Broiler Meat: Emerging Challenges and Food Safety Measures. *Foods* **2020**, *9*, 776. [CrossRef]

131. Bonardi, S.; Bruini, I.; Bolzoni, L.; Cozzolino, P.; Pierantoni, M.; Brindani, F.; Bellotti, P.; Renzi, M.; Pongolini, S. Assessment of *Salmonella* survival in dry-cured Italian salami. *Int. J. Food Microbiol.* **2017**, *262*, 99–106. [CrossRef]

132. Morales-Partera, Á.M.; Cardoso-Toset, F.; Jurado-Martos, F.; Astorga, R.J.; Huerta, B.; Luque, I.; Tarradas, C.; Gómez-Laguna, J. Survival of selected foodborne pathogens on dry cured pork loins. *Int. J. Food Microbiol.* **2017**, *258*, 68–72. [CrossRef]

133. Shiroda, M.; Pratt, Z.L.; Döpfer, D.; Wong, A.C.; Kaspar, C.W. RpoS impacts the lag phase of *Salmonella enterica* during osmotic stress. *FEMS Microbiol. Lett.* **2014**, *357*, 195–200. [CrossRef]

134. Heir, E.; Holck, A.; Omer, M.; Alvseike, O.; Høy, M.; Mage, I.; Axelsson, L. Reduction of verotoxigenic *Escherichia coli* by process and recipe optimisation in dry-fermented sausages. *Int. J. Food Microbiol.* **2010**, *141*, 195–202. [CrossRef]

135. Sartz, L.; De Jong, B.; Hjertqvist, M.; Plym-Forshell, L.; Alsterlund, R.; Löfdahl, S.; Osterman, B.; Ståhl, A.; Eriksson, E.; Hansson, H.B.; et al. An Outbreak of *Escherichia coli* O157:H7 Infection in Southern Sweden Associated with Consumption of Fermented Sausage; Aspects of Sausage Production That Increase the Risk of Contamination. *Epidemiol. Infect.* **2008**, *136*, 370. [CrossRef]

136. European Food Safety Authority (EFSA); European Centre for Disease Prevention and Control (ECDC). The European Union summary report on trends and sources of zoonoses, zoonotic agents and food-borne outbreaks in 2017. *EFSA J.* **2018**, *16*, e05500. [CrossRef]

137. Raimann, E.; Schmid, B.; Stephan, R.; Tasara, T. The Alternative Sigma Factor Sigma(L) of *L. monocytogenes* Promotes Growth under Diverse Environmental Stresses. *Foodborne Pathog. Dis.* **2009**, *6*, 583–591. [CrossRef]

138. Medveov, A.; Valk, L. Ubomr *Staphylococcus aureus*: Characterisation and Quantitative Growth Description in Milk and Artisanal Raw Milk Cheese Production. In *Structure and Function of Food Engineering*; Eissa, A., Ed.; InTech: London, UK, 2012; pp. 71–102. [CrossRef]

139. Sihto, H.-M.; Stephan, R.; Engl, C.; Chen, J.; Johler, S. Effect of food-related stress conditions and loss of agr and sigB on seb promoter activity in *S. aureus*. *Food Microbiol.* **2017**, *65*, 205–212. [CrossRef]

140. Peck, M.W.; van Vliet, A.H. Impact of *Clostridium botulinum* genomic diversity on food safety. *Curr. Opin. Food Sci.* **2016**, *10*, 52–59. [CrossRef]

141. Courtney-Long, E.; Carrol, D.; Zhang, Q.; Stevens, A.; Griffin-Blake, S.; Armour, B.; Campbell, V. Prevalence of Disability and Disability Type Among Adults—United States 2013. *MMWR Morb. Mortal. Wkly. Rep.* **2015**, *64*, 777. [CrossRef] [PubMed]

142. Laranjo, M.; Gomes, A.; Agulheiro-Santos, A.C.; Potes, M.E.; Cabrita, M.J.; Garcia, R.; Rocha, J.M.; Roseiro, L.C.; Fernandes, M.J.; Fraqueza, M.H.; et al. Characterisation of "Catalão" and "Salsichão" Portuguese traditional sausages with salt reduction. *Meat Sci.* **2016**, *116*, 34–42. [CrossRef]

143. Cittadini, A.; Domínguez, R.; Gómez, B.; Pateiro, M.; Pérez-Santaescolástica, C.; López-Fernández, O.; Sarriés, M.V.; Lorenzo, J.M. Effect of NaCl replacement by other chloride salts on physicochemical parameters, proteolysis and lipolysis of dry-cured foal "cecina". *J. Food Sci. Technol.* **2020**, *57*, 1628–1635. [CrossRef] [PubMed]

144. Khan, I.; Tango, C.N.; Miskeen, S.; Lee, B.H.; Oh, D.H. Hurdle Technology: A Novel Approach for Enhanced Food Quality and Safety—A Review. *Food Control* **2017**, *73*, 1426–1444. [CrossRef]

145. Shan, L.C.; Regan, A.; Monahan, F.J.; Li, C.; Murrin, C.; Lalor, F.; Wall, P.G.; McConnon, A. Consumer views on "healthier" processed meat. *Br. Food J.* **2016**, *118*, 1712–1730. [CrossRef]

146. Devlieghere, F.; Vermeiren, L.; Debevere, J. New preservation technologies: Possibilities and limitations. *Int. Dairy J.* **2004**, *14*, 273–285. [CrossRef]

147. Sadaghiani, S.K.; Aliakbarlu, J.; Tajik, H.; Mahmoudian, A. Anti-listeria activity and shelf life extension effects of *Lactobacillus* along with garlic extract in ground beef. *J. Food Saf.* **2019**, *39*, e12709. [CrossRef]

148. Castellano, P.; Peña, N.; Ibarreche, M.P.; Carduza, F.; Soteras, T.; Vignolo, G. Antilisterial efficacy of *Lactobacillus* bacteriocins and organic acids on frankfurters. Impact on sensory characteristics. *J. Food Sci. Technol.* **2018**, *55*, 689–697. [CrossRef]

149. Hu, Z.Y.; Balay, D.; Hu, Y.; McMullen, L.M.; Gänzle, M.G. Effect of chitosan, and bacteriocin—Producing *Carnobacterium maltaromaticum* on survival of *Escherichia coli* and *Salmonella Typhimurium* on beef. *Int. J. Food Microbiol.* **2019**, *290*, 68–75. [CrossRef]

150. Kumar, Y.; Kaur, K.; Shahi, A.K.; Kairam, N.; Tyagi, S.K. Antilisterial, antimicrobial and antioxidant effects of pediocin and *Murraya koenigii* berry extract in refrigerated goat meat emulsion. *LWT Food Sci. Technol.* **2017**, *79*, 135–144. [CrossRef]

151. Yang, H.-J.; Wang, H.-F.; Tao, F.; Li, W.-X.; Cao, G.-T.; Yang, Y.-Y.; Xu, X.-L.; Zhou, G.-H.; Shen, Q. Structural basis for high-pressure improvement in depolymerization of interfacial protein from RFRS meat batters in relation to their solubility. *Food Res. Int.* **2021**, *139*, 109834. [CrossRef]

152. Xiao, Q.; Xu, M.; Xu, B.; Chen, C.; Deng, J.; Li, P. Combined Effect of High-Pressure Processing with Spice Extracts on Quality of Low-Salt Sausage during Refrigerated Storage. *Foods* **2021**, *10*, 2610. [CrossRef]

153. Horita, C.N.; Baptista, R.C.; Caturla, M.Y.; Lorenzo, J.M.; Barba, F.J.; Sant'Ana, A.S. Combining reformulation, active packaging and non-thermal post-packaging decontamination technologies to increase the microbiological quality and safety of cooked ready-to-eat meat products. *Trends Food Sci. Technol.* **2018**, *72*, 45–61. [CrossRef]

154. Grossi, A.; Søltoft-Jensen, J.; Knudsen, J.C.; Christensen, M.; Orlien, V. Reduction of salt in pork sausages by the addition of carrot fibre or potato starch and high pressure treatment. *Meat Sci.* **2012**, *92*, 481–489. [CrossRef]

155. Hygreeva, D.; Pandey, M. Novel approaches in improving the quality and safety aspects of processed meat products through high pressure processing technology—A review. *Trends Food Sci. Technol.* **2016**, *54*, 175–185. [CrossRef]

156. Duranton, F.; Guillou, S.; Simonin, H.; Chéret, R.; de Lamballerie, M. Combined use of high pressure and salt or sodium nitrite to control the growth of endogenous microflora in raw pork meat. *Innov. Food Sci. Emerg. Technol.* **2012**, *16*, 373–380. [CrossRef]

157. Stollewerk, K.; Jofré, A.; Comaposada, J.; Arnau, J.; Garriga, M. The effect of NaCl-free processing and high pressure on the fate of *Listeria monocytogenes* and *Salmonella* on sliced smoked dry-cured ham. *Meat Sci.* **2012**, *90*, 472–477. [CrossRef] [PubMed]

158. Hammou, F.B.; Skali, S.N.; Idaomar, M.; Abrini, J. Combinations of Nisin with Salt (NaCl) to Control *Listeria monocytogenes* on Sheep Natural Sausage Casings Stored at 6 °C. *African J. Biotechnol.* **2010**, *9*, 1191–1195.

159. Ojha, K.S.; Keenan, D.F.; Bright, A.; Kerry, J.P.; Tiwari, B. Ultrasound assisted diffusion of sodium salt replacer and effect on physicochemical properties of pork meat. *Int. J. Food Sci. Technol.* **2016**, *51*, 37–45. [CrossRef]

160. Barros, J.C.; Gois, T.S.; Pires, M.A.; Rodrigues, I.; Trindade, M.A. Sodium reduction in enrobed restructured chicken nuggets through replacement of NaCl with $CaCl_2$. *J. Food Sci. Technol.* **2019**, *56*, 3587–3596. [CrossRef] [PubMed]

161. Pandya, J.K.; Decker, K.E.; Goulette, T.; Kinchla, A.J. Sodium reduction in Turkey breast meat by using sodium anion species. *LWT Food Sci. Technol.* **2020**, *124*, 109110. [CrossRef]

162. Aliño, M.; Grau, R.; Toldrá, F.; Blesa, E.; Pagán, M.J.; Barat, J.M. Physicochemical properties and microbiology of dry-cured loins obtained by partial sodium replacement with potassium, calcium and magnesium. *Meat Sci.* **2010**, *85*, 580–588. [CrossRef]

163. Blesa, E.; Aliño, M.; Barat, J.M.; Grau, R.; Toldrá, F.; Pagán, M. Microbiology and physico-chemical changes of dry-cured ham during the post-salting stage as affected by partial replacement of NaCl by other salts. *Meat Sci.* **2008**, *78*, 135–142. [CrossRef]

164. Harper, N.M.; Getty, K.J. Effect of Salt Reduction on Growth of *Listeria monocytogenes* in Meat and Poultry Systems. *J. Food Sci.* **2012**, *77*, M669–M674. [CrossRef]

165. Shadbolt, C.; Ross, T.; McMeekin, T. Differentiation of the effects of lethal pH and water activity: Food safety implications. *Lett. Appl. Microbiol.* **2001**, *32*, 99–102. [CrossRef]

166. Devlieghere, F.; Vermeiren, L.; Bontenbal, E.; Lamers, P.-P.; Debevere, J. Reducing salt intake from meat products by combined use of lactate and diacetate salts without affecting microbial stability. *Int. J. Food Sci. Technol.* **2009**, *44*, 337–341. [CrossRef]

167. Lorenzo, J.M.; Cittadini, A.; Bermúdez, R.; Munekata, P.E.; Domínguez, R. Influence of partial replacement of NaCl with KCl, $CaCl_2$ and $MgCl_2$ on proteolysis, lipolysis and sensory properties during the manufacture of dry-cured lacón. *Food Control* **2015**, *55*, 90–96. [CrossRef]

168. Samapundo, S.; Ampofo-Asiama, J.; Anthierens, T.; Xhaferi, R.; Van Bree, I.; Szczepaniak, S.; Goemaere, O.; Steen, L.; Dhooge, M.; Paelinck, H. Influence of NaCl reduction and replacement on the growth of *Lactobacillus* sakei in broth, cooked ham and white sauce. *Int. J. Food Microbiol.* **2010**, *143*, 9–16. [CrossRef]

169. Rios-Mera, J.D.; Saldaña, E.; Patinho, I.; Selani, M.M.; Contreras-Castillo, C.J. Enrichment of NaCl-reduced burger with long-chain polyunsaturated fatty acids: Effects on physicochemical, technological, nutritional, and sensory characteristics. *Meat Sci.* **2021**, *177*, 108497. [CrossRef]

170. Vidal, V.A.; Biachi, J.P.; Paglarini, C.; Pinton, M.B.; Campagnol, P.C.; Esmerino, E.A.; da Cruz, A.G.; Morgano, M.A.; Pollonio, M.A. Reducing 50% sodium chloride in healthier jerked beef: An efficient design to ensure suitable stability, technological and sensory properties. *Meat Sci.* **2019**, *152*, 49–57. [CrossRef]

171. França, F.; Harada-Padermo, S.D.S.; Frasceto, R.A.; Saldaña, E.; Lorenzo, J.M.; Vieira, T.M.F.D.S.; Selani, M.M. Umami ingredient from shiitake (*Lentinula edodes*) by-products as a flavor enhancer in low-salt beef burgers: Effects on physicochemical and technological properties. *LWT* **2022**, *154*, 112724. [CrossRef]

172. Gaudette, N.J.; Pietrasik, Z. The sensory impact of salt replacers and flavor enhancer in reduced sodium processed meats is matrix dependent. *J. Sens. Stud.* **2017**, *32*, e12247. [CrossRef]

173. Horita, C.; Morgano, M.A.; Celeghini, R.; Pollonio, M. Physico-chemical and sensory properties of reduced-fat mortadella prepared with blends of calcium, magnesium and potassium chloride as partial substitutes for sodium chloride. *Meat Sci.* **2011**, *89*, 426–433. [CrossRef]

174. García-Lomillo, J.; González-SanJosé, M.A.L.; Del Pino-García, R.; Rivero-Pérez, M.A.D.; Muñiz-Rodríguez, P. Alternative natural seasoning to improve the microbial stability of low-salt beef patties. *Food Chem.* **2017**, *227*, 122–128. [CrossRef]

175. Omer, M.; Prieto, B.; Rendueles, E.; Alvarez-Ordóñez, A.; Lunde, K.; Alvseike, O.; Prieto, M. Microbiological, physicochemical and sensory parameters of dry fermented sausages manufactured with high hydrostatic pressure processed raw meat. *Meat Sci.* **2015**, *108*, 115–119. [CrossRef]

176. Rendueles, E.; Omer, M.; Alvseike, O.; Alonso-Calleja, C.; Capita, R.; Prieto, M. Microbiological food safety assessment of high hydrostatic pressure processing: A review. *LWT* **2011**, *5*, 1251–1260. [CrossRef]

177. Aymerich, T.; Picouet, P.; Monfort, J. Decontamination technologies for meat products. *Meat Sci.* **2008**, *78*, 114–129. [CrossRef]

178. Bajovic, B.; Bolumar, T.; Heinz, V. Quality considerations with high pressure processing of fresh and value added meat products. *Meat Sci.* **2012**, *92*, 280–289. [CrossRef]

179. Hereu, A.; Dalgaard, P.; Garriga, M.; Aymerich, T.; Bover-Cid, S. Analysing and modelling the growth behaviour of *Listeria monocytogenes* on RTE cooked meat products after a high pressure treatment at 400 MPa. *Int. J. Food Microbiol.* **2014**, *186*, 84–94. [CrossRef]

180. Bover-Cid, S.; Belletti, N.; Garriga, M.; Aymerich, T. Model for *Listeria monocytogenes* inactivation on dry-cured ham by high hydrostatic pressure processing. *Food Microbiol.* **2011**, *28*, 804–809. [CrossRef]

181. Rubio, B.; Possas, A.; Rincón, F.; García-Gimeno, R.M.; Martínez, B. Model for *Listeria monocytogenes* inactivation by high hydrostatic pressure processing in Spanish chorizo sausage. *Food Microbiol.* **2018**, *69*, 18–24. [CrossRef]

182. Luckose, F.; Pandey, M.C.; Chauhan, O.P.; Sultana, K.; Abhishek, V. Effect of High Pressure Processing on the Quality Characteristics and Shelf Life of Low-Sodium Re-Structured Chicken Nuggets. *J. Food Nutr. Res.* **2015**, *54*, 334–345.

183. EURL Lm. *EURL Lm Technical Guidance Document on Challenge Tests and Durability Studies for Assessing Shelf-Life of Ready-to-Eat Foods Related to Listeria monocytogenes*; EURL Lm: Maisons-alfort, France, 2021; pp. 1–60.

184. Nikodinoska, I.; Baffoni, L.; Di Gioia, D.; Manso, B.; García-Sánchez, L.; Melero, B.; Rovira, J. Protective cultures against foodborne pathogens in a nitrite reduced fermented meat product. *LWT* **2019**, *101*, 293–299. [CrossRef]

185. Bernardo, R.; Barreto, A.S.; Nunes, T.; Henriques, A.R. Estimating *Listeria monocytogenes* Growth in Ready-to-Eat Chicken Salad Using a Challenge Test for Quantitative Microbial Risk Assessment. *Risk Anal.* **2020**, *40*, 2427–2441. [CrossRef] [PubMed]

186. Di Gioia, D.; Mazzola, G.; Nikodinoska, I.; Aloisio, I.; Langerholc, T.; Rossi, M.; Raimondi, S.; Melero, B.; Rovira, J. Lactic acid bacteria as protective cultures in fermented pork meat to prevent *Clostridium* spp. growth. *Int. J. Food Microbiol.* **2016**, *235*, 53–59. [CrossRef] [PubMed]

187. Bonilauri, P.; Merialdi, G.; Ramini, M.; Bardasi, L.; Taddei, R.; Grisenti, M.S.; Daminelli, P.; Cosciani-Cunico, E.; Dalzini, E.; Frustoli, M.A.; et al. Modeling the behavior of *Listeria innocua* in Italian salami during the production and high-pressure validation of processes for exportation to the U.S. *Meat Sci.* **2021**, *172*, 108315. [CrossRef]

188. Borges, A.F.; Cózar, A.; Patarata, L.; Gama, L.T.; Alfaia, C.M.; Fernandes, M.J.; Pérez, H.V.; Fraqueza, M.J. Effect of high hydrostatic pressure challenge on biogenic amines, microbiota, and sensory profile in traditional poultry- and pork-based semidried fermented sausage. *J. Food Sci.* **2020**, *85*, 1256–1264. [CrossRef]

189. Álvarez-Ordóñez, A.; Leong, D.; Hickey, B.; Beaufort, A.; Jordan, K. The challenge of challenge testing to monitor *Listeria monocytogenes* growth on ready-to-eat foods in Europe by following the European Commission (2014) Technical Guidance document. *Food Res. Int.* **2015**, *75*, 233–243. [CrossRef]

190. Palanichamy, A.; Jayas, D.S.; Holley, R.A. Predicting Survival of *Escherichia coli* O157:H7 in Dry Fermented Sausage Using Artificial Neural Networks. *J. Food Prot.* **2008**, *71*, 6–12. [CrossRef]

191. Skjerdal, T.; Gangsei, L.E.; Alvseike, O.; Kausrud, K.; De Cesare, A.; Alexa, E.-A.; Alvarez-Ordóñez, A.; Moen, L.H.; Osland, A.M.; From, C.; et al. Development and validation of a regression model for *Listeria monocytogenes* growth in roast beefs. *Food Microbiol.* **2021**, *98*, 103770. [CrossRef]

192. Hereu, A.; Dalgaard, P.; Garriga, M.; Aymerich, T.; Bover-Cid, S. Modeling the high pressure inactivation kinetics of *Listeria monocytogenes* on RTE cooked meat products. *Innov. Food Sci. Emerg. Technol.* **2012**, *16*, 305–315. [CrossRef]

193. Huang, L. Evaluating the Performance of a New Model for Predicting the Growth of *Clostridium perfringens* in Cooked, Uncured Meat and Poultry Products under Isothermal, Heating, and Dynamically Cooling Conditions. *J. Food Sci.* **2016**, *81*, M1754–M1765. [CrossRef]

194. Bover-Cid, S.; Belletti, N.; Aymerich, T.; Garriga, M. Modeling the protective effect of a w and fat content on the high pressure resistance of *Listeria monocytogenes* in dry-cured ham. *Food Res. Int.* **2015**, *75*, 194–199. [CrossRef]

195. Gunvig, A.; Borggaard, C.; Hansen, F.; Hansen, T.B.; Aabo, S. ConFerm—A tool to predict the reduction of pathogens during the production of fermented and matured sausages. *Food Control* **2016**, *67*, 9–17. [CrossRef]

196. Hansen, T.B.; Abdalas, S.; Al-Hilali, I.; Hansen, L.T. Predicting the effect of salt on heat tolerance of *Listeria monocytogenes* in meat and fish products. *Int. J. Food Microbiol.* **2021**, *352*, 109265. [CrossRef]

197. Alía, A.; Rodríguez, A.; Andrade, M.J.; Gómez, F.M.; Córdoba, J.J. Combined effect of temperature, water activity and salt content on the growth and gene expression of *Listeria monocytogenes* in a dry-cured ham model system. *Meat Sci.* **2019**, *155*, 16–19. [CrossRef]

198. Qu, K.; Guo, F.; Liu, X.; Lin, Y.; Zou, Q. Application of Machine Learning in Microbiology. *Front. Microbiol.* **2019**, *10*, 1–10. [CrossRef]

 *foods*

Review

Treatment of Fresh Meat, Fish and Products Thereof with Cold Atmospheric Plasma to Inactivate Microbial Pathogens and Extend Shelf Life

Peter Paulsen [1], Isabella Csadek [1], Alexandra Bauer [2], Kathrine H. Bak [1], Pia Weidinger [3], Karin Schwaiger [1], Norbert Nowotny [3,4], James Walsh [5], Emilio Martines [6] and Frans J. M. Smulders [1,*]

[1] Unit of Food Hygiene and Technology, Institute of Food Safety, Food Technology and Veterinary Public Health, University of Veterinary Medicine, Veterinaerplatz 1, 1210 Vienna, Austria
[2] Canfelis, Kienmayergasse 5, 1140 Vienna, Austria
[3] Viral Zoonoses, Emerging and Vector-Borne Infections Group, Institute of Virology, University of Veterinary Medicine, Veterinaerplatz 1, 1210 Vienna, Austria
[4] Department of Basic Medical Sciences, College of Medicine, Mohammed Bin Rashid University of Medicine and Health Sciences, Dubai P.O. Box 505055, United Arab Emirates
[5] Centre for Plasma Microbiology, University of Liverpool, Liverpool L69 3BX, UK
[6] Department of Physics "G. Occhialini", University of Milano—Bicocca, Piazza della Scienza 3, 20126 Milano, Italy
* Correspondence: frans.smulders@vetmeduni.ac.at; Tel.: +43-1-25077-3318

Abstract: Assuring the safety of muscle foods and seafood is based on prerequisites and specific measures targeted against defined hazards. This concept is augmented by 'interventions', which are chemical or physical treatments, not genuinely part of the production process, but rather implemented in the framework of a safety assurance system. The present paper focuses on 'Cold Atmospheric pressure Plasma' (CAP) as an emerging non-thermal intervention for microbial decontamination. Over the past decade, a vast number of studies have explored the antimicrobial potential of different CAP systems against a plethora of different foodborne microorganisms. This contribution aims at providing a comprehensive reference and appraisal of the latest literature in the area, with a specific focus on the use of CAP for the treatment of fresh meat, fish and associated products to inactivate microbial pathogens and extend shelf life. Aspects such as changes to organoleptic and nutritional value alongside other matrix effects are considered, so as to provide the reader with a clear insight into the advantages and disadvantages of CAP-based decontamination strategies.

Keywords: cold atmospheric plasma; antimicrobial effects; physical-chemical properties; foodborne pathogens management; longitudinally integrated safety assurance; shelf-life extension

Citation: Paulsen, P.; Csadek, I.; Bauer, A.; Bak, K.H.; Weidinger, P.; Schwaiger, K.; Nowotny, N.; Walsh, J.; Martines, E.; Smulders, F.J.M. Treatment of Fresh Meat, Fish and Products Thereof with Cold Atmospheric Plasma to Inactivate Microbial Pathogens and Extend Shelf Life. *Foods* **2022**, *11*, 3865. https://doi.org/10.3390/foods11233865

Academic Editor: Arun K. Bhunia

Received: 13 November 2022
Accepted: 24 November 2022
Published: 30 November 2022

Publisher's Note: MDPI stays neutral with regard to jurisdictional claims in published maps and institutional affiliations.

1. Introduction

In the food production sector, 'shelf life' is one of the most essential quality parameters. Even when microbial contamination and subsequent growth of pathogenic organisms are successfully counteracted, microbial and/or chemical spoilage will cause foods of animal origin to be withdrawn from the market. The latter is one of the main worries of the United Nations (UN), as such represents the main reason for food waste. Approximately one-third of the world's foods of animal origin is lost through waste and this markedly reduces food security [1]. According to estimates of the UN's Food and Agricultural Organisation (FAO), published in 2015, it results in approximately USD 940 billion per year in economic losses. It also results in significant environmental impacts. For example, food loss and waste are responsible for 8% of the world's greenhouse gas emissions [2]. In fact, if food loss and waste were contained to one country, that country would be the world's third-largest emitter after the United States and China [1].

Arguably, food security is not only dependent on minimising food waste, but also on the enhancement of the efficacy of meat production. Thus, farm-animal species play crucial roles in satisfying demands for meat on a global scale, and environmental as well as genetic factors [3] need to be optimised. In particular, one of the important aims is to increase skeletal muscle growth in farm animals [4,5]. The enhancement of muscle development and growth is crucial to meet consumers' demands for meat [5,6].

The veterinary medical curriculum includes enough elements of biology and physiology to allow graduates to function as 'doctors', yet many of them end up working in the food industry as 'veterinary public health' (VPH) professionals. These have the legal responsibility to remain aware of the latest technologies and techniques applied by industry to assure their products are safe, nutritious and have the desirable physical-chemical and sensory properties to appeal to the customer. Over the past decades, VPH officials have gradually shifted their attention from 'end-product oriented inspection' towards 'longitudinally integrated safety assurance' (LISA; [7]) and as health officials they concentrate on assuring the absence of pathogenic microorganisms in foods of animal origin as evidence for 'quality'. However, in the current political climate, public health authorities need to assure that besides 'food safety' (the first, apparently most significant parameter), also 'food security' and 'sustainability' issues are adequately addressed. 'Food security' has been defined by the UN's FAO as: *'assuring that all people, at all times, have physical and economic access to sufficient, safe and nutritious food, that meets their dietary needs and food preferences for an active and healthy life'*, and 'sustainability' as: *'meeting the needs of the present without compromising the ability of the future generations to meet their own needs'* [8]. The paramount importance of food security was emphasised in the UN World Commission on Environment and Development Report in 1987 [9].

Concentrating on safety, security and sustainability is 'part and parcel' of the EU's current pathogen management strategy, which has been summarised in its May 2020 strategy paper 'From Farm to Fork' [10]. For this contribution, this means that the authors take an approach beyond merely judging the antimicrobial efficacy of risk management strategies against pathogens, but rather additionally consider effects on variables such as shelf life and physical-chemical and sensory attributes. In particular, the exposure of foods to reactive species produced in an ionised gas—'plasma'—and their effects on microbial contaminants and on the food matrix (in terms of sensory quality and alterations of proteins and lipids) will be discussed. To this end, we provide (i) an introduction on the composition and generation of plasma, and (ii) an overview of the application of plasma technology for the microbial decontamination of selected food commodities of animal origin, with (iii) special consideration of the effects of plasma on the sensory quality of meat products, in particular those related to oxidative reactions. The antibacterial potential of the application of plasma on meat-based products was emphasised recently [11], but the oxidation of lipids [12] and proteins [11] has been identified as a potential drawback. Although protein modification can have positive side effects (e.g., improving gelling quality, [13]), protein oxidation has been identified as a major cause for quality loss in muscle-based foods [14,15].

2. What Is 'Cold Atmospheric Plasma' and How Is It Generated?

2.1. Generation of Plasma

The term 'plasma', originally coined by Nobel prize winner Irving Langmuir, designates in physical sciences a gas where a fraction of the particles is ionised; that is, stripped of one electron and converted into an electron–ion couple. The plasma state, which is considered as the fourth state of matter, is thus a mixture of electrons, ions and neutral particles [16]. The fraction of charged to total particles, called 'degree of ionization', is a function of several factors, among which the most notable one is the power density used to produce the plasma, and can range from very low values (weakly ionised plasma) to 1 (fully ionised plasma). The plasmas typically used for the treatment of food are considered weakly ionised, so most of the gas particles are electrically neutral.

In almost all practical situations, plasma is produced by the application of an electric field. The electric field accelerates free electrons to the energy required to ionise neutral atoms and molecules in the gas (typically between 5 and 25 eV). This is, however, opposed by collisions that electrons undergo in their motion through the gas: in particular, inelastic collisions cause electrons to lose energy, which is transferred to the neutral gas in the form of the excitation of bound electrons to higher energy levels, or the excitation of the rotation or vibrational states of molecules. Incidentally, the emission of photons when the excited bound electrons return to their original state is the source of visible light, which gives plasmas their typical glowing appearance, and can also be a source of biocidal ultraviolet (UV) radiation. Given this balance between acceleration and collisional energy loss, a minimum applied voltage is required for plasma ignition, a phenomenon called 'breakdown', because the presence of free charged particles converts the previously dielectric neutral gas into an electrical conductor.

To achieve breakdown, it is required that the few free electrons naturally present in the gas are increased in number. This occurs through a process, by which collisions between neutral species and sufficiently energetic electrons result in ionisation events, causing the number of free electrons in the gas to grow exponentially. This process is known as an 'electron avalanche'. Subsequent processes strongly depend on the manner in which the electrical energy is applied. In the case of a stationary (direct current, DC) or slowly varying electric field, the positive ions produced in ionisation events are accelerated towards the negative electrode, called the cathode (assuming that the electric field is obtained by applying a potential difference between two electrodes), and have a certain probability of extracting an electron called the 'secondary electron' from its surface. When this process is sufficiently intense to provide a new electron for each electron lost to the anode, the process becomes self-sustaining. A second approach is to vary the electric field fast enough that electrons perform an oscillation with an amplitude smaller than the electrode gap. In this case, there is no electron flux to the anode, and each electron can produce others in its oscillatory motion (the electron number does not grow indefinitely because of diffusion losses). For typical system sizes, ranging from mm to cm, the required oscillation frequency is in the order of a few MHz or larger: this defines the so-called 'radio frequency (RF) plasmas', or, when frequencies are in the GHz range, 'microwave plasmas'.

Since the electric field readily transfers energy to electrons, they typically have high temperatures, in the order of 1 eV or higher (in plasma physics, temperatures are given through their equivalent mean kinetic energy: 1 eV corresponds to a temperature of 11,600 K). Still, unless very high power is used, and the plasma is very well confined (for example, by magnetic fields, as is the case in thermonuclear fusion studies), most electrons do not have the time to transfer their energy to the ions and to the neutral gas, which thus remain at relatively low temperatures. In this case, which is the one of interest in the following, the plasma can be described as 'non-thermal'. If the ions and the neutral gas remain at, or very near room temperature, the notion of 'cold plasma' arises. This is exactly the concept of interest here since the use of plasma for food decontamination must avoid thermal effects. For a detailed description of the underlying principles of plasma generation and applications in materials modification, the reader is referred to Lieberman and Lichtenberg [17].

The interest of plasmas in the context of disinfection and decontamination (as in many other fields) stems from the fact that the free hot electrons shift chemical equilibria, giving rise to a wealth of reactive species, which interact in a destructive way with microbes. In particular, reactive oxygen and nitrogen species (ROS and RNS) are of relevance for this application, due to their interaction with the cell membrane [18,19]. Furthermore, the plasma is responsible for generating other agents, namely UV radiation, intense electric fields and charged species, which may also play a role in the decontamination process [20,21].

2.2. Low Pressure Plasma vs. Atmospheric-Pressure Plasma

Plasmas are most easily produced at low pressure (several orders of magnitude below atmospheric pressure) because the electrons do not undergo too many collisions and are easily accelerated to the energy required for ionisation. This results in a reduced breakdown voltage, typically of the order of a few hundred volts. However, low-pressure plasmas are not well suited for food treatment. On the contrary, generating plasma at atmospheric pressure requires higher voltages, typically a few kV [22]. Consequently, the requirement of keeping the neutral component at or near room temperature results in the need to limit the current achieved after breakdown, otherwise it is very easy to obtain high power levels. There are different solutions for this problem, which can be summarised into two main approaches. One is the dielectric barrier discharge (DBD), where the high-voltage electrode is separated from the grounded electrode by at least one layer of dielectric material. After the breakdown, charge quickly accumulates on the dielectric surface, and extinguishes the current. These devices are typically operated at a frequency of a few kHz, although lower frequencies can be used, or with pulsed voltages [23]. The second one is the use of radio frequency (RF) voltage, which changes polarity so quickly that the peak current is limited. In the context of plasma food treatment, RF is seldom used, as such plasmas are typically 'hotter' than those generated using the DBD approach.

2.3. Application of Atmospheric-Pressure Plasma to Tissues, Foods and Food Contact Surfaces

The potential of atmospheric-pressure plasma to inactivate bacteria and other pathogens has been known for a long time [24]. However, only in the last twenty years has the development of low-power plasma sources enabled the treatment of organic substrates without thermal damage, leading to its use in the emerging discipline of 'plasma medicine' [25]. This includes not only disinfection [26], but also the use of the plasma-produced chemical species to manipulate cellular processes or structures [19] giving rise to therapeutic effects. For example, the stimulation of wound healing is a well-studied process [27–30], which has found application also in the context of veterinary medicine [31]. A field where the use of plasma-based decontamination may prove to be a game changer is that of food decontamination [32]. A particular mode of plasma generation has been studied more recently, with a view to assess its suitability as a means of antimicrobial intervention during food production [33]. The scientific principles on which the generation of reactive gas species is based, and the modes of antimicrobial action of cold atmospheric plasma (CAP) when applied to various foods of animal origin, with special reference to meat and meat products, have been reviewed recently [34].

It is essential to realise that CAP exerts its antimicrobial action primarily on the surface of a treated food item. Hence, bacterial (or viral) surface contamination resulting from slaughter and subsequent processing could therefore—at least partly—be inactivated before further processing/packaging. Obviously, the chosen method of applying plasma determines the antimicrobial efficacy of exposing microbial contaminants to plasma.

The specific mix of reactive species produced by a plasma source depends on several interlinked factors. First is the gas used for the process, which is typically either air or a noble gas (helium or argon) mixed with small fractions of air or air constituents. Second is the amplitude, frequency and waveform of the voltage used to produce the plasma, the applied power and the gas flow (if applicable). For example, in the case of an air DBD, the prevalence of ROS or RNS will be dictated by the power level [35]. As another example, humidity will affect ozone production [36], and this may lead to the loss of bactericidal effect [37]. The closer the target is to the plasma source, the shorter the time needed for reactive species to reach the target. In settings with a distance between electrodes and the sample, the presence of long-lived radicals such as NO_2, O_3 and N_2O is an important factor co-determining the array of reactive compounds.

In recent years, a plethora of different plasma systems have been developed for the decontamination of food contact surfaces [38] and food products [39]. Typically, but not exclusively, these prototype systems have been based on the DBD family of discharges. Two distinct modes of application have arisen, and the first involves plasma interacting directly with food products, typically achieved by placing the product between the electrodes of a parallel plate reactor or in the effluent of a plasma jet (Figure 1a,b, respectively). Direct contact systems are highly efficient as reactive and short-lived chemical species, such as O, N and OH, directly impinge on, and interact with, the food matrix. Despite their efficiency, a direct contact between plasma and food poses a number of technical challenges; for example, the plasma characteristics are inevitably and inextricably linked to the electrical characteristics of the food product, a situation that can compromise repeatability. Another challenge relates to the complexity of the discharge chemistry reaching the product and its impact on the food matrix, a process that could potentially involve over 1000 complex biochemical reactions. Without a clear understanding of the underpinning processes that give rise to the intriguing antimicrobial effects associated with plasma treatment, the regulatory approval necessary for the commercial application of direct-contact plasma technology may not be forthcoming.

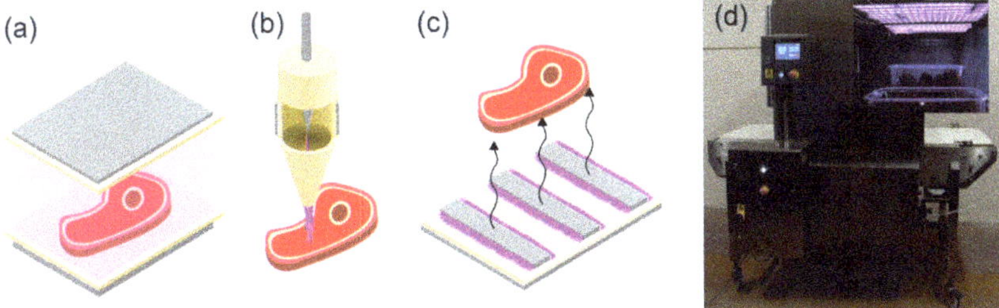

Figure 1. Typical DBD systems used for the treatment of food products: (**a**) direct-contact parallel plate reactor, (**b**) direct-contact plasma jet, (**c**) indirect contact surface barrier discharge and (**d**) pilot-scale surface barrier discharge system developed at the University of Liverpool.

The second mode of application relates to the indirect exposure of food products to plasma, typically achieved by generating plasma in close proximity to a product and relying on the diffusion and/or convection of chemical species to its surface (Figure 1c). In this scenario, short-lived chemical species react before reaching the food product, yielding a number of longer-lived intermediaries; for example, in the case of air plasma O_3, NO, N_2O and NO_2 [40]. Due to the absence of highly reactive chemical species, indirect approaches are often considered less effective for microbial inactivation compared to their direct-contact counterparts. Conversely, a vast reduction in the variety of chemical species reaching the product is conducive when attempting to elucidate the underpinning mode of action. A further benefit of many indirect treatment approaches is their ability to be easily scaled to cover large areas (Figure 1d), and they remain unaffected by the electrical properties of the food product, enhancing repeatability.

2.4. In-Package Cold Plasma Treatment

Direct exposure to CAP can also be achieved for already packed products. In this case, the packaging material itself is used as the dielectric barrier and external electrodes are used to apply a high voltage, resulting in plasma formation directly within the sealed pack. This efficacy of 'in-pack' plasma treatment of foods has been demonstrated against foodborne pathogens [41,42] and spoilage microorganisms to extend the shelf life of end products [43–47]. As products are already sealed within the package prior to plasma disinfection, there is little opportunity for further contamination, which is considered a major advantage of the approach.

A drawback of the approach is the requirement that the packaging material can withstand plasma treatment without degradation, which could potentially contaminate foods sealed within. Very few studies have considered the impact of in-package plasma on the packaging material, or how the material influences the production of plasma species [46,47]. Previous studies have shown that the physical-chemical and microbiological condition of fresh beef, packaged in a polyethylene–polyamide–polyethylene (PE/PA/PE) film after it had been inoculated with *S. aureus*, *L. monocytogenes* and *E. coli*, is in no way affected by subjecting it to treatment with atmospheric-pressure cold plasma ([48]; some details in Section 3).

Using in-package plasma treatment (2 to 60 s), statistically significant reductions (0.8–1.6 log cycles) in *Listeria innocua* contamination of 'Bresaola' (a dried, ready-to-eat beef ham product) were recorded [49]. Using 'dielectric barrier' electrodes and air as gas, a 3 min plasma treatment of packaged chicken cubes allowed reductions in *Salmonella*, *E. coli* O157:H7 and *L. monocytogenes* of up to 3.7 log [50]. Similar findings were recorded by Jayasena et al. [51], who established antimicrobial effects of up to 2.6 log *S. typhimurium* and *L. monocytogenes* in PE-packaged beef and pork using a DBD system.

British studies have shown that at retail level a sizeable portion of the cross-contamination of fresh meat occurs via both the external and internal surfaces of the packaging film [52]. Until the CAP exposure of food and food products has gained the necessary regulatory approval, the application of CAP to inactivate pathogens on the external surfaces of packaging materials is a viable way forward, provided one can demonstrate that the packaging matrix is not breached by cold plasma and hence that the packaged product does not have to be classified as a 'novel food' according to EU Regulation 2015/2283 [53]. According to the latter regulation, any lasting change incurred beyond what could be expected naturally as a result of plasma exposure would require '*novel food*' certification [53]. As regards plasma generated from ambient air, it is debatable if the action of ROS and RNS would qualify plasma-treated food as 'novel food' (i.e., food with intentionally modified molecular structures that were not present in foods within the Union before 15th May 1997, or '*food resulting from a production process not used for food production within the Union before 15 May 1997, which gives rise to significant changes in the composition or structure of a food, affecting its nutritional value, metabolism or level of undesirable substances*'; Article 3, paragraphs (i) and (vii) of Regulation (EU) 2015/2283; [53]). European partner countries initiated a COST project (013/20) with the purpose of creating a database that will ultimately allow 'understanding plasma's most important processes including aspects of (Novel food) legislation, energy consumption, food safety and quality'. However, provided a CAP treatment does NOT cause a lasting change, it can be considered a processing aid and can be applied without extra labelling or consumer information [54]. A case-by-case evaluation of plasma-matrix combinations has been suggested [55]. In 2017, Ekezie et al. [56] reported that CAP had not been implemented in food industry settings because of uncertainties about matrix modifications and subsequent legal issues. In the last years, this knowledge gap has been at least partially filled; see the following sections.

3. Effects of CAP Treatment of Fresh Meat and Meat Products

3.1. Fresh Meat

Microbial contamination of meat occurs at slaughter and numerous contamination scenarios are possible along the fresh-meat chain [57,58]. Control of such contamination is essential to prevent food from becoming 'unsafe', i.e., either hazardous to human health or unfit for human consumption due to spoilage (Regulation (EC) No 178/2002) [59]. The prospects and limitations of 'Good Hygiene Practice' and, more specifically, of 'Hazard Analysis Critical Control Point' systems in safeguarding fresh meat have been extensively discussed [60,61] and the usefulness of interventions as additional tools has been proposed and studied [62–68].

Typically, such methods would be preservation or processing, but these usually alter product appearance or other characteristics. Thus, the array of intervention methods for fresh meat is rather limited. Treatment with cold or hot water or with steam and dilute organic acids may be applied by rinsing, spraying or immersion, with no or negligible effects on the appearance of fresh meat and without leaving residues [69–76]). Since nearly a decade, the application of dilute lactic acid has been allowed in the EU, albeit only as part of the pre-chill treatment of beef carcasses (Commission Regulation (EU) No 101/2013) [77].

The possibilities of exposing meat surfaces to further decontaminating treatments that aim to eliminate both pathogens and spoilage flora has remained a relevant topic. Treatment with CAP has been suggested as a promising option [33,48]. Three major questions need to be addressed, i.e., 'What level of microbial reduction can be achieved?', 'Are there significant 'side-effects' on the meat matrix' and, in in-pack exposure settings: 'How do packaging material and headspace in the package influence the effect of CAP?' The latter issue has already been addressed in this paper (see Section 2.4). A number of studies have demonstrated the antibacterial activity of CAP on meat surfaces, considering the possibility of changes in the meat matrix due to lipid oxidation, protein denaturation and the state of haem pigments.

The magnitude of the reduction in contaminant bacteria by CAP technology is influenced by the nature and abundance of plasma gas species, which, in turn, is different depending on the medium in which the plasma is generated and on the way the gas species get in contact with the food matrix, e.g., direct exposure or circulating gases or liquids. Thus, it has been suggested to consider key parameters when comparing the outcomes of different studies [11]. Since our aim was not to identify the 'best treatment protocol', we refrained from presenting all experimental details in the various studies discussed in this paper. Still, the matrix in which plasma was generated (gases or liquids) and exposure conditions are given.

There are also differences in susceptibility between bacterial genera and the physical state of the meat samples (e.g., chilled or deep-frozen). For example, Choi et al. [78] contaminated samples of frozen and fresh pork with *L. monocytogenes* and *E. coli*. Samples were then exposed to CAP generated by a corona discharge plasma system (20 kV DC, 58 kHz) with a fan delivering air to the plasma source. A 120 s exposure to CAP effectuated a reduction in the numbers of *E. coli* by about 1.6 log CFU in fresh, but significantly more (about 2.7 log) in frozen pork. Reductions in *Listeria* were ca. 1.1 log CFU, with no significant difference between chilled and deep-frozen pork. Arguably, all contaminated surfaces must be exposed to CAP to achieve optimum results. Thus, Yong et al. [79] found that *E. coli* on raw chicken breast was reduced by 1.14 log CFU when one side was exposed to CAP (generated from an O_2:N_2 mix) for 5 min, but by 1.44 log CFU, when both sides of the fillet were exposed for 2.5 min each.

Depending on the distance from the plasma source to the target surface, either a wide array of reactive gas species may reach and interact with the target, or only long-lived species may arrive; see Section 2.3. The gas species will react with compounds of the cell wall (including the cell membrane) and with cytoplasmic components and nucleic acids [80]. Most compounds (with the exception of cell wall components) are not exclusive for bacteria, but also prevail in eukaryotic cells. Thus, it can be expected that the majority of the arriving reactive species will react with the more abundant food matrix rather than with the bacterial cells. Since penetration depth is low and the underlying meat parts will act as buffers, the 'average' immediate effect on a piece of meat with several cm diameter can be expected to be negligible. A somewhat different situation may exist with respect to triggering the auto-oxidation of lipids. Since (apart from water) protein and lipids are the major constituents of meat [81], numerous studies have focused on the consequences of CAP exposure to lipids [12,82] and proteins, including haem proteins [83]. In addition to chemical tests, colour measurement has proven to be able to assess sarcoplasmic denaturation (indicated by an increase in L*) [48] or oxygenation or oxidation of myo- and haemoglobin (indicated as changes in a* values). Finally, nitrate generated in water exposed to CAP can be ultimately reduced to NO, resulting in the development of NO-myoglobin, which, after heat treatment or other type of denaturation, turns into the pink NO-haemochrome [83]. Notably, in water exposed to CAP generated in air, nitrate will accumulate, which would allow the use of such plasma-activated water as a curing agent (see Section 4). RNS are typically produced in atmospheric pressure air plasmas with a high-power density at higher voltage, whereas plasma generated at lower powers are dominated by ROS [52]. In principle, voltage adjustment would allow fine-tuning if antibacterial action with the side-effect of oxidation (low power) or curing (high power)—with an antioxidative side-effect—is aimed at [84,85].

Bauer et al. [48] studied the effects of plasma treatment of packaged fresh beef. Vacuum packaged and non-packaged beef longissimus samples were treated with CAP (generated from ambient air, at different powers) over a 10-day period of vacuum, and a subsequent 3-day period of aerobic storage. It is important to realise that their approach was fundamentally different from treating foods 'in-pack' (i.e., after packaging) as described above under Section 2.4. Exposure of 'non-covered' beef samples to high-power CAP conditions resulted in increased a*, b*, Chroma and Hue values, but CAP treatment of packaged loins did not impact colour (L*, a*, b*, Chroma, Hue), lipid peroxidation, sarcoplasmic protein denaturation, nitrate/nitrite uptake or myoglobin isoform distribution [48]. Colour values measured after 3 days of aerobic storage following un-packaging (i.e., at 20 days post-mortem) were similar and all compliant with consumer acceptability standards. Exposure to CAP of the polyamide-polyethylene packaging film inoculated with *Staphylococcus aureus*, *Listeria monocytogenes* and two *Escherichia coli* strains resulted in a >2 log reduction without affecting the integrity of the packaging matrix. Results indicate that CAP can reduce microbial numbers on the surfaces of beef packages without affecting the characteristics of the packaged beef.

3.2. Meat Products

The potential use of CAP as a decontamination technology has also been studied in meat products. This included dried—'jerky style' [84–86]—products as well as dried-cured meat products such as dry ham [37,49].

In cooked/cured meat products, CAP has been used mainly as a curing agent—either by direct curing or by making use of plasma-treated water (PTW)—through the formation of reactive nitrogen species (RNS) when relying on N_2 as part of the carrier gas mixture leading to the formation of nitrite, and hence producing a curing effect [87–91].

Cooked/cured meat products are generally microbiologically safe due to the combination of heat and nitrite [92,93]. However, post-processing handling such as cutting, slicing and packaging may lead to recontamination of the surface [94,95]. Thus, it is advantageous that CAP is effective on the product surface due to the nature of plasma [48]. As regards the post-processing and pre-packaging contamination of ready-to-eat cooked meats, *Listeria monocytogenes* is the pathogen of concern [96], the more so as pH and water activity are often not low enough [97] to prevent the multiplication of *L. monocytogenes* during the shelf life of the product. In a study on typical Austrian cooked ready-to-eat meat products [98], this issue was addressed in detail, and a decision tool was developed to estimate to what extent pH or water activity of a given product need to be adjusted, albeit the authors concluded that there are limited possibilities to do so without altering sensory product characteristics and impacting consumers´ acceptance. This issue has been studied in detail in typical Austrian cured-cooked meats, see Csadek et al. [99], with respect to *Listeria* and *E. coli* and to colour changes after exposure to CAP generated from ambient air. The authors found that *E. coli* was more readily reduced than *Listeria*. It has been speculated that Gram-negative bacteria (*E. coli*) are more susceptible for CAP than are Gram-positives (*Listeria*), since the membrane lipids in Gram-negative organisms are directly exposed to CAP molecules, in particular ozone, whereas the cell wall of Gram-positive organisms would protect the cell membrane. However, experimental data are inconclusive [37,39,48]. Differences were also observed between the high and low power settings of the CAP device, but also between similar products from different manufacturers. Since the composition of the samples was—according to the information provided on the label—practically the same, it remains to be explored why different results were obtained.

4. Effects of CAP Treatment of Aquatic Foods of Animal Origin: Review of Recent Model Experiments

'Aquatic food' means food grown in or harvested from water (including all types of fish, reptiles and amphibians) and mixtures containing aquatic foods and synthetic foods, such as surimi. It is important to realise that the terms 'aquatic foods' and 'seafoods' are not necessarily considered to be synonymous. Generally, the term 'seafood' is understood to stand for ´any form of sea life regarded as foods by humans, prominently (but not exclusively) including fish and shellfish´ [100]. Shellfish include various species of molluscs (e.g., bivalve molluscs such as clams, oysters and mussels and cephalopods such as octopus and squid), crustaceans (e.g., shrimp, crabs and lobster) and echinoderms (e.g., sea cucumbers and sea urchins).

In recent years, a considerable number of scientific studies have been dedicated to analysing the microbiological (and sensory) effects of the CAP treatment of fish and 'seafoods'. In the following, we will restrict ourselves to seafoods of animal origin.

From a nutritional viewpoint, fish species may be conveniently divided into oily fish (i.e., fish in which lipids in the soft tissues and the coelom are present as oil) and whitefish [101]. Fish oil from 'oily fish' species is valued for its in vitamin and omega-3-fatty acid contents [102]. Arguably, studies on the antibacterial action of CAP in oily fish species need to consider lipid oxidation as well.

Rathod et al. [103] concluded that CAP treatment would retard bacterial spoilage (i.e., protein degradation and lipid oxidation) and, thus, CAP could be recommended as a minimal processing intervention for preserving the quality of seafood of animal origin.

4.1. Oily Fish

4.1.1. Atlantic Mackerel

Atlantic mackerel [(*Scomber scombrus*), a swarm fish caught in coastal waters, is one of the most abundant fish species in Europe, containing high levels of long-chain polyunsaturated fatty acids (PUFAs), which, consequently, are highly susceptible to oxidation, and, thus, may cause the production of off-flavours and -odours. The effects of CAP (generated from ambient air) on fillets of fresh mackerel were investigated in 2017 [104]. When fresh mackerel fillets were stored in packages and subjected to CAP using a 'large-gap' (i.e., the target is placed between the electrodes, see Section 2.4) DBD (70–80 kV for 1, 3 and 5 min), microbiological and quality characteristics were improved significantly. The spoilage bacteria (i.e., psychrotrophic aerobic flora, *Pseudomonas* and Lactic acid bacteria) of mackerel fillets were reduced by ca. 1 log cycle through CAP within 24 h of post-CAP treatment. A significant increase in lipid oxidation parameters (i.e., peroxide values, dienes) was observed in CAP-treated samples. The intensity and duration of CAP treatment of mackerel fillets also have a great impact on their microbiological condition. Nevertheless, no changes in pH and colour (with the exception of L* values) were recorded as a result of CAP treatment. These results imply that CAP could be considered as a means of reducing spoilage bacteria and thus extending the shelf life of mackerel, provided an antioxidant is added during storage to keep lipid oxidation in check. Recently, Trevisani et al. [105] confirmed that mackerel subjected to CAP using a DBD (3.8 kV at 12.7 Hz) did not, with the exception of slightly higher lightness (L*) values attributed to the oxidation of haemoproteins [106], appreciably change sensory traits when stored for 5 days at 4 °C. CAP had been generated from ambient air, and exposure was under wet conditions. Trevisani et al. [105] conclude that the treatment of fish fillets before long distance transportation (under challenging environmental conditions) may contribute to safety and extend their shelf life.

4.1.2. Tuna

The latest update on global fish consumption shows that Tuna (*Thunnus obesus*) is the world's most popular fish food and Pan et al. [107] recently investigated the effects of CAP on its quality. Tuna slices of 10 g (2.5 cm × 5 cm) were subjected to 40 kV CAP, generated from ambient air, in a ´large gap´ DBD design (i.e., the electrodes are at a distance which allows the sample to be placed between them, see also Section 2.4). No changes in sensory effects on tuna sashimi (i.e., raw, fresh, finely filleted tuna) were reported, but significantly different levels of a volatile compound ('1-hexanol', a chemical known to indicate the level of protection against flavour changes that negatively affect shelf life) indicate a superior 'freshness' of CAP-treated tuna [107].

4.1.3. Herring

Albertos et al. [108] investigated the use of a 'large gap' DBD design to generate a CAP discharge within the headspace of packaged herring (*Clupea harengus*) fillets, and its effects on microbiological and quality markers after 11 days storage at 4 °C. DBD plasma treatment conditions were 70 kV or 80 kV for 5 min treatment time. The results showed that the microbial load (total aerobic mesophilic-/total aerobic psychrotrophic bacteria, *Pseudomonas*, lactic acid bacteria and *Enterobacteriaceae*) was significantly ($p < 0.05$) lower in the treated samples, compared with untreated controls. Samples exposed to the lowest applied voltage better retained key quality factors (i.e., lower oxidation and less colour modification). DBD-treatment caused a reduction in 'trapped water' in the myofibrillar network, as assessed by the 'low-field nuclear magnetic resonance of protons' technique. The results indicate that in-package DBD plasma treatment could be employed as an effective treatment for reducing spoilage bacteria in highly perishable fish products.

4.2. Whitefish

4.2.1. Alaska Pollock

Choi et al. [109] investigated the effect of a corona discharge plasma jet (CDPJ) using ambient air on microbial reduction and the physical-chemical and sensory characteristics of dried Alaska Pollock (*Pollachius pollachius*, a cod species) shreds. All of the spoilage or pathogenic bacteria, moulds and yeasts researched were significantly reduced by 1–2.3 log units. A 3 min exposure reduced the water content from 15 to 8.6%, and a significant increase in Thiobarbituric Acid Reactive Substances (TBARS) was observed. Although individual colour coordinates (L*, a*, b*) remained unchanged, a significant increase in ΔE (up to 2.2 units) was noted. This change, although rated as 'distinct' [110], would not necessarily be perceived by consumers. Delta-E [$(\Delta E = (\Delta L^*)^2 + (\Delta a^*)^2 + (\Delta b^*)^2)^{0.5}$] [111] was used as a proxy for visually perceived colour changes. ΔE is a single number that represents the 'distance' between two colours, the idea being that a ΔE of 1 is the smallest colour difference the human eye can perceive [112]. More specifically, $\Delta E < 2$ indicates a colour change visible to an experienced observer only and $\Delta E > 5$ indicates the impression of two different colours [111].

Likewise, there was a change in texture, with CAP-exposed samples rated as 'more crispy'. Other sensory quality parameters were not affected. Both oxidation and drying may have contributed to the changes in ΔE, and crispiness was most likely affected by drying. The authors concluded that 2 min exposure time would yield an optimal condition in terms of quality traits and the inactivation of bacteria.

4.2.2. Hairtail Fish

Koddy et al. [113] studied the effect of CAP at 50 kV with different treatment times on the crude protease extract and muscle protein from Hairtail fish ((*Trichiurus Lepturus*), a saltwater species ('daiyu' in Chinese), one of the most popular fish species on the Chinese table). The results suggest that implementing CAP at 50 kV can inhibit the activity of crude protease extract to the lowest value of 0.035 units/mg protein after 240 s. Protein oxidation indices (carbonyls and sulfhydryl) varied significantly after crude protease enzyme was exposed to plasma active species. An enhancement in the colour and water-holding capacity properties was observed in hairtail samples treated with CAP. Therefore, CAP treatment could be used as an effective non-thermal method to maintain the quality of hairtail fish and extend the shelf life. Supplementary research is needed to provide knowledge concerning the effect of CAP treatment on the lipid oxidation of hairtail muscle.

4.3. Shrimps

4.3.1. Pacific White Shrimp and Greasyback Shrimp

In recent years, some CAP research has been conducted on Pacific white shrimp ((*Lito-)Penaeus vannamei*), primarily with a view to investigate if 'melanosis' (the enzymatic oxidation of shrimps leading to 'black spots', a condition associated with serious economic losses) can be counteracted by a 10 min CAP treatment (DBD configuration, plasma generated from air, 500 Hz, 40 kV; [114]). Although, immediately after CAP exposure, microbiological variables (i.e., counts of mesophilic/psychrotrophic bacteria, *Staphylococcus*) were lower in the treated group than in the control group, no significant differences were noted after 3 and 6 days of storage. However, treated samples had significantly lower pH values, higher water-binding capacity and (for a storage period of up to 9 days), a lower cooking loss than controls, and ΔE values were lower. The assessment of overall sensory quality (index composed from six factors) indicated that the shelf life of plasma-treated shrimps was >4 days longer than that of the controls (14.1 vs. 9.8 days), which would increase marketability enormously.

Recently, Elliot et al. [115] suggested to incorporate CAP in the traditional processing chain of fresh shrimp (*Penaeus vannamei*), i.e., immediately after the traditional double wash preceding refrigerated storage at 4 °C for 12 days—a minimal treatment with cold plasma (DBD, 60 kV, 69/90/120 or 150 s). This treatment results in more desirable quality outcomes. The latter are characterised by low malondialdehyde concentration, low volatile nitrogen products content and comparable proximate composition as compared with the traditional approach without CAP. Texture, pH and colour are remarkably retained at 120 and 150 s of CAP pre-treatment and protein degradation is negligible up to 90 s, as opposed to 120 and 150 s of pre-treatment [115].

Whereas most trials subjecting shrimps to CAP rely on direct treatment of the surface with the gases, and positive effects, particularly on some reported physical-chemical effects, there are few data on microbiological effects. Several years ago, a group of Chinese researchers [116] reported the superior quality of Greasyback Shrimps (*Metapenaeus ensis*) that had been stored in ice prepared from plasma-activated water (PAW), as compared with tap water. The former treatment group exhibited less microbial growth, thus extending the storage life 4 to 8 days. During storage, pH values remained < 7.7 and less off-colours and surface 'hardness' were observed. The total volatile basic nitrogen (TVBN) values remained at levels < 20 mg/100 g, i.e., at levels significantly lower ($p < 0.05$) than the controls not stored in ice prepared from CAP-treated water.

4.3.2. A Short Note on Freshwater Shrimps' Role in Spreading Antibiotics Resistance

Recently, serious concerns have been raised about veterinary drug residues in imported shrimp from Asia [117]. The Chinese freshwater grass-shrimp (*Paleomonetes sinensis*, generally used as an aquarium 'cleaner' rather than as food for human consumption) is a shrimp species that eats dead/decaying plants and animals). It has recently been indicated that this shrimp could play a significant role in spreading antibiotic resistance.

In the USA, laboratory tests have shown that most frozen freshwater shrimp samples imported from Asia (e.g., Thailand, China, India, Indonesia, Bangladesh) and Ecuador contain residues of, e.g., oxytetracyclin, nitrofurantoin, fluoroquinolone and malachite green, i.e., antibiotics that are restricted or banned under US food standards. Apparently, existing screening protocols and enforcement measures are insufficient to prevent this from happening. There are also serious doubts if, currently, adequate labelling rules are followed [117].

The use of freshwater shrimp as a human food is generally advised against [118]. Although freshwater shrimp is entirely edible, its reputation is that it is 'not worth the effort' (as there is hardly any meat to eat, and it is classified as a 'gooey' (syrupy, viscous, sticky) substance that is better left for 'monster fishes' to eat) [118,119].

Arguably, alternative antibacterial interventions could help to reduce the use of antimicrobials in shrimp production. This might include the application of CAP to sanitise the water in the breeding/holding pens.

4.4. Squid

Choi et al. [120,121] investigated the effect of a CDPJ, with plasma generated from air, on microbial reduction and the physical-chemical and sensory characteristics of semi-dried and dried squid *(Todarodes pacificus)* shreds. All the spoilage or pathogenic bacteria, moulds and yeasts researched were significantly reduced. In semi-dried squid shreds, exposure times > 3 min resulted in a change of flavour and significant increase in TBARS and L*and b* values, with ΔE up to 6.6 after 10 min exposure. The levels of TVBN and trimethylamine were not affected. The overall sensory acceptance was not impaired. Likewise, a 3 min. exposure of dried shreds resulted in increases in L* (ΔE maximum 2.5) and TBARS, indicative for oxidative changes. Again, the overall acceptance was not impaired. The authors concluded that a 2 min. exposure would warrant a sufficient reduction in bacteria without compromising product characteristics.

4.5. Molluscs (Mussels and Oysters)

Mussels and oysters have been implicated in foodborne poisoning, either by contaminant bacteria or viruses [122–131]. Unlike fish, this type of seafood is often traded alive and some species are even consumed raw (e.g., oysters). Thus, no traditional food processing techniques with antimicrobial or antiviral effect (e.g., heat treatment; [132]) can be applied. Choi et al. [133] exposed an oyster *(Crassostrea gigas)* slurry to plasma generated by a jet-type CAP device for 30 min and could demonstrate a reduction (>1 log) in human norovirus without compromising the colour and pH of the oyster [133]. Although the authors selected a highly relevant viral pathogen, biosafety concerns make the use of surrogate viruses more feasible, which is an issue discussed in more detail in the following section.

Csadek et al. [134] studied the inactivation of surrogate viruses on an oyster slurry, but employed a DBD instead of a jet-type plasma generator. Plasma was generated from ambient air. The authors observed a higher antiviral effect towards a double-stranded DNA virus (Equid Alphaherpesvirus 1, EHV-1; 2.3–2.8 log) than against a single-stranded RNA virus (Bovine Coronavirus, BCoV; 1.4–1.0 log) in Dulbecco's Modified Eagle's Medium (DMEM). Plasma generated at low power (ozone dominated) had a higher virus inactivation effect than was the case at high power (nitrogen dioxide dominated). Plasma exposure caused a decline of glucose contents in DMEM, which might have been caused by a reaction of carbohydrates with amino acids (Maillard reaction). The exposure of an oyster slurry to CAP did not result in any change of pH and colour, corroborating the findings of Choi et al. [133]. However, the oyster matrix resembles a buffered medium, and, thus, pH changes were not really expected. Regarding colour, different mechanisms may apply than in muscle foods, since the electron acceptor is haemocyanin instead of haemoglobin [135]. Thus, the absence of colour changes is maybe less suitable for assessing CAP-induced changes. The authors observed an accumulation of nitrogen in the oyster slurry due to CAP exposure, which could only in part be explained by higher nitrate and nitrite contents. The (entirely plausible) assumption that nitrate from the plasma reacted with the compounds of the oyster matrix remains to be substantiated.

Both Choi et al. [133] and Csadek et al. [134] studied oyster tissue and not live oyster, and it was a contamination scenario, not an infection. However, the antiviral effects of CAP exposure can also be expected for live oysters, whereas the significance of the observed nitrogen accumulation in oyster slurry for live animals is not entirely clear. It can be assumed that the exposure of the cells to nitrate is a stressor for cell homoeostasis. Studies on exposure to ACP of oyster cell monolayers with or without viral contaminants might allow the obtainment of an estimate if the extent and the benefit of antiviral action are counteracted by oyster cell damage.

Admittedly, the direct exposure of oysters to CAP is not very practical, compared to sanitising the (salt) water in the holding tanks. Although the concept of applying CAP to reduce viral contamination in water is promising [136], it will largely depend on the composition of the plasma species. When NO_x are generated, they will be dissolved in water and lower the pH [134]. This pH drop is an additional stressor for the oysters.

A CAP-based system for sanitising water in holding tanks for live oysters could allow the water to circulate from the holding tank via a compartment for CAP exposure to a station adjusting the pH (either by adding alkali or by denitrification) and again back to the holding tank. Denitrification is usually a biological process, but there are also chemical systems available [137,138]. Based on these references, the authors of this article suggest studies on a circulating system (Figure 2), in which CAP treatment would prevent re-infections or new infections and eventually reduce the pathogen load of already-infected oysters and mussels.

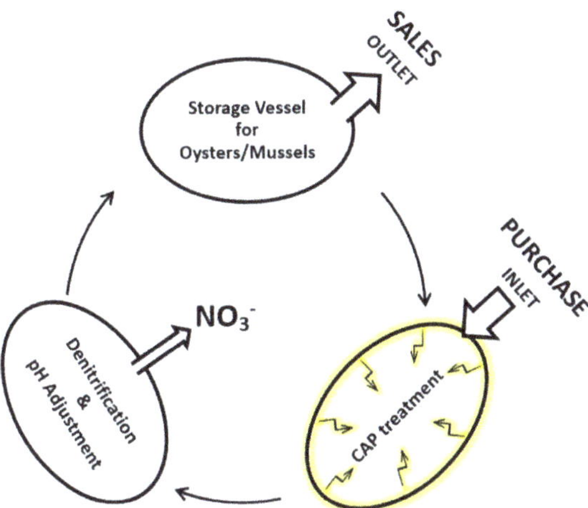

Figure 2. Concept of a circulation system for denitrification and cold atmospheric plasma treatment of stored molluscs (source: authors).

5. A Note on Food-Contaminating Viruses; Why Surrogate Viruses Are Used for Studying the Virucidal Effect of Cold Atmospheric Plasma

While food-transmitted pathogenic viruses are of concern for food safety, food security is threatened by viruses causing disease in production animals, e.g., Newcastle disease virus (NDV) and porcine reproductive and respiratory syndrome virus (PRRSV), sometimes with zoonotic potential (highly pathogenic avian influenza virus, HPAIV). It can be assumed that CAP would effectuate virus inactivation on the surface of biological materials the same way as it acts on food surfaces.

The most recent authoritative review on cold plasma effects on viruses stems from Filipić et al. [139], who acknowledge that, so far, insufficient data are available that would allow selecting the correct treatment options. Unfortunately, the literature on plasma effects against foodborne viruses is rather scarce and relevant parameters (including a treatment duration that would allow optimal interaction with contaminated material) have not yet been studied sufficiently. The biosecurity problems associated with studying pathogenic viruses have been mentioned above and might also be a reason why so few studies have been conducted on viruses. Thus, it is worth addressing the use of surrogate viruses in more detail.

The literature on foodborne viruses focuses on noroviruses, enteric adenoviruses and hepatitis A virus, which are the leading causes of acute gastroenteritis, the second most infectious disease worldwide, responsible for high levels of hospitalisation and mortality [139,140]. The infectivity of these agents explains why research on the effects of CAP on these viruses is rather limited and usually relies on the use of 'surrogate' viruses that are relatively easy to culture/propagate and are safe to work with [141]. By the same token, it has been proposed to use bacteriophages as surrogates or indicators for the presence of enteric infectious viruses in wastewater [140].

Pathogenic viruses have always posed a great risk to humans and animals alike, and pandemics can quickly wreak havoc on the livelihood of millions of people, as currently seen due to SARS-CoV-2. Apart from human-to-human transmission, contaminated surfaces are another source of viral infections, and the ingestion of dangerous pathogens on food surfaces frequently results in significant morbidities and mortalities [142]. Apart from considerable evidence of the microbicidal effect of atmospheric-pressure cold plasma on bacteria and fungi, several studies indicate that CAP treatment also induces virus inactivation and is therefore considered a promising tool to combat human pathogenic viruses [143].

An evaluation of the efficacy of inactivation methods for pathogenic human viruses may imply significant health hazards, which may require specialised buildings and equipment, such as biosafety level 3 (BSL-3) laboratories, which are not generally available. Furthermore, such resources are not only expensive, but also depend on specially trained personnel; thus, extensive testing of such viruses is often not economically justified. As a consequence, surrogate viruses that can be handled in BSL-1 or BSL-2 facilities are usually favoured for testing and optimising plasma technology [142]. In addition, most food-related viruses cannot be propagated in cell culture (e.g., hepatitis viruses and caliciviruses), and/or do not cause cytopathic effects, which would be required to directly assess a viral reduction caused by, e.g., cold plasma in cell culture. Thus, ways to inactivate infectious viruses are most commonly investigated by the use of cultivable surrogate viruses, or by the detection of viral nucleic acids by real-time polymerase chain reaction (RT-)PCR, which, however, does not provide information about infectivity [141], unless free DNA or RNA molecules released from destroyed viral particles are removed prior to PCR by DNase or RNase treatment, respectively [144].

Surrogate viruses should be closely related or have very similar traits to mimic the pathogen of interest as well as possible, including biological, biophysical and biochemical characteristics [141]. For instance, noroviruses are single-stranded, non-enveloped RNA viruses belonging to the family *Caliciviridae*. Thus, viruses from the same family are the best surrogate choice, such as feline calicivirus, which is cultivable (but does not cause a cytopathic effect) and has been used as a surrogate for norovirus in several studies since the 1970s [141]. When murine norovirus, which is even more closely related to food-contaminating noroviruses, was discovered in 2003, researchers turned to this virus, as it is also resistant to low pH values. However, when it was found that the murine norovirus is highly sensitive to alcohols, the search expanded to other cultivable caliciviruses [141], which shows that, aside from genetic and morphologic similarity, there are several other features of a surrogate that have to be considered.

6. Conclusions and Future Perspectives

Cold plasma technology is a cornerstone of modern society given its ubiquitous use in materials manufacturing applications (e.g., semiconductor fabrication and polymer treatment). Recently, its action on biological material has been studied extensively, with three major fields of application, i.e., medical treatments (e.g., cell regeneration and wound healing), non-thermal surface disinfection and food science. Food science applications are more complex, since they aim at inactivating contaminant bacteria and viruses on food surfaces, but need to take into account that plasma species will also react with the food matrix.

In this contribution, we have shown that cold atmospheric-pressure plasma can be generated with low energy consumption simply using air as the precursor for the generation of reactive chemical species. A large number of studies have shown that not only the gas composition, but also the mode of plasma generation and the spatial-temporal distance from the plasma source to the target govern which chemical species will reach the surface of the target, which in turn affects the ability of the CAP to inactivate microorganisms. In this contribution, we have shown that, on the surfaces of meat and seafood treated with indirect CAP systems, ozone and nitrate are the major reactive gas species, and their inactivating action on bacteria and viruses is known. By the same token, their effects on the food matrix are well described (i.e., lipid oxidation, the action of NO on haem pigments and eventually the oxidation of (haem) proteins, oyster tissue) but are not plasma-specific.

In summary, this contribution has shown that atmospheric-pressure cold plasma is capable of effectively reducing the load of bacteria and viruses on the surfaces of fresh as well as processed meat and seafood. Available data suggest, in most cases, no or only negligible effects of plasma species on food matrices in terms of chemical composition, colour and physical-chemical properties; nitrogen accumulation on oyster tissue being an exception.

Going forward, the insight gained from an expert opinion published in 2012 must be considered, where the diversity of plasma-generation devices, conditions and treatment protocols was identified as a drawback in conducting a detailed risk assessment with respect to the applicability of CAP in sanitising fresh meat and thus could not clearly answer if such treated food items would be 'novel foods' according to EU legislation [55]. Given the multitude of data generated in the last 10 years, the application of atmospheric-pressure cold plasma on meat and fish (products) deserves a formal re-assessment in this respect.

Author Contributions: Conceptualisation, F.J.M.S., P.P., J.W. and N.N.; Methodology, J.W. and A.B.; Formal Analysis, I.C.; Investigation, I.C., A.B. and P.P.; Resources, K.S.; Data Curation, I.C. and P.W.; Writing—Original Draft Preparation, P.P., F.J.M.S., J.W., N.N., P.W. and K.H.B.; Writing—Review and Editing, F.J.M.S., I.C., E.M., P.P. and K.S.; Supervision, F.J.M.S. All authors have read and agreed to the published version of the manuscript.

Funding: This research received no external funding.

Data Availability Statement: Not applicable.

Conflicts of Interest: The authors declare no conflict of interest.

References

1. Lipinski, B. Why do animal-based food loss and waste matter? *Anim. Front.* **2020**, *10*, 48–52. [CrossRef] [PubMed]
2. FAO. Food and Agriculture Organization of the United Nations. Food Wastage Footprint and Climate Change, Rome Italy, 2015. Available online: https://www.fao.org/3/bb144e/bb144e.pdf (accessed on 19 August 2022).
3. Warner, R.D.; Greenwood, P.L.; Pethick, D.W.; Ferguson, D.M. Genetic and environmental effects on meat quality. *Meat Sci.* **2010**, *86*, 171–183. [CrossRef] [PubMed]
4. Jan, A.T.; Lee, E.J.; Ahmad, S.; Choi, I. Meeting the meat: Delineating the molecular machinery of muscle development. *J. Anim. Sci. Technol.* **2016**, *58*, 18. [CrossRef] [PubMed]
5. Mohammadabadi, M.; Bordbar, F.; Jensen, J.; Du, M.; Guo, W. Key genes regulating skeletal muscle development and growth in farm animals. *Animals* **2021**, *11*, 835. [CrossRef] [PubMed]
6. Mohammadi, A.; Nassiry, M.R.; Mosafer, J.; Mohammadabadi, M.R. Distribution of BoLA-DRB3 allelic frequencies and identification of a new allele in the Iranian cattle breed Sistani (*Bos indicus*). *Russ. J. Genet.* **2009**, *45*, 198–202. [CrossRef]
7. Mossel, D.A.A. Adequate protection of the public against food-transmitted diseases of microbial aetiology. Achievements and challenges half a century after the introduction of the Prescott-Meyer-Wilson strategy of active intervention. *Int. J. Food Microbiol.* **1989**, *9*, 271–294. [CrossRef]
8. FAO. Food and Agriculture Organization of the United Nations. Rome Declaration on World Food Security. World Food Summit, 13–17 November 1996, Rome, Italy. 1996. Available online: https://www.fao.org/3/w3613e/w3613e00.htm (accessed on 2 September 2022).
9. UN World Commission on Environment and Development. Report of the World Commission on Environment and Development: Our Common Future. 1987. Available online: https://sustainabledevelopment.un.org/content/documents/5987our-common-future.pdf (accessed on 21 August 2022).
10. European Commission. Farm to Fork Strategy; for a Fair, Healthy and Environmentally-Friendly System. Available online: https://food.ec.europa.eu (accessed on 22 September 2020).

11. Warne, G.R.; Williams, P.M.; Pho, H.Q.; Tran, N.N.; Hessel, V.; Fisk, I.D. Impact of cold plasma on the biomolecules and organoleptic properties of foods: A review. *Food Sci.* **2021**, *86*, 3762–3777. [CrossRef]

12. Jadhav, H.B.; Annapure, U. Consequences of non-thermal cold plasma treatment on meat and dairy lipids—A review. *Future Foods* **2021**, *4*, 100095. [CrossRef]

13. Olatunde, O.O.; Singh, A.; Shiekh, K.A.; Nuthong, P.; Benjakul, S. Effect of High Voltage Cold Plasma on Oxidation, Physiochemical, and Gelling Properties of Myofibrillar Protein Isolate from Asian Sea Bass (*Lates calcarifer*). *Foods* **2021**, *10*, 326. [CrossRef]

14. Domínguez, R.; Pateiro, M.; Munekata, P.E.S.; Zhang, W.; Garcia-Oliveira, P.; Carpena, M.; Prieto, M.A.; Bohrer, B.; Lorenzo, J.M. Protein Oxidation in Muscle Foods: A Comprehensive Review. *Antioxidants* **2022**, *11*, 60. [CrossRef]

15. Nawaz, A.; Irshad, S.; Ali Khan, I.; Khalifa, I.; Walayat, N.; Aadil, R.M.; Kumar, M.; Wang, M.; Chen, F.; Cheng, K.-W.; et al. Protein oxidation in muscle-based products: Effects on physicochemical properties, quality concerns, and challenges to food industry. *Food Res. Int.* **2022**, *157*, 111322. [CrossRef] [PubMed]

16. Raizer, Y.P. *Gas Discharge Physics*; Springer: Berlin/Heidelberg, Germany; New York, NY, USA, 1997.

17. Lieberman, M.A.; Lichtenberg, A.J. *Principles of Plasma Discharges and Materials Processing*, 2nd ed.; Wiley & Sons: Hoboken, NJ, USA, 2005.

18. Bourke, P.; Ziuzina, D.; Han, L.; Cullen, P.J.; Gilmore, B.F. Microbiological Interactions with Cold Plasma. *J. Appl. Microbiol.* **2017**, *123*, 308–324. [CrossRef] [PubMed]

19. Martines, E. 2020. Interaction of Cold Atmospheric Plasmas with Cell Membranes in Plasma Medicine Studies. *Jpn. J. Appl. Phys.* **2020**, *59*, SA0803. [CrossRef]

20. Mendis, D.A.; Rosenberg, M.; Azam, F. A Note on the Possible Electrostatic Disruption of Bacteria. *IEEE Trans. Plasma Sci.* **2000**, *28*, 1304–1306. [CrossRef]

21. Lackmann, J.-W.; Schneider, S.; Edengeiser, E.; Jarzina, F.; Brinckmann, S.; Steinborn, E.; Havenith, M.; Benedikt, J.; Bandow, J.E. Photons and Particles Emitted from Cold Atmospheric-Pressure Plasma Inactivate Bacteria and Biomolecules Independently and Synergistically. *J. R. Soc. Interface.* **2013**, *10*, 20130591. [CrossRef] [PubMed]

22. Bruggeman, P.J.; Iza, F.; Brandenburg, R. Foundations of Atmospheric Pressure Non-Equilibrium Plasmas. *Plasma Sources Sci. Technol.* **2017**, *26*, 123002. [CrossRef]

23. Brandenburg, R. Dielectric Barrier Discharges: Progress on Plasma Sources and on the Understanding of Regimes and Single Filaments. *Plasma Sources Sci. Technol.* **2017**, *26*, 053001. [CrossRef]

24. Laroussi, M. Nonthermal Decontamination of Biological Media by Atmospheric-Pressure Plasmas: Review, Analysis, and Prospects. *IEEE Trans. Plasma Sci.* **2002**, *30*, 1409–1415. [CrossRef]

25. Laroussi, M.; Bekeschus, S.; Bogaerts, A.; Fridman, A.; Lu, X.; Miller, V.; Laux, C.; Walsh, J.; Jiang, C.; Liu, D.; et al. Low-Temperature Plasma for Biology, Hygiene, and Medicine: Perspective and Roadmap. *IEEE Trans. Radiat. Plasma Med. Sci.* **2022**, *6*, 127–157. [CrossRef]

26. Brun, P.; Vono, M.; Venier, P.; Tarricone, E.; Deligianni, V.; Martines, E.; Zuin, M.; Spagnolo, S.; Cavazzana, R.; Cardin, R.; et al. Disinfection of Ocular Cells and Tissues by Atmospheric-Pressure Cold Plasma. *PLoS ONE* **2012**, *7*, e33245. [CrossRef]

27. Diwan, R.; Debta, F.M.; Deoghare, A.; Ghom, S.; Khandelwhal, A.; Sikdar, S.; Ghom, A. Plasma therapy; An overview. *J. Indian Acad. Oral Med. Radiol.* **2011**, *23*, 120–123. [CrossRef]

28. Lackmann, J.-W.; Bandow, J.E. Inactivation of microbes and macromolecules by atmospheric-pressure plasma jets. *Appl. Microbiol. Biotechnol.* **2014**, *98*, 6205–6213. [CrossRef] [PubMed]

29. Zimmermann, J. Cold Atmospheric Plasma in Medicine—From Basic Research to Application, Max Planck Institute of Pathology, Technical University Munich, Habilitation Treatise; 2013. Available online: http://mediatum.ub.tum.de/?id=1219297 (accessed on 2 September 2022).

30. Bekeschus, S.; von Woedtke, T.; Emmert, S.; Schmidt, A. Medical Gas Plasma-Stimulated Wound Healing: Evidence and Mechanisms. *Redox Biol.* **2021**, *46*, 102116. [CrossRef] [PubMed]

31. Melotti, L.; Martinello, T.; Perazzi, A.; Martines, E.; Zuin, M.; Modenese, D.; Cordaro, L.; Ferro, S.; Maccatrozzo, L.; Iacopetti, I.; et al. Could cold plasma act synergistically with allogeneic mesenchymal stem cells to improve wound skin regeneration in a large size animal model? *Res. Vet. Sci.* **2021**, *136*, 97–110. [CrossRef]

32. Scholtz, V.; Pazlarova, J.; Souskova, H.; Julak, J. Nonthermal Plasma—A Tool for Decontamination and Disinfection. *Biotechnol. Adv.* **2015**, *33*, 1108–1119. [CrossRef]

33. Misra, N.N.; Schlüter, O.; Cullen, P.J. Chapter 1—Plasma in Food and Agriculture. In *Cold Plasma in Food and Agriculture*; Misra, N.N., Schlüter, O., Cullen, P.J., Eds.; Academic Press: San Diego, CA, USA, 2016; pp. 1–16.

34. Csadek, I.; Paulsen, P.; Bak, K.H.; Smulders, F.J.M. Application of atmospheric pressure cold plasma (ACP) on meat and meat products. Part 1. Effects on bacterial surface flora. *Fleischwirtschaft* **2021**, *101*, 96–104.

35. Shimizu, T.; Sakiyama, Y.; Graves, D.; Zimmerman, J.; Morfill, G. The dynamics of ozone generation and mode transition in air surface micro-discharge plasma at atmospheric pressure. *New J. Phys.* **2012**, *14*, 103028. [CrossRef]

36. Reuter, S.; Winter, J.; Iseni, S.; Schmidt-Bleker, A.; Dünnbier, M.; Masur, K.; Wende, K.; Weltmann, K.-D. The Influence of Feed Gas Humidity Versus Ambient Humidity on Atmospheric Pressure Plasma Jet-Effluent Chemistry and Skin Cell Viability. *IEEE Trans. Plasma Sci.* **2015**, *43*, 3185–3192. [CrossRef]

37. Lis, K.A.; Boulaaba, A.; Binder, S.; Li, Y.; Kehrenberg, C.; Zimmermann, J.L.; Klein, G.; Ahlfeld, B. Inactivation of *Salmonella* Typhimurium and *Listeria monocytogenes* on Ham with Nonthermal Atmospheric Pressure Plasma. *PLoS ONE* **2018**, *13*, e0197773. [CrossRef]

38. Katsigiannis, A.S.; Bayliss, D.L.; Walsh, J.L. Cold Plasma for the Disinfection of Industrial Food-contact Surfaces: An Overview of Current Status and Opportunities. *Comp. Rev. Food Sci. Food Saf.* **2022**, *21*, 1026–1124. [CrossRef]

39. López, M.; Calvo, T.; Prieto, M.; Múgica-Vidal, R.; Muro-Fraguas, I.; Alba-Elías, F.; Alvarez-Ordóñez, A. A Review on Non-Thermal Atmospheric Plasma for Food Preservation: Mode of Action, Determinants of Effectiveness, and Applications. *Front. Microbiol.* **2019**, *10*, 622. [CrossRef] [PubMed]

40. Hasan, M.I.; Walsh, J.L. Influence of Gas Flow Velocity on the Transport of Chemical Species in an Atmospheric Pressure Air Plasma Discharge. *Appl. Phys. Lett.* **2017**, *110*, 134102. [CrossRef]

41. Misra, N.N.; Yepez, X.; Xu, L.; Keener, K. In-Package Cold Plasma Technologies. *J. Food Eng.* **2019**, *244*, 21–31. [CrossRef]

42. Ziuzina, D.; Patil, S.; Cullen, P.J.; Keener, K.M.; Bourke, P. Atmospheric Cold Plasma Inactivation of *Escherichia coli* in Liquid Media inside a Sealed Package. *J. Appl. Microbiol.* **2013**, *114*, 778–787. [CrossRef]

43. Huang, M.; Wang, J.; Zhuang, H.; Yan, W.; Zhao, J.; Zhang, J. Effect of In-Package High Voltage Dielectric Barrier Discharge on Microbiological, Color and Oxidation Properties of Pork in Modified Atmosphere Packaging during Storage. *Meat Sci.* **2019**, *149*, 107–113. [CrossRef] [PubMed]

44. Patange, A.; Boehm, D.; Bueno-Ferrer, C.; Cullen, P.J.; Bourke, P. Controlling *Brochothrix thermosphacta* as a Spoilage Risk Using In-Package Atmospheric Cold Plasma. *Food Microbiol.* **2017**, *66*, 48–54. [CrossRef]

45. Wang, J.; Zhuang, H.; Hinton, A.; Zhang, J. Influence of in-package cold plasma treatment on microbiological shelf life and appearance of fresh chicken breast fillets. *Food Microbiol.* **2016**, *60*, 142–146. [CrossRef]

46. Levif, P.; Séguin, J.; Moisan, M.; Soum-Glaude, A.; Barbeau, J. Packaging materials for plasma sterilization with the flowing afterglow of an N_2—O_2 discharge: Damage assessment and inactivation efficiency of enclosed bacterial spores. *J. Phys. D Appl. Phys.* **2011**, *44*, 405201. [CrossRef]

47. Brayfield, R.S.; Jassem, A.; Lauria, M.V.; Fairbanks, A.J.; Keener, K.M.; Garner, A.L. Characterization of High Voltage Cold Atmospheric Plasma Generation in Sealed Packages as a Function of Container Material and Fill Gas. *Plasma Chem. Plasma Process* **2018**, *38*, 379–395. [CrossRef]

48. Bauer, A.; Ni, Y.; Bauer, S.; Paulsen, P.; Modic, M.; Walsh, J.L.; Smulders, F.J.M. The effects of atmospheric pressure cold plasma treatment on microbiological, physical-chemical and sensory characteristics of vacuum packaged beef loin. *Meat Sci.* **2017**, *128*, 77–87. [CrossRef]

49. Rød, S.K.; Hansen, F.; Leipold, F.; Knøchel, S. Cold atmospheric pressure plasma treatment of ready-to-eat meat: Inactivation of *Listeria innocua* and changes in product quality. *Food Microbiol.* **2012**, *30*, 233–238. [CrossRef] [PubMed]

50. Roh, S.; Oh, Y.; Le, S.; Kann, J.; Min, S. Inactivation of *Escherichia coli* O157:H7, *Salmonella*, *Listeria monocytogenes* and Tulane virus in processed chicken breast via atmospheric in-package cold plasma treatment. *LWT Food Sci. Technol.* **2020**, *127*, 109429. [CrossRef]

51. Jayasena, D.; Kim, H.; Yong, H.; Park, S.; Kim, K.; Choe, W.; Jo, C. Flexible thin-layer dielectric barrier discharge plasma treatment of pork butt and beef loin: Effects on pathogen inactivation and meat-quality attributes. *Food Microbiol.* **2015**, *46*, 51–57. [CrossRef] [PubMed]

52. Harrison, W.A.; Griffith, C.J.; Tennant, D.; Peters, A.C. Incidence of *Campylobacter* and *Salmonella* isolated from retail chicken and associated packaging in South Wales. *Lett. Appl. Microbiol.* **2001**, *33*, 450–454. [CrossRef] [PubMed]

53. Regulation (EU) 2015/2283 of the European Parliament and of the Council of 25 November 2015 on novel foods, amending Regulation (EU) No 1169/2011 of the European Parliament and of the Council and repealing Regulation (EC) No 258/97 of the European Parliament and of the Council and Commission Regulation (EC) No 1852/2001. *Off. J. Eur. Union* **2015**, *L 327*, 1–22. Available online: https://eur-lex.europa.eu/legal-content/EN/TXT/?uri=celex%3A32015R2283 (accessed on 13 November 2022).

54. Cooperation in Science and Technology (COST 013/20) Decision 24 March; Memorandum of Understanding for the Implementation of the COST Action "Plasma Application for Smart and Sustainable Agriculture (P/Agri) CA 19110", Brussels, 16 pp., COST, Avenue Boulevard 21, Brussels, Belgium, 2020. Available online: https://plagri.eu/wp-content/uploads/2021/02/CA19110-e.pdf (accessed on 2 September 2022).

55. 55. German Research Community (Deutsche Forschungsgemeinschaft). Stellungnahme Zum Einsatz von Plasmaverfahren zur Behandlung von Lebensmitteln [Opinion on the Application of Plasma-technology for Treatment of Foods], 2012. Available online: www.dfg.de/sklm (accessed on 19 August 2022).

56. Ekezie, C.; Suan, D.; Cheng, J. Review on recent advances in cold plasma technology for the food industry: Current applications and future trends. *Trends Food Sci. Technol.* **2017**, *69*, 46–58. [CrossRef]

57. Nørrung, B.; Buncic, S. Microbial safety of meat in the European Union. *Meat Sci.* **2008**, *78*, 14–24. [CrossRef]

58. Das, A.K.; Nanda, P.K.; Das, A.; Biswas, S. Hazards and Safety Issues of Meat and Meat Products. In *Food Safety and Human Health*; Singh, R.L., Mondal, S., Eds.; Academic Press: Cambridge, MA, USA, 2019; pp. 145–168.

59. Regulation (EC) No 178/2002 of the European Parliament and of the Council of 28 January 2002 laying down the general principles and requirements of food law, establishing the European Food Safety Authority and laying down procedures in matters of food safety. *Off. J. Eur. Union* **2002**, *L 31*, 1–24. Available online: https://eur-lex.europa.eu/legal-content/EN/ALL/?uri=celex%3A32002R0178 (accessed on 13 November 2022).
60. Pearce, R.A.; Bolton, D.J.; Sheridan, J.J.; McDowell, D.A.; Blair, I.S.; Harrington, D. Studies to determine the critical control points in pork slaughter hazard analysis and critical control point systems. *Int. J. Microbiol.* **2004**, *90*, 331–339. [CrossRef]
61. Antic, D.; Houf, K.; Michalopoulou, E.; Blagojevic, B. Beef abattoir interventions in a risk-based meat safety assurance system. *Meat Sci.* **2021**, *182*, 108622. [CrossRef]
62. Paulsen, P.; Smulders, F.J.M. Combining natural antimicrobial systems with other preservation techniques: The case of meat. In *Food Preservation Techniques*; Zeuthen, P., Bogh-Sorensen, L., Eds.; Woodhead Publishing: Abington, UK, 2003; pp. 71–89.
63. Aymerich, T.; Picouet, T.A.; Monfort, J.M. Decontamination technologies for meat products. *Meat Sci.* **2008**, *78*, 114–129. [CrossRef] [PubMed]
64. Loretz, M.; Stephan, R.; Zweifel, C. Antibacterial activity of decontamination treatments for cattle hides and beef carcasses. *Food Control* **2011**, *22*, 347–359. [CrossRef]
65. Buncic, S.; Sofos, J. Interventions to Control *Salmonella* Contamination during Poultry, Cattle and Pig Slaughter. *Food Res. Int.* **2012**, *45*, 641–655. [CrossRef]
66. Buncic, S.; Nychas, G.-J.; Lee, M.R.F.; Koutsoumanis, K.; Hébraud, M.; Desvaux, M.; Chorianopoulos, N.; Bolton, D.; Blagojevic, B.; Antic, D. Microbial pathogen control in the beef chain: Recent research advances. *Meat Sci.* **2014**, *97*, 288–297. [CrossRef]
67. Albert, T.; Braun, P.G.; Saffaf, J.; Wiacek, C. Physical Methods for the Decontamination of Meat Surfaces. *Curr. Clin. Microbiol. Rep.* **2021**, *8*, 9–20. [CrossRef]
68. Zdolec, N.; Kotsiri, A.; Houf, K.; Alvarez-Ordóñez, A.; Blagojevic, B.; Karabasil, N.; Salines, M.; Antic, D. Systematic Review and Meta-Analysis of the Efficacy of Interventions Applied during Primary Processing to Reduce Microbial Contamination on Pig Carcasses. *Foods* **2022**, *11*, 2110. [CrossRef]
69. Gill, C.O.; Bedard, D.; Jones, T. The decontaminating performance of a commercial apparatus for pasteurizing polished pig carcasses. *Food Microbiol.* **1997**, *14*, 71–79. [CrossRef]
70. Gill, C.O.; Jones, T.; Badoni, M. The effects of hot water pasteurizing treatments on the microbiological conditions and appearances of pig and sheep carcasses. *Food Res. Int.* **1998**, *31*, 273–278. [CrossRef]
71. Alban, L.; Sørensen, L.L. Hot-water decontamination is an effective way of reducing risk of *Salmonella* in pork. *Fleischwirtschaft* **2010**, *90*, 109–113.
72. Paulsen, P.; Smulders, F.J.M. Reduction of the microbial contamination of carcasses and meat cuts with particular reference to the application of organic acids. In *Food Safety and Veterinary Public Health, Volume 2: Safety Assurance during Food Processing*; Smulders, F.J.M., Collins, J.D., Eds.; Wageningen Academic Publishers: Wageningen, The Netherlands, 2004; pp. 177–199.
73. Smulders, F.J.M.; Wellm, G.; Hiesberger, J.; Rohrbacher, I.; Bauer, A.; Paulsen, P. Microbiological and sensory effects of the combined application of hot-cold organic acid sprays and steam condensation at subatmospheric pressure for decontamination of inoculated pig tissue surfaces. *J. Food Prot.* **2011**, *74*, 1338–1344. [CrossRef]
74. Smulders, F.J.M.; Wellm, G.; Hiesberger, J.; Bauer, A.; Paulsen, P. The potential of the combined application of hot water sprays and steam condensation at subatmospheric pressure for decontaminating inoculated pig skin and muscle surfaces. *Food Control* **2012**, *24*, 154–159. [CrossRef]
75. EFSA Panel on Biological Hazards (BIOHAZ). Scientific Opinion on the evaluation of the safety and efficacy of lactic acid for the removal of microbial surface contamination of beef carcasses, cuts and trimmings. *EFSA J.* **2011**, *9*, 2317.
76. EFSA Panel on Food Contact Materials, Enzymes and Processing Aids (CEP); Silano, V.; Barat Baviera, J.M.; Bolognesi, C.; Brüschweiler, B.J.; Chesson, A.; Cocconcelli, P.S.; Crebelli, R.; Gott, D.M.; Grob, K.; et al. Evaluation of the safety and efficacy of the organic acids lactic and acetic acids to reduce microbiological surface contamination on pork carcasses and pork cuts. *EFSA J.* **2018**, *16*, e05482.
77. Commission Regulation (EU) No 101/2013 of 4 February 2013 concerning the use of lactic acid to reduce microbiological surface contamination on bovine carcasses. *Off. J. Eur. Union* **2013**, *L34*, 1–3. Available online: https://eur-lex.europa.eu/LexUriServ/LexUriServ.do?uri=OJ:L:2013:034:0001:0003:EN:PDF (accessed on 13 November 2022).
78. Choi, S.; Puligundla, P.; Mok, C. Corona discharge plasma jet for inactivation of *Escherichia coli* O157:H157 and *Listeria monocytogenes* on inoculated pork and its impact on meat quality attributes. *Ann. Microbiol.* **2016**, *66*, 685–694. [CrossRef]
79. Yong, J.; Kim, H.; Park, S.; Choe, W.; Oh, M.; Jo, C. Evaluation of the Treatment of Both Sides of Raw Chicken Breasts with an Atmospheric Pressure Plasma Jet for the Inactivation of *Escherichia Coli*. *Foodborne Pathog. Dis.* **2014**, *11*, 652–657. [CrossRef]
80. Misra, N.N.; Jo, C. Applications of cold plasma technology for microbiological safety in meat industry. *Trends Food Sci. Tech.* **2017**, *64*, 74–86. [CrossRef]
81. Lawrie, R.A.; Ledward, D.A. *Lawrie's Meat Science*, 7th ed.; Woodhead: Cambridge, UK, 2006.
82. Pérez-Andrés, J.M.; Cropotova, J.; Harrison, S.M.; Brunton, N.P.; Cullen, P.J.; Rustad, T.; Tiwari, B.K. Effect of Cold Plasma on Meat Cholesterol and Lipid Oxidation. *Foods* **2020**, *9*, 1786. [CrossRef]
83. Bak, K.H.; Csadek, I.; Paulsen, P.; Smulders, F.J.M. Application of atmospheric pressure cold plasma (ACP) on meat and meat products. Part 2. Effects on the sensory quality with special focus on meat colour and lipid oxidation. *Fleischwirtschaft* **2021**, *101*, 100–105.

84. Kim, J.-S.; Lee, E.-J.; Choi, E.H.; Kim, Y.-J. Inactivation of *Staphylococcus aureus* on the beef jerky by radio-frequency atmospheric pressure plasma discharge treatment. *Innov. Food Sci. Emerg. Technol.* **2014**, *22*, 124–130. [CrossRef]

85. Yong, H.I.; Lee, H.; Park, S.; Park, J.; Choe, W.; Jung, S.; Jo, C. Flexible thin-layer plasma inactivation of bacteria and mold survival in beef jerky packaging and its effects on the meat's physicochemical properties. *Meat Sci.* **2017**, *123*, 151–156. [CrossRef] [PubMed]

86. Yong, H.I.; Lee, S.H.; Kim, S.Y.; Park, S.; Park, J.; Choe, W.; Jo, C. Color development, physiochemical properties, and microbiological safety of pork jerky processed with atmospheric pressure plasma. *Innov. Food Sci. Emerg. Technol.* **2019**, *53*, 78–84. [CrossRef]

87. Jung, S.; Kim, H.J.; Park, S.; Yong, H.I.; Choe, J.H.; Jeon, H.-J.; Choe, W.; Jo, C. The use of atmospheric pressure plasma-treated water as a source of nitrite for emulsion-type sausage. *Meat Sci.* **2015**, *108*, 132–137. [CrossRef] [PubMed]

88. Jung, S.; Kim, H.J.; Park, S.; Yong, H.I.; Choe, J.H.; Jeon, H.J.; Choe, W.; Jo, C. Color Developing Capacity of Plasma-treated Water as a Source of Nitrite for Meat Curing. *Korean J. Food Sci. Anim. Resour.* **2015**, *35*, 703–706. [CrossRef] [PubMed]

89. Lee, J.; Jo, K.; Lim, Y.; Jeon, H.J.; Choe, J.H.; Jo, C.; Jung, S. The use of atmospheric pressure plasma as a curing process for canned ground ham. *Food Chem.* **2018**, *240*, 430–436. [CrossRef]

90. Yong, H.I.; Park, J.; Kim, H.-J.; Jung, S.; Park, S.; Lee, H.J.; Choe, W.; Jo, C. An innovative curing process with plasma-treated water for production of loin ham and for its quality and safety. *Plasma Process. Polym.* **2018**, *15*, 1700050. [CrossRef]

91. Jo, K.; Lee, S.; Yong, H.I.; Choi, Y.-S.; Jung, S. Nitrite sources for cured meat products. *LWT-Food Sci. Technol.* **2020**, *129*, 109583. [CrossRef]

92. Alahakoon, A.U.; Jayasena, D.D.; Ramachandra, S.; Jo, C. Alternatives to nitrite in processed meat: Up to date. *Trends Food Sci. Technol.* **2015**, *45*, 37–49. [CrossRef]

93. Guerrero-Legarreta, I. Meat and poultry—Spoilage of Cooked Meat and Meat Products. In *Encyclopedia of Food Microbiology*, 2nd ed.; Batt, C.A., Tortorello, M.L., Eds.; Academic Press: Oxford, UK, 2014; pp. 508–513.

94. Ganan, M.; Hierro, E.; Hospital, X.F.; Barroso, E.; Fernández, M. Use of pulsed light to increase the safety of ready-to-eat cured meat products. *Food Control* **2013**, *32*, 512–517. [CrossRef]

95. Possas, A.; Valdramidis, V.; García-Gimeno, R.M.; Pérez-Rodríguez, F. High hydrostatic pressure processing of sliced fermented sausages: A quantitative exposure assessment for *Listeria monocytogenes*. *Innov. Food Sci. Emerg. Technol.* **2019**, *52*, 406–419. [CrossRef]

96. Hadjicharalambous, C.; Grispoldi, L.; Goga, B.C. Quantitative risk assessment of *Listeria monocytogenes* in a traditional RTE product. *EFSA J.* **2019**, *17* (Suppl. S2), e170906. [CrossRef] [PubMed]

97. European Commission. Commission Regulation (EC) No 2073/2005 of 15 November 2005 on microbiological criteria for foodstuffs. *Off. J. Eur. Union* **2005**, *L338*, 1–26.

98. Awaiwanont, N.; Smulders, F.J.M.; Paulsen, P. Growth potential of *Listeria monocytogenes* in traditional Austrian cooked-cured meat products. *Food Control* **2017**, *50*, 150–156. [CrossRef]

99. Csadek, I.; Vankat, U.; Schrei, J.; Graf, M.; Bauer, S.; Pilz, B.; Schwaiger, K.; Smulders, F.J.M.; Paulsen, P. Treatment of Ready-to-Eat Cooked Meat Products with Cold Atmospheric Plasma to Inactivate *Listeria* and *Escherichia coli*. *Foods. submitted*.

100. Wikipedia. Seafood. Available online: https://en.wikipedia.org/wiki/Seafood (accessed on 19 August 2022).

101. FSA (Food Standards Agency). What Is an Oily Fish? Available online: https://webarchive.nationalarchives.gov.uk/ukgwa/20101210005807/http://www.food.gov.uk/news/newsarchive/2004/jun/oilyfishdefinition (accessed on 19 August 2022).

102. NHS (National Health Services). 2018. Available online: https://www.nhs.uk/live-well/eat-well/food-types/fish-and-shellfish-nutrition (accessed on 19 August 2022).

103. Rathod, N.B.; Chudaman, R.; Bhagwat, P.K.; Ozugul, F.; Benjakul, S.; Sottawat, B.; Pilai, S.; Annapure, U.S. Cold Plasma for the preservation of aquatic food products: An overview. *Compr. Rev. Food Sci. Food Saf.* **2021**, *20*, 4407–4425. [CrossRef] [PubMed]

104. Albertos, I.; Martin Diana, A.B.; Cullen, P.J.; Tiwari, B.K.; Ojha, S.K.; Bourke, P.; Alvarez, C.; Rico, D. Effects of dielectric barrier discharge (DBD) generated plasma on microbial reduction and quality parameters of fresh mackerel (*Scomber scombrus*) fillets. *Innov. Food Sci. Emerg. Technol.* **2017**, *44*, 117–122. [CrossRef]

105. Trevisani, M.; Cavoli, C.; Ragni, L.; Cecchini, M.; Berardinelli, A. Effect of non-thermal atmospheric plasma on viability and histamine-producing activity of psychrotrophic bacteria in mackerel fillets. *Front. Microbiol.* **2021**, *12*, 653597. [CrossRef]

106. Kulawik, P.; Kumar Tiwari, B. Recent advancements in the application of non-thermal plasma technology for the seafood industry. *Crit. Rev. Food Sci. Nutr.* **2018**, *59*, 3199–3210. [CrossRef]

107. Pan, W.; Benjakul, S.; Sanmartin, C.; Guidi, A.; Ying, X.; Ma, L.; Weng, X.; Yu, J.; Deng, S. Characterization of the Flavor Profile of Bigeye Tuna Slices Treated by Cold Plasma Using E-Nose and GC-IMS. *Fishes* **2022**, *7*, 13. [CrossRef]

108. Albertos, I.; Martin Diana, A.B.; Cullen, P.J.; Tiwari, B.K.; Ojha, S.K.; Bourke, P.; Rico, D. Shelf life extension of herring (*Clupea harengus*) using in-package atmospheric plasma technology. *Innov. Food Sci. Emerg. Technol.* **2019**, *53*, 85–91. [CrossRef]

109. Choi, S.; Puligundla, P.; Mok, C. Microbial decontamination of dried Alaska pollock shreds using corona discharge plasma jet: Effects on physico-chemical and sensory characteristics. *J. Food Sci.* **2016**, *81*, M952. [CrossRef]

110. Tiwari, B.K.; Muthukumarappan, K.; O'Donnell, C.P.; Cullen, P.J. Effects of sonication on the kinetics of orange juice quality parameters. *J. Agric. Food Chem.* **2008**, *56*, 2423–2428. [CrossRef]

111. CIE (Commission Internationale de L'Éclairage (International Commission on Illumination). *Recommendations on Uniform Color Spaces, Color-Difference Equations, Psychometric Color Terms*; Commission Internationale de l'Eclairage: Paris, France, 1978.

112. Upton, S. Delta E: The Color Difference. Delta E, CHROMIX Colornews, 17. Available online: http://www.chromix.com/ColorNews/ (accessed on 13 November 2022).
113. Koddy, J.K.; Miao, W.; Hatab, S.; Tang, L.; Xu, H.; Nyaisaba, B.M.; Chen, M.; Deng, S. Understanding the role of atmospheric cold plasma (CAP) in maintaining the quality of hairtail (*Trichiurus lepturus*). *Food Chem.* **2020**, *343*, 128418. [CrossRef]
114. de Souza Silva, D.A.; da Silva Campelo, M.C.; de Oliviera Soares Rebouças, L.; de Oliveira Vitoriano, J.; Alves, C., Jr.; Alves da Silva, J.B.; de Oliveira Lima, P. Use of cold atmospheric plasma to preserve the quality of white shrimp (*Litopenaeus vannamei*). *J. Food Prot.* **2019**, *82*, 1217–1223. [CrossRef]
115. Elliot, M.; Chen, J.; Chen, D.-Z.; Ekaterina, N.; Deng, S.G. Effect of cold plasma-assisted shrimp processing chain in biochemical and sensory quality alterations in Pacific White Shrimps (*Penaeus vannamei*). *Food Sci. Techn. Internat.* **2022**, *28*, 683–693. [CrossRef]
116. Liao, X.; Su, V.; Liu, D.; Chen, S.; Hu, Y.; Ye, X.; Wang, J.; Ding, T. Application of atmospheric cold plasma-activated water (PAW) ice for preservation of shrimps (*Metapenaeus ensis*). *Food Control* **2018**, *94*, 307–314. [CrossRef]
117. Kahn, M.; Lively, J.A. Determination of sulphite and antimicrobial residues in important shrimps to the USA. *Aquac. Rep.* **2020**, *18*, 100529. [CrossRef]
118. Rahman, M. Are Ghost Shrimps Edible? Available online: https://acuariopets.com/are-ghost-shrimps-edible/ (accessed on 21 August 2022).
119. Roberts, J. Ghost Shrimp (Complete Care, Diet, Set-up and Breeding Guide). Available online: http://www.buildyouraquarium.com/ghost-shrimp (accessed on 21 August 2022).
120. Choi, S.; Puligundla, P.; Mok, C. Impact of corona discharge plasma treatment on microbial load and physicochemical and sensory characteristics of semi-dried squid (*Todarodes pacificus*). *Food Sci. Biotechnol.* **2017**, *26*, 1137–1144. [CrossRef]
121. Choi, S.; Puligundla, P.; Mok, C. Effect of corona discharge plasma on microbial decontamination of dried squid shreds including physico-chemical and sensory evaluation. *LWT-Food Sci. Technol.* **2017**, *75*, 323–328. [CrossRef]
122. Dumen, E.; Ekici, G.; Ergin, S.; Bayrakal, G.M. Presence of Foodborne Pathogens in Seafood and Risk Ranking for Pathogens. *Foodborne Pathog. Dis.* **2020**, *17*, 541–546. [CrossRef]
123. Hall, A.J.; Wikswo, M.E.; Pringle, K.; Gould, L.H.; Parashar, U.D. Vital Signs: Foodborne Norovirus Outbreaks—United States, 2009–2012. *Morb. Mortal. Wkly. Rep.* **2014**, *63*, 491–495.
124. Koopmans, M. Foodborne Viruses and Seafood Safety in an Environmental Health Perspective. *Epidemiology* **2009**, *20*, S233. [CrossRef]
125. Mizan, F.R.; Jahid, I.K.; Ha, S.D. Microbial biofilms in seafood: A food-hygiene challenge. *Food Microbiol.* **2015**, *49*, 41–55. [CrossRef]
126. Sala, M.R.; Arias, C.; Dominguez, A.; Bartolomé, R.; Muntada, J.M. Foodborne outbreak of gastroenteritis due to Norovirus and *Vibrio parahaemolyticus*. *Epidemol. Infect.* **2008**, *137*, 626–629. [CrossRef]
127. Elbashir, S.; Parveen, S.; Schwarz, J.; Rippen, T.; Jahncke, M.; DePaola, A. Seafood pathogens and information on antimicrobial resistance: A review. *Food Microbiol.* **2018**, *70*, 85–93. [CrossRef]
128. Alfano-Sobsey, M.; Davies, M.; Ledford, S.I. Norovirus Outbreak associated with undercooked oysters and secondary household transmission. *Epidemiol. Infect.* **2012**, *140*, 276–282. [CrossRef]
129. Brucker, R.; Bui, T.; Kwan-Gett, T.; Stewart, L. Centers for Disease Control and Prevention, Notes from the Field: Norovirus Infections Associated with Frozen Raw Oysters, Washington. *Morb. Mortal. Wkly. Rep.* **2011**, *307*, 1480. Available online: http://www.cdc.gov/mmwr/preview/mmwrhtmL/mm6106a3.htm (accessed on 29 September 2021).
130. Chironna, M.; Germinario, C.; De Medici, D.; Fiore, A.; Di Pasquale, S.; Quartoa, M.; Barbuti, S. Detection of hepatitis A virus in mussels from different sources marketed in Puglia region (South Italy). *Int. J. Food Microbiol.* **2002**, *75*, 11–18. [CrossRef]
131. Centers for Disease Control and Prevention. Increase in Vibrio parahaemolyticus Illnesses Associated with Consumption of Shellfish from Several Atlantic Coast Harvest Areas, United States. 2013. Available online: https://www.cdc.gov/vibrio/investigations/vibriop-09-13/index.html (accessed on 12 October 2021).
132. Zuber, S.C.; Butot, S.; Baert, L. Effects of treatments used in food processing on viruses. In *Food Safety Assurance and Veterinary Public Health, Volume 6: Foodborne Viruses and Prions and Their Significance for Public Health; ECVPH (European College of Veterinary Public Health/European Food Safety Authority)*; Smulders, F.J.M., Nørrung, B., Budka, H., Eds.; Wageningen Academic Publishers: Wageningen, The Netherlands, 2013; pp. 113–136.
133. Choi, M.S.; Jeon, E.B.; Kim, J.; Chou, E.H.; Lim, J.S.; Choi, J.; Ha, K.S.; Kwon, J.Y.; Jeong, S.H.; Park, S.Y. Virucidal Effects of Dielectric Barrier Discharge Plasma on Human Norovirus Infectivity in Fresh Oysters (*Crassostrea gigas*). *Foods* **2020**, *9*, 1731. [CrossRef]
134. Csadek, I.; Paulsen, P.; Weidinger, P.; Bak, K.H.; Bauer, S.; Pilz, B.; Nowotny, N.; Smulders, F.J.M. Nitrogen Accumulation in Oyster (*Crassostrea gigas*) Slurry Exposed to Virucidal Cold Atmospheric Plasma Treatment. *Life* **2021**, *11*, 1333. [CrossRef]
135. van Holde, K.E.; Miller, K.I.; Decker, H. Hemocyanins and Invertebrate Evolution. *J. Biol. Chem.* **2001**, *19*, 15563–15566. [CrossRef] [PubMed]
136. Guo, L.; Xu, R.; Gou, L.; Liu, Z.; Zhao, Y.; Liu, D.; Zhang, L.; Chen, H. Mechanism of Virus Inactivation by Cold Atmospheric-Pressure Plasma and Plasma-Activated Water. *Appl. Environ. Microbiol.* **2018**, *84*, e00726-18. [CrossRef] [PubMed]
137. Wu, Y.; Wang, Y.; Wang, J.; Xu, S.; Yu, L.; Corvini, P.; Wintgens, T. Nitrate removal from water by new polymeric adsorbent modified with amino and quaternary ammonium groups: Batch and column adsorption study. *J. Taiwan Inst. Chem. Eng.* **2016**, *66*, 191–199. [CrossRef]

138. El-Hanache, L.; Sundermann, L.; Lebeau, B.; Toufaily, J.; Hamieh, T.; Daou, T.J. Surfactant-modified MFI-type nanozeolites: Super-adsorbents for nitrate removal from contaminated water. *Microporous Mesoporous Mater.* **2019**, *283*, 1–13. [CrossRef]
139. Filipić, A.; Gutierrez-Aguirre, I.; Primc, G.; Mozetič, M.; Dobnik, D. Cold Plasma, a New Hope in the Field of Virus Inactivation. *Trends Biotechnol.* **2020**, *38*, 1278–1291. [CrossRef]
140. McMinn, B.R.; Ashbolt, N.J.; Korajkic, A. Bacteriophages as indicators of faecal pollution and enteric virus removal. *Lett. Appl. Microbiol.* **2017**, *65*, 11–26. [CrossRef]
141. Cromeans, T.; Park, G.W.; Costantini, V.; Lee, D.; Wang, Q.; Farkas, T.; Lee, A.; Vinjé, J. Comprehensive comparison of cultivable norovirus surrogates in response to different inactivation and disinfection treatments. *Appl. Environ. Microbiol.* **2014**, *80*, 5743–5751. [CrossRef]
142. Mohamed, H.; Nayak, G.; Rendine, N.; Wigdahl, B.; Krebs, F.C.; Bruggeman, P.J.; Miller, V. Non-thermal plasma as a novel strategy for treating or preventing viral infection and associated disease. *Front. Phys.* **2021**, *9*, 683118. [CrossRef]
143. Weiss, M.; Daeschlein, G.; Kramer, A.; Burchardt, M.; Brucker, S.; Wallwiener, D.; Stope, M.B. Virucide properties of cold atmospheric plasma for future clinical applications. *J. Med. Virol.* **2017**, *89*, 952–959. [CrossRef]
144. Mormann, S.; Dabisch, M.; Becker, B. Effects of technological processes on the tenacity and inactivation of norovirus genogroup II in experimentally contaminated foods. *Appl. Environ. Microbiol.* **2010**, *76*, 536–545. [CrossRef]

 *foods*

Article

Treatment of Ready-To-Eat Cooked Meat Products with Cold Atmospheric Plasma to Inactivate *Listeria* and *Escherichia coli*

Isabella Csadek, Ute Vankat, Julia Schrei, Michelle Graf, Susanne Bauer, Brigitte Pilz, Karin Schwaiger, Frans J. M. Smulders * and Peter Paulsen

Unit of Food Hygiene and Technology, Institute of Food Safety, Food Technology and Veterinary Public Health, University of Veterinary Medicine, Veterinaerplatz 1, 1210 Vienna, Austria
* Correspondence: frans.smulders@vetmeduni.ac.at; Tel.: +43-1-25077-3325

Abstract: Ready-to-eat meat products have been identified as a potential vehicle for *Listeria monocytogenes*. Postprocessing contamination (i.e., handling during portioning and packaging) can occur, and subsequent cold storage together with a demand for products with long shelf life can create a hazardous scenario. Good hygienic practice is augmented by intervention measures in controlling post-processing contamination. Among these interventions, the application of 'cold atmospheric plasma' (CAP) has gained interest. The reactive plasma species exert some antibacterial effect, but can also alter the food matrix. We studied the effect of CAP generated from air in a surface barrier discharge system (power densities 0.48 and 0.67 W/cm^2) with an electrode-sample distance of 15 mm on sliced, cured, cooked ham and sausage (two brands each), veal pie, and calf liver pâté. Colour of samples was tested immediately before and after CAP exposure. CAP exposure for 5 min effectuated only minor colour changes (ΔE max. 2.7), due to a decrease in redness (a*), and in some cases, an increase in b*. A second set of samples was contaminated with *Listeria (L.) monocytogenes*, *L. innocua* and *E. coli* and then exposed to CAP for 5 min. In cooked cured meats, CAP was more effective in inactivating *E. coli* (1 to 3 log cycles) than *Listeria* (from 0.2 to max. 1.5 log cycles). In (non-cured) veal pie and calf liver pâté that had been stored 24 h after CAP exposure, numbers of *E. coli* were not significantly reduced. Levels of *Listeria* were significantly reduced in veal pie that had been stored for 24 h (at a level of ca. 0.5 log cycles), but not in calf liver pâté. Antibacterial activity differed between but also within sample types, which requires further studies.

Keywords: cold atmospheric plasma; dielectric barrier discharge; antimicrobial effects; *Listeria*; *Escherichia coli*; cooked cured meat products; colour

Citation: Csadek, I.; Vankat, U.; Schrei, J.; Graf, M.; Bauer, S.; Pilz, B.; Schwaiger, K.; Smulders, F.J.M.; Paulsen, P. Treatment of Ready-To-Eat Cooked Meat Products with Cold Atmospheric Plasma to Inactivate *Listeria* and *Escherichia coli*. *Foods* **2023**, *12*, 685. https://doi.org/10.3390/foods12040685

Academic Editor: Paulo Eduardo Sichetti Munekata

Received: 22 December 2022
Revised: 24 January 2023
Accepted: 28 January 2023
Published: 4 February 2023

1. Introduction

Listeria (L.) monocytogenes is an important food-borne pathogen and can thrive and persist in a wide range of environmental conditions, even under industrial conditions in food processing companies [1]. Asymptomatic 'healthy' animals and humans may carry and shed the pathogen [2,3]. However, clinical symptoms may develop and range from mild fever to severe diarrhoeal disease, fatalities or even miscarriages, with young, old, and immunocompromised consumers at particular risk. Predominant symptoms are not necessarily specific, e.g., chills, headache, arthralgia, prostration, malaise, swollen lymph nodes [4].

Food intended for human consumption can be contaminated with *L. monocytogenes* at virtually any level in the food chain, i.e., primary production at the farm level, during processing or at the retail or consumer level due to insufficient hygiene precautions [3]. As early as 1983, Schlech et al. reported transmission of the bacterium via food [5]. *L. monocytogenes* is considered the most important food-borne pathogen in ready-to-eat (RTE) foods due to its ability to survive and multiply under cold storage conditions, in vacuum or modified atmosphere packed foods and due to its persistence in food processing premises [6]. Although thermal treatment at temperatures > 65 °C is effective in killing *L. monocytogenes*, all cooked meats can become contaminated with listeriae during slicing and further handling. Thus, it is not surprising that not only unheated RTE foods (e.g., dry-cured or cold-smoked foods) have been identified as source for food-borne listeriosis [7] but also pasteurized products that are portioned and packed. Post-processing contamination of an otherwise nearly sterile product and prolonged shelf life under refrigerated conditions contribute to a risk scenario for introduction and multiplication of *L. monocytogenes* [8]. RTE foods implicated in food-borne listeriosis outbreaks are often of traditional type and manufactured by small local producers [9,10]. In 1993, for example, listeriosis outbreaks in France were associated with the consumption of rillettes (an RTE delicatessen food with ham cooked in grease) [11]. Besides the direct negative consequences for the health of consumers, contamination with *Listeria* requires ceasing delivery or recalls of food batches, which impairs development of domestic producers [12].

Ferreira et al. [13] reported that 50% of human listeriosis cases in the US were linked to the consumption of ready meals and that contamination was found at the retail level. In Europe, the number of listeriosis cases was found to be alarming [14]. In the European Union (EU), *L. monocytogenes* was the most serious zoonotic food-related disease with the highest fatality rate [15]. Out of 1876 cases of listeriosis, 780 were hospitalized and 167 died [15]. In November 2022, an RTE product (fish cake) from Denmark caused listeriosis in seven people (up to the publication of this manuscript, there was no further follow-up information available) [16].

In 2021, a food-borne outbreak caused by *L. monocytogenes* was reported in Austria. Five people were affected and two fatalities were noted caused by contaminated meat and meat products. Due to such cases and given the fact that every year up to two outbreaks of food-borne listeriosis are reported in Austria, 3835 samples were examined for the presence of listeriae in the year 2021, including 1300 samples of RTE food. Two of these were harmful to health and three were judged unfit for human consumption [17].

According to EU legislation [18], levels of *L. monocytogenes* in RTE foods must not exceed 100 cfu/g throughout the product's shelf life. At the end of the manufacturing process, before the food item leaves the processing plant, the food business operator has to assure that *L. monocytogenes* is not detectable in 5×25 g food. For RTE foods that are considered not to support growth of *L. monocytogenes*, a limit of 100 cfu/g applies. This latter category comprises products (i) with pH $\leq$ 4.4 or $a_w \leq$ 0.92, (ii) with pH $\leq$ 5.0 and $a_w \leq$ 0.94 or (iii) with a shelf life of less than 5 days [18].

In Austria, there is a large number of RTE traditional specialties made from cured, boiled, chopped meat [19]. The standards of identity in the Austrian Food Codex [20] give no requirements in terms of pH or a_w for those dishes. Meat is very popular in Austria and often finds a place on the dining table at home. The Agricultural Marketing Agency (AMA) reported that in 2020, the per capita consumption of meat (including poultry) in this country was 90.8 kg, ca. 50% of which was consumed as sausages and other specialties [21].

2. Rationale for Application of CAP to Cooked and/or Cured Meat Products

'Plasma' designates a gas where a fraction of the particles is in an ionized state. This can be accomplished under various conditions, e.g., by exposing gases to an electrical field under atmospheric pressure [22]. The array of plasma species is, among others, dependent on the gases used. When ambient air is used, reactive oxygen species dominate al lower electrical voltage [23], whereas at higher voltages (10 kV), more reactive oxygen/nitrogen species (RNS, RONS) are formed [24]. The antibacterial effect of cold, atmospheric plasma (CAP) on cured meats has already been documented in a number of studies [25–28].

Generation of NO_x in plasma-treated water has received much attention, since nitrate will accumulate and such treated water allows curing of meats without the addition of nitrite salt [29–33]. Nitrite/nitrate curing of foods serves (besides other effects) as a protection against microorganisms [34,35]. However, recontamination can occur during further processing, such as shredding, portioning and packaging. Unless pH and/or water activity are sufficiently low, *L. monocytogenes* will be able to thrive in these contaminated foods [18]. A study on typical Austrian cooked ready-to-eat meat products addressed this issue in detail [36] and presented a decision tool to estimate to what extent pH or water activity of a given product need to be lowered to render a food that does not favour the multiplication of *L. monocytogenes*. The authors concluded that few, if any, options exist to lower water activity or pH without changing the sensory characteristics and impacting on acceptance of consumers.

Given the abovementioned constraints in post-processing control of *L. monocytogenes* and in consideration of the mode of action of CAP on contaminant bacteria on food surfaces, we studied the potential of CAP for reducing numbers of contaminant bacteria on two brands ('A', 'B') of sliced, cooked, cured ham (ham 'A', ham 'B'), on two brands of sliced/pasteurized emulsified sausages (sausage 'A', sausage 'B') and on veal pie and calf liver pâté. We used not only *L. monocytogenes* isolates but also *L. innocua* as a surrogate [25] and *E. coli* as an established marker of (faecal) contamination [37].

3. Materials and Methods

3.1. Characterisation of the Samples and Exposure to CAP

Cured sliced meats and non-cured meat pies and pâtés were obtained pre-packed and had a shelf life > 5 days.

Samples were exposed to CAP generated by a surface barrier discharge (SBD) plasma generator (described in [38]), with two power settings (see Table 1), placed at a distance of 15 mm from the product and an exposure time of 2 and 5 min (cooked cured ham and sausages) or 3 and 5 min (non-cured meat pie and pâté). For technical reasons, 15 mm was the nearest distance we could go without running the risk of the CAP device being contaminated by contact with samples containing listeriae or *E. coli*. Selection of exposure times was based on assessment of sample colour changes during CAP exposure.

Table 1. Settings of the CAP device.

		Comment
CAP device	SBD-type, 9 kHz frequency	Device described in Bauer et al. [38]
CAP settings	low power	Power input 20.7 W Output voltage 8.16 kV Power density 0.48 W/cm^2
	high power	Power input 29.9 W Output voltage 9.44 kV Power density 0.67 W/cm^2
Exposure time	2 or 5 min	for cooked cured ham and cooked cured sausage
	3 or 5 min	for veal pie and calf liver pâté
Distance sample to electrode	15 mm	for all samples

3.2. Measurement of Water Activity and pH

Water activity (a_w) (Lab-Swift, Novasina, Lachen, Switzerland) and pH (penetrating electrode LoT 406-M6-DXK-S7/25; Mettler-Toledo, Urdorf, Switzerland, and pH-Meter Testo 230; Testo AG, Lenzkirch, Germany) were measured and the average of five such measurements reported.

3.3. Colour Measurement

Colour (L*, a*, b*) was measured in the centre of the sample's surface using a double-beam spectrophotometer with an aperture size of 8 mm and a D65 illuminant and an observer angle of 10° (Phyma Codec 400, Phyma, Gießhübl, Austria). Surface colour was measured immediately before and after CAP treatment. Control samples (i.e., stored at ambient air under ambient light without CAP exposure) were measured at the same time intervals. Each measurement was the average of five scans and the number of replicates was 4–5. The total number of samples per product was 28 for ham and sausage, and 30 for pie and pâté.

Delta-E [($\Delta E = (\Delta L^*)^2 + (\Delta a^*)^2 + (\Delta b^*)^2)^{0.5}$] [39] was used as a proxy for visually perceived colour changes. ΔE is a single number that represents the 'distance' between two colours, the idea being that a ΔE of 1 is the smallest colour difference the human eye can perceive [40]. More specifically, $\Delta E < 2$ indicates a colour change visible to an experienced observer only and $\Delta E > 5$ indicates the impression of two different colours [41].

3.4. Preparation of the Inoculum and the Samples

A second set of samples was contaminated with *E. coli* (mix of NCTC 9001 and ATCC 11303), *Listeria monocytogenes* (NCTC 11994 and in-house isolate 17001) and *L. innocua* (in-house isolates 16777 and 16908-2). *E. coli* had been stored on slant agar and was activated by overnight incubation in buffered peptone water (Oxoid CM1049; Oxoid, Basingstoke, UK) at 37 °C. Likewise, freeze-dried pellets of *L. monocytogenes* and *L. innocua* were separately inoculated into brain–heart infusion broth (Merck 110493; Merck KG, Darmstadt, Germany) and incubated overnight at 30 °C. Serial decimal dilutions from the overnight cultures were prepared in 0.89% sterile saline. Aliquots from the dilutions were streaked onto plate count agar (PCA; Merck 105463), colonies were counted after 24 h incubation at 37 °C, and cell concentration/mL was calculated. In the meantime, 1:10 dilutions of the overnight cultures were maintained at 0–2 °C. This dilution was then adjusted to 7 and 6 log cfu/mL for *E. coli* and *Listeria* species, respectively. Adjusted dilutions were mixed and used within 3 h.

Samples were cut using a sterile 30 mm cork borer. On each sample surface, 100 µL or 20 µL of the mix was evenly spread. It was observed that 20 µL inoculum was easily spread on the surface (i.e., the area of the sample facing the mesh electrode of the CAP generator, taking care that there was no drip to the unexposed sides of the sample), whereas for the 100 µL inoculum, a moisture film remained. After a period of 5 min, samples were either directly vacuum-packed (control group) or exposed to CAP and then vacuum-packed (treatment group). Ham and sausage samples were stored for 24 h in the dark at 2 ± 2 °C. Veal pie and calf liver pâté samples stored for 1 and 7 days. Subsequently, the entirety of the samples was suspended in 9 parts of maximum recovery diluent (Oxoid CM0733) and macerated in a Stomacher lab blender (Seward Medical, Worthing, UK) for 3 min. Serial decimal dilutions were plated onto Listeria-selective agar (OCLA; Oxoid CM1080; incubation 72 h at 37 °C; with a turbid halo around a colony indicative of *L. monocytogenes*) and on Chrom ID E. coli agar (BioMerieux 42017; BioMerieux, Marcy l'Etoile, F; incubation 24 h at 42 °C). After incubation, typical colonies were counted and results given as log cfu/g. Experiments were done in triplicate, with a total number of samples of n = 18 per product.

3.5. Statistical Analysis

The colour values before and after treatment were analysed by pairwise comparison (paired *t*-test), with a level of significance set to $p < 0.05$. Within each product group (cured ham; cured sausage; non-cured meats), water activity and pH of the two samples each were compared by t-test. For each storage day, numbers of bacteria in CAP-exposed samples were compared to those in the control samples (multiple sample comparison procedure; Statgraphics 3.0, Statistical Graphics Corp., Warrenton, VA, USA), with a level of significance set to $p < 0.05$.

4. Results

4.1. Water Activity and pH of Samples

Water activity and pH of samples are reported in Table 2. According to current EU legislation [18], samples were considered to be able to support growth of *L. monocytogenes*.

Table 2. Physicochemical sample characteristics.

Product	Code	Characteristics	
		pH (n = 5)	Water Activity (a_w) (n = 5)
Sliced cooked cured ham	Ham 'A *'	6.28 ± 0.03	0.96 ± 0.01
	Ham 'B *'	6.32 ± 0.02	0.96 ± 0.01
Sliced cooked cured sausage	Sausage 'A *'	6.27 [a] ** ± 0.02	0.95 ± 0.01
	Sausage 'B *'	6.33 [b] ± 0.02	0.96 ± 0.01
Sliced cooked meats	Veal pie	5.95 [a] ± 0.01	0.92 [c] ± 0.01
	Calf liver pâté	5.48 [b] ± 0.04	0.94 [d] ± 0.01

* 'A' and 'B' indicate the manufacturer. ** Figures with different superscripts differ significantly ($p < 0.05$).

Statistically significant, yet small differences were observed for pH between the two sausage samples, and for pH and water activity between the two non-cured meats.

4.2. Changes in Colour

After CAP treatment of cooked cured ham 'A', a statistically significant decrease in a* values was observed immediately following 2 min CAP exposure at low power, and after 5 min exposure to high and low power. An increase of b* was observed after 5 min exposure. ΔE values were in the range of 1.4 to 2.1. No statistically significant changes in colour parameters were observed in controls (Table 3). In ham 'B' (high power), statistically significant changes in a* were observed only after exposure to high power CAP. As for ham 'A', an increase of b* was observed after 5 min exposure, and no statistically significant changes in colour parameters were observed in controls. ΔE values were in the range of 1.2 to 1.9 (Table 3).

For the two sliced cured sausage samples, a statistically significant decrease in a* values was observed immediately following CAP exposure, regardless of the mode of CAP exposure protocol and of sample type (Table 4), whereas no significant decrease was observed in non–CAP-exposed controls. Likewise, a significant, yet small increase in b* values was observed in CAP-exposed sausage 'A' samples and in sausage 'B' at 5 min. However, the differences were small, and did not exceed 1 for a* and b* at 2 min exposure or 3 at 5 min. exposure. Average ΔE values were in the range of 0.9 to 1.3 for 2 min exposure, and slightly higher after 5 min exposure (1.6–2.7).

Similar findings were found for CAP-exposed veal pie and calf liver pâté, with average ΔE values in the range of 0.7–1.7 and 1.1–2.0, respectively (Table 5). In all treatments, a small, yet significant increase was found for redness (a*). In liver pâté, lightness (L*) decreased significantly. In control samples exposed to air and ambient light, no significant differences were observed.

Although ΔE values > 2 indicate changes in colour visible also for inexperienced observers, it is assumed that a change in colour is perceived by the majority of consumers at higher ΔE values of >3 [42]. Thus, we used the 5 min exposure protocol for subsequent experiments with bacterial contaminants.

Table 3. Colour (L*, a* and b*) of cooked cured ham before and after CAP exposure, with colour difference expressed as ΔE.

Brand	Power	Time (min)	Prior to or after Treatment	L*	a*	b*	ΔE
A	low	2	P	72.8 ± 0.9	4.0 [a] ± 0.4	8.3 ± 0.8	
			a	74.2 ± 1.9	2.4 [b] ± 0.8	8.4 ± 0.5	2.11
	high	2	P	71.9 ± 2.8	4.7 ± 1.0	8.9 ± 0.5	
			a	70.2 ± 1.0	4.4 ± 0.7	8.8 ± 0.4	1.75
	no (control)	2	P	72.0 ± 1.1	4.2 ± 0.5	8.5 ± 0.9	
			a	71.6 ± 1.4	4.1 ± 0.6	8.6 ± 0.8	0.42
	low	5	P	71.0 ± 3.3	5.3 [a] ± 1.3	7.3 [c] ± 0.3	
			a	70.8 ± 2.4	4.2 [b] ± 1.1	8.5 [d] ± 0.2	1.61
	high	5	P	71.3 ± 2.4	4.7 [a] ± 0.8	8.7 [c] ± 1.1	
			a	71.5 ± 2.6	3.9 [b] ± 0.8	9.9 [d] ± 1.3	1.39
	no (control)	5	P	72.0 ± 1.1	4.2 ± 0.5	8.5 ± 0.9	
			a	71.3 ± 1.5	4.0 ± 0.8	8.36 ± 0.7	0.74
B	low	2	P	68.2 ± 2.3	8.0 ± 1.0	8.6 ± 0.3	
			a	67.5 ± 2.8	7.0 ± 1.3	8.6 ± 1.0	1.21
	high	2	P	68.6 ± 2.1	6.8 [a] ± 1.0	8.5 ± 1.0	
			a	69.4 ± 2.0	5.9 [b] ± 0.9	8.9 ± 1.0	1.18
	no (control)	2	P	71.0 ± 1.6	7.3 ± 1.3	7.1 ± 0.5	
			a	71.1 ± 1,9	6.9 ± 0.9	7.3 ± 0.3	0.46
	low	5	P	70.3 ± 3.5	7.0 ± 1.5	7.0 [c] ± 0.3	
			a	70.2 ± 3.4	6.2 ± 1.3	8.1 [d] ± 0.6	1.35
	high	5	P	71.0 ± 2.8	7.3 [a] ± 1.2	7.5 [c] ± 0.6	
			a	71.5 ± 2.0	6.0 [b] ± 0.9	8.8 [d] ± 0.3	1.85
	no (control)	5	P	71.0 ± 1.6	7.3 ± 1.3	7.1 ± 0.5	
			a	71.2 ± 1,5	6.7 ± 1.5	7.2 ± 0.6	0.64

Note: $n = 5$ for low- and $n = 4$ for high-power treatment. Within-sample treatment combinations, different superscripts indicate statistically significant ($p < 0.05$) differences between colour parameters before and after treatment.

Table 4. Colour (L*, a* and b*) of cooked cured sausage before and after CAP exposure, with colour difference expressed as ΔE.

Brand	Power	Time (min)	Prior to or after Treatment	L*	a*	b*	ΔE
A	low	2	P	74.8 ± 0.6	6.4 [a] ± 0.3	9.7 [c] ± 0.5	
			a	74.9 ± 1.5	5.6 [b] ± 0.2	10.2 [d] ± 0.5	0.89
	high	2	P	74.8 ± 1.0	6.5 [a] ± 0.3	9.8 [c] ± 0.6	
			a	73.5 ± 1.9	6.0 [b] ± 0.2	10.7 [d] ± 0.6	1.64
	no (control)	2	P	74.5 ± 0.6	6.7 ± 0.5	9.0 ± 0.5	
			a	74.1 ± 0.9	6.8 ± 0.6	9.3 ± 0.4	0.51
	low	5	P	74.1 ± 0.7	6.5 [a] ± 0.4	9.7 [c] ± 0.1	
			a	74.3 ± 0.6	5.5 [b] ± 0.3	10.3 [d] ± 0.1	1.19
	high	5	P	74.1 ± 1.6	7.0 [a] ± 0.1	8.9 [c] ± 0.2	
			a	73.7 ± 0.8	5.3 [b] ± 0.1	10.1 [d] ± 0.4	2.16
	no (control)	2	P	74.5 ± 0.6	6.7 ± 0.5	9.0 ± 0.5	
			a	73.9 ± 1.0	7.0 ± 0.8	9.2 ± 0.6	0.71
B	low	2	P	66.2 ± 1.0	11.9 [a] ± 0.7	9.6 ± 0.8	
			a	65.4 ± 1.1	10.8 [b] ± 0.8	9.7 ± 0.6	1.33
	high	2	P	66.1 ± 0.8	12.4 [a] ± 0.5	9.6 ± 0.6	
			a	65.5 ± 1.3	11.8 [b] ± 0.7	9.7 ± 0.8	0.89
	no (control)	5	P	64.0 ± 1.2	13.2 [a] ± 0.7	8.7 ± 0.4	
			a	64.6 ± 1.2	12.9 [b] ± 0.4	9.0 ± 0.5	0.73
	low	5	P	64.3 ± 1.0	13.3 [a] ± 0.6	8.5 [c] ± 0.2	
			a	64.8 ± 1.3	10.6 [b] ± 0.2	9.1 [d] ± 0.3	2.73
	high	5	P	65.4 ± 1.1	13.3 [a] ± 0.2	8.6 [c] ± 0.4	
			a	65.2 ± 0.5	10.7 [b] ± 0.4	9.4 [d] ± 0.3	2.72
	no (control)	5	P	64.0 ± 1.2	13.2 ± 0.7	8.7 ± 0.4	
			a	63.1 ± 1.7	12.7 ± 1.0	8.9 ± 0.7	0.81

Note: $n = 5$ for low- and $n = 4$ for high-power treatment. Within-sample treatment combinations, different superscripts indicate statistically significant ($p < 0.05$) differences between colour parameters before and after treatment.

Table 5. Colour (L*, a* and b*) of (non-cured) veal pie and liver pâté before and after CAP exposure, with colour difference expressed as ΔE.

Product	Power	Time (min)	Prior to or after Treatment	L*	a*	b*	ΔE
Veal pie	low	3	P	57.5 ± 0.5	12.9 [c] ± 0.2	15.0 ± 0.2	
			a	56.8 ± 0.7	13.8 [d] ± 0.2	15.3 ± 0.2	1.21
	high	3	P	57.9 ± 0.5	13.1 [c] ± 0.2	15.2 ± 0.2	
			a	57.5 ± 0.4	14.8 [d] ± 0.2	15.5 ± 0.3	1.68
	no (control)	3	P	57.9 ± 0.5	13.1 ± 0.2	15.2 ± 0.3	
			a	57.8 ± 0.5	13.6 ± 0.7	15.5 ± 0.3	0.58
	low	5	P	56.6 ± 0.9	13.5 [c] ± 0.4	15.2 ± 0.2	
			a	56.1 ± 0.7	14.5 [d] ± 0.1	14.9 ± 0.1	1.12
	high	5	P	57.0 ± 0.8	13.1 [c] ± 0.2	14.7 ± 0.4	
			a	57.2 ± 0.8	13.7 [d] ± 0.4	15.0 ± 0.2	0.74
	no (control)	5	P	57.9 ± 0.5	13.1 ± 0.2	15.2 ± 0.3	
			a	57.7 ± 0.4	13.5 ± 0.7	15.5 ± 0.3	0.52
Liver pâté	low	3	P	63.9 [a] ± 1.6	12.2 [c] ± 0.5	15.5 ± 0.5	
			a	62.9 [b] ± 1.2	12.6 [d] ± 0.6	15.9 ± 0.7	1.10
	high	3	P	63.3 [a] ± 0.6	11.9 [c] ± 0.2	15.8 ± 0.3	
			a	62.5 [b] ± 0.5	12.7 [d] ± 0.5	16.4 ± 0.3	1.28
	no (control)	3	P	63.9 ± 0.6	12.5 ± 0.3	15.6 ± 0.4	
			a	63.5 ± 0.7	12.6 ± 0.3	16.0 ± 0.6	0.58
	low	5	P	64.0 [a] ± 0.7	12.1 [c] ± 0.3	15.6 ± 0.4	
			a	63.1 [b] ± 0.4	13.1 [d] ± 0.3	16.1 ± 0.6	1,39
	high	5	P	63.0 [a] ± 1.0	12.5 [c] ± 0.3	16.0 ± 0.4	
			a	61.4 [b] ± 1.1	13.6 [d] ± 0.3	16.4 ± 0.5	1.96
	no (control)	5	P	63.9 ± 0.6	12.5 ± 0.3	15.6 ± 0.4	
			a	63.5 ± 0.5	12.6 ± 0.4	16.1 ± 0.6	0.61

Note: n = 5 for low- and high-power treatment. Within-sample treatment combinations, different superscripts indicate statistically significant ($p < 0.05$) differences between colour parameters before and after treatment.

4.3. Changes in Bacterial Load

Numbers of *E. coli* were significantly lower ($p < 0.05$) in CAP-exposed cured sausage and ham than in controls (Figure 1). Significant reductions of numbers of listeriae in CAP-exposed samples were found in cured ham 'B', and in sliced sausage samples 'A' and 'B' (only 20 µL inoculum). For the sake of simplicity, only *L. monocytogenes* will be reported in the following, since we observed the same ratio between *L. monocytogenes* and *L. innocua* in the inoculum as well as on controls and CAP-exposed samples.

CAP was obviously more effective in inactivating *E. coli* (1 to 3 log cycles) than *Listeria* (from 0.2 to max. 1.5 log cycles; sliced sausage 'A'). There was no consistent pattern as regards the effect of inoculum size and plasma type (low power or high power).

In veal pie and calf liver pâté, no significant reductions were observed for *E. coli* 24 h after CAP exposure, whereas after 7 days' storage, a significant reduction was observed only for samples exposed to low-power CAP (up to 1 log cycle). In veal pie, levels of *Listeria* were significantly reduced in samples tested at 24 h (at level of ca. 0.6 log cycles), but not in calf liver pâté. Significant reductions in listeriae were observed only in liver pâté 7 days after low-power CAP exposure (ca. 0.4 log cycles; Figure 2).

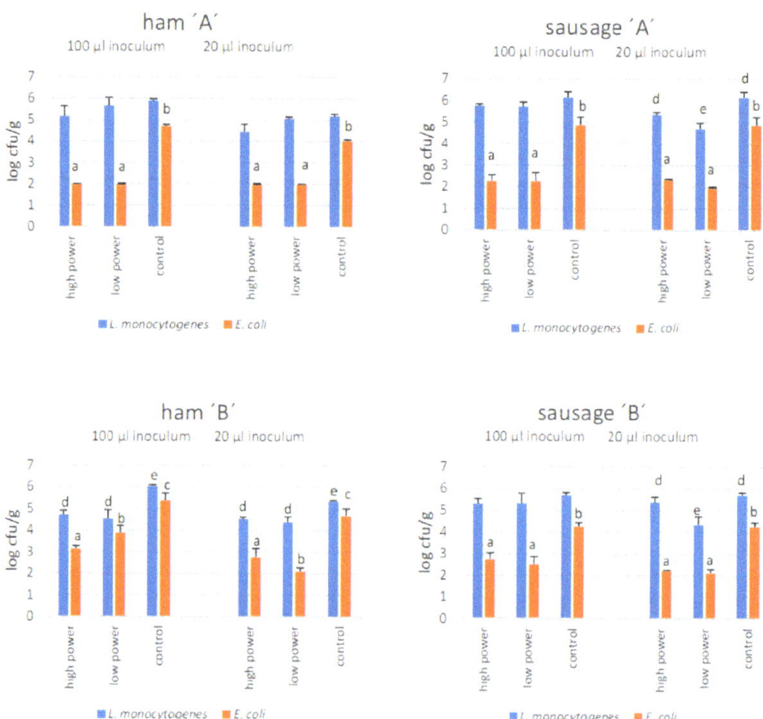

Figure 1. The effects of a 5 min exposure to high-power vs. low-power cold atmospheric pressure plasma treatment on the survival of *L(isteria)* and *E(scherichia) coli*, inoculated on the surface of cured/cooked/sliced meat products (n = 3) manufactured in two different enterprises ('A' and 'B'). Note that the limit of detection is 2.0 log cfu/g, i.e., bars at 2.0 log with a standard deviation of 0 indicate that bacterial counts were actually <2 log. Within the 100 µL or 20 µL inoculum groups, statistically significant differences ($p < 0.05$) between numbers of bacteria on treated and control samples are indicated by different superscripts (a,b,c for *E. coli* and d,e, for *L. monocytogenes*) above the columns.

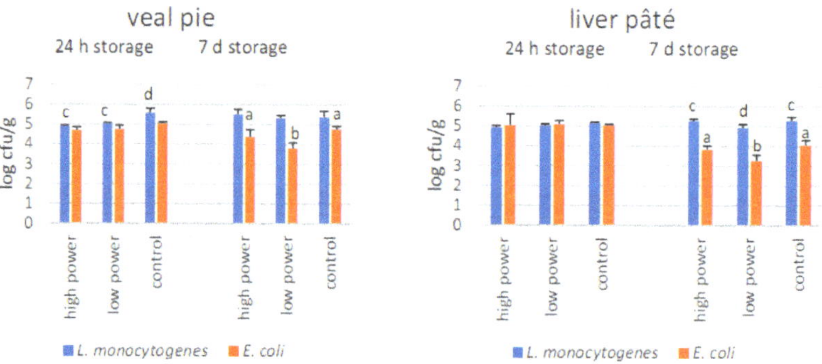

Figure 2. Numbers of *L(isteria) monocytogenes* and *E(scherichia) coli* inoculated on veal pie and calf liver pâté (n = 3) and subjected to 5 min exposure to high- vs. low-power cold atmospheric pressure plasma, followed by 1 to 7 days' storage at 2 ± 2 °C. For the 1- and 7-day storage groups, statistically significant differences ($p < 0.05$) between numbers of bacteria on treated and control samples are indicated by different superscripts (a,b for *E. coli* and c,d for *L. monocytogenes*) above the columns.

5. Discussion

5.1. Effect on Contaminant Bacteria

We considered typical cured and non-cured, heat-treated, ready-to-eat meat pro-ducts that can easily be contaminated during portioning and slicing. Physicochemical characteristics indicated that the products can favour the multiplication of *Listeria monocytogenes* during the shelf life of these products [18]. For such high-risk products, strict adherence to good hygiene practices is a prerequisite, and the establishment of operation prerequisite programs should be considered [43].

The implementation of additional antibacterial measures/interventions has been suggested repeatedly, but the magnitude of the effect of biological agents is not always certain (e.g., anti-listerial bacteriophages [44,45]) and limitations may apply to physicochemical treatments in terms of residues or changes of organoleptic properties of properties (see EFSA series of scientific opinions).

Surface pasteurization of vacuum-packed cooked ready-to-eat meat products requires temperatures of 96 °C and holding times of 10 min to effectuate a 2 to 4 log reduction in *Listeria monocytogenes* [46], but such conditions are not feasible for all meat products. A 1% cetylpyridinium chloride (CPC) spray applied on Polish-style sausage before vacuum-packing was highly effective against *L. monocytogenes* (depending on the inoculation level, an immediate reduction of 1–3 log cycles was observed, and after 42 days of storage it was 2–4 log units) [47], but such additives are not accepted by all consumers.

Cold atmospheric plasma has demonstrated its ability to reduce numbers of bacteria on food surfaces and is thus well suited for managing bacterial contamination post-processing and pre-packaging [25]. With regard to ready-to-eat meats, the reductions of *Listeria* we observed (up to 1.5 log) are in the range as reported in other studies [25–28], albeit differences in experimental design make detailed comparisons difficult.

Our results support the assumption that Gram-negative bacteria (*E. coli*) are more susceptible to CAP than are Gram-positives (*Listeria*). This assumption would be logical since the membrane lipids in Gram-negative organisms are directly exposed to CAP molecules in particular ozone, whereas the cell wall of Gram-positive organisms would protect the cell membrane [48]. However, experimental data are inconclusive [26,37]. While the higher susceptibility of *E. coli* could be explained, it is unclear why *Listeria* reduction differed between similar products ('A', 'B') from different producers, the more so as the ingredient list was nearly identical. The lower reductions observed in pâté and pie compared to cured meats deserves attention and warrants further studies, particularly as on the labels of these products, no antioxidants were declared. We observed no consistent pattern as regards the effect of inoculum size (with respect of the moisture film on the sample surface) or plasma type (low power or high power), although it has been established that humidity or water films influence plasma composition [26,49,50] and that ROS and RONS act differently on bacterial cells [48]. Since our studies were designed as pilots, further experiments are envisaged to study these issues in detail.

A limitation of the methodology we applied for enumeration of bacteria after CAP exposure is that direct plating onto selective agar media for enumeration of bacteria does not consider the possibility of sublethal injury or a viable but not culturable (VBNC) state of the contaminant bacteria [51] post-CAP exposure. The VBNC issue has been studied for thermal and acidic stress in *Listeria* [52–55], but specific studies on CAP are still lacking.

Likewise, instead of the pre-packaging CAP exposure we studied, an in-package CAP treatment with formation of the plasma species in the headspace of the package could be more feasible, since the product is then already sealed and protected from contamination [25,56].

5.2. Effect on the Food Matrix

The role of the food matrix in the CAP–bacterium interplay is poorly studied. It can be expected that-given the abundance of meat protein, fat and water in the food matrix compared to that in the bacterial cells, the majority of CAP species react with the food matrix. As regards plasma generated from ambient air, it is debatable if reactive oxygen substances and reactive nitrogen substances in the plasma would react with the food matrix in a way that would result in a 'novel food', i.e., in molecular structures that were not present in foods within the EU before 15 May 1997 (Article 3 of Regulation (EU) 2015/2283 [57]). With respect to muscle foods, we recently reviewed the effect of CAP on myoglobin forms and thus on colour [58]. In the cured meat products, veal pie and calf liver pâté, a decrease in redness (a*) was most frequently observed, indicating some effect on the myoglobin forms present in the meat products, and in fewer cases a significant, yet small increase in b*. Lightness (L*) was not affected at all, indicating that CAP exposure had no effect on water-binding capacity.

The magnitude of changes in a* and b* was moderate: ΔE values of up to 2.2 were observed for some 2 min exposure protocols, and values up to 2.7 for 5 min exposure. Notably, in the control samples, ΔE values were consistently <1, whereas in all treatment groups, it was >1. ΔE values < 1 are not likely to be recognised as differences, and values of >2 indicate changes in colour also visible to inexperienced observers, and it is assumed that a change in colour is perceived by the majority of consumers at ΔE values of >3 [39–42]. Further experiments should explore if or to what extent colour differences are observed in CAP-exposed samples after cold storage.

The small increase in redness (a*) in non-cured products is most probably not due to a curing reaction, since the myoglobin is already heat-denaturated. A decrease in lightness in CAP-exposed liver pâté might simply indicate that the product is more sensitive to drying [59] than other products under study.

6. Conclusions

CAP treatment of sliced, cured, cooked ham and sausage effectuated significant reductions in *E. coli* (up to 3 log units), but less pronounced reductions in listeriae. In traditional non-cured cooked meats, CAP was less effective. Colour changes (expressed as ΔE values) were in an acceptable range, although changes in redness (a*) indicated some effect of CAP on the myoglobin present in cured foods. Differences were observed between cured and non-cured meats, but also between products of similar type, which warrants further studies.

Author Contributions: Conceptualization, F.J.M.S., P.P. and I.C.; methodology, P.P.; formal analysis, I.C.; investigation, U.V., J.S., M.G., I.C. and P.P.; resources, K.S.; data curation, U.V., J.S. and M.G.; writing—original draft preparation, P.P. and F.J.M.S.; writing—review and editing, P.P., F.J.M.S., I.C., S.B., K.S. and B.P.; supervision, P.P. All authors have read and agreed to the published version of the manuscript.

Funding: This study received no funding.

Data Availability Statement: Data are contained in the DVM Diploma Theses from Ute Vankat, Julia Schrei and Michelle Graf. These are available at https://permalink.obvsg.at/AC15668870 (accessed on 24 January 2023), https://permalink.obvsg.at/AC15615153 (accessed on 24 January 2023), and https://permalink.obvsg.at/AC15615235 (accessed on 24 January 2023).

Acknowledgments: The authors thank J. Walsh for his comments on the manuscript draft and S. Vali for his technical support.

Conflicts of Interest: The authors declare no conflict of interest.

References

1. Camargo, A.C.; Woodward, J.J.; Call, D.; Nero, L.A. *Listeria monocytogenes* in Food-Processing Facilities, Food Contamination, and Human Listeriosis: The Brazilian Scenario. *Foodborne Pathog. Dis.* **2017**, *14*, 623–636. [CrossRef] [PubMed]
2. Schoder, D. Chapter 425. Listeria: Listeriosis. In *The Encyclopedia of Food and Health*, 1st ed.; Caballero, B., Finglas, P., Toldrá, F., Eds.; Academic Press: Oxford, UK, 2015; pp. 561–566.
3. Schoder, D.; Guldimann, C.; Märtlbauer, E. Asymptomatic Carriage of *Listeria monocytogenes* by Animals and Humans and Its Impact on the Food Chain. *Foods* **2022**, *11*, 3472. [CrossRef] [PubMed]
4. Bintsis, T. Foodborne pathogens. *AIMS Microbiol.* **2017**, *3*, 529–563. [CrossRef] [PubMed]

5. Schlech, W.; Lavigne, M.; Bortolussi, R.; Allen, A.; Haldane, V.; Wort, J.; Hightower, A.; Johnson, S.; King, S.; Nicholls, E. Epidemic Listeriosis—Evidence for Transmission by Food. *N. Engl. J. Med.* **1983**, *308*, 203–206. [CrossRef]
6. Kerry, J.P.; Kerry, J.F. *Processed Meats: Improving Safety, Nutrition and Quality*; Woodhead Publishing Series in Food Science, Technology and Nutrition 1st Edition; Woodhead Publishing: Sawston, UK, 2011.
7. Kurpas, M.; Wieczorek, K.; Osek, J. Ready-to-eat Meat Products as a Source of *Listeria monocytogenes*. *J. Vet. Res.* **2018**, *62*, 49–55. [CrossRef]
8. Nesbakken, T.; Kapperud, G.; Caugant, D.A. Pathways of *Listeria monocytogenes* contamination in the meat processing industry. *Int. J. Food Microbiol.* **1996**, *31*, 161–171. [CrossRef] [PubMed]
9. Samelis, J.; Metaxopoulos, J. Incidence and principal sources of *Listeria* spp. and *Listeria monocytogenes* contamination in processed meats and a meat processing plant. *Food Microbiol.* **1999**, *16*, 465–477. [CrossRef]
10. Vitas, A.I.; Aguado, V.; Garcia-Jalon, I. Occurrence of *Listeria monocytogenes* in fresh and processed foods in Navarra (Spain). *Int. J. Food Micro.* **2004**, *90*, 349–356. [CrossRef] [PubMed]
11. Goulet, V.; Rocourt, J.; Rebiere, I.; Jacquet, C.; Moyse, C.; Dehaumont, P.; Salvat, G.; Veit, P. Listeriosis Outbreak Associated with the Consumption of Rillettes in France in 1993. *J. Infect. Dis.* **1998**, *177*, 155–160. [CrossRef]
12. Farber, J.M.; Zwietering, M.; Wiedmann, M.; Schaffner, D.; Hedberg, C.W.; Harrison, M.A.; Hartnett, E.; Chapman, B.; Donnelly, C.W.; Goodburn, K.E.; et al. Alternative approaches to the risk management of *Listeria monocytogenes* in low risk foods. *Food Control* **2021**, *123*, 150–170. [CrossRef]
13. Ferreira, V.; Wiedmann, M.; Teixeira, P.; Stasiewicz, M.J. *Listeria monocytogenes* Persistence in Food-Associated Environments: Epidemiology, Strain Characteristics, and Implications for Public Health. *J. Food Prot.* **2014**, *77*, 150–170. [CrossRef]
14. EFSA Panel on Biological Hazards (BIOHAZ); Ricci, A.; Allende, A.; Bolton, D.; Chemaly, M.; Davies, R.; Fernández Escámez, P.S.; Girones, R.; Herman, L.; Koutsoumanis, K. Listeria monocytogenes contamination of ready-to-eat foods and the risk for human health in the EU. *EFSA J.* **2018**, *16*, e05134.
15. EFSA; ECDC. The European Union One Health 2020 Zoonoses Report. *EFSA J.* **2021**, *19*, 6971. [CrossRef]
16. Rapid Alert System for Food and Feed (RASFF). Notification 2022.6589 Listeria Monocytogenes in Fishcakes. Available online: https://webgate.ec.europa.eu/rasff-window/screen/notification/579466 (accessed on 30 November 2022).
17. Agency for Health for Humans, Animals & Plants (AGES). Pathogen Listeria. Available online: https://www.ages.at/mensch/krankheit/krankheitserreger-von-a-bis-z/listerien (accessed on 26 November 2022).
18. European Commission. Commission Regulation (EC) No 2073/2005 of 15 November 2005 on microbiological criteria for foodstuffs. *OJ* **2005**, *L338*, 1–26.
19. Traditional Foods in Austria (Meat). Available online: https://info.bml.gv.at/themen/lebensmittel/trad-lebensmittel/Fleisch.html (accessed on 30 November 2022).
20. Codex Alimentarius Austriacus. Available online: https://www.lebensmittelbuch.at/ (accessed on 27 November 2022).
21. Agricultural Market Analysis (AMA). Available online: https://www.ama.at/marktinformationen/vieh-und-fleisch/konsumverhalten (accessed on 28 November 2022).
22. Paulsen, P.; Csadek, I.; Bauer, A.; Bak, K.H.; Weidinger, P.; Schwaiger, K.; Nowotny, N.; Walsh, J.; Martines, E.; Smulders, F.J.M. Treatment of Fresh Meat, Fish and Products Thereof with Cold Atmospheric Plasma to Inactivate Microbial Pathogens and Extend Shelf Life. *Foods* **2022**, *11*, 3865. [CrossRef]
23. Shimizu, T.; Sakiyama, Y.; Graves, D.; Zimmerman, J.; Morfill, G. The dynamics of ozone generation and mode transition in air surface micro-discharge plasma at atmospheric pressure. *New J. Phys.* **2012**, *14*, 103028. [CrossRef]
24. Zimmermann, J. Cold Atmospheric Plasma in Medicine—From Basic Research to Application, Max Planck Institute of Pathology, Technical University Munich, Habilitation Treatise. 2013. Available online: http://mediatum.ub.tum.de/?id=1219297 (accessed on 2 September 2022).
25. Rød, S.K.; Hansen, F.; Leipold, F.; Knøchel, S. Cold atmospheric pressure plasma treatment of ready-to-eat-meat: Inactivation of *Listeria innocua* and changes in product quality. *Food Microbiol.* **2012**, *30*, 233–238. [CrossRef] [PubMed]
26. Lis, K.A.; Boulaaba, A.; Binder, S.; Li, Y.; Kehrenberg, C.; Zimmermann, J.L.; Klein, G.; Ahlfeld, B. Inactivation of *Salmonella Typhimurium* and *Listeria Monocytogenes* on Ham with Nonthermal Atmospheric Pressure Plasma. *PLoS ONE* **2018**, *13*, e0197773. [CrossRef]
27. Zeraatpisheh, F.; Tabatabaei Yazdl, F.; Shahidi, F. Investigation of effect of cold plasma on microbial load and physicochemical properties of ready-to-eat sliced chicken sausage. *J. Food Sci. Technol.* **2022**, *59*, 3928–3937. [CrossRef]
28. Yadav, B.; Spinelli, A.C.; Misra, N.N.; Tsui, Y.Y.; McMullen, L.M.; Roopesh, M.S. Effect of in-package atmospheric cold plasma discharge on microbial safety and quality of ready-to-eat ham in modified atmospheric packaging during storage. *J. Food Sci.* **2020**, *85*, 1203–1212. [CrossRef]
29. Jung, S.; Kim, H.J.; Park, S.; Yong, H.I.; Choe, J.H.; Jeon, H.-J.; Choe, W.; Jo, C. The use of atmospheric pressure plasma-treated water as a source of nitrite for emulsion-type sausage. *Meat Sci.* **2015**, *108*, 132–137. [CrossRef] [PubMed]
30. Jung, S.; Kim, H.J.; Park, S.; Yong, H.I.; Choe, J.H.; Jeon, H.J.; Choe, W.; Jo, C. Color Developing Capacity of Plasma-treated Water as a Source of Nitrite for Meat Curing. *Korean J. Food Sci. Anim. Resour.* **2015**, *35*, 703–706. [CrossRef] [PubMed]
31. Lee, J.; Jo, K.; Lim, Y.; Jeon, H.J.; Choe, J.H.; Jo, C.; Jung, S. The use of atmospheric pressure plasma as a curing process for canned ground ham. *Food Chem.* **2018**, *240*, 430–436. [CrossRef] [PubMed]

32. Yong, H.I.; Park, J.; Kim, H.-J.; Jung, S.; Park, S.; Lee, H.J.; Choe, W.; Jo, C. An innovative curing process with plasma-treated water for production of loin ham and for its quality and safety. *Plasma Process. Polym.* **2018**, *15*, 1700050. [CrossRef]
33. Jo, K.; Lee, S.; Yong, H.I.; Choi, Y.-S.; Jung, S. Nitrite sources for cured meat products. *LWT-Food Sci. Technol.* **2020**, *129*, 109583. [CrossRef]
34. Guerrero-Legarreta, I. Meat and poultry—Spoilage of Cooked Meat and Meat Products. In *Encyclopedia of Food Microbiology*, 2nd ed.; Batt, C.A., Tortorello, M.L., Eds.; Academic Press: Oxford, UK, 2014; pp. 508–513.
35. Honikel, K.-O. The use and control of nitrate and nitrite for the processing of meat products. *Meat Sci.* **2008**, *78*, 68–76. [CrossRef]
36. Awaiwanont, N.; Smulders, F.J.M.; Paulsen, P. Growth potential of *Listeria monocytogenes* in traditional Austrian cooked-cured meat products. *Food Control* **2017**, *50*, 150–156. [CrossRef]
37. Ekici, G.; Dümen, E. Escherichia coli and Food Safety. In *The Universe of Escherichia coli*; Starčič Erjavec, M., Ed.; IntechOpen: London, UK, 2019. [CrossRef]
38. Bauer, A.; Ni, Y.; Bauer, S.; Paulsen, P.; Modic, M.; Walsh, J.L.; Smulders, F.J.M. The effects of atmospheric pressure cold plasma treatment on microbiological, physical-chemical and sensory characteristics of vacuum packaged beef loin. *Meat Sci.* **2017**, *128*, 77–87. [CrossRef]
39. CIE (Commission Internationale de L'Éclairage (International Commission on Illumination). *Recommendations on Uniform Color Spaces, Color-Difference Equations, Psychometric Color Terms*; Commission Internationale de l'Eclairage: Paris, France, 1978.
40. Upton, S. Delta E: The Color Difference. CHROMIX Colornews, 17. Available online: http://www.colorwiki.com/wiki/Delta_E (accessed on 18 February 2005).
41. Altmann, B.A.; Gertheiss, J.; Tomasevic, I.; Engelkes, C.; Glaesener, T.; Meyer, J. Differences using computer vision system measurements of raw pork loin. *Meat Sci.* **2022**, *183*, 108766. [CrossRef]
42. Upton, S. Delta E: The Color Difference. Delta E, CHROMIX Colornews, 17. Available online: http://www.chromix.com/ColorNews/ (accessed on 27 November 2022).
43. European Commission. Commission Notice on the implementation of food safety management systems covering prerequisite programs (PRPs) and procedures based on the HACCP principles, including the facilitation/flexibility of the implementation in certain food businesses. *OJ* **2016**, *C28*, 1–32.
44. European Food Safety Authority (EFSA) (EFSA Panel on Biological Hazards). Scientific opinion on the evaluation of the safety and efficacy of ListexTMP100 for reduction of pathogens on different ready-to-eat (RTE) food products. *EFSA J.* **2016**, *14*, 4565. [CrossRef]
45. Kawacka, I.; Olejnik-Schmidt, A.; Schmidt, M.; Sip, A. Effectiveness of Phage-Based Inhibition of *Listeria monocytogenes* in Food Products and Food Processing Environments. *Microorganisms* **2020**, *8*, 1764. [CrossRef] [PubMed]
46. Houben, J.H.; Eckenhausen, F. Surface pasteurization of vacuum-sealed precooked ready-to-eat meat products. *J. Food Prot.* **2006**, *69*, 459–468. [CrossRef] [PubMed]
47. Singh, M.; Gill, V.S.; Thippareddi, H.; Phebus, R.K.; Marsden, J.L.; Herald, T.J.; Nutsch, A.L. Cetylpyridinium chloride treatment of ready-to-eat Polish sausages: Effects on *Listeria monocytogenes* populations and quality attributes. *Foodborne Pathog. Dis.* **2005**, *2*, 233–241. [CrossRef]
48. Bourke, P.; Ziuzina, D.; Han, L.; Cullen, P.J.; Gilmore, B.F. Microbiological interactions with cold plasma. *J. Appl. Microbiol.* **2017**, *123*, 308–324. [CrossRef]
49. Reuter, S.; Winter, J.; Iseni, S.; Schmidt-Bleker, A.; Dünnbier, M.; Masur, K.; Wende, K.; Weltmann, K.-D. The Influence of Feed Gas Humidity versus Ambient Humidity on Atmospheric Pressure Plasma Jet-Effluent Chemistry and Skin Cell Viability. *IEEE Trans. Plasma Sci.* **2015**, *43*, 3185–3192. [CrossRef]
50. Trevisani, M.; Cavoli, C.; Ragni, L.; Cecchini, M.; Berardinelli, A. Effect of non-thermal atmospheric plasma on viability and histamine-producing activity of psychrotrophic bacteria in mackerel fillets. *Front. Microbiol.* **2021**, *12*, 653597. [CrossRef]
51. Donnelly, C.W. Detection and Isolation of *Listeria monocytogenes* from Food Samples: Implications of Sublethal Injury. *J. AOAC Int.* **2002**, *85*, 495–500. [CrossRef]
52. Arvaniti, M.; Tsakanikas, P.; Papadopoulou, V.; Giannakopoulou, A.; Skandamis, P. *Listeria monocytogenes* Sublethal Injury and Viable-but-Nonculturable State Induced by Acidic Conditions and Disinfectants. *Microbiol. Spectr.* **2021**, *9*, e01377-21. [CrossRef]
53. Arvaniti, M.; Tsakanikas, P.; Paramithiotis, S.; Papadopoulou, V.; Balomenos, A.; Giannakopoulou, A.; Skandamis, P. Deciphering the induction of *Listeria monocytogenes* into sublethal injury using fluorescence microscopy and RT-qPCR. *Int. J. Food Microbiol.* **2022**, *385*, 109983. [CrossRef]
54. Siderakou, D.; Zilelidou, E.; Poimenidou, S.; Tsipra, I.; Ouranou, E.; Papadimitriou, K.; Skandamis, P. Assessing the survival and sublethal injury kinetics of *Listeria monocytogenes* under different food processing-related stresses. *Int. J. Food Microbiol.* **2021**, *346*, 109159. [CrossRef]
55. Koutsoumanis, K.P.; Kendall, P.A.; Sofos, J.N. Effect of food processing-related stresses on acid tolerance of *Listeria monocytogenes*. *Appl. Environ. Microbiol.* **2003**, *69*, 7514–7516. [CrossRef]
56. Roh, S.; Oh, Y.; Le, S.; Kann, J.; Min, S. Inactivation of *Escherichia coli* O157, H7, *Salmonella*, *Listeria monocytogenes* and Tulane virus in processed chicken breast via atmospheric in-package cold plasma treatment. *LWT Food Sci. Technol.* **2020**, *127*, 109429. [CrossRef]

57. Regulation (EU) 2015/2283 of the European Parliament and of the Council of 25 November 2015 on novel foods, amending Regulation (EU) No 1169/2011 of the European Parliament and of the Council and repealing Regulation (EC) No 258/97 of the European Parliament and of the Council and Commission Regulation (EC) No 1852/2001. *Off. J. Eur. Union* **2015**, *L 327*, 1–22.
58. Bak, K.H.; Csadek, I.; Paulsen, P.; Smulders, F.J.M. Application of atmospheric pressure cold plasma (ACP) on meat and meat products. Part 2. Effects on the sensory quality with special focus on meat colour and lipid oxidation. *Fleischwirtschaft* **2021**, *101*, 100–105.
59. Jadhay, H.B.; Annapure, U. Consequences of non-thermal cold plasma treatment on meat and dairy lipids—A review. *Future Foods* **2021**, *4*, 100095. [CrossRef]

MDPI

St. Alban-Anlage 66

4052 Basel

Switzerland

Tel. +41 61 683 77 34

Fax +41 61 302 89 18

www.mdpi.com

Foods Editorial Office

E-mail: foods@mdpi.com

www.mdpi.com/journal/foods